*Now you can learn
the abstract concepts of algebra
through concrete modeling!*

Computer Interactive Algeblocks, Volumes 1 and 2

by Anita Johnston, Jackson Community College

Taking algebra from frustrating to fun!

Learning abstract concepts from a lecture or a book is one of the more difficult hurdles many students face when they enter the algebra curriculum. Frustration may lead some to give up before ever building the strong algebraic foundation necessary for them to move on to higher-level courses. Now, with *Computer Interactive Algeblocks*, you actually learn by doing *and* seeing algebra—giving you the skills you need to continue with confidence.

Three easy steps to mathematical success!

1. *Follow the examples:* On-screen examples are automatic and guide you every step of the way. Each example begins with an explanation of how to solve the problem at hand, and then proceeds with an animated demonstration of how to use *Algeblocks* to solve the problem.

2. *Practice the concepts:* Once comfortable with the examples, you're ready to perform practice exercises. The program gives you the option to solve equations with or without the help of *Algeblocks*.

3. *Self-test:* The self-test allows you to review the concepts presented in the *Algeblocks* lab without the guidance of the program. If you have any difficulty, simply click on "Examples" or "Practice" to review, then return to the self-test when you're ready.

System Requirements

Macintosh®: System 7.0 or higher, 68030 Processor or higher, 8MB RAM, color monitor.
Windows®: Windows 3.1 or higher, VGA, 386/40 MHz processor, 8MB RAM, 640 x 480 screen resolution.

To order a copy of *Computer Interactive Algeblocks, Volume 1* or *2*, please contact your college store or fill out the form on the back and return with your payment to Brooks/Cole.

ORDER FORM

_____Yes! Send me a copy of *Computer Interactive Algeblocks, Volume 1*
(Macintosh® ISBN: 0-534-95144-9)

_____Yes! Send me a copy of *Computer Interactive Algeblocks, Volume 1*
(Windows® ISBN: 0-534-95146-5)

_____Yes! Send me a copy of *Computer Interactive Algeblocks, Volume 2*
(Macintosh® ISBN: 0-534-95443-X)

_____Yes! Send me a copy of *Computer Interactive Algeblocks, Volume 2*
(Windows® ISBN: 0-534-95443-X)

_____Copies x $20.95* =_____

Residents of: AL, AZ, CA, CT, CO, FL, GA, IL, IN, KS, KY, LA,
MA, MD, MI, MN, MO, NC, NJ, NY, OH, PA, RI, SC, TN, TX,
UT, VA, WA, WI must add appropriate state sales tax.

Subtotal _____
Tax _____
Handling __$4.00__
Total Due _____

* Call after 12/1/97 for current prices: 1-800-487-3575

Payment Options

_____ Check or money order enclosed

Bill my _____VISA _____MasterCard _____American Express

Card Number: _____ Expiration Date: _____

Signature: _____

Note: Credit card billing and shipping addresses must be the same.

Please ship my order to: (Please print.)

Name _____

Institution _____

Street Address_____

City _____ State _____ Zip+4_____

Telephone ( )_____ e-mail _____

Your credit card will not be billed until your order is shipped. Prices subject to change without
notice. We will refund payment for unshipped out-of-stock titles after 120 days and for not-yet-
published titles after 180 days unless an earlier date is requested in writing from you.

Mail to:

Brooks/Cole Publishing Company
Source Code 8BCTC041
511 Forest Lodge Road
Pacific Grove, California 93950-5098
Phone: (408) 373-0728; Fax: (408) 375-6414
e-mail: info@brookscole.com

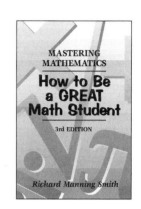

*Mastering Mathematics:
How to Be a Great Math Student,
Third Edition*
by Richard Manning Smith

A complete guide to improving your learning and performance in math courses!

Providing solid tips for every stage of study, *Mastering Mathematics: How to Be a Great Math Student, Third Edition* stresses the importance of a positive attitude and gives you the tools to succeed in your math course. This practical guide will help you:

- avoid mental blocks during math exams
- identify and improve your areas of weakness
- get the most out of class time
- study more effectively
- overcome a perceived "low math ability"
- be successful on math tests
- get back on track when you are feeling "lost"
  …and much more!

To order a copy of *Mastering Mathematics*, please contact your college store or fill out the form on the back and return with your payment to Brooks/Cole. You can also order on-line at http://www.brookscole.com

ORDER FORM

_____Yes! Send me a copy of *Mastering Mathematics: How to be a Great Math Student*   (ISBN: 0-534-34864-5)

_____Copies x $15.95* =_____

Residents of: AL, AZ, CA, CT, CO, FL, GA, IL, IN, KS, KY, LA, MA, MD, MI, MN, MO, NC, NJ, NY, OH, PA, RI, SC, TN, TX, UT, VA, WA, WI must add appropriate state sales tax.

Subtotal _____
Tax _____
Handling ___$4.00___
Total Due _____

Payment Options

_____ Check or money order enclosed

Bill my   ____VISA ____MasterCard ____American Express

Card Number: _____

Expiration Date: _____

Signature: _____

Note: Credit card billing and shipping addresses must be the same.

Please ship my order to: (Please print)

Name _____

Institution _____

_____

Street Address_____

_____

City _____ State _____ Zip+4_____

Telephone (   )_____ e-mail _____

Your credit card will not be billed until your order is shipped. Prices subject to change without notice. We will refund payment for unshipped out-of-stock titles after 120 days and for not-yet-published titles after 180 days unless an earlier date is requested in writing from you.

Mail to:

Brooks/Cole Publishing Company
Source Code 8BCMA139
511 Forest Lodge Road
Pacific Grove, California 93950-5098
Phone: (408) 373-0728; Fax: (408) 375-6414
e-mail: info@brookscole.com

10/97

* Call after 12/1/97 for current prices: 1-800-487-3575

Beginning Algebra

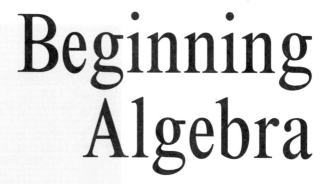

Beginning Algebra

Bill E. Jordan
Seminole Community College

William P. Palow
Miami-Dade Community College

BROOKS/COLE PUBLISHING COMPANY

I(T)P® An International Thomson Publishing Company

Pacific Grove • Albany • Belmont • Bonn • Boston • Cincinnati • Detroit • London • Madrid • Melbourne
Mexico City • New York • Paris • San Francisco • Singapore • Tokyo • Toronto • Washington

Sponsoring Editor: Bob Pirtle

Marketing Team: Laura Caldwell, Jennifer Huber, Romy Taormina

Editorial Assistant: Melissa Duge

Production Editor: Timothy Wardell

Editorial/Production Supervision: Kelli Jauron, Carlisle Publishers Services

Interior Design: The Davis Group, Inc.

Cover Design: Craig Hanson

Typesetting: Carlisle Communications, Ltd.

Cover Printing: Courier Westford

Printing and Binding: Courier Westford

COPYRIGHT © 1998 by Brooks/Cole Publishing Company

A division of International Thomson Publishing, Inc.

I(T)P The ITP logo is a registered trademark under license.

For more information, contact:

BROOKS/COLE PUBLISHING COMPANY
511 Forest Lodge Road
Pacific Grove, CA 93950
USA

International Thomson Publishing Europe
Berkshire House 168-173
High Holborn
London WC1V 7AA
England

Thomas Nelson Australia
102 Dodds Street
South Melbourne, 3205
Victoria, Australia

Nelson Canada
1120 Birchmount Road
Scarborough, Ontario
Canada M1K 5G4

International Thomson Editores
Seneca 53
Col. Polanco
11560 México, D. F., México

International Thomson Publishing GmbH
Königswinterer Strasse 418
53227 Bonn
Germany

International Thomson Publishing Asia
221 Henderson Road
#05-10 Henderson Building
Singapore 0315

International Thomson Publishing Japan
Hirakawacho Kyowa Building, 3F
2-2-1 Hirakawacho
Chiyoda-ku, Tokyo 102
Japan

Printed in the United States of America

10 9 8 7 6 5 4 3 2 1

Library of Congress Cataloging-in-Publication Data

Jordan, Bill E.
　　Beginning algebra / Bill E. Jordan, William P. Palow.
　　　　p.　　cm.
　　Includes index.
　　ISBN 0–534–95010–8
　　1. Algebra.　I. Palow, William P.　II. Title.
QA152.2.J67　1998
512.9—dc21　　　　　　　　　　　97–41800
　　　　　　　　　　　　　　　　　　　　CIP

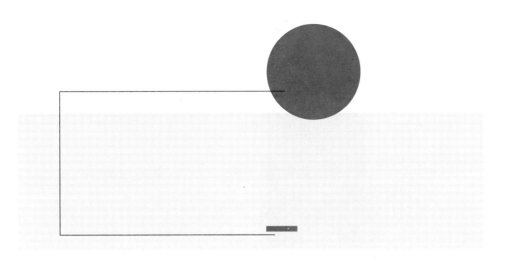

Preface

Purpose and Goals

This text was written for students who have little or no previous experience with algebra. Upon completion of this text, students will have a solid foundation from which they can pursue more advanced courses in algebra. Every attempt was made to present all the skills necessary for success in future classes in a coherent and understandable manner.

Much of this text was taken from *Integrated Arithmetic and Algebra* by Jordan and Palow and, consequently, shares many of the characteristics of that text. Since Jordan wrote the majority of the integrated text and Palow was not directly involved with the development of this text, over 85% was written by Jordan. One of the issues addressed in Standards for Introductory College Mathematics adopted by AMATYC is more integration of the various fields of mathematics. Consequently, there is more emphasis on the relationships between arithmetic and algebra than is found in most beginning algebra texts. Much of elementary algebra involves performing the same operations on polynomials that are performed on numbers. Consequently, a review of an operation on numbers is given just before or at the time the corresponding operation is performed on polynomials. This approach is used extensively in the sections on performing operations on rational expressions. This allows students to see the relationships between arithmetic and algebra rather than viewing them as entirely separate subjects and makes algebra a lot "friendlier." There is also more geometry incorporated throughout the text than in most texts on this level.

The major goals of *Beginning Algebra* are:

1) To prepare students whose entry level is Beginning Algebra for Intermediate Algebra.

2) To show the close relationships between arithmetic and algebra.

3) To present the material in an organized, easy to read and understandable manner.

4) To have students think critically about mathematics.

5) To show how mathematics relates to other fields and to real life situations.

6) To have the student understand the basic principles of Beginning Algebra.

Beginning Algebra accomplishes these goals through the following.

Concept Versus Algorithm

Success in mathematics and being able to use mathematics is dependent upon an understanding of why mathematical principles are true and not just on how to do an exercise and get the right answer. Too often students view mathematics as a "bag of tricks" and the way to learn mathematics is to memorize the trick that applies to a given situation. Nothing could be further from the truth, so this text places heavy emphasis on concept development throughout. If we understand something we not only remember it longer, but we also remember more of it and can apply it to other situations. Hence, the concept (the "why") is introduced and developed before the process or algorithm (the "how") for performing the operation. The student is never told how to do something without explaining why it is done. This approach is demonstrated in virtually all sections of the text. For specific concepts, see Section 1.7 (Subtraction of Integers), 1.9 (Combining Like Terms), and 2.1 (Multiplication of Integers).

Critical Thinking/Problem Solving

We must emphasize teaching thinking skills and problem-solving. We believe that these important skills must start on the very first page and be carried throughout the book. We teach thinking skills and problem-solving by:

1) The first stage of problem solving (classification) is introduced in Chapter 1 and used in subsequent chapters.

2) The second and third stages of problem solving (representation and solution) are introduced and developed in Chapters 1, 2, and 3. The fourth stage (verification) is introduced and emphasized in Chapter 3. These are then continued throughout the book.

3) Almost every exercise set contains writing exercises that are designed to help students formulate and understand mathematical ideas. Usually these can not be answered by simply looking in the text, but require an understanding of a mathematical concept and critical thinking.

4) Group Projects are also found in many exercise sets. These give students additional opportunities to formulate and understand mathematical ideas and often require critical thinking. Sometimes they show how mathematics relates to careers and other fields of study.

5) Almost all exercise sets contain some problems that are not just skills-oriented, but require the student to go beyond the skill to an application of that skill. For example, in the sections on products of polynomials students are asked to represent the area of geometric figures where the lengths of the sides are polynomials. Also, in all sections where appropriate, students are asked to solve applications problems that involve the concepts presented in that section.

6) Special attention has been given to the language of mathematics. Without the proper vocabulary and grammar, mathematics cannot be meaningful.

Features of the Text

Worktext Format

This text introduces a concept, develops and explains the appropriate algorithm, and demonstrates the algorithm with carefully chosen examples. The student is then asked to practice and apply the concept through the algorithm. This is accomplished through the use of practice exercises embedded into the flow of the text and additional practice exercises located in the margin. In this manner each topic is introduced, illustrated, and then practiced until the section exercises are reached where there is an ample opportunity to practice and draw ideas together. The practice exercises and additional practice exercises are written so that the first practice exercise is like the "a" part of the preceding example, the second is like the "b" part, etc. In this manner the student can get immediate help with any practice exercise (or additional practice exercise) that is being troublesome. The answers to the practice and additional practice exercises are located in the margin at the top of the next page for easy reference. Answers to the odd numbered exercises and all the answers to chapter tests are located in the back of the book.

Examples

There is an extremely large number of carefully selected examples that are designed to cover all the variations of any particular concept that are appropriate to a text at this level. The examples are carefully done in a step-by-step manner in a two-column format with the steps on the left and the explanations on the right. The two-column format is not unusual, but the way we did it is somewhat different. In most texts the explanations to the right of each step is what was done to get that step. This is in the reverse order of the way we think. We do not do something and then wonder why we did it. In this text the step is given and the explanation to the right is what is to be done to get the next step. This is more consistent with the way we think when solving a problem. We determine what has to be done next and then we do it. This teaches students good problem solving techniques.

Approach to Technology

The use of technology should be a natural part of mathematics and should be incorporated as such. The amount of calculator usage at this level varies among colleges from none to total immersion. We have incorporated some specific uses of calculators in the early chapters where appropriate and many exercise sets have problems that are specifically designated as calculator exercises. We have also included some Calculator Exploration Activities that lead students to discover a relationship. Answers to these activities are in the margin at the top of the next page. These activities are optional and followed by a complete mathematical justification of the relationships the students were asked to discover. We feel that, on the whole, instructors should incorporate the amount of technology with which they are comfortable. With the great number of calculators on the market today and the tremendous differences in the way they operate, it is difficult to specify key strokes. We played it safe and gave key strokes for scientific calculators—both with and without parentheses. The key strokes would be different for graphing calculators.

Order of Topics

The order of topics in the early chapters is somewhat different than the order in most texts. As previously mentioned, much of a beginning algebra course involves performing the same operations on polynomials that are performed on numbers. Rather than present all the operations on integers and then somewhere later in the text present the same operation on polynomials, we present an operation on integers and then the same operation on polynomials. This reinforces the operation on integers, since they have to perform this operation when performing the same operation on polynomials, and gives students an immediate reason for learning this operation on integers. Too often students learn mathematical principles in isolation of similar principles (like adding integers and adding like terms) and never see how seemingly isolated principles fit together in an algebraic system. This order of topics shows the "big picture" of how similar topics are related rather than presenting algebra as a series of isolated topics. This approach has been used very successfully for the past several years in *Integrated Arithmetic and Algebra* and in the class testing of this text.

Exercise Sets

Each exercise set contains an abundant number of exercises that are written so that consecutive odd and even exercises are alike. Great care has been taken to be certain that all cases for each concept are included. Almost all exercise sets contain problems that require the student to go beyond the basic skills level and apply the basic skill(s) of that section to a situation. Often this is done using geometric concepts or solving applications problems. Almost all exercise sets contain writing exercises that require the student to understand and think about the concept(s) of that section. Usually the answers to writing exercises can not be found directly in the text. Many exercise sets have some writing exercises labeled as "Critical Thinking Exercises" because they require more thought and synthesis than the usual writing exercise. Almost all exercise sets also contain group projects many of which can be assigned as writing exercises.

Applications Problems

Applications problems are found in many sections of the text. Applications involving fractions are in Section 1.2 and those involving order of operations on whole numbers are located in Section 1.3 Applications involving perimeter, area, and volume are in Section 1.5. Applications involving addition and subtraction of integers are in Sections 1.6 and 1.7 respectively while applications involving multiplication and division of integers are in Sections 2.1 and 2.5. Applications with integers and the order of operations are in Section 2.6. Many real-world applications are found in each of Sections 3.1–3.3 and 3.5 where solving particular types of linear equations are discussed. Geometric applications problems are in Section 3.8. In addition, Sections 3.6 and 3.7 contain the types of applications problems that are traditionally found in virtually all algebra texts and are even called traditional applications problems. If some of the purposes of a beginning algebra course are to teach critical thinking, to promote problem solving skills, and to promote an understanding of an algebraic system, then these traditional problems serve this purpose well. They are grouped by types so it is easy to pick and choose the type(s) that you wish without any loss of continuity in the remainder of the text. Other applications problems are found in Sections 4.9, 5.8, 5.9, 6.2, 6.3, 6.5, 6.6, 7.4, 8.6, 8.7, 9.1–9.3, and 9.5.

Geometry

Beginning Algebra emphasizes geometry more than most texts at this level through the following.

* In Section 1.4 students are asked to apply the skills of substituting for a variable and simplifying using the order of operations learned in Section 1.3 to find perimeters, areas, and volumes of geometric figures.

- In Section 1.5 students are asked to solve real-world applications problems involving geometry that do not require the use of variables. These problems involve perimeters, areas, and volumes of geometric figures.
- In Section 1.9 students are asked to represent the perimeters of geometric figures with sides that are polynomials and in Sections 2.3 and 2.4 to represent areas and volumes in the same manner. This is in preparation for solving applications problems where it is necessary to represent sides of geometric figures as polynomials.
- In Section 3.5 students are asked to solve real-world applications problems that involve the use of variables. These problems involve perimeter, angle relationships of parallel lines, and angle relationships of triangles. In Sections 4.9 and 9.1 and 9.3 there are applications problems that involve area and volume.
- In Section 8.7 students are asked to solve right triangles, rectangles, squares, and real-world problems involving the Pythagorean Theorem.
- In many sections throughout the text students are asked to apply geometric formulas in the exercise sets.

Chapter Summaries

Each chapter summary contains the important ideas, definitions, algorithms and procedures from the chapter and serves as a quick overview and review of the chapter.

Chapter Review Exercises

Each chapter review contains additional exercises that are keyed to the section of the text that explains the concepts and algorithms needed to do these exercises. These allow the student to quickly test his/her understanding of the chapter before attempting the chapter test.

Chapter Tests

Each chapter contains a sample test that students can use to evaluate their understanding of the material found in the chapter. These questions are in a free response format and space to work the exercises is provided. The answers to the questions in the chapter tests are in the back of the book.

Writing Exercises and Study Tips to Reduce Math Anxiety

Writing about difficulties has been shown to be an effective means to understand our problems and reduce anxiety. We suggest informal writing like prewriting and/or journals to allow students to write about their feelings and difficulties. We have included extensive writing exercises throughout the text to give students the opportunity to write about mathematics. Since mathematics is not learned the same way as other subjects, study tips have been dispersed throughout Chapters 1, 2, and 3 to assist and encourage students as they attempt to learn a subject about which many of them are very anxious. Attention to concept and understanding has been shown to reduce math anxiety by as much as 40%.

Definitions and Algorithms

All definitions, algorithms, and other material of importance are clearly labeled and boxed for easy reference. The first time a mathematical term is used, it is printed in bold-face type.

Be Careful and Note

Throughout the text students are warned about common errors by paragraphs that begin with **Be Careful**! Paragraphs that begin with **Note** provide additional information, explanations or observations about the concept just discussed.

Class Testing

This text has been extensively class tested for four semesters in manuscript and custom published form at Seminole Community College by five instructors. As a result of the class testing, we have been able to eliminate many of the errors usually associated with a first edition book and improve the manuscript through student input.

Ancillaries

At the present time, we are planning to have an instructor's manual with even answers and a test bank (hard copies), a computerized test generator, audio and video tapes, and a students' solution manual.

We would like to thank the following reviewers for their valuable suggestions: Carole Bauer, Triton College; Paul Foutz, Temple College; Curtis Herink, Mercer University; Timothy Klinger, Delta College; Timothy Magnavita, Bucks County Community College; Robert McCarthy, Community College of Allegheny County; Dan Schapiro, Yakima Valley Community College; Don Shriner, Frostburg State University; and Muserref Wiggins, University of Akron. (There were lots of other reviewers when the text was with PWS.)

We would also like to thank the instructors and students at Seminole Community College who participated in the class testing of the text. Their suggestions and corrections helped make it a better text.

Many thanks also to all the folks at Brooks/Cole especially our editor, Bob Pirtle, for his advice and guidance through this project; Melissa Duge, Editorial Assistant; Jennifer Huber, Marketing Manager; Linda Row, Ancillary Editor; Timothy Wardell, Production Editor; and Roy Neuhaus for the beautiful and striking cover design.

This book has been written with the characteristics and needs of the beginning mathematics student in mind. It is based on the research and experience of authors who have taught over a quarter of a century each and who have received numerous awards for teaching excellence. We have tried to combine our experience with current learning theory and practice to produce a book that we hope will make a difference. It is our thought that the beginning mathematics student needs a fresh way of viewing mathematics so that he/she is not looking at the same old thing again. We hope that your experiences with the text are pleasant ones.

Sincerely,
B. E. Jordan
W. P. Palow

Table of Contents

CHAPTER ONE

Addition and Subtraction of Real Numbers and Polynomials

1.1	Number Systems, Order, and Absolute Value	2
1.2	A Review of Fractions	12
1.3	Exponents, Variables, and Order of Operations	27
1.4	Evaluating Geometric Formulas	37
1.5	Geometric Applications	53
1.6	Addition of Real Numbers	59
1.7	Subtraction of Real Numbers	70
1.8	Terms and Polynomials	79
1.9	Combining Like Terms and Polynomials	86

CHAPTER SUMMARY   97
CHAPTER REVIEW EXERCISES   101
CHAPTER TEST   106

CHAPTER TWO

Multiplication and Division of Integers and Polynomials

2.1	Multiplication of Real Numbers	109
2.2	Multiplication Laws of Exponents	119
2.3	Products of Polynomials	126
2.4	Special Products	132
2.5	Division of Integers	139
2.6	Order of Operations on Real Numbers	146
2.7	Quotient Rule and Integer Exponents	152
2.8	An Application of Exponents: Scientific Notation	161
2.9	Power Rule for Quotients and Using Combined Laws of Exponents	170
2.10	Division of Polynomials by Monomials	177

CHAPTER SUMMARY   182

CHAPTER REVIEW EXERCISES   184

CHAPTER TEST   187

CHAPTER THREE

Linear Equations and Inequalities

3.1	Addition Property of Equality	190
3.2	Multiplication Property of Equality	200
3.3	Combining Properties in Solving Linear Equations	209
3.4	Solving Linear Inequalities	219
3.5	Percent Equations	233
3.6	Traditional Applications Problems	237
3.7	More Traditional Applications Problems	248
3.8	Geometric Applications Problems	260
3.9	Using and Solving Formulas	268

CHAPTER SUMMARY   273

CHAPTER REVIEW EXERCISES   276

CHAPTER TEST   280

CHAPTER FOUR

Factoring and Solving Quadratic Equations

4.1	Prime Factorization and Greatest Common Factor	283
4.2	Factoring Polynomials with Common Factors and by Grouping	288
4.3	Factoring Binomials	294
4.4	Factoring Perfect Square Trinomials	299
4.5	Factoring General Trinomials with Leading Coefficient of One	304
4.6	Factoring General Trinomials with First Coefficient Other Than One	313
4.7	Mixed Factoring	323
4.8	Solving Quadratic Equations by Factoring	326

CHAPTER SUMMARY   335

CHAPTER REVIEW EXERCISES   337

CHAPTER TEST   339

CHAPTER FIVE

Operations with Rational Expressions

5.1	Reducing Rational Expressions to Lowest Terms	341
5.2	Multiplication of Rational Expressions	351
5.3	Division of Rational Expressions	358
5.4	Least Common Multiple, Least Common Denominator and Equivalent Rational Expressions	368
5.5	Addition and Subtraction of Rational Expressions	375
5.6	Complex Fractions	386

5.7	Solving Equations Containing Rational Numbers and Expressions	394
5.8	Applications with Rational Expressions	401
5.9	Ratio and Proportion	414

CHAPTER SUMMARY 422
CHAPTER REVIEW EXERCISES 425
CHAPTER TEST 430

CHAPTER SIX

Graphing Linear Equations and Inequalities

6.1	Ordered Pairs and Solutions of Linear Equations with Two Variables	434
6.2	Graphing Linear Equations with Two Variables	441
6.3	Graphing Linear Inequalities with Two Variables	457
6.4	Slope of a Line	468
6.5	Writing Equations of Lines	586
6.6	Relations and Functions	503

CHAPTER SUMMARY 511
REVIEW EXERCISES 514
CHAPTER TEST 519

CHAPTER SEVEN

Systems of Linear Equations and Inequalities

7.1	Defining Linear Systems and Solving by Graphing	524
7.2	Solving Linear Systems Using Elimination by Addition	535
7.3	Solving Linear Systems Using Elimination by Substitution	546
7.4	Applications Using Systems of Linear Equations	553
7.5	Systems of Linear Inequalities	564

CHAPTER SUMMARY 573
REVIEW EXERCISES 574
CHAPTER TEST 578

CHAPTER EIGHT

Roots and Radicals

8.1	Defining and Finding Roots	582
8.2	Simplifying Radicals	689
8.3	Products and Quotients of Radicals	597
8.4	Addition, Subtraction, and Mixed Operations with Radicals	504
8.5	Rationalizing the Denominator	613
8.6	Solving Equations with Radicals	623
8.7	Pythagorean Theorem	631

CHAPTER SUMMARY 641
REVIEW EXERCISES 643
CHAPTER TEST 648

CHAPTER NINE

Solving Quadratic
Equations

9.1	Solving Incomplete Quadratic Equations	651
9.2	Solving Quadratic Equations by Completing the Square	659
9.3	Solving Quadratic Equations by the Quadratic Formula	668
9.4	Solving Mixed Equations	675
9.5	Applications Involving Quadratic Equations	683

CHAPTER SUMMARY   692
REVIEW EXERCISES   694
CHAPTER TEST   697

APPENDIX

Table 1   631

Answers   633

Index   663

1.1   Number Systems, Order, and Absolute Value

1.2   A Review of Fractions

1.3   Variables, Exponents, and Order of Operations

1.4   Evaluating Geometric Formulas

1.5   Geometric Applications

1.6   Addition of Real Numbers

1.7   Subtraction of Real Numbers

1.8   Terms and Polynomials

1.9   Combining Like Terms and Polynomials

Addition and Subtraction of Real Numbers and Polynomials

CHAPTER OVERVIEW

We begin the chapter by reviewing number systems and introducing exponents and variables which may be your first exposure to algebra. Variables allow us to generalize the concepts of arithmetic and then expand on these concepts. We follow this with the order of operations on whole numbers that you will need in order to perform operations on the next topic, signed numbers. Operations on signed numbers are spread over Chapters 1 and 2. We next discuss addition and subtraction of signed numbers and then show how these operations are used to combine like terms and polynomials.

Much of arithmetic is the study of performing operations on numbers. Much of algebra is performing these same operations on polynomials.

As you work through the chapter, note how the topics later in the chapter depend upon topics that came earlier in the chapter. This is typical of the structure of mathematics and is the reason you must learn one topic before proceeding to the next. Failure to do this is one of the major reasons some students are not successful in mathematics.

SECTION 1.1

Number Systems, Order, and Absolute Value

OBJECTIVES

When you complete this section, you will be able to:

Ⓐ   Identify natural numbers, whole numbers, and integers.

Ⓑ   Draw the graphs of natural numbers, whole numbers, and integers.

Ⓒ   Determine order relations for integers.

Ⓓ   Find the negative (opposite) of an integer.

Ⓔ   Find the absolute value of an integer.

A **set** is any collection of objects. The objects that make up the set are called the **elements** of the set. In mathematics, the sets that we are most often concerned with are sets whose elements are numbers. Sets are usually named by using a capital letter and are indicated by enclosing its elements in braces. If the elements of a set are listed individually, then the set is said to be listed by **roster**. For example, {a, b, c} is read "The set whose elements are a, b, and c." and the elements are listed by roster. If all the elements of a set A are also elements of a set B, then set A is a **subset** of set B. For example, the set {1, 2} is a subset of the set {1, 2, 3, 4}.

Historically, numbers have been invented as they were needed. Our ancestors probably kept track of "how many" by using tally marks or stones, one for each object being "counted." No doubt the first numbers needed were those that indicated how many. We call these the **natural** or **counting** numbers and they are listed by roster below.

SET OF NATURAL NUMBERS

$$N = \{1, 2, 3, 4, 5, ...\}$$

The "..." is used to indicate that the set continues in the same manner with no last element.

If the set of natural numbers is expanded to include 0, we get another set of numbers called the whole numbers. Note that the natural numbers are a subset of the whole numbers since all the elements of the set of natural numbers are also elements of the set of whole numbers.

SET OF WHOLE NUMBERS

$$W = \{0, 1, 2, 3, 4, ...\}$$

If limited to just the whole numbers, we would have no way of representing 10° below zero or a loss of $150. Consequently, negative numbers were invented to help us make these representations. If $+10°$ represents 10° above 0° then $-10°$ represents 10° below 0°. Likewise, if $+45$ feet represents 45 feet above sea level, then -45 feet would represent 45 feet below sea level.

ADDITIONAL PRACTICE

a) If $+23$ represents going up 23 floors in an elevator, how would you represent going down 23 floors?

b) If $+\$250$ means receiving $250, what would $-\$250$ represent?

c) If $+33$ represents going 33 miles north, how would you represent going 33 miles south?

PRACTICE EXERCISES

Answer the following.

1) If $+3$ yards represents a gain of 3 yards, then _____ represents a loss of 3 yards.

2) If $-\$100$ represents a withdrawal of $100, then _____ represents a deposit of $100.

3) If $+12°$ represents a rise of 12° in temperature, then _____ represents a drop of 12° in temperature.

If additional practice is needed, do the Additional Practice Exercises in the margin.

If the set of whole numbers is expanded to include the negatives of the natural numbers, we get a new set of numbers called the integers. Note that both the natural numbers and the whole numbers are subsets of the integers.

SET OF INTEGERS

$$I = \{ \ldots, -3, -2, -1, 0, +1, +2, +3, \ldots \}$$

Positive and negative numbers are often referred to as signed numbers with 0 being neither positive nor negative. Another way of representing the integers is by using the number line. The number line is like a thermometer that is horizontal instead of vertical. In order to construct a number line, draw a horizontal line, choose an arbitrary point and label it 0. Choose a unit of length and mark off the line both to the left and right of 0 in terms of this unit. Label the units to the right of 0 with the positive integers and the units to the left with the negative integers. Positive numbers are usually written without the + sign. For example, +3 is written as 3 and +6 is written as 6. The resulting number line should look something like the one following.

In this manner any integer can be paired with a point on the number line. This procedure is called **graphing**.

- **Graphing the integers**

EXAMPLE 1
Graph the integers {−4, −2, 0, 3}.
We put a dot on the number line at the location of each of the integers −4, −2, 0, 3.

ADDITIONAL PRACTICE

Draw a number line and graph each of the following.

d) {−3, −1, 0, 2, 4}

e) {−4, −2, 5, 6}

PRACTICE EXERCISES

Graph the following sets of integers.

4) {−5, −2, 3, 6}

5) {−4, −3, −1, 1, 2, 3}

If more practice is needed, do the Additional Practice Exercises in the margin.

Stating definitions and general rules in mathematics can be very cumbersome in words and are most often stated using variables.

- **Graphing Rational Numbers**

DEFINITION OF VARIABLE

A variable is a symbol, usually a letter, that is used to represent a number.

Variables will be discussed in more detail in Section 1.3.

There are points on the number line that do not have integers assigned to them. Consequently, other types of numbers will be needed in order to assign a number to every point on the number line. Some of these numbers are **rational numbers** that are defined as follows using **set builder notation**. The braces are read as "the set", the | is read as "such that" and then a rule is given.

SET OF RATIONAL NUMBERS

$$\mathbf{Q} = \left\{ \frac{a}{b} \,\middle|\, a \text{ and } b \text{ are integers and } b \text{ does not equal } 0. \right\}$$

This is read "The set of numbers a divided by b such that a and b are integers and b does not equal 0." These numbers are called *ratio*nal because they can be written as a *ratio* of integers and the letter **Q** is used to identify the rationals because they are represented as a *q*uotient. Examples of rational numbers are $\frac{2}{3}$, $\frac{-4}{5}$, and $\frac{6}{1}$. A number like $\frac{\pi}{3}$ is not a rational number since π is not an integer. The natural numbers, whole numbers, and integers are all subsets of the rational numbers because each can be written as a ratio whose denominator is 1. For example, $3 = \frac{3}{1}$, and $-4 = \frac{-4}{1}$.

The manner that we graph rational numbers on the number line is similar to the way we graph integers. If the units of the number line are divided into equal parts, then we can locate points on the number line that have distances from 0 represented by rational numbers.

EXAMPLE 2
Graph the following rational numbers.

a) $\dfrac{3}{4}$

Solution:

The denominator, 4, means divide each unit into 4 equal parts. Since the number is positive go to the right of 0. The numerator, 3, means put the dot on the third division point.

b) $-\dfrac{5}{2}$

Solution:

The denominator, 2, means divide each unit into 2 equal parts. The negative sign means go to the left of 0. The numerator, 5, means put the dot on the fifth division point. This is the same as $-2\frac{1}{2}$.

Graph the following rational numbers on the number line.

6) $-\dfrac{5}{3}$

7) $\dfrac{13}{4}$

- **Representing rational numbers as division and converting rational numbers into decimals**

Any rational number represents division where the numerator is divided by the denominator. For example, $\frac{3}{4} = 3 \div 4$. In general, $\frac{a}{b} = a \div b$. This interpretation allows us to write a rational number as a decimal and to change an improper fraction to a mixed number.

EXAMPLE 3

Change the following rational numbers into decimals.

 a) $\dfrac{3}{4}$

Solution:

$\frac{3}{4}$ means $3 \div 4$. Since $4 > 3$, it is necessary to put a decimal and some zeros after the 3.

$$
\begin{array}{r}
0.75 \\
4\overline{)3.00} \\
-28 \\
\hline
20 \\
-20 \\
\hline
0
\end{array}
$$

Therefore, $\frac{3}{4} = 0.75$. Since the division comes to an end, this is a **terminating** decimal.

 b) $\dfrac{5}{11}$

Solution:

$\frac{5}{11}$ means $5 \div 11$. We will need to put a decimal after the 5 followed by as many 0s as needed.

$$
\begin{array}{r}
0.4545 \\
11\overline{)5.00000} \\
-44 \\
\hline
60 \\
-55 \\
\hline
50 \\
-44 \\
\hline
60 \\
-55 \\
\hline
5
\end{array}
$$

6)

$-\frac{5}{3}$

-3 -2 -1 0 1 2 3 4

7)

$\frac{13}{4}$

-3 -2 -1 0 1 2 3 4

The digits 4 and 5 will continue to repeat. This is indicated by using ... (called ellipsis) as in 0.454545... or by placing a bar over the block of repeating digits as in $0.\overline{45}$. Since a block of digits repeat indefinitely, this is a **repeating** decimal.

Such divisions are most easily done using a calculator. Using long division or a calculator we find that $\frac{12}{5} = 2.4$ (a terminating decimal) and $\frac{3}{7} = 0.\overline{428571}$ (a repeating decimal).

Note that when written as decimals, all rational numbers either terminate ($\frac{3}{4} = 0.75$) or repeat ($\frac{2}{3} = 0.666...$).

PRACTICE EXERCISES

Write each of the following as a decimal.

8) $\frac{3}{8}$ 9) $\frac{4}{11}$ 10) $\frac{13}{4}$ 11) $\frac{17}{6}$

Another type of number, the **irrational (Ir)** numbers, will be discussed in Chapter 8, but essentially a number is irrational if its decimal representation neither terminates nor repeats. Examples of irrational numbers are $\sqrt{2}$, $\sqrt[3]{4}$, and π. The set that is made up of the natural numbers, whole numbers, integers, rational numbers, and irrational numbers is the set of **real numbers** (R). The relationship between the different number systems is illustrated in the following diagram. The numbers are examples of numbers that are in that set that are not in the next smaller subset.

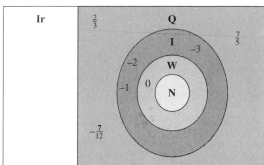

Real Numbers

The real numbers possess the property of being **ordered**. By this, we mean that when given two real numbers, it is possible to determine whether they are equal or which are the larger and smaller of the two.

To show the order relationships of numbers, we use the following symbols.

ORDER SYMBOLS

$a = b$ Read "a equals b." This means that the number represented by a has the same value as the number represented by b. For example, $5 = 5$.

$a \neq b$ Read "a is not equal to b." This means that the number represented by a does not have the same value as the number represented by b. For example, $-3 \neq 4$.

$a < b$ Read "a is less than b." This means that the number represented by a is to the left of the number represented by b on the number line. For example, $2 < 3$ because 2 is to the left of 3 on the number line.

$a \le b$ Read "a is less than or equal to b." This means that the number represented by a is either less than the number represented by b or the number represented by a is equal to the number represented by b. Examples: $3 \le 4$, $5 \le 5$.

$a > b$ Read "a is greater than b." This means that the number represented by a is to the right of the number represented by b on the number line. For example, $5 > 1$ because 5 is to the right of 1 on the number line.

$a \ge b$ Read "a is greater than or equal to b." This means that the number represented by a is either greater than the number represented by b or the number represented by a is equal to the number represented by b. Examples: $4 \ge 2$, $7 \ge 7$.

Remember, the smaller number is always to the left of the larger on the number line. Or equivalently, the larger is always to the right of the smaller.

• Determining order relations for real numbers

EXAMPLE 4

a) 0 is to the left of 3. Therefore, $0 < 3$ or equivalently $3 > 0$. We could also write $0 \le 3$ or $3 \ge 0$.

b) -2 is to the right of -4. Therefore, $-2 > -4$ or $-4 < -2$.

c) -5 is to the left of 2. Therefore, $-5 < 2$ or $2 > -5$.

d) $\frac{2}{3}$ is to the left of $\frac{7}{4}$. Therefore, $\frac{2}{3} < \frac{7}{4}$.

e) -4.2 is to the right of -6.1. Therefore, $-4.2 > -6.1$.

PRACTICE EXERCISES

Insert the proper symbol ($=$, $<$, or $>$) in order to make each of the following true.

12) a) -2 ____ 3   **b)** 4 ____ -5 **c)** -5 ____ -8

d) 0 ____ -6 **e)** 8 ____ 8   **f)** $-\frac{1}{2}$ ____ $\frac{3}{4}$

g) $\frac{3}{5}$ ____ $\frac{7}{6}$ **h)** -4.3 ____ -2.5

From the diagram below, we see that each number, except 0, can be paired with another number the same distance from 0, but on the opposite side. Each such number is called the opposite or negative of the other. In the next section we will also call them additive inverses.

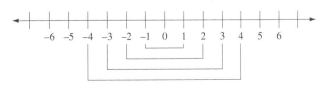

The negative or opposite of 3 is -3, the negative or opposite of -4 is 4, etc. The negative of a number is usually indicated by use of the negative sign. Consequently, the negative or opposite of -4 is written $-(-4)$ and equals 4. This leads us to an important characteristic of signed numbers.

OBSERVATION

$$-(-x) = x \text{ for all values of } x$$

Another important concept associated with signed numbers is that of absolute value. The absolute value of a number is the distance from 0 to the number. Absolute value is indicated by placing the number whose absolute value you wish to find between vertical bars. Since absolute value is a distance, it can never be negative.

EXAMPLE 5
Find the indicated absolute values.

a) $|4| = 4$ The distance from 0 to 4 is 4.

b) $|-4| = 4$ The distance from 0 to -4 is 4.

c) $|0| = 0$ The distance from 0 to 0 is 0.

d) $|-8| = 8$ The distance from 0 to -8 is 8.

e) $|8| = 8$ The distance from 0 to 8 is 8.

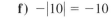 **f)** $-|10| = -10$ The "$-$" is outside the absolute value marks, so first find the absolute value of 10 and then take its negative, which is -10.

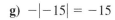 **g)** $-|-15| = -15$ The "$-$" is outside the absolute value marks, so first find the absolute value of -15 and then take its negative, which is -15.

h) $\left|-\dfrac{5}{6}\right| = \dfrac{5}{6}$ The distance from 0 to $-\frac{5}{6}$ is $\frac{5}{6}$.

PRACTICE EXERCISES
Find the indicated absolute values.

13) $|9| = $ __   **14)** $|-10| = $ __   **15)** $|13| = $ __

16) $|-5| = $ __   **17)** $-|18| = $ __   **18)** $-|-23| = $ __

19) $\left|-\dfrac{3}{4}\right| = $__   **20)** $|-1.5| = $ __

STUDY TIP #1
Developing a Positive Approach

A positive approach can help ensure your success in your math courses. Your approach should include the following:

1) Recognize that your degree of success depends upon you and you alone.

2) Make an all-out effort to do well in the course.

3) Work hard enough to do much better than just pass. Set high, but realistic goals for yourself.

4) Make a commitment to overcome *any* setbacks, personal or otherwise, and work hard until the very end of the course.

    Commitment and determination can go a long way toward guaranteeing that you will be successful. You **can** do it!

Answer each of the following using signed numbers.

1) If +8 represents 8 units to the right, how do you represent 8 units to the left?

2) If −80 feet represents a scuba diver descending 80 feet, how do you represent a scuba diver ascending 50 feet?

3) If +12 pounds represents gaining 12 pounds, what would −10 pounds represent?

4) If −$25 represents losing $25, what would +$55 represent?

5) If 250 miles represents 250 miles east, what would −250 miles represent?

6) If +5 represents gaining 5 points on an exam, what would −5 represent?

Draw a number line for each of the following and graph the given set of numbers.

7) $\{-5, -4, -1, 0\}$ **8)** $\{-3, -1, 2, 3\}$

9) $\{-10, -7, -2, 0, 4\}$ **10)** $\{-8, -3, 0, 2, 6\}$

11) $\left\{-\frac{11}{4}, -\frac{3}{2}, -\frac{2}{3}, \frac{3}{5}, 2\right\}$

12) $\left\{-\frac{17}{5}, -3, -\frac{5}{2}, -\frac{2}{3}, \frac{15}{4}\right\}$

Change the following rational numbers into decimals.

13) $\frac{7}{8}$ **14)** $\frac{5}{8}$

15) $\frac{2}{9}$ **16)** $\frac{5}{9}$

17) $\frac{5}{7}$ **18)** $\frac{7}{11}$

Given the set $\{-5, -\frac{4}{3}, 0, \sqrt{2}, 2.3, 4.\overline{12}, 6\}$. List the elements that belong to the following sets.

19) Rational Numbers   **20)** Irrational

21) Whole Numbers   **22)** Integers

23) Rationals but not Integers

24) Integers but not Whole Numbers

Answer the following as always true, sometimes true, or never true.

25) A rational number is an integer.

26) A whole number is a rational number.

27) An integer is a rational number.

28) An integer is a whole number.

29) A natural number is an integer.

30) A whole number is a natural number.

31) A whole number is an integer.

32) An integer is a natural number.

33) A natural number is a whole number.

Find the negative (opposite) of each of the following.

34) 19   **35)** 17   **36)** −12   **37)** −15

38) $\frac{3}{4}$ **39)** $\frac{2}{3}$ **40)** $-\frac{5}{6}$ **41)** $-\frac{3}{7}$

42) 2.3   **43)** −3.4

Simplify the following involving absolute values.

44) $|-2|$ **45)** $|-5|$ **46)** $|7|$ **47)** $|9|$

48) $-|3|$ **49)** $-|11|$ **50)** $\left|\frac{2}{3}\right|$ **51)** $\left|\frac{4}{7}\right|$

52) $\left|-\frac{4}{9}\right|$ **53)** $\left|-\frac{7}{10}\right|$ **54)** $-|-4|$ **55)** $-|-9|$

56) $|5-2|$ **57)** $|7-1|$

Insert the proper symbol (=, <, or >) in order to make each of the following true.

58) 8 ___ 10   **59)** 29 ___ 12   **60)** 13 ___ −3

61) −10 ___ 8   **62)** −8 ___ −4   **63)** −3 ___ −7

64) $\frac{2}{3}$ ___ $\frac{7}{3}$ **65)** $\frac{2}{9}$ ___ $\frac{5}{9}$ **66)** $\frac{1}{2}$ ___ $-\frac{3}{4}$

67) $-\frac{2}{3}$ ___ $\frac{5}{6}$ **68)** $-\frac{7}{4}$ ___ $-\frac{1}{4}$ **69)** $-\frac{5}{3}$ ___ $-\frac{2}{3}$

70) 2.1 ___ 4.5   **71)** 4.7 ___ 1.6

72) 2.5 ___ −1.6   **73)** 3.1 ___ −2.2

74) −6.3 ___ −4.2   **75)** −1.5 ___ −3.7

76) $|6|$ ___ 6   **77)** $|-5|$ ___ −5

78) $-|-4|$ ___ 4   **79)** −7 ___ $-|-7|$

80) $|10|$ ___ $|-10|$ **81)** $-|-10|$ ___ $-(-10)$

82) $|24-18|$ ___ $|8-1|$ **83)** $|32-15|$ ___ $|13-5|$

84) $-(-5)$ ___ 5 **85)** 8 ___ $-(-8)$

86) -10 ___ $-(-10)$

WRITING EXERCISES

87) Give three examples of numbers that are integers but that are not whole numbers. Explain why they are not whole numbers.

88) Why is the absolute value of a number always non-negative?

89) Why is any positive number always greater than any negative number?

90) Is saying a number is positive the same as saying the number is non-negative? Why or why not?

CRITICAL THINKING EXERCISES

91) Does $-(-x)$ always represent a positive number? Why or why not?

92) If the absolute value of a number is the same as the number, what kind of number must it be?

93) If x is a negative number, what type of number does $-x$ represent?

94) If $x > y$ is $|x| > |y|$ always true, sometimes true, or never true? Explain.

95) Explain why there are two values of x that make $|x| = 6$ a true statement.

GROUP PROJECT

96) In this section, examples of the occurrence of negative numbers were given, including the withdrawal of money, a drop in temperature, and so on. Give at least three more examples of where negative numbers occur other than those given in the text.

SECTION 1.2

A Review of Fractions

OBJECTIVES

When you complete this section, you will be able to:

A Write a composite number in terms of prime factors.
B Reduce fractions to lowest terms.
C Multiply two or more fractions and/or mixed numbers.
D Divide two fractions or mixed numbers.
E Add and/or subtract fractions and/or mixed numbers.
F Solve real world problems that involve fractions or mixed numbers.

INTRODUCTION

In this section we give a brief review of fractions. We discuss how to reduce, multiply, divide, add, and subtract fractions. A thorough discussion of fractions will occur in Chapter 5.

Before discussing fractions, we need some concepts that will assist us in working with fractions. If two or more numbers are multiplied, the result is called a **product** and the numbers being multiplied are called **factors**. For example, 18 can be written as $3 \cdot 6$ so 18 is the product of 3 and 6, thus 3 and 6 are factors of 18. The expression $3 \cdot 6$ is called a **factorization** of 18. Other factorizations of 18 are $1 \cdot 18$ and $2 \cdot 9$. Consequently, other factors of 18 are 1, 2, 9, and 18. Note than any number is **divisible** by each of its factors. This means that if a number is divided by one of its factors, the remainder will be 0.

A natural number is **prime** if it has exactly two factors that are the number itself and 1. Consequently, a prime number is divisible only by 1 and itself. The first few prime numbers are:

$$2, 3, 5, 7, 11, 13, 17, 19, 23, 29, 31, 37, 41, \text{etc.}$$

If a natural number, other than 1, is not prime, then it is called **composite**. Note that 1 is neither prime nor composite. The first few composite numbers are:

$$4, 6, 8, 9, 10, 12, 14, 15, 16, 18, 20, 21, 22, 24, 25, \text{etc.}$$

EXAMPLE 1

Determine if the following numbers are prime or composite.

a) 45

Solution:

45 is divisible by 5 and 9 in addition to 1 and 45, so 45 is composite.

b) 53

Solution:

If we try to divide 53 by consecutive prime numbers, we will find that it is not divisible by any of them until we get to 53. Consequently, the only factors of 53 are 1 and 53, so 53 is prime.

c) 1,936

Solution:

We can divide 1,936 by 2, so 1,936 is composite.

PRACTICE EXERCISES

Determine if the following numbers are prime or composite.

1) 63   **2)** 97   **3)** 4,798

In order to work with fractions, it is necessary to write each of the numerators and denominators as the product of its prime factors. To write a number in terms of prime factors, we use the following procedure.

• **Writing composite numbers in terms of prime factors**

PROCEDURE FOR FINDING THE PRIME FACTORIZATION

1) Divide the composite number by the smallest prime number by which it is divisible.

2) Divide the resulting quotient by the smallest prime number by which it is divisible.

3) Continue this process until the quotient is prime.

4) The prime factorization is the product of all the divisors and the last quotient.

To assist us, we repeat the first few prime numbers: {2, 3, 5, 7, 11, 13, 17, 19, 23, 29, 31, ...}. We illustrate the procedure with some examples.

EXAMPLE 2

Find the prime factorization for each of the following.

a) 24   2 is the smallest prime number which divides evenly into 24.

$$\frac{12}{2\overline{)24}}$$ 12 is also divisible by 2.

$$\frac{6}{2\overline{)12}}$$ 6 is also divisible by 2.
$$2\overline{)24}$$

$$\begin{array}{r} 3 \\ 2\overline{)6} \\ 2\overline{)12} \\ 2\overline{)24} \end{array}$$ 3 is prime.

Therefore, $24 = 2 \cdot 2 \cdot 2 \cdot 3$.

b) 105   3 is the smallest prime which divides evenly into 105.

$$\begin{array}{r} 35 \\ 3\overline{)105} \end{array}$$ 5 is the smallest prime which divides evenly into 35.

$$\begin{array}{r} 7 \\ 5\overline{)35} \\ 3\overline{)105} \end{array}$$ 7 is prime.

Therefore, $105 = 3 \cdot 5 \cdot 7$ is the prime factorization of 105.

c) 89   89 is not divisible by 2, 3, 5, 7, 11, 13, etc. We suspect that 89 might be prime but how do we know? Do we have to try dividing 89 by consecutive prime numbers until we get to 89? Fortunately, we do not. If we begin by trying to divide by the smallest prime number 2, then 3, then 5 and so on, then **we can stop looking for prime factors when we arrive at a prime number whose square is greater than the number we are trying to factor**. In this case, there are no prime factors smaller than 11 and 11^2 is greater than 89, so there will be no prime factors greater than 11. Therefore, we conclude that 89 is prime.

PRACTICE EXERCISES

Write each of the following in terms of prime factors.

4) 84   **5)** 90   **6)** 73

A **fraction** is a number in the form of $\frac{a}{b}$ where a and b represent any numbers with $b \neq 0$. The number a is called the numerator and b the denominator. The rational numbers are special fractions whose numerators and denominators are integers. If the numerator is smaller than the denominator the fraction is called a **proper** fraction and if the numerator is greater than or equal to the denominator the fraction is called an **improper** fraction. A number like $2\frac{3}{4}$ is called a **mixed number**. To change a mixed number to an improper fraction, multiply the denominator of the fractional part by whole number part and then add the numerator of the fractional part. This result is the numerator of the improper fraction and the denominator is the denominator of the fractional part. For example, $2\frac{3}{4} = \frac{4 \cdot 2 + 3}{4} = \frac{11}{4}$. To change an improper fraction to a mixed number, divide the denominator into the numerator. The number of times that it goes into the numerator is the whole part and the remainder is put over the denominator. For example, $\frac{13}{5} = 2\frac{3}{5}$.

A fraction is reduced to lowest terms if the numerator and denominator have no common factors other than 1. Although you probably already know one method of reducing fractions to lowest terms, the following procedure is recommended because it closely resembles the procedure used to reduce algebraic fractions in Chapter 5.

• **Reducing fractions to lowest terms**

REDUCING FRACTIONS TO LOWEST TERMS

1) Write the numerator and denominator in terms of prime factors.

2) Divide the numerator and denominator by the factor(s) that are common to both.

3) Multiply the remaining factors in the numerator and denominator.

EXAMPLE 3

Reduce the following to lowest terms.

a) $\dfrac{10}{25} =$ Write the numerator and denominator as the product of prime factors.

$\dfrac{2 \cdot 5}{5 \cdot 5} =$ Divide by the factors common to both.

$\dfrac{2 \cdot \cancel{5}}{5 \cdot \cancel{5}} =$ $\dfrac{5}{5} = 1.$

$\dfrac{2}{5}$ Therefore, $\dfrac{10}{25} = \dfrac{2}{5}$ reduced to lowest terms.

b) $\dfrac{24}{36} =$ Write the numerator and denominator as the product of prime factors.

$\dfrac{2 \cdot 2 \cdot 2 \cdot 3}{2 \cdot 2 \cdot 3 \cdot 3} =$ Divide by the factors common to both.

$\dfrac{\cancel{2} \cdot \cancel{2} \cdot 2 \cdot \cancel{3}}{\cancel{2} \cdot \cancel{2} \cdot 3 \cdot \cancel{3}} =$ $\dfrac{2}{2}$ and $\dfrac{3}{3} = 1.$

$\dfrac{2}{3}$ Therefore, $\dfrac{24}{36} = \dfrac{2}{3}.$

c) $\dfrac{21}{40} =$ Write the numerator and denominator as the product of prime factors.

$\dfrac{3 \cdot 7}{2 \cdot 2 \cdot 2 \cdot 5} =$ There are no factors common to the numerator and denominator, so $\frac{21}{40}$ is already reduced to lowest terms.

PRACTICE EXERCISES

Reduce the following to lowest terms.

7) $\dfrac{18}{45}$ 8) $\dfrac{60}{126}$ 9) $\dfrac{33}{56}$

• **Multiplying fractions**

When multiplying fractions, we multiply numerator times numerator and denominator times denominator. Following is the rule for multiplication of fractions using variables. Remember, multiplication is often indicated by using the times sign, $\times$, the raised dot, $\cdot$, or parentheses, ( ). Recall, the result of multiplication is called a product.

MULTIPLICATION OF FRACTIONS

For all numbers a, b, c, and d with b and $d \neq 0$, $\dfrac{a}{b} \cdot \dfrac{c}{d} = \dfrac{a \cdot c}{b \cdot d}$. In words, to multiply two fractions find the product of the numerators divided by the product of the denominators.

Remember, when performing any operation on fractions be sure the answer is reduced to lowest terms.

EXAMPLE 4
Find the following products.

a) $\dfrac{3}{5} \cdot \dfrac{2}{7} =$ Put the product of the numerators over the product of the denominators.

$\dfrac{3 \cdot 2}{5 \cdot 7} =$ Multiply.

$\dfrac{6}{35}$ Product.

b) $\dfrac{4}{9} \cdot \dfrac{3}{10} =$ Put the product of the numerators over the product of the denominators.

$\dfrac{4 \cdot 3}{9 \cdot 10} =$ Write each factor of the numerator and denominator as the product of prime factors.

$\dfrac{2 \cdot 2 \cdot 3}{3 \cdot 3 \cdot 2 \cdot 5} =$ Divide by the factors common to both.

$\dfrac{\cancel{2} \cdot 2 \cdot \cancel{3}}{\cancel{3} \cdot 3 \cdot \cancel{2} \cdot 5} =$ Multiply the remaining factors.

$\dfrac{2}{15}$ Product.

c) $8 \cdot \dfrac{1}{2} =$ Think of 8 as $\dfrac{8}{1}$.

$\dfrac{8}{1} \cdot \dfrac{1}{2} =$ Put the product of the numerators over the product of the denominators.

$\dfrac{8 \cdot 1}{1 \cdot 2} =$ Multiply.

$\dfrac{8}{2} =$ $8 \div 2 = 4$.

4 Product.

d) $2\dfrac{2}{7} \cdot 3\dfrac{1}{2} =$ Change the mixed numbers into improper fractions.

$\dfrac{7 \cdot 2 + 2}{7} \cdot \dfrac{2 \cdot 3 + 1}{2} =$ Simplify.

$\dfrac{16}{7} \cdot \dfrac{7}{2} =$ Put the product of the numerators over the product of the denominators.

$\dfrac{16 \cdot 7}{7 \cdot 2} =$ Write each factor as the product of prime factors.

$$\frac{2 \cdot 2 \cdot 2 \cdot 2 \cdot 7}{7 \cdot 2} =$$ Divide by the factors common to both.

$$\frac{\cancel{2} \cdot 2 \cdot 2 \cdot 2 \cdot \cancel{7}}{\cancel{7} \cdot \cancel{2}} =$$ Multiply the remaining factors.

8 Product.

PRACTICE EXERCISES

Find the following products.

10) $\dfrac{2}{3} \cdot \dfrac{4}{9}$ **11)** $\dfrac{9}{25} \cdot \dfrac{5}{6}$ **12)** $9 \cdot \dfrac{1}{3}$ **13)** $3\dfrac{1}{3} \cdot 6\dfrac{3}{5}$

Before we discuss division of fractions, we need to discuss the **multiplicative inverse**, also called the **reciprocal**. Two numbers are multiplicative inverses if their product is 1. To find the multiplicative inverse of a number written as a fraction, we interchange the numerator and denominator. This procedure is sometimes referred to as "inverting the fraction." Consider the following.

Number	Multiplicative Inverse	Product
$\dfrac{2}{3}$	$\dfrac{3}{2}$	$\dfrac{2}{3} \cdot \dfrac{3}{2} = \dfrac{6}{6} = 1$
$\dfrac{4}{7}$	$\dfrac{7}{4}$	$\dfrac{4}{7} \cdot \dfrac{7}{4} = \dfrac{28}{28} = 1$
6	$\dfrac{1}{6}$	$6 \cdot \dfrac{1}{6} = \dfrac{6}{1} \cdot \dfrac{1}{6} = \dfrac{6}{6} = 1$

• **Division by fractions**

Recall to divide by a fraction, we "invert the fraction on the right and multiply." For example, $\dfrac{2}{3} \div \dfrac{5}{4} = \dfrac{2}{3} \cdot \dfrac{4}{5}$. The result of division is the **quotient**. This property is stated more formally as follows.

DIVISION OF FRACTIONS

For all numbers *a, b, c,* and *d* with *b, c,* and $d \neq 0$, $\dfrac{a}{b} \div \dfrac{c}{d} = \dfrac{a}{b} \cdot \dfrac{d}{c}$. In words, to divide by a number, multiply by its multiplicative inverse (reciprocal).

EXAMPLE 5

Find the following quotients.

a) $\dfrac{2}{3} \div \dfrac{3}{5} =$ Multiply by the reciprocal of $\dfrac{3}{5}$.

$\dfrac{2}{3} \cdot \dfrac{5}{3} =$ Put the product of the numerators over the product of the denominators.

$\dfrac{2 \cdot 5}{3 \cdot 3} =$ Multiply.

$\dfrac{10}{9}$ Quotient.

b) $\dfrac{4}{7} \div \dfrac{20}{21} =$ Multiply by the reciprocal of $\dfrac{20}{21}$.

$\dfrac{4}{7} \cdot \dfrac{21}{20} =$ Put the product of the numerators over the product of the denominators.

$\dfrac{4 \cdot 21}{7 \cdot 20} =$ Write each factor as products of prime factors.

$\dfrac{2 \cdot 2 \cdot 3 \cdot 7}{7 \cdot 2 \cdot 2 \cdot 5} =$ Divide by the factors common to the numerator and denominator.

$\dfrac{\cancel{2} \cdot \cancel{2} \cdot 3 \cdot \cancel{7}}{\cancel{7} \cdot \cancel{2} \cdot \cancel{2} \cdot 5} =$ Multiply the remaining factors.

$\dfrac{3}{5}$ Quotient.

c) $\dfrac{3}{7} \div 8 =$ $8 = \dfrac{8}{1}$.

$\dfrac{3}{7} \div \dfrac{8}{1} =$ Multiply by the reciprocal of $\dfrac{8}{1}$.

$\dfrac{3}{7} \cdot \dfrac{1}{8} =$ Put the product of the numerators over the product of the denominators.

$\dfrac{3 \cdot 1}{7 \cdot 8} =$ Multiply.

$\dfrac{3}{56}$ Quotient.

d) $3\dfrac{1}{3} \div 2\dfrac{1}{7} =$ Change the mixed numbers to improper fractions.

$\dfrac{10}{3} \div \dfrac{15}{7} =$ Multiply by the reciprocal of $\dfrac{15}{7}$.

$\dfrac{10}{3} \cdot \dfrac{7}{15} =$ Put the product of the numerators over the product of the denominators.

$\dfrac{10 \cdot 7}{3 \cdot 15} =$ Write each factor as products of prime factors.

$\dfrac{2 \cdot 5 \cdot 7}{3 \cdot 3 \cdot 5} =$ Divide by the factors common to both.

$\dfrac{2 \cdot \cancel{5} \cdot 7}{3 \cdot 3 \cdot \cancel{5}} =$ Multiply the remaining factors.

$\dfrac{14}{9}$ or $1\dfrac{5}{9}$ Quotient.

PRACTICE EXERCISES

Find the following quotients.

14) $\dfrac{3}{4} \div \dfrac{2}{7}$ **15)** $\dfrac{14}{15} \div \dfrac{21}{20}$ **16)** $\dfrac{2}{9} \div 5$ **17)** $5\dfrac{1}{3} \div 2\dfrac{2}{5}$

- **Addition of fractions with the same denominators**

Shaded regions may be used to illustrate addition of fractions with the same denominators. Suppose $\frac{3}{8}$ of a region is shaded.

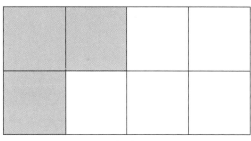

1 unit

Now shade another $\frac{4}{8}$ of the same region.

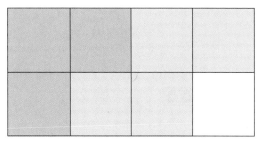

1 unit

Observe that $\frac{7}{8}$ of the region is now shaded. In other words, $\frac{3}{8} + \frac{4}{8} = \frac{3+4}{8} = \frac{7}{8}$. Note that the denominators of the fractions are the same and only the numerators are added. This leads to the following procedure for adding fractions with like denominators. The result of addition is the **sum**.

ADDITION OF FRACTIONS WITH THE SAME DENOMINATORS

For all numbers a, b, and c, with $c \neq 0$, $\frac{a}{c} + \frac{b}{c} = \frac{a+b}{c}$. In words, to add two or more fractions with the same denominators, add the numerators and place the sum over the same denominator.

It can easily be shown that a similar procedure is used to subtract fractions. The result of subtraction is the **difference**. The procedure is given below.

SUBTRACTION OF FRACTIONS WITH THE SAME DENOMINATORS

For all numbers a, b, and c, with $c \neq 0$, $\frac{a}{c} - \frac{b}{c} = \frac{a-b}{c}$. In words, to subtract two or more fractions with the same denominators, subtract the numerators and place the difference over the same denominator.

EXAMPLE 6

Find the following sums or differences.

a) $\dfrac{1}{5} + \dfrac{3}{5} =$ The denominators are the same, so add the numerators and place the sum over the same denominator.

$\dfrac{1 + 3}{5} =$ Add 1 and 3.

$\dfrac{4}{5}$ Sum.

b) $\dfrac{7}{16} - \dfrac{5}{16} =$ The denominators are the same, so subtract the numerators and place the difference over the same denominator.

$\dfrac{7-5}{16} =$ Subtract 7 and 5.

$\dfrac{2}{16} =$ Reduce to lowest terms.

$\dfrac{1}{8}$ Difference.

c) $\dfrac{2}{13} + \dfrac{4}{13} + \dfrac{1}{13} =$ The denominators are the same, so add the numerators and place the sum over the same denominator.

$\dfrac{2 + 4 + 1}{13} =$ Add 2, 4, and 1.

$\dfrac{7}{13}$ Sum.

d) $\dfrac{3}{11} + \dfrac{5}{11} - \dfrac{4}{11} =$ The denominators are the same, so add and subtract the numerators and place the sum and/or difference over the same denominator.

$\dfrac{3 + 5 - 4}{11} =$ Add 3 and 5 then subtract 4.

$\dfrac{4}{11}$ Answer.

PRACTICE EXERCISES

Find the following sums.

18) $\dfrac{4}{11} + \dfrac{2}{11}$ **19)** $\dfrac{11}{18} - \dfrac{5}{18}$

20) $\dfrac{5}{17} + \dfrac{3}{17} + \dfrac{6}{17}$ **21)** $\dfrac{3}{10} + \dfrac{5}{10} - \dfrac{1}{10}$

Suppose we want to add two fractions that do not have a common denominator like $\frac{1}{3}$ and $\frac{1}{4}$ using shaded regions. We would first shade $\frac{1}{3}$ of the region.

Now we shade $\frac{1}{4}$ of the region.

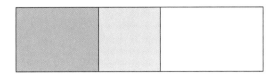

What part of the region has been shaded? Looking at the figure it is difficult to tell. Let's divide the figure into 12 equal regions.

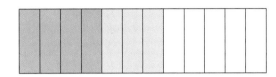

From the figure we see that $\frac{1}{3}$ became $\frac{4}{12}$ and $\frac{1}{4}$ became $\frac{3}{12}$ so $\frac{1}{3} + \frac{1}{4} = \frac{3}{12} + \frac{4}{12} = \frac{7}{12}$ of the region has been shaded. In other words, $\frac{1}{3} + \frac{1}{4} = \frac{4}{12} + \frac{3}{12} = \frac{7}{12}$. The number 12 is called the **least common denominator** of the other denominators 3 and 4. In general the least common denominator is the smallest number that is divisible by all the denominators. A method of finding the least common denominators using prime factorization is presented in Chapter 5. The denominators of the fractions in this section are small enough that this procedure is not needed.

If you have difficulty finding the least common denominator, try the following. Write the multiples of each denominator until you find one or more numbers that are common multiples of both denominators. The smallest of these common multiples is the least common denominator. For example, for $\frac{1}{3}$ and $\frac{1}{4}$ we write the multiples of 3 which are 3, 6, 9, 12, 15, 18, 21, 24,... and the multiples of 4 which are 4, 8, 12, 16, 24, 28,... . Note that 12 and 24 are multiples of both 3 and 4. Since 12 is smaller than 24, 12 is the least common denominator for $\frac{1}{3}$ and $\frac{1}{4}$. Following is the procedure for adding fractions with unlike denominators.

ADDING FRACTIONS WITH UNLIKE DENOMINATORS

To add two or more fractions with unlike denominators:

1) Determine the least common denominator.

2) Rewrite each fraction as an equivalent fraction with the least common denominator as its denominator.

3) Add the fractions using the procedure for adding fractions with like denominators.

In order to rewrite a fraction as an equivalent fraction with the least common denominator as its denominator, it is necessary to multiply the fraction by 1 in the form of an improper fraction with equal numerator and denominator. For example, to rewrite $\frac{4}{7}$ as a fraction with a denominator of 21, multiply $\frac{4}{7}$ by $\frac{3}{3}$ to get $\frac{4}{7} \cdot \frac{3}{3} = \frac{12}{21}$. To rewrite $\frac{2}{3}$ as an equivalent fraction with denominator of 12, multiply $\frac{2}{3}$ by $\frac{4}{4}$ to get $\frac{2}{3} \cdot \frac{4}{4} = \frac{8}{12}$.

EXAMPLE 7

Find the following sums and/or differences.

 a) $\dfrac{5}{6} + \dfrac{7}{8} =$

The least common denominator is 24, so rewrite each fraction as an equivalent fraction with a denominator of 24. Since $6 \cdot 4 = 24$, multiply $\frac{5}{6}$ by $\frac{4}{4}$. Since $8 \cdot 3 = 24$, multiply $\frac{7}{8}$ by $\frac{3}{3}$.

$\dfrac{5}{6} \cdot \dfrac{4}{4} + \dfrac{7}{8} \cdot \dfrac{3}{3} =$ Multiply.

$\dfrac{20}{24} + \dfrac{21}{24} =$ Add the fractions.

$\dfrac{41}{24}$ Sum.

b) $\dfrac{7}{12} + \dfrac{3}{4} =$

The least common denominator is 12, so rewrite each fraction as an equivalent fraction with a denominator of 12. $\frac{7}{12}$ already has 12 as its denominator. Since $4 \cdot 3 = 12$, multiply $\frac{3}{4}$ by $\frac{3}{3}$.

$\dfrac{7}{12} + \dfrac{3}{4} \cdot \dfrac{3}{3} =$ Multiply.

$\dfrac{7}{12} + \dfrac{9}{12} =$ Add the fractions.

$\dfrac{16}{12} =$ Reduce to lowest terms.

$\dfrac{4}{3}$ Sum.

 c) $6 - \dfrac{5}{12} =$

Think of 6 as $\frac{6}{1}$, so the least common denominator is 12. Consequently, multiply $\frac{6}{1}$ by $\frac{12}{12}$ and leave $\frac{5}{12}$ as is.

$\dfrac{6}{1} \cdot \dfrac{12}{12} - \dfrac{5}{12} =$ Multiply.

$\dfrac{72}{12} - \dfrac{5}{12} =$ Subtract the fractions.

$\dfrac{67}{12}$ Difference.

d) $6\dfrac{1}{8} - 4\dfrac{5}{6} =$

Change each mixed number into an improper fraction.

$\dfrac{49}{8} - \dfrac{29}{6} =$

The least common denominator is 24, so multiply $\frac{49}{8}$ by $\frac{3}{3}$ and multiply $\frac{29}{6}$ by $\frac{4}{4}$.

$\dfrac{49}{8} \cdot \dfrac{3}{3} - \dfrac{29}{6} \cdot \dfrac{4}{4} =$ Multiply the fractions.

$\dfrac{147}{24} - \dfrac{116}{24} =$ Subtract the fractions.

$\dfrac{31}{24}$ or $1\dfrac{7}{24}$ Difference.

Find the following sums or differences.

22) $\dfrac{5}{12} - \dfrac{7}{18}$ **23)** $\dfrac{11}{18} - \dfrac{1}{9}$ **24)** $\dfrac{5}{6} + 3$ **25)** $6\dfrac{3}{8} - 4\dfrac{1}{3}$

Fractions frequently occur in everyday situations.

EXAMPLE 8
Solve the following.

a) Each fill-up of a six-gallon tank of gasoline for an outboard motor requires $\frac{1}{2}$ quart of oil. If a container has 5 quarts of oil, there is enough oil in the container for how many fill-ups of the six-gallon tank?

Solution:

In order to find the number of fill-ups, we divide the number of quarts of oil in the container by the number of quarts needed per fill-up. Therefore,

$$5 \div \frac{1}{2} =$$ Multiply by the reciprocal of $\frac{1}{2}$.

$$5 \cdot 2 =$$ Multiply.

$$10$$ Therefore, there is enough oil for 10 fill-ups.

b) A recipe for a cake calls for $\frac{2}{3}$ of a cup of chopped nuts. Find the number of cups of nuts needed for 9 such cakes.

Solution:

In order to find the number of cups of nuts needed for 9 cakes, we multiply the number of cups per cake by the number of cakes. Hence,

$$\frac{2}{3} \cdot 9 =$$ Think of 9 as $\frac{9}{1}$.

$$\frac{2}{3} \cdot \frac{9}{1} =$$ Multiply the fractions.

$$\frac{18}{3} =$$ Divide.

$$6$$ Therefore, 6 cups of nuts are required for 9 cakes.

c) From Jenny's house it is $1\frac{1}{4}$ miles to the shopping center and $2\frac{2}{3}$ miles beyond the shopping center to the post office. How far is it from her house to the post office?

Solution:

In order to find the distance from Jenny's house to the post office, we add the distance from her house to the shopping center and the distance from the shopping center to the post office.

$$1\frac{1}{4} + 2\frac{2}{3} =$$ Change the mixed numbers to improper fractions.

$$\frac{5}{4} + \frac{8}{3} =$$ Change each fraction into an equivalent fraction whose denominator is the least common denominator of 12.

$$\frac{5}{4} \cdot \frac{3}{3} + \frac{8}{3} \cdot \frac{4}{4} = \qquad \text{Multiply the fractions.}$$

$$\frac{15}{12} + \frac{32}{12} = \qquad \text{Add the fractions.}$$

$$\frac{47}{12} \text{ or } 3\frac{11}{12} \qquad \text{Therefore, the distance from Jenny's house to the post office is } 3\frac{11}{12} \text{ miles.}$$

PRACTICE EXERCISES

Solve the following.

26) A baby bottle holds $\frac{1}{8}$ of a quart of formula. If a case of formula contains 16 quarts, how many times can the bottle be filled from a case of formula?

27) A pharmacist fills a prescription for a medication for high blood pressure. She uses $\frac{2}{3}$ of the pills in a bottle that is $\frac{2}{5}$ full. What part of a full bottle did she use?

28) On Monday, Sandy rode her bike $3\frac{3}{4}$ miles and on Tuesday she rode $2\frac{1}{5}$ miles. What is the total distance she rode for the two days?

STUDY TIP #2

Motivation—The Key to Success

The attitude with which you approach a course is the single most important factor in determining your success in that course. Begin with a positive attitude and a belief that you can be successful. Mathematics can be learned and understood. We have done our very best to take the mysteries out of mathematics. But for you to reach your goals, you need to make a commitment right now to attend class, to study as we suggest and to give this course your very best effort. With the proper determination, you can master this course.

E X E R C I S E   S E T   1.2

Classify each of the following as prime or composite.

1) 17   **2)** 23   **3)** 28   **4)** 35

5) 51   **6)** 69   **7)** 71   **8)** 83

Find the prime factorization of each of the following.

9) 28   **10)** 20   **11)** 45   **12)** 63

13) 100   **14)** 225   **15)** 154   **16)** 232

17) 180   **18)** 126   **19)** 315   **20)** 525

Reduce the following to lowest terms.

21) $\frac{12}{18}$ **22)** $\frac{24}{36}$ **23)** $\frac{30}{45}$ **24)** $\frac{63}{70}$

25) $\frac{42}{28}$ **26)** $\frac{72}{64}$

Find the following products.

27) $\frac{2}{5} \cdot \frac{4}{9}$ **28)** $\frac{3}{7} \cdot \frac{1}{4}$ **29)** $\frac{4}{7} \cdot \frac{3}{5}$

30) $\frac{5}{8} \cdot \frac{3}{7}$ **31)** $\frac{14}{15} \cdot \frac{33}{35}$ **32)** $\frac{5}{12} \cdot \frac{18}{25}$

33) $\frac{35}{39} \cdot \frac{26}{42}$ **34)** $\frac{14}{30} \cdot \frac{45}{44}$ **35)** $6 \cdot \frac{2}{3}$

36) $8 \cdot \frac{3}{4}$ **37)** $\frac{3}{10} \cdot 15$ **38)** $\frac{5}{12} \cdot 18$

39) $4\frac{4}{5} \cdot \frac{3}{4}$ **40)** $\frac{2}{5} \cdot 2\frac{2}{3}$ **41)** $2\frac{2}{3} \cdot 3\frac{3}{5}$

42) $2\frac{1}{2} \cdot 1\frac{9}{21}$

Find the following quotients.

43) $\dfrac{3}{5} \div \dfrac{7}{4}$ 44) $\dfrac{4}{9} \div \dfrac{3}{2}$ 45) $\dfrac{7}{11} \div \dfrac{3}{8}$

46) $\dfrac{7}{15} \div \dfrac{3}{5}$ 47) $\dfrac{3}{7} \div \dfrac{3}{5}$ 48) $\dfrac{4}{5} \div \dfrac{4}{11}$

49) $\dfrac{3}{7} \div \dfrac{9}{21}$ 50) $\dfrac{14}{35} \div \dfrac{3}{28}$ 51) $\dfrac{2}{5} \div 3$

52) $\dfrac{5}{8} \div 6$ 53) $7 \div \dfrac{3}{4}$ 54) $8 \div \dfrac{3}{7}$

55) $\dfrac{26}{8} \div 13$ 56) $\dfrac{42}{30} \div 14$ 57) $\dfrac{5}{8} \div 3\dfrac{3}{4}$

58) $\dfrac{10}{7} \div 2\dfrac{1}{2}$ 59) $3\dfrac{1}{2} \div 3\dfrac{3}{8}$ 60) $4\dfrac{2}{3} \div 1\dfrac{2}{5}$

Find the following sums and differences.

61) $\dfrac{6}{17} + \dfrac{9}{17}$ 62) $\dfrac{3}{19} + \dfrac{11}{19}$ 63) $\dfrac{9}{13} - \dfrac{4}{13}$

64) $\dfrac{9}{11} - \dfrac{5}{11}$ 65) $\dfrac{9}{16} + \dfrac{3}{16}$ 66) $\dfrac{5}{24} + \dfrac{7}{24}$

67) $\dfrac{2}{7} + \dfrac{1}{7} + \dfrac{3}{7}$ 68) $\dfrac{4}{23} + \dfrac{8}{23} + \dfrac{3}{23}$

69) $\dfrac{7}{13} + \dfrac{4}{13} - \dfrac{2}{13}$ 70) $\dfrac{11}{15} + \dfrac{4}{15} - \dfrac{7}{15}$

71) $\dfrac{7}{18} + \dfrac{13}{18} + \dfrac{1}{18}$ 72) $\dfrac{8}{15} + \dfrac{2}{15} + \dfrac{11}{15}$

73) $3\dfrac{9}{10} + 6\dfrac{3}{10}$ 74) $8\dfrac{2}{7} + 4\dfrac{3}{7}$

75) $12\dfrac{5}{8} - 8\dfrac{3}{8}$ 76) $10\dfrac{7}{9} - 5\dfrac{4}{9}$

77) $\dfrac{1}{4} + \dfrac{5}{12}$ 78) $\dfrac{1}{5} + \dfrac{7}{15}$

79) $\dfrac{2}{3} - \dfrac{2}{5}$ 80) $\dfrac{3}{7} + \dfrac{4}{5}$

81) $\dfrac{11}{15} - \dfrac{4}{9}$ 82) $\dfrac{9}{10} - \dfrac{3}{14}$

83) $\dfrac{7}{10} + \dfrac{8}{25}$ 84) $\dfrac{9}{14} + \dfrac{8}{21}$

85) $5 - \dfrac{5}{6}$ 86) $7 + \dfrac{3}{4}$

87) $\dfrac{2}{3} + \dfrac{1}{4} - \dfrac{5}{6}$ 88) $\dfrac{3}{4} - \dfrac{2}{5} + \dfrac{1}{6}$

89) $\dfrac{2}{3} + \dfrac{5}{6} + \dfrac{1}{4}$ 90) $\dfrac{5}{8} + \dfrac{3}{4} + \dfrac{1}{2}$

91) $3\dfrac{3}{8} - 1\dfrac{1}{2}$ 92) $4\dfrac{5}{6} - 2\dfrac{2}{3}$

93) $7\dfrac{5}{8} + 4\dfrac{1}{6}$ 94) $4\dfrac{5}{12} - 2\dfrac{1}{8}$

CHALLENGE EXERCISES (95-96)

95) $\dfrac{3}{5} \cdot \dfrac{1}{7} \cdot \dfrac{9}{10}$ 96) $\dfrac{3}{11} \cdot \dfrac{5}{13} \div \dfrac{2}{5}$

Answer the following.

97) If ten sugar bowls hold $\frac{1}{2}$ cup sugar each, how much sugar does it take to fill them all?

98) If a tank holds $\frac{1}{4}$ gallon of gasoline, how many times can the tank be filled from a 5 gallon can?

99) A pizza is cut into eight equal pieces. If Sue eats one piece, John eats two pieces, and Ellen eats two pieces, what part of the pizza has been eaten?

100) Allen, Jose, and Ismel take a trip. If Allen has driven $\frac{4}{13}$ of the trip, Jose has driven $\frac{2}{13}$ of the trip, and Ismel has driven $\frac{5}{13}$ of the trip, what part of the trip has been completed?

101) John needs 7 pieces of lumber each of which is $\frac{7}{10}$ of a foot long. How many feet of lumber does he need?

102) If Candido spends $1\frac{3}{8}$ hours on his algebra homework each day, how many hours does he spend on his algebra homework per week?

103) A board 12 feet long is to be cut into pieces that are $\frac{2}{3}$ of a foot each. How many pieces will there be?

104) Tia works part time and works $\frac{2}{3}$ of a day. How many days will she have to work in order to work the equivalent of 8 full days?

105) A mini-marathon is approximately 13 miles long. If Susy has completed $8\frac{2}{3}$ miles, how much further must she run?

106) John is building a deck. He cuts a piece $2\frac{2}{5}$ ft long from a board 10 ft long. What is the length of the remaining piece?

107) In 1 hour, Paul can paint $\frac{1}{6}$ of a room, his wife Jenny can paint $\frac{1}{4}$ of the room, and Paul's brother Mark can paint $\frac{1}{12}$ of the room. What part of the room could they paint in one hour if they worked together?

108) One pipe can fill $\frac{1}{4}$ of a storage tank in one hour, another pipe can fill $\frac{1}{6}$ of the tank in one hour while a third pipe can empty $\frac{1}{3}$ of the tank in one hour. If all three pipes are open, what part of the tank is being filled in one hour?

109) A tailor needs $3\frac{3}{4}$ square yards of material to make a suit. How many square yards of material are needed to make 16 suits?

110) It takes $4\frac{5}{12}$ feet of ribbon to wrap a birthday gift. How many feet of ribbon would it take to wrap 18 of these gifts?

111) A chemist has a beaker that is $\frac{2}{3}$ full of hydrochloric acid. In an experiment she uses $\frac{1}{3}$ of the acid in the beaker. What fractional part of a full beaker did she use?

112) A pharmacist fills a prescription for a medication for high blood pressure. He uses $\frac{3}{4}$ of a bottle that is $\frac{2}{3}$ full. What part of a full bottle did he use?

113) If $2\frac{3}{4}$ ounces of silver cost $44, what is the cost of one ounce?

114) If a developer paid $286,000 for $4\frac{2}{5}$ acres of land, how much did she pay for one acre?

115) How many $\frac{1}{4}$-acre lots can a developer get from a $3\frac{1}{2}$-acre tract of land?

116) A bag of cookies that is $\frac{2}{3}$ full is divided equally among 4 children. What part of a full bag did each receive?

117) It takes 10 gallons of water to fill an aquarium $\frac{2}{5}$ full. How many gallons does the aquarium hold?

WRITING EXERCISE

118) When multiplying fractions you multiply both the numerators and denominators, but when adding you add only the numerators and not the denominators. Explain.

WRITING EXERCISES OR GROUP PROJECTS

If done as a group project, each group should write at least one exercise of each type. Then groups should exchange exercises with other groups and solve. If done as writing exercises, each student should both write and solve.

119) Write an application problem that requires the multiplication of fractions.

120) Write an application problem that requires the division of fractions.

121) Write an application problem that requires the addition of fractions with the same denominators.

SECTION 1.3

Variables, Exponents, and Order of Operations

OBJECTIVES

When you complete this section, you will be able to:

Ⓐ Identify variables and constants.
Ⓑ Read and evaluate expressions raised to powers.
Ⓒ Simplify expressions involving more than one operation.
Ⓓ Evaluate expressions containing variables when given the value(s) of the variable(s).

INTRODUCTION

We now expand our discussion of variables that were introduced in Section 1.1. In this book we will discuss some algebra, much of which is arithmetic from a more general point of view. This more general point of view is made possible by the use of variables introduced in Section 1.1. You have previously used variables in many formulas for the perimeters, areas, or volumes for several geometric figures. For example, in the formula for the circumference of a circle, $C = 2\pi r$, both C and r are variables. When given a value for r, we can find a value for C. Remember π is a **constant** whose approximate value is 3.14.

DEFINITION OF VARIABLE AND CONSTANT

A variable is a symbol, usually a letter, that is used to represent a number.
A constant is a symbol (a numeral) whose value does not change.

Variables are to mathematics as pronouns are to a language. Pronouns take the place of nouns and variables take the place of numbers.

Constants are used in a variety of ways in mathematics. Often they are given special names depending on how they are used. In this text we will use constants as numerical coefficients, exponents, terms of polynomials, and several other ways.

EXAMPLE 1

Identify the constant and the variable(s) in each of the following:

a) $3x$ The constant is 3 and the variable is x.

b) $4xy$ The constant is 4 and the variables are x and y.

c) n This means $1 \cdot n$, so the constant is 1 and the variable is n.

d) 5 The constant is 5 and there is no variable.

PRACTICE EXERCISES

Identify the constant and the variable(s) in each of the following.

1) $6y$ Constant = ___ , variable = ___ .

2) $5xy$ Constant = ___ , variables = ___ and ___ .

3) z Constant = ___ , variable = ___ .

4) 8 Constant = ___ , variable = ___ .

In arithmetic, multiplication is a short way of performing repeated addition of the same number. For example $4 \cdot 3$ means $3 + 3 + 3 + 3$. The 4 tells us how many 3s are being added. Likewise, there is a short way of indicating repeated multiplication of the same number using **exponents**. For example, $2 \cdot 2 \cdot 2$ is written as 2^3 where 3 is the **exponent** and 2 is the **base** of the exponent. Since each of the numbers being multiplied is called a **factor** the exponent tells us how many times the base is used as a factor.

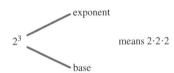

In reading expressions that have exponents, the base is read first and then the exponent is read as a power. The expression above, 2^3, is read "two to the third power." The power of 2 is most often read as "squared" and the power of 3 is most often read as "cubed." Consequently, 2^3 could also be read as "two cubed." There are no special names for powers greater than three.

Since exponents indicate multiplication, constants raised to powers can be evaluated.

EXAMPLE 2

	Exponential Expression	Read as	Meaning	Value
1)	5^2	5 squared or 5 to the second power	$5 \cdot 5$	25
2)	6^3	6 cubed or 6 to the third power	$6 \cdot 6 \cdot 6$	216
3)	8^5	8 to the fifth power	$8 \cdot 8 \cdot 8 \cdot 8 \cdot 8$	32,768
4)	$\left(\dfrac{2}{3}\right)^4$	Two-thirds to the fourth power	$\dfrac{2}{3} \cdot \dfrac{2}{3} \cdot \dfrac{2}{3} \cdot \dfrac{2}{3}$	$\dfrac{16}{81}$

PRACTICE EXERCISES

Complete the following chart.

Exponential Expression	Read as	Meaning	Value
5) 4^2			
6) 5^3			
7) 3^6			
8) $\left(\dfrac{1}{2}\right)^5$			

Order of Operations

Suppose a problem has more than one operation. Which one do we do first? For example, if we add first $2 + 3 \cdot 4 = 5 \cdot 4 = 20$, but if we multiply first $2 + 3 \cdot 4 = 2 + 12 = 14$. Both answers can not be correct, so which is correct?
 Consider the following situations.

a) You are shopping for some clothes. You buy a shirt for $30 and four pairs of socks for $2 per pair. The amount you would pay is $30 + 4 \cdot 2 = \$38$. Would you add or multiply first?

b) An item is marked three for $12. How much would you pay for two of these items? The amount you would pay is $12 \div 3 \times 2 = \$8$. Do you divide or multiply first?

c) Three friends are buying soft drinks for a party. They purchase five cartons at $6 per carton. If they pay equal amounts, how much does each pay? The amount each pays is $5 \cdot 6 \div 3 = \$10$. Would you multiply or divide first?

d) You purchase a jacket for $30 and two shirts marked at $20 each but with a tag for $5 off on each. You also return a pair of pants originally marked $15 but purchased with a $3 discount. How much money do you owe? You owe $30 + 2(20 - 5) - (15 - 3) = \$48$. In what order would you perform these calculations?

 In situation (a) you multiply before adding. In situation (b) you divide before multiplying, but in situation (c) you multiply before dividing. In situation (d) you simplify inside the parentheses first, then multiply by 2 and finally add and subtract in order from left to right.
 It appears that we need some rules as to which order to perform operations in problems that involve more than one operation. Hence we have the following order of operations.

• **Order of Operations**

ORDER OF OPERATIONS

If an expression contains more than one operation, they are to be performed in the following order:

1) If parentheses or other inclusion symbols (braces or brackets) are present, begin within the innermost and work outward, using the order in steps 3–5 below in doing so.

2) If a fraction bar is present, simplify above and below the fraction bar separately in the order given by steps 3–5 below.

3) First evaluate all indicated powers.

---

4) Then perform all multiplication or division in the order in which they occur as you work from left to right.

5) Then perform all additions or subtractions in the order in which they occur as you work from left to right.

EXAMPLE 3

Simplify the following using the order of operations.

a) $8 - 6 + 5 =$ Add or subtract left to right, so subtract 6 from 8.

    $2 + 5 =$ Add 2 and 5.

    7   Answer.

b) $4 \cdot 5 - 8 =$ Multiply before subtracting so multiply 4 and 5.

    $20 - 8 =$ Subtract 8 from 20.

    12   Answer.

c) $5 + 12 \div 4 =$ Divide before adding, so divide 12 by 4.

    $5 + 3 =$ Add 5 and 3.

    8   Answer.

d) $2 + 3 \cdot 4^2 =$ Raise to powers first, so square 4.

    $2 + 3 \cdot 16 =$ Multiply before adding, so multiply 3 and 16.

    $2 + 48 =$ Add 2 and 48.

    50   Answer.

e) $2^3 \cdot 3^2 =$ Raise to powers first.

    $8 \cdot 9 =$ Multiply.

    72   Answer.

f) $5(6 - 3) + 2 =$ Inside parentheses first, subtract 3 from 6.

    $5(3) + 2 =$ Multiply before adding, so multiply 5 and 3.

    $15 + 2 =$ Add 15 and 2.

    17   Answer.

g) $15 - 2(3 + 2^2) =$ Inside parentheses first, so square 2.

    $15 - 2(3 + 4) =$ Still inside parentheses, so add 3 and 4.

    $15 - 2(7) =$ Multiply before adding, so multiply 2 and 7.

    $15 - 14 =$ Subtract 14 from 15.

    1   Answer.

h) $15 + 3 \cdot 8 \div 12 =$ Multiply 3 and 8.

    $15 + 24 \div 12 =$ Divide 24 by 12.

    $15 + 2 =$ Add 15 and 2.

    17   Answer.

i) $\dfrac{4 + 3 \cdot 2^2}{2 + 3 \cdot 2} =$ Simplify above and below the fraction bar separately. Square 2 in the numerator and multiply in the denominator.

    $\dfrac{4 + 3 \cdot 4}{2 + 6} =$ Multiply in the numerator and add in the denominator.

$$\frac{4 + 12}{8} =$$ Add 4 and 12.

$$\frac{16}{8} =$$ Divide 8 into 16.

2   Answer.

ADDITIONAL PRACTICE

Simplify the following using the order of operations.

a) $6 + 8 - 3$

b) $5 + 4(6)$

c) $4 + 2 \cdot 3^3$

d) $5^2 \cdot 4^2$

e) $7(25 - 15 \div 3) - 8$

f) $6 + 2(9 + 3 \cdot 7)$

g) $24 - 15 \div 3 \cdot 2^2$

h) $\dfrac{6 + 12 \div 3}{2 \cdot 3 - 1}$

PRACTICE EXERCISES

Simplify the following using the order of operations.

9) $10 - 8 + 5$ **10)** $9 + 3 \cdot 6$

11) $12 - 10 \div 5 \cdot 3$ **12)** $5 + 4 \cdot 3^3$

13) $6(12 - 2^2) + 4$ **14)** $15 - 20 \div 4 \cdot 2 - 3$

15) $4 + 3(6 + 5 \cdot 3^2) - 10$ **16)** $\dfrac{8 + 15 \cdot 2}{2 \cdot 6 + 7}$

If more practice is needed, do the Additional Practice Exercises in the margin.

Calculators (Optional)

Most scientific calculators using algebraic logic have the order of operations agreement built in. Consequently, expressions that do not involve parentheses may be simplified by entering the expression in the order in which it is written. Let us look back at some previous examples and see how they would be done on a scientific calculator.

EXAMPLE C1
$4 \cdot 5 - 8$

Key strokes: 4, ×, 5, −, 8, =
    (Some calculators have an "enter" key instead of =.)

EXAMPLE C2
$2 + 3 \cdot 4^2$

Key strokes: 2, +, 3, ×, 4, x^y, 2, =
    (Some calculators have a special key for squaring.)

EXAMPLE C3
$15 + 3 \cdot 8 \div 12$

Key strokes: 15, +, 3, ×, 8, ÷, 12, =

If parentheses are involved and your calculator does not have this capability, it becomes a little tricky. The expression in the parentheses must be evaluated first.

EXAMPLE C4
$5(6 - 3) + 2$

Key strokes: 6, −, 3, =, ×, 5, +, 2, =. The extra = after the 3 is necessary in order to obtain a value for the expression inside the parentheses before multiplying by 5. If your calculator has parentheses, the key strokes are 5, ×, (, 6, −, 3,), +, 2, =.

- Evaluating expressions containing variables

EXAMPLE C5

$$15 + 2(4 + 2^2)$$

Key strokes: 4, +, 2, x^y, 2, =, ×, 2, +, 15, = or if your calculator has parentheses: 15, +, 2, ×, (, 4, +, 2, x^y, 2,), =.

The same order of operations applies when working with expressions containing variables. If a variable and a constant or two variables are written together with no operation sign, the operation is understood to be multiplication. For example, $3x$ means $3 \cdot x$ and xy means $x \cdot y$. We substitute the value(s) for the variable(s) and simplify by performing the correct order of operations.

EXAMPLE 4

Let $x = 2$ and $y = 5$ and evaluate.

a) $3x + 4y =$ Substitute 2 for x and 5 for y.

$\quad 3(2) + 4(5) =$ Multiply in order left to right, so multiply 3(2) and 4(5).

$\quad 6 + 20 =$ Add 6 and 20.

$\quad 26$ Answer.

b) $2x^2 + 3y^3 =$ Substitute 2 for x and 5 for y.

$\quad 2(2)^2 + 3(5)^3 =$ First raise to powers.

$\quad 2(4) + 3(125) =$ Multiply in order from left to right.

$\quad 8 + 375 =$ Add.

$\quad 383$ Answer.

c) $5x^2y^2 =$ Substitute 2 for x and 5 for y.

$\quad 5(2)^2(5)^2 =$ Raise to powers.

$\quad 5 \cdot 4 \cdot 25 =$ Multiply in order left to right.

$\quad 20 \cdot 25 =$ Multiply.

$\quad 500$ Answer.

d) $10x^2 \div y =$ Substitute 2 for x and 5 for y.

$\quad 10(2)^2 \div 5 =$ Raise to powers.

$\quad 10(4) \div 5 =$ Multiply and divide in order from left to right, so multiply 10 and 4.

$\quad 40 \div 5 =$ Divide.

$\quad 8$ Answer.

ADDITIONAL PRACTICE

Let $x = 4$ and $y = 5$ and evaluate.

i) $2x^2 + y^2$

j) $3x^2y^2$

k) $3x^3y + 4y^2$

l) $(5x^3) \div (2y)$

PRACTICE EXERCISES

Let $x = 3$ and $y = 4$ and evaluate.

17) $4x + 3y$	**18)** $2x^2 + 5y$	**19)** $2x^3y^2$
20) $5x^3y - 2y^2$	**21)** $(8x^3) \div (2y)$	

If more practice is needed, do the Additional Practice Exercises in the margin.

Order of operations occur in everyday situations.

EXAMPLE 5
Answer the following.

a) Tara, who has expensive taste, buys a living room suite by paying $500 down and $150 for 18 months. Find the total amount that she pays.

Solution:

The amount that she pays can be represented by $500 + 150 \cdot 18$.

$500 + 150 \cdot 18$	Multiply before adding.
$500 + 2700$	Add.
3200	Therefore, she pays $3,200 for the living room suite.

b) Jonathon leased a car with the following terms. He paid $1000 down, $199 per month for 2 years, and 12 cents per mile for all mileage over 24,000. When he returned the car it had 32,000 miles. Find the total amount that he paid for the lease.

Solution:

The amount that he paid can be represented by $1000 + 199 \cdot 2 \cdot 12 + 0.12(32000 - 24000)$.

$1000 + 199 \cdot 2 \cdot 12 + 0.12(32000 - 24000) =$	Simplify inside the parentheses.
$1000 + 199 \cdot 2 \cdot 12 + 0.12(8000) =$	Multiply 199 and 2.
$1000 + 398 \cdot 12 + 0.12(8000) =$	Multiply $398 \cdot 12$.
$1000 + 4776 + 0.12(8000) =$	Multiply .12 and 8000.
$1000 + 4776 + 960 =$	Add.
6736	Therefore, the lease cost $6736.

PRACTICE EXERCISES

22) Wolfgang bought a car by paying $1,200 down and $350 per month for 4 years. What is the total amount that he paid for the car?

23) Five people agree to split the cost of some take-out food. Three of the orders cost $4.99 each and two orders cost $6.99 each. They also purchased two two-liter soft drinks that cost $1.29 each. To the nearest cent, how much did each pay?

STUDY TIP #3
Mathematics Is Sequential Learning

Mathematical knowledge is sequential. New concepts and principles build upon previously learned concepts and principles. The material you learn each day will depend upon material that you should have learned prior to thatday. You must do all of your homework BEFORE each class meeting.

If you are absent from class you can not begin learning again at the same point as the rest of the class. You must first learn the material that was covered while you were absent.

Write a statement indicating how each of the following would be read and evaluate.

1) 8^2
2) 4^2
3) 5^3
4) 8^3
5) 9^4
6) 2^4
7) 2^5
8) 10^5
9) 10^6
10) 3^4
11) $\left(\frac{2}{3}\right)^2$
12) $\left(\frac{3}{4}\right)^2$
13) $(1.3)^3$
14) $(2.1)^3$

Simplify the following using the order of operations.

15) $7 - 3 + 5$
16) $8 + 6 - 4$
17) $5 - 6 \div 2$
18) $8 + 15 \div 3$
19) $7 + 2 \cdot 6$
20) $10 - 2 \cdot 4$
21) $8 + 3 \cdot 4^2$
22) $25 - 3 \cdot 2^3$
23) $4^2 + 5 \cdot 6$
24) $3^3 + 12 \div 3$
25) $24 - 3(8 - 2)$
26) $25 - 5(12 - 7)$
27) $10 + (13 - 9) \div 2$
28) $15 + (24 - 6) \div 3$
29) $2^3 \cdot 6^2$
30) $2^5 \cdot 4^2$
31) $3^2 \cdot 4^3 \cdot 2^2$
32) $5^2 \cdot 7^2 \cdot 2^3$
33) $\left(\frac{1}{2}\right)^3 \cdot \left(\frac{2}{3}\right)^2$
34) $\left(\frac{4}{3}\right)^2 \cdot \left(\frac{1}{4}\right)^3$
35) $2^3 + 3 \cdot 4^2$
36) $3^4 - 4 \cdot 2^3$
37) $26 - 18 \div 3^2$
38) $32 - 24 \div 2^3$
39) $3^4 - 36 \div 3^2$
40) $5^3 - 4 \cdot 3^3$
41) $22 + 18 \div 3 - 12$
42) $31 - 24 \div 6 + 8$
43) $32 - 4^2 + (8 - 3)$
44) $9 - 2^2 + (11 - 7)$
45) $45 \div (18 - 3^2)$
46) $36 \div (22 - 2^2)$
47) $3(11 - 8) \div 3$
48) $4(21 - 6) \div 20$
49) $4(2^4) - 32$
50) $5(3^3) - 45$
51) $6(4^2 - 8) \div 12$
52) $5(6^2 - 28) \div 10$
53) $17 - 9 + 8 \cdot 4 \div 16$
54) $29 - 12 \cdot 5 \div 30$
55) $48 \div 6 \cdot 5 - 16$
56) $64 \div 8 \cdot 4 - 17$
57) $5 + 35 \div 7 \cdot 3 - 16$
58) $14 + 81 \div 9 \cdot 2 - 26$
59) $200 - 84 \div 4 \cdot 2^3 + 29$
60) $225 - 48 \div 8 \cdot 3^2 - 17$

61) $\dfrac{8 + 2 \cdot 3}{7}$
62) $\dfrac{6 + 5 \cdot 4}{13}$
63) $\dfrac{9 - 12 \div 4 \cdot}{8 \div 4}$
64) $\dfrac{12 - 16 \div 8}{15 \div 3}$
65) $\dfrac{6^2 + 8 \cdot 2^2}{4^2 + 1}$
66) $\dfrac{8^2 + 6 \cdot 4^2}{6^2 + 4 \cdot 11}$

CALCULATOR EXERCISES

Evaluate the following using a calculator.

C1) 3^7
C2) 12^3
C3) $(7.32)^2$
C4) $4^4 \cdot 6^5$
C5) $(4.13)(8.95) + (2.6)^2$
C6) $20.48 \div 5.12 - 3.15$
C7) $51.23(6.14^2 - 23.46) - 17.3^2$
C8) $8.1^3 \div 2.7^2 - 4.7 - 3.1^3$

Let x = 2 and y = 3 and evaluate.

67) $3x + 4y$
68) $5x + 3y$
69) $8x^2$
70) $5y^3$
71) $x^2 y$
72) xy^4
73) $5x^2 y^2$
74) $7x^3 y$
75) $3x^2 + 4y^2$
76) $6x^3 - 2y^2$
77) $6x^2 \div y$
78) $18x^3 \div y^2$
79) $\dfrac{x^2 + 2y^2}{2x^2 + 3}$
80) $\dfrac{3x^2 + y^2}{2x^2 - 1}$

Find the value of each of the following when $x = \frac{1}{3}$ and $y = \frac{2}{5}$.

81) xy^2
82) $x^3 y$
83) $9x^2 y^2$
84) $25x^2 y^3$

CALCULATOR EXERCISES

Find the value of the following for x = 2.4 and y = 3.6.

C9) $1.7x + 2.6y$
C10) $4.2y - 1.2x$
C11) $1.6x^2 + 3.7y^3$
C12) $3.3x^3 + 8.7y^2$
C13) $6x^3 \div 3y$
C14) $1.2x^4 \div 0.6y^2$

Solve each of the following.

85) Tara bought a sofa by paying $150 down and $28 per month for eighteen months. The total cost to Tara is $150 + 28(18)$. Find the cost.

86) Diane bought a refrigerator by paying $225 down and $48 per month for twelve months. The total cost to Diane is $225 + 48(12)$. Find the cost.

87) Deon buys a used car by paying $600 down and $190 per month for twenty-four months. What is the total amount Deon paid for the car?

88) An automobile lease company offers a plan where the customer pays $1,000 down and $399 per month for twenty-four months. Find the total cost of leasing the car.

89) Under the property settlement terms of his divorce, Manuel is to pay his ex-wife $3,000 per month for the first four years and then $3,500 per month for the next six years. How much will Manual pay in property settlement?

90) Harold buys a lot with the following terms. He is to pay $250 per month for the first five years and $350 per month for the next five years. Find the total amount Harold will pay for the lot.

91) Five fraternity brothers agree to split the cost equally for three pizzas that cost $10.99 each and five soft drinks that cost 79 cents each. Find the amount (to the nearest cent) that each pays.

92) Six people agree to split the cost equally for some Chinese take-out. If two orders cost $3.99 each, three orders cost $3.50 each, one order costs $4.50, and they also purchase three bottles of soft drinks that cost $1.29 each, find the amount (to the nearest cent) that each pays.

93) Ezra leased a car with the following terms. He agreed to pay $800 down, $299 per month for three years, and 15 cents per mile for all mileage in excess of 45,000. When Ezra returned the car at the end of the lease period, it had 58,000 miles. The total cost of the lease can be represented as $800 + 299(3)(12) + 0.15(58,000 - 45,000)$. Find the total amount Ezra paid for the lease.

94) Candace leased a car with the following terms. She agreed to pay $1,500 down, $499 per month for 24 months, and 20 cents per mile for all mileage in excess of 24,000. When she returned the car at the end of the lease period, it had 32,000 miles. Find the total amount Candace paid for the lease.

CHALLENGE EXERCISES (95–104)

Simplify the following using the order of operations.

95) $6^2 + 2^3(4 \cdot 5^2 - 4 \cdot 5) \div 20 + 2$

96) $4^3 + 3 \cdot 5(36 \div 3^2 + 7 \cdot 4) \div 8 - 3^2$

97) $\dfrac{3}{4} + \left(\dfrac{2}{3}\right)^2 \cdot \dfrac{3}{8} - \dfrac{5}{16}$

98) $\dfrac{5}{6} - \dfrac{9}{16} \div \left(\dfrac{3}{4}\right)^3 \cdot \dfrac{5}{9} + \dfrac{2}{3}$

99) $\dfrac{1}{2} + \dfrac{3}{4}\left(\dfrac{5}{6} - \dfrac{2}{3}\right)^2 - \dfrac{3}{2} \div 4$

100) $\dfrac{3}{8} - \dfrac{1}{2}\left(\dfrac{3}{4} - \dfrac{5}{8}\right) + \dfrac{5}{16} \div \left(\dfrac{3}{2}\right)^2$

101) $2|-4| + 3|-5| - 2 \cdot 3^2$

102) $4|5| - 2|-3| + 3 \cdot 2^3$

103) $5|4^2 + 3 \cdot 2| - 4^2 \cdot 2$

104) $6|5 \cdot 6^2 - 8 \cdot 7| - 8 \cdot 2^3$

Evaluate each of the following for x = 3, y = 4 and z = 6.

105) $x^2y \div 2z + 3y^3 \div xy^2 - z^2$

106) $y^2z^2 \div x^2y \div (2y)(4x^2y)$

WRITING EXERCISES

107) How do constants and variables differ in meaning?

108) Give three different ways of indicating multiplication.

109) When a number is written with an exponent, what is the meaning of the exponent?

110) Why is the second power often called squared and the third power is often called cubed?

111) Why is an agreement on order of operations necessary?

CRITICAL THINKING EXERCISES

112) Compare the meaning of multiplication to the meaning of raising a number to a power.

113) Write a word problem which could represent the expression $3 \cdot 2 + 2(20 - 5)$.

WRITING EXERCISES OR GROUP ACTIVITIES

If these activities are done as a group project, each group should write at least three exercises of each type. Then groups should exchange exercises with other groups and solve. If done as writing exercises, each student should both write and solve one of each.

114) Write an application problem similar to 85–94 above that involves at least two operations.

115) Write an application problem similar to 85–94 above that involves at least three operations.

Evaluating Geometric Formulas

OBJECTIVES

When you complete this section, you will be able to:

Ⓐ Find the perimeters of geometric figures including triangles, squares, and rectangles, and the circumference of a circle.

Ⓑ Find the areas of rectangles, squares, triangles, parallelograms, trapezoids, and circles.

Ⓒ Find the volumes of rectangular solids, cubes, right circular cylinders, and cones.

INTRODUCTION

One concept that involves both the evaluation of expressions for specific values of the variable(s) and the order of operations is the evaluation of geometric formulas.

The **perimeter** of a geometric figure is the distance around it. Consequently, to find the perimeter of a figure, we find the sum of the lengths of the sides. Since perimeter is distance, it is measured in linear units such as inches, feet, miles, centimeters, meters, and so on. If the sides are not given in the same units (feet, inches, etc.), then you must convert to the same units before doing any calculations.

• **Finding the perimeter of geometric figures**

PERIMETER OF GEOMETRIC FIGURES

If $a, b, c, \ldots$ are the lengths of the sides of a geometric figure, then the perimeter of the geometric figure is the sum of the lengths of the sides. That is,

$$P = a + b + c + \cdots$$

Be careful: When substituting into geometric formulas, be sure all the measurements are in the same units. For example, one side cannot be in feet and another in inches.

EXAMPLE 1

Find the perimeter of each of the following.

a)

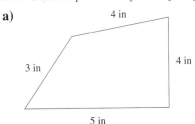

To find the perimeter, add the lengths of the sides. Hence, $P = 3$ in $+ 5$ in $+ 4$ in $+ 4$ in $= 16$ in.

b)

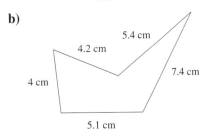

To find the perimeter, add the lengths of the sides. Hence, $P = 4$ cm $+ 5.1$ cm $+ 7.4$ cm $+ 5.4$ cm $+ 4.2$ cm $= 26.1$ cm.

c)

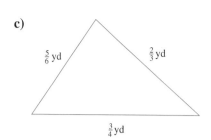

$\frac{5}{6}$ yd

$\frac{2}{3}$ yd

$\frac{3}{4}$ yd

To find the perimeter, add the lengths of the sides. So, $P = \frac{5}{6}$ yd $+ \frac{3}{4}$ yd $+ \frac{2}{3}$ yd. Write each with the least common denominator of 12. $P = \frac{10}{12}$ yd $+ \frac{9}{12}$ yd $+ \frac{8}{12}$ yd $= \frac{27}{12}$ or $2\frac{1}{4}$ yd.

PRACTICE EXERCISES

Find the perimeter of each of the following.

1)

9 mm

6 mm   7 mm

7 mm

2)

7.3 in

5 in   4.1 in

3.2 in

7.6 in

3)

$\frac{7}{12}$ ft $\frac{2}{3}$ ft

$\frac{3}{5}$ ft

• Perimeters of triangles, rectangles, squares, and the circumference of a circle

Some special geometric figures have formulas for finding their perimeter. Following are examples of some common geometric figures, their characteristics, and the formula for calculating the perimeter (circumference in the case of the circle) of each.

Figure	Characteristics	Formula for Perimeter
Triangle	Three sides of length a, b, and c.	$P = a + b + c$
Rectangle	Four sides, all angles are right angles (90°), opposite sides are equal in length. The two longest sides are usually called the length and the two shortest sides the width.	$P = 2L + 2W$

Figure	Characteristics	Formula for Perimeter
Square	Four sides of equal length and four right angles. The length of each side is s.	$P = 4s$

Figure	Characteristics	Formula for Circumference
Circle	The set of all points in a plane that are an equal distance from a given point in the plane. The given point is called the **center** of the circle and the given distance is the **radius**. The distance around the circle is called the **circumference** rather than the perimeter. A line segment with end points on the circle and passing through the center is called a diameter. Hence, $d = 2r$ where d is the diameter and r is the radius.	$C = \pi d$ or $C = 2\pi r$ $\pi \approx 3.14$ or $\frac{22}{7}$ The symbol "$\approx$" means "is approximately equal to."

EXAMPLE 2

Find the perimeter of each of the following.

a)

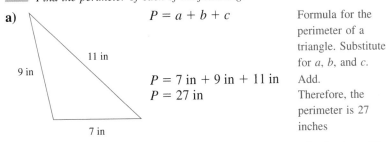

$P = a + b + c$ — Formula for the perimeter of a triangle. Substitute for a, b, and c.

$P = 7 \text{ in} + 9 \text{ in} + 11 \text{ in}$ — Add.

$P = 27 \text{ in}$ — Therefore, the perimeter is 27 inches

b) A rectangle with length of 2 feet and with width of 8 inches. Give the answer in inches.

Since the sides are given in different units, we must express both in the same units, which are inches in this case. Note that 1 foot = 12 inches and 2 feet = 2 · 12 inches = 24 inches.

$$P = 2L + 2W$$ Formula for the perimeter of a rectangle. Substitute for L and W.

$$P = 2(24 \text{ in}) + 2(8 \text{ in})$$ Multiply before adding.

$$P = 48 \text{ in} + 16 \text{ in}$$ Add.

$$P = 64 \text{ in}$$ Therefore, the perimeter is 64 inches.

c)

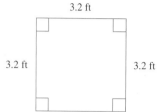

$$P = 4s$$ Formula for the perimeter of a square. Substitute for s.

$$P = 4(3.2 \text{ ft})$$ Multiply.

$$P = 12.8 \text{ ft}$$ Therefore, the perimeter is 12.8 feet.

d) Find the circumference of the circle below. Use $\pi \approx 3.14$.

$$C = \pi d$$ Formula for the circumference of a circle. Substitute for π and d.

$$C = (3.14)(8 \text{ ft})$$ Multiply.

$$C = 25.12 \text{ ft}$$ Therefore, the circumference is 25.12 feet.

e) Find the circumference of a circle with radius $\frac{9}{5}$ feet. Use $\pi \approx \frac{22}{7}$.

$$C = 2\pi r$$ Formula for the circumference of a circle. Substitute for π and r.

$$C = 2\left(\frac{22}{7}\right)\left(\frac{9}{5} \text{ ft}\right)$$ Multiply.

$$C = \frac{396}{35} \text{ ft}$$ Therefore, the circumference is $\frac{396}{35}$ feet.

Note: In the formula for the circumference of a circle the constant π is used. The ratio of the circumference to the diameter of any circle is π. Since π is a constant, exact answers are often left in terms of π rather than substituting a value and multiplying. For example, the answer for (2d) above could have been expressed as: $C = \pi d = \pi(8 \text{ ft}) = 8\pi$ ft and the answer for (2e) would be $C = 2\pi r = 2\pi\left(\frac{9}{5}\right)$ ft $= \frac{18}{5}\pi$ ft.

ADDITIONAL PRACTICE

Find the perimeter of each of the following:

a) A triangle whose sides are 2 ft, 18 in, and 15 in. Give the answer in inches.

b)

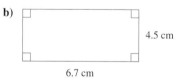

c)

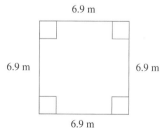

PRACTICE EXERCISES

Find the perimeter of each of the following.

4) A triangle whose sides are 3 inches, 5 inches, and 6 inches.

5)

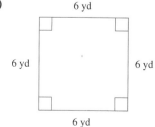

6)

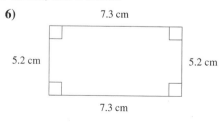

Find the circumference of the following circles.

d) Use $\pi \approx 3.14$ and round the answer to the nearest meter.

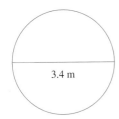

3.4 m

e) Radius of $\frac{5}{3}$ ft. Use $\pi \approx \frac{22}{7}$.

• **Developing the concept of area**
• **Area units of measure**

ANSWERS:
Practice 4–6

4) 14 in   **5)** 24 yd

6) 25 cm

Additional Practice a–c

a) 57 in   **b)** 22.4 cm

c) 27.6 m

• **Area of a rectangle**

Find the circumference of the following circles.

7) Diameter of 5.6 yards. Use $\pi \approx 3.14$ and round the answer to the nearest tenth of a yard.

8) Use $\pi \approx \frac{22}{7}$.

$\frac{11}{3}$ in

If more practice is needed, do the Additional Practice Exercises in the margin.

The units used for measuring perimeter are linear units that measure distances. Suppose we want to carpet the floor of a room that is 14 feet long and 10 feet wide. We could not measure the amount of carpet needed using linear units since the size of the floor cannot be expressed as a distance. The floor of the room is a surface. The measure of the size of a surface is called its **area**. The units used to measure area are **square units**. A square unit is a square with sides each of which measure one linear unit. Below are some examples of square units.

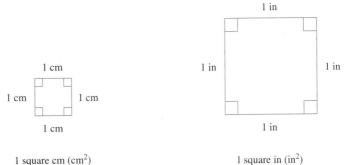

1 square cm (cm²)   1 square in (in²)

Instead of writing square cm or square in, we will use the notation cm^2 and in^2. Thus, ft^2 is read "square foot" and m^2 is read "square meter."

In finding the area of a surface, we find the number of square units that are contained within the surface. For example, in the 10 feet by 14 feet room mentioned above, the area of the floor is the same as the number of ft^2 of carpet needed to cover the floor. We will develop the formula for the area of a rectangle to reinforce the concept of area, and then give the formulas for other geometric figures.

EXAMPLE 3

a) Find the area of a rectangle 5 centimeters long and 3 centimeters wide.

Solution:
We begin by drawing the rectangle.

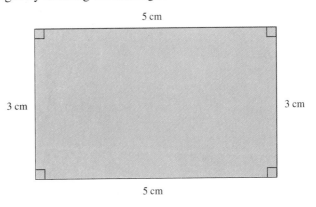

5 cm

3 cm   3 cm

5 cm

We now draw horizontal and vertical lines one centimeter apart.

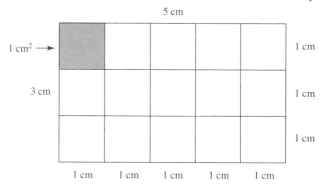

We have created a number of squares within the rectangle. The length of each side of each square is 1 centimeter, so each square is 1 cm^2. Each row contains 5 squares and there are 3 rows of squares. Therefore, there are $5 \times 3 = 15$ squares. Consequently, the area of the rectangle is 15 cm^2. Notice that this is the same as multiplying the length (5 cm) times the width (3 cm). Therefore, the area of a rectangle with length L and width W is $A = LW$.

One advantage of the unit^2 notation is that we will automatically get the correct unit of measure if we will write the unit as well as the numerical value when performing computations. In the rectangle above, $A = (5 \text{ cm})(3 \text{ cm}) = (5 \cdot 3)(\text{cm} \cdot \text{cm}) = 15 \text{ cm}^2$ (think of cm·cm as cm^2).

Below we give formulas for the areas of some common geometric figures. As was the case with perimeters, the dimensions must be given in the same units.

Figure	Characteristics	Formula for Area
Rectangle	Four sides, all angles are right angles (90°), opposite sides are equal in length. The two longest sides are usually called the length and the two shortest sides the width.	$A = LW$
Square	Four sides of equal length and four right angles. The length of each side is s.	$A = s^2$
Triangle	Any side can be used as the base, b, and the perpendicular (forms a right angle) from the opposite vertex (point where two sides intersect) to that side is the height, h.	$A = \dfrac{bh}{2}$

Figure	Characteristics	Formula for Area
Parallelogram 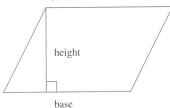	Has four sides and opposite sides are parallel (do not intersect and remain the same distance apart) and are equal in length. Any side can be used as the base, b, and the perpendicular to that side is the height, h.	$A = bh$
Trapezoid 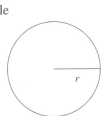	Has four sides in which one pair is parallel. The parallel sides are called the bases. The longer base is designated as B and the shorter base, b. The perpendicular distance between the parallel sides is the height, h.	$A = \dfrac{h(B+b)}{2}$
Circle 	Given earlier when perimeter was discussed.	$A = \pi r^2$

EXAMPLE 4

Find the area of each of the following.

a)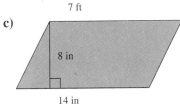

2 in

4 in

$A = LW$ Substitute 4 in for L and 2 in for W.

$A = (4\text{ in})(2\text{ in})$ Multiply.

$A = 8\text{ in}^2$ Therefore, the area is 8 square inches.

b)

7 ft

7 ft

$A = s^2$ Substitute 7 ft for s.

$A = (7\text{ ft})^2$ Raise to the power.

$A = 49\text{ ft}^2$ Therefore, the area is 49 square feet.

c)

8 in

14 in

$A = bh$ Substitute for 14 in for b and 8 in for h.

$A = (14\text{ in})(8\text{ in})$ Multiply.

$A = 112\text{ in}^2$ Therefore, the area is 112 square inches.

d)

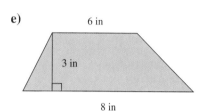

$A = \dfrac{bh}{2}$	Substitute for b and h.
$A = \dfrac{(10 \text{ in})(8 \text{ in})}{2}$	Multiply in the numerator.
$A = \dfrac{80 \text{ in}^2}{2}$	Divide.
$A = 40 \text{ in}^2$	Therefore, the area is 40 square inches.

e)

$A = \dfrac{h(B + b)}{2}$	Substitute for h, B, and b.
$A = \dfrac{(3 \text{ in})(8 \text{ in} + 6 \text{ in})}{2}$	Add inside parentheses.
$A = \dfrac{(3 \text{ in})(14 \text{ in})}{2}$	Multiply in the numerator.
$A = \dfrac{42 \text{ in}^2}{2}$	Divide.
$A = 21 \text{ in}^2$	Therefore, the area is 21 square inches.

f) Circle with a diameter of 8.2 inches. Round the answer to the nearest hundredth of an in². Use $\pi \approx 3.14$.

Since the diameter is 8.2 inches, the radius is $\dfrac{8.2}{2} = 4.1$ inches.

$A = \pi r^2$	Substitute for r.
$A = \pi(4.1 \text{ in})^2$	Square 4.1 in and substitute for π.
$A = (3.14)(16.81 \text{ in}^2)$	Multiply.
$A = 52.7834 \text{ in}^2$	Round to the nearest hundredth of an in².
$A = 52.78 \text{ in}^2$	Therefore, the area is 52.78 square inches.

Note: The exact answer for (4f) is $A = \pi r^2 = \pi(4.1)^2 = 16.81\pi \text{ in}^2$.

ADDITIONAL PRACTICE

Find the area of the following parallelogram.

f)

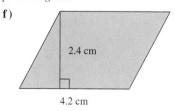

2.4 cm

4.2 cm

PRACTICE EXERCISES

Find the area of each of the following.

9) Find the area of the following rectangle.

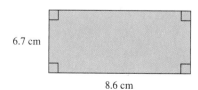

6.7 cm

8.6 cm

10) Find the area of the following square.

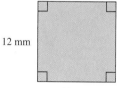

12 mm

12 mm

Find the area of the following triangle.

g) Base of 4.2 cm and height of 5 cm. Round the answer to the nearest tenth of a cm², if necessary.

Find the area of the following trapezoid.

h)

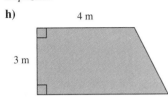

Find the area of the following circle.

i) Diameter of 12.8 ft. Round the answer to the nearest hundredth.

• **Finding volumes of geometric solids**

• **The word "cube" is often misused in everyday English. Have you ever seen an ice cube that is really a cube?**

11) Find the area of the following parallelogram.

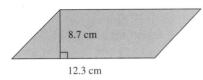

12) Find the area of the following triangle.

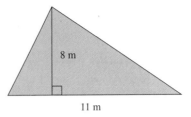

13) Find the area of the following trapezoid. Give answer in square inches.

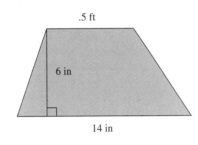

14) Find the area of the following circle. Use $\pi \approx 3.14$.

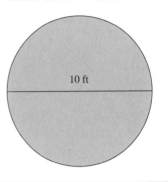

If more practice is needed, do the Additional Practice Exercises in the margin.

If we wanted to find how much cypress mulch it would take to fill a box 4 feet long, 2 feet wide, and 3 feet high, we could not use linear units since we are not finding a distance, and we could not use square units since we are not finding the size of a surface. In order to find the volume of a solid object, we must use another type of unit called a **cubic unit**.

Imagine a box, including the top, each of whose faces is a square. This type of figure is a **cube**. A cube is a three dimensional geometric solid with six faces each of which is a square. A cubic unit is a cube each of whose faces is a square whose sides are one unit each. For example, the figure below is a cubic inch (in³).

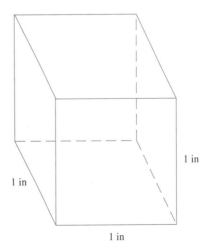

When finding the volume of a solid, we are finding the number of cubic units it takes to fill the solid. This may be a little confusing since the volumes we encounter in everyday situations are usually liquid measures of volume like ounces and gallons.

A **rectangular solid** is a geometric figure with six faces each of which is a rectangle. A box is an example of a rectangular solid. See the figure below.

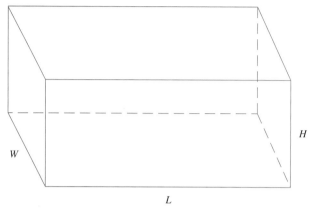

• **Volume of a rectangular solid**

We will develop the formula for the volume of a rectangular solid and then give the formulas for the volumes of other types of solids. Let us find the volume of a rectangular solid that is 3 feet long, 4 feet wide, and 2 feet high. We are finding the number of cubic feet necessary to fill the solid. First we calculate how many cubic feet (ft³) it would take to cover the bottom of the rectangular solid.

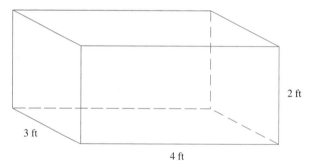

We mark off units of one foot each along the length and width of the base of the solid and draw horizontal lines at these units.

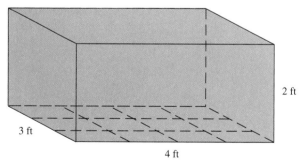

The lines divide the base into $3 \cdot 4 = 12$ unit squares each with sides of one foot. Since each face of a cube is a square, we can place one cubic foot on top of each of these squares, forming one layer of 12 cubic feet on the base.

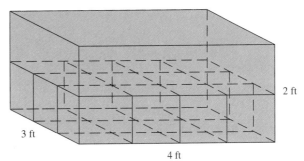

Since the height is 2 feet, it would take another layer of cubes identical to this layer to completely fill the rectangular solid. Therefore, it would take $12 \cdot 2 = 24$ ft^3 to fill the rectangular solid. Notice the length (3) times the width (4) gave us the number of cubes in the first layer (12). The height (2) gave us the number of layers. Therefore, the number of cubic feet in the solid is $12 \cdot 2 = 24$ ft^3. Therefore, the volume is $3 \cdot 4 \cdot 2 = 24$ ft^3. Consequently, the volume of a rectangular solid is $V = LWH$. Following are some common geometric solids and the formulas for their volumes.

Type of Solid	Characteristics	Formula for Volume
Rectangular Solid	Six faces all of which are rectangles.	$V = LWH$
Cube	A rectangular solid in which each face is a square. The intersection of any two faces is called an edge and is denoted by e. All edges are of the same length.	$V = e^3$
Right Circular Cylinder 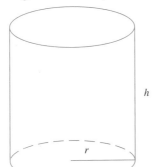	Looks like a tin can. The top and bottom are circles and the "sides" are perpendicular to the top and bottom. The radius of the base is denoted by r and the distance between the top and bottom is the height, h.	$V = \pi r^2 h$

Type of Solid	Characteristics	Formula for Volume
Cone 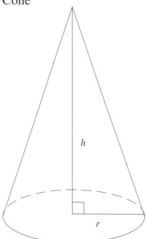	The bottom, called the base, is a circle and the figure comes to a single point at the top, called the vertex. The radius of the base is denoted by r. The perpendicular distance between the vertex and the base is called the height and is denoted by h.	$V = \dfrac{1}{3}\pi r^2 h$ or $\dfrac{\pi r^2 h}{3}$

EXAMPLE 5

Find the volumes of the following geometric solids. If necessary, round answers to the nearest hundredth.

a)

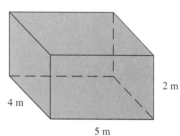

Solution:

$V = LWH$ — Formula for the volume of a rectangular solid. Substitute for L, W, and H.

$V = (5\text{ m})(4\text{ m})(2\text{ m})$ — Multiply.

$V = 40\text{ m}^3$ — Therefore, the volume is 40 m³

b) Find the volume of the following cube.

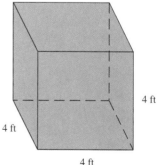

$V = e^3$ — Formula for the volume of a cube. Substitute for e.

$V = (4\text{ ft})^3$ — Raise to the power.

$V = 64\text{ ft}^3$ — Therefore, the volume is 64 cubic feet.

Find the volumes of the following right circular cylinders.

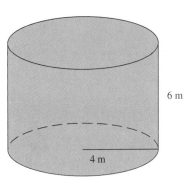

c)

$$V = \pi r^2 h \qquad \text{Substitute for } \pi, r, \text{ and } h.$$
$$V = (3.14)(4 \text{ m})^2(6 \text{ m}) \qquad \text{Raise to powers first.}$$
$$V = (3.14)(16 \text{ m}^2)(6 \text{ m}) \qquad \text{Multiply.}$$
$$V = 301.44 \text{ m}^3 \qquad \text{Therefore, the volume is 301.44 cubic meters.}$$

d) Radius of the base is 2.1 meters and the height is 3.2 meters. Round the answer to the nearest hundredth of a cubic meter. Use $\pi \approx 3.14$.

$$V = \pi r^2 h \qquad \text{Substitute for } \pi, r, \text{ and } h.$$
$$V = (3.14)(2.1 \text{ m})^2(3.2 \text{ m}) \qquad \text{Raise to powers first.}$$
$$V = (3.14)(4.41 \text{ m}^2)(3.2 \text{ m}) \qquad \text{Multiply.}$$
$$V = 44.31168 \text{ m}^3 \qquad \text{Round to the nearest hundredth.}$$
$$V = 44.31 \text{ m}^3 \qquad \text{Therefore, the volume is 44.31 cubic meters to the nearest hundredth.}$$

Note: Since π has a constant value, we often leave the results in terms of π rather than substituting and multiplying out. For example, we could have expressed the answers to (5c) and (5d) above as:

$$V = \pi r^2 h = \pi (4 \text{ m})^2(6 \text{ m}) = \pi (16 \text{ m}^2)(6 \text{ m}) = 96\pi \text{ m}^3.$$
$$V = \pi r^2 h = \pi (2.1 \text{ m})^2(3.2 \text{ m}) = \pi (4.41 \text{ m}^2)(3.2 \text{ m}) = 14.112\pi \text{ m}^3.$$

e) Find the volume of the following cone. Leave answer in terms of π.

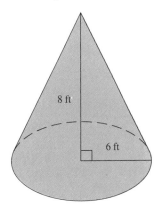

$$V = \frac{\pi r^2 h}{3}$$ Substitute for r and h.

$$V = \frac{\pi (6 \text{ ft})^2 (8 \text{ ft})}{3}$$ Raise to powers first.

$$V = \frac{\pi (36 \text{ ft}^2)(8 \text{ ft})}{3}$$ Multiply in the numerator.

$$V = \frac{\pi (288 \text{ ft}^3)}{3}$$ Divide.

$$V = 96\pi \text{ ft}^3$$ Therefore, the volume is 96π cubic feet.

ADDITIONAL PRACTICE

Find the volume of the following rectangular solid.

j) Length of 8 in, width of 4 in, and height of 5 in.

Find the volume of the following cube.

k)

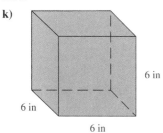

6 in
6 in
6 in

Find the volume of the following right circular cylinder.

l) Radius of the base is 5.1 ft and the height is 2.4 ft. Leave answer in terms of π.

Find the volume of the following cone.

m) Leave answer in terms of π.

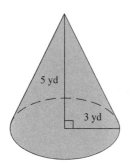

5 yd
3 yd

PRACTICE EXERCISES

Find the volume of the following geometric solids.

15) Rectangular solid with length of 13 mm, width of 2 mm, and height of 24 mm.

16) Find the volume of the cube below.

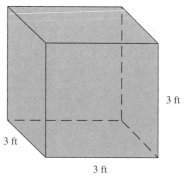

3 ft
3 ft
3 ft

17) Find the volume of the following right circular cylinder. Radius of the base is $\frac{3}{2}$ cm and the height is $\frac{11}{4}$ cm. Use $\pi \approx \frac{22}{7}$.

18) Find the volume of the cone below. Leave answer in terms of π.

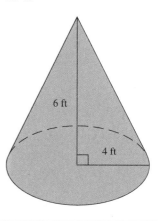

6 ft
4 ft

If more practice is needed, do the Additional Practice Exercises in the margin.

We summarize as follows.

FINDING PERIMETERS, AREAS, OR VOLUMES

1) If necessary, identify the type of figure.

2) Select the appropriate formula.

3) Substitute for the variables in the formula.

4) Simplify using the order of operations. Be careful with the units!

STUDY TIP #4

Reading Mathematics Critically

Mathematics is read differently. In reading a novel or a story, a whole chapter may contribute a single idea to the plot of the story, but in mathematics every single word or symbol has a precise meaning and each may contribute one or more ideas. Consequently, you must read slowly and carefully so that you understand the meaning of each word and symbol.

You must be able to tell the difference between expressions that look very much alike but have different meanings. For example, "the difference of 8 and 5, written $8 - 5$" is very different from "the difference of 5 and 8, written $5 - 8$."

Read with pencil and paper available. If you do not understand what the author did in getting from one step to the next, take the time to work it out for yourself.

EXERCISE SET   **1.4**

Find the perimeter of each of the following.

1)

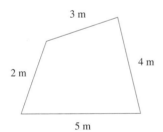

3)

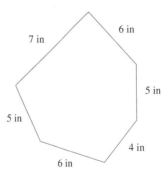

2)

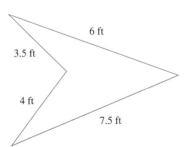

4)

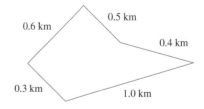

Find the perimeter and area of each of the following rectangles.

5)

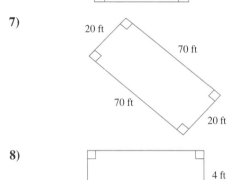

18 m

3 m

6)

4 in

10 in

7)

20 ft

70 ft

70 ft

20 ft

8)

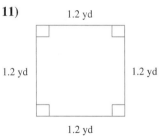

4 ft

2 yd

Give answers in terms of feet.

9) Length of $2\frac{3}{4}$ yds and width of $1\frac{2}{3}$ yds.

10) Length of $6\frac{1}{6}$ ft and width of $2\frac{1}{2}$ ft.

Find the area and perimeter of each of the following squares.

11)

1.2 yd

1.2 yd   1.2 yd

1.2 yd

12) Each of whose sides is 9 decimeters long.

13) Each side is $\frac{5}{7}$ cm.

14) Each side is $3\frac{1}{2}$ ft.

Find the perimeter and area of the following triangles.

15)

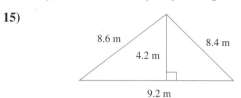

8.6 m   8.4 m

4.2 m

9.2 m

16)

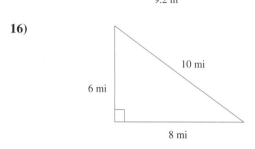

10 mi

6 mi

8 mi

17) A triangle whose sides are of length 6 inches, 7 inches, and 9 inches, and the height to the 6-inch side is 5 inches.

18) A triangle whose sides are 12 km, 14 km, and 18 km long, and the height to the 14-km side is 10 km.

Find the exact value of the circumference and area of each of the following circles. (Leave answers in terms of π.)

19)

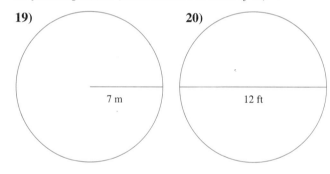

7 m

20)

12 ft

Find the circumference and area of each of the following circles. Use $\pi \approx 3.14$ and round the answers to the nearest tenth of a unit.

21) A circle of radius 5.5 feet.

22) A circle with diameter 9.8 centimeters.

Find the circumference and area of each of the following circles. Use $\pi \approx \frac{22}{7}$.

23) A circle of radius $\frac{7}{3}$ in.

24) A circle with diameter of $\frac{3}{5}$ ft.

Find the area of each of the following parallelograms.

25)

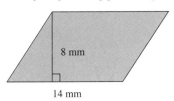

8 mm

14 mm

26)

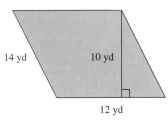

14 yd   10 yd

12 yd

27)

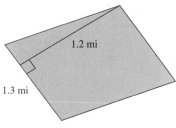

1.2 mi

1.3 mi

28) A parallelogram with base 18.4 feet and altitude 12.6 feet.

Find the areas of the following trapezoids.

29)

6 mi

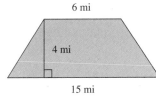

4 mi

15 mi

30)

11 ft

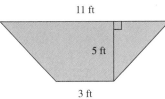

5 ft

3 ft

31)

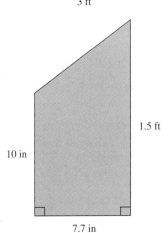

1.5 ft

10 in

7.7 in

32) A trapezoid with large base 11.3 centimeters, small base 9.7 cm, and altitude 6.5 cm.

Find the volume of the following rectangular solids.

33)

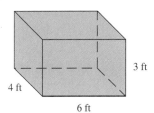

3 ft

4 ft

6 ft

34)

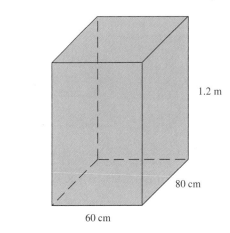

1.2 m

80 cm

60 cm

35) Length of 8 meters, width of 5.2 meters, and height of 6.5 meters.

36) Length of 8 feet, width of 7 feet, and height of 4 yards.

Find the volume of the following cubes.

37)

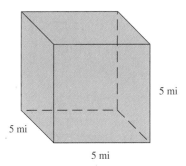

5 mi

5 mi

5 mi

38) Length of each edge is 2.3 kilometers.

39) Each edge is $\frac{3}{5}$ cm.

40) Each edge is $1\frac{2}{3}$ ft.

Find the volumes of the following right circular cylinders. Leave answers in terms of π.

41)

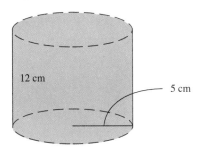

12 cm

5 cm

42)

8 ft

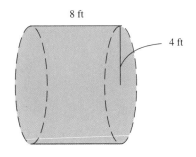

4 ft

43) Radius of the base is 4.2 meters and the height is 6 meters.

44) Radius of the base is 3 centimeters and the height is 8.5 centimeters.

Find the volumes of the following cones. Round the answers to the nearest tenth.

45) Use π ≈ 3.14.   **46)**

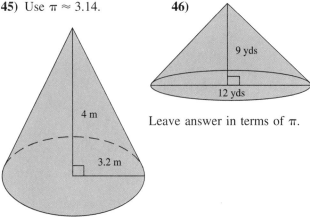

9 yds

12 yds

4 m

Leave answer in terms of π.

3.2 m

47) Base radius of $\frac{5}{7}$ in and height of 14 in. Use π ≈ $\frac{22}{7}$.

48) Base diameter of 14 ft and height of $\frac{23}{11}$ ft. Use π ≈ $\frac{22}{7}$.

GROUP PROJECT

49) Look around your campus or local community and find examples of at least five of the geometric figures that we studied in this section. If the figure is two-dimensional (square, rectangle, circle and so on), find the perimeter and area of each. If the figure is three dimensional (rectangular solid, right circular cylinder or other solid) find the volume of each.

SECTION 1.5

Geometric Applications

OBJECTIVE

When you complete this section, you will be able to:

Ⓐ   Solve application problems involving perimeter, area, and volume.

In the previous section we learned how to compute the perimeters, areas, and volumes of some of the most common geometric figures. The use of these geometric formulas occurs in many everyday situations. In order to solve these problems, we must determine the appropriate formula, the value(s) of the variable(s), substitute for the variables, and simplify. Often there are additional factors that must be taken into consideration. We illustrate with some examples.

EXAMPLE 1

a) A carpenter needs to put molding around the ceiling of a rectangular room with length 15 feet and with width 12 feet. How many feet of molding does he need? See the figure in the margin.

Solution:
    This is the same as finding the perimeter of a rectangle with length 15 feet and width 12 feet.

12 ft

15 ft

$$P = 2L + 2W$$ Formula for the perimeter of a rectangle.

$$P = 2(15 \text{ ft}) + 2(12 \text{ ft})$$ Substituting for L and W.

$$P = 30 \text{ ft} + 24 \text{ ft}$$ Multiply before adding.

$$P = 54 \text{ ft}$$ Therefore, the carpenter needs 54 feet of molding.

b) A gardener wishes to put a border around a circular flower bed with radius 3 feet. How much of the border material will she need? See the figure in the margin.

Solution:

This is the same as finding the circumference of a circle with radius 3 feet.

$$C = 2\pi r$$ Formula for the circumference of a circle.

$$C = 2(3.14)(3 \text{ ft})$$ Substituting for π and r.

$$C = 18.84 \text{ ft}$$ Therefore, the gardener needs 18.84 feet of border material.

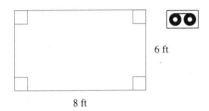

c) A piece of wood is in the shape of a rectangle 8 feet long and 6 feet wide. If the wood costs $1.70 per square foot, what is the cost of the piece of wood? See the figure in the margin.

Strategy:

We must first find the number of square feet of wood in the piece. This is the same as the area of a rectangle 6 feet long and 8 feet wide.

Solution:

$$A = LW$$ Substitute for L and W.

$$A = (8 \text{ ft})(6 \text{ ft})$$ Multiply.

$$A = 48 \text{ ft}^2$$ Therefore, there are 48 square feet of wood.

Since the wood costs $1.70 per square foot and there are 48 ft², multiply 48 times $1.70. Therefore, cost = (48)(1.70) = $81.60.

d) An irrigation pipe 150 feet long is anchored at one end and the other end is mounted on wheels. Consequently, it sweeps out a circle when in operation. Suppose the field is planted in potatoes. The potatoes can be fertilized by putting liquid fertilizer into the irrigation system. When diluted, one gallon of fertilizer will fertilize 900 ft² of potatoes and costs $8.50. How much does it cost to fertilize the field? See figure.

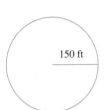

Strategy:

We must determine how many gallons of fertilizer are needed to fertilize a circular field 150 feet in radius. Since one gallon will fertilize 900 ft², the number of gallons needed is equal to the area of the field divided by 900. Therefore, we must first determine the area of the field.

Solution:

$$A = \pi r^2$$ Substitute for π and r.

$$A = (3.14)(150 \text{ ft})^2$$ Square 150 feet.

$$A = (3.14)(22{,}500 \text{ ft}^2)$$ Multiply.

$$A = 70{,}650 \text{ ft}^2$$ Therefore, the area of the field is 70,650 square feet.

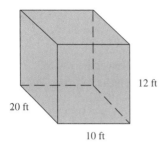

Since one gallon will fertilize 900 ft², the number of gallons of fertilizer needed is $70650 \div 900 = 78.5$ gallons. Each gallon costs \$8.50, so the cost of the fertilizer is $(78.5)(8.50) = \$667.25$.

e) A truck has an enclosed compartment that is 20 feet long, 10 feet wide and 12 feet high. A product comes in boxes which are 5 feet long, 2 feet wide, and 3 feet high. How many of these boxes will the truck hold if there is no empty space around the boxes? See the figure.

Strategy:
We will first calculate the volume of the truck compartment. Then we will calculate the volume of one box. If we divide the volume of the truck compartment by the volume of one box, we will get the number of boxes the truck will hold.

Solution:
Volume of the truck compartment.

$$V = LWH = (20 \text{ ft})(10 \text{ ft})(12 \text{ ft}) = 2400 \text{ ft}^3.$$

Volume of one box.

$$V = LWH = (5 \text{ ft})(2 \text{ ft})(3 \text{ ft}) = 30 \text{ ft}^3.$$

Divide the volume of the truck container by the volume of one box.
Number of boxes = $2400 \text{ ft}^3 \div 30 \text{ ft}^3 = 80$ boxes.

f) Sand falls from a conveyer belt into the shape of a cone. The pile of sand has a base radius of 6 yards and a height of 5 yards. A dump truck can hold 7 cubic yards of sand and the sand sells for \$20 per load. What is the value of the sand in the pile? Round the number of truck loads to the nearest whole load. See the figure.

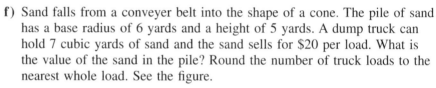

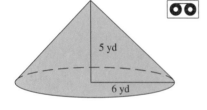

Strategy:
First we need to find the volume of the pile of sand. Since the truck holds 7 cubic yards, we divide the volume of sand by 7 in order to find the number of truck loads in the pile. Since each load costs \$20, multiply the number of loads by \$20 to get the value of the sand.

Solution:
Volume of the sand:

$$V = \frac{1}{3}\pi r^2 h = \frac{1}{3}(3.14)(6 \text{ yd})^2(5 \text{ yd}) = \frac{1}{3}(3.14)(36 \text{ yd}^2)(5 \text{ yd})$$
$$= \frac{1}{3}(565.2 \text{ yd}^3)$$
$$= 188.4 \text{ yd}^3.$$

Number of truck loads:
Number of truckloads = (volume of sand) $\div 7 = 188.4 \div 7 = 26.9$ truck loads which we round off to 27.

Value of the sand:
Value of the sand = (number of truck loads)(\$20) = $(27)(20) = \$540$.

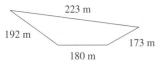

223 m
192 m   **173 m**
180 m

PROBLEM 1

180 ft

PROBLEM 2

50 ft
75 ft

PROBLEM 3

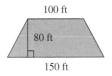

100 ft
80 ft
150 ft

PROBLEM 4

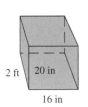

2 ft **20 in**
16 in

PROBLEM 5

PRACTICE EXERCISES

Solve each of the following. Figures for some are in the margin.

1) A surveyor finds the lengths of the sides of a field with four sides to be 180 meters, 192 meters, 223 meters, and 173 meters. Find the perimeter of the field.

2) A drainage ditch is dug around a square field each of whose sides is 180 feet long. If Charles is paid $.60 per foot for digging the ditch, how much money does he earn?

3) How many square feet of sod will it take to cover a rectangular yard 75 feet long and 50 feet wide?

4) A field is in the shape of a trapezoid whose parallel sides are 150 feet and 100 feet long. The height of the trapezoid is 80 feet. A bag of grass seed costs $6.00 and will plant 500 square feet. How much will the seed to plant the field cost?

5) An aquarium is 2 feet long, 16 inches wide, and 20 inches high. If one gallon of water contains 231 in^3, how many gallons of water are in the aquarium? Round the answer to the nearest gallon.

6) A hole must be dug for a basement. The hole is to be in the shape of a rectangular solid that is 28 yards long, 20 yards wide, and 4 yards deep. The truck holds 14 yd^3 of dirt, and it costs $15 per load to haul away the dirt. How much will it cost to haul the dirt away from the hole?

7) A walkway is to be poured with concrete. The walkway is to be 48 feet long, 4 feet wide, and 4 inches thick. How many cubic yards of concrete will it take? Round the answer off to the nearest tenth of a cubic yard.

E X E R C I S E   S E T   1.5

Solve each of the following. If necessary, use π ≈ 3.14 and round your answers to the nearest hundredth of a unit. Figures are drawn for exercises 1–10.

1) A carpenter wants to make a square window frame. If the window measures 52 inches on each of the four sides, how much material does he need to make the frame?

52 in

2) A rectangular football field measures 150 feet long by 120 feet wide. If the school band is to march all the way around the field, how far will they march?

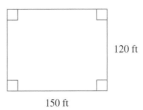

120 ft
150 ft

3) A police officer wishes to tape off an irregular shaped crime scene. From point A to point B is 12 feet, from point B to point C is 15 feet, from point C to point D is 10 feet, from point D to point E is 6 feet, and from point E to point A is 26 feet. How much tape will the officer need to go all the way around the scene?

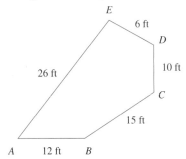

4) Magna has a square garden plot which measures 18 feet on each side. If fencing cost $.50 a linear foot, how much does it cost to fence her garden?

5) A local sandwich shop wishes to put up a sign with a neon light border. The shape and size of the sign is given in the diagram below. If the neon tubing costs $5.00 per foot, how much does it cost for the border around the sign?

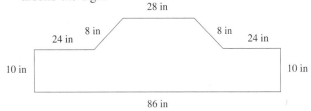

6) A landscape architect is designing a triangular flower bed to be put in the city's new park. If the flower bed is to be 8 yards by 12 yards by 8 yards, how many landscape timbers, each of which is 4 feet long, are needed to put around the flower bed? If the timbers cost $4.95 each, find the cost of the timbers.

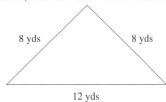

7) A lawn sprinkler rotates on a shaft so that it wets a circular region of the lawn. If the water sprays out a

distance of 50 feet from the sprinkler, what is the circumference of the circle of wet grass?

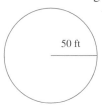

8) Plumbers use diameter to measure water pipes. What is the circumference of a pipe 2.5 inches in diameter?

9) The diameter of the barrels of the largest guns used on a battleship is 16 inches. What is the circumference of these barrels?

10) The floor of a rectangular shower stall measures 52 inches long and 32 inches wide. What is the area of the floor of the shower stall?

11) A certain type of plant needs 1 square yard of ground to grow in. If a field is 200 yards long and 150 yards wide, how many plants can be grown in the field? If each plant costs $2.95, find the cost of the plants.

12) A circular fountain has a diameter of 10 meters. How many square shaped tiles are needed to cover the bottom of the fountain if each tile is 1 square decimeter? Assume no waste. If each tile costs $.69, find the cost of the tile.

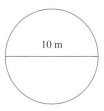

13) A university wants to install artificial turf in its football stadium. The football field, including the end zones, is rectangular and measures 150 yards long and 120 yards wide. How many square yards of artificial turf are needed?

14) The airport authority wants to design a triangular caution sign. How many square centimeters of reflective material will be needed for the sign if the height of the triangle is 90 centimeters and the base is 100 centimeters?

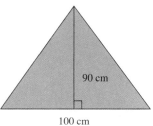

15) A business concern wants to build a decorative wall in the shape of a trapezoid that measures 25 feet on the

top base and 40 feet on the bottom base. If the wall is 8 feet high, how many 1 foot square ceramic pieces of tile are necessary to cover the wall? Assume no waste. If each piece of tile costs $2.59, find the cost of the tile.

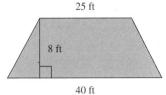

16) Acoustic ceiling tile with an area of 1 square foot costs $.49 a tile. What would tile cost for an acoustical ceiling measuring 20 feet long by 17 feet wide?

17) A family wishes to put an addition on their house. The roof of the addition is rectangular and measures 25 feet long and 12 feet wide. If roofing costs $3.50 per square foot, what will be the total cost of materials for the new roof?

18) A cereal box is 11 inches high, 7 inches long, and 2 inches wide. How many cubic inches of cereal can it hold?

19) The body of a dump truck used for hauling dirt and sand is in the shape of a rectangular solid and measures 8 feet by 12 feet by 4 feet. How many cubic yards of sand will it hold?

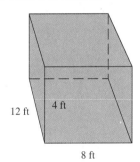

20) A building contractor wishes to put a basement in a new office building on campus. The basement is to measure 12 feet deep, and the building is to measure 90 feet long by 40 feet wide. How many cubic yards of dirt must be removed before the basement can be built?

21) A storage tank is in the shape of a right circular cylinder that has a base diameter of 42 inches and a height of 2 yards. Find the volume in cubic inches and leave the answer in terms of π.

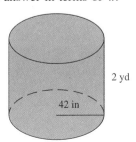

22) John and Kay wish to fill their swimming pool. Their pool is shaped like a rectangular solid measuring 15 feet wide by 40 feet long with the water at a depth of 6 feet. If water costs $0.75 per 100 cubic feet, how much would it cost to fill the pool?

23) Alfredo is estimating the cost of an addition to his home. He will build two rectangular rooms with a combined area that measures 60 feet by 18 feet. If the concrete slab (foundation) must be 8 inches thick and the concrete costs $27.00 a cubic yard, what will be the total cost of concrete for the slab?

24) It costs the department of transportation $0.43 per square foot for the material used in making traffic signs. Each sign is rectangular with length of 2 feet and width of $1\frac{1}{2}$ feet. Find the cost of the material to make 400 such signs.

25) A room is $12\frac{2}{3}$ feet long and $18\frac{3}{4}$ feet wide. If carpet costs $18.95 per square yard, find the total cost of the carpet needed to carpet the floor of the room.

CHALLENGE EXERCISES (26–29)

26) The runway at a certain airport is 500 yards long and 60 yards wide. How many runway lights are needed if they are to be placed every 20 yards at all edges of the runway?

27) An automobile tire has a diameter of 14 inches. How far will the tire roll in five revolutions?

28) An artist wishes to make one wall of his studio out of glass block to capture the natural light. The wall measures 26 feet long and 12 feet high. If the glass blocks are 6 inch squares and cost $1.25 each, how much will the blocks for the wall cost?

29) A gasoline storage tank at a service station is in the shape of a right circular cylinder with a diameter of 6 feet and a height of 12 feet. If there are 231 cubic inches in a gallon and the gasoline costs the service station owner $0.89 per gallon, how much will it cost to fill the tank? Use $\pi \approx 3.14$ and round the number of gallons to the nearest whole gallon.

WRITING EXERCISES OR GROUP PROJECTS

If done as a group project, each group should write at least one exercise of each type. Then groups should exchange exercises with other groups and solve. If done as writing exercises, each student should both write and solve one of each type.

30) Write an application problem involving the perimeter of a triangle.

31) Write an application problem involving the circumference of a circle.

32) Write an application problem involving the area of a rectangle.

33) Write an application problem involving the area of a circle.

34) Write an application problem involving the volume of a rectangular solid.

GROUP PROJECT

35) Interview someone from campus maintenance, the campus carpentry shop, a construction worker/foreman, a landscaping company or other work area and make a list of at least five instances in which they have been required to solve geometric application problems on the job.

SECTION <u>1.6</u>

Addition of Real Numbers

OBJECTIVES

When you complete this section, you will be able to:

Ⓐ Add integers with the same and opposite signs.

Ⓑ Translate verbal expressions into mathematical expressions and simplify.

Ⓒ Identify properties of addition.

INTRODUCTION

The addition of integers occurs in many everyday situations. If the temperature rises 8° from 9:00 to 10:00 and then rises 5° from 10:00 to 11:00, there has been a total change in temperature of $8° + 5° = 13°$ from 9:00 to 11:00. If a football team gains 8 yards ($+8$) on first down and loses 3 yards (-3) on second down, then there is a net gain on the two downs of $(+8) + (-3) = 5$ yards. If a gambler loses \$20 (-20) on the first hand and wins \$8 ($+8$) on the second hand, there is a total loss of $(-\$20) + \$8 = -\$12$ on the two hands. If a scuba diver descends 30 feet (-30) and then descends another 20 feet (-20), then she has descended a total of $(-30) + (-20) = -50$ feet.

CALCULATOR EXPLORATION ACTIVITY—OPTIONAL

Using a calculator, complete the following.

Column A	Column B
$3 + 3 =$	$6 + (-2) =$
$4 + 5 =$	$-6 + 2 =$
$6 + 2 =$	$4 + (-6) =$
$(-3) + (-4) =$	$5 + (-2) =$
$(-5) + (-7) =$	$-8 + 5 =$
$(-2) + (-6) =$	$-4 + 9 =$

Based on your answers to Column A, answer the following.

1) When adding two numbers with the same sign, is the absolute value of the answer the sum or difference of the absolute values of the numbers being added?

2) If the signs of the numbers being added are the same, how does the sign of the answer compare with the signs of the numbers being added?

3) Based on your answers to 1 and 2, write a rule for adding two numbers with the same sign.

(continued)

• **Adding real numbers**

When adding integers, it is convenient to think of them as directed
numbers. In the following examples the number line will be used to show the
addition of integers with direction to the right as positive and direction to the
left as negative. We will always begin at 0. The operation sign, $+$, is like a verb
since it is telling you what action to perform. Think of the action as "and then
go." Do not confuse the addition sign with the $+$ sign indicating a positive
number.

EXAMPLE 1
Find the following sums using the number line.

a) $(+3) + (+2)$ (This is usually written as $3 + 2$ since a number written without a
sign is assumed to be positive.)

This means begin at 0 and go 3 units to the right (because 3 is positive).
Then go 2 more to the right (because 2 is also positive). You should now be
at $+5$, showing that $(+3) + (+2) = +5$ or $3 + 2 = 5$. Using the number
line it appears as follows.

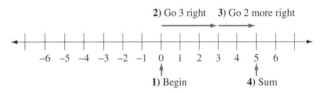

b) $(+6) + (-2)$ (Usually written as $6 + (-2)$ or $6 - 2$.)

This means begin at 0 and go 6 units to the right. Then go 2 units to the left.
You should now be at 4. Therefore, $6 + (-2) = 4$ as illustrated below.

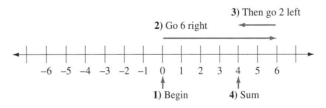

c) $-6 + (+4)$ (Usually written as $-6 + 4$.)

This means begin at 0 and go 6 units to the left. Then go 4 units to the right. You should now be at -2. Therefore, $-6 + 4 = -2$ as the following illustrates.

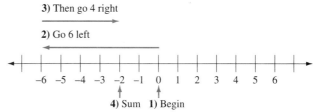

d) $(-4) + (-2)$

This means begin at 0 and go 4 units left. Then go 2 more units left. You should now be at -6. Therefore, $(-4) + (-2) = -6$ as illustrated below.

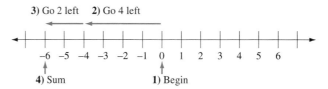

PRACTICE EXERCISES

Use the number line to find the following sums.

1) $2 + 4$

2) $5 + (-4)$

3) $-6 + 3$

4) $-2 + (-3)$

Following is a table summarizing the results of Example 1.

Problem	Signs of Addends	Add or Subtract the Absolute Values of the Addends	Sign of the Answer
$(+3) + (+2) = 5$	same (both $+$)	add	positive—the same as sign of the addends
$(+6) + (-2) = 4$	opposite	subtract	positive—the same as addend with larger absolute value
$-6 + (+4) = -2$	opposite	subtract	negative—the same as addend with larger absolute value
$(-4) + (-2) = -6$	same (both $-$)	add	negative—the same as sign of the addends

• One way to remember whether to add or subtract the absolute values is to remember what is happening on the number line. If you go in the same direction, add the absolute values. If you go in opposite directions, subtract the absolute values.

Look back at the situations in the introduction and see that the results are consistent. This brings us to a general rule for adding signed numbers.

ADDING SIGNED NUMBERS

1) To add numbers with the same sign, add the absolute values of the numbers and give the answer the same sign as the numbers being added.

2) To add numbers with opposite signs, subtract the absolute values of the numbers and give the answer the sign of the number with the larger absolute value.

EXAMPLE 2

Find the following sums.

a) $7 + 5 =$ Signs are the same, so add the absolute values.

  12   Both numbers are positive, so sum is positive.

b) $6 + (-13) =$ Signs are opposite, so subtract the absolute values.

  -7 -13 has the larger absolute value, so the sum is negative.

c) $-6 + (-7) =$ Signs are the same, so add the absolute values.

  -13 Both numbers are negative, so the sum is negative.

d) $-14 + 5 =$ Signs are opposite, so subtract the absolute values.

  -9 -14 has the larger absolute value, so the sum is negative.

e) $-\dfrac{3}{5} + \dfrac{4}{5} =$ Think of $-\dfrac{3}{5}$ as $\dfrac{-3}{5}$ and add the numerators.

  $\dfrac{-3 + 4}{5} =$ Add.

  $\dfrac{1}{5}$ Sum.

ADDITIONAL PRACTICE

Find the following sums.

a) $7 + 12$

b) $-15 + 9$

c) $-12 + (-15)$

d) $13 + (-13)$

e) $-\dfrac{2}{5} + \dfrac{4}{5}$

PRACTICE EXERCISES

Find the following sums.

5) $9 + 7$ 6) $-12 + 7$ 7) $-8 + (-5)$

8) $9 + (-16)$ 9) $\dfrac{4}{9} + \left(-\dfrac{2}{9}\right)$

If more practice is needed, do the Additional Practice Exercises in the margin.

If we are adding more than two numbers apply the order of operations by adding in order from left to right. If parentheses are present, simplify inside the parentheses first. To avoid confusion, brackets are often used to show grouping when parentheses are already present.

ADDITIONAL PRACTICE

Find the following sums.

f) −14 + 9 + (−16)

g) 24 + (−9) + (−22)

h) 18 + 27 + (−17)

i) (−23 + 52) + (−45 + 33)

 EXAMPLE 3

Find the following sums.

a) 4 + (−8) + 7 =   First add 4 and −8.

   −4 + 7 =   Now add −4 and 7.

   3   Sum.

b) −24 + 16 + (−18) =   First add −24 and 16.

   −8 + (−18) =   Now add −8 and −18.

   −26   Sum.

 c) 26 + [17 + (−13)] =   Inside brackets first, so add 17 and −13.

   26 + 4 =   Add 26 and 4.

   30   Sum.

d) (−52 + 34) + [−21 + (−8)] =   Inside parentheses first.

   −18 + (−29) =   Add −18 and −29.

   −47   Sum.

PRACTICE EXERCISES

Find the following sums.

10) −10 + 8 + (−15)   11) 32 + [−18 + (−21)]

12) −9 + [15 + (−6)]   13) (−45 + 34) + [36 + (−22)]

If more practice is needed, do the Additional Practice Exercises in the margin.

Often it is necessary to translate expressions written in English into mathematical expressions. There are many expressions in English that indicate addition. For example, "sum," "more than," and "increased by" all imply addition.

 EXAMPLE 4

Write a numerical expression for each of the following and evaluate.

a) The sum of −7 and −2.

   Solution:

   Sum means add, so this means add −7 and −2.

      −7 + (−2) =   Signs are the same, so add the absolute values.

      −9   Both signs are negative, so the sum is negative.

 b) −10 increased by 4

   Solution:

   This implies that you begin with −10 and increase it by 4 which means add 4 to −10.

      −10 + 4 =   Signs are opposite, so subtract the absolute values.

      −6   −10 has the larger absolute value, so the sum is negative.

c) 6 more than −2

Solution:

This implies that you begin with −2 and you need 6 more which means add −2 and 6.

$$-2 + 6 =$$ Signs are opposite, so subtract the absolute values.

$$4$$ 6 has the larger absolute value, so the sum is positive.

PRACTICE EXERCISES

Write a numerical expression for each of the following and evaluate.

14) the sum of −12 and −5   **15)** −8 increased by 14

16) 5 more than −13   **17)** $\frac{2}{3}$ added to $-\frac{3}{5}$

- One way to remember the commutative property is that each day you commute from home to school, you take the same route, but in opposite directions. The commutative property involves the same numbers, but in the opposite order.

The operation of addition has several properties that are of such importance that they are given names for easy reference. You may have already noticed that the order in which the numbers are added makes no difference. For example, $2 + 4 = 4 + 2$, and $5 + 9 = 9 + 5$. The same is true for the addition of integers. This relationship is known as the **commutative property of addition**. It says that the order in which the numbers are added does not matter since the sum is the same.

EXAMPLE 5

a) $-4 + 7 = 7 + (-4)$ Simplify each side.

$\quad\quad 3 = 3$

b) $-5 + (-8) = -8 + (-5)$ Simplify each side.

$\quad\quad -13 = -13$

If three or more numbers are added, the manner in which they are grouped makes no difference. For example, $2 + (3 + 4) = (2 + 3) + 4$. So it does not matter which we add first, the 3 and the 4 or the 2 and the 3. This relationship is called the **associative** property of addition. It says the manner in which the numbers are grouped does not matter since the sum is the same.

EXAMPLE 6

a) $4 + (5 + 8) = (4 + 5) + 8$ Simplify each side by adding inside the parentheses.

$\quad\quad 4 + 13 = 9 + 8$ Add.

$\quad\quad 17 = 17$ Therefore, both sides are equal.

b) $-6 + [(-4) + 9] = [-6 + (-4)] + 9$ Simplify inside brackets first.

$\quad\quad -6 + 5 = -10 + 9$ Add.

$\quad\quad -1 = -1$ Therefore, both sides are equal.

The number 0 is called the **additive identity** or the identity element for addition. Adding 0 to any number gives a result identical to the original number. If the sum of two numbers is 0, the two numbers are **additive inverses** of each other. In Section 1.1 these were called negatives or opposites.

EXAMPLE 7

a) The additive inverse of 5 is −5 because $5 + (−5) = 0$.

b) The additive inverse of −14 is 14 because $−14 + 14 = 0$.

c) 0 is its own additive inverse because $0 + 0 = 0$.

Below is a summary of the properties of addition.

PROPERTIES OF ADDITION

All real numbers *a, b*, and *c* satisfy the following properties.

Commutative Property of Addition

$a + b = b + a$ Addition may be performed in any order.

Associative Property of Addition

$a + (b + c) = (a + b) + c$ Addition may be grouped in any manner.

Additive Inverse

Every real number, *a*, has an additive inverse, −*a*, so that $a + (−a) = (−a) + a = 0$.

Additive Identity

The real number 0 is called the additive identity because $a + 0 = 0 + a = a$. The sum of any number and 0 is identical to the number.

EXAMPLE 8

Give the name of the property illustrated by each of the following.

a) $3 + (−5) = −5 + 3$ Order is different. Therefore, commutative.

b) $4 + (−6 + 2) = [4 + (−6)] + 2$ Order is same and grouping is different. Therefore, associative.

c) $5 + (8 + 4) = 5 + (4 + 8)$ Grouping is same and order is different. Therefore, commutative.

d) $5[6 + (−6)] = 5(0)$ $6 + (−6) = 0$. Therefore, additive inverse.

e) $−6 + 0 = −6$ Additive identity.

PRACTICE EXERCISES

Give the name of the property illustrated by each of the following.

18) $−8 + 2 = 2 + (−8)$ _____

19) $7 + (−6 + 3) = [7 + (−6)] + 3$ _____

20) $4(6 + 3) = 4(3 + 6)$ _____

21) $−9(−3 + 3) = −9(0)$ _____

22) $5 + 0 = 5$ _____

Memorizing and Understanding Mathematics

Some things in mathematics must be memorized. Symbols, definitions, rules, and algorithms have to be both memorized and understood. We all forget, so those things which must be memorized also need to be reviewed periodically. We can do this by writing definitions, rules, etc. on 3 × 5 cards or 8 × 5 cards and reading them while we wait for a bus or train or between classes at the college or university. However, you cannot memorize everything in a mathematics course. You may be able to memorize enough for one unit exam, but there is always a final exam.

Psychologists tell us that we learn best when what we are learning has meaning for us. That is, we can learn more and keep it longer when we understand the material to be learned. The more we review, think about, and see how things fit together, the more meaning these things have for us and the better we understand them and can apply them.

In mathematics there is a reason why we do everything that we do and we must know and understand that reason. Mathematics is not learned just by doing problems. It is learned by doing the problems and understanding why we did them the way we did. The "why" of mathematics is as important, if not more important, than the "how" of mathematics. In general, mathematics is to be *understood*, not memorized.

EXERCISE SET 1.6

ANSWERS:
Practice 18–22

18) commutative   **19)** associative

20) commutative   **21)** additive inverse

22) additive identity

Find the following sums.

1) $2 + 16$

2) $9 + 5$

3) $-2 + 13$

4) $-5 + 165$

5) $7 + (-14)$

6) $8 + (-18)$

7) $17 + (-8)$

8) $14 + (-5)$

9) $-17 + 6$

10) $-21 + 3$

11) $-7 + (-4)$

12) $-6 + (-3)$

13) $13 + 35$

14) $32 + 43$

15) $-24 + 35$

16) $-54 + 67$

17) $-25 + 19$

18) $-42 + 37$

19) $63 + (-34)$

20) $72 + (-54)$

21) $\frac{7}{11} + \frac{-3}{11}$

22) $\frac{-5}{9} + \frac{2}{9}$

23) $-\frac{3}{4} + \frac{5}{6}$

24) $\frac{3}{4} + \left(-\frac{2}{3}\right)$

25) $-\frac{2}{5} + \left(-\frac{5}{6}\right)$

26) $-\frac{4}{7} + \left(-\frac{2}{3}\right)$

27) $4.3 + (-2.5)$

28) $-6.7 + 5.3$

29) $-9.2 + 4.5$

30) $7.6 + (-4.1)$

31) $8 + (-5) + 4$

32) $9 + (-6) + (-6)$

33) $-8 + 10 + (-7)$

34) $-6 + 12 + (-4)$

35) $13 + (-17) + (-21)$

36) $21 + (-16) + (-32)$

37) $-16 + (-23) + (-31)$

38) $-25 + (-32) + (-47)$

39) $[9 + (-3)] + (-8 + 5)$

40) $[7 + (-4)] + (-6 + 9)$

41) $(-13 + 9) + (-21 + 14)$

42) $(-17 + 8) + (-27 + 17)$

43) $[-4 + (-7)] + (-10 + 4)$

44) $[-8 + (-5)] + (-13 + 7)$

45) $[16 + (-32)] + [17 + (-6)]$

46) $[-5 + (-43)] + [-37 + (-21)]$

CALCULATOR EXERCISES

Find the following sums using a calculator.

C1) $14.62 - 33.21$ **C2)** $-67.43 + 49.62$

C3) $-896.456 + (-45.7689)$

C4) $-985.35 + (-83678.587)$

C5) $(-765.45 + 65.78) + [473.21 + (-860.32)]$

C6) $[-75849.359 + (-94327.432)] + [79235.937 + (-27807.857)]$

Give the name of the property illustrated by each of the following.

47) $6 + (-2) = -2 + 6$ _____

48) $5 + (6 + 4) = (5 + 6) + 4$ _____

49) $-8 + 8 = 0$ _____

50) $57 + 0 = 57$ _____

51) $7(5 + 3) = 7(3 + 5)$ _____

52) $6[4 + (-2)] = 6[-2 + 4]$ _____

53) $-5(-4 + 4) = -5(0)$ _____

54) $0 + 8 = 8 + 0$ _____

55) $12 + (-4 + 9) = 12 + [9 + (-4)]$ _____

56) $0 + (6 + 4) = 6 + 4$ _____

57) $(3 + 5) + (-7) = 3 + [5 + (-7)]$ _____

58) $(-10 + 10) + 7 = 0 + 7$ _____

Complete each of the following using the given property.

59) $-5 + 6 =$ _____ Commutative for addition

60) $-4 + (9 + 12) =$ _____ Associative for addition

61) $-4 + (9 + 12) =$ _____ Commutative for addition

62) $0 + 34 =$ _____ Additive identity

63) $-14 + 14 =$ _____ Additive inverse

64) $45 + 0 =$ _____ Commutative for addition

65) $7(5 + 8) =$ _____ Commutative for addition

Write a numerical expression for each of the following and evaluate.

66) The sum of -9 and 5

67) The sum of -11 and 8

68) The sum of -3, 6 and -7

69) The sum of -6, -4 and 10

70) 6 increased by 9   **71)** 8 increased by 11

72) -7 increased by 14   **73)** -12 increased by 7

74) 23 more than 36   **75)** 35 more than 49

76) 17 more than -30 **77)** 36 more than -18

Problems 78–85 involve changes in temperature. Write each as an addition problem and solve.

	Current Temp	Change	Addition	New Temperature
78)	78°	Drops 15°	_____	_____
79)	92°	Drops 25°	_____	_____
80)	18°	Drops 28°	_____	_____
81)	12°	Drops 19°	_____	_____
82)	−5°	Rises 15°	_____	_____
83)	−12°	Rises 24°	_____	_____
84)	−14°	Drops 12°	_____	_____
85)	−21°	Drops 7°	_____	_____

Problems 86–89 involve a scuba diver. Express each as an addition problem and solve.

	Current Depth	Change in Depth	Addition	New Depth
86)	−43 feet	Ascends 27 feet	_____	_____
87)	−68 feet	Ascends 39 feet	_____	_____
88)	−45 feet	Descends 22 feet	_____	_____
89)	−65 feet	Descends 43 feet	_____	_____

Solve each of the following word problems using addition of integers.

Use the following information for Exercises 90–93. In Europe the first floor of a building is called ground floor and what we call the second floor is called the first floor and so on. Consequently, we could think of the ground floor as being the "0" floor. Suppose a European hotel has a 30th floor and 4 levels of underground parking garages. Represent each of the following as an addition problem and solve.

90) If an elevator in a European hotel starts on the 27th floor and goes down 29 levels, where does it stop?

91) Suppose the elevator in Exercise 90 begins on the 16th floor and goes down 19 levels. Where does it stop?

92) Suppose the elevator in Exercise 90 begins on the third parking level below ground and goes up 23 levels. Where does it stop?

93) Suppose the elevator in Exercise 90 begins on the 2nd parking level below ground and goes up 14 levels. Where does it stop?

94) Hernando has a balance of $130 in his checking account. He makes a deposit of $230. What is his new balance?

95) Farshad has a balance of $289 in his checking account. He writes a check for $146. What is his new balance?

96) Sara has a balance of $319 in her checking account. She makes a deposit of $132 and then writes a check for $350. What is her new balance?

97) Mira has a balance of $293 in her checking account. She writes a check for $167 and later makes a deposit of $250. What is her new balance?

98) At the opening bell a stock was selling for $53\frac{1}{2}$. During the day it gained $3\frac{1}{8}$ points. What was the closing price?

99) At the opening bell, a stock was selling for $37\frac{3}{8}$ and gained $1\frac{7}{8}$ before the final bell. What was the closing price?

100) A hiker begins at the rim of the Grand Canyon and descends 1500 feet into the canyon. He then turns around and ascends 850 feet. How far is the hiker from the rim?

101) The balance in a credit card account is $1396 before a payment of $450 is made. Find the new balance.

102) A group of explorers is located in Death Valley and are 32 feet below sea level. An airplane drops supplies from a height of 1800 feet. How far did the supplies fall?

103) On successive plays a running back gained 5 yards, lost 2 yards, and gained 8 yards. If a net gain of 10 yards is needed to earn a first down (a) did the team earn a first down? b) Why?

CHALLENGE EXERCISES (104–107)

Evaluate each of the following.

104) $4^2(-7 + 2 \cdot 5) \div 3 \cdot 4$

105) $(-3 + 6 \cdot 3^2) \div 17 \cdot 6$

106) $3^2 \cdot 4(-12 + 4^2) \div 2^2 \cdot 3^2$

107) $48 \div 2^2 \cdot 3(-18 + 2 \cdot 4^2)$

WRITING EXERCISE

108) Why is 0 its own negative (opposite)?

CRITICAL THINKING

109) In your own words explain why you add the absolute values of two numbers with like signs when finding their sum.

110) In your own words explain why you subtract the absolute values of two numbers with unlike signs when finding their sum.

GROUP PROJECTS

111) In the introduction, several examples were given where the addition of integers occur in everyday situations. Give three additional examples not found in the text.

112) Interview an engineer, physics or chemistry instructor, or other professional and make a list of at least three instances in which they have used addition of signed numbers.

SECTION <u>1.7</u>

Subtraction of Real Numbers

OBJECTIVES

When you complete this section, you will be able to:

Ⓐ Subtract integers.

Ⓑ Determine if a signed number is a solution of an equation.

In this section, we will see that we already know how to subtract signed numbers.

CALCULATOR EXPLORATION ACTIVITY—OPTIONAL

Complete the following using a calculator.

Column A	Column B
$4 - (+3) =$	$4 + (-3) =$
$-6 - (+5) =$	$-6 + (-5) =$
$7 - (-3) =$	$7 + (+3) =$
$-8 - (-2) =$	$-8 + (+2) =$

By looking at the corresponding lines in Columns A and B, answer the following.

1) Subtracting a number is the same as _____ its _____.

Following are some real world situations that support the conclusion reached above.

As with the addition of integers, the subtraction of integers occurs in many everyday situations. Look for a pattern in the following examples using temperature.

EXAMPLE 1

a) If the temperature changes from 50° to 70°, what is the change in temperature? We know the temperature has risen 20° which can be represented as $+20°$, but how do we find it mathematically? The change is (the new temperature) $-$ (the old temperature). Therefore, $70° - (+50°) = +20°$ represents the change in temperature. The positive indicates the temperature rose by 20°. Note that $70° - (+50°) = 20°$ which is the same as $70° + (-50°)$.

b) If the temperature changes from 60° to 50°, what is the change in temperature? We know the temperature has dropped 10° which can be represented by $-10°$. The change can be found mathematically by taking (the new temperature) $-$ (the old temperature). Therefore, the change would be $50° - (+60°) = -10°$ where the negative indicates that the temperature has dropped. Note that $50° - (+60°) = 50° + (-60°) = -10°$.

c) If the temperature changes from $-10°$ to 20°, what is the change in temperature? We know the temperature has risen 30° which can be represented by $+30°$. The change can be found mathematically by taking (the new temperature) $-$ (the old temperature). Therefore, the change would be $20° - (-10°) = +30°$ where the positive indicates that the temperature has risen. Note that $20° - (-10°) = 20° + (+10°) = +30°$.

d) Likewise, if the temperature changes from $-20°$ to $-10°$, how much did it change? We know it has risen 10°. The change in temperature is again found mathematically by taking (the new temperature) $-$ (the old temperature). Therefore, the change would be $(-10°) - (-20°) = +10°$. The positive indicates the temperature has risen by 10°. Notice $(-10°) - (-20°) = -10° + (+20°) = +10°$. Following is a table summarizing the results of Example 1.

Temperature Change	Written as Subtraction	Written as Addition
From 50° to 70°	$70° - (+50°) = 20°$	$70° + (-50°) = 20°$
From 60° to 50°	$50° - (+60°) = -10°$	$50° + (-60°) = -10°$
From $-10°$ to 20°	$20° - (-10°) = 30°$	$20° + (+10°) = 30°$
From $-20°$ to $-10°$	$-10° - (-20°) = 10°$	$-10° + (+20°) = 10°$

From the examples above it seems there is a very close relationship between subtraction and addition since each subtraction problem can be rewritten as addition. You will notice that the procedure is to change the sign of the number being subtracted and change the operation from subtraction to addition. This leads to the following procedure for subtracting integers.

PROCEDURE FOR SUBTRACTING

$a - b = a + (-b)$.   To subtract a number, add its negative (opposite). Remember, a and b can represent either positive or negative numbers.

EXAMPLE 2

Rewrite each of the following as addition and evaluate.

a) $7 - 5 =$ $7 - 5$ means $7 - (+5)$, so change the subtraction to addition and $+5$ to -5.

$7 + (-5) =$ Add.

2 Difference.

b) $-6 - 2 =$ $-6 - 2$ means $-6 - (+2)$, so change the subtraction to addition and $+2$ to -2.

$-6 + (-2) =$ Add.

-8 Difference.

c) $8 - (-5) =$ Change the subtraction to addition and -5 to $+5$.

$8 + 5 =$ It is not necessary to write $+ (+5)$ since 5 means $+5$.

13 Difference.

d) $-12 - (-8) =$ Change the subtraction to addition and -8 to $+8$.

$-12 + 8 =$ Add.

-4 Difference.

ADDITIONAL PRACTICE

Evaluate each of the following.

a) $8 - 13$

b) $-3 - 9$

c) $5 - (-12)$

d) $-16 - (-9)$

PRACTICE EXERCISES

Rewrite each of the following as addition and evaluate.

1) $12 - 8$ **2)** $-7 - 5$

3) $9 - (-4)$ **4)** $-14 - (-6)$

If more practice is needed, do the Additional Practice Exercises in the margin.

Note: Since $a - b = a + (-b)$, we will no longer write two signs when adding or subtracting, except in the case of subtracting a negative number. We will assume the operation is addition and the sign is the sign of the number following it. For example, $6 - 4 = 6 + (-4) = 2$ and $-8 - 5 = -8 + (-5) = -13$.

If there are more than two numbers, simplify inside any parentheses or brackets first from the innermost to the outermost, then add or subtract in order from left to right as in the following examples.

EXAMPLE 3

Evaluate the following. Remember the order of operations.

a) $8 - 10 + 5 =$ Add 8 and -10. $(8 - 10 = 8 + (-10))$

$-2 + 5 =$ Add -2 and 5. $(-2 + 5 = -2 + (+5))$

3 Sum.

b) $-14 - 7 + 6 - (-4) =$ Add -14 and -7. $(-14 - 7 = -14 + (-7))$

$-21 + 6 - (-4) =$ Add -21 and 6. $(-21 + 6 = -21 + (+6))$

$-15 - (-4) =$ Change $- (-4)$ to $+ 4$.

$-15 + 4 =$ Add -15 and 4. $(-15 + 4 = -15 + (+4))$

-11 Sum.

c) $14 - (9 - 21) =$ Inside parentheses first, so add 9 and -21.

$14 - (-12) =$ Change $- (-12)$ to $+ 12$.

$14 + 12 =$ Add 14 and 12.

26 Sum.

d) $(-16 - 8) - (18 - 9) =$ Inside parentheses first.

$-24 - 9 =$ Add -24 and -9.

-33 Sum.

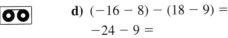

ADDITIONAL PRACTICE

Simplify the following.

e) −9 + 7 − 3

f) 14 − 6 − (−7) + 5

g) 15 − (−11 + 7)

h) (12 − 19) − (−6 + 13)

i) 5 − [6 − (−15 + 9)]

---

e) 5 − [7 − (−6 − 5)] =	Inside parentheses first.
5 − [7 − (−11)] =	Change − (−11) to + 11.
5 − [7 + 11] =	Inside brackets next.
5 − 18 =	Add 5 and −18.
−13	Sum.

Note: In Example 3e above, we used brackets and parentheses to show grouping since it is confusing to have parentheses inside parentheses.

PRACTICE EXERCISES

Evaluate the following.

5) −7 + 10 − 8   **6)** 10 − (−6) + 4 − 12

7) 17 − (−5 − 8)   **8)** (8 − 16) − (12 − 21)

9) 8 − [−9 − (−24 + 18)]

If more practice is needed, do the Additional Practice Exercises in the margin.

Often problems present themselves to us in the form of words rather than symbols. When this happens it is necessary to translate from English to mathematics. Remember, sum means add and difference means subtract.

EXAMPLE 4

Write each of the following as a subtraction problem and evaluate.

a) The difference of 7 and 12

Solution:

The difference of 7 and 12 means subtract 12 from 7 and is written as 7 − 12.

7 − 12 =	Add 7 and −12.
−5	Therefore, the difference of 7 and 12 is −5.

b) 8 less than 2

Solution:

8 less than 2 means subtract 8 from 2 and is written as 2 − 8.

2 − 8 =	Add 2 and −8.
−6	Therefore, 8 less than 2 is −6.

c) The difference of 4 and −5 added to 10

Solution:

The difference of 4 and −5 means subtract −5 from 4 which is 4 − (−5). This entire difference is added to 10 which is written as 10 + [4 − (−5)]. The brackets around 4 − (−5) are necessary to show that we are adding the difference of 4 and −5 to 10. Without the brackets we would have 10 + 4 − (−5) which means we are adding 10 and 4 and then finding the difference between that sum and −5. So we have:

$10 + [4 − (−5)] =$ Rewrite $4 − (−5)$ as $4 + 5$.

$10 + [4 + 5] =$ Add 4 and 5. Drop the brackets.

$10 + 9 =$ Add 10 and 9.

19 Therefore, the difference of 4 and −5 added to 10 is 19.

Solution:

The sum of −8 and 6 means add −8 and 6 and is written as $−8 + 6$. The difference of −3 and 2 means subtract 2 from −3 and is written as $−3 − 2$. We need to subtract $−8 + 6$ from $−3 − 2$. This is written as $(−3 − 2) − (−8 + 6)$. The parentheses are necessary to show we are subtracting the *sum* of −8 and 6 (in parentheses) from the *difference* of −3 and 2 (in parentheses). So we have:

$(−3 − 2) − (−8 + 6) =$ Simplify inside the parentheses first.

$(−5) − (−2) =$ Rewrite $(−5) − (−2)$ as $−5 + 2$.

$−5 + 2 =$ Add.

$−3$ Therefore, the sum of −8 and 6 subtracted from the difference of −3 and 2 is −3.

PRACTICE EXERCISES

Write each of the following as a subtraction problem and evaluate.

10) The difference of −5 and −9.

11) 10 less than −3.

12) The difference of 3 and −1 added to −3.

13) The sum of 5 and −8 subtracted from the difference of −4 and 6.

• **Finding solutions of equations**

An **equation** is a statement that two expressions are equal. For example, $2 − 8 = −6$, $5 = 12 − 7$, $P = 2L + 2W$, and $x − 4 = 7$ are all equations. The equations in which we will be interested are those involving variables like $x − 4 = 7$. A number is a solution of an equation if the resulting statement is true when the variable is replaced by that number. An equation is said to be **solved** by any number that is a solution of the equation.

EXAMPLE 5
Determine if the given value of the variable is a solution of the equation.

a) $x − 4 = 7$: $x = 11$ Substitute 11 for x.

$11 − 4 = 7$ Add 11 and −4.

$7 = 7$ This is a true statement. Therefore, 11 is a solution.

Note: Example (5a) can be thought of as, "If $x − 4 = 7$, can $x = 11$?"

b) $3x − 5 = 1$: $x = 2$ Substitute 2 for x.

$3 \cdot 2 − 5 = 1$ Multiply 3 and 2.

$6 − 5 = 1$ Add 6 and −5.

$1 = 1$ This is a true statement. Therefore, 1 is a solution.

c) $4x - 12 = -3 - x: x = 5$ Substitute 5 for x.

$4 \cdot 5 - 12 = -3 - 5$ Multiply 4 and 5 and add −3 and −5.

$20 - 12 = -8$ Add 20 and −12.

$8 = -8$ This is a false statement. Therefore, 5 is not a solution.

ADDITIONAL PRACTICE

Determine if the given number is a solution of the equation.

j) $x - 3 = 7: x = 10$

k) $3x - 5 = -1: x = 2$

l) $4x - 3 = -15 + 3x: x = 3$

PRACTICE EXERCISES

Determine if the given value of the variable is a solution of the equation.

14) $x - 7 = 5: x = 12$ **15)** $3x - 10 = -2: x = 4$

16) $4x - 3 = 7 - x: x = 2$

If more practice is needed, do the Additional Practice Exercises in the margin.

The solutions to an equation must come from a specified set of numbers which is called the **domain** or **replacement set**.

EXAMPLE 6

Find the one solution of each of the following equations from the given domain {−3, −2, −1, 0, 1, 2, 3}.

a) $x + 2 = 1$ Substitute values for x from the domain until you find the value which solves the equation. Try −3.

$-3 + 2 = 1$

$-1 \neq 1$ Therefore, −3 is not a solution. Try −2.

$-2 + 2 = 1$

$0 \neq 1$ Therefore, −2 is not a solution. Try −1.

$-1 + 2 = 1$

$1 = 1$ True statement. Therefore, −1 is a solution.

b) $x - 3 = -5$ Try −3.

$-3 - 3 = -5$

$-6 \neq -5$ Therefore, −3 is not a solution. Try −2.

$-2 - 3 = -5$

$-5 = -5$ True statement. Therefore, −2 is a solution.

ADDITIONAL PRACTICE

Find the one solution of each of the following from the given domain {−2, −1, 0, 1, 2}.

m) $x - 6 = -7$

n) $x + 5 = 7$

PRACTICE EXERCISES

Find the one solution of each of the following from the given domain {−3, −2, −1, 0, 1, 2, 3}.

17) $x + 5 = 4$ **18)** $x - 2 = -4$

If more practice is needed, do the Additional Practice Exercises in the margin.

Evaluate each of the following.

1) $9 - 3$ **2)** $11 - 7$ **3)** $6 - 12$

4) $9 - 15$ **5)** $8 - (-4)$ **6)** $6 - (-7)$

7) $-5 - (-9)$ **8)** $-7 - (-8)$ **9)** $-14 - (-6)$

10) $-9 - (-3)$ **11)** $\dfrac{7}{5} - \dfrac{4}{5}$ **12)** $\dfrac{12}{7} - \dfrac{6}{7}$

13) $-\dfrac{4}{3} + \dfrac{5}{6}$ **14)** $-\dfrac{7}{4} + \dfrac{3}{8}$ **15)** $-\dfrac{3}{4} - \dfrac{2}{5}$

16) $-\dfrac{2}{3} - \dfrac{4}{7}$ **17)** $\dfrac{5}{6} - \left(-\dfrac{3}{8}\right)$ **18)** $\dfrac{7}{8} - \left(-\dfrac{5}{12}\right)$

19) $7.5 - 9.2$ **20)** $4.6 - 10.3$ **21)** $-7 + 4 - 6$

22) $-5 + 9 - 8$ **23)** $-13 - 6 + 7$

24) $-17 - 7 + 4$ **25)** $21 - (-8) - 4$

26) $14 - 6 - 9$ **27)** $\dfrac{7}{6} - \dfrac{2}{3} + \dfrac{3}{4}$

28) $\dfrac{5}{8} - \dfrac{1}{4} - \dfrac{1}{6}$ **29)** $10.6 - 5.1 - 3.3$

30) $11.4 - 2.1 + 6.3$ **31)** $8 - (6 - 2)$

32) $12 - (7 - 5)$ **33)** $9 - (5 - 8)$

34) $7 - (3 - 8)$ **35)** $(16 - 12) - (8 - 5)$

36) $(26 - 17) - (7 - 3)$

37) $(14 - 23) - (31 - 26)$

38) $(16 - 28) - (34 - 27)$

39) $(-9 + 24) - (-19 - 5)$

40) $(-14 + 29) - (-24 - 9)$

41) $\left(\dfrac{7}{12} - \dfrac{5}{12}\right) - \left(\dfrac{5}{6} - \dfrac{1}{6}\right)$ **42)** $\left(\dfrac{5}{9} - \dfrac{7}{9}\right) - \left(\dfrac{7}{6} - \dfrac{4}{6}\right)$

43) $8 - [3 - (6 - 9)]$ **44)** $6 - [4 - (3 - 9)]$

45) $-13 + [24 - (-9 + 3)]$

46) $-25 + [37 - (-8 + 7)]$

47) $[(33 - 24) - 12] - 23$

48) $[(46 - 39) - 32] - 44$

CALCULATOR EXERCISES

Find the following differences using a calculator when necessary.

C1) $7564.968 - (-8573.3547)$

C2) $6503.954 - (-2759.8735)$

C3) $56.45 - [45.93 - (45.98 - 65.43)]$

C4) $87.4 - [8.94 - (5.7 - 87.64)]$

Write each of the following as a subtraction problem and evaluate. Remember, the difference of a and b means subtract b from a and is written $a - b$.

49) The difference of 7 and −5

50) The difference of 9 and −5

51) The difference of −10 and 6

52) The difference of −14 and 9

53) 5 less than 8

54) 7 less than 12

55) 8 less than −5

56) 13 less than −8

57) 6 less 9

58) 7 less 12

59) −13 subtracted from 9

60) −21 subtracted from 11

61) 32 subtracted from 26

62) 29 subtracted from 21

63) The difference of 6 and −5 added to 14

64) The difference of 8 and −3 added to 23

65) 15 added to the difference of −4 and −6

66) 7 added to the difference of 9 and −5

67) The sum of −5 and 6 subtracted from the difference of 9 and −4

68) The sum of 7 and −3 subtracted from the sum of −7 and 5

69) The difference of −5 and −2 added to the difference of 7 and −3

70) The difference of 4 and 9 added to the difference of −5 and 8

Determine if the given value of the variable is a solution of the equation.

71) $x + 5 = 9$; $x = 4$ **72)** $x - 2 = -4$; $x = -2$

73) $4x - 6 = -2$; $x = 2$ **74)** $2x - 12 = -2$; $x = 5$

75) $3x - 3 = 5 - x$; $x = 2$

76) $4x - x = -4 + x$; $x = 1$

Find the solution of each of the following from the given domain $\{-3, -2, -1, 0, 1, 2, 3\}$.

77) $x + 7 = 6$ **78)** $x - 5 = -7$ **79)** $4 - x = 6$

80) $-3 - x = -5$ **81)** $x - 9 = -6$ **82)** $x + 12 = 12$

Problems 83–87 involve changes in temperature. Write as subtraction and complete the table.

	Old Temperature	New Temperature	Subtraction	Change in Temperature
83)	78°	53°	_____	_____
84)	57°	82°	_____	_____
85)	12°	−15°	_____	_____
86)	−13°	−5°	_____	_____
87)	−23°	18°	_____	_____

Problems 88–90 involve a scuba diver. Write as subtraction and complete the table.

	Previous Depth	Present Depth	Subtraction	Change in Depth
88)	−22 ft	−37 ft	_____	_____
89)	−34 ft	−15 ft	_____	_____
90)	−84 ft	−49 ft	_____	_____

Write each as subtraction and answer the following.

91) A submarine is submerged at a depth of 145 feet and an airplane searching for the sub flies over at a height of 350 feet. What is the distance between them?

92) If the submarine in Exercise 91 is submerged at a depth of 245 feet and the airplane is at an altitude of 435 feet, what is the distance between them?

93) If a company loses $5000 in its first year of operation and has a profit of $12,000 the following year, how much greater were the company's earnings the second year?

94) If a company has a profit of $6500 dollars one year and a loss of $7300 the next, how much less did the company earn the second year than the first?

95) A mine shaft that began at 500 feet above sea level ended at 300 feet below sea level. Suppose a miner is 250 feet above sea level and descends to 125 feet below sea level. What is his change in altitude? Remember, change = new altitude − old altitude.

96) Suppose the miner in Exercise 95 is 210 feet below sea level and ascends to 109 feet below sea level. What is his change in altitude?

97) An airplane is flying over Death Valley at a height of 2500 feet when it flies over a hiker whose elevation is 187 feet below sea level. What is the distance between the airplane and the hiker?

98) For the first half of the fiscal year a company had a profit of $123,000 and during the second half of the year had a loss of $105,000. Find the difference in the company's earnings for the first and second halves of the year.

99) A gauge initially had a reading of $17\frac{1}{2}$ and later had a reading of $-6\frac{3}{4}$. By how much did the reading change?

100) On the same day in February, the low temperature in East Yellowstone, Montana was −36°F and the low temperature in Orlando, Florida was 52°F. How much higher was the low temperature in Orlando than in East Yellowstone?

Marsha tracked the progress of one of her stocks for a week and recorded the changes in the following chart. Use the chart to answer Exercises 101–102.

Monday	Tuesday	Wednesday	Thursday	Friday
+3	$-2\frac{1}{8}$	$+2\frac{3}{4}$	$-3\frac{1}{2}$	$-1\frac{7}{8}$

101) What was the net gain or loss for the week?

102) How much higher was the stock on Monday than on Tuesday?

John conducts a small part-time business from his home. The graph below shows the daily profit or loss for his business. Use the graph to answer Exercises 103–104.

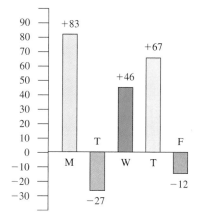

103) What was the total profit/loss for the week?

104) How much more did the company earn on Monday than on Tuesday?

CHALLENGE EXERCISES (105–108)

Evaluate the following.

105) $5^2[6^2 \cdot 2 - (-4)] \div 19$

106) $2^4[48 - (-12) - 32] \div 7 \cdot 4$

107) $5^2\{6 - 2[10 - 3(-9 + 12)]\}$

108) $2^3\{24 - 3[12 - 2(-7 + 11)]\}$

WRITING EXERCISES

109) Is subtraction commutative? Why or why not? (Hint: Try some examples.)

110) Is subtraction associative? Why or why not? (Hint: Try some examples.)

WRITING EXERCISE OR GROUP PROJECT

If done as a group project, each group should write at least two exercises. Then groups should exchange with other groups and solve. If done as a writing exercise, each student should write and solve one exercise.

111) Write an application problem involving the subtraction of integers, at least one of which is negative.

SECTION **1.8**

Terms and Polynomials

OBJECTIVES

When you complete this section, you will be able to:

Ⓐ Identify terms.
Ⓑ Identify the numerical coefficient and the variable(s) of a term.
Ⓒ Find the degree of a term.
Ⓓ Identify types of polynomials.
Ⓔ Write polynomials in descending order and find their degree.
Ⓕ Evaluate polynomials for a specified value of the variable.

INTRODUCTION

In arithmetic, operations are performed on numbers. In algebra, operations are performed on algebraic expressions. An **algebraic expression** is any constant, variable, or combination of constants and variables using the operations of addition, subtraction, multiplication, division, raising to powers, or taking of roots (which will be discussed later). Examples of algebraic expressions are:

$$5, -3, x, 3x, x^3, -4x^4y, 4x^2 - 5x + 2, \frac{2x}{y}, \frac{2x-5}{3x+5}, \frac{2x^3 - 4y^2}{4y^4 + 3x^2}$$

• **Identifying terms**

A term is a number, variable, or product and/or quotient of numbers and variables raised to powers. We have not used 0 or negative integers as powers, but we will in Chapter 2 where we will show that $x^0 = 1$ if $x \neq 0$. We mention this now because we need it when we discuss the degree of a polynomial later in this section. Examples of terms are 5, x, $5x$, $-3x^2$, $2x^3y^2$, $-7x^2y^3z^4$, $\frac{x}{y}$ and $4x^{-2}$. An expression like $3x + 2$ is not a term since addition or subtraction is not allowed in a term. This is actually the sum of two terms.

• **Definition of polynomial**

A **polynomial** is the sum or difference of a **finite** number of terms in which there is no variable as a divisor and the exponents on the variables are whole numbers. A set is finite if it is possible to represent the number of elements in the set with a whole number. Examples of polynomials are x, $2x + 3$, $5x^2 + 2x - 4$, $2x^3 + 2x^2 - 4x + 6$, $\frac{2}{3}x^2 + \frac{4}{5}x - 4$, and 3.

Expressions like $3x^{-1} + 2$ and $\frac{3}{x} + 6$ are not polynomials because negative exponents on variables or division by variables is not allowed in a polynomial.

• **Identifying types of polynomials**

Polynomials can also be classified by the number of terms they contain as illustrated in the following table.

Name	Number of Terms	Examples	Terms
Monomial	One term	4	4
		x	x
		$-5x^2$	$-5x^2$
Binomial	Two terms	$3x - 2$	$3x, -2$
		$4x^3 - 3x$	$4x^3, -3x$
Trinomial	Three terms	$y^2 - 3y + 5$	$y^2, -3y, 5$
		$6x^3 + 2x^2 - 4x$	$6x^3, 2x^2, -4x$

Note: Polynomials with more than three terms are not given special names. You will be asked to suggest names for some of these in the writing exercises.

EXAMPLE 1

Classify each of the following as a monomial, binomial, trinomial, or none of these.

a) $2x + 4$ — There are two terms, $2x$ and 4. Therefore, this is a binomial.

b) $3x^3$ — There is only one term. Therefore, this is a monomial.

c) $2y^3 + 4y - 2$ — There are three terms, $2y^3$, $4y$, and -2. Therefore, this is a trinomial.

d) $6x^2 - 4x^4 + 5x - 4$ — There are four terms. Therefore, this is none of these.

e) -4 — There is one term. Therefore, this is a monomial.

f) $\dfrac{5}{y} - 10$ — We have division by a variable. Therefore, this is not a polynomial.

PRACTICE EXERCISES

Classify each of the following as monomial, binomial, trinomial, or none of these.

1) $2y^2 + 4$ 　　　　　　　　　　　**2)** $8x^4$

3) $-3x^2 - 6x^5 - 7x$ 　　　　　　**4)** $2x^4 - 2x - x^2 + 9$

5) 7 　　　　　　　　　　　　　　**6)** $\dfrac{3}{x} + 5$

• **Coefficients, variables, and degree of monomials**

When writing a term, the constant that is usually written before the variable(s) is called the numerical coefficient, or simply, the coefficient. For example, in the term $3x$, the numerical coefficient is 3. If a monomial has only one variable, the exponent on the variable is called the degree of the monomial. If there is more than one variable, the degree of the monomial is the sum of all the exponents on all the variables.

EXAMPLE 2

Give the coefficient, variable, and degree of each of the following monomials.

	Monomial	Coefficient	Variable(s)	Degree
a)	$4x^3$	4	x	3
b)	$-8z^6$	-8	z	6
c)	y	$1(y = 1y^1)$	y	1
d)	$2m^3n^4$	2	m, n	$3 + 4 = 7$
e)	-3	-3	none	0 since $-3 = -3x^0$

Be careful: In Example (2e) the degree of -3 is 0 since $-3 = -3x^0$. It is often mistakenly given a degree of 1. Remember, the degree of a term is the exponent of the *variable*, so $-3x$ has a degree of 1. Thus, -3 and $-3x$ could not have the same degree.

PRACTICE EXERCISES

Give the coefficient, variable(s), and degree of each of the following.

7) $-8x^3$ coefficient = ___ variable(s) = ___ degree = ___

8) y coefficient = ___ variable(s) = ___ degree = ___

9) $2x^2y^4$ coefficient = ___ variable(s) = ___ degree = ___

10) -5 coefficient = ___ variable(s) = ___ degree = ___

• **Writing polynomials in descending order and determining degree**

    A polynomial of one variable is written in **descending order** when the term of highest degree is written first, the term of next highest degree second, etc. For example, $3x^2 - 2x + 4$ is in descending order, but $2x - 3x^3 - 5$ is not because the term of highest degree is not written first. The **degree of a polynomial** is the same as the degree of the term of the polynomial with the highest degree. See the chart in Example 3.

EXAMPLE 3

Write each polynomial in descending order and find the degree of the polynomial.

Polynomial	Descending Order	Terms	Degree of Terms	Degree of Polynomial
a) $3 + 4x$	$4x + 3$	$4x, 3$	1, 0	1
b) $2x - 4 + x^2$	$x^2 + 2x - 4$	$x^2, 2x, -4$	2, 1, 0	2
c) $4x - 5x^3 - 7x^4 + 2$	$-7x^4 - 5x^3 + 4x + 2$	$-7x^4, -5x^3, 4x, 2$	4, 3, 1, 0	4

ADDITIONAL PRACTICE

Write each of the following in descending order and give its degree.

a) $3 - 7x$

b) $7 - 3x + 5x^2$

c) $5y - 4y^2 + 6y^4 - y^3$

PRACTICE EXERCISES

Write each of the following polynomials in descending order and give its degree.

Polynomial	Descending Order	Degree
11) $-5 + 3x$	_____	_____
12) $-5 + 2x^2 + 8x$	_____	_____
13) $6x - 7x^4 - 8x^2 + x^3$	_____	_____

    If more practice is needed, do the Additional Practice Exercises in the margin.

• **Evaluating polynomials**

    Since polynomials contain variables and variables represent numbers, it is possible to evaluate a polynomial when given a value for its variable. Remember to keep the order of operations in mind when evaluating the polynomial. For convenience in evaluating, polynomials are often written using a special notation. The polynomial is given a name, usually a letter, and then its variable is specified in parentheses. For example, we might write $P(x) = x^2 + 2x - 3$ where P is the name we have given the polynomial and x is the variable. When we wish to assign the variable a value, the value replaces the variable inside the parentheses and is substituted for the variable throughout the expression. If we wish to evaluate $P(x) = x^2 + 2x - 3$ for $x = 3$, we write $P(3) = 3^2 + 2(3) - 3$ and then simplify. So, $P(3) = 9 + 6 - 3 = 12$.

ADDITIONAL PRACTICE

Evaluate each of the following for the given value of the variable.

d) $P(x) = 3x - 7$; $x = 2$

e) $P(t) = t^2 + 4t - 4$; $t = 6$

EXAMPLE 4

Evaluate the following polynomials for the given value of the variable.

a) $P(x) = 2x + 4$; $x = 3$ Substitute 3 for x.

    $P(3) = 2(3) + 4$ Multiply 2 and 3.

    $P(3) = 6 + 4$ Add 6 and 4.

    $P(3) = 10$ Therefore, the value of P when x is 3 is 10.

 b) $P(t) = t^2 + 3t - 4$; $t = 2$ Substitute 2 for t.

    $P(2) = (2)^2 - 3(2) - 4$ Raise to powers first, so square 2.

    $P(2) = 4 + 3(2) - 4$ Multiply 3 and 2.

    $P(2) = 4 + 6 - 4$ Add in order left to right.

    $P(2) = 10 - 4$ Add.

    $P(2) = 6$ Therefore, the value of P when t is 2 is 6.

PRACTICE EXERCISES

Evaluate each of the following polynomials for the given value of the variable.

14) $P(x) = 2x + 5$; $x = 2$ **15)** $P(y) = y^2 + 3y - 6$; $y = 4$

If more practice is needed, do the Additional Practice Exercises in the margin.

Evaluations of polynomials occur in many situations in the sciences and the real world.

EXAMPLE 5

Solve the following:

A particle moves along a straight line such that the distance from the starting point is given by $s(t) = t^2 + 3t + 2$ where $s(t)$ is the distance after t seconds. Find the location of the particle after 3 seconds by finding $s(3)$.

    $s(t) = t^2 + 3t + 2$ To find $s(3)$ replace t with 3.

    $s(3) = 3^2 + 3(3) + 2$ Square 3.

    $s(3) = 9 + 3(3) + 2$ Multiply 3(3).

    $s(3) = 9 + 9 + 2$ Add in order left to right.

    $s(3) = 20$ Therefore, the particle is 20 units from the starting point after 3 seconds.

PRACTICE EXERCISE

16) If x units of a product are sold, the revenue is given by
$R(x) = 1,000x + 0.2x^2$. Find the revenue when 25 units are sold by finding $R(25)$.

The polynomial plays a central role in elementary algebra. A great deal of the remainder of this book will be spent performing operations on polynomials.

Making Sure: Confidence Building

We feel good about what we have done when we know that it is right. We can tell if we are right by checking our answers. Sometimes the answer is in the back of the book, but most of the time we need to perform our own check. The check should be different from and shorter than reworking the exercise. For example, the check for subtraction is addition. $8 - 5 = 3$, if $3 + 5 = 8$. Also, the check for division is multiplication. $54 \div 9 = 6$, if $6 \cdot 9 = 54$. Always check your work!

EXERCISE SET **1.8**

ANSWERS:
Practice 14–15

14) 9   **15)** 22

Additional Practice d–e

d) −1   **e)** 56

Practice 16

16) $25,125

Classify each of the following as a monomial, binomial, trinomial, or none of these.

1) $2x^2$ _____

2) $3y^3$ _____

3) $x^2 - 3x + 4$ _____

4) $x^2 - 5x + 2$ _____

5) $5 + 8x$ _____

6) $3 - 7x$ _____

7) $2x - 2x^2 + 4$ _____

8) $4 - 3x + 5x^2$ _____

9) $\dfrac{2x^2}{y^2} - 4x$ _____

10) $\dfrac{3y^3}{2x^2} - 2y$ _____

11) 5 _____

12) -7 _____

13) $x^3 + 3x - 4x^2 - 3$ _____

14) $x^4 - 3x + 2x^3 + 4$ _____

Give the coefficient, variable(s), and degree of each of the following.

	Coefficient	Variable(s)	Degree
15) $3x$	_____	_____	_____
16) $2y$	_____	_____	_____
17) x^2	_____	_____	_____
18) y^4	_____	_____	_____
19) $-5x$	_____	_____	_____
20) $-7z$	_____	_____	_____

	Coefficient	Variable(s)	Degree
21) $4x^3$	_____	_____	_____
22) $7y^5$	_____	_____	_____
23) $-x^3$	_____	_____	_____
24) $-y^5$	_____	_____	_____
25) $10x^2y^4$	_____	_____	_____
26) $12x^5y^2$	_____	_____	_____
27) $-13x^3z^3$	_____	_____	_____
28) $-16y^4z^6$	_____	_____	_____

Write each of the following in descending order and give the degree of the polynomial.

	Descending Order	Degree
29) $4 - 5x$	_____	_____
30) $5 + 2y$	_____	_____
31) $3x - 4 + 2x^2$	_____	_____
32) $5x - 6 + 4x^2$	_____	_____
33) $3x - 4x^2 - x^4 + x^3$	_____	_____
34) $2y + 3y^3 - y^4 + 5$	_____	_____
35) $x + 3x^3$	_____	_____
36) $4 + 3x^2$	_____	_____

Evaluate each of the following for the indicated value of the variable.

37) $P(x) = 3x - 9; x = 3$

38) $P(y) = 4y - 3; y = 2$

39) $P(x) = 2x^2 - 6; x = 4$

40) $P(x) = 2x^2 - 5; x = 3$

41) $P(x) = x^2 + 3x - 5; x = 1$

42) $P(x) = x^2 + 4x - 3; x = 5$

43) $P(s) = 2s^2 + 5s - 7; s = 4$

44) $P(s) = 3s^2 + 6s - 12; s = 5$

45) $P(x) = x^3 + 2x^2 + 3x - 3; x = 2$

46) $P(x) = x^3 + 4x^2 + 2x + 5; x = 3$

47) $P(x) = 2x + 3; x = \dfrac{1}{2}$

48) $P(x) = 4x - 5; x = \dfrac{3}{4}$

49) $P(x) = x^2 + 2x + 1; x = \dfrac{1}{3}$

50) $P(x) = x^2 - 3x + 3; x = \dfrac{1}{4}$

If an object is dropped from rest, the distance it falls is given by $s(t) = 16t^2$. Use this to answer Exercises 51–52.

51) While hunting in rural Georgia a hunter finds the location of an old homesite with an open well. In order to estimate the depth of the well, he drops a stone into it and observes that it took $t = 2.5$ seconds for the stone to hit the water. Find the depth of the well by finding $s(2.5)$.

52) If a rock is dropped from the top of the Sears Tower, find how far it has fallen after 3 seconds by finding $s(3)$.

A particle moves along a line such that the distance from the starting point is given by $s(t) = t^3 + 2t^2 + 3t + 5$ with t measured in seconds. Use this to answer Exercises 53–54.

53) Find the location of the particle after 2 seconds by finding $s(2)$.

54) Find the location of the particle after 5 seconds by finding $s(5)$.

A furniture manufacturing company finds that the cost of manufacturing x sofas is given by $C(x) = 3500 + 23x + 0.4x^2$. Use this to answer Exercises 55–56.

55) Find the cost of producing 100 sofas by finding $C(100)$.

56) Find the cost of producing 500 sofas by finding $C(500)$.

A farmer has 200 feet of fencing and wishes to enclose a rectangular field. If the length of the field is x, then the area is $A(x) = 200x - 2x^2$. Use this to answer Exercises 57–58.

57) Find the area when the length is 50 feet by finding $A(50)$.

58) Find the area when the length is 30 feet by finding $A(30)$.

CALCULATOR EXERCISES

C1) Use a calculator to evaluate $P(x) = x^3 - 5.1x^2 + 4.3x - 6.7$ for $x = 4.7$.

C2) Use a calculator to evaluate $P(x) = 3.2x^4 - 6.5x^3 + 8.5x - 9.8$ for $x = 2.3$.

WRITING EXERCISES

59) How does an algebraic expression differ from a polynomial?

60) Give three examples of algebraic expressions that are not polynomials.

61) If special names were given to polynomials of more than three terms, what suggestions would you have for the name of a polynomial of four terms? five terms?

62) How do you find the degree of a monomial? a polynomial with two or more terms?

63) What is meant by writing a polynomial in descending order?

64) Why does 2^4y^2 have degree 2 and x^4y^2 have degree 6?

65) How does a term differ from a monomial?

CRITICAL THINKING

66) Are all terms monomials? Explain. Are all monomials terms? Explain.

SECTION 1.9

Combining Like Terms and Polynomials

OBJECTIVES

When you complete this section, you will be able to:

Ⓐ Recognize and apply the distributive property.
Ⓑ Add and subtract like terms.
Ⓒ Add and subtract polynomials.

INTRODUCTION

If we had three apples and two oranges and purchased four apples and six oranges we would now have seven apples and eight oranges. We add the number of apples to the number of apples and the number of oranges to the number of oranges, but not the number of apples with the number of oranges. Though very simplistic, this is basically the idea behind adding **like terms** which is one of the topics of this section.

Before we can discuss adding terms, we need another number property. Let us simplify $3(5 + 2)$ in two different ways. If we follow the order of operations, we get $3(5 + 2) = 3(7) = 21$. If we find $3 \cdot 5 + 3 \cdot 2$, we get $15 + 6 = 21$. Therefore, $3(5 + 2) = 3 \cdot 5 + 3 \cdot 2$. This example can be generalized into the **distributive property** stated below.

- **Distributive property**

DISTRIBUTIVE PROPERTY OF MULTIPLICATION OVER ADDITION

For all numbers a, b, and c, $a(b + c) = a \cdot b + a \cdot c$.

The distributive property means that it does not matter whether we add inside the parentheses and then multiply or multiply each term first and then add. We will not discuss multiplication of signed numbers until Chapter 2, so Example 1 will be limited to natural numbers only. The distributive property holds, however, for all real numbers.

EXAMPLE 1

Verify the distributive property for each of the following by evaluating both sides and showing that they are equal.

a) $2(3 + 5) = 2 \cdot 3 + 2 \cdot 5$ $a(b + c) = ab + ac$ and use the order of operations
$2(8) = 6 + 10$ on each side.
$16 = 16$

b) $5(2 + 8) = 5 \cdot 2 + 5 \cdot 8$ $a(b + c) = ab + ac$ and use the order of operations
$5(10) = 10 + 40$ on each side.
$50 = 50$

PRACTICE EXERCISES

Verify the distributive property for each of the following. Rewrite each expression as in Example 1, then evaluate each side and show they are equal.

1) $3(4 + 2)$ **2)** $5(6 + 3)$

We need to be able to recognize the distributive property when we see it in various forms. Instead of writing it as $a(b + c) = ab + ac$, another form of the distributive property is $(b + c)a = ba + ca$. This equation may be "turned around" and written as $ba + ca = (b + c)a$. For example, $3 \cdot 4 + 5 \cdot 4 = (3 + 5) \cdot 4$. It is this form that we will use in Example 2.

Terms that have the same variables with the same exponents on these variables are called **like terms**. Examples of like terms are $3x$ and $5x$, $6y$ and $9y$, $11z^2$ and $9z^2$, $3x^2y$ and $7x^2y$. The terms $4x$ and $3y$ are not like terms because the variables are not the same. The terms $3x$ and $3x^2$ are not like terms because the exponents on the variables are not the same.

Now we will use the distributive property in the form $ba + ca = (b + c)a$ to add like terms. Study the following examples carefully.

EXAMPLE 2
Find the following sums.

a) $2 \cdot x + 4 \cdot x =$ Apply $ba + ca = (b + c)a$.

$\quad (2 + 4)x =$ Now add the 2 and 4.

$\quad 6x$ Therefore, $2x + 4x = 6x$.

b) $5xy - 3xy =$ Apply $ba + ca = (b + c)a$.

$\quad (5 - 3)xy =$ Now add 5 and -3.

$\quad 2xy$ Therefore, $5xy - 3xy = 2xy$.

c) $3x^2 + 8x^2 =$ Apply $ba + ca = (b + c)a$.

$\quad (3 + 8)x^2 =$ Now add 3 and 8.

$\quad 11x^2$ Therefore, $3x^2 + 8x^2 = 11x^2$.

d) $3x + 7y$ The distributive property does not apply, so this expression cannot be simplified.

In Examples (2a–c) we were adding like terms. In each case the variable was placed outside the parentheses with the coefficients inside and the coefficients were then added. This leads to the following procedure for adding like terms.

ADDITION OF LIKE TERMS

To add like terms, add the numerical coefficients and leave the variable portion unchanged.

The above rule greatly simplifies the addition of like terms since we no longer have to apply the distributive property each time we add.

EXAMPLE 3
Simplify the following, if possible, by adding like terms.

a) $6x + 2x =$ Like terms, so add the coefficients, 6 and 2.

$\quad 8x$ Leave the variable portion unchanged.

b) $4x - 7x =$ Like terms, so add the coefficients, 4 and -7.

$\quad -3x$ Leave the variable portion unchanged.

c) $5x^2 + 4x^2 =$ Like terms, so add the coefficients, 5 and 4.

$\quad 9x^2$ Leave the variable portion unchanged.

d) $6xy - 2xy + 8 =$ $6xy$ and $-2xy$ are like terms, so add them.

$\quad 4xy + 8$

e) $5x + 4y - 2x + 6y =$ $5x$ and $-2x$ are like terms and $4y$ and $6y$ are also like terms, so add each pair.

$\quad 3x + 10y$

Note: Some people prefer to use the commutative and associative properties to rearrange the terms so the like terms are together before adding. Example (3e) could be rewritten as follows: $5x + 4y - 2x + 6y = 5x - 2x + 4y + 6y = 3x + 10y$.

---

f) $3x^2 + 4x - 7x^2 + 5 + 3x =$ $3x^2$ and $-7x^2$ are like terms and so are $4x$ and $3x$.
$-4x^2 + 7x + 5$ Add each pair.

As in Example (3e), the terms could be rearranged before adding. So, $3x^2 + 4x - 7x^2 + 5 + 3x = 3x^2 - 7x^2 + 4x + 3x + 5 = -4x^2 + 7x + 5$.

g) $2x - (-5x) =$ Change to addition.
$2x + 5x =$ Add like terms.
$7x$

h) $3x + 4y - 3x^2$ There are no like terms, so this expression cannot be simplified.

ADDITIONAL PRACTICE

Simplify the following, if possible.

a) $4y + 6y$

b) $5ab + 9ab$

c) $7x - 4x + 5x$

d) $5z^2 + 9z^2$

e) $9x^2y - 5x + 3x^2y$

f) $6y + 2z - 9y - 7z$

g) $7x - 6x^2 - 9x + 6 + 7x^2$

h) $7y - (-5y)$

i) $7r^2 + 3s - 5t$

PRACTICE EXERCISES

Simplify the following, if possible.

3) $7x + 5x$ **4)** $4xy + 8xy$ **5)** $6y^2 + 9y^2$

6) $5x - 9x + 3x$ **7)** $4x^2y - 5 + 8x^2y$ **8)** $7z + 6x - 3z + 9x$

9) $4x - 5x^2 - 6x + 4 - 3x$ **10)** $4x - (-7x)$

11) $6a + 7b + 5a^2$

If more practice is needed, do the Additional Practice Exercises in the margin.

Since polynomials are made of terms, adding or subtracting polynomials is very similar to adding like terms. Parentheses are used to show which polynomials are being added or subtracted, so we must remove the parentheses and combine like terms. If there is no sign preceding the parentheses, there is an understood positive sign and we do not change any signs. We found the negative of an integer by changing its sign. In the same manner we will find the negative of a polynomial by changing *all* its signs.

EXAMPLE 4

Remove the parentheses from each of the following.

a) $(2x - 3) = 2x - 3$ There is an understood $+$ sign before the parentheses, so do not change any signs.

b) $-(3x - 6) = -3x + 6$ There is a $-$ sign before the parentheses, so change all the signs of the polynomial.

c) $(-4x + 2) = -4x + 2$ There is an understood $+$ sign before the parentheses, so do not change any signs.

d) $-(2x^2 + 4x - 5) =$ There is a $-$ sign before the parentheses, so change all
$-2x^2 - 4x + 5$ the signs of the polynomial.

To add or subtract polynomials, we remove the parentheses and add like terms.

EXAMPLE 5

Find the following sums or differences.

a) $(2x + 4) + (3x - 6) =$ Remove the parentheses. Do not change signs.
$2x + 4 + 3x - 6 =$ Add like terms.
$5x - 2$ Sum.

---

b) $(2x^2 - 6x + 4) + (x^2 - 3x - 7) =$ Remove parentheses.

$2x^2 - 6x + 4 + x^2 - 3x - 7 =$ Add like terms.

$3x^2 - 9x - 3$ Sum.

 **c)** $(2x^2 - 6x + 3) - (x^2 - 2x + 7) =$ Remove parentheses and change all the signs of the second polynomial.

$2x^2 - 6x + 3 - x^2 + 2x - 7 =$ Add like terms.

$x^2 - 4x - 4$ Difference.

d) $(x^2 + 6x - 5) - (-2x^2 + 7x - 3) =$ Remove parentheses and change all the signs of the second polynomial.

$x^2 + 6x - 5 + 2x^2 - 7x + 3 =$ Add like terms.

$3x^2 - x - 2$ Difference.

ADDITIONAL PRACTICE

Add or subtract the following polynomials.

j) $(5x - 3) + (-3x + 6)$

k) $(x^2 - 4x + 2) + (x^2 + 6x - 5)$

l) $(x^2 + x - 7) - (x^2 - 8x - 8)$

m) $(3x^2 + 2x - 9) - (-x^2 + 2x + 3)$

PRACTICE EXERCISES

Add or subtract the following polynomials.

12) $(4x + 6) + (3x + 5)$ **13)** $(x^2 - 5x + 4) + (2x^2 - 3x - 6)$

14) $(3x^2 - 7x - 4) - (x^2 + 5x - 6)$

15) $(2x^2 + 4x - 7) - (-3x^2 - 8x + 3)$

If more practice is needed, do the Additional Practice Exercises in the margin.

As was the case with integers, addition and subtraction of polynomials is often expressed in words.

EXAMPLE 6

Write each of the following as an addition and/or subtraction problem and simplify.

a) Find the sum of $2x^2 + 3x - 6$ and $4x^2 - 5x + 2$.

Solution:

Since sum means add, this is written as $(2x^2 + 3x - 6) + (2x^2 + 3x - 6) + (4x^2 - 5x + 2)$.

$(2x^2 + 3x - 6) + (4x^2 - 5x + 2) =$ Remove the parentheses.

$2x^2 + 3x - 6 + 4x^2 - 5x + 2 =$ Add like terms.

$6x^2 - 2x - 4$ Sum.

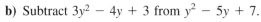 **b)** Subtract $3y^2 - 4y + 3$ from $y^2 - 5y + 7$.

Solution:

This implies that we begin with $y^2 - 5y + 7$ and subtract $3y^2 - 4y + 3$ from it. This is written as $(y^2 - 5y + 7) - (3y^2 - 4y + 3)$.

$(y^2 - 5y + 7) - (3y^2 - 4y + 3) =$ Remove parentheses.

$y^2 - 5y + 7 - 3y^2 + 4y - 3 =$ Add like terms.

$-2y^2 - y + 4$ Difference.

c) From the sum of $4x + 5$ and $2x - 8$ subtract $7x - 1$.

Solution:

This means that we first add $4x + 5$ and $2x - 8$ and from this sum subtract $7x - 1$. This is written as $[(4x + 5) + (2x - 8)] - (7x - 1)$.

$[(4x + 5) + (2x - 8)] - (7x - 1) =$	Remove the parentheses inside the brackets.
$[4x + 5 + 2x - 8] - (7x - 1) =$	Add like terms inside the brackets.
$[6x - 3] - (7x - 1) =$	Remove the parentheses and brackets.
$6x - 3 - 7x + 1 =$	Add like terms.
$-x - 2$	Answer.

PRACTICE EXERCISES

Write each of the following as addition and/or subtraction and simplify.

16) Find the sum of $5z^2 - 7z + 4$ and $2z^2 + 4z - 6$.

17) Subtract $4a^2 - 5a - 5$ from $2a^2 + 7a - 8$.

18) From the sum of $5a + 2$ and $-2a - 5$, subtract $8a - 1$.

Optional

Addition and subtraction of polynomials may be done vertically as well as horizontally. The procedure is similar to the way whole numbers and decimals are added. We place digits with the same place value underneath each other and add. In adding polynomials vertically, we put them in descending order, then place like terms underneath each other and add. Of course, there is no carry when adding polynomials.

EXAMPLE 7

Add the following vertically.

a) 374 and 68

Solution:

Place the digits with the same place value underneath each other and add.

$$
\begin{array}{r}
374 \\
68 \\
\hline
442
\end{array}
$$

b) Add $13p^2 + 3p - 4$ and $7p^2 - 4p - 4$.

Solution:

Place like terms underneath each other and add.

$$
\begin{array}{r}
13p^2 + 3p - 4 \\
7p^2 - 4p - 4 \\
\hline
20p^2 - p - 8
\end{array}
$$

c) Add $9w^2 + 12w - 14$ and $-5w^2 + 12$.

Solution:
Place like terms underneath each other and add.

$$\begin{array}{r} 9w^2 + 12w - 14 \\ -5w^2 \qquad + 12 \\ \hline 4w^2 + 12w - 2 \end{array}$$

d) Add $3x^3 - 5x + 7$ and $2x^2 + 4x - 9$.

Solution:
Place like terms underneath each other and add.

$$\begin{array}{r} 3x^3 \qquad\quad - 5x + 7 \\ 2x^2 + 4x - 9 \\ \hline 3x^3 + 2x^2 - x - 2 \end{array}$$

Remember, when subtracting polynomials we add the negative of the polynomial being subtracted. Consequently, all its signs are changed. When subtracting vertically, the polynomial being subtracted is the one on the bottom, so we change the signs of the polynomial on the bottom and add.

EXAMPLE 8
Subtract the following vertically.

a) Subtract $3y^2 - 8y - 10$ from $5y^2 - 9y + 3$.

Solution:
Written horizontally, this is $(5y^2 - 9y + 3) - (3y^2 - 8y - 10)$, so it is necessary to change the signs of $3y^2 - 8y - 10$. Written vertically, it appears as follows.

Subtract:
$$\begin{array}{r} 5y^2 - 9y + 3 \\ 3y^2 - 8y - 10 \end{array}$$

Change the signs of the polynomial on the bottom since it is the polynomial being subtracted and change the operation to addition.

Add:
$$\begin{array}{r} 5y^2 - 9y + 3 \\ -3y^2 + 8y + 10 \\ \hline 2y^2 - y + 13 \end{array}$$

b) Find the difference of $6x^2 - 11x + 9$ and $3x^2 + 6$.

Solution:
Just like with whole numbers, difference means subtract the second expression from the first.

Subtract:
$$\begin{array}{r} 6x^2 - 11x + 9 \\ 3x^2 \qquad\quad + 6 \end{array}$$

Change the signs of the bottom polynomial and add.

Add:
$$6x^2 - 11x + 9$$
$$\underline{-3x^2 \qquad - 6}$$
$$3x^2 - 11x + 3$$

c) Subtract $4x^2 - 6x - 8$ from $3x^3 - 7x^2 + 5$. (This is the same as finding the difference of $3x^3 - 7x^2 + 5$ and $4x^2 - 6x - 8$.)

Solution:

Subtract:
$$3x^3 - 7x^2 \qquad + 5$$
$$\underline{\qquad 4x^2 - 6x - 8}$$

Change the signs of the bottom polynomial and add.

Add:
$$3x^3 - \ 7x^2 \qquad + \ 5$$
$$\underline{\quad - \ 4x^2 + 6x + \ 8}$$
$$3x^3 - 11x^2 + 6x + 13$$

PRACTICE EXERCISES

Add or subtract the following vertically.

19) Add $5x^2 + 7x - 3$ and $3x^2 - 8x + 5$.

20) Add $10x^2 + 9x - 6$ and $-4x^2 - 9$.

21) Add $3x^3 - 4x + 5$ and $2x^2 + 6x - 9$.

22) Subtract $6x^2 + 7x - 9$ from $2x^2 + 5x + 7$.

23) Find the difference of $8y^2 - 9y + 3$ and $5y^2 + 9$.

24) Subtract $7x^2 - 8x + 2$ from $8x^3 - 9x^2 + 4$.

EXERCISE SET   **1.9**

Simplify the following, if possible, by combining like terms.

1) $4x + 7x$

2) $6x + 9x$

3) $-3y + 9y$

4) $-7y + 4y$

5) $\dfrac{2}{7}x + \dfrac{3}{7}x$

6) $\dfrac{2}{9}z + \dfrac{5}{9}z$

7) $\dfrac{3}{5}x - \dfrac{2}{3}x$

8) $\dfrac{3}{7}y - \dfrac{3}{5}y$

9) $12z - 8z + 3$

10) $15z - 9z + 7$

11) $3x - 8x - 6x$

12) $4z - 5z - 2z$

13) $\dfrac{9}{11}x - \dfrac{3}{11}x - \dfrac{2}{11}x$

14) $\dfrac{7}{9}b - \dfrac{4}{9}b - \dfrac{2}{9}b$

15) $\dfrac{1}{4}x + \dfrac{2}{3}x - \dfrac{5}{6}x$

16) $\dfrac{4}{7}x - \dfrac{1}{3}x + \dfrac{3}{14}x$

17) $5xy + 6x - 9xy + 2x$

18) $9xy + 5x - 3xy + 8x$

19) $3x^2 - 2y^2 - 6x^2 - 5x^2$

20) $4x^2 + 5y^2 - 6x^2 + 3y^2$

21) $9z - 5 + 3x + 3z - 2x - 7$

22) $8y - 4k + 2 - 4y + 3 - 9k$

23) $9b^2 - 8b + 2 - 2b^2 - 2b + 10$

24) $4d^2 - 4 + 4d + 4d^2 - 9 - 3d$

25) $m^2r + 3r - 5mr^2 + 3r + 5mr^2$

26) $n^2t - 8t + 6tn^2 - 2t - n^2t$

27) $\dfrac{4}{5}x^2y + \dfrac{3}{7}xy^2 - \dfrac{2}{5}x^2y + \dfrac{2}{7}xy^2$

28) $\dfrac{7}{9}ab^2 + \dfrac{3}{11}a^2b^2 - \dfrac{5}{9}ab^2 - \dfrac{5}{11}a^2b^2$

29) $\dfrac{3}{13}m^3n^2 - \dfrac{5}{9}mn^2 - \dfrac{7}{13}m^3n^2 - \dfrac{2}{9}mn^2$

30) $\dfrac{4}{17}a^4b^3 - \dfrac{5}{19}a^2b - \dfrac{8}{17}a^4b^3 - \dfrac{3}{19}a^2b$

31) $\dfrac{1}{2}x^3y - \dfrac{1}{3}xy^2 + \dfrac{2}{9}x^3y + \dfrac{1}{2}xy^2$

32) $\dfrac{7}{9}ab^2 + \dfrac{5}{3}a^2b - \dfrac{5}{18}ab^2 - \dfrac{5}{9}a^2b$

Add or subtract the following polynomials.

33) $(7x - 5) + (3x - 2)$

34) $(5y + 5) + (-4y - 3)$

35) $(6x - 5y) + (3x + 2y)$

36) $(3x + 8y) + (7x - 2y)$

37) $(5x - 4) - (3x - 6)$

38) $(4y + 7) - (6y - 2)$

39) $(12x + 8y) - (9x - 5y)$

40) $(15x - 7y) - (4x + 7y)$

41) $(2x^2 - 7x + 9) + (3x^2 + 2x + 5)$

42) $(3x^2 - 5x + 4) + (4x^2 - 5x + 8)$

43) $(4xy - 6y + 5) + (6xy + 9y - 7)$

44) $(2xz - 7x + 4) + (4xz + 6x - 7)$

45) $(2m^2c^2 + 5mc - 9c^2 + 5) + (3m^2c^2 - mc - 2)$

46) $(3q^2d^2 + 6qd - 4d^2 + 4) + (3q^2d^2 - qd - 3)$

47) $(4p^3 - 7p^2k - 4p) + (2p^2k - 9p + 11)$

48) $(7b^3 - 5b^2 + 4) + (b^2 + 9b - 1)$

49) $(4r^2 + 3r - 5) - (9r^2 - 3r + 5)$

50) $(2k^2 + 9k - 5) - (9k^2 - 9k + 5)$

51) $(8p^2m^2 + 8mp - 3m^2 + 3) - (3p^2m^2 - mp - 2)$

52) $(7n^2r^2 + 7nr - 3r^2 + 4) - (3n^2r^2 - nr - 5)$

53) $(8rp^2 - 5rp) - (r^3 + 4rp) + (4rp - 7rp^2)$

54) $(-5m^2t + 9) - (4mt^2 + 9m - 9) + (6mt^2 + 5m^2t)$

55) $(4h^2 + 7h - 4) - (3h^2 - 4h - 5) + (-11h - 10)$

56) $(3p^3 - 9p + 7) + (3p^2 - 5p + 6) - (8p^2 - 3)$

57) $\left(\dfrac{2}{5}x^2 - \dfrac{2}{3}x + \dfrac{1}{9}\right) + \left(\dfrac{1}{5}x^2 + \dfrac{1}{3}x + \dfrac{4}{9}\right)$

58) $\left(\dfrac{5}{7}a^2 - \dfrac{2}{5}a + \dfrac{2}{3}\right) + \left(\dfrac{1}{7}a^2 - \dfrac{2}{5}a - \dfrac{1}{3}\right)$

59) $\left(\dfrac{3}{11}r^2 + \dfrac{2}{7}r - \dfrac{4}{5}\right) - \left(\dfrac{5}{11}r^2 - \dfrac{4}{7}r - \dfrac{2}{5}\right)$

60) $\left(\dfrac{5}{13}n^2 + \dfrac{3}{5}n - \dfrac{2}{9}\right) - \left(\dfrac{11}{13}n^2 - \dfrac{1}{5}n - \dfrac{2}{9}\right)$

CHALLENGE EXERCISES (61–62)

61) $\left(\dfrac{7}{10}x^2 - \dfrac{3}{8}x + \dfrac{1}{5}\right) - \left(\dfrac{4}{5}x^2 - \dfrac{1}{4}x - \dfrac{3}{10}\right)$

62) $\left(\dfrac{5}{8}x^2 + \dfrac{6}{7}x - \dfrac{2}{3}\right) - \left(\dfrac{5}{16}x^2 - \dfrac{4}{21}x + \dfrac{7}{8}\right)$

Represent each of the following using addition and/or subtraction and simplify.

63) Add $7x^2 + 9x - 1$ and $4x^2 - 3x + 4$.

64) Add $4x^2 + 9x - 2$ and $9x^2 - 3x - 7$.

65) Add $6x^3 - 5x^2 - 3$ and $5x^2 - 10x - 9$.

66) Add $-8x^3 + 7x - 6$ and $4x^2 - 8x + 3$.

67) Subtract $6x^2 + 7x - 3$ from $3x^2 - 3x + 1$.

68) Subtract $2z^2 + z - 3$ from $4z^2 + 7z - 4$.

69) Subtract $6x^2 - 8x + 4$ from $2x^3 - 8x^2 + 5$.

70) Subtract $9x^2 - 6$ from $5x^3 - 6x^2 + 7$.

71) Find the sum of $3x + 13$ and $5x - 8$.

72) Find the sum of $5x + 16$ and $7x - 3$.

73) Find the difference of $4x - 7$ and $2x - 12$.

74) Find the difference of $5y + 7$ and $2y - 6$.

75) Find the sum of $x^2 - 3x + 4$ and $2x^2 + 4x - 6$.

76) Find the sum of $y^2 - 7y + 3$ and $3y^2 + 6y - 8$.

77) Find the difference of $z^2 + 7z - 9$ and $z^2 - 5z + 2$.

78) Find the difference of $w^2 - 8w + 4$ and $w^2 - 7w - 9$.

79) From the sum of $2x^2 - 7x - 2$ and $x^2 + 5$, subtract $2x^2 + 4x - 3$.

80) From the sum of $3y^2 + 8y - 6$ and $2y^2 - 9$, subtract $4y^2 - 7y - 4$.

81) To the difference of $3x - 7$ and $4x + 2$ add $5x - 9$.

82) To the difference of $4x + 2$ and $6x - 3$ add $9x + 5$.

Recall that the perimeter of a triangle is found by adding the lengths of all three sides. For example, if the sides of a triangle are x inches, 4x inches, and 2x inches, the perimeter is x + 4x + 2x = 7x inches. Find the following.

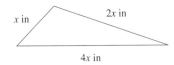

83) Find the perimeter of a triangle with sides $2x$ cm, $5x$ cm, and x cm.

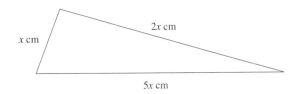

84) Find the perimeter of a triangle with sides x ft, $6x$ ft, and $4x$ ft.

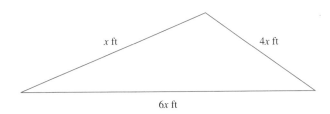

85) The lengths of a triangular sign are x inches, $2x - 3$ inches, and $5x + 4$ inches. What is the perimeter of the sign?

86) The lengths of the sides of a triangular piece of land are x feet, $3x - 2$ feet, and $7x + 1$ feet. How many feet of fencing would it take to enclose it?

A quadrilateral is a figure with four sides. The perimeter of a quadrilateral is found by adding the lengths of all four sides. For example, if the sides are 2x ft, 3x ft, x ft and 2x ft, the perimeter is 2x + 3x + x + 2x = 8x ft.

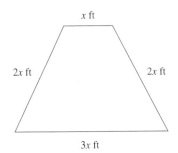

87) Find the perimeter of a quadrilateral whose sides are x in, $3x$ in, $5x$ in, and $7x$ in.

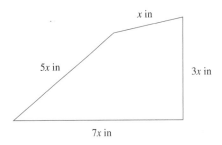

88) Find the perimeter of a quadrilateral whose sides are x m, $4x$ m, $2x$ m, and $8x$ m.

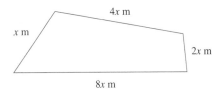

89) If the dimensions of a yard are x feet, $2x + 3$ feet, $x + 4$ feet, and $3x + 2$ feet, how many feet of fencing would it take to enclose it?

90) If the dimensions of a patio are x feet, $3x - 1$ feet, $4x + 2$ feet, and $5x - 7$ feet, how many feet of screen would it take to enclose it? Assume that each piece of screen will reach to the ceiling.

Represent each of the following using variables.

91) John is building a piece of furniture. He needs a board that is x feet long and another that is $2x - 2$ feet long. What is the total length of the two boards needed?

92) If John has x dollars and Mary has $3x - 5$ dollars, how much money do they have combined?

93) Christine is making a shelf for a closet and has a board $2x - 5$ feet long. She has to cut off a piece $x - 3$ feet long. How long is the remaining piece?

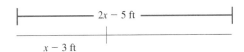

94) A carpenter is installing baseboard and has a board $3y + 2$ feet long. He has to cut off a piece $y + 4$ feet long. How long is the remaining piece?

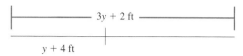

95) If Chan climbs $3x + 3$ meters and then climbs $2x - 5$ meters more, how high has he climbed all together?

96) If Yolanda has driven $5x - 6$ miles and then drives $3x + 2$ miles farther, what is the total distance she has driven?

WRITING EXERCISES

97) Explain why $3x \cdot x$ and $-4x^2$ are like terms.

98) Explain why $15x^2y$ and $-6xy^2$ are not like terms.

99) When subtracting polynomials, why is it necessary to change all the signs of the subtrahend (the second polynomial), not just the sign of the first term?

GROUP PROJECT (CRITICAL THINKING)

100) Do the following. (1) Choose any number. (2) Triple the number. (3) Add 1. (4) Subtract the original number. (5) Repeat steps 3 and 4. (6) Repeat steps 3 and 4 again. (7) The result is always 3. Write a paragraph explaining why.

101) Write a "riddle" similar to the one in Exercise 100.

Summary

Types of Numbers: [Section 1.1]

Natural Numbers = {1, 2, 3, ...}

Whole Numbers = {0, 1, 2, 3, ...}

Integers = {...,−3, −2, −1, 0, 1, 2, 3, ...}

Rational Numbers = {$\frac{a}{b}$ | a and b are integers and $b \neq 0$}

Irrational Numbers = {non-repeating and non-terminating decimals}

Real Numbers = {all numbers that are rational or irrational}

Order Symbols: [Section 1.1]

=	equals
≠	not equal
<	less than
≤	less than or equal to
>	greater than
≥	greater than or equal to

Negatives or Opposites: [Section 1.1]

Two numbers are opposites if they are the same distance from 0, but on opposite sides.

Negative of a Negative: [Section 1.1]

$-(-x) = x$ for all x.

Absolute Value: [Section 1.1]

The absolute value of a number is its distance from 0. The symbol for absolute value of x is $|x|$.

Definition of Prime and Composite Numbers: [Section 1.2]

A number is prime if it has only two factors, 1 and the number itself. Equivalently, a prime number is divisible only by 1 and itself.

A number is composite if it has at least one factor other than 1 and itself. Equivalently, a composite number is divisible by at least one number other than 1 and itself.

Prime Factorization: [Section 1.2]

Every real integer has one and only one prime factorization. To write a number in terms of prime factors

1) divide the composite number by the smallest prime number by which it is divisible,

2) divide the resulting quotient by the smallest prime number by which it is divisible,

3) continue this process until the quotient is prime,

4) the prime factorization is the product of all the divisors and the last quotient.

Definition of Fraction: [Section 1.2]

$\frac{a}{b}, b \neq 0$ The top number, a, is the numerator and the bottom number, b, is the denominator (not zero). A fraction is a form of division. If $a < b$, the fraction is proper, if $a \geq b$, the fraction is improper.

Mixed Number: [Section 1.2]

A mixed number is any number of the form $a\frac{b}{c}$ with $c \neq 0$. To change a mixed number to a fraction use the form $\frac{c \cdot a + b}{c}$.

Lowest Terms: [Section 1.2]

A fraction is reduced to lowest terms if the numerator and denominator have no factors in common.

Reducing a Fraction to Lowest Terms: [Section 1.2]

1) Write the numerator and denominator in terms of prime factors.

2) Divide the numerator and denominator by all factors common to both.

3) Multiply any remaining factors in the numerator and/or denominator.

Multiplication of Fractions: [Section 1.2]

$\frac{a}{b} \cdot \frac{c}{d} = \frac{a \cdot c}{b \cdot d}$ if $b, d \neq 0$ To multiply two fractions, find the product of the numerators divided by the product of the denominators.

Quotient of Fractions: [Section 1.2]

$\frac{a}{b} \div \frac{c}{d} = \frac{a}{b} \cdot \frac{d}{c}$ if $b, d, c \neq 0$ To divide a number by a number, multiply by its multiplicative inverse (reciprocal).

Sum of Fractions with the Same Denominators: [Section 1.2]

$\dfrac{a}{c} + \dfrac{b}{c} = \dfrac{a+b}{c}$ if $c \neq 0$ To add two or more fractions with the same denominator, add the numerators and place the sum over the same denominator.

Difference of Fractions with the Same Denominators: [Section 1.2]

$\dfrac{a}{c} - \dfrac{b}{c} = \dfrac{a-b}{c}$ if $c \neq 0$ To subtract two or more fractions with the same denominator, subtract the numerators and place the difference over the same denominator.

Least Common Denominator: [Section 1.2]

The least common denominator for a set of fractions is the smallest number that is divisible by all the denominators.

Combining Fractions with Unlike Denominators: [Section 1.2]

1) Determine the least common denominator.

2) Rewrite each fraction as an equivalent fraction with the least common denominator as its denominator.

3) Add the fractions using the procedure for adding fractions with common denominators.

Definition of Natural Number Exponents: [Section 1.3]

a^n means multiply n factors of a.

Definition of Variables and Constants: [Section 1.3]

A variable is a symbol, usually a letter, which is used to represent a number.

A constant is a symbol, a number, whose value does not change.

Order of Operations: [Section 1.3]

1) If inclusion symbols are present, simplify within the inclusion symbols first. Begin with the innermost inclusion symbol and work outward using the order below.

2) Evaluate all indicated powers.

3) Perform all multiplications and divisions in the order in which they occur as you work from left to right.

4) Perform all additions and subtractions in the order in which they occur as you work from left to right.

5) If a fraction is present, simplify above and below the fraction bar using the order given in steps 2–4.

Perimeter, Area, and Volume: [Section 1.4]

The perimeter of a geometric figure is the distance around the figure. The distance around a circle is its circumference. Perimeter and circumference are expressed in linear units. The area of a geometric figure is the measure of the size of a surface and is expressed in square units. The volume of a geometric solid is a measure of its interior or capacity and is expressed in cubic units.

Perimeter Formulas: [Section 1.4]

Figure	Formula
Triangle	$P = a + b + c$
Rectangle	$P = 2L + 2W$
Square	$P = 4s$
Circle	$C = 2\pi r$ or $C = \pi d$

Area Formulas: [Section 1.4]

Figure	Formula
Rectangle	$A = LW$
Square	$A = s^2$
Triangle	$A = \dfrac{bh}{2}$
Parallelogram	$A = bh$
Trapezoid	$A = \dfrac{h(B+b)}{2}$
Circle	$A = \pi r^2$

Volume Formulas: [Section 1.4]

Solid	Formula
Rectangular Solid	$V = LWH$
Cube	$V = e^3$
Right Circular Cylinder	$V = \pi r^2 h$
Cone	$V = \dfrac{\pi r^2 h}{3}$

Rules for Addition of Integers: [Section 1.6]

To add integers with the same sign, add their absolute values and give the answer the same sign as the numbers being added.

To add integers with opposite signs, subtract their absolute values and give the answer the sign of the number with the larger absolute value.

Properties of Addition: [Section 1.6]

Commutative:

$a + b = b + a$ Addition may be performed in any order.

Associative:

$$a + (b + c) = (a + b) + c$$ Addition may be grouped in any manner.

Inverse:

For every integer a, there exists an integer $-a$ such that $a + (-a) = -a + a = 0$.

Identity:

The integer 0 is the additive identity since $0 + a = a + 0 = a$.

Rule for Subtraction: [Section 1.7]

$$a - b = a + (-b)$$ To subtract a number, add its inverse (opposite).

Definition of an Equation: [Section 1.7]

An equation is a statement indicating that two expressions are equal.

Solutions of Equations: [Section 1.7]

A number is a solution of an equation if the equation is a true statement when the variable is replaced by that number.

Definition of Term: [Section 1.8]

A term is a number, variable, or product and/or quotient of numbers and variables that are raised to powers.

Numerical Coefficient: [Section 1.8]

The numerical coefficient of a term is the number factor of the term.

Degree of a Monomial: [Section 1.8]

The degree of a monomial is the exponent on the variable if there is only one variable and is the sum of the exponents on the variables if there is more than one variable.

Definition of Polynomial: [Section 1.8]

A polynomial is the sum of a finite number of terms that do not have a variable as a divisor and exponents on variables are whole numbers.

Types of Polynomials: [Section 1.8]

A monomial is a polynomial with one term.

A binomial is a polynomial with two terms.

A trinomial is a polynomial with three terms.

Descending Order: [Section 1.8]

A polynomial is written in descending order if the term of highest degree is written first, the term of second highest degree is written second, etc.

Degree of a Polynomial: [Section 1.8]

The degree of a polynomial is the same as the degree of the term of the polynomial with the highest degree.

Distributive Property of Multiplication over Addition: [Section 1.9]

For all numbers a, b, and c, $a(b + c) = ab + ac$.

Definition of Like Terms: [Section 1.9]

Like terms are terms which have the same variables with the same exponents on these variables.

Addition of Like Terms: [Section 1.9]

To add like terms, add the numerical coefficients and leave the variable portion unchanged.

Addition of Polynomials: [Section 1.9]

To add polynomials, add the like terms and leave the sum in descending order.

Subtraction of Polynomials: [Section 1.9]

To subtract a polynomial, add the negative of the polynomial.

Review Exercises

Answer each of the following using a signed number (integer). [Section 1.1]

1) If $+25$ represents winning \$25, how would you represent losing \$50?

2) If -5 feet represents the depth of a hole, how would you represent the height of a pile of dirt 7 feet high?

Give the negative (opposite) of each of the following. [Section 1.1]

3) 23   **4)** -35

5) $\dfrac{7}{16}$ **6)** $-\dfrac{2}{4}$

7) -8.3 **8)** 4.7

Simplify the following involving absolute values. [Section 1.1]

9) $|-5|$ **10)** $-|-5|$

Insert the proper symbol (=, <, or >) to make each of the following true. [Section 1.1]

11) $5 \underline{} 9$ **12)** $-28 \underline{} 9$

13) $-8 \underline{} -10$ **14)** $|12| \underline{} |-12|$

15) $-|-16| \underline{} |16|$ **16)** $|-32| \underline{} -|32|$.

17) $\dfrac{5}{11} \underline{} \dfrac{2}{11}$ **18)** $-\dfrac{3}{5} \underline{} \dfrac{5}{9}$

19) $\dfrac{4}{3} \underline{} \dfrac{7}{6}$ **20)** $-\dfrac{7}{12} \underline{} -\dfrac{5}{6}$

21) $-4.1 \underline{} -6.2$ **22)** $4.5 \underline{} 3.2$

Reduce the following to lowest terms. [Section 1.2]

23) $\dfrac{15}{45}$ **24)** $\dfrac{48}{36}$

Evaluate the following. [Section 1.2]

25) $\dfrac{3}{4} \cdot \dfrac{5}{2}$ **26)** $\dfrac{6}{7} \div \dfrac{11}{9}$ **27)** $\dfrac{3}{11} + \dfrac{4}{11}$

28) $\dfrac{8}{9} - \dfrac{4}{9}$ **29)** $\dfrac{12}{13} \cdot 4\dfrac{5}{6}$ **30)** $\dfrac{4}{9} \div \dfrac{12}{13}$

31) $\dfrac{5}{9} + \dfrac{7}{15}$ **32)** $\dfrac{3}{10} - \dfrac{7}{45}$ **33)** $8\dfrac{5}{12} + 3\dfrac{3}{8}$

34) $4\dfrac{9}{20} - 2\dfrac{3}{8}$

35) How many shelves each of which is $1\frac{2}{3}$ feet long can be cut from a board 15 feet long?

36) In 1 hour, Frank can mow $\frac{1}{12}$ of the camp ground using a self-propelled lawn mower and Suzy can

mow $\frac{3}{8}$ of the campground using a riding lawnmower. What part of the campground can they mow in one hour working together?

37) If it takes $4\frac{2}{3}$ yards of material to make a dress, how much material would it take to make 12 such dresses?

38) For a chemistry experiment, Cheryl used $\frac{3}{5}$ of a beaker of sulfuric acid that was $\frac{2}{3}$ full. What part of a full beaker did she use?

Write each of the following using exponents. [Section 1.3]

39) $9 \cdot 9 \cdot 9 \cdot 9 \cdot 9$ **40)** $(-2)(-2)(-2)(-2)$

41) $3 \cdot 3 \cdot 3 \cdot 5 \cdot 5 \cdot 5 \cdot 5$ **42)** $12 \cdot 9 \cdot 9 \cdot 9 \cdot 2 \cdot 2$

43) $3 \cdot 3 \cdot 3 \cdot a \cdot a \cdot a$

44) $7 \cdot 7 \cdot 7 \cdot r \cdot r \cdot r \cdot r \cdot s \cdot s$

Evaluate each of the following. [Section 1.3]

45) $2^2 \cdot 3^4$ **46)** $5 \cdot 2^4$ **47)** $3^2 \cdot 5^2 \cdot 2^2$

48) $10^2 \cdot 2^2 \cdot 5^3$ **49)** $\left(\dfrac{1}{2}\right)^2\left(\dfrac{4}{5}\right)^2$ **50)** $\left(\dfrac{3}{2}\right)^3\left(\dfrac{8}{9}\right)^2$

Find the value of each of the following when $x = 3$ and $y = 2$. [Section 1.3]

51) $x^3 y$ **52)** $x^2 y^2$ **53)** $5x^2 y$ **54)** $4xy^5$

Evaluate each of the following using the order of operations. [Section 1.3]

55) $8 - 2 \cdot 3$ **56)** $2^4 + 24 \div 6$

57) $29 - 4(7 - 2)$ **58)** $5 + 2 \cdot 3^2$

59) $4^2 - 24 \div 2^2$ **60)** $64 \div (3^3 - 11)$

61) $8 + 48 \div 8 \cdot 3 + 5$ **62)** $12 + 8 \cdot 9 \div 3 - 9$

Find the value of each of the following when $x = 6$ and $y = 3$. [Section 1.3]

63) $5x - 2y^2$ **64)** $2x^2 \div 3y$ **65)** $5xy^2$

Solve the following. [Section 1.3]

66) Don went to the hardware store and purchased three bolts that cost four cents each and five nuts that cost two cents each. How much did he pay for the bolts and nuts?

67) Four sorority sisters share equally in the cost of two pizzas that cost \$8.99 each and four soft drinks that cost 79 cents each. To the nearest cent, how much did each pay?

Find the perimeter and area of each of the following.
[Section 1.4]

68)

3 in

7 in

69)

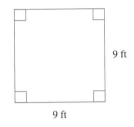

9 ft

9 ft

70)

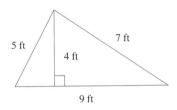

5 ft

7 ft

4 ft

9 ft

71) Find the circumference and area. Use π ≈ 3.14.

5 m

Find the areas of the following. [Section 1.4]

72)

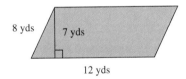

8 yds

7 yds

12 yds

73)

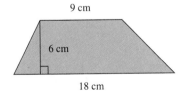

9 cm

6 cm

18 cm

Find the volumes of the following. [Section 1.4]

74)

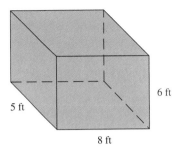

5 ft

6 ft

8 ft

75)

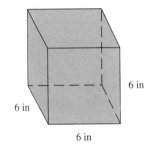

6 in

6 in

6 in

76) Leave answer in terms of π.

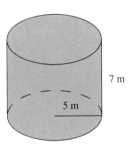

7 m

5 m

77) Leave answer in terms of π.

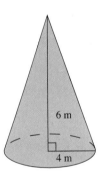

6 m

4 m

Answer the following. [Section 1.5]

78) A rancher is planning to fence a holding pen for his cattle. If the pen is to be in the shape of a square 250 feet on each side, how much fencing will he need?

79) An irrigation system consists of a pipe which is anchored on one end to remain stationary and the other end is on wheels. When the system is turned on, the pipe rotates about the stationary end and the end on wheels sweeps out a circle. If the pipe is 150 feet long, how far do the wheels travel in one rotation.

80) A rectangular bedspread is 120 inches long and 90 inches wide. Pat is going to sew a lace border around the bedspread. If the lace costs $1.50 per yard, how much will the lace cost?

81) Find the cost of the carpet needed to cover the floor of a room that is 14 yards long and 10 yards wide if the carpet costs $13.00 per square yard?

82) A window is in the shape of a square each of whose sides is 52 inches long. What is the area of the window?

83) A sign for an advertisement is in the shape of a trapezoid. If the parallel sides are 10 inches and 6 inches long and the height of the sign is 8 inches, find the area of the sign.

84) Find the area watered in one revolution by the irrigation system in Exercise 79.

85) A table top is in the shape of a triangle two of whose sides are 17 inches long and the third side is 16 inches long. The height to the 16 inch side is 15 inches. Find the area of the table top.

86) A box of rice is in the shape of a rectangular solid that is six inches long, one inch wide, and eight inches high. What is the volume of the box?

87) A can of soup is in the shape of a right circular cylinder whose diameter is three inches and whose height is five inches. What is the volume of the can?

88) A drinking cup is in the shape of a right circular cylinder whose radius is two inches and whose height is six inches. What is the volume of the cup?

Find the following sums and/or differences. [Sections 1.6 and 1.7]

89) $19 - 13$

90) $9 - (-6)$

91) $-16 - (-9)$

92) $8 - 12 + 17$

93) $-24 - (-33) + 14$

94) $(-9 + 5) + (5 - 8)$

95) $(-53 - 37) - (32 - 27)$

96) $(-13 - 32) - (41 - 56)$

97) $-\dfrac{7}{16} + \dfrac{5}{16}$

98) $\dfrac{5}{18} - \dfrac{13}{18}$

99) $-\dfrac{5}{16} - \dfrac{7}{12}$

100) $-\dfrac{8}{15} - \dfrac{7}{20}$

Give the name of the property illustrated by each of the following. [Section 1.6]

101) $9 + 12 = 12 + 9$

102) $6 + (5 + 1) = (6 + 5) + 1$

103) $6 + (5 + 1) = 6 + (1 + 5)$

104) $(3 + 8) + 5 = (8 + 3) + 5$

105) $9 + (-4 + 4) = 9 + 0$

106) $0 + (-3 + 2) = -3 + 2$

Write an expression for each of the following and evaluate. [Sections 1.6 and 1.7]

107) The sum of -9 and 6

108) The difference of 15 and -8

109) 16 more than -5

110) -6 less than 3

111) -9 increased by 8

112) -3 decreased by -7

113) The sum of 18 and -8 added to the sum of -5 and 13

114) The difference of -5 and 9 added to the difference of -9 and -2

115) The difference of 4 and -5 subtracted from the sum of 14 and -8

116) The product of 5 and the sum of -6 and 10

Give the coefficient, variable(s), and degree of each of the following. [Section 1.8]

	Coefficient	Variable(s)	Degree
117) $9x^2$	_____	_____	_____
118) $-13a^6$	_____	_____	_____
119) $-9x^5y^4$	_____	_____	_____

Write each of the following in descending order and classify each as a monomial, binomial, trinomial, or none of these. Give the degree of those that are polynomials. [Section 1.9]

120) $4x^3 + 6x$

121) $7y^2 - 8y + 2$

122) $5x^7$

123) $9x^4 - 7x^5 + 2x$

124) $2a^3 - 5a^2 + 14a^6 + 19$

125) $\dfrac{3x^6}{y} + 9x - 8$

Evaluate each of the following for the indicated value of the variable. [Section 1.8]

126) $P(x) = 3x^2 + 2x - 8; x = 1$

127) $P(t) = t^2 + 6t - 9; t = 3$

128) $P(a) = 2a^3 + 5a; a = 2$

129) $P(b) = 3b^4 + 5b - 5; b = 3$

Simplify the following by adding like terms. [Section 1.9]

130) $2x - 5x$

131) $8x + 4y - 6x - 3y$

132) $-13w^3 + 7u^2 - 5 + 8w^3 - 13u^2 + 8$

133) $14x^2y - 23xy^2 - 19x^2y + 29xy^2 + 4$

Find the following sums and/or differences of polynomials. Leave the answers in descending powers. [Section 1.9]

134) $(4x + 3) + (6x - 8)$

135) $(7a + 3b) - (2a - 8b)$

136) $(u^4 - 2u^3 + u) + (4u^4 - 3u^3 + u)$

137) $(4a^2 - 7a + 2) - (2a^2 + 7a - 3)$

138) $(z^4 - 3z^3 + 6z) + (2 - 4z + 4z^3)$

139) $(x^4 - 1 - 2x^2 + 3x^3) - (3x^3 - 4x^4 + 2x^2)$

140) $(3z^2 + 5x - 1) - (7z^2 - 7z + 8) + (5z^2 - 6z + 1)$

Write each of the following as addition and/or subtraction and simplify. [Section 1.9]

141) Find the sum of $5x^2 - 9x + 5$ and $6x^2 - 2x - 1$.

142) Find the sum of $3x^3 + 4x^2 - 4$ and $2x^2 - 4x - 7$.

143) Find the difference of $4x^2 + 7x - 2$ and $6x^2 - 6x + 3$.

144) Subtract $3u^4 - 3u^3 + 4u^2 - 7$ from $u^4 + u^2 - 3u + 1$.

145) From the sum of $-x^3 - 4x^2 + 2x + 2$ and $2x^3 - 3x^2 - 2x + 1$, subtract $3x^3 + 6x^2 - 2x - 2$.

146) Find the perimeter of a triangle whose sides are $3x - 5$ ft, $4x + 1$ ft, and $2x - 2$ ft.

147) If a board is $2x + 6$ feet long and a piece $x - 5$ feet long is cut off, what is the length of the remaining piece?

148) If John climbs $x + 5$ feet and then climbs $3x - 8$ feet farther, what is the total distance he has climbed?

149) If Clem has $4x - 6$ dollars and spends $2x + 3$ dollars, how much money does he have left?

Chapter 1 Test

Evaluate the following.

1) $\dfrac{5}{6} \cdot \dfrac{7}{9}$

2) $\dfrac{4}{17} + \dfrac{7}{17}$

3) $\dfrac{15}{8} \div \dfrac{5}{32}$

4) $\dfrac{13}{4} + \dfrac{12}{25}$

5) A developer purchases a 51-acre tract of land and plans to divide it into lots that are $\frac{3}{4}$ of an acre each. How many lots will there be?

Evaluate each of the following involving exponents.

6) 4^3

7) $(-2)^4(3^2)$

Evaluate each of the following.

8) $36 - 2 \cdot 3^2 \div 6$

9) $5(8^2 - 16) \div 12 + 3$

10) Find the value of $8xy \div 2y^3$ if $x = 4$ and $y = 2$.

11) Find the perimeter and area of the following.

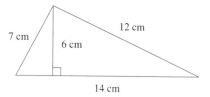

12) Find the circumference and area of the given circle. Use $\pi \approx 3.14$.

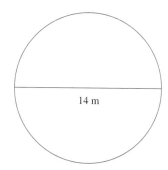

13) Find the volume of the following rectangular solid.

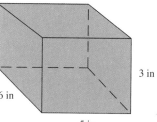

14) Magna plans to fence in a square garden plot that measures 18 feet on each side. If fencing costs $.50 per linear foot, how much will all the fencing cost?

15) A rectangular wall is 15 feet long and 12 feet high. How much will it cost in total to paint the wall if it costs $.40 per square foot?

Insert the proper symbol (<, >, or =) in order to make each of the following true.

16) -3 _____ $|-6|$

17) $-|-10|$ _____ $|10|$

Find the following sums or differences.

18) $-16 + 5$

19) $-9 + (-6) + 8$

20) $(-11 + 5) - (7 - 12)$

21) $-12 - [22 - (-8 - 2)]$

22) $\dfrac{8}{15} - \dfrac{7}{20}$

23) $-\dfrac{5}{9} - \dfrac{7}{15}$

Name the property illustrated by each of the following.

24) $(-5 + 7) + 4 = [7 + (-5)] + 4$

25) $8[4 + (-4)] = 8(0)$

Write an expression for each of the following and evaluate.

26) The sum of -7 and 11

27) The difference of 9 and -6 added to -12

28) The sum of -3 and 9 subtracted from the sum of 4 and -13

29) The difference of $x^2 - 6x + 2$ and $2x^2 + 5x - 3$

30) The sum of $3x + 6$ and $2x - 3$ subtracted from $4x + 5$

31) A board is $3x + 6$ meters long and a piece $2x - 2$ meters long is cut off. How long is the remaining piece?

32) Answer the following for the polynomial
$3x^2 + 4x^3 + 6x$.

a) What is the numerical coefficient of the second term?

b) What is the degree of the polynomial?

c) Write the polynomial in descending order.

d) Is the polynomial a monomial, binomial, trinomial, or none of these?

33) If a scuba diver is at -22 feet and descends another 18 feet, what is her present depth?

34) A submarine is submerged at a depth of 152 feet, and an airplane flies over the sub at an elevation of 345 feet. What is the distance between them?

35) Evaluate $P(x) = 3x^2 - 2x + 8$ for $x = 2$.

Simplify the following.

36) $9xy + 6yz - 4xy - 10yz$

37) $(2x^2 - 5x + 7) + (3x^2 + 2x - 11)$

38) $(6n^3 + 7n^2 - 8) - (9n^2 - 7n - 6)$

39) Albert leased a car with the following terms. He agreed to pay $900 down, $249 per month for three years, and 18 cents per mile in excess of 45,000. When he returned the car at the end of the lease period it had 54,000 miles. What was the total cost of the lease?

40) Is $-x$ always a negative number? Explain.

2

2.1 Multiplication of Real Numbers
2.2 Multiplication Laws of Exponents
2.3 Products of Polynomials
2.4 Special Products
2.5 Division of Integers
2.6 Order of Operations on Real Numbers
2.7 Quotient Rule and Integer Exponents
2.8 An Application of Exponents: Scientific Notation
2.9 Power Rule for Quotients and Using Combined Laws of Exponents
2.10 Division of Polynomials by Monomials

Multiplication and Division of Integers and Polynomials

I n this chapter we will continue our discussion of operations on integers and polynomials. In Chapter 1 we found the sums and differences of integers and polynomials. In this chapter we will find the products and quotients of integers, the products of polynomials, the quotients of monomials, and the quotients of polynomials and monomials. We will not discuss the division of polynomials by polynomials other than monomials until Chapter 5.

CHAPTER OVERVIEW

SECTION 2.1

Multiplication of Real Numbers

OBJECTIVES

When you complete this section, you will be able to:

Ⓐ Multiply signed numbers.
Ⓑ Raise signed numbers to powers.
Ⓒ Identify properties of multiplication.

INTRODUCTION

In Chapter 1 we learned to add and subtract signed numbers. In this section we will use our knowledge of how signed numbers are added to develop rules for multiplication of signed numbers.

CALCULATOR EXPLORATION ACTIVITY 1—OPTIONAL

Using a calculator, complete the following.

Column A	Column B
3(3) =	−3(3) =
3(2) =	−3(2) =

(continued)

Recall that multiplication is a shortcut for repeatedly adding the same number. For example, $3 \cdot (5) = 5 + 5 + 5$. The first number, 3, tells us how many of the second number, 5, are being added. Likewise, $4 \cdot (6) = 6 + 6 + 6 + 6$. In the following examples, we will develop rules for multiplying a positive number by a positive number, a positive number by a negative number, a negative number by a positive number, and a negative number by a negative number.

• **Finding products of signed numbers**

EXAMPLE 1

The multiplication of a positive number by a positive number is already familiar to you from arithmetic.

a) $3 \cdot 5 = 5 + 5 + 5 = 15$ Positive 3 times positive 5 equals positive 15.

b) $4 \cdot 6 = 6 + 6 + 6 + 6 = 24$ Positive 4 times positive 6 equals positive 24.

It is apparent from the above example, and your prior knowledge, that a positive number multiplied by a positive number results in a positive number. Example 2 illustrates the product of a positive number and a negative number.

EXAMPLE 2

Find the following products.

a) $+3(−5) = (−5) + (−5) + (−5) =$ Definition of multiplication.

-15 Therefore, $+3$ times $−5$ equals $−15$.

b) $+4(−6) = (−6) + (−6) + (−6) + (−6) =$ Definition of multiplication.

-24 Therefore, $+4$ times $−6$ equals $−24$.

The above example suggests that a positive number multiplied by a negative number results in a negative number.

In order to discuss a negative number multiplied by a positive number, we need the **commutative property** for multiplication. We know from arithmetic that $3(4) = 4(3)$ and $6(2) = 2(6)$. These products suggest that the order in which two numbers are multiplied makes no difference. We can generalize this to the statement of the commutative property for multiplication.

COMMUTATIVE PROPERTY FOR MULTIPLICATION

For any two integers a and b, $a \cdot b = b \cdot a$.

EXAMPLE 3

Find the following products.

a) $(-3)(3) =$ Apply the Commutative Property.

    $(3)(-3) =$ Multiply.

    -9 Therefore, -3 times $+3$ equals -9.

b) $(-6)(4) =$ Apply the Commutative Property.

    $(4)(-6)$ Multiply.

    -24 Therefore, -6 times $+4$ equals -24.

From Example 3, we can conclude that a negative number multiplied by a positive number results in a negative number. Unfortunately, we cannot use the definition of multiplication to develop a rule for multiplying a negative number by a negative number. For example, $(-3)(-5)$ cannot mean add -3, -5's.

EXAMPLE 4

Look for a pattern in the following.

$(3)(-4) = -12$ } Increase of 4
$(2)(-4) = -8$ } Increase of 4
$(1)(-4) = -4$ } Increase of 4
$(0)(-4) = 0$

Notice that the number on the left is decreasing by one and the product is increasing by 4 in each case. Let's continue the pattern.

$(3)(-4) = -12$

$(2)(-4) = -8$

$(1)(-4) = -4$

$(0)(-4) = 0$ } Increase of 4
$(-1)(-4) = 4$ } Increase of 4
$(-2)(-4) = 8$

The above pattern suggests a negative number multiplied by a negative number results in a positive number.

Looking back at Examples 1 and 4, we notice that if the signs are both positive or both negative the product is positive. From Examples 2 and 3 we see that if the signs are different (one positive and one negative) the product is negative. Based on these examples we make the following rules for multiplying signed numbers.

PRODUCTS OF SIGNED NUMBERS

The product of two numbers with the same (like) signs is positive. The product of two numbers with different (unlike) signs is negative.
Using symbols, $(+)(+) = +$, $(-)(-) = +$, $(+)(-) = -$, and $(-)(+) = -$.

EXAMPLE 5
Find the following products.

a) $4 \cdot 6 = 24$ Same signs. Therefore, the product is positive.

b) $(-5)(5) = -25$ Different signs. Therefore, the product is negative.

c) $(-6)(-7) = 42$ Same signs. Therefore, the product is positive.

d) $\left(\frac{3}{5}\right)\left(-\frac{2}{7}\right) = -\frac{6}{35}$ Different signs. Therefore, the product is negative.

ADDITIONAL PRACTICE

Find the following products.

a) $5 \cdot 3$ **b)** $(-6)(4)$

c) $(-5)(-9)$ **d)** $\left(-\frac{3}{4}\right)\left(-\frac{5}{7}\right)$

PRACTICE EXERCISES

Find the following products.

1) $6 \cdot 3$ **2)** $(-8)(4)$

3) $(-8)(-3)$ **4)** $\left(\frac{1}{3}\right)\left(-\frac{4}{5}\right)$

If more practice is needed, do the Additional Practice Exercises in the margin.

If more than two numbers are multiplied, remember to multiply in order from left to right unless the numbers are grouped using the associative property.

EXAMPLE 6
Find the following products.

a) $4(-3)(5) =$ Multiply 4 and -3.

     $(-12)(5) =$ Multiply -12 and 5.

     -60 Product

b) $(-5)(-2)(4) =$ Multiply -5 and -2.

     $(10)(4) =$ Multiply 10 and 4.

     40 Product.

c) $6(-4)(-5)(-3) =$ Multiply 6 and -4.

     $(-24)(-5)(-3) =$ Multiply -24 and -5.

     $(120)(-3) =$ Multiply 120 and -3.

     -360 Product.

d) $(-2)(-3)(-5)(-1) =$ Multiply -2 and -3.

$6(-5)(-1) =$ Multiply 6 and -5.

$(-30)(-1) =$ Multiply -30 and -1.

30 Product.

Be careful: Remember that we multiply in order from left to right. A common error when multiplying signed numbers is to "distribute" the first factor over the remaining factors. For instance, in Example (6a), $4(-3)(5) \neq (-12)(20)$.

By looking closely at Example 6, you may discover an easy way to determine the sign of the product in advance.

PRODUCTS WITH NEGATIVE NUMBERS

If there is an even number of negative signs, the product is positive. If there is an odd number of negative signs, the product is negative.

Remember, the even numbers are 2, 4, 6, ... and the odd numbers are 1, 3, 5, ...

ADDITIONAL PRACTICE

Find the following products.

e) $(4)(-3)(6)$

f) $(-2)(7)(-3)$

g) $(-3)(-2)(-5)(4)$

h) $(-3)(-1)(-3)(-5)$

PRACTICE EXERCISES

Find the following products.

5) $(6) \cdot (4) \cdot (-3)$ **6)** $(5) \cdot (-4) \cdot (-3)$

7) $(-3) \cdot (-4) \cdot (6) \cdot (-2)$ **8)** $(-3) \cdot (-1) \cdot (-4) \cdot (-5)$

If more practice is needed, do the Additional Practice Exercises in the margin.

Now that we know how to multiply signed numbers, we can raise signed numbers to powers. Recall from Section 1.1, an exponent tells us how many times the base is used as a factor.

CALCULATOR EXPLORATION ACTIVITY 2—OPTIONAL

Use a calculator to evaluate the following. If your calculator does not have parentheses, be very careful.

Column A	Column B	Column C*
$(-5)^2 =$	$(-5)^3 =$	$-5^2 =$
$(-3)^4 =$	$(-3)^5 =$	$-3^4 =$
$(-2)^6 =$	$(-2)^7 =$	$-2^5 =$
$(-1)^8 =$	$(-1)^9 =$	$-1^9 =$

*Some calculators without parentheses give incorrect answers to some of these.

(continued)

CALCULATOR EXPLORATION ACTIVITY 2—OPTIONAL *(continued)*

6) Based on your answers to Column A, is the result of raising a negative number to an even power positive or negative?

7) Based on your answers to Column B is the result of raising a negative number to an odd power positive or negative?

8) Based on your answers to Column C is the negative of a positive number raised to any positive integer power positive or negative?

We will now confirm these results mathematically.

In Example 7 look for a relationship between the exponent and the sign of the answer.

EXAMPLE 7
Evaluate the following.

a) $(-3)^2 = (-3)(-3) = 9$

b) $(-5)^3 = (-5)(-5)(-5) = (25)(-5) = -125$

c) $(-1)^4 = (-1)(-1)(-1)(-1) = 1(-1)(-1) = (-1)(-1) = 1$

d) $(-2)^5 = (-2)(-2)(-2)(-2)(-2) = 4(-2)(-2)(-2) = (-8)(-2)(-2) = (16)(-2) = -32$

Since the exponent indicates how many times the base is used as a factor, the procedure for finding the sign of a product applies to finding the sign of a negative number raised to a power.

RAISING A NEGATIVE NUMBER TO A POWER

If a negative number is raised to an even power, the answer is positive. If a negative number is raised to an odd power, the answer is negative.

Be careful: There is a great deal of difference between $(-3)^2$ and -3^2. The exponent applies only to the symbol immediately preceding it so the "−" is not part of the base unless we use parentheses. Consequently, $(-3)^2 = (-3)(-3) = 9$. But in the expression -3^2, which is the negative of 3^2, the base of the exponent is 3 and not −3. You may think of -3^2 as $-1 \cdot 3^2$, so the meaning is $-1 \cdot 3 \cdot 3 = -3 \cdot 3 = -9$. The only way to raise a negative number to a power is to put the negative number inside parentheses and the power outside the parentheses as in $(-3)^2$.

EXAMPLE 8
Evaluate the following.

a) $(-2)^2 = (-2)(-2) = 4$

b) $-2^2 = -1 \cdot 2 \cdot 2 = -2 \cdot 2 = -4$

c) $-8^2 = -1 \cdot 8 \cdot 8 = -8 \cdot 8 = -64$

d) $(-8)^2 = (-8)(-8) = 64$

e) $(-3)^3 = (-3)(-3)(-3) = 9(-3) = -27$

f) $-3^3 = -1 \cdot 3^3 = -1 \cdot 3 \cdot 3 \cdot 3 = -3 \cdot 3 \cdot 3 = -9 \cdot 3 = -27$

ADDITIONAL PRACTICE

Evaluate the following.

i) $(-6)^2$ **j)** $(-4)^3$

k) $(-3)^4$ **l)** -2^5

m) $(-4)^2$ **n)** -4^2

ANSWERS:
Calculator Exploration
Activity 2

1) positive

2) negative

3) negative

ADDITIONAL PRACTICE

Find the following products.

o) $(-2)^2(4)^2$

p) $(-4)^2(-5)^2$

PRACTICE EXERCISES

Evaluate the following.

9) $(-4)^2$ **10)** $(-2)^3$ **11)** $(-4)^4$

12) -2^4 **13)** $(-5)^2$ **14)** -5^2

If more practice is needed, do the Additional Practice Exercises in the margin.

In problems involving both multiplication and powers, remember to raise to powers first.

EXAMPLE 9

Find the following products.

a) $(-2)^2(3)^3 =$ Square -2 and cube 3.

     $(4)(27) =$ Multiply 4 and 27.

     108   Product.

b) $(-3)^3(-1)^2 =$ Cube -3 and square -1.

     $(-27)(1) =$ Multiply -27 and 1.

     -27 Product.

c) $(-5)^2(-4)^3 =$ Square -5 and cube -4.

     $(25)(-64) =$ Multiply 25 and -64.

     -1600 Product.

PRACTICE EXERCISES

Find the following products.

15) $(-3)^2(2)^2$ **16)** $(-1)^3(-2)^3$

17) $(-4)^2(-5)^3$ **18)** $(4)^2(-3)^3$

If more practice is needed, do the Additional Practice Exercises in the margin.

Earlier the commutative property of multiplication was mentioned. Like addition, multiplication has a number of properties which are summarized below.

PROPERTIES OF MULTIPLICATION

Commutative

     $ab = ba$ Multiplication may be performed in any order.

Associative

     $a(bc) = (ab)c$ Multiplication may be grouped in any manner.

Identity

The number 1 is the identity for multiplication since $a \cdot 1 = 1 \cdot a = a$, for any number a.

PROPERTIES OF MULTIPLICATION *(continued)*

Inverse

For any number $a \neq 0$, there exists a number $\frac{1}{a}$ such that $a \cdot \frac{1}{a} = 1$.

Multiplication by 0

$a \cdot 0 = 0 \cdot a = 0$ for all numbers a.

EXAMPLE 10

Identify the property illustrated by each of the following.

a) $(-3)(2) = (2)(-3)$ Order is different. Therefore, commutative.

b) $2[(-3)(-5)] = [2(-3)](-5)$ Grouping is different. Therefore, associative.

c) $2(-3 + 5) = (-3 + 5)2$ Order is different. Therefore, commutative.

d) $(-4) \cdot 1 = -4$ Identity for multiplication.

e) $5 \cdot \dfrac{1}{5} = 1$ Inverse for multiplication.

f) $5 \cdot 0 = 0$ Multiplication by 0.

PRACTICE EXERCISES

Identify the property illustrated by each of the following.

19) $(5)(-6) = (-6)(5)$ 20) $-3[2(-4)] = [(-3)2](-4)$

21) $-3(5 + 6) = (5 + 6)(-3)$ 22) $1 \cdot (-7) = -7$

23) $6 \cdot \dfrac{1}{6} = 1$ 24) $8 \cdot 0 = 0$

EXERCISE SET 2.1

Find the following products.

1) $6(-4)$ 2) $(-12)(-8)$ 3) $(-6)(4)$

4) $(14)(-5)$ 5) $(-6)(-4)$ 6) $(-8)(-9)$

7) $(-7)(3)$ 8) $(-5)(8)$ 9) $\left(\dfrac{2}{3}\right)\left(-\dfrac{4}{5}\right)$

10) $\left(-\dfrac{3}{4}\right)\left(\dfrac{5}{7}\right)$ 11) $\left(-\dfrac{8}{15}\right)\left(-\dfrac{25}{6}\right)$

12) $\left(-\dfrac{10}{21}\right)\left(-\dfrac{14}{15}\right)$ 13) $(-1.6)(5.7)$

14) $(-2.1)(-7.6)$ 15) $(-1)(3)(-5)$

16) $(-4)(-1)(6)$ 17) $(4)(-3)(4)$

18) $(5)(-4)(6)$ 19) $(-5)(-3)(-6)$

20) $(-4)(-7)(-10)$ 21) $(-2)(4)(-3)(-5)$

22) $(-6)(4)(-3)(-1)$ 23) $\left(\dfrac{3}{4}\right)\left(-\dfrac{1}{5}\right)\left(-\dfrac{3}{2}\right)$

24) $\left(-\dfrac{1}{3}\right)\left(-\dfrac{2}{5}\right)\left(\dfrac{4}{7}\right)$ 25) $\left(-\dfrac{3}{4}\right)\left(-\dfrac{5}{6}\right)\left(-\dfrac{12}{25}\right)$

26) $\left(-\dfrac{5}{6}\right)\left(-\dfrac{9}{10}\right)\left(-\dfrac{5}{18}\right)$ 27) $(-8)^2$

28) -8^2

29) -2^4

30) $(-2)^4$ 31) $\left(-\dfrac{2}{3}\right)^2$

32) $-\left(\dfrac{2}{3}\right)^2$ 33) $(-2)^2(3)^3$

34) $(-4)^2(3)^2$ 35) $(-1)^5(-5)^3$

36) $(-6)^2(-2)^3$

CALCULATOR EXERCISES

Find the following products using a calculator. The calculator used determines how to enter a negative number. On many calculators, first enter the number and then press the $\boxed{+/-}$ *key. This key changes the sign of the number currently showing on the display. Do not attempt to use the subtraction key.*

C1) $(-5.74)(8.65)$ C2) $(98.67)(-6.45)$

C3) $(-89.54)(-67.34)$ C4) $(-8.546)(-6.768)$

Evaluate each of the following for $x = -2$ and $y = -4$.

37) $3x$

38) $4y$

39) $-5x$

40) $-2y$

41) $6x^2$

42) $3y^2$

43) $-5x^3$

44) $-4y^3$

45) x^2y^3

46) x^2y^2

47) x^3y^2

48) x^3y^3

Find the solution(s) of each of the following from the domain $\{-3, -2, -1, 0, 1, 2, 3\}$.

49) $3x = -6$

50) $4x = -4$

51) $-2x = -4$

52) $-5x = -15$

53) $6x = -12$

54) $9x = -18$

55) $2x^2 = 2$

56) $2x^2 = 18$

57) $2x - 3 = -5$

58) $3x + 1 = -5$

59) $-4x + 3 = -5$

60) $-5x - 4 = 1$

Write an expression for each of the following and evaluate.

61) The product of 5 and -3 increased by 6.

62) The product of -4 and 7 increased by 9.

63) 5 decreased by the product of 4 and -7.

64) -3 decreased by the product of -5 and 3.

65) The product of 3 and the square of -4.

66) The product of -5 and the square of 5.

67) The product of -6 and -3 decreased by 8.

68) The product of -3 and -5 decreased by 4.

69) The product of 4 and -2 added to the product of -5 and 4.

70) The product of 8 and -3 added to the product of -5 and 7.

Identify the property illustrated by each of the following.

71) $(-2)[(3)(-4)] = [(-2)(3)](-4)$

72) $[(-4)(9)](-5) = (-4)[(9)(-5)]$

73) $3(-4 + 6) = (-4 + 6)(3)$

74) $(-4)(5 - 7) = (5 - 7)(-4)$

75) $(-9) \cdot 1 = -9$

76) $1 \cdot (-6) = -6$

77) $(-4)(-7) + 2 = (-7)(-4) + 2$

78) $5 + (-5)(8) = 5 + (8)(-5)$

79) $-3 \cdot \dfrac{-1}{3} = 1$

80) $5 \cdot \dfrac{1}{5} = 1$

81) $0 \cdot 7 = 0$

82) $9 \cdot 0 = 0$

Complete each of the following using the given property of multiplication.

83) $3[(-9)(-7)] =$ _____ Associative property

84) $6[(-2)(3)] =$ _____ Associative property

85) $3(-7) =$ _____ Commutative property

86) $(-5)(8) =$ _____ Commutative property

87) $4(-9 + 2) =$ _____ Distributive property

88) $5(-2 + 9) =$ _____ Distributive property

89) $4(-9 + 2) =$ _____ Commutative property

90) $5(-2 + 9) =$ _____ Commutative property

91) $(-2) \cdot 1 =$ _____ Identity for multiplication

92) $(-5) \cdot 1 =$ _____ Identity for multiplication

93) $8 \cdot \dfrac{1}{8} =$ _____ Inverse for multiplication

94) $4 \cdot \dfrac{1}{4} =$ _____ Inverse for multiplication

95) $2 \cdot 0 =$ _____ Multiplication property of 0

96) $0 \cdot 5 =$ _____ Multiplication property of 0

CHALLENGE EXERCISES (97–100)

Evaluate each of the following.

97) $-2 \cdot 3^2 + 5 \cdot (-4)^2 \div (5)(2)$

98) $4 \cdot (-6)^2 \div (3)(2)^3$

99) $-4[-5 - (-6)]^2 + (-7)$

100) $-6 - 5[-3^2 - (-6)]^2$

Answer the following.

101) When a cold front moved through, the temperature dropped at an average rate of 4° per hour. Using a signed number, find the change in temperature after three hours.

102) On a sunny day in Florida, the temperature rose at an average rate of 3.5° per hour from 10:00 A.M. to 2:00 P.M. Using a signed number, find the change in temperature from 10:00 A.M. to 2:00 P.M.

103) On a steep mountain road the elevation dropped seven feet per 100 feet. Using a signed number, find the change in elevation after 1000 feet.

104) On a steep roof there is a vertical drop of two feet for every horizontal change of ten feet. Using a signed number, find the vertical change for a horizontal change of 25 feet.

105) On Black Monday a certain stock dropped at an average rate of $\frac{3}{4}$ points per hour. Using a signed number, find the drop after eight hours.

106) A parachutist is descending at a rate of 8 feet per second. Using a signed number, find the number of feet that she descends in 15 seconds.

107) If a product has sixteen positive factors and twenty-three negative factors, is the product positive or negative? Why?

108) Is $(-2)^{48}$ positive or negative? Why?

109) Why is a product that contains an even number of negative signs positive?

110) Why is a product that contains an odd number of negative signs negative?

111) Compare and contrast the commutative and associative properties of multiplication.

112) Why is there no multiplicative inverse for 0?

113) Why is $-x^2$ always nonpositive?

114) Write an application exercise similar to 101–106 above. Exchange your exercise with another group and solve their exercise.

SECTION 2.2

Multiplication Laws of Exponents

OBJECTIVES

When you complete this section, you will be able to:

Ⓐ Simplify expressions of the form $a^m \cdot a^n$.
Ⓑ Simplify expressions of the form $(ab)^n$.
Ⓒ Simplify expressions of the form $(a^m)^n$.
Ⓓ Simplify expressions involving two or more of the forms A through C above.

INTRODUCTION

Previously, we raised whole numbers and integers to powers. In this section we will develop rules to simplify products involving exponents whose bases are the same, to raise the product of two numbers to a power, and to raise a number to a power to another power,. All of these rules depend on the definition of a positive integral exponent. Remember, if n is a positive integer, a^n means the product of n factors of a.

CALCULATOR EXPLORATION ACTIVITY 1—OPTIONAL

Use a calculator to evaluate each of the following columns.

Column A	Column B
$2^2 \cdot 2^3 =$	$2^5 =$
$3^3 \cdot 3^4 =$	$3^7 =$
$4^2 \cdot 4^4 =$	$4^6 =$

By looking at the corresponding lines in Columns A and B, answer the following.

1) If two exponential expressions with the same base are multiplied, leave the _____ unchanged and _____ the exponents.

We now give a mathematical justification for the above observation.

By the definition of exponents $(a^2)(a^3) = (a \cdot a)(a \cdot a \cdot a) = a \cdot a \cdot a \cdot a \cdot a = a^5$.

$$\uparrow \qquad \uparrow \qquad\qquad \uparrow$$

$$2 \text{ factors of } a \ + \ 3 \text{ factors of } a \ = \ 5 \text{ factors of } a$$

Also, $(a^4)(a^3) = (a \cdot a \cdot a \cdot a)(a \cdot a \cdot a) = a \cdot a \cdot a \cdot a \cdot a \cdot a \cdot a = a^7$

$$\uparrow \qquad\qquad \uparrow \qquad\qquad \uparrow$$

$$4 \text{ factors of } a + 3 \text{ factors of } a = 7 \text{ factors of } a$$

Based on the two examples above, we generalize to the first law of exponents for multiplication.

FIRST LAW OF EXPONENTS: PRODUCT RULE

$a^m \cdot a^n = a^{m+n}$ To multiply two expressions with the same base, leave the base unchanged and add the exponents.

The above property is easily extended to more than two expressions. For example, $x^3 \cdot x^4 \cdot x^2 = x^{3+4+2} = x^9$. Remember, if no exponent is given, it is understood to be 1.

EXAMPLE 1

Find the following products. Leave the answers in exponential form.

a) $x^3 \cdot x^5 =$ Apply $a^m \cdot a^n = a^{m+n}$.

    $x^{3+5} =$ Add the exponents.

    x^8 Therefore, $x^3 \cdot x^5 = x^8$.

b) $2^4 \cdot 2^2 =$ Apply $a^m \cdot a^n = a^{m+n}$.

    $2^{4+2} =$ Add the exponents. Note: The base remains 2.

    2^6 Therefore, $2^4 \cdot 2^2 = 2^6$.

c) $3 \cdot 3^4 =$ Apply $a^m \cdot a^n = a^{m+n}$.

    $3^{1+4} =$ Add the exponents. Remember, 3 means 3^1.

    3^5 Therefore, $3 \cdot 3^4 = 3^5$.

d) $a^5 \cdot a^4 \cdot a =$ Apply $a^m \cdot a^n = a^{m+n}$.

    $a^{5+4+1} =$ Add the exponents.

    a^{10} Therefore, $a^5 \cdot a^4 \cdot a = a^{10}$.

Be careful: When multiplying exponential expressions with constant bases as in parts b and c above, a common error is to multiply the bases as well as adding the powers. A product of the type $3^2 \cdot 3^4$ means multiply 2 factors of 3 by 4 factors of 3 for a total of 6 factors of 3, which is written as 3^6. A common error is to write the product $3^2 \cdot 3^4$ as 9^6, which means multiply 6 factors of 9 when, in fact, 9 is not used as a factor.

If the terms have numerical coefficients, we use the commutative and associative properties to rearrange the order of the factors and multiply the coefficients and variable factors with the same bases separately.

EXAMPLE 2

Find the following products.

a) $(2x^2)(3x^3)$ Rearrange the order of the terms.

    $(2 \cdot 3)(x^2 \cdot x^3)$ Multiply 2 and 3. Apply $a^m \cdot a^n = a^{m+n}$.

    $6x^{2+3}$ Add the exponents.

    $6x^5$ Therefore, $(2x^2)(3x^3) = 6x^5$.

b) $(-4x^4y^2)(6x^3y^5)$ Rearrange the order of the terms.

    $(-4 \cdot 6)(x^4 \cdot x^3)(y^2 \cdot y^5)$ Multiply -4 and 6. Apply $a^m \cdot a^n = a^{m+n}$.

    $-24x^{4+3}y^{2+5}$ Add the exponents.

    $-24x^7y^7$ Therefore, $(-4x^4y^2)(6x^3y^5) = -24x^7y^7$.

c) $(3^2 \cdot 4^4)(3^5 \cdot 4^2)$ Rearrange the order of the terms.

    $(3^2 \cdot 3^5)(4^4 \cdot 4^2)$ Apply $a^m \cdot a^n = a^{m+n}$.

    $3^{2+5} \cdot 4^{4+2}$ Add the exponents.

    $3^7 \cdot 4^6$ Therefore, $(3^2 \cdot 4^4)(3^5 \cdot 4^2) = 3^7 \cdot 4^6$.

ADDITIONAL PRACTICE

Find the following products.

a) z^4z^3

b) 2^42^2

c) $c^3c^2c^5$

d) $(5x^3)(2x^5)$

e) $(-4w^3z)(3w^5z^3)$

f) $(3^4 \cdot 6^3)(3^3 \cdot 6^4)$

PRACTICE EXERCISES

Find the following products.

1) $x^2 \cdot x^4$ **2)** $3^2 \cdot 3^6$ **3)** $b^4 \cdot b^3 \cdot b^5$

4) $(4x^4)(3x^5)$ **5)** $(5y^2z^3)(-6y^3z^5)$ **6)** $(2^4 \cdot 5^3)(2^2 \cdot 5^4)$

If more practice is needed, do the Additional Practice Exercises in the margin.

CALCULATOR EXPLORATION ACTIVITY 2—OPTIONAL

Evaluate each of the following using a calculator. Be careful if your calculator does not have parentheses.

Column A	Column B
$(2 \cdot 3)^3 =$	$2^3 \cdot 3^3 =$
$(3 \cdot 5)^3 =$	$3^3 \cdot 5^3 =$
$(3 \cdot 4)^2 =$	$3^2 \cdot 4^2 =$

By looking at corresponding lines of Columns A and B, answer the following.

2) Raising a product to a power is the same as raising each _____ to the _____.

We now provide a mathematical justification for the above observation.

- **Simplifying expressions of the form $(ab)^n$**

In an expression of the form $(ab)^3$, the base of the exponent 3 is the product ab. Consequently, $(ab)^3 = (ab)(ab)(ab)$. Using the commutative and associative properties of multiplication, $(ab)(ab)(ab) = (a \cdot a \cdot a)(b \cdot b \cdot b) = a^3b^3$.

Therefore, $(ab)^3 = a^3b^3$. Likewise, $(2y)^4 = (2y)(2y)(2y)(2y) = (2 \cdot 2 \cdot 2 \cdot 2)(y \cdot y \cdot y \cdot y) = 2^4y^4 = 16y^4$. These examples suggest the following generalization.

SECOND LAW OF EXPONENTS: POWER OF A PRODUCT

$(a \cdot b)^n = a^nb^n$.   To raise a product to a power, raise each factor to the power.

The above property can be extended to include more than two factors. For example,

$$(2ab)^4 = (2ab)(2ab)(2ab)(2ab)$$
$$= (2 \cdot 2 \cdot 2 \cdot 2)(a \cdot a \cdot a \cdot a)(b \cdot b \cdot b \cdot b)$$
$$= 2^4a^4b^4 = 16a^4b^4.$$

EXAMPLE 3

Find the following products.

a) $(xy)^2 = x^2y^2$

b) $(3y)^4 = 3^4y^4 = 81y^4$ Do not forget to raise 3 to the 4th power!

c) $(-4z)^2 = (-4)^2z^2 = 16z^2$

d) $(abc)^5 = a^5b^5c^5$

e) $(-4ab)^3 = (-4)^3a^3b^3 = -64a^3b^3$

ADDITIONAL PRACTICE

Find the following products.

g) $(ef)^7$ **h)** $(4w)^3$

i) $(-5y)^4$ **j)** $(rst)^8$

k) $(-5st)^3$

PRACTICE EXERCISES

Find the following products.

7) $(ab)^5$ **8)** $(5x)^3$ **9)** $(-6x)^2$

10) $(xyz)^6$ **11)** $(3ab)^4$

If more practice is needed, do the Additional Practice Exercises in the margin.

CALCULATOR EXPLORATION ACTIVITY 3—OPTIONAL

Evaluate each of the following columns using a calculator.

Column A	Column B
$(2^2)^3 =$	$2^6 =$
$(3^4)^2 =$	$3^8 =$
$(2^3)^3 =$	$2^9 =$

By comparing the corresponding lines in Columns A and B, answer the following.

3) To raise a number to a power to another power, leave the _____ unchanged and _____ the exponents.

Following is justification for the above observation.

- Simplifying expressions of the form $(a^m)^n$

In the expression $(a^2)^3$, the base of the exponent 3 is a^2. Consequently, $(a^2)^3 = a^2 \cdot a^2 \cdot a^2$. From the Product Rule, we know $a^2 \cdot a^2 \cdot a^2 = a^{2+2+2} = a^6$. Since multiplication is a shortcut for repeated additions of the same number, $2 + 2 + 2 = 3 \cdot 2$. Therefore, $(a^2)^3 = a^{(3)(2)} = a^6$. In the same manner, $(b^3)^4 = b^3 \cdot b^3 \cdot b^3 \cdot b^3 = b^{3+3+3+3} = b^{(4)(3)} = b^{12}$. These examples generalize into the third law of exponents.

> **THIRD LAW OF EXPONENTS: POWER TO A POWER**
>
> $(a^m)^n = a^{mn}$.   To raise a number to a power to another power, leave the base unchanged and multiply the powers.

EXAMPLE 4

Simplify the following.

a) $(x^2)^4 =$ Apply $(a^m)^n = a^{mn}$.

  $x^{(2)(4)} =$ Multiply the exponents.

  x^8 Therefore, $(x^2)^4 = x^8$.

b) $(2^3)^4 =$ Apply $(a^m)^n = a^{mn}$.

  $2^{(3)(4)} =$ Multiply the exponents.

  2^{12} Therefore, $(2^3)^4 = 2^{12}$.

c) $(y^4)^2 =$ Apply $(a^m)^n = a^{mn}$.

  $y^{(4)(2)} =$ Multiply the exponents.

  y^8 Therefore, $(y^4)^2 = y^8$.

ADDITIONAL PRACTICE

Simplify the following.

l) $(q^2)^5$

m) $(5^4)^4$

n) $(w^3)^6$

PRACTICE EXERCISES

Simplify the following.

12) $(a^7)^2$ 13) $(4^3)^5$ 14) $(z^4)^5$

If more practice is needed, do the Additional Practice Exercises in the margin.

Often it is necessary to simplify expressions that involve more than one of the laws of exponents. Be careful in distinguishing which rule applies to each situation.

- Simplifying expressions involving more than one law of exponents

EXAMPLE 5

Simplify each of the following. Leave answers in exponential form.

a) $(x^2 y^3)^3 =$ Apply $(ab)^n = a^n b^n$.

  $(x^2)^3 (y^3)^3 =$ Apply $(a^m)^n = a^{mn}$.

  $x^6 y^9$ Answer.

b) $(2x^3 y)^4 =$ Apply $(ab)^n = a^n b^n$.

  $2^4 (x^3)^4 (y^1)^4 =$ Apply definition of exponents and $(a^m)^n = a^{mn}$.

  $16x^{12} y^4$ Answer.

c) $(2xy)^2(3xy)^3 =$ Apply $(ab)^n = a^n b^n$.

    $2^2 \cdot x^2 \cdot y^2 \cdot 3^3 \cdot x^3 \cdot y^3 =$ Regroup and apply the definition of exponent.

    $(4 \cdot 27)(x^2 \cdot x^3)(y^2 \cdot y^3) =$ Apply $a^n \cdot a^m = a^{m+n}$.

    $108x^5y^5$ Answer.

 d) $(x^3y^2)^2(x^4y^3)^2 =$ Apply $(ab)^n = a^n b^n$.

    $(x^3)^2(y^2)^2(x^4)^2(y^3)^2 =$ Apply $(a^m)^n = a^{mn}$.

    $x^6y^4 \cdot x^8y^6 =$ Regroup.

    $(x^6 \cdot x^8)(y^4 \cdot y^6) =$ Apply $a^m \cdot a^n = a^{m+n}$.

    $x^{14}y^{10}$ Answer.

ADDITIONAL PRACTICE

Simplify the following.

o) $(w^4z^3)^2$ **p)** $(-2x^4y^4)^4$

q) $(cd)^4(cd)^3$ **r)** $(xy^5)^3(x^3y^5)^4$

PRACTICE EXERCISES

Simplify the following.

15) $(x^4y^2)^4$ **16)** $(-4a^3b)^2$

17) $(3ab)^3(2ab)^4$ **18)** $(a^3b^4)^2(a^2b^4)^3$

If more practice is needed, do the Additional Practice Exercises in the margin.

Sometimes it is necessary to add like terms after applying one or more laws of exponents. Remember, when adding like terms we add the coefficients only. We *do not add* the exponents.

EXAMPLE 6
Simplify the following.

a) $(2x)(4x) + (6x)(2x) =$ Multiply before adding. Use $a^m \cdot a^n = a^{m+n}$.

    $8x^2 + 12x^2 =$ Add like terms. Add coefficients only.

    $20x^2$ Answer.

b) $(2x)^3 + (3x)(-2x^2) =$ Simplify $(2x)^3$ and $(3x)(-2x^2)$.

    $8x^3 - 6x^3 =$ Add like terms. Add coefficients only.

    $2x^3$ Answer.

c) $(-3x^2y)(2xy) + (4xy^2)(3x^2) =$ Multiply before adding. Use $a^m \cdot a^n = a^{m+n}$.

    $-6x^3y^2 + 12x^3y^2 =$ Add like terms. Add coefficients only.

    $6x^3y^2$ Answer.

ADDITIONAL PRACTICE

Simplify the following.

s) $(-2x)(4x) + (4x)(3x)$

t) $(4x)^2 + (5x)(-2x)$

u) $(6pq)(4p^2q) + (5p^2q^2)(3p)$

PRACTICE EXERCISES

Simplify the following.

19) $(4y)(2y) + (3y)(-5y)$ **20)** $(3r)^4 + (4r^2)(-10r^2)$

21) $(5a^2b^2)(-2ab^2) + (4ab^3)(3a^2b)$

If more practice is needed, do the Additional Practice Exercises in the margin.

STUDY TIP #7

Preparing for Class

How well you prepare for class determines how much you will get from that class. Organize your day and specify at least one hour a day to study mathematics. It is best to study in concentrated short intervals of approximately thirty minutes with five to ten minute breaks between. *The following are some things you should do.*

1) Read your class notes as soon after class as possible, preferably the same day. Highlight important formulas, statements, and so on.

2) Write any definitions, rules, formulas, or important statements on your review cards. If your instructor went over these things in class, they are important. Check the chapter summary.

3) Read the textbook slowly and carefully with pencil and paper available. Mark those things that you do not understand and come back to them after you have read all of the material. Check the previous material for something similar.

4) If necessary, use other textbooks and study guides. Many texts have supplemental study guides that may

be purchased in the college bookstore. Many math learning centers, or the library, have texts that may be checked out. Your instructor may also have additional texts or materials.

5) Do your homework as soon as possible after class is over and most definitely before the next class meeting. Do not skip steps. The reason for a step is often something that you learned previously. Putting in each step will help reinforce these principles and help you remember them. Careless errors often occur when steps are skipped. Review your homework immediately prior to attending the next class meeting. Make a list of things that you are unsure of and ask your instructor.

6) Preview the material to be covered in class the next day by reading the material with pencil in hand. Mark those parts you are unsure of and make a list of questions to ask your instructor if his/her explanation does not answer your question.

7) If your instructor does not have time in class to answer all of your questions, make an appointment to see him/her in his/her office during office hours.

EXERCISE SET 2.2

Simplify the following using the laws of exponents.

1) $r^2 \cdot r^6$

2) $s^3 \cdot s^5$

3) $4^3 \cdot 4^4$

4) $5^4 \cdot 5^5$

5) $x^3 \cdot x^4 \cdot x^2$

6) $y^3 \cdot y^6 \cdot y^5$

7) $(4x^5)(5x^4)$

8) $(3y^5)(6y^6)$

9) $(-7a^4)(-3a^4)$

10) $(-6b^6)(5b^6)$

11) $(x^5y^3)(x^3y^4)$

12) $(a^3b^6)(a^7b^3)$

13) $(4a^5b^4)(2a^5b^7)$

14) $(5y^6z^2)(7y^5z^5)$

15) $(ab)^5$

16) $(xy)^7$

17) $(2y)^4$

18) $(3a)^5$

19) $(-4x)^2$

20) $(-7x)^2$

21) $-(5x)^2$

22) $-(8b)^4$

23) $-(-3c)^3$

24) $(-5a)^3$

25) $(-3b)^3$

26) $(x^4)^5$

27) $(y^3)^6$

28) $(3^4)^2$

29) $(4^5)^3$

30) $(ab)^5(ab)^4$

31) $(xy)^2(xy)^5$

32) $(2d)^4(2d)^5$

33) $(4x)^5(4x)^3$

34) $(a^3b^3)^5$

35) $(c^4d^3)^6$

36) $(3^4x^5)^3$

37) $(4^5a^3)^4$

38) $-(2^2a^4)^3$

39) $-(4^4b^5)^5$

40) $(-5^3x^2)^4$

41) $(-6^2b^3)^3$

42) $(2x^3y^4)^3(3xy^3)^2$

43) $(3a^5b^2)^3(2ab^3)^2$

44) $(3^2a^4)^3(3^4a^5)^3$

45) $(4^4b^4)^3(4^2b^4)^4$

46) $(x^2y^3z^2)^3(x^4y^4z^2)^3$

47) $(a^3b^2c^3)^2(a^2b^3c^4)^4$

Simplify the following.

48) $(5y)(4y) + (3y)(-6y)$

49) $(6a)(-2a) + (7a)(4a)$

50) $(2z)^3 + (3z)(-5z^2)$

51) $(3x)^3 + (-4x)(6x^2)$

52) $(-2n)^4 + (2n^2)^2$

53) $(2x)^6 + (5x^3)^2$

54) $(3x^2)(-2x) + (-3x)(-5x^2)$

55) $(-6a^3)(2a) - (-4a^2)(-3a^2)$

56) $(2x^2y)^2 - (3x^3y)(-4xy)$

57) $(4m^2n^3)^2 + (-3mn^5)(6m^3n)$

WRITING EXERCISES

58) Are $(-2)^2$ and -2^2 equal? Explain why or why not.

59) Are $(-2)^3$ and -2^3 equal? Are their meanings the same? Explain why or why not.

60) Are 3^2 and $3 \cdot 2$ equal? Explain why or why not.

61) If we simplify $(2x^3)(3x^3)$ we multiply the coefficients and add the exponents. If we simplify $2x^3 + 3x^3$ we add the coefficients only and do not change the exponents. Why?

S E C T I O N 2.3

Products of Polynomials

OBJECTIVES

When you complete this section, you will be able to:

Ⓐ Find the product of two or more monomials.

Ⓑ Find the product of a monomial and a polynomial.

Ⓒ Find the product of two polynomials.

INTRODUCTION

The laws of exponents developed in the previous section are used extensively in finding the products of polynomials. In showing that $a^m \cdot a^n = a^{m+n}$ and in subsequent examples, we were multiplying monomials by monomials. For example, $(2x^3y^2)(3x^3y^4) = 6x^6y^6$ is the product of two monomials.

• **Product of a monomial and a polynomial**

In multiplying a monomial and a polynomial, we use the distributive property which you will recall is $a(b + c) = ab + ac$. The distributive property is easily extended to include multiplication of a monomial over polynomials of more than two terms. For example, $a(b + c + d) = ab + ac + ad$ and $a(b + c + d + e) = ab + ac + ad + ae$, and so on.

EXAMPLE 1

Find the following products.

a) $2(x + 4) =$ ⟶ Apply the distributive property.

$2 \cdot x + 2 \cdot 4 =$ ⟶ Multiply 2 and 4.

$2x + 8$ ⟶ These are not like terms. Therefore, this is the product.

b) $2x^2(x - 5) =$ ⟶ Apply the distributive property.

$2x^2 \cdot x - 2x^2 \cdot 5 =$ ⟶ $2x^2 \cdot x = 2x^3$ and $2x^2 \cdot 5 = 10x^2$.

$2x^3 - 10x^2$ ⟶ These are not like terms. Therefore, this is the product.

c) $3x^2y(2xy^2 + 4x^3y^2) =$ ⟶ Apply the distributive property.

$(3x^2y)(2xy^2) + (3x^2y)(4x^3y^2) =$ ⟶ Apply $a^m \cdot a^n = a^{m+n}$.

$6x^3y^3 + 12x^5y^3$ ⟶ These are not like terms. Therefore, this is the product.

d) $-2x^2(3x^2 - 4x + 6) =$ Apply the distributive property.

$(-2x^2)(3x^2) + (2x^2)(4x) - (2x^2)(6) =$ Apply $a^m \cdot a^n = a^{m+n}$.

$-6x^4 + 8x^3 - 12x^2$ These are not like terms. Therefore, this is the product.

ADDITIONAL PRACTICE

Find the following products.

a) $4(a + 2)$

b) $a(3a - 6)$

c) $3yz^2(3yz - 4y^3z^2)$

d) $-4a^2(3a^3 - 5a^2 + 4)$

PRACTICE EXERCISES

Find the following products.

1) $3(x + 5)$

2) $3x^3(x - 6)$

3) $2ab^2(4a^2b - 3a^2b^2)$

4) $-2y^2(4y^3 - 3y - 5)$

If more practice is needed, do the Additional Practice Exercises in the margin.

In multiplying a binomial and a polynomial of two or more terms, we need to multiply each term of the polynomial by each term of the binomial. The easiest way to do this is to use the distributive property twice. We distribute each term of the binomial over the polynomial as illustrated by the following examples.

EXAMPLE 2

Find the following products.

a) $(x + 2)(x + 3) =$ Rewrite using the distributive property.

$x(x + 3) + 2(x + 3) =$ Apply the distributive property.

$x^2 + 3x + 2x + 6 =$ Combine like terms.

$x^2 + 5x + 6$ Product.

b) $(2x - 4)(3x - 2) =$ Rewrite using the distributive property.

$2x(3x - 2) - 4(3x - 2) =$ Apply the distributive property.

$6x^2 - 4x - 12x + 8 =$ Combine like terms.

$6x^2 - 16x + 8$ Product.

c) $(2a - 3)(2a^2 - 4a + 5) =$ Rewrite using the distributive property.

$2a(2a^2 - 4a + 5) - 3(2a^2 - 4a + 5) =$ Apply the distributive property.

$4a^3 - 8a^2 + 10a - 6a^2 + 12a - 15 =$ Combine like terms.

$4a^3 - 14a^2 + 22a - 15$ Product.

d) $(a - 5)^2 =$ Rewrite as a product.

$(a - 5)(a - 5) =$ Rewrite using the distributive property.

$a(a - 5) - 5(a - 5) =$ Apply the distributive property.

$a^2 - 5a - 5a + 25 =$ Combine like terms.

$a^2 - 10a + 25$ Product.

ADDITIONAL PRACTICE

ADDITIONAL PRACTICE

Find the following products.

e) $(b + 4)(b + 1)$

f) $(2a - 4)(4a + 3)$

g) $(2b + 3)(b^2 + 4b - 1)$

h) $(a - 6)^2$

ANSWERS:
Practice 1–4

1) $3x + 15$ **2)** $3x^4 - 18x^3$

3) $8a^3b^3 - 6a^3b^4$

4) $-8y^5 + 6y^3 + 10y^2$

Additional Practice a–d

a) $4a + 8$

b) $3a^2 - 6a$

c) $9y^2z^3 - 12y^4z^4$

d) $-12a^5 + 20a^4 - 16a^2$

PRACTICE EXERCISES

Find the following products.

5) $(y + 4)(y + 6)$ **6)** $(3x - 2)(4x - 1)$

7) $(3a + 2)(a^2 - 3a + 2)$ **8)** $(x - 4)^2$

If more practice is needed, do the Additional Practice Exercises in the margin.

Multiplication of polynomials can also be done vertically in much the same manner whole numbers are multiplied. Since any two terms can be multiplied, it is not necessary to align like terms before multiplying as was necessary with addition. When multiplying whole numbers, it is necessary to line up digits with the same place value before adding to get final answer. When multiplying polynomials vertically, it is necessary to align like terms before adding. Study the following examples.

EXAMPLE 3

a) *Find the product of 46 and 67 vertically.*

$$\begin{array}{r} 46 \\ \underline{67} \\ 322 \\ \underline{276} \\ 3082 \end{array}$$

     Product of 7 and 46.

     Product of 60 and 46. Align place values.

     Product.

b) *Find the product of 3x + 2 and 2x − 7 vertically.*

$$\begin{array}{r} 3x + 2 \\ \underline{2x - 7} \\ -21x - 14 \\ \underline{6x^2 + 4x} \\ 6x^2 - 17x - 14 \end{array}$$

     Product of $3x + 2$ *and* -7.

     Product of $3x + 2$ and $2x$. Align like terms and add.

     Product.

c) *Find the product of $2x^2 - 4x + 2$ and 3x + 4 vertically.*

$$\begin{array}{r} 2x^2 - 4x + 2 \\ \underline{3x + 4} \\ 8x^2 - 16x + 8 \\ \underline{6x^3 - 12x^2 + 6x} \\ 6x^3 - 4x^2 - 10x + 8 \end{array}$$

     Product of 4 and $2x^2 - 4x + 2$.

     Product of $3x$ and $2x^2 - 4x + 2$. Align like terms and add.

     Product.

ADDITIONAL PRACTICE

Find the following products vertically.

i) $(2x - 1)(3x - 2)$

j) $(4x^2 - 5x + 1)(x - 4)$

PRACTICE EXERCISES

Find the following products.

9) $\begin{array}{r} 3x - 5 \\ \underline{2x + 3} \end{array}$ **10)** $\begin{array}{r} 3x^2 - 5x + 2 \\ \underline{3x - 1} \end{array}$

If more practice is needed, do the Additional Practice Exercises in the margin.

If the dimensions of a geometric figure are given in terms of variables, it is possible to represent the area in terms of these same variables. Remember, the formula for the area of a rectangle is $A = LW$ and the formula for the area of a triangle is $A = \frac{bh}{2}$.

EXAMPLE 4

Write an expression for the area of each of the following using the given dimensions.

a) *A rectangle with $L = 2x - 3$ and $W = 3x - 4$.*

$A = LW$	Substitute for L and W.
$A = (2x - 3)(3x - 4)$	Rewrite using the distributive property.
$A = 2x(3x - 4) - 3(3x - 4)$	Apply the distributive property.
$A = 6x^2 - 8x - 9x + 12$	Add like terms.
$A = 6x^2 - 17x + 12$	Therefore, the area is represented as $6x^2 - 17x + 12$.

b) *A triangle with $b = 4x$ and $h = 5x$.*

$A = \dfrac{bh}{2}$	Substitute for b and h.
$A = \dfrac{(4x)(5x)}{2}$	Multiply $4x$ and $5x$.
$A = \dfrac{20x^2}{2}$	Divide 2 into 20.
$A = 10x^2$	Therefore, the area is represented as $10x^2$.

ADDITIONAL PRACTICE

Write an expression for the area of each of the following using the given dimensions.

k) Rectangle with $L = x - 5$ and $W = x + 3$.

l) Triangle with $b = 7a$ and $h = 4a$.

PRACTICE EXERCISES

Write an expression for the area of each of the following using the given dimensions.

11) A rectangle with $L = x - 3$ and $W = 2x + 2$.

12) A triangle with $b = 5x$ and $h = 6x$.

If more practice is needed, do the Additional Practice Exercises in the margin.

EXERCISE SET 2.3

Find the products of the following monomials.

1) $(7ax^2)(-a^2x)$

2) $(5rq^3)(-3r^4q)$

3) $(6fkb^5)(-7f)(-f^5k)$

4) $(-5ab^2c)(2b)(-3ab^3c^2)$

5) $(-3p^3r)(4p^2qr^4)(-2q^4r^3)$

6) $(5p^4r^2)(-4p^3q^2)(-q^3r^4)$

CALCULATOR EXERCISES

Find the following products using a calculator as needed.

C1) $(-6.78x^3y^5)(7.93x^6y^8)$

C2) $(8.52a^6b^2)(-9.97a^2b)$

C3) $(7.68x^4y^5)(-4.5x^7y^4)$

C4) $(-67.4m^4n^8)(-56.9m^2n^8)$

Find the products of the following monomials and polynomials.

7) $2h^3(h^2 + 8h - 5)$

8) $3y^2(y^2 + 5y - 9)$

9) $2b^4(b^2 + 7b - 6)$

10) $2b^5(b^2 + 6b - 5)$

11) $-2r^3(3r^3 - 4r^2 + 2r - 3)$

12) $-3s^2(4s^4 - 5s^2 + 6s - 5)$

13) $5x^2y(2x^2y^2 - 4xy^3)$

14) $3ab^2(6a^3b - 2a^2b^2)$

15) $-4p^2q^3(3p^4q - p^2q^2 + q)$

16) $-3u^2v^2(4u + 3u^4v^2 - u^2v^4)$

CALCULATOR EXERCISES

Find the following products using a calculator as needed.

C5) $4.3x^3(5.3x^2 - 7.8x - 7.2)$

C6) $-6.4y^2(-4.1y^2 - 9.3y + 9.2)$

Find the products of the following polynomials.

17) $(a + 4)(a + 5)$ **18)** $(q + 7)(q + 2)$

19) $(z + 5)(z - 8)$ **20)** $(w - 1)(w + 5)$

21) $(x + 4)(x - 4)$ **22)** $(x - 5)(x + 5)$

23) $(2x - y)(2x + y)$ **24)** $(x + 4y)(x - 2y)$

25) $(x + y)(x - 2y)$ **26)** $(r - d)(r + 3d)$

27) $(4w + 3)(5w - 8)$ **28)** $(2c - 1)(9c + 5)$

29) $(3z - 4)(2z - 6)$ **30)** $(7a - 5)(2a - 3)$

31) $(2a - b)(3a + 4b)$ **32)** $(4x - 3y)(2x + 5y)$

33) $(x - 1)(x^2 + x + 3)$ **34)** $(x - 2)(x^2 + x - 4)$

35) $(x + 3)(2x^2 - 4x + 3)$ **36)** $(x + 4)(3x^2 - 2x + 5)$

37) $(x + 2)(x^2 - 2x + 4)$ **38)** $(x - 3)(x^2 + 3x + 9)$

39) $(2x - 3)(3x^2 + 4x - 3)$

40) $(3x - 4)(3x^2 - 5x - 2)$

CALCULATOR EXERCISES

Find the following products using a calculator.

C7) $(5.3x - 4.8)(7.1x + 3.6)$

C8) $(6.2x + 3.8)(2.6x - 7.2)$

Find the following products.

41) $2x + 4$
   $x + 3$

42) $3x + 2$
   $x + 5$

43) $3x - 5$
   $2x - 5$

44) $4x - 3$
   $3x - 2$

45) $a^2 - 5a + 6$
   $2a + 4$

46) $b^2 - 4b + 7$
   $3b + 5$

47) $2y^2 + 3y - 2$
   $4y - 2$

48) $3t^2 - 5t + 4$
   $5t - 1$

Recall that the formula for the area of a rectangle is $A = LW$. Write an expression for the area of each of the following rectangles with the given length and width.

49) $L = x, W = 2x$ **50)** $L = 2x, W = 3x$

51) $L = 4x, W = x + 6$ **52)** $L = 5x, W = 3x - 5$

53) $L = x + 2, W = 2x - 3$

54) $L = x - 3, W = 3x + 1$

55) $L = x^2 - x + 2, W = 3x + 2$

56) $L = x^2 + 2x + 9, W = 4x - 3$

Recall that the formula for the area of a triangle is $A = \frac{bh}{2}$. Write an expression for the area of each of the following.

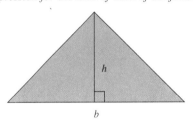

57) $b = 2y, h = 4y$ **58)** $b = 4x, h = 6x$

59) $b = 6z, h = 3z$ **60)** $b = 4a, h = 5a$

61) Fran, who is an architect, has a strange habit. In all the houses that she designs, the width of the living room is 8 feet less than twice the length. If y represents the length of a living room, how would you represent the amount of carpet needed to carpet the living room?

62) The Martinez family wants to sod their backyard. They found that the length is 3 feet more than twice the width. If x represents the width of their backyard, how would you represent the amount of sod they need to sod their backyard?

63) A rectangular pasture is $(3x + 5)$ feet long and $(3x - 2)$ feet wide. How many square feet of land are in the pasture?

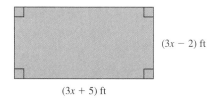

(3x − 2) ft

(3x + 5) ft

64) A rectangular bathroom wall is $(2x + 8)$ feet long and $(3x - 9)$ feet high. How many square feet of tile would it take to cover it?

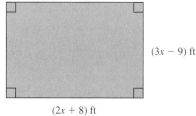

(3x − 9) ft

(2x + 8) ft

CHALLENGE EXERCISES (65–66)

65) One brand of pile carpet sells for $12.50 per square yard. How would you represent the cost to carpet a room $(x + 3)$ yards long and x yards wide?

66) A certain type of sod sells for $0.50 per square foot. How would you represent the cost to sod a rectangular yard that is $(2x + 4)$ feet long and $(x + 2)$ feet wide?

WRITING EXERCISE

67) Is $(a + 3)^2 = a^2 + 3^2$? Why or why not?

68) Is the product of a monomial and a trinomial always a trinomial? Explain.

69) Can the product of two binomials be a binomial? Explain.

SECTION 2.4

Special Products

OBJECTIVES

When you complete this section, you will be able to:

Ⓐ Find the product of two binomials.
Ⓑ Find products of the form $(a + b)(a - b)$.
Ⓒ Square binomials.
Ⓓ Optional: Recognize products of the form $(a + b)(a^2 - ab + b^2)$ and $(a - b)(a^2 + ab + b^2)$.

INTRODUCTION

• **Multiplying binomials using FOIL**

The abilities to multiply binomials quickly and to recognize special products are very important in Chapter 5. In the previous section we found the products of binomials by applying the distributive property twice. We need to be able to multiply binomials more quickly and easily. Consider the following.

$$(x + 3)(2x + 4) = x(2x + 4) + 3(2x + 4) = 2x^2 + 4x + 6x + 12$$

The term, $2x^2$, is the product of the *F*irst terms of the two binomials, x and $2x$. The term, $4x$, is the product of the *O*utside terms (terms farthest apart), x and 4. To find the term, $6x$, multiply the *I*nside terms (terms closest together), 3 and $2x$. The term, 12, is the product of the *L*ast terms of the two binomials, 3 and 4. The underlined letters spell the word FOIL which is an easy way to remember which terms to multiply and allows us to multiply binomials mentally. The only thing left to do is add like terms and get the final product of $2x^2 + 10x + 12$. The diagram below illustrates the use of FOIL.

First   Last

$$(x + 3)(2x + 4) = 2x^2 + 4x + 6x + 12$$

F   O   I   L

Inside

Outside

Now combine the like terms $4x$ and $6x$ to get the product $2x^2 + 10x + 12$.

EXAMPLE 1

Find the following products using FOIL.

a) $(x - 4)(x + 2) = x^2 + 2x - 4x - 8 = x^2 - 2x - 8$

with labels: *Outside*, *Last*, *First*, *Inside*

b) $(2x - 5)(2x + 3) =$ — Apply FOIL.

$2x \cdot 2x + 2x \cdot 3 - 5 \cdot 2x - 5 \cdot 3 =$ — Multiply.

$4x^2 + 6x - 10x - 15 =$ — Add like terms.

$4x^2 - 4x - 15$ — Product.

c) $(4a - b)(3a - 2b) =$ — Apply FOIL.

$4a \cdot 3a - 4a \cdot 2b - b \cdot 3a + b \cdot 2b =$ — Multiply.

$12a^2 - 8ab - 3ab + 2b^2 =$ — Add like terms.

$12a^2 - 11ab + 2b^2$ — Product.

d) $(x + 2y)(3a + b) =$ — Apply FOIL.

$3ax + bx + 6ay + 2by$ — Product since there are no like terms.

ADDITIONAL PRACTICE

Find the following products using FOIL.

a) $(r + 5)(r - 3)$

b) $(5x - 2)(3x + 4)$

c) $(4y + 3z)(2y + 7z)$

d) $(2a + 4b)(3c - 5d)$

- **Products of the form**
  $(a + b)(a - b)$

PRACTICE EXERCISES

Find the following products using FOIL.

1) $(q + 2)(q - 4)$

2) $(3a - 2)(5a - 3)$

3) $(2w + 5t)(3w - 2t)$

4) $(3a + 2b)(4c + 5d)$

If more practice is needed, do the Additional Practice Exercises in the margin.

Products of the form $(a + b)(a - b)$ are of particular interest since the answer is a binomial instead of the usual trinomial. Notice the factors are the sum and difference of the exact same two terms. Products of this form are often referred to as **conjugate pairs**. As you will see, the sum of the outer and inner products is always zero which leaves the difference of two squares.

EXAMPLE 2

Find the following product.

a) $(a + b)(a - b) =$ — Apply FOIL.

$a^2 - ab + ab - b^2 =$ — Combine like terms.

$a^2 - b^2$ — Sum of outer and inner products is 0. The product is the difference of squares.

This example leads to the following observation.

PRODUCTS OF THE FORM $(a + b)(a - b)$ (CONJUGATE PAIRS)

The product of two binomials that are the sum and difference of the same terms results in a binomial that is a difference of squares. In symbols, $(a + b)(a - b) = a^2 - b^2$.

ADDITIONAL PRACTICE

Find the following products.

e) $(a + 1)(a - 1)$

f) $(2p + 6q)(2p - 6q)$

From the example above, we see that in the product of conjugate pairs, the outside and inside products always have a sum of zero. Therefore, it is not necessary to compute them.

 EXAMPLE 3

Find the following products.

a) $(3x - 4)(3x + 4) =$ Form of $(a + b)(a - b)$.

   $9x^2 - 16$ Multiply first by first and last by last.

 b) $(3b + 4c)(3b - 4c) =$ Form of $(a + b)(a - b)$.

   $9b^2 - 16c^2$ Multiply first by first and last by last.

Note that in each case the answer is the difference of squares.

PRACTICE EXERCISES

Find the following products.

5) $(x + 6)(x - 6)$ 6) $(2a - 6b)(2a + 6b)$

If more practice is needed, do the Additional Practice Exercises in the margin.

Another special product is the square of a binomial. As illustrated below, the outside and inside products are always the same. This leads to a quick way of squaring a binomial mentally.

EXAMPLE 4

Square the following binomials.

a) $(a + b)^2 =$ Apply the definition of exponent.

   $(a + b)(a + b) =$ Apply FOIL.

   $a^2 + ab + ab + b^2 =$ Add like terms. Note the outside and inside products are equal.

   $a^2 + 2ab + b^2$ Product. Note the middle term is twice the product ab.

b) $(a - b)^2 =$ Apply the definition of exponent.

   $(a - b)(a - b) =$ Apply FOIL.

   $a^2 - ab - ab + b^2 =$ Add like terms. Note the outside and inside products are equal.

   $a^2 - 2ab + b^2$ Product. Note the middle term is twice the product $-ab$.

This leads to the following procedure for squaring a binomial.

SQUARE OF A BINOMIAL

To square a binomial, square the first term, add two times the product of the two terms, and then add the square of the last term. In symbols, $(a + b)^2 = a^2 + 2ab + b^2$ and $(a - b)^2 = a^2 - 2ab + b^2$.

This procedure is diagrammed below for the sum of two terms.

$$\underset{\downarrow}{\overset{\textit{First term squared}}{}} \quad \underset{\downarrow}{\overset{\textit{Last term squared}}{}}$$
$$(a + b)^2 = a^2 + 2ab + b^2$$
$$\underset{\textit{Two times the product of the terms}}{\overset{\uparrow}{}}$$

EXAMPLE 5
Square the following binomials.

a) $(x + 3)^2 =$ Apply $(a + b)^2 = a^2 + 2ab + b^2$.

   $x^2 + 2(x)(3) + 3^2 =$ Simplify.

   $x^2 + 6x + 9$ Answer.

b) $(y - 4)^2$ Apply $(a - b)^2 = a^2 - 2ab + b^2$.

   $y^2 - 2(y)(4) + (-4)^2$ Simplify.

   $y^2 - 8y + 16$ Answer.

c) $(w - 2q)^2$ Apply $(a - b)^2 = a^2 - 2ab + b^2$.

   $w^2 - 2(w)(2q) + (-2q)^2$ Simplify.

   $w^2 - 4wq + 4q^2$ Answer.

d) $(2a + 3b)^2$ Apply $(a + b)^2 = a^2 + 2ab + b^2$.

   $(2a)^2 + 2(2a)(3b) + (3b)^2$ Multiply and raise to powers.

   $4a^2 + 12ab + 9b^2$ Answer.

ADDITIONAL PRACTICE

Square the following.

g) $(q + 5)^2$ h) $(r - 2)^2$

i) $(c - 2d)^2$ j) $(4a + 2b)^2$

PRACTICE EXERCISES

Square the following binomials.

7) $(a + 2)^2$ 8) $(b - 5)^2$

9) $(c + 3d)^2$ 10) $(3x - 2y)^2$

If more practice is needed, do the Additional Practice Exercises in the margin.

Optional

Another special product is that of a binomial (two terms) multiplied by a trinomial (three terms) where the trinomial has the following properties.

1) The first term of the trinomial is the square of the first term of the binomial.

2) The second term of the trinomial is the negative of the product of the terms of the binomial.

3) The third term of the trinomial is the square of the last term of the binomial.

The following illustrates the form of this product.

$$\underset{\downarrow}{\overset{\textit{Square of first term}}{}} \quad \underset{\downarrow}{\overset{\textit{Square of last term}}{}}$$
$$(a + b)(a^2 - ab + b^2)$$
$$\underset{\textit{Negative of the product of the two terms}}{\overset{\uparrow}{}}$$

A product of this form always results in the sum of two cubes if the binomial is a sum. The product will be the difference of two cubes if the binomial is a difference. The terms of the product are the cubes of the terms of the binomial.

EXAMPLE 6
Find the following products.

a) $(a + b)(a^2 - ab + b^2) =$ Rewrite using distributive property.

$a(a^2 - ab + b^2) + b(a^2 - ab + b^2) =$ Apply distributive property.

$a^3 - a^2b + ab^2 + a^2b - ab^2 + b^3 =$ Add like terms.

$a^3 + b^3$ Sum of the cubes of the terms of the binomial.

b) $(x - y)(x^2 + xy + y^2) =$ Rewrite using distributive property.

$x(x^2 + xy + y^2) - y(x^2 + xy + y^2) =$ Apply distributive property.

$x^3 + x^2y + xy^2 - x^2y - xy^2 - y^3 =$ Add like terms.

$x^3 - y^3$ Difference of the cubes of the terms of the binomial.

c) $(x + 2)(x^2 - 2x + 4) =$ Form of $(a + b)(a^2 - ab + b^2)$.

$x^3 + 2^3 =$ Sum of cubes of terms of binomial.

$x^3 + 8$ Answer.

d) $(2x - 3y)(4x^2 + 6xy + 9y^2) =$ Form of $(a - b)(a^2 + ab + b^2)$.

$(2x)^3 - (3y)^3 =$ Difference of cubes. Apply $(ab)^n = a^n b^n$.

$8x^3 - 27y^3$ Answer.

ADDITIONAL PRACTICE

Find the following products.

k) $(r + s)(r^2 - rs + s^2)$

l) $(a - b)(a^2 + ab + b^2)$

m) $(a + 4)(a^2 - 4a + 16)$

n) $(2a + 3b)(4a^2 - 6ab + 9b^2)$

PRACTICE EXERCISES

Find the following products.

11) $(c + d)(c^2 - cd + d^2)$ 12) $(w - z)(w^2 + wz + z^2)$

13) $(y + 3)(y^2 - 3y + 9)$ 14) $(3a - b)(9a^2 + 3ab + b^2)$

If more practice is needed, do the Additional Practice Exercises in the margin.

STUDY TIP #8
Making Effective Use of Class Time

You need to get as much out of your time in class as possible. It is here that you have the benefit of having an expert on the subject at your disposal—your teacher. The following are some things you should do in order to get the most out of class time.

1) Attend all class meetings. Be on time so that you do not miss anything. You cannot learn from the teacher unless you are there. If you do miss a class, be sure to find out what the assignment was and complete it *before* the next class meeting.

2) Participate in class. Do not just listen to the instructor. Think along with the instructor and anticipate the next step when he/she is doing an example. Answer questions that the instructor asks. Ask questions when you do not understand what the teacher has said or why he/she said it.

3) Understand as much as possible before you leave class.

4) Sit as close to the front of the classroom as possible so you can see and hear everything. In front you will also find yourself more involved with classroom activities.

5) Keep a mathematics notebook with your class notes in one section and your homework in another. When taking notes, leave extra space so you will have room to add additional comments as you review your notes.

6) If you do not understand an example the instructor is doing, ask the instructor to explain the first step that you do not understand.

7) Get help outside of class either from the instructor, a classmate, or, if available, the math learning center. Forming a study group with others in the class is an excellent idea.

8) Try tape recording the class and use the recording to supplement your notes. Listen to the recording as you review your notes. Don't forget to ask the instructor's permission to record the class.

ANSWERS:
Practice 11–14

11) $c^3 + d^3$ **12)** $w^3 - z^3$

13) $y^3 + 27$ **14)** $27a^3 - b^3$

Additional Practice k–n

k) $r^3 + s^3$ **l)** $a^3 - b^3$

m) $a^3 + 64$ **n)** $8a^3 + 27b^3$

EXERCISE SET 2.4

Find the product of the following binomials using FOIL.

1) $(x + 3)(x + 2)$ **2)** $(y + 4)(y + 5)$

3) $(a + 3)(a - 4)$ **4)** $(b - 4)(b + 6)$

5) $(c - 5)(c - 2)$ **6)** $(q - 5)(q - 1)$

7) $(a + 3b)(a + 4b)$ **8)** $(t + 5s)(t + 2s)$

9) $(r - 5s)(r + 3s)$ **10)** $(x + 4y)(x - 6y)$

11) $(2a + 3b)(3a + 2b)$ **12)** $(3s + 5t)(2s + 3t)$

13) $(3x - 5y)(2x + 3y)$ **14)** $(4x + 3y)(5x - 2y)$

15) $(2x - 5y)(3x - 2y)$ **16)** $(3a - 5b)(2a - 7b)$

CALCULATOR EXERCISES

Find the following products using a calculator.

C1) $(3.1x + 6.3y)(8.2x - 5.2y)$

C2) $(8.3x - 3.1y)(9.5x + 4.8y)$

Find the following products of the form $(a + b)(a - b)$.

17) $(x + 3)(x - 3)$ **18)** $(x + 5)(x - 5)$

19) $(p - q)(p + q)$ **20)** $(r - s)(r + s)$

21) $(a + 4b)(a - 4b)$ **22)** $(w + 3q)(w - 3q)$

23) $(6a - b)(6a + b)$ **24)** $(3x + y)(3x - y)$

25) $(4a + 5)(4a - 5)$ **26)** $(5a + 2)(5a - 2)$

27) $(2x + 5y)(2x - 5y)$ **28)** $(3x + 5y)(3x - 5y)$

CALCULATOR EXERCISES

Find the following products using a calculator.

C3) $(7.2a + 6.7)(7.2a - 6.7)$

C4) $(2.8x - 8.3)(2.8x + 8.3)$

Square the following binomials.

29) $(x + 2)^2$ **30)** $(y + 6)^2$ **31)** $(a - 4)^2$

32) $(b - 7)^2$ **33)** $(r + s)^2$ **34)** $(p + q)^2$

35) $(t - s)^2$ **36)** $(r - s)^2$ **37)** $(2x + 3)^2$

38) $(3x + 4)^2$ **39)** $(3a - 1)^2$ **40)** $(5a - 2)^2$

41) $(2x + 5y)^2$ **42)** $(3a + 4b)^2$ **43)** $(4x - 3y)^2$

44) $(5x - 6y)^2$ **45)** $(5t^2 - w)^2$ **46)** $(b + 2z^2)^2$

CALCULATOR EXERCISES

Square the following binomials using a calculator.

C5) $(4.7x + 3.7y)^2$ **C6)** $(5.7a - 7.4b)^2$

Find the following products.

47) $(2a + 5b)(3a - 2b)$ **48)** $(4x - 3y)(2x + 5y)$

49) $(4a - 3b)(4a + 3b)$ **50)** $(3s + 5)(3s - 5)$

51) $(3a + 5)^2$ **52)** $(2a - 7)^2$

53) $(2 - 3b)(4 - 3b)$ **54)** $(3 + 5c)(3 + 2c)$

55) $(6 + y)(6 - y)$ **56)** $(7 + x)(7 - x)$

57) $(4 + 3w)^2$ **58)** $(5 + 4c)^2$

59) $(4c + 7d)(4c - 7d)$ **60)** $(5x - 8y)(5x + 8y)$

Write an expression or equation representing each of the following and simplify when possible.

61) The square of the sum of 3 and x.

62) The square of the difference of y and 4.

63) 3 more than the square of x.

64) 5 less than the square of y.

65) The quantity 5 less than x squared is equal to 25.

66) The quantity 7 more than x squared is equal to 16.

Recall that the formula for the area of a rectangle is $A = LW$. Represent the area of the following rectangles.

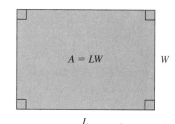

$A = LW$ W

L

67) $L = x + 5, W = x - 8$ **68)** $L = y - 7, W = y + 9$

69) $L = 4a - 5, W = 3a + 9$ **70)** $L = 5x - 6, W = x + 5$

Recall that the area of a square is $A = s^2$ where s is the length of a side. For example, the area of a square each of whose sides is $x + 7$ is $A = (x + 7)^2 = x^2 + 14x + 49$. Represent the area of each of the following squares.

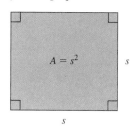

$A = s^2$ s

s

71) $s = x + 8$ **72)** $s = y - 9$

73) $s = 3a - 5b$ **74)** $s = 4y + 3z$

Find the following products.

75) $(p + q)(p^2 - pq + q^2)$

76) $(t - s)(t^2 + ts + s^2)$

77) $(x + 3)(x^2 - 3x + 9)$

78) $(a + 5)(a^2 - 5a + 25)$

79) $(b - 4)(b^2 + 4b + 16)$

80) $(r - 2)(r^2 + 2r + 4)$

81) $(3a + 2b)(9a^2 - 6ab + 4b^2)$

82) $(2z - 4w)(4z^2 + 8wz + 16w^2)$

WRITING EXERCISES

83) In finding products of the form $(a + b)(a - b)$ there is no middle term. Why?

84) When squaring a binomial, why is it necessary to multiply the product of the two terms by 2 to get the middle term?

85) By assigning values to a and b, show that $(a + b)^2 \neq a^2 + b^2$. In general, show that $(a + b)^n \neq a^n + b^n$ for $n = 2, 3,$ and 5.

GROUP PROJECT

86) Using the formula for the area of a rectangle ($A = LW$) and the area of a square ($A = s^2$), construct a geometric figure that illustrates that $(a + b)^2 = a^2 + 2ab + b^2$. (Hint. Begin with a square.)

SECTION **2.5**

Division of Integers

OBJECTIVES

When you complete this section, you will be able to:

Ⓐ Divide integers.

Ⓑ Simplify quotients whose numerators/denominators first require simplification.

INTRODUCTION

In Section 2.1 we discussed the multiplication of integers and found that $(+)(+) = +, (+)(-) = -, (-)(+) = -,$ and $(-)(-) = +$. In other words, we found the product of two numbers with the same sign is positive, and the product of two numbers with opposite signs is negative. In this section we will develop similar rules for division, but first we need to define division.

We know that $\frac{24}{3} = 8$, because $3 \cdot 8 = 24$. Likewise, $\frac{12}{4} = 3$ because $4 \cdot 3 = 12$. Division can always be checked by multiplication. Generalizing on these examples, we state the following definition of division.

DEFINITION OF DIVISION

For all numbers a, b, and c with $b \neq 0$, if $\frac{a}{b} = c$, then $b \cdot c = a$.

In the above definition $b \neq 0$. In other words, you can not divide by 0. This is because of the definition of division. If $\frac{a}{0} = c$, then $0 \cdot c = a$ which is impossible since 0 times any real number is 0.

 CALCULATOR EXPLORATION ACTIVITY—OPTIONAL

Evaluate each of the following columns using a calculator.

Column A	Column B
$\frac{12}{6} =$	$\frac{-12}{6} =$
$\frac{18}{2} =$	$\frac{18}{-3} =$
$\frac{-16}{-4} =$	$\frac{-16}{4} =$
$\frac{-21}{-3} =$	$\frac{21}{-3} =$

1) Based on your answers to Column A, is the quotient of two numbers with the same sign positive or negative?

2) Based on your answers to Column B is the quotient of two numbers with different signs positive or negative?

We now give justification for the preceding observations.

- **Dividing a positive number by a positive number**

From our prior knowledge of arithmetic, we know a positive number divided by a positive number is a positive number. For example, $\frac{18}{6} = 3$ and $\frac{28}{7} = 4$. Stated in symbols, $\frac{(+)}{(+)} = (+)$, because $(+)(+) = (+)$.

Using the definition of division stated above, we will develop rules for dividing a positive number by a negative number, a negative number by a positive number, and a negative number by another negative number.

- **Dividing a positive number by a negative number**

 EXAMPLE 1

a) Find $\frac{12}{-3}$.

If $\frac{12}{-3} = \underline{\quad}$, then $-3 \cdot \underline{\quad} = 12$. Since $(-3)(\underline{-4}) = 12$, the blank would be filled by -4. Therefore, $\frac{12}{-3} = -4$. Check: $(-3)(-4) = 12$.

• Dividing a negative number by a positive number

• Dividing a negative number by a negative number

b) Find $\frac{32}{-4}$.

If $\frac{32}{-4} =$ ___ , then $-4 \cdot$ ___ $= 32$. Since $(-4)(\underline{-8}) = 32$, the blank would be filled by -8. Therefore, $\frac{32}{-4} = -8$. Check: $(-4)(-8) = 32$.

From Example 1, we can conclude that a positive number divided by a negative number results in a negative number. Stated in symbols, $\frac{+}{-} = (-)$, because $(-)(-) = (+)$.

EXAMPLE 2

a) Find $\frac{-18}{3}$.

If $\frac{-18}{3} =$ ___ , then $3 \cdot$ ___ $= -18$. Since $(3)(\underline{-6}) = -18$, the blank would be filled with -6. Therefore, $\frac{-18}{3} = -6$. Check: $(3)(-6) = -18$.

b) Find $\frac{-12}{4}$.

If $\frac{-12}{4} =$ ___ , then $4 \cdot$ ___ $= -12$. Since $(4)(\underline{-3}) = -12$, the blank would be filled with -3. Therefore, $\frac{-12}{4} = -3$. Check: $(4)(-3) = -12$.

From Example 2, we can conclude that a negative number divided by a positive number results in a negative number. Written symbolically, $\frac{(-)}{(+)} = (-)$, because $(+)(-) = (-)$.

EXAMPLE 3

a) Find $\frac{-8}{-2}$.

If $\frac{-8}{-2} =$ ___ , then $-2 \cdot$ ___ $= -8$. Since $(-2)(\underline{4}) = -8$, the blank would be filled by 4. Therefore, $\frac{-8}{-2} = 4$. Check: $(-2)(4) = -8$.

b) Find $\frac{-27}{-3}$.

If $\frac{-27}{-3} =$ ___ , then $-3 \cdot =$ ___ -27. Since $(-3)(\underline{9}) = -27$, the blank would be filled by 9. Therefore, $\frac{-27}{-3} = 9$. Check: $(-3)(9) = -27$.

From Example 3, we can conclude that a negative number divided by a negative number results in a positive number. Stated symbolically, $\frac{(-)}{(-)} = (+)$ because $(-)(+) = (-)$.

Note that the rules for division of signed numbers are exactly the same as the rules for multiplication and are summarized in the following box.

DIVISION OF SIGNED NUMBERS

The quotient of two numbers with the same sign is positive and the quotient of two numbers with opposite signs is negative. In symbols: $\frac{(+)}{(+)} = (+)$, $\frac{(+)}{(-)} = (-)$, $\frac{(-)}{(+)} = (-)$, and $\frac{(-)}{(-)} = (+)$.

EXAMPLE 4
Find the following quotients.

a) $\dfrac{10}{-2} = -5$ \qquad A positive divided by a negative is a negative.

 b) $\dfrac{-14}{-7} = 2$ \qquad A negative divided by a negative is a positive.

---

c) $\dfrac{24}{8} = 3$ A positive divided by a positive is a positive.

d) $\dfrac{-26}{13} = -2$ A negative divided by a positive is a negative.

ADDITIONAL PRACTICE

Find the following quotients.

a) $\dfrac{-16}{8}$ **b)** $\dfrac{-24}{-6}$

c) $\dfrac{15}{5}$ **d)** $\dfrac{36}{-4}$

- **Simplifying quotients**

PRACTICE EXERCISES

Find the following quotients.

1) $\dfrac{24}{-3}$ **2)** $\dfrac{-21}{-7}$ **3)** $\dfrac{32}{8}$ **4)** $\dfrac{-36}{4}$

If more practice is needed, do the Additional Practice Exercises in the margin.

Remember, when quotients are involved, the numerator and the denominator are simplified separately using the order of operations agreement as illustrated by the following examples.

EXAMPLE 5
Simplify the following.

a) $\dfrac{2 \cdot 3^2 - 4}{4^2 - 9} =$ Raise to powers in both the numerator and the denominator.

$\dfrac{2 \cdot 9 - 4}{16 - 9} =$ Multiply in the numerator, subtract in the denominator.

$\dfrac{18 - 4}{7} =$ Subtract in the numerator.

$\dfrac{14}{7} =$ Divide.

2 Quotient.

b) $\dfrac{(-8)(-3) + 4}{2^2 + 3} =$ Multiply in the numerator and raise to the power in the denominator.

$\dfrac{24 + 4}{4 + 3} =$ Add in both the numerator and the denominator.

$\dfrac{28}{7} =$ Divide.

4 Quotient.

c) $\dfrac{5^2 - 4^2}{-3(4^2 - 13)} =$ Powers in the numerator and power in parentheses in the denominator.

$\dfrac{25 - 16}{-3(16 - 13)} =$ Subtract in the numerator, subtract inside parentheses in the denominator.

$\dfrac{9}{-3(3)} =$ Multiply in the denominator.

$\dfrac{9}{-9} =$ Divide.

-1 Quotient.

Evaluate each of the following.

e) $\dfrac{(-6)(7) - 6}{4^2 + 8}$

f) $\dfrac{3(-4)^2 + 2}{3(-5) + 5}$

g) $\dfrac{6^2 - 4^2}{-2(3^2 - 4)}$

ANSWERS:
Practice 1–4

1) -8	**2)** 3
3) 4	**4)** -9

Additional Practice a–d

a) -2	**b)** 4
c) 3	**d)** -9

PRACTICE EXERCISES

Evaluate each of the following.

5) $\dfrac{3 \cdot 4^2 - 16}{5^2 - 9}$ **6)** $\dfrac{4(-2)^2 + 10}{-2(-4) + 5}$ **7)** $\dfrac{4^2 + 6^2}{-2(3^2 - 7)}$

If more practice is needed, do the Additional Practice Exercises in the margin.

When translating from English to mathematics, one way of indicating division is by using the word "quotient." When using quotient, look for the words "of" and "and." If division is indicated by using a fraction, the expression following "of" is the numerator and the expression following "and" is the denominator.

EXAMPLE 6

Write an expression for each of the following and simplify.

a) The quotient of 24 and -6 increased by 7.

Solution:

The quotient of 24 and -6 means $\frac{24}{-6}$. The quotient is increased by 7 which means add 7 to the quotient. Consequently, the quotient of 24 and -6 increased by 7 is expressed as:

$\dfrac{24}{-6} + 7 =$	Divide before adding.
$-4 + 7 =$	Add.
3	Answer.

b) The quotient of -15 and -3 subtracted from the quotient of 36 and -9.

Solution:

The quotient of -15 and -3 means $\frac{-15}{-3}$ and the quotient of 36 and -9 means $\frac{36}{-9}$. The first quotient is subtracted from the second which is expressed as:

$\dfrac{36}{-9} - \dfrac{-15}{-3} =$	Find each quotient.
$-4 - 5 =$	Add.
-9	Answer.

PRACTICE EXERCISES

Write an expression for each of the following and simplify.

8) The quotient of -35 and 7 decreased by 8.

9) The quotient of 48 and -12 subtracted from the quotient of -56 and 8.

EXERCISE SET **2.5**

Find the following quotients.

1) $\dfrac{36}{-9}$ **2)** $\dfrac{-36}{-6}$ **3)** $\dfrac{-42}{7}$

4) $\dfrac{63}{9}$ **5)** $\dfrac{-48}{-8}$ **6)** $\dfrac{56}{-7}$

7) $\dfrac{-121}{-11}$ **8)** $\dfrac{108}{9}$ **9)** $\dfrac{144}{-18}$

10) $\dfrac{-392}{98}$ **11)** $\dfrac{15}{5-2}$ **12)** $\dfrac{24}{-5-3}$

13) $\dfrac{35-7}{-4}$ **14)** $\dfrac{17-(-5)}{-11}$

15) $\dfrac{(-8)(-7)}{14}$ **16)** $\dfrac{18(-6)}{-27}$

17) $\dfrac{-140}{(35)(-2)}$ **18)** $\dfrac{180}{(-45)(-2)}$

19) $\dfrac{(-12)(18)}{(-6)(-9)}$ **20)** $\dfrac{(-28)(-13)}{-91}$

21) $\dfrac{(4)(-3)-4}{(3)(3)-1}$ **22)** $\dfrac{(-2)(8)+6}{(-3)(4)+10}$

23) $\dfrac{(-4)(2)+(-3)(6)}{(-4)(3)-(-5)(5)}$ **24)** $\dfrac{(-4)(-5)-(-2)(8)}{(-3)(-5)+(-4)(6)}$

25) $\dfrac{(4)(4)+3^2}{(-3)(3)+2^2}$ **26)** $\dfrac{(3)(6)-2^3}{(8)(4)-3^3}$

CALCULATOR EXERCISES

Round answers to the nearest hundredth.

C1) $\dfrac{-9.18}{2.7}$ **C2)** $\dfrac{32.745}{-9.25}$

C3) $\dfrac{-14.65+42.6268}{-4.16}$ **C4)** $\dfrac{-2103.19-1468.73}{-72.6}$

Determine if the given number is a solution to the given equation.

27) $\dfrac{x}{-3} = -6$: $x = 18$

28) $\dfrac{x}{-4} = 7$: $x = 28$

29) $\dfrac{x}{3} + 4 = 1$: $x = -9$

30) $\dfrac{x}{-5} - 6 = -4$: $x = -10$

31) $\dfrac{x+3}{4} = -1$: $x = 1$

32) $\dfrac{x-6}{3} = -2$: $x = 12$

33) $\dfrac{2x-6}{5} = 2$: $x = 8$

34) $\dfrac{3x+4}{-7} = 2$: $x = -6$

35) $\dfrac{x+6}{-4} - 5 = -7$: $x = -2$

36) $\dfrac{x-3}{-5} + 7 = 9$: $x = -12$

Write an expression for each of the following and simplify.

37) The quotient of 12 and −4 increased by −8.

38) The quotient of −15 and 5 increased by 9.

39) The quotient of −18 and −9 decreased by −6.

40) The quotient of −24 and −4 decreased by −9.

41) 12 less the quotient of 36 and −9.

42) −5 less the quotient of 45 and −5.

43) The quotient of 48 and −16 subtracted from 7.

44) The quotient of −28 and −7 subtracted from −8.

45) The quotient of −9 and 3 added to the quotient of 14 and −7.

46) The quotient of −24 and 6 added to the quotient of −32 and 8.

47) The quotient of −36 and 18 subtracted from the quotient of 48 and −8.

48) The quotient of −42 and 6 subtracted from the quotient of 54 and −9.

CHALLENGE EXERCISES (49–52)

Evaluate the following.

49) $(-4)^2(6 + 6 \cdot 3) \div (-2)^2(-3)$

50) $(-6)^2(42 - 3 \cdot 2) \div (-3)^2(-2)$

51) $12 - 3 \cdot 6^2 \div (-3)^3 \div (-2)$

52) $32 - 4 \cdot 3^3 \div (-6)^2 \div (-3)$

Answer the following.

53) When a cold front moved through, the temperature dropped 28° in four hours. Using a signed number, find the average drop per hour.

54) Paula went on a diet and lost ten pounds in four weeks. Using a signed number, find the average number of pounds per week that Paula lost.

55) On a steep mountain road the elevation dropped at the rate of nine feet per 150 feet. Using a signed number, find the rate at which the road dropped per foot.

56) A football team lost twelve yards in three plays. Using a signed number, find the average loss per play.

57) A parachutist descends 54 feet in six seconds. Using a signed number, find the rate at which she is descending per second.

58) A stock dropped five points in four hours. Using a signed number, find the average drop per hour.

WRITING EXERCISES

59) Why is the quotient of a negative number and a positive number equal to a negative number?

60) Why is the quotient of a negative number and a negative number equal to a positive number?

61) Why is $\frac{2}{0}$ undefined? (Do not say because you can not divide by 0!)

62) If the temperature drops 12° in three hours, how would you represent the average drop per hour using a signed number? Explain how you arrived at your answer and what the answer means.

63) If an airliner descends 2000 feet in 10 minutes, how would you represent the average drop in altitude per minute as a signed number? Explain how you arrived at your answer and what the answer means.

64) Is division commutative? Why or why not?

SECTION 2.6

Order of Operations on Real Numbers

OBJECTIVES

When you complete this section, you will be able to:

Ⓐ Evaluate expressions containing integers which involve combinations of addition, subtraction, multiplication, division, and raising to powers.

Ⓑ Evaluate expressions containing variables when given the value(s) of the variables.

INTRODUCTION

Because integers had not been discussed in Section 1.3 when we did order of operations, we were limited to operations with whole numbers only. Now that we know how to perform operations on integers, we will revisit the order of operations. We repeat the order of operations.

ORDER OF OPERATIONS

If an expression contains more than one operation, they are to be performed in the following order:

1) If parentheses or other inclusion symbols (braces or brackets) are present, begin within the innermost and work outward, using the order in steps 3–5 below in doing so.

2) If a fraction bar is present, simplify above and below the fraction bar separately in the order given by steps 3–5 below.

3) First evaluate all indicated powers.

4) Then perform all multiplication or division in the order in which they occur as you work from left to right.

5) Then perform all additions or subtractions in the order in which they occur as you work from left to right.

EXAMPLE 1

Evaluate the following using the order of operations.

a) $-5 - 6 + 8 =$ Add and subtract in order from left to right.

$\quad$ $-11 + 8 =$ Continue adding.

$\quad$ -3 Answer.

b) $8 - 15 \div (-3) =$ Divide before adding. $-15 \div (-3) = 5$.

$\quad$ $8 + 5 =$ Add.

$\quad$ 13 Answer.

c) $-7 - 3(-2)^2 =$ Raise to powers first. $(-2)^2 = 4$.

$\quad$ $-7 - 3 \cdot 4 =$ Multiplication before subtraction.

$\quad$ $-7 - 12 =$ Add.

$\quad$ -19 Answer.

d) $6 - (12 - 18) \div 3 =$ Add inside parentheses first.

$\quad$ $6 - (-6) \div 3 =$ Division before subtraction.

$\quad$ $6 - (-2) =$ $-(-x) = x$.

$\quad$ $6 + 2 =$ Add.

$\quad$ 8 Answer.

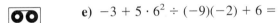

e) $-3 + 5 \cdot 6^2 \div (-9)(-2) + 6 =$ Raise to powers first.

$\quad$ $-3 + 5 \cdot 36 \div (-9)(-2) + 6 =$ Multiply 5 and 36.

$\quad$ $-3 + 180 \div (-9)(-2) + 6 =$ Divide 180 by -9.

$\quad$ $-3 + (-20)(-2) + 6 =$ Multiply -20 and -2.

$\quad$ $-3 + 40 + 6 =$ Add -3 and 40.

$\quad$ $37 + 6 =$ Add 37 and 6.

$\quad$ 43 Answer.

f) $-3(5^2 - 4) \div (-7) =$ Power inside parentheses first.

$\quad$ $-3(25 - 4) \div (-7) =$ Add inside parentheses.

$\quad$ $-3(21) \div (-7) =$ Multiply before dividing in order from left to right.

$\quad$ $-63 \div (-7) =$ Divide

$\quad$ 9 Answer.

g) $\dfrac{4 - 6^2}{4^2 - 8} =$ Simplify the numerator and the denominator separately. Raise to powers.

$\quad$ $\dfrac{4 - 36}{16 - 8} =$ Add in the numerator and the denominator.

$\quad$ $\dfrac{-32}{8} =$ Divide.

$\quad$ -4 Answer.

h) $\dfrac{-5(-2) - [(-4)(-5) + 6^2]}{(-2)^2(6) - 3} =$ Power inside the brackets in the numerator and the power in the denominator first.

$\quad$ $\dfrac{-5(-2) - [(-4)(-5) + 36]}{(4)(6) - 3} =$ Multiply inside the brackets in the numerator and multiply in the denominator.

$$\frac{-5(-2) - [20 + 36]}{24 - 3} =$$
Add inside the brackets in the numerator and add in the denominator.

$$\frac{-5(-2) - 56}{21} =$$
Multiply in the numerator.

$$\frac{10 - 56}{21} =$$
Add in the numerator.

$$\frac{-46}{21}$$
Answer.

ADDITIONAL PRACTICE

Evaluate the following using the order of operations agreement.

a) $9 - 12 - 8$

b) $-6 - (-8) \div 2$

c) $9 - 4(-3)^3$

d) $-7 - (7 - 19) \div (-3)$

e) $-2(8^2 - 22) \div (-14)$

f) $-4 + 2 \cdot 4^3 \div (-16) - 7$

g) $\dfrac{3^3 - 2^4}{4^2 - 9}$

h) $\dfrac{4(-3) - [(-8)(3) - 4]}{(-4)^2(-1) - 9}$

PRACTICE EXERCISES

Evaluate the following using the order of operations agreement.

1) $-12 + 5 - 8$

2) $-9 - 18 \div (-9)$

3) $6 - (-3)(-4)^2$

4) $8 - (5 - 13) \div (-4)$

5) $6(7^2 - 13) \div (-9)$

6) $7 - 3 \cdot 8^2 \div (-8)(-3) + 5$

7) $\dfrac{6 - 6^2}{5^2 - 8}$

8) $\dfrac{4(-2) - [(-5)(3) - 21]}{(-3)^2(-2) + 15}$

If more practice is needed, do the Additional Practice Exercises in the margin.

In Sections 1.1 and 1.2 variables were introduced, and we evaluated expressions containing variables. The values used for the variables were whole numbers since operations with integers had not been discussed. We can now use integer values as illustrated by the following examples.

EXAMPLE 2

Evaluate each of the following for $x = -2$, $y = -3$, and $z = 6$.

a) $5x^2 - 2y =$
Substitute -2 for x and -3 for y.

$5(-2)^2 - 2(-3) =$
First raise to powers.

$5(4) - 2(-3) =$
Multiply in order from left to right.

$20 + 6 =$
Add.

26
Answer.

b) $-3xy^2z =$
Substitute for x, y, and z.

$-3(-2)(-3)^2(6) =$
First raise to powers.

$-3(-2)(9)(6) =$
Multiply in order from left to right.

$6 \cdot 9 \cdot 6 =$
Continue multiplying.

$54 \cdot 6 =$
Continue multiplying.

324
Answer.

c) $x - y(x^4 - 2z) =$
Substitute for x, y, and z.

$-2 - (-3)[(-2)^4 - 2 \cdot 6] =$
Raise to powers inside the parentheses.

$-2 - (-3)(16 - 2 \cdot 6) =$
Multiply inside the parentheses.

$-2 - (-3)(16 - 12) =$
Add inside the parentheses.

$-2 - (-3)(4) =$
Multiply before adding.

$-2 - (-12) =$
$-(-x) = x$.

$-2 + 12 =$
Add.

10
Answer.

d) $\dfrac{2x^2 - 3y}{2z^2 - 3x^4} =$ Substitute for x, y, and z.

$\dfrac{2(-2)^2 - 3(-3)}{2(6)^2 - 3(-2)^4} =$ First raise to powers.

$\dfrac{2 \cdot 4 - 3(-3)}{2 \cdot 36 - 3 \cdot 16} =$ Multiply in order, left to right.

$\dfrac{8 + 9}{72 - 48} =$ Add.

$\dfrac{17}{24}$ Answer.

ADDITIONAL PRACTICE

Evaluate each of the following for $x = -4$, $y = 3$, and $z = -1$.

i) $4z^2 - 2x^2$ **j)** $-5x^2yz^2$

k) $-3y + 2x(2z^2 - 2x^2)$

l) $\dfrac{3y^2 - 5z}{3x^2 - 5y}$

PRACTICE EXERCISES

Evaluate each of the following for $x = -3$, $y = 2$, and $z = -4$.

9) $3x^2 - 4z$ **10)** $-2xy^2z^2$

11) $-2y - x(3y^2 - 2z^2)$ **12)** $\dfrac{4y - 5x^2}{3y^3 - z^2}$

If more practice is needed, do the Additional Practice Exercises in the margin.

STUDY TIP #9

Preparing for a Test

Proper test preparation is certainly one of the most important factors in determining how well you do in a course. The following should help.

1) Set your goal high. Try for 100% rather than trying just to pass the test. With your goal set higher, your preparation will be more complete and you should score higher.

2) Avoid getting a mental block. Adequately prepare for the test. Inadequate preparation causes a loss of confidence and this loss of confidence is what produces the mental block. So be prepared.

3) Begin your test preparation early. Do not wait until the night before the test to begin studying. Begin your preparation at least a week in advance and study at least an hour a day for the exam.

4) Be organized. Make a list of specific topics to be covered on the test. Find and solve specific problems for each topic. Be sure to include all types of problems that could be contained within each topic.

5) Start at the beginning of the material to be tested and work through each section in turn. Master each section before going to the next.

6) Practice by doing the chapter review in your textbook. Do all of the exercises if you can. If you can't do an exercise, go back to the indicated section and study the examples. If you still can't do an exercise, ask your instructor.

7) In your review, answer each problem, confirm that the answer is correct (Check your work!), and examine your understanding of the problem. Do not allow yourself to get "stuck." If you can't do an exercise after 10 minutes, go to the next exercise.

8) Review your notes and the text and clear up any questions that you might have. Think about the material as you review. How does this relate to previous material? How do different parts of this material relate to other parts of the material?

9) Be able to distinguish between the different types of problems that might be on the test.

10) Try to find or construct a practice test. The practice test might be one of your teacher's previous tests, tests from the text, or from a study guide.

E X E R C I S E   S E T   **2.6**

Evaluate the following.

1) $-3 + 7 - 6$ **2)** $-7 + 9 - 5$

3) $12 - 5 + 15$ **4)** $14 - 9 + 6$

5) $8 - 4 \cdot 6$ **6)** $5 - 7 \cdot 3$

7) $-5 + 4(-7)$ **8)** $-3 + 7(-6)$

9) $8 - 12 \div 3$ **10)** $6 - 18 \div 3$

11) $9 - 21 \div (-7)$ **12)** $4 - 15 \div (-5)$

13) $10 - 2(-4)^2$

14) $14 - 4(-2)^2$

15) $9 - (8 - 16) \div 4$

16) $5 - (3 - 12) \div 3$

17) $-8 - (15 - 5) \div (-5)$

18) $-4 - (13 - 7) \div (-3)$

19) $6 + 4 \cdot 5^2 \div 25(-2) + 3$

20) $3 + 3 \cdot 8^2 \div 32(-3) + 5$

21) $-5 - 2 \cdot 6^2 \div (-36)(-4) + 8$

22) $-9 - 6 \cdot 4^2 \div (-16)(-5) + 7$

23) $4(6^2 - 4) \div (-8)$

24) $6(4^2 + 12) \div (-7)$

25) $-3^2(6 - 3^2)$

26) $-4^2(8 - 4^2)$

27) $-36 \div (4^2 - 7)$

28) $-48 \div (5^2 - 9)$

29) $\dfrac{7^2 - 13}{5^2 - 7}$ **30)** $\dfrac{9^2 - 17}{6^2 - 4}$

31) $\dfrac{2(-3)^2 - 6}{-3(-2)^2 + 6}$ **32)** $\dfrac{2(-3)^2 + 6}{3(-3)^2 - 3}$

33) $\dfrac{2(4^2) - 7^2}{5^2 - 3(-3)^2}$

34) $\dfrac{3(-2)^2 - 5^2}{9^2 - 4(-5)^2}$

35) $\dfrac{(-3)(-5) - [5(-4) + 7^2]}{(-3)^2(4) - 7}$

36) $\dfrac{(-2)(-6) - [6(-3) + 5^2]}{(-4)^2(3) - 21}$

CALCULATOR EXERCISES

Evaluate each of the following using a calculator. If necessary, round off to the nearest hundredth.

C1) $7.02 + (5.72)(9.43)$

C2) $9.40 + (7.34)(6.92)$

C3) $18.2 - 208.32 \div 16.8$

C4) $14.9 - 325.26 \div (-23.4)$

C5) $(56.2)^2 + (-82.4)^2$

C6) $(-12.4)^3 - (58.9)^2$

C7) $68.23 - 34.7(56.1^2 - 67.2) \div 39.1$

C8) $95.35 + 62.5(9.8^2 - 68.9) \div 62.8$

C9) $\dfrac{(13.2)^2 + (32.1)(45.3)^2}{(12.6)^3 - (76.3)(18.5)^2}$

C10) $\dfrac{(5.7)^4 - (19.7)^2(96.3)}{(26.7)^2 + (14.6)^2(31.1)^2}$

Evaluate each of the following for x = −2, y = 3, and z = −3.

37) $4x^2 - 2y$ **38)** $5z^2 - 3x$

39) $-3y^3 + 4x$ **40)** $-5x^2 + 3z$

41) $5x^2 - 3y^2$ **42)** $2y^2 - 3x^2$

43) $3x^2y$ **44)** $4y^2z$

45) $-5xz^2$ **46)** $-6yz^2$

47) $3x^2y^2z$ **48)** $2xy^2z^2$

49) $3y - 2(3x + z)$ **50)** $3x - 4(2y - x)$

51) $-3x + z(3x - 4y)$ **52)** $-3z + x(4y - 3x)$

53) $-y^2 - (2x^2 - 3z)$ **54)** $-x^2 - (3z^2 - 4x)$

55) $2x^3y - x^2(3z^2 - 4x^2)$ **56)** $3y^3x - z^2(5x^2 - y^2)$

57) $\dfrac{2x + 3y}{3y + 2z}$ **58)** $\dfrac{3x + 3z}{4y + 2x}$

59) $\dfrac{x^2 - y^2}{y^2 + z^2}$ **60)** $\dfrac{y^2 + x^2}{x^2 - z^2}$

61) $\dfrac{2x^3 - 3y^2}{3x^3 - 2z^2}$ **62)** $\dfrac{4x^3 - 2y^2}{5x^3 - 3z^3}$

Answer the following.

63) The owner of a produce stand sold 55 pounds of tomatoes at a profit of 15 cents per pound. As the tomatoes began to age, she sold 22 pounds at a loss of 8 cents per pound. What was her net profit/loss on the sale of the tomatoes?

64) The produce stand owner in Exercise 63 also sold 20 pounds of onions at a profit of 12 cents per pound and later sold eight pounds at a loss of fifteen cents per pound. What was her net profit/loss on the sale of the onions?

65) During a recent price war among the airlines, 30 seats on one flight were sold at a profit of $40 per seat and 60 seats were sold at a loss of $25 per seat. Find the net profit/loss for this flight.

66) During the same price war as in Exercise 65, another flight sold 45 seats at a profit of $25 per seat and 50 seats at a loss of $25 per seat. Find the net profit/loss for this flight.

67) Ms. Jones recently sold some of her stock. She made a profit of $30 per share on 45 shares, had a loss of $12 per share on 20 shares, and a profit of $8 per share on 36 shares. What was her net profit/loss on the sale of the stock?

68) Mr. Lopez sold 20 shares of stock at a loss of $12 per share, 32 shares at a profit of $20 per share, and 40 shares at a loss of $7 per share. What was his net profit/loss from the sale of the stock?

WRITING EXERCISES

69) What is the error in the following?
$8 - 3(2^2 + 4) = 8 - 3(4 + 4) = 8 - 3(8) = 5(8) = 40$.
Rework the problem correctly.

70) What is the error in the following?
$5 - 2 \cdot 4^2 = 5 - 8^2 = 5 - 64 = -59$. Rework the problem correctly.

GROUP PROJECT

71) Write a problem involving the order of operations on integers similar to 63–68 above. Exchange your problem with another group and solve the problem that you received.

S E C T I O N 2.7

Quotient Rule and Integer Exponents

OBJECTIVES

When you complete this section, you will be able to:

Ⓐ Simplify exponential expressions using the property, $\frac{a^m}{a^n} = a^{m-n}$.

Ⓑ Simplify expressions with zero and negative integer exponents.

INTRODUCTION

In Section 2.2 we developed laws of exponents that involved products of expressions with exponents. For reference, these were:

1) $a^m \cdot a^n = a^{m+n}$,

2) $(a^m)^n = a^{mn}$, and

3) $(ab)^n = a^n b^n$.

In this section we will develop similar laws for quotients of expressions with exponents.

CALCULATOR EXPLORATION ACTIVITY 1—OPTIONAL

Evaluate each of the following columns using a calculator.

Column A	Column B
$\dfrac{2^5}{2^2} =$	$2^3 =$
$\dfrac{3^7}{3^4} =$	$3^3 =$
$\dfrac{5^6}{5^4} =$	$5^2 =$

(continued)

Remember, the exponent indicates how many times the base is to be used as a factor. Another fact that we will need is that any number (other than 0) divided by itself is equal to 1. We also need the procedure for multiplying fractions. Recall, we multiply numerator times numerator and denominator times denominator. For example, $\frac{2}{3} \cdot \frac{5}{7} = \frac{2 \cdot 5}{3 \cdot 7} = \frac{10}{21}$.

- Developing $\dfrac{a^m}{a^n} = a^{m-n}$

EXAMPLE 1

Simplify the following using the definition of exponents. Leave the answer in exponential form.

a) $\dfrac{2^6}{2^3} =$ Rewrite using the definition of exponent.

$\dfrac{2 \cdot 2 \cdot 2 \cdot 2 \cdot 2 \cdot 2}{2 \cdot 2 \cdot 2} =$ Rewrite as the multiplication of fractions.

$\dfrac{2}{2} \cdot \dfrac{2}{2} \cdot \dfrac{2}{2} \cdot \dfrac{2}{1} \cdot \dfrac{2}{1} \cdot \dfrac{2}{1} =$ $\dfrac{2}{2} = 1$ and $\dfrac{2}{1} = 2$.

$1 \cdot 1 \cdot 1 \cdot 2 \cdot 2 \cdot 2 =$ Apply the identity for multiplication.

$2 \cdot 2 \cdot 2 =$ Rewrite using definition of exponent.

2^3 Answer.

Notice that three of the six factors of 2 in the numerator were divided by the three factors of 2 in the denominator leaving $6 - 3 = 3$ factors of 2 in the numerator. Therefore, $\frac{2^6}{2^3} = 2^{6-3} = 2^3$. Notice that the base remained as 2.

b) $\dfrac{x^5}{x^2} =$ Rewrite using the definition of exponent.

$\dfrac{x \cdot x \cdot x \cdot x \cdot x}{x \cdot x} =$ Rewrite as multiplication of fractions.

$\dfrac{x}{x} \cdot \dfrac{x}{x} \cdot \dfrac{x}{1} \cdot \dfrac{x}{1} \cdot \dfrac{x}{1} =$ $\dfrac{x}{x} = 1$, if $x \neq 0$ and $\dfrac{x}{1} = x$.

$1 \cdot 1 \cdot x \cdot x \cdot x =$ Apply the identity for multiplication.

$x \cdot x \cdot x =$ Rewrite using definition of exponent.

x^3 Answer.

Again, notice that two of the factors of x in the numerator were divided by the two factors of x in the denominator, leaving $5 - 2 = 3$ factors of x in the numerator. Therefore, $\frac{x^5}{x^2} = x^{5-2} = x^3$.

Note that in each of the preceding examples, the exponent in the numerator was larger than the exponent in the denominator. Based on these examples, we generalize to the following law of exponents for division of expressions with the same base but different exponents.

FOURTH LAW OF EXPONENTS: QUOTIENT RULE

For any two positive integers m and n with $m > n$ and $a \neq 0$, $\frac{a^m}{a^n} = a^{m-n}$. To divide two numbers with the same base, subtract the bottom exponent from the top exponent leaving the base unchanged.

Be careful: A common error is to divide the bases as well as subtracting the exponents. The base does not change when using this law of exponents. For example, $\frac{3^5}{3^3} \neq 1^{5-3}$. Using the Quotient Rule, $\frac{3^5}{3^3} = 3^{5-3} = 3^2$.

EXAMPLE 2

Simplify the following using the Quotient Rule. Leave the quotients in exponential form. All variables represent nonzero quantities.

a) $\dfrac{3^5}{3^2} =$ \qquad Apply $\dfrac{a^m}{a^n} = a^{m-n}$.

$3^{5-2} =$ \qquad Subtract the exponents.

3^3 \qquad Quotient.

b) $\dfrac{5^6}{5^2} =$ \qquad Apply $\dfrac{a^m}{a^n} = a^{m-n}$.

$5^{6-2} =$ \qquad Subtract the exponents.

5^4 \qquad Quotient.

c) $\dfrac{z^7}{z^4} =$ \qquad Apply $\dfrac{a^m}{a^n} = a^{m-n}$.

$z^{7-4} =$ \qquad Subtract the exponents.

z^3 \qquad Quotient.

ADDITIONAL PRACTICE

Simplify each of the following. Leave answers in exponential form.

a) $\dfrac{5^6}{5^4}$ \qquad **b)** $\dfrac{9^7}{9^4}$

c) $\dfrac{r^5}{r^2}$

PRACTICE EXERCISES

Simplify the following using the Quotient Rule. Leave the answers in exponential form. All variables represent nonzero quantities.

1) $\dfrac{4^5}{4^3} =$ \qquad **2)** $\dfrac{6^8}{6^4} =$ \qquad **3)** $\dfrac{x^6}{x^3} =$

If more practice is needed, do the Additional Practice Exercises in the margin.

All the exponents that we have discussed to this point have been positive integers. If the restriction that $m > n$ is removed from the Quotient Rule, it would be possible to have zero or negative exponents. Using the Quotient Rule, $\frac{2^3}{2^3} = 2^{3-3} = 2^0$. What does 2^0 mean? It surely cannot mean use 2 as a factor 0 times. Example 3 will give meaning to 0 as an exponent.

• Defining 0 exponents

EXAMPLE 3

Simplify the following.

a) $\dfrac{2^4}{2^4} = \dfrac{2 \cdot 2 \cdot 2 \cdot 2}{2 \cdot 2 \cdot 2 \cdot 2} = \dfrac{2}{2} \cdot \dfrac{2}{2} \cdot \dfrac{2}{2} \cdot \dfrac{2}{2} = 1 \cdot 1 \cdot 1 \cdot 1 = 1$. However, if we use the

Quotient Rule $\dfrac{2^4}{2^4} = 2^{4-4} = 2^0$. Since $\dfrac{2^4}{2^4} = 1$ and $\dfrac{2^4}{2^4} = 2^0$, we conclude that $2^0 = 1$.

b) $\dfrac{x^3}{x^3} = \dfrac{x \cdot x \cdot x}{x \cdot x \cdot x} = \dfrac{x}{x} \cdot \dfrac{x}{x} \cdot \dfrac{x}{x} = 1 \cdot 1 \cdot 1 = 1$. If we use the Quotient Rule,

$\dfrac{x^3}{x^3} = x^{3-3} = x^0$. Since $\dfrac{x^3}{x^3} = 1$ and $\dfrac{x^3}{x^3} = x^0$, we conclude that $x^0 = 1$ for

$x \neq 0$.

Based on Example 3, we make the following definition for 0 exponents.

DEFINITION OF 0 EXPONENTS

For all $a \neq 0$, $a^0 = 1$. Therefore, any nonzero number raised to the 0 power is equal to 1.

In using 0 as an exponent, great care must be taken in determining the base of the exponent, as illustrated by the following. Remember, the base is the symbol immediately preceding the exponent unless parentheses are used in which case the base is everything inside the parentheses.

EXAMPLE 4

Evaluate each of the following.

a) $5^0 = 1$	Definition of 0 exponents.
b) $5x^0 = 5 \cdot 1 = 5$	The base of the 0 exponent is x only, not $5x$. Remember, $x \neq 0$.
c) $(5x)^0 = 1$	The base of the 0 exponent is $5x$.
d) $-3y^0 + 7x^0 =$	The bases of the 0 exponents are y and x only. $y^0 = 1$, $x^0 = 1$.
$-3 \cdot 1 + 7 \cdot 1 =$	Apply the identity for multiplication.
$-3 + 7 =$	Add.
4	Answer.
e) $(4x)^0 - 6x^0 =$	The bases of the 0 exponents are $4x$ and x. $(4x)^0 = 1$, $x^0 = 1$.
$1 - 6 \cdot 1 =$	Apply the identity for multiplication.
$1 - 6 =$	Add.
-5	Answer.

ADDITIONAL PRACTICE

Evaluate each of the following.

d) 10^0 **e)** $10y^0$

f) $(10x)^0$ **g)** $3a^0 - 7a^0$

h) $(-5x)^0 - 8x^0$

PRACTICE EXERCISES

Evaluate each of the following. Assume $x \neq 0$ and $z \neq 0$.

4) $8^0 =$ **5)** $8x^0 =$ **6)** $(8x)^0 =$

7) $4z^0 - 9z^0 =$ **8)** $6x^0 - (-3x)^0 =$

• **Defining negative exponents**

If more practice is needed, do the Additional Practice Exercises in the margin.

Now we will consider negative exponents.

CALCULATOR EXPLORATION ACTIVITY 2—OPTIONAL

Evaluate each of the following columns using a calculator.

Column A	Column B
$2^{-3} =$	$\dfrac{1}{2^3} =$
$5^{-2} =$	$\dfrac{1}{5^2} =$
$3^{-4} =$	$\dfrac{1}{3^4} =$

By looking at the corresponding lines of Columns A and B answer the following.

2) Raising a number to a negative power is the same as _____
_____ .

Following is justification for the previous observation.

Using the Quotient Rule, $\frac{3^2}{3^5} = 3^{2-5} = 3^{-3}$. What does 3^{-3} mean? It cannot mean use 3 as a factor -3 times. The following examples will give meaning to these types of expressions.

EXAMPLE 5

Simplify the following.

a) $\dfrac{3^2}{3^4} = \dfrac{3 \cdot 3}{3 \cdot 3 \cdot 3 \cdot 3} = \dfrac{3}{3} \cdot \dfrac{3}{3} \cdot \dfrac{1}{3} \cdot \dfrac{1}{3} = 1 \cdot 1 \cdot \dfrac{1 \cdot 1}{3 \cdot 3} = \dfrac{1}{3^2}$. If we use the Quotient

Rule, we get $\dfrac{3^2}{3^4} = 3^{2-4} = 3^{-2}$. Since $\dfrac{3^2}{3^4} = \dfrac{1}{3^2}$ and $\dfrac{3^2}{3^4} = 3^{-2}$, then $3^{-2} = \dfrac{1}{3^2}$.

b) $\dfrac{x^3}{x^4} = \dfrac{x \cdot x \cdot x}{x \cdot x \cdot x \cdot x} = \dfrac{x}{x} \cdot \dfrac{x}{x} \cdot \dfrac{x}{x} \cdot \dfrac{1}{x} = 1 \cdot 1 \cdot 1 \cdot \dfrac{1}{x} = \dfrac{1}{x}$. If we use the Quotient

Rule, we get $\dfrac{x^3}{x^4} = x^{3-4} = x^{-1}$. Since $\dfrac{x^3}{x^4} = \dfrac{1}{x}$ and $\dfrac{x^3}{x^4} = x^{-1}$, then

$x^{-1} = \dfrac{1}{x}$.

From the above example, we generalize to the following definition for negative integer exponents.

DEFINITION OF NEGATIVE EXPONENTS

For all $x \neq 0$, $x^{-n} = \dfrac{1}{x^n}$.

It is interesting to note that $x^n \cdot x^{-n} = x^{n + (-n)} = x^0 = 1$, $x \neq 0$. This means that x^n and x^{-n} are multiplicative inverses (reciprocals). For example, $2^1 = 2$ and $2^{-1} = \frac{1}{2}$ so $2 \cdot 2^{-1} = 2 \cdot \frac{1}{2} = 1$. Another way of stating it is that x^n denotes multiplication and x^{-n} denotes division.

EXAMPLE 6

Rewrite each of the following with positive exponents only. Assume all variables represent nonzero quantities.

a) $4^{-3} = \dfrac{1}{4^3} = \dfrac{1}{64}$ ⟶ Definition of negative exponents and $4^3 = 64$.

b) $x^{-5} = \dfrac{1}{x^5}$ ⟶ Definition of negative exponents.

c) $3a^{-4} = 3 \cdot \dfrac{1}{a^4} = \dfrac{3}{a^4}$ ⟶ The base of the exponent is a only, not $3a$.

d) $(4x)^{-2} = \dfrac{1}{(4x)^2} =$ ⟶ The base of the exponent is $(4x)$.

$\dfrac{1}{4^2 x^2} = \dfrac{1}{16x^2}$ ⟶ Apply $(ab)^n = a^n b^n$ and $4^2 = 16$.

Be careful: A common mistake is to confuse negative exponents with negative numbers. The value of $3^{-2} \neq -9$ and also $3^{-2} \neq (-2)(3)$. The value of $3^{-2} = \frac{1}{3^2} = \frac{1}{9}$.

PRACTICE EXERCISES

Rewrite each of the following with positive exponents only. Assume all variables represent nonzero quantities.

9) $5^{-3} =$ **10)** $y^{-6} =$ **11)** $6y^{-4}$ **12)** $(7z)^{-3}$

There is also an easy way to simplify expressions of the form $\frac{1}{x^{-n}}$ as is illustrated by the following examples.

EXAMPLE 7

a) In the expression $(2^{-3})^{-1}$, the base of the exponent -1 is (2^{-3}). Therefore, by the definition of a negative exponent, $(2^{-3})^{-1} = \frac{1}{2^{-3}}$. However, using the $(a^m)^n = a^{mn}$ which can be shown applies to negative exponents also, $(2^{-3})^{-1} = 2^{(-3)(-1)} = 2^3$. Since $(2^{-3})^{-1} = \frac{1}{2^{-3}}$ and $(2^{-3})^{-1} = 2^3$, then $\frac{1}{2^{-3}} = 2^3$.

b) In the expression $(x^{-4})^{-1}$, the base of the exponent -1 is (x^{-4}). Therefore, by the definition of a negative exponent, $(x^{-4})^{-1} = \frac{1}{x^{-4}}$. But, if we use the Power to a Power law of exponents, $(x^{-4})^{-1} = x^4$. Therefore, $\frac{1}{x^{-4}} = x^4$ since they both equal $(x^{-4})^{-1}$.

From the preceding examples, we make the following observation.

> **OBSERVATION**
>
> For all $x \neq 0$, $\dfrac{1}{x^{-n}} = x^n$.

EXAMPLE 8

Write the following with positive exponents only. Assume all variables represent nonzero quantities.

a) $\dfrac{1}{2^{-4}} = 2^4$ Apply $\dfrac{1}{x^{-n}} = x^n$.

b) $\dfrac{1}{x^{-6}} = x^6$ Apply $\dfrac{1}{x^{-n}} = x^n$.

c) $\dfrac{2}{x^{-2}} = 2x^2$ Apply $\dfrac{1}{x^{-n}} = x^n$.

PRACTICE EXERCISES

Write the following with positive exponents only. Assume all variables represent nonzero quantities.

13) $\dfrac{1}{5^{-7}} =$ 14) $\dfrac{1}{x^{-8}} =$ 15) $\dfrac{4}{y^{-3}} =$

Since 0 and negative exponents now have meaning, it is no longer necessary to restrict $m > n$ in the Quotient Rule. Therefore, $\dfrac{a^m}{a^n} = a^{m-n}$ for all integer values of m and n.

GENERALIZED QUOTIENT RULE

For all integer values of m and n, $\dfrac{a^m}{a^n} = a^{m-n}$ for $a \neq 0$.

It can be shown that all the laws of exponents hold for zero and negative exponents. The following examples make that assumption. Remember, $a^m \cdot a^n = a^{m+n}$, $(ab)^n = a^n b^n$, and $(a^m)^n = a^{mn}$.

EXAMPLE 9

Simplify each of the following. Leave answers in exponential form with positive exponents only. Assume all variables represent nonzero quantities.

a) $3^{-3}3^5 =$ Apply $a^m a^n = a^{m+n}$.

$3^{-3+5} =$ Add exponents.

3^2 Answer.

b) $a^4 a^{-6} =$ Apply $a^m a^n = a^{m+n}$.

$a^{4+(-6)} =$ Add exponents.

$a^{-2} =$ Apply definition of negative exponents.

$\dfrac{1}{a^2}$ Answer.

c) $\dfrac{x^{-4}}{x^2} =$ Apply $\dfrac{a^m}{a^n} = a^{m-n}$.

$x^{-4-2} =$ Subtract exponents.

$x^{-6} =$ Apply definition of negative exponents.

$\dfrac{1}{x^6}$ Answer.

Example 9c could also be done as follows: $\dfrac{x^{-4}}{x^2} = x^{-4} \cdot \dfrac{1}{x^2} = \dfrac{1}{x^4} \cdot \dfrac{1}{x^2} = \dfrac{1}{x^6}$

d) $\dfrac{a^{-3}}{a^{-5}} =$ Apply $\dfrac{a^m}{a^n} = a^{m-n}$.

$a^{-3-(-5)} =$ $-(-x) = x$.

$a^{-3+5} =$ Add the exponents.

a^2 Answer.

e) $(w^{-3})^2 =$ Apply $(a^m)^n = a^{mn}$.

$w^{(-3)(2)} =$ Multiply the exponents.

$w^{-6} =$ Apply the definition of negative exponents.

$\dfrac{1}{w^6}$ Answer.

ADDITIONAL PRACTICE

Simplify each of the following. Leave answers in exponential form with positive exponents only.

i) 2^{-5} **j)** d^{-6}

k) $r^{-2}r^5$ **l)** q^3q^{-6}

m) $a^{-5}a$ **n)** $\dfrac{z^{-3}}{z^3}$

o) $\dfrac{r^2}{r^{-5}}$ **p)** $(a^{-4})^2$

PRACTICE EXERCISES

Simplify each of the following. Leave answers in exponential form with positive exponents only. Assume all variables represent nonzero quantities.

16) $4^{-5}4^7 =$ **17)** $w^3w^{-5} =$ **18)** $\dfrac{b^{-4}}{b^2} =$

19) $\dfrac{h^4}{h^{-3}} =$ **20)** $(a^{-4})^{-2} =$

If more practice is needed, do the Additional Practice Exercises in the margin.

EXERCISE SET   2.7

Simplify each of the following. Leave answers in exponential form with positive exponents only. Assume all variables represent nonzero quantities.

1) $\dfrac{2^6}{2^2}$ **2)** $\dfrac{3^8}{3^3}$ **3)** $\dfrac{c^5}{c}$ **4)** $\dfrac{d^7}{d}$

5) $\dfrac{z^9}{z^5}$ **6)** $\dfrac{a^8}{a^6}$ **7)** 2^{-4} **8)** 3^{-2}

9) a^{-6} **10)** b^{-7} **11)** $\dfrac{1}{5^{-3}}$ **12)** $\dfrac{1}{8^{-4}}$

13) $\dfrac{1}{y^{-5}}$ **14)** $\dfrac{1}{z^{-9}}$ **15)** $\dfrac{x^2}{x^5}$ **16)** $\dfrac{y^4}{y^5}$

17) $\dfrac{5^3}{5^6}$ **18)** $\dfrac{4^2}{4^6}$ **19)** 8^0 **20)** 6^0

21) $3a^0$ **22)** $7z^0$ **23)** $(-4s)^0$ **24)** $(-7h)^0$

25) $(3x)^0 + (2y)^0$ **26)** $(6w)^0 - (2z)^0$ **27)** $3x^0 - 4z^0$

28) $2x^{-3}$ **29)** $5a^{-2}$ **30)** $-7b^{-4}$

31) $-9c^{-6}$ **32)** $(2x)^{-3}$ **33)** $(5a)^{-2}$

34) $8q^0 - 2q^0$ **35)** $2^{-3} \cdot 2^5$ **36)** $4^5 \cdot 4^{-2}$

37) $5^{-6} \cdot 5^4$ **38)** $6^3 \cdot 6^{-7}$ **39)** $x^{-3}x^{-2}$

40) $y^{-4}y^{-5}$ **41)** $z^{-4}z^2$ **42)** q^5q^{-2}

43) $\dfrac{3^{-2}}{3^2}$ **44)** $\dfrac{2^{-4}}{2^5}$ **45)** $\dfrac{y^{-4}}{y^5}$

46) $\dfrac{4^4}{4^{-5}}$ **47)** $\dfrac{6^5}{6^{-2}}$ **48)** $\dfrac{x^4}{x^{-3}}$

49) $\dfrac{r^2}{r^{-5}}$ **50)** $\dfrac{5^{-3}}{5^{-5}}$ **51)** $\dfrac{6^{-2}}{6^{-3}}$

52) $\dfrac{x^{-6}}{x^{-3}}$ **53)** $\dfrac{z^{-8}}{z^{-4}}$ **54)** $(5^2)^5$

16) 4^2 **17)** $\dfrac{1}{w^2}$ **18)** $\dfrac{1}{b^6}$

19) h^7 **20)** a^8

Additional Practice i–p

i) $\dfrac{1}{2^5}$ **j)** $\dfrac{1}{d^6}$ **k)** r^3

l) $\dfrac{1}{q^3}$ **m)** $\dfrac{1}{a^4}$ **n)** $\dfrac{1}{z^6}$

o) r^7 **p)** $\dfrac{1}{a^8}$

Simplify each of the following. Leave answers with positive exponents only. Assume all variables represent nonzero quantities.

69) $(2x^{-3}y^2)^2(-3x^3y^{-3})^3$

70) $(-4a^2b^{-3})^3(5a^{-3}b^{-3})^2$

71) $(2m^{-4}n^2)^{-2}(-2m^2n^{-4})^3$

72) $(8c^{-5}d^3)^{-1}(8c^3d^{-5})^2$

73) $\dfrac{(4a^{-4}b^2)^3}{(2a^3b^{-4})^3}$

74) $\dfrac{(6m^{-5}n^2)^2}{(3m^2n^{-6})^2}$

WRITING EXERCISES

75) How does 4^{-3} differ from $(-3)(4)$?

76) Why is $3^{-3} \neq -27$?

77) What is wrong with the following? $\dfrac{4^5}{4^3} = 1^{5-3} = 1^2 = 1.$

CRITICAL THINKING

78) Is it possible to raise a positive number to a power and get a negative answer? Why or why not?

79) We know that $x^0 = 1$ if $x \neq 0$. Why is $x = 0$ not allowed?

55) $(8^3)^6$ **56)** $(p^4)^4$ **57)** $(r^5)^4$

58) $(4^5)^{-4}$ **59)** $(6^3)^{-5}$ **60)** $(t^6)^{-2}$

61) $(y^7)^{-3}$ **62)** $(6^{-1})^4$ **63)** $(8^{-3})^6$

64) $(b^4)^{-2}$ **65)** $(z^3)^{-5}$ **66)** $(6^{-2})^{-5}$

67) $(9^{-5})^{-2}$ **68)** $(a^{-2})^{-4}$

SECTION 2.8

An Application of Exponents: Scientific Notation

OBJECTIVES

When you complete this section you will be able to:

A Change numbers written in scientific notation into standard notation.

B Write numbers in scientific notation.

C Multiply and divide very large and/or very small numbers using scientific notation.

INTRODUCTION

A certain computer can perform 15,000 computations per second. Therefore, in one day, it can perform $(15{,}000)(60)(60)(24)$ computations. When I performed this calculation on my calculator, I got 1.296E9. What does this mean?

Very large and very small numbers occur frequently in the sciences and in values encountered everyday, like the population of the United States, the national debt, and the gross national product. Frequently, (as in the computer example) these numbers have so many digits that a calculator cannot display all of them. We use an alternative method of writing extremely large and small numbers called **scientific notation**.

DEFINITION OF SCIENTIFIC NOTATION

A number is written in scientific notation if it is in the form of $a \times 10^n$ with $1 \leq a < 10$ and n is an integer.

To say $1 \leq a < 10$ means that there is one nonzero digit to the left of the decimal point. Remember that the integers are $\{... -3, -2, -1, 0, 1, 2, 3, ...\}$. Examples of numbers written in scientific notation are:

$$2.3 \times 10^4, \ 4.06 \times 10^7, \ 9.23 \times 10^{-5}, \text{ and } 5.34 \times 10^{-3}.$$

In performing the calculation $(15,000)(60)(60)(24)$, the calcualtor gave the answer in scientific notation with the number following the "E" is the exponent of 10. Consequently, this computer can perforn 1.296×10^9 calculations per day. What does this number represent?

Before we write numbers in scientific notation, we need to make some observations regarding powers of 10 and how multiplying and dividing by powers of 10 affects the movement of the decimal point. Study the table below.

Power of 10	Value
10^0	1
10^1	10
10^2	100
10^3	1,000
10^4	10,000

Observe that the exponent of 10 is the same as the number of zeros following the 1 in the value of the power of 10. Consequently, 10^7 equals 1 followed by seven 0's or 10,000,000. Study the following table.

Power of 10	Value		
10^{-1}	$\dfrac{1}{10}$		$= 0.1$
10^{-2}	$\dfrac{1}{10^2}$	$= \dfrac{1}{100}$	$= 0.01$
10^{-3}	$\dfrac{1}{10^3}$	$= \dfrac{1}{1000}$	$= 0.001$
10^{-4}	$\dfrac{1}{10^4}$	$= \dfrac{1}{10000}$	$= 0.0001$

Notice the total number of decimal places is the same as the absolute value of the exponent of 10. Consequently, 10^{-6} has a total of 6 decimal places and equals 0.000001.

Study the following examples and look for a relationship between the exponent of 10 and the movement of the decimal point. When graphing integers on the number line, movement to the right is positive and movement to the left is negative. We will use the same idea in moving the decimal point.

EXAMPLE 1

Find the following products.

a) 3.2×10^3 $10^3 = 1,000$. Write vertically and multiply.

$$3.2$$
$$\underline{\times \ 1000}$$ Place three zeros behind 32 and mark off one decimal place.
$$3200.0$$ Notice that the decimal has been moved three places to the right which is the same as the exponent of 10.

b) 4.67×10^5 $10^5 = 100{,}000.$ Write vertically and multiply.

$$4.67$$
$$\underline{\times\ 100000}$$
$$467000.00$$

Place five zeros behind 467 and mark off two decimal places.

Notice that the decimal has been moved five places to the right which is the same as the exponent of 10.

c) 2.452×10^{-4} $10^{-4} = 0.0001.$ Write vertically and multiply.

$$2.452$$
$$\underline{\times\ 0.0001}$$
$$0.0002452$$

$1 \times 2452 = 2452$ and mark off seven decimal places.

Notice that the decimal has been moved four places to the left which is the same as the absolute value of the exponent of 10.

d) 9.6×10^{-2} $10^{-2} = 0.01.$ Write vertically and multiply.

$$9.6$$
$$\underline{\times\ 0.01}$$
$$0.096$$

$1 \times 96 = 96$ and mark off three decimal places.

Notice that the decimal has been moved two places to the left which is the same as the absolute value of the exponent of 10.

Based on the results of Example 1, we make the following observation.

MULTIPLYING BY POWERS OF 10

To find a product of the form $a \times 10^n$, move the decimal n places to the right if n is positive or move the decimal $|n|$ places to the left if n is negative.

An easy way to remember this rule is a positive exponent on 10 makes the product larger, so move the decimal point to the right and a negative exponent makes the product smaller, so move the decimal point to the left.

Since all the numbers in Example 1 were given in scientific notation, the rule above is used to change a number from scientific notation into standard notation.

EXAMPLE 2

Change each of the following from scientific notation into standard notation.

a) 6.89×10^4 Since 4 is positive, make the product larger by moving the decimal point four places to the right.

    68,900   Standard notation.

b)  3.01×10^{-3} Since -3 is negative, make the product smaller by moving the decimal point three places to the left.

    0.00301   Standard notation.

PRACTICE EXERCISES

Change each of the following from scientific notation into standard notation.

1) 9.42×10^5 **2)** 1.72×10^{-5}

3) 7.82×10^2 **4)** 2.05×10^{-6}

Now that we know how to multiply numbers by powers of 10 and change numbers from scientific notation into standard notation, we are ready to write numbers in scientific notation. Remember that in order for a number to be written in scientific notation, $a \times 10^n$, the "a" must have one nonzero digit to the left of the decimal. Consequently, our first task is to properly place the decimal. Our second task is to determine the exponent of 10 so that the number written in scientific notation is equal to the original number.

EXAMPLE 3

Write the following numbers in scientific notation.

a) 91,000

Solution:

Since we need one nonzero digit to the left of the decimal, place the decimal between 9 and 1. Consequently, 91,000 written in scientific notation is of the form 9.1×10^n. Since $9.1 < 91{,}000$, we need to multiply 9.1 by a power of 10 that will make the product larger by moving the decimal point four places to the right. Therefore, $n = 4$. Consequently, $91{,}000 = 9.1 \times 10^4$.

b) 0.000091

Solution:

Again, the decimal is placed between 9 and 1. Consequently, 0.000091 written in scientific notation is of the form 9.1×10^n. Since $9.1 > 0.000091$ we need to multiply 9.1 by a power of 10 that will make the product smaller by moving the decimal point five places to the left. Therefore, $n = -5$. Consequently, $0.000091 = 9.1 \times 10^{-5}$.

c) 43,600,000

Solution:

Since we need one nonzero digit to the left of the decimal, place the decimal between 4 and 3. Consequently, 43,600,000 written in scientific notation is of the form 4.36×10^n. Since $4.36 < 43{,}600{,}000$, we need to multiply 4.36 by a power of 10 that will make the product larger by moving the decimal point seven places to the right. Therefore, $n = 7$. Consequently, $43{,}600{,}000 = 4.36 \times 10^7$.

d) 0.00361

Solution:

Since we need one nonzero digit to the left of the decimal, place the decimal between 3 and 6. Consequently, 0.00361 written in scientific notation is of the form 3.61×10^n. Since $3.61 > 0.00361$ we need to multiply 3.61 by a power of 10 that will make the product smaller by moving the decimal point three places to the left. Therefore, $n = -3$. Consequently, $0.00361 = 3.61 \times 10^{-3}$.

PRACTICE EXERCISES

Write the following numbers in scientific notation.

5) 470,000   **6)** 0.00000056

7) 5,630   **8)** 0.000972

The need for scientific notation often occurs in the various branches of the natural sciences, such as chemistry, physics, biology, and astronomy.

• A joule is a unit of measure of energy and 4.184 J = 1 calorie. Hence, 4.184 J is the amount of heat required to raise the temperature of 1 gram of water by 1 degree Celsius.

 EXAMPLE 4

a) The average distance between the earth and the sun is 93,000,000 miles. Express this distance using scientific notation.

Solution:
   The decimal is placed between 9 and 3. So 93,000,000 = 9.3×10^n. If $9.3 \times 10^n = 93,000,000$, the decimal must be moved seven places to the right. Therefore, $n = 7$. Consequently, the average distance between the earth and sun is 9.3×10^7 miles.

b) Einstein stated that the energy, E, equivalent to the mass, m, can be calculated by the equation $E = mc^2$. According to this equation, 9.0×10^7 Joules of energy are equivalent to 1.0×10^{-6} grams of mass. Write this number of grams of mass in standard form and give the fractional part of a gram that it represents.

Solution:
   Since the exponent of 10 is negative, move the decimal six places to the left. Therefore, 1.0×10^{-6} grams = 0.000001 grams. As a fraction, 0.000001 = $\frac{1}{1000000}$. Hence, according to Einstein's equation, one, one-millionth of a gram of mass is converted into 90,000,000 Joules of energy. This accounts for the tremendous amount of energy released during a nuclear explosion.

PRACTICE EXERCISES

9) The age of the earth is estimated at about 4,500,000,000 years. Express the age of the earth in scientific notation.

10) The radius of an atom of a certain substance is 1.42×10^{-7} millimeters. Write the radius of this atom in standard form.

In order to perform operations on numbers written in scientific notation, we need to recall two laws of exponents: $a^m \cdot a^n = a^{m+n}$ and $\frac{a^m}{a^n} = a^{m-n}$. The method that we use to multiply and divide numbers written in scientific notation is very much like the method used to multiply and divide monomials. We will refer to the "a" of $a \times 10^n$ as the coefficient and the "10^n" as the exponential part. Consider the following.

Monomials	Scientific Notation
$(3.2a^2)(1.6a^4)$	$(3.2 \times 10^2)(1.6 \times 10^4)$
Regroup the factors so the coefficients and variable parts are together.	Regroup the factors so the coefficients and exponential parts are together.
$(3.2 \cdot 1.6)(a^2 \cdot a^4)$	$(3.2 \cdot 1.6) \times (10^2 \cdot 10^4)$
Multiply the coefficients and the variable parts.	Multiply the coefficients and the exponential parts.
$5.12a^6$	5.12×10^6
Therefore, $(3.2a^2)(1.6a^4) = 5.12a^6$.	Therefore, $(3.2 \times 10^2)(1.6 \times 10^4) = 5.12 \times 10^6$.

$$\frac{2.6a^5}{1.3a^3}$$

We think of this as coefficient divided by coefficient, and variable parts divided by variable parts.

$$\frac{2.6}{1.3} \cdot \frac{a^5}{a^3}$$

Divide the coefficients and the variables separately.

$2a^2$

Therefore, $\frac{2.6a^5}{1.3a^3} = 2a^2$.

$$\frac{2.6 \times 10^5}{1.3 \times 10^3}$$

We think of this as coefficient divided by coefficient, and exponential part divided by exponential part.

$$\frac{2.6}{1.3} \times \frac{10^5}{10^3}$$

Divide the coefficients and the exponentials separately.

2×10^2

Therefore, $\frac{2.6 \times 10^5}{1.3 \times 10^3} = 2 \times 10^2$.

ANSWERS:
Practice 9–10
9) 4.5×10^9 years
10) 0.000000142 mm

EXAMPLE 5

Simplify the following using scientific notation. Leave answers in standard notation.

a) $(4.1 \times 10^3)(2.3 \times 10^{-6}) =$ Regroup the factors so the coefficients and exponential parts are together.

$(4.1 \cdot 2.3) \times (10^3 \cdot 10^{-6}) =$ Multiply coefficients and $10^3 \cdot 10^{-6} = 10^{3 + (-6)} = 10^{-3}$.

$9.43 \times 10^{-3} =$ Move the decimal three places to the left.

0.00943 Product.

b) $\dfrac{7.44 \times 10^{-2}}{3.1 \times 10^4} =$ Divide coefficient by coefficient and exponential part by exponential part.

$\dfrac{7.44}{3.1} \times \dfrac{10^{-2}}{10^4} =$ Divide coefficients and $\dfrac{10^{-2}}{10^4} = 10^{-2-4} = 10^{-6}$.

$2.4 \times 10^{-6} =$ Move the decimal six places to the left.

0.0000024 Quotient.

c) $(4,100)(56,000) =$ Rewrite each number in scientific notation.

$(4.1 \times 10^3)(5.6 \times 10^4) =$ Regroup the factors so the coefficients and exponential parts are together.

$(4.1 \cdot 5.6) \times (10^3 \cdot 10^4) =$ Multiply coefficients and $10^3 \cdot 10^4 = 10^{3 + 4} = 10^7$.

$22.96 \times 10^7 =$ Move the decimal seven places to the right.

$229,600,000$ Product.

d) $\dfrac{(120,000)(0.0018)}{3600} =$ Rewrite each number in scientific notation.

$\dfrac{(1.2 \times 10^5)(1.8 \times 10^{-3})}{3.6 \times 10^3} =$ Regroup in the numerator.

$\dfrac{(1.2 \cdot 1.8) \times (10^5 \cdot 10^{-3})}{3.6 \times 10^3} =$ Multiply coefficients and $10^5 \cdot 10^{-3} = 10^{5 - 3} = 10^2$.

$\dfrac{2.16 \times 10^2}{3.6 \times 10^3} =$ Divide coefficient by coefficient and exponential part by exponential part.

$\dfrac{2.16}{3.6} \times \dfrac{10^2}{10^3} =$ Divide. $\dfrac{10^2}{10^3} = 10^{2 - 3} = 10^{-1}$.

$0.6 \times 10^{-1} =$ Move the decimal one place to the left.

0.06 Answer.

PRACTICE EXERCISES

Simplify each of the following using scientific notation. Leave answers in standard notation.

11) $(4.4 \times 10^4)(2.3 \times 10^3)$

12) $\dfrac{9.3 \times 10^7}{3.0 \times 10^3}$

13) $(5,300)(0.000025)$

14) $\dfrac{0.00345}{15000}$

Computations involving scientific notation often occur in application problems.

EXAMPLE 6

Solve the following using scientific notation. Leave the answers in standard notation.

a) A very slow computer can perform a calculation in 0.000003 of a second. How long would it take this computer to perform 5 billion (5,000,000,000) calculations?

Solution:

The time required to perform 5,000,000,000 calculations is equal to:

$(5,000,000,000)(0.000003)$	Write each in scientific notation.
$(5 \times 10^9)(3 \times 10^{-6})$ 15×10^3	Regroup the factors.
$(5 \cdot 3) \times (10^9 \cdot 10^{-6})$	Multiply coefficients and $10^9 \cdot 10^{-6} = 10^{9-6} = 10^3$.
15×10^3	Move the decimal three places to the right.
15000	Therefore, it takes the computer 15000 seconds to perform 5 billion computations.

b) The solubility constant of barium sulfate is 1.5×10^{-9} and the solubility constant of silver bromide is 5×10^{-13}. How many times greater is the solubility constant of barium sulfate than silver bromide?

Solution:

The number of times greater the solubility constant of barium sulfate than silver bromide is equal to the solubility constant of barium sulfate divided by the solubility constant of silver bromide.

$\dfrac{1.5 \times 10^{-9}}{5 \times 10^{-13}}$	Divide coefficient by coefficient and exponential part by exponential part.
$\dfrac{1.5}{5} \times \dfrac{10^{-9}}{10^{-13}}$	Divide. $\dfrac{10^{-9}}{10^{-13}} = 10^{-9-(-13)} = 10^{-9+13} = 10^4$.
0.3×10^4	Move the decimal four places to the right.
3000	Therefore, the solubility constant of barium sulfate is 3000 times greater than the solubility constant of silver bromide.

PRACTICE EXERCISES

Solve the following using scientific notation. Leave the answers in standard notation.

15) The mass of a helium atom is 6.65×10^{-24} grams. Find the mass of a sample of helium that contains 3.0×10^{26} atoms.

16) The approximate distance from earth to the planet Saturn is 7.942×10^8 miles. How many hours would it take a spacecraft traveling at 20,000 miles per hour to reach Saturn?

E X E R C I S E   S E T   2.8

Write the following in standard notation.

1) 3.5×10^4 **2)** 4.7×10^2 **3)** 9.5×10^{-3}

4) 6.3×10^{-5} **5)** 4.79×10^6 **6)** 3.07×10^8

7) 9.24×10^{-6} **8)** 2.19×10^{-2} **9)** 1×10^2

10) 7×10^5 **11)** 1×10^{-8} **12)** 9×10^{-4}

Write the following in scientific notation.

13) 7600   **14)** 83000   **15)** 0.00035

16) 0.0026   **17)** 857000   **18)** 13400000

19) 0.000000498   **20)** 0.000913   **21)** 600000

22) 10000000   **23)** 0.00000001   **24)** 0.00006

Perform the following operations using scientific notation. Leave answers in standard notation.

25) $(1.2 \times 10^3)(2.5 \times 10^5)$

26) $(3.4 \times 10^4)(1.7 \times 10^2)$

27) $(4.3 \times 10^6)(3.2 \times 10^{-4})$

28) $(8.3 \times 10^7)(5.1 \times 10^{-5})$

29) $(3.12 \times 10^{-6})(4.23 \times 10^3)$

30) $(7.21 \times 10^{-8})(4.82 \times 10^4)$

31) $(6.28 \times 10^{-3})(4.21 \times 10^{-5})$

32) $(3.04 \times 10^{-2})(2.5 \times 10^{-4})$

33) $\dfrac{4.2 \times 10^7}{2.1 \times 10^4}$ **34)** $\dfrac{3.9 \times 10^6}{1.3 \times 10^2}$ **35)** $\dfrac{3.6 \times 10^{-3}}{2.4 \times 10^4}$

36) $\dfrac{4.5 \times 10^{-5}}{1.8 \times 10^2}$ **37)** $\dfrac{1.8 \times 10^3}{1.5 \times 10^{-3}}$ **38)** $\dfrac{1.065 \times 10^5}{4.26 \times 10^{-5}}$

39) $\dfrac{1.28 \times 10^{-3}}{2.56 \times 10^{-5}}$ **40)** $\dfrac{7.29 \times 10^{-5}}{2.43 \times 10^{-2}}$

41) $\dfrac{(1.25 \times 10^{-5})(5 \times 10^7)}{2.5 \times 10^{-2}}$

42) $\dfrac{(3.6 \times 10^6)(4.8 \times 10^{-4})}{1.44 \times 10^{-3}}$

43) $\dfrac{(5.4 \times 10^{-3})(7.2 \times 10^6)}{(2.7 \times 10^{-2})(1.6 \times 10^{-1})}$

44) $\dfrac{(7.5 \times 10^5)(1.0 \times 10^{-3})}{(1.5 \times 10^{-2})(2.5 \times 10^8)}$

Perform the following operations using scientific notation. Leave answers in standard notation.

45) $(55000)(2300)$ **46)** $(350000)(2600000)$

47) $(5600)(0.000045)$ **48)** $(7200000)(0.00036)$

49) $(0.0000025)(0.000000036)$ **50)** $(0.00000034)(0.0031)$

51) $\dfrac{345000000}{15000}$ **52)** $\dfrac{1280000}{1600}$

53) $\dfrac{1800}{72000000}$ **54)** $\dfrac{22500}{112500000}$

55) $\dfrac{37400}{0.0017}$ **56)** $\dfrac{9240000}{0.0000021}$

57) $\dfrac{0.000612}{0.0034}$ **58)** $\dfrac{0.0084}{0.00000056}$

59) $\dfrac{(120000)(0.0018)}{3600}$ **60)** $\dfrac{(24000)(0.00000028)}{560000}$

61) $\dfrac{(8000)(0.000252)}{(0.00063)(400)}$ **62)** $\dfrac{(112000)(0.0015)}{(0.00002)(1400)}$

63) The astronomical unit is 150,000,000 kilometers. Write the astronomical unit in scientific notation.

64) The net assets of a company are $1,250,000,000. Write the assets in scientific notation.

65) The density of the element mercury is 13,600 kilograms per cubic meter. Express the density of mercury in scientific notation.

66) The coefficient of linear expansion of aluminum is 0.000024. Express the linear expansion coefficient of aluminum in scientific notation.

67) The wavelength of an x-ray is 0.0000013 meters. Express the wavelength of this x-ray in scientific notation.

68) The solubility product constant of lead sulfate is 0.000000013. Express the solubility constant of lead sulfate in scientific notation.

69) The speed of light is approximately 2.997925×10^8 meters per second. Express the velocity of light in standard notation.

70) The half-life of a radioactive substance is the time required for one-half of the amount present to decay. The half-life of actinium is 7.04×10^8 years. Express the half-life of actinium in standard notation.

71) The density of gold is 1.93×10^4 kilograms per cubic meter. Express the density of gold in standard notation.

72) The density of hydrogen is $8.99 - 10^{-2}$ kilograms per cubic meter. Express the density of hydrogen in standard notation.

73) The mass of a hydrogen atom is 1.673×10^{-24} grams. Find the mass of the atoms in a sample of hydrogen that contains 1,000,000,000 hydrogen atoms.

74) If a computer can execute a command in 0.00005 seconds, how long will it take the computer to execute 5 million commands?

75) The distance from earth to the nearest star, Alpha Centauri, is approximately 25,200,000,000,000 miles. A light-year is the distance that light can travel in one year and is approximately 6,000,000,000,000 miles. Find the distance to Alpha Centauri to the nearest tenth of a light-year.

76) Density is defined as $\frac{mass}{volume}$. If the mass of the earth is 5.98×10^{27} grams and the volume of the earth is 1.08×10^{27} cubic centimeters, find the density of the earth to the nearest tenth of a gram per cubic centimeter.

77) In chemistry the number of moles (a measure of the number of particles of a substance) is equal to $\frac{\text{number of molecules of the substance}}{6.02 \times 10^{23}}$. Find the number of moles of a particular substance if 9.03×10^{23} molecules are present.

CHALLENGE EXERCISES (78–80)

78) The mass of a hydrogen atom is 1.673×10^{-24} grams. How many atoms are in a 10-gram sample of hydrogen?

79) In Exercise 75, the distance from earth to Alpha Centauri was given as approximately 25,200,000,000,000 miles. Currently spacecraft travel at approximately 25,000 miles per hour. Approximately how many years would it take one of our current spacecraft to reach Alpha Centauri? Is this a problem for manned space travel?

80) The mass of a helium atom is 6.65×10^{-24} grams. How many atoms are in a 4-gram sample of helium?

WRITING EXERCISE

81) Why is scientific notation preferable when performing operations on extremely large or extremely small numbers?

GROUP PROJECT

82) Interview some science instructors, engineers, astronomers, or other technical professionals, and find at least five examples in which they have used scientific notation.

SECTION 2.9

Power Rule for Quotients and Using Combined Laws of Exponents

OBJECTIVES

When you complete this section, you will be able to:

Ⓐ Simplify exponential expressions using the property $\left(\frac{a}{b}\right)^n = \frac{a^n}{b^n}$.

Ⓑ Simplify expressions which involve integer exponents using more than one law of exponents.

INTRODUCTION

One of the laws of exponents for products is $(ab)^n = a^n b^n$. This means that to raise a product to a power, you raise each of the factors to the power. There is a similar property for quotients.

Remember, to multiply fractions you multiply numerator times numerator and divide by denominator times denominator. In general, $\frac{a}{b} \times \frac{c}{d} = \frac{a \cdot c}{b \cdot d}$.

• Developing $\left(\dfrac{a}{b}\right)^n = \dfrac{a^n}{b^n}$

EXAMPLE 1

Simplify each of the following using the definition of exponents. Leave the answer in exponential form. Assume all variables represent nonzero quantities.

a) $\left(\dfrac{2}{3}\right)^3 =$ Rewrite using the definition of exponent.

$\dfrac{2}{3} \cdot \dfrac{2}{3} \cdot \dfrac{2}{3} =$ Multiply the fractions.

$\dfrac{2 \cdot 2 \cdot 2}{3 \cdot 3 \cdot 3} =$ Rewrite using the definition of exponent.

$\dfrac{2^3}{3^3}$ Answer.

Notice that both the numerator and the denominator are raised to the third power.

b) $\left(\dfrac{x}{y}\right)^4 =$ Rewrite using the definition of exponent.

$\dfrac{x}{y} \cdot \dfrac{x}{y} \cdot \dfrac{x}{y} \cdot \dfrac{x}{y} =$ Multiply the fractions.

$\dfrac{x \cdot x \cdot x \cdot x}{y \cdot y \cdot y \cdot y} =$ Rewrite using the definition of exponent.

$\dfrac{x^4}{y^4}$ Answer.

Notice that both the numerator and the denominator were raised to the fourth power.

Based on the examples above, we generalize to the following law for raising a quotient of a power.

FIFTH LAW OF EXPONENTS: POWER RULE FOR QUOTIENTS

For any positive integer n and $b \neq 0$, $\left(\frac{a}{b}\right)^n = \frac{a^n}{b^n}$. To raise a quotient to a power, raise both the numerator and the denominator to the power.

EXAMPLE 2

Simplify the following using the Power Rule for Quotients. Assume all variables represent nonzero quantities.

a) $\left(\frac{2}{5}\right)^4 = \frac{2^4}{5^4} = \frac{16}{625}$ $\qquad$ $\left(\frac{a}{b}\right)^n = \frac{a^n}{b^n}.$

b) $\left(\frac{a}{b}\right)^5 = \frac{a^5}{b^5}$ $\qquad$ $\left(\frac{a}{b}\right)^n = \frac{a^n}{b^n}.$

c) $\left(\frac{3}{y}\right)^2 = \frac{3^2}{y^2} = \frac{9}{y^2}$ $\qquad$ $\left(\frac{a}{b}\right)^n = \frac{a^n}{b^n}.$

PRACTICE EXERCISES

Simplify the following using the Power Rule for Quotients. Leave the answer in exponential form. Assume all variables represent nonzero quantities.

1) $\left(\frac{5}{3}\right)^6$ $\qquad\qquad$ **2)** $\left(\frac{r}{s}\right)^8$ $\qquad\qquad$ **3)** $\left(\frac{z}{5}\right)^3$

Another interesting property occurs when a quotient is raised to a negative power. We need to recall the rule for dividing fractions. We multiply by the reciprocal of the fraction on the right and change the operation to multiplication. For example, $\frac{1}{2} \div \frac{3}{5} = \frac{1}{2} \cdot \frac{5}{3}$.

EXAMPLE 3

Simplify the following leaving the quotient with positive exponents only.

a) $\left(\frac{a}{b}\right)^{-3} =$ $\qquad$ Apply $\left(\frac{a}{b}\right)^n = \frac{a^n}{b^n}.$

$\dfrac{a^{-3}}{b^{-3}} =$ $\qquad$ Apply the definition of negative exponents.

$\dfrac{\frac{1}{a^3}}{\frac{1}{b^3}} =$ $\qquad$ $\frac{a}{b} = a \div b.$

$\dfrac{1}{a^3} \div \dfrac{1}{b^3} =$ $\qquad$ Rewrite as multiplication of fractions.

$\dfrac{1}{a^3} \cdot \dfrac{b^3}{1} =$ $\qquad$ Multiply the fractions.

$\dfrac{b^3}{a^3} =$ Apply $\dfrac{a^n}{b^n} = \left(\dfrac{a}{b}\right)^n$.

$\left(\dfrac{b}{a}\right)^3$ Answer.

Therefore, $\left(\dfrac{a}{b}\right)^{-3} = \left(\dfrac{b}{a}\right)^3$. Notice the fraction has been inverted and the exponent has been changed to positive.

Therefore, we have the following rule.

QUOTIENTS RAISED TO NEGATIVE POWERS

For $a \neq 0$ and $b \neq 0$, $\left(\dfrac{a}{b}\right)^{-n} = \left(\dfrac{b}{a}\right)^n$. To raise a fraction to a negative power, invert the fraction and change the power to positive.

EXAMPLE 4

Simplify the following leaving answers with positive exponents only. Assume that all variables represent nonzero quantities.

a) $\left(\dfrac{3}{5}\right)^{-2} =$ Apply $\left(\dfrac{a}{b}\right)^{-n} = \left(\dfrac{b}{a}\right)^n$.

$\left(\dfrac{5}{3}\right)^2 =$ Apply $\left(\dfrac{a}{b}\right)^n = \dfrac{a^n}{b^n}$.

$\dfrac{5^2}{3^2} =$ Simplify the powers.

$\dfrac{25}{9}$ Answer.

b) $\left(\dfrac{x}{y}\right)^{-3} =$ Apply $\left(\dfrac{a}{b}\right)^{-n} = \left(\dfrac{b}{a}\right)^n$.

$\left(\dfrac{y}{x}\right)^3 =$ Apply $\left(\dfrac{a}{b}\right)^n = \dfrac{a^n}{b^n}$.

$\dfrac{y^3}{x^3}$ Answer.

PRACTICE EXERCISES

Simplify the following, leaving answers with positive exponents only. Assume that all variables represent nonzero quantities.

4) $\left(\dfrac{3}{5}\right)^{-3} =$ 5) $\left(\dfrac{a}{b}\right)^{-5} =$

Often it is necessary to use two or more of the laws of exponents in order to simplify an expression. For reference, these laws and the properties of integer exponents are listed in the following box.

LAWS OF EXPONENTS

For any integers m and n and any real numbers a and b:

1) $a^m a^n = a^{m+n}$ Product Rule

2) $(ab)^n = a^n b^n$ Power Rule for Products

3) $(a^m)^n = a^{mn}$ Power to a Power

4) $\dfrac{a^m}{a^n} = a^{m-n},\ a \neq 0$ Quotient Rule

5) $\left(\dfrac{a}{b}\right)^n = \dfrac{a^n}{b^n},\ b \neq 0$ Power Rule for Quotients

6) $\left(\dfrac{a}{b}\right)^{-n} = \left(\dfrac{b}{a}\right)^n$ if $a, b, \neq 0$ Quotients to Negative Powers

7) $a^0 = 1$ if $a \neq 0$ Definition of Zero Exponents

8) $a^{-n} = \dfrac{1}{a^n}$ if $a \neq 0$ Definition of Negative Exponents

9) $\dfrac{1}{a^{-n}} = a^n$ if $a \neq 0$ Definition of Negative Exponents

---

• Simplifying expressions using more than one law of exponents

EXAMPLE 5

Simplify each of the following. Leave answers in exponential form with positive exponents only. Assume all variables represent nonzero quantities.

a) $\dfrac{x^2 x^4}{x^3} =$ Apply $a^m a^n = a^{m+n}$ in the numerator.

$\dfrac{x^6}{x^3} =$ Apply $\dfrac{a^m}{a^n} = a^{m-n}$.

$x^{6-3} =$ Subtract the exponents.

x^3 Answer.

b) $(3x^{-2}y^3)^{-3} =$ Apply $(ab)^n = a^n b^n$.

$(3^{-3})(x^{-2})^{-3}(y^3)^{-3} =$ Apply definition of negative exponent and $(a^m)^n = a^{mn}$.

$\dfrac{1}{3^3} x^6 y^{-9} =$ Apply definition of negative exponents.

$\dfrac{1}{27} x^6 \cdot \dfrac{1}{y^9} =$ Multiply the fractions.

$\dfrac{x^6}{27y^9}$ Answer.

c) $\dfrac{(4x^{-3})^2}{x^2 x^{-4}} =$ Apply $(a^m)^n = a^{mn}$ in the numerator and $a^m a^n = a^{m+n}$ in the denominator.

$\dfrac{16x^{-6}}{x^{-2}} =$ Apply $\dfrac{a^m}{a^n} = a^{m-n}$.

$$16x^{-6-(-2)} =$$ $-(-2) = +2.$

$$16x^{-6+2} =$$ Add exponents.

$$16x^{-4} =$$ $x^{-4} = \dfrac{1}{x^4}.$

$$\dfrac{16}{x^4}$$ Answer.

d) $\left(\dfrac{x^4}{y^2}\right)^3 =$ Apply $\left(\dfrac{a}{b}\right)^n = \dfrac{a^n}{b^n}.$

$$\dfrac{(x^4)^3}{(y^2)^3} =$$ Apply $(a^m)^n = a^{mn}.$

$$\dfrac{x^{12}}{y^6}$$ Answer.

e) $\dfrac{(x^{-2})^3(x^3)^4}{(x^{-3})^3} =$ Apply $(a^m)^n = a^{mn}.$

$$\dfrac{x^{-6}x^{12}}{x^{-9}} =$$ Apply $a^m a^n = a^{m+n}$ in the numerator.

$$\dfrac{x^6}{x^{-9}} =$$ Apply $\dfrac{a^m}{a^n} = a^{m-n}.$

$$x^{6-(-9)} =$$ $-(-9) = +9.$

$$x^{6+9} =$$ Add exponents.

$$x^{15}$$ Answer.

ADDITIONAL PRACTICE

Simplify each of the following. Leave answers in exponential form with positive exponents only.

a) $\dfrac{x^5 x^3}{x^6}$

b) $(4p^4 q^{-3})^{-2}$

c) $(4z^{-4})^{-3}$

d) $\dfrac{(3x^{-1})^4}{x^{-3} x^{-2}}$

e) $\left(\dfrac{p^3}{q^2}\right)^5$

f) $\dfrac{(b^{-1})^4 (b^{-3})^{-2}}{(b^3)^{-1}}$

PRACTICE EXERCISES

Simplify each of the following. Leave answers in exponential form with positive exponents only. Assume all variables represent nonzero quantities.

6) $\dfrac{x^3 x^5}{x^4}$

7) $(2a^{-3} b^2)^{-4}$

8) $(3y^{-2})^{-2}$

9) $\dfrac{(3x^4)^{-2}}{x^{-4} x^3}$

10) $\left(\dfrac{a^4}{b^2}\right)^4$

11) $\dfrac{(a^3)^{-3} (a^4)^{-2}}{(a^{-4})^2}$

If more practice is needed, do the Additional Practice Exercises in the margin.

STUDY TIP #10

Taking a Math Test

The type of test in which you show all work and write the answers is called an open-ended test. The type of test in which you choose the correct answer from those presented is called a multiple choice test. There are many things you can do before and during the test that can improve your score.

1) Arrive early so you will be ready when the test is passed out.

2) As soon as you get the test, write down any formulas or definitions that you might need.

3) Read the test over from front to back. Mark the questions that you know you can answer with a check mark. Mark the ones you are unsure about with a question mark and the ones you cannot do with an "x."

4) Answer the questions in this order.

   a) Those you know how to do starting with the ones you think are easiest.

   b) Those you are unsure about.

   c) Those you thought you did not know how to do. Sometimes doing the ones that you know reminds you of how to do others.

5) Estimate the amount of time needed for each question by dividing the number of minutes for the test by the number of items on the test. If you are spending more than this amount of time on an item, go to the next item and come back to it later if you have time.

6) Read the directions to each question carefully. Underline the key words, such as *not* equal.

7) On an open-ended test, write down the information given, what you are asked to find,

and any relevant formulas, definitions, or theorems. Sometimes an estimate of the answer will give you a clue as to how to solve the problem.

8) On an open-ended test, show all your work in a neat and organized manner. Box in your answers.

9) Check all your answers, time permitting.

10) If you are taking a multiple choice test, on which there *is* a penalty for incorrect answers and you can narrow the answer to two choices, guess. If there is *no* penalty, guess and leave nothing blank.

11) Take all of the time permitted for the test. Never worry about being the last one to leave a test. By far the first to leave a test are the ones who know everything or those who know very little.

Remember, the idea in taking a test is to show what you know, so attempt what you know first.

ANSWERS:
Practice 6–11

6) x^4
7) $\dfrac{a^{12}}{16b^8}$
8) $\dfrac{y^4}{9}$

9) $\dfrac{1}{9x^7}$
10) $\dfrac{a^{16}}{b^8}$
11) $\dfrac{1}{a^9}$

Additional Practice a–f

a) x^2
b) $\dfrac{q^6}{16p^8}$
c) $\dfrac{z^{12}}{4^3}$

d) $81x$
e) $\dfrac{p^{15}}{q^{10}}$
f) b^5

E X E R C I S E   S E T   2.9

Write each of the following without exponents and evaluate.

1) $\left(\dfrac{2}{3}\right)^2$
2) $\left(\dfrac{3}{4}\right)^3$
3) $\left(\dfrac{3}{5}\right)^3$

4) $\left(\dfrac{4}{3}\right)^2$
5) $\left(\dfrac{5}{2}\right)^3$
6) $\left(\dfrac{2}{5}\right)^{-2}$

7) $\left(\dfrac{3}{2}\right)^{-2}$
8) $\left(\dfrac{5}{4}\right)^{-3}$
9) $\left(\dfrac{3}{7}\right)^{-4}$

Write each of the following with positive exponents only.

10) $\left(\dfrac{r}{s}\right)^2$
11) $\left(\dfrac{a}{b}\right)^3$
12) $\left(\dfrac{x}{z}\right)^{-4}$

13) $\left(\dfrac{p}{q}\right)^{-4}$
14) $\left(\dfrac{3a}{b}\right)^4$
15) $\left(\dfrac{2x}{y}\right)^3$

16) $\left(\dfrac{3x}{y}\right)^{-4}$
17) $\left(\dfrac{p}{3q}\right)^{-3}$
18) $(r^{-2}s^2)^3$

19) $(a^3b^{-5})^4$
20) $(z^{-2}w^3)^{-4}$
21) $(x^4y^{-2})^{-6}$

22) $(2x^3)^{-2}$
23) $(4x^2)^{-3}$
24) $(-4x^{-3})^{-4}$

25) $(-2y^{-5})^{-2}$
26) $(2x^{-3}y^{-2})^3$
27) $(5a^{-2}b^{-5})^2$

28) $(3x^3y^{-4})^{-4}$
29) $(2w^3z^{-2})^{-2}$
30) $\left(\dfrac{x^3}{y^5}\right)^3$

31) $\left(\dfrac{w^3}{z^6}\right)^3$
32) $\left(\dfrac{a^5}{b^3}\right)^{-5}$
33) $\left(\dfrac{x^4}{z^6}\right)^{-5}$

34) $\dfrac{(2x^2)(3x^7)}{(x^3)(x^3)}$
35) $\dfrac{(4w^3)(6w^4)}{(w^2)(w^2)}$
36) $\dfrac{(-5a^2)(3a^5)}{(a^6)(a^4)}$

37) $\dfrac{(x^4)(3x^6)}{(5x^8)(2x^3)}$
38) $\dfrac{(7x^3y^2)(6x^4y^4)}{(5x^2y^3)(x^5y^5)}$
39) $\dfrac{(x^4)^2}{(x^2)^3}$

40) $\dfrac{(y^3)^4}{(y^5)^2}$
41) $\dfrac{(a^{-3})^3}{(a^3)^2}$
42) $\dfrac{(z^{-2})^4}{(z^5)^2}$

43) $\dfrac{(a^{-2})^{-3}}{(a^{-4})^2}$
44) $\dfrac{(p^{-4})^{-2}}{(p^{-5})^3}$
45) $\dfrac{(2y^2)^4}{(y^4)^4}$

46) $\dfrac{(3q^3)^3}{(q^2)^4}$

47) $\dfrac{(x^2)^5}{x^2x^4}$

58) $\dfrac{(3b^5)^{-3}(2b^3)^{-3}}{(36b^4)^{-2}}$

59) $\dfrac{(2x^{-3})^3}{x^2x}$

48) $\dfrac{(y^4)^3}{y^3y^2}$

49) $\dfrac{(3x^4)^0}{(2x^2)^2}$

60) $\dfrac{(2y^{-4})^4}{y^5y^2}$

61) $\dfrac{(-3a^{-3})^4}{(a^{-3})^5}$

50) $\dfrac{(5c^3)^2}{(8c^6)^0}$

51) $\dfrac{(7^0g^4)^3}{(2g^2)^4}$

CHALLENGE EXERCISES (62–63)

Simplify the following. Leave answers with positive exponents only. Assume all variables represent nonzero quantities.

52) $\dfrac{(4a^4)^3}{(5^0a^2)^4}$

53) $\dfrac{(x^{-2})^4}{x^3x^4}$

54) $\dfrac{(z^{-1})^5}{z^5z^2}$

55) $\dfrac{(q^2)^3(q^4)^2}{(q^3)^5}$

62) $\dfrac{(2x^{-3}y^2)^4(3xy^{-3})^{-2}}{(3x^{-4}y^{-2})^{-4}}$

56) $\dfrac{(p^4)^2(p^5)^3}{(p^2)^7}$

57) $\dfrac{(4a^{-4})^3(2a^2)^{-3}}{(2a^{-2})^3}$

63) $\dfrac{(4a^{-3}b^{-2})^{-2}(6a^3b^{-5})^3}{(2a^{-2}b^{-2})^{-4}}$

SECTION 2.10

Division of Polynomials by Monomials

OBJECTIVES

When you complete this section, you will be able to:

Ⓐ   Divide monomials by monomials.

Ⓑ   Divide polynomials of more than one term by monomials.

INTRODUCTION

Recall from Section 1.8 that a monomial is a polynomial with only one term. Examples are 4, $5x^2$, $6xy$, $-4z^2y^3$, and $8p^3q^2r^5$. Remember, the exponent on any variable must be a whole number. This means the exponent on a variable cannot be negative, and there cannot be a variable in the denominator. Recall also that the number in front of the variable is called the coefficient. In order to divide a monomial by another monomial, we will use the Quotient Law discussed in Section 2.7.

• **Dividing a monomial by a monomial**

EXAMPLE 1

Find the quotients of the following monomials. Leave answers with positive exponents only. Assume all variables represent nonzero quantities only.

a) $\dfrac{4x^4}{2x^3} =$ Divide the coefficients and variables separately.

$\dfrac{4}{2} \cdot \dfrac{x^4}{x^3} =$ Divide the coefficients and apply $\dfrac{a^m}{a^n} = a^{m-n}$ to the variables.

$2x$ Answer.

b) $\dfrac{-12x^5y^3}{3x^2y^2} =$ Divide the coefficients and like variables separately.

$\dfrac{-12}{3} \cdot \dfrac{x^5}{x^2} \cdot \dfrac{y^3}{y^2} =$ Divide the coefficients and apply $\dfrac{a^m}{a^n} = a^{m-n}$ to the variables.

$-4x^3y$ Answer.

---

c) $\dfrac{6x^3}{-2x^5} =$ Divide the coefficients and the variables separately.

$\dfrac{6}{-2} \cdot \dfrac{x^3}{x^5} =$ Divide the coefficients and apply $\dfrac{a^m}{a^n} = a^{m-n}$ to the variables.

$-3x^{-2} =$ Apply the definition of negative exponents.

$\dfrac{-3}{x^2}$ Answer.

Note: The answer to Example 1c is not a monomial since there is division by a variable. Consequently, the quotient of two monomials may not be a monomial just as the quotient of two integers may not be an integer.

d) $\dfrac{24a^4b^3z^6}{8a^2b^3z^8} =$ Divide the coefficients and like variables separately.

$\dfrac{24}{8} \cdot \dfrac{a^4}{a^2} \cdot \dfrac{b^3}{b^3} \cdot \dfrac{z^6}{z^8} =$ Divide the coefficients and apply $\dfrac{a^m}{a^n} = a^{m-n}$ to the variables.

$3a^2b^0z^{-2} =$ Apply the definition of zero and negative exponents.

$\dfrac{3a^2}{z^2}$ Answer.

e) $\dfrac{(4p^2q^3)^2}{8p^3q^4} =$ Apply $(ab)^n = a^n b^n$ in the numerator.

$\dfrac{(4)^2(p^2)^2(q^3)^2}{8p^3q^4} =$ Apply $(a^m)^n$ in the numerator.

$\dfrac{16p^4q^6}{8p^3q^4} =$ Divide the coefficients and apply $\dfrac{a^m}{a^n} = a^{m-n}$ to the variables.

$2pq^2$ Answer.

ADDITIONAL PRACTICE

Find the quotients of the following monomials. Leave answers with positive exponents only.

a) $\dfrac{18a^6}{9a^4}$ b) $\dfrac{-24p^6q^6}{6p^4q^5}$

c) $\dfrac{-32r^5}{-4r^7}$ d) $\dfrac{-12a^4b^5c^3}{2a^3b^5c^6}$

e) $\dfrac{(4a^4b^4)^3}{8a^6b^8}$

PRACTICE EXERCISES

Find the quotients of the following monomials. Leave answers with positive exponents only. Assume all variables represent nonzero quantities only.

1) $\dfrac{8y^6}{2y^3}$ 2) $\dfrac{-18x^8y^7}{6x^3y^2}$ 3) $\dfrac{16p^3}{-4p^5}$

4) $\dfrac{-15x^3y^5z^2}{3x^3y^3z^2}$ 5) $\dfrac{(6x^3y^4)^2}{9x^5y^4}$

If more practice is needed, do the Additional Practice Exercises in the margin.

Now we will divide polynomials by monomials. From Section 1.2 we know how to add fractions with a common denominator, $\dfrac{a}{c} + \dfrac{b}{c} = \dfrac{a+b}{c}$. If we "turn this expression around", it gives us the method for dividing a polynomial by a monomial, $\dfrac{a+b}{c} = \dfrac{a}{c} + \dfrac{b}{c}$. We summarize in the following rule.

ANSWERS:

Practice 1–5

1) $4y^3$
2) $-3x^5y^5$

3) $\dfrac{-4}{p^2}$
4) $-5y^2$

5) $4xy^4$

Additional Practice a–e

a) $2a^2$
b) $-4p^2q$

c) $\dfrac{8}{r^2}$
d) $\dfrac{-6a}{c^3}$

e) $8a^6b^4$

• **Dividing polynomials by monomials**

DIVISION OF A POLYNOMIAL BY A MONOMIAL

$$\frac{a+b}{c} = \frac{a}{c} + \frac{b}{c}.$$ To divide a polynomial by a monomial, divide each term of the polynomial by the monomial.

EXAMPLE 2

Find the following quotients of polynomials and monomials. Assume all variables represent nonzero quantities only.

a) $\dfrac{4x + 6}{2} =$ Divide each term of the polynomial by the monomial.

$\dfrac{4x}{2} + \dfrac{6}{2} =$ Divide the coefficients.

$2x + 3$ Quotient.

b) $\dfrac{6r^4 + 9r^5}{3r^2} =$ Divide each term of the polynomial by the monomial.

$\dfrac{6r^4}{3r^2} + \dfrac{9r^5}{3r^2} =$ Divide the coefficients and apply $\dfrac{a^m}{a^n} = a^{m-n}$ to the variables.

$2r^2 + 3r^3$ Quotient.

c) $\dfrac{8y^4 - 12y^2}{4y^3} =$ Divide each term of the polynomial by the monomial.

$\dfrac{8y^4}{4y^3} - \dfrac{12y^2}{4y^3} =$ Divide the coefficients and apply $\dfrac{a^m}{a^n} = a^{m-n}$ to the variables.

$2y - 3y^{-1} =$ Apply the definition of negative exponents.

$2y - \dfrac{3}{y}$ Quotient.

Note: The answer to Example 2c is not a polynomial. Therefore, the quotient of a polynomial with a monomial need not be a polynomial.

d) $\dfrac{10n^5 + 15n^4 - 5n^3}{5n^2} =$ Divide each term of the polynomial by the monomial.

$\dfrac{10n^5}{5n^2} + \dfrac{15n^4}{5n^2} - \dfrac{5n^3}{5n^2} =$ Divide the coefficients and apply $\dfrac{a^m}{a^n} = a^{m-n}$ to the variables.

$2n^3 + 3n^2 - n$ Answer.

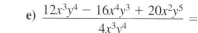

e) $\dfrac{12x^3y^4 - 16x^4y^3 + 20x^2y^5}{4x^3y^4} =$ Divide each term of the polynomial by the monomial.

$\dfrac{12x^3y^4}{4x^3y^4} - \dfrac{16x^4y^3}{4x^3y^4} + \dfrac{20x^2y^5}{4x^3y^4} =$ Divide the coefficients and apply the Quotient Rule to the variables.

$3x^0y^0 - 4xy^{-1} + 5x^{-1}y =$ Apply the definition of zero and negative exponents.

$3 - \dfrac{4x}{y} + \dfrac{5y}{x}$ Answer.

Find the following quotients of polynomials and monomials.

f) $\dfrac{8x - 12}{4}$ **g)** $\dfrac{21x^6 + 15x^3}{3x^2}$

h) $\dfrac{15b^5 - 25b^2}{5b^4}$

i) $\dfrac{8x^6 + 16x^5 - 24x^3}{8x^3}$

j) $\dfrac{9r^4s^5 - 15r^2s^3 + 21r^5s^4}{3r^3s^4}$

PRACTICE EXERCISES

Find the following quotients of polynomials and monomials. Assume all variables represent nonzero quantities only.

6) $\dfrac{6x + 12}{2}$ **7)** $\dfrac{10a^5 + 25a^3}{5a^2}$ **8)** $\dfrac{12r^4 - 20r^2}{4r^3}$

9) $\dfrac{24p^6 - 12p^4 + 18p^7}{6p^3}$ **10)** $\dfrac{21p^2q^4 - 14p^5q^3 + 7p^3q^7}{7p^3q^5}$

If more practice is needed, do the Additional Practice Exercises in the margin.

EXERCISE SET 2.10

Find the quotients of the following monomials.

1) $\dfrac{5x^2}{x}$ **2)** $\dfrac{6y^3}{y}$ **3)** $\dfrac{-3z^4}{z^2}$ **4)** $\dfrac{-5a^5}{a^3}$

5) $\dfrac{6x^4}{3x^2}$ **6)** $\dfrac{14b^5}{b^2}$ **7)** $\dfrac{-15a^4}{3a^7}$ **8)** $\dfrac{-24w^5}{8w^9}$

9) $\dfrac{18n^6}{-3n^4}$ **10)** $\dfrac{28u^8}{-4u^2}$ **11)** $\dfrac{30c^4}{15c^4}$ **12)** $\dfrac{32w^6}{16w^6}$

13) $\dfrac{x^4y^5}{x^2y^3}$ **14)** $\dfrac{a^6b^4}{a^4b^3}$ **15)** $\dfrac{r^3s^6}{-r^5s^2}$ **16)** $\dfrac{-q^5r^2}{qr^5}$

17) $\dfrac{18x^5y^3}{6x^3y}$ **18)** $\dfrac{28y^6z^4}{7y^3z^2}$ **19)** $\dfrac{-32x^3y^5}{8x^6y^2}$

20) $\dfrac{-16a^2b^6}{8a^4b^3}$ **21)** $\dfrac{x^3y^6z^4}{x^2y^4z^6}$ **22)** $\dfrac{a^5b^3c^8}{a^3b^7c^4}$

23) $\dfrac{48x^4y^6z^8}{-12x^6y^3z^{10}}$ **24)** $\dfrac{36a^3b^5c^2}{-18a^5b^2c}$ **25)** $\dfrac{-28m^2n^5p^3}{-14m^2n^3p^6}$

26) $\dfrac{-32x^4y^2z^6}{-8x^4y^6z^3}$ **27)** $\dfrac{(6x^3y^5)^2}{4x^4y^8}$ **28)** $\dfrac{(8a^4b^3)^2}{16a^6b^4}$

29) $\dfrac{-(5x^5y^2)^2}{5x^{10}y^6}$

Find the following quotients of polynomials and monomials.

30) $\dfrac{2x + 8}{2}$ **31)** $\dfrac{3x + 9}{3}$ **32)** $\dfrac{x^2 + x}{x}$

33) $\dfrac{a^2 - a}{a}$ **34)** $\dfrac{2x + 4}{x}$ **35)** $\dfrac{3y + 7}{y}$

36) $\dfrac{4x^2 - 2x}{2x}$ **37)** $\dfrac{12z^2 - 6z}{6z}$

38) $\dfrac{12x^3 + 8x}{4x^2}$ **39)** $\dfrac{24y^3 + 16y}{8y^2}$

40) $\dfrac{6x^3y^4 - 12x^4y^5}{3x^2y^2}$ **41)** $\dfrac{20a^5b^3 - 10a^3b^6}{5a^2b^2}$

42) $\dfrac{12x^2y^4 + 24x^5y^2}{6x^4y^3}$ **43)** $\dfrac{18mn^4 + 27m^4n^2}{9m^2n^3}$

44) $\dfrac{4x^2 - 6x + 2}{2}$ **45)** $\dfrac{6x^2 + 9x - 3}{3}$

46) $\dfrac{x^3 - x^2 - x}{x}$ **47)** $\dfrac{y^4 - y^3 - y^2}{y}$

48) $\dfrac{9x^4 - 3x^3 + 6x^2}{3x}$ **49)** $\dfrac{15m^4 - 5m^3 + 10m^2}{5m}$

50) $\dfrac{6x^3 - 12x^2 + 9x}{3x^2}$ **51)** $\dfrac{12y^4 - 16y^3 + 8y}{4y^2}$

52) $\dfrac{21x^2y^4 - 12x^3y^5 + 18x^5y^3}{3xy^2}$

53) $\dfrac{24m^3n^2 - 16m^4n^3 + 32m^5n^4}{8m^2n^2}$

Write an expression for each of the following and simplify.

54) Find the quotient of $18x^6y^9$ and $9x^4y^3$.
55) Find the quotient of $14m^4n^8$ and $7m^3n^6$.
56) Divide $-24a^3b^2$ by $-12a^5b$.
57) Divide $-32m^4n^2$ by $-16m^7n$.
58) Find the quotient of $32x^5 - 36x^4$ and $4x^2$.
59) Find the quotient of $16a^7 - 24a^5$ and $8a^3$.
60) Divide $16x^4 - 32x^2 + 24$ by $8x^2$.
61) Divide $12y^5 - 16y^3 - 8y$ by $4y^3$.

6) $3x + 6$

7) $2a^3 + 5a$

8) $3r - \dfrac{5}{r}$

9) $4p^3 - 2p + 3p^4$

10) $\dfrac{3}{pq} - \dfrac{2p^2}{q^2} + q^2$

Additional Practice f–j

f) $2x - 3$

g) $7x^4 + 5x$

h) $3b - \dfrac{5}{b^2}$

i) $x^3 + 2x^2 - 3$

j) $3rs - \dfrac{5}{rs} + 7r^2$

62) Divide the sum of $6x^2y^3$ and $-2x^2y^3$ by $2xy$.

63) Divide the sum of $8x^3y^3$ and $-2x^3y^3$ by $3xy$.

64) Divide the sum of $2x^4 - 7x^3 + 3x^2$ and $4x^4 - 2x^3 - 6x^2$ by $-3x^2$.

65) Divide the sum of $4y^5 - 6y^3 + y^2$ and $2y^5 - 4y^3 + 7y^2$ by $2y^2$.

66) Is the quotient of a monomial with a monomial always a monomial? Why or why not? In not, give examples.

67) Is the quotient of a polynomial with a monomial always a polynomial? Why or why not? If not, give examples.

68) Does the quotient of a trinomial with a monomial always have three terms?

69) What is wrong with the following?
$$\frac{4x^2 + 5x + 6}{6} = \frac{4x^2 + 5x + \cancel{6}}{\cancel{6}} = 4x^2 + 5x + 1.$$

70) Without using a calculator, evaluate
$\dfrac{x^{1000} - x^{999}}{x^{999}}$ when $x = 1000$.

71) The quotient of a polynomial and $3xy$ is $2x^2y - 6x^3$. What is the polynomial? Why?

72) If $\dfrac{12x^a - 18x^b + 9x^c + 6x^d}{3x} = 4x^3 - 6x^2 + 3x + 2$, find a, b, c, and d.

73) If the area of a rectangle is $8x^3 + 12x^2 + 4$ and the width is x^2, find the length.

Summary

Rules for Multiplying Signed Numbers: [Section 2.1]

The product of two numbers with the same sign is positive. $(+)(+) = (+)$ and $(-)(-) = (+)$

The product of two numbers with opposite signs is negative. $(+)(-) = (-)$ and $(-)(+) = (-)$

Signed Numbers to Powers: [Section 2.1]

A negative number raised to an even power is positive.

A negative number raised to an odd power is negative.

The base of $(-a)^n$ is $-a$, so $(-a)^n$ means multiply n, $-a$'s.

The base of $-a^n$ is a, so $-a^n$ means $-1 \cdot a^n$.

Properties of Multiplication: [Section 2.1]

Commutative:

$ab = ba$ Multiplication may be performed in any order.

Associative:

$a(bc) = (ab)c$ Factors may be grouped in any manner.

Identity:

The number 1 is the identity for multiplication since $a \cdot 1 = 1 \cdot a = a$.

Inverse:

For any number $a \neq 0$, there exists a number $\frac{1}{a}$ such that $a \cdot \frac{1}{a} = 1$.

Multiplication Laws of Exponents: [Section 2.2]

Product Rule: $a^m \cdot a^n = a^{m+n}$

Power of a Product: $(ab)^n = a^n b^n$

Power to a Power: $(a^m)^n = a^{mn}$

Products of Polynomials: [Section 2.3]

Monomial by monomial:

Multiply coefficients and multiply variables with the same base using the Product Rule.

Polynomial by monomial:

Use the distributive property to multiply each term of the polynomial by the monomial.

Polynomial by polynomial:

Use the distributive property to multiply the second polynomial by each term of the first polynomial.

$(a + b)(c + d + e) = a(c + d + e) + b(c + d + e)$

Special Products: [Section 2.4]

Binomial by binomial—FOIL:

Sum and difference of the same terms:

Multiply the first and last only. The product is the difference of squares.

$$(a + b)(a - b) = a^2 - b^2$$

Square of a binomial:

Square the first term, add twice the product of the terms, add the square of the last term.

$$(a + b)^2 = a^2 + 2ab + b^2$$
$$(a - b)^2 = a^2 - 2ab + b^2$$

Products resulting in the sum of cubes:

$$(a + b)(a^2 - ab + b^2) = a^3 + b^3$$

Products resulting in the difference of cubes:

$$(a - b)(a^2 + ab + b^2) = a^3 - b^3$$

Definition of Division: [Section 2.5]

If $\frac{a}{b} = c$, then $bc = a$. $b \neq 0$.

Rules for Dividing Signed Numbers: [Section 2.5]

The quotient of two numbers with the same sign is positive. $\frac{(+)}{(+)} = (+)$ and $\frac{(-)}{(-)} = (+)$

The quotient of two numbers with opposite signs is negative. $\frac{(+)}{(-)} = (-)$ and $\frac{(-)}{(+)} = (-)$

Order of Operations on Integers: [Section 2.6]

The order of operations on integers is the same as with whole numbers. If needed, see Chapter 1 Summary.

Laws of Exponents for Quotients: [Sections 2.7 and 2.9]

Quotient Rule: $\dfrac{a^m}{a^n} = a^{m-n}$, $a \neq 0$

Power Rule for Quotients: $\left(\dfrac{a}{b}\right)^n = \dfrac{a^n}{b^n}$, $b \neq 0$

Integer Exponents: [Section 2.7 and 2.9]

1) $x^0 = 1$, if $x \neq 0$

2) $a^{-n} = \dfrac{1}{a^n}$, $a \neq 0$

3) $\dfrac{1}{a^{-n}} = a^n$, $a \neq 0$

4) $\left(\dfrac{a}{b}\right)^{-n} = \left(\dfrac{b}{a}\right)^n$, $a, b \neq 0$

Scientific Notation: [Section 2.8]

1) A number is written in scientific notation if it is in the form $a \times 10^n$ with $1 \leq a < 10$ and n is an integer.

2) To change a number from scientific notation to standard notation, move the decimal n places to the right if n is positive and $|n|$ places to the left if n is negative.

3) To write a number in standard notation in scientific notation, place the decimal so there is one nonzero digit to the left of the decimal. Then determine n by the number of places the decimal would have to be moved so the number in scientific notation will be equal to the number in standard notation.

4) To multiply two numbers in scientific notation, multiply the coefficients and the exponential parts separately.

5) To divide two numbers in scientific notation, divide the coefficients and exponential parts separately.

Division of a Polynomial by a Monomial: [Section 2.10]

To divide a polynomial by a monomial, divide each term of the polynomial by the monomial. In symbols, $\frac{a+b}{c} = \frac{a}{b} + \frac{b}{c}, c \neq 0$.

Review Exercises

Find the following products. [Section 2.1]

1) $(-4)(2)$

2) $(-5)(-3)$

3) $(4)(-6)(3)$

4) $(-4)(5)(-3)(-2)$

5) $(-3)^2(4)^2$

6) $(-1)^3(-4)^3$

Evaluate each of the following for $x = -3$ and $y = -2$. [Section 2.1]

7) $5x$　　　**8)** $-3y^2$　　　**9)** x^2y^3　　　**10)** $-4xy^3$

Write an expression for each of the following and evaluate. [Section 2.1]

11) The product of -4 and 8 increased by 7

12) -6 decreased by the product of -5 and 5

13) The sum of the product of -6 and 3 and the product of 6 and -4

Give the name of the property illustrated by each of the following. [Section 2.1]

14) $(-4)(5) = (5)(-4)$

15) $(-2 \cdot 5)(6) = -2(5 \cdot 6)$

16) $(-5)(-3)(7) = (-3)(-5)(7)$

17) $-5(4 + 7) = (-5)(4) + (-5)(7)$

18) $9 \cdot \dfrac{1}{9} = 1$

19) $1 \cdot 13 = 13$

Simplify the following using the multiplication laws of exponents. [Section 2.2]

20) $w^5 \cdot w^8$

21) $4^3 \cdot 4^5 \cdot 4^6$

22) $(8x^3)(-5x^7)$

23) $(6a^4b^6)(-5a^2b^6)$

24) $(xy)^7$

25) $(5d)^3$

26) $(-2ab)^4$

27) $(xy)^5(xy)^4$

28) $(4x)^3(-2x)^5$

29) $(x^5)^4$

30) $(4^3)^8$

31) $(a^4b^6)^4$

32) $(4a^3b^7)^4$

33) $(a^4b^3)^3(a^5b^2)^5$

34) $(2m^3n^3)^3(3m^2n^5)^2$

Find the products of the following polynomials. [Sections 2.3 and 2.4]

35) $(-3x^4y^3)(-5x^6y^2)$

36) $6y(3y^2 - 2)$

37) $-5x^2(4x^2 - 5)$

38) $7c^3(3c^2 - 7c + 2)$

39) $-2x^2y^3(3x^4y - 8xy^5 + 4x^3)$

40) $(x + 2)(x + 5)$

41) $(4m + 5)(3m - 6)$

42) $(5m - 6n)(4m + 3n)$

43) $(x - 9)(x + 9)$

44) $(5x - 6)(5x + 6)$

45) $(2s + 5t)(2s - 5t)$

46) $(m + n)^2$

47) $(3a - 2)^2$

48) $(6a - 2b)^2$

49) $(a + 4)(a^2 - 5a + 4)$

50) $(2m - 5)(3m^2 + 2m - 4)$

51) $(4x - 2y)(3x^2 - 5xy + 2y^2)$

Find the following products. (Optional 52–53): [Section 2.3]

52) $(x - 4)(x^2 + 4x + 16)$

53) $(3x + 5)(9x^2 - 15x + 25)$

54) Find the area of the rectangle with length $5a - 7$ and width $3a + 2$. [Section 2.3]

55) Find the area of the triangle with base $5x$ and height $4x$. [Section 2.3]

56) Find the area of the square each of whose sides is $3z + 8$. [Section 2.4]

Write an expression for each of the following and simplify. [Section 2.4]

57) The square of the quantity 4 less than the product of 6 and x

58) The square of the quantity 2 more than the product of 3 and z

Find the following quotients. [Section 2.5]

59) $\dfrac{48}{-6}$

60) $\dfrac{-28}{-14}$

61) $\dfrac{-18}{2}$

62) $\dfrac{16}{13 - 17}$

63) $\dfrac{23 - 35}{-6}$

64) $\dfrac{-120}{(15)(-4)}$

65) $\dfrac{(-16)(6)}{(-8)(3)}$

Simplify the following using the Order of Operations Agreement. [Section 2.6]

66) $-6 + 13 - 9$

67) $8 - 4 \cdot 5$

68) $17 + 36 \div (-4)$

69) $14 - 3(-5)^2$

70) $7 + 2(8) \div (-4)$

71) $6 - 2(9 + 4)$

72) $15 + 4 \cdot 6^2 \div (-9)(3) - 7$

73) $-4(6^2 - 2^2) \div (-8)$

74) $\dfrac{(-3)(-6) + 6}{(-6)(3) + 6}$

75) $\dfrac{2 \cdot 4^2 + 16}{-3 \cdot 2^2 + 4}$

76) $\dfrac{8^2 - 4^2}{-2(3^2 - 1)}$

77) $\dfrac{(-4)(5) + [6(-4) + 6^2]}{(-3)^2(4) - 28}$

Find the value of each of the following for $x = -3$, $y = 2$, and $z = -2$. [Section 2.6]

78) $3x^2 - 4z$

79) $-4x^3y^2$

80) $3xy^2z$

81) $3x - 4(y - 3z)$

82) $-y^2 - (2x^2 - 3z^2)$

83) $\dfrac{3x - 2y}{4y - 2z}$

84) $\dfrac{2x^2 - 2z^3}{5x^2 + 3z^2}$

Simplify each of the following. Leave answers in exponential form with positive exponents only. Assume all variables represent nonzero quantities. [Section 2.7]

85) $\dfrac{2^8}{2^3}$

86) $\dfrac{m^8}{m^6}$

87) 3^{-4}

88) x^{-6}

89) $\dfrac{5^4}{5^6}$

90) $3x^0 + (5x)^0$

91) $5^{-5} \cdot 5^4$

92) z^4z^{-2}

93) $x^{-4}x^{-2}$

94) $\dfrac{5^{-4}}{5^2}$

95) $\dfrac{a^{-3}}{a^{-6}}$

96) $\dfrac{7^3}{7^{-7}}$

Write the following numbers in standard notation. [Section 2.8]

97) 4.6×10^5

98) 3.06×10^6

99) 7.03×10^{-5}

100) 7.123×10^{-3}

Write the following numbers in scientific notation. [Section 2.8]

101) 97,000,000,000

102) 46,700

103) 0.00000478

104) 0.000307

Simplify each of the following using scientific notation. [Section 2.8]

105) $(4.5 \times 10^4)(3.1 \times 10^{-7})$

106) $(5.07 \times 10^{-5})(4.06 \times 10^{-2})$

107) $\dfrac{4.8 \times 10^4}{1.6 \times 10^6}$

108) $\dfrac{3.9 \times 10^{-5}}{1.3 \times 10^3}$

109) $(4,600)(0.0000012)$

110) $(0.000047)(0.00000034)$

111) $\dfrac{48000}{0.00096}$

112) $\dfrac{0.00033}{0.00000022}$

Write each of the following without exponents and evaluate. [Section 2.9]

113) $\left(\dfrac{3}{5}\right)^2$

114) $\left(\dfrac{3}{2}\right)^{-2}$

Write each of the following in simplest form with positive exponents only. [Section 2.9]

115) $(3^{-5})^2$

116) $(r^{-5})^{-4}$

117) $(x^{-4}y^4)^{-3}$

118) $(4x^{-2})^{-3}$

119) $(5a^4)^{-3}$

120) $(2x^{-4}y^6)^{-3}$

121) $\left(\dfrac{m^4}{n^3}\right)^{-2}$

122) $\dfrac{(m^{-4})^{-5}}{(m^3)^{-4}}$

123) $\dfrac{-24n^6}{-3n^3}$

124) $\dfrac{m^4n^6}{m^3n^2}$

125) $\dfrac{-48m^8n^5}{-8m^{10}n^5}$

126) $\dfrac{(-6x^3)(-5x^5)}{(2x^2)(-3x^4)}$

127) $\dfrac{(3m^{-4}n^3)(-8m^3n^{-6})}{(4m^5n^{-3})(-2m^{-3}n^6)}$

128) $\dfrac{(m^3)^4(m^{-6})^2}{(m^8)^{-2}(m^{-3})^{-3}}$

Find the following quotients of polynomials by monomials. [Section 2.10]

129) $\dfrac{9x - 18}{9}$

130) $\dfrac{5y^3 + 3y}{y}$

131) $\dfrac{28a^2 - 16a}{4a}$

132) $\dfrac{36m^7n^3 + 18m^4n^7}{9m^5n^5}$

133) $\dfrac{12x^4 - 9x^3 - 21x^2}{3x^2}$

134) $\dfrac{8m^5n^3 - 32m^2n^7 + 24m^6n^4}{8m^6n^5}$

Write an expression for each of the following and simplify. [Section 2.9]

135) The quotient of $48m^6n^3$ and $-16m^4n^5$.

136) The quotient of $24a^6 - 16a^4$ and $8a^3$.

137) Divide the sum of $4x^4 - 2x^3 - 3x^2$ and $2x^4 - 7x^3 + 2x^2$ by $3x^2$.

138) The quotient of $28x^6 - 40x^3$ and $4x^4$.

139) Divide the sum of $5x^2 - 7x + 5$ and $3x^2 + 3x - 9$ by $2x$.

Chapter 2 Test

Find the following products.

1) $(-3)(5)(-4)$

2) $(-2)^3(3)^2$

3) $(3x^4y^3)(-2xy^2)$

4) $(3x^2y)(-4x^2y)^2$

5) $2x^2(3x - 4)$

6) $p^2q(2pq^3 - 4p^2q - 2)$

7) $(3p + 4)(2p - 1)$

8) $(3a - 4b)(2a - b)$

9) $(3y - 1)(y^2 + 2y + 5)$

10) $(5x - 3y)(5x + 3y)$

11) $(2a - 7)^2$

Find the following quotients. If the exercise involves exponents, leave the answer with positive exponents only.

12) $\dfrac{-24 - 8}{8}$

13) $\dfrac{4 \cdot 5^2 - 25}{(-3)(-4) + 13}$

14) $\dfrac{36x^4y^3z^5}{-18x^7y^3z^2}$

15) $\dfrac{(4a^6b^{-3})^2}{(2a^3b^4)^3}$

16) $\dfrac{(6a^{-2}b^{-4})(-8a^{-5}b^8)}{(4a^{-3}b^4)(-6a^6b^{-6})}$

17) $\left(\dfrac{2x^2}{y^3}\right)^{-2}$

18) $\dfrac{8x^3 - 10x^2 + 6x}{2x^2}$

Evaluate each of the following.

19) $-8 - (12 - 36) \div (-4)$

20) $\dfrac{4(-2)^3 + 2}{8^2 + 2(-3)^3}$

21) Find the value of $2y - (3x^2 - 3y) \div x^3$ if $x = -2$ and $y = -4$.

22) Write each of the following in scientific notation.
   a) 4,790,000,000   b) 0.0000000749

23) Evaluate $(43,000,000)(0.000025)$ using scientific notation. Leave the answer in standard notation.

Write an expression for each of the following and simplify.

24) The product of 7 and -3 subtracted from the quotient of 12 and -4

25) The product of $x + 3$ and $2x - 5$ added to $3x^2 - 7x + 5$

3

3.1 | Addition Property of Equality
3.2 | Multiplication Property of Equality
3.3 | Combining Properties in Solving Linear Equations
3.4 | Solving Linear Inequalities
3.5 | Percent Equations
3.6 | Traditional Applications Problems
3.7 | More Traditional Applications Problems
3.8 | Geometric Applications Problems
3.9 | Using and Solving Formulas

Linear Equations and Inequalities

CHAPTER OVERVIEW

One of the primary uses of addition, subtraction, multiplication, and division of signed numbers is in the solution of linear equations and inequalities. The notions of equations and inequalities were previously mentioned in Chapter 1. In this chapter we will expand on these notions and use them to solve various types of real-world and not so real-world applications problems.

SECTION 3.1

Addition Property of Equality

OBJECTIVE

When you complete this section, you will be able to:

Ⓐ Solve linear equations using the Addition Property of Equality.
Ⓑ Translate mathematical expressions into English and English expressions into mathematics.
Ⓒ Solve "real-world" problems.

INTRODUCTION

Recall that an equation is a mathematical sentence indicating that two expressions are equal. For example, $3 + 2 = 5$ and $x + 4 = 9$ are equations. Recall also that a solution of an equation is any value of the variable that makes the equation true. For example, $x = 5$ is a solution of $x + 4 = 9$ because if we replace x with 5 we get $5 + 4 = 9$, which is a true statement. When the solutions are written as a set, we refer to it as the solution set. So, the solution set of $x + 4 = 9$ is $\{5\}$.

In Chapters 1 and 2 you were given a number and asked to determine whether or not this number was a solution of a given equation. In this and following sections we will discuss techniques that allow us to solve **linear**

equations. A linear equation is any equation that can be put in the form of $ax + b = c$ where a, b, and c are constants and $a \neq 0$.

Suppose we begin with an equation and add the same number to both sides of the equation. Is the resulting sentence still an equation? Consider the following.

$4 + 5 = 9$	Add 6 to both sides.
$4 + 5 + 6 \overset{?}{=} 9 + 6$	Simplify both sides.
$15 = 15$	Adding 6 to both sides of the equation did not change the equality.
$4 + 5 = 9$	Add -8 to both sides.
$4 + 5 - 8 \overset{?}{=} 9 - 8$	Simplify both sides.
$1 = 1$	Adding -8 to both sides of the equation did not change the equality.

The above examples lead us to the following generalization.

ADDITION PROPERTY OF EQUALITY

If $a = b$, then $a + c = b + c$ for any number c. In words, if the same number is added to both sides of an equation the result is still an equation.

Equations that have the same solutions are called **equivalent equations.** If the addition property is applied to an equation containing variables, the resulting equation is equivalent to the original equation.

Before continuing, we need to recall two principles. The first is the additive inverse property. It states that every number, a, has an additive inverse, $-a$, such that $a + (-a) = 0$. The second property is the additive property of zero. It states that $0 + a = a + 0 = a$ for any number a.

The goal of equation solving will be to get the variable on one side of the equation and a constant on the other. By doing so we will get the simplest equivalent equation in the form of (variable) = (constant). In the following examples we isolate the variable by adding the additive inverse of the constant that is on the same side of the equation as the variable to both sides of the equation and simplifying. After solving an equation, we can verify that our solution is correct by substituting that value into the original equation and simplifying.

EXAMPLE 1
Solve the following.

a) $y - 3 = 9$ Since 3 is the additive inverse of -3, add 3 to both sides of the equation.

$y - 3 + 3 = 9 + 3$	Simplify both sides of the equation.
$y + 0 = 12$	$y + 0 = y$.
$y = 12$	Therefore, $y = 12$ is the solution and $\{12\}$ is the solution set.

Check:

$y - 3 = 9$	Substitute 12 for y.
$12 - 3 = 9$	Simplify the left side.
$9 = 9$	Therefore, 12 is the correct solution.

b) $6 = 8 + z$

Since -8 is the additive inverse of 8, add -8 to both sides of the equation.

$-8 + 6 = -8 + 8 + z$ Simplify both sides of the equation.

$-2 = 0 + z$ $0 + z = z$.

$-2 = z$ or $z = -2$ Therefore, -2 is the solution and $\{-2\}$ is the solution set.

Check:

$6 = 8 + z$ Substitute -2 for z.

$6 = 8 + (-2)$ Simplify the right side.

$6 = 6$ Therefore, -2 is the correct solution.

c) $x + \dfrac{1}{2} = \dfrac{3}{4}$

Since $-\frac{1}{2}$ is the additive inverse of $\frac{1}{2}$, add $-\frac{1}{2}$ to both sides of the equation.

$x + \dfrac{1}{2} - \dfrac{1}{2} = \dfrac{3}{4} - \dfrac{1}{2}$ Simplify both sides of the equation.

$x = \dfrac{1}{4}$ Therefore, $\frac{1}{4}$ is the solution and $\{\frac{1}{4}\}$ is the solution set.

Check:

$x + \dfrac{1}{2} = \dfrac{3}{4}$ Substitute $\dfrac{1}{4}$ for x.

$\dfrac{1}{4} + \dfrac{1}{2} = \dfrac{3}{4}$ Rewrite the left side with LCD of 4.

$\dfrac{1}{4} + \dfrac{2}{4} = \dfrac{3}{4}$ Add $\dfrac{1}{4}$ and $\dfrac{2}{4}$.

$\dfrac{3}{4} = \dfrac{3}{4}$ Therefore, $\dfrac{1}{4}$ is the correct solution.

Note: We will no longer write the solution sets of equations except in the case there are no solutions. If there are no solutions, then the solution set is the empty set indicated by $\varnothing$.

ADDITIONAL PRACTICE

Solve the following.

a) $x - 3 = 6$

b) $r + 4 = -1$

c) $x + \dfrac{1}{4} = \dfrac{3}{8}$

PRACTICE EXERCISES

Solve the following.

1) $v - 8 = 7$ **2)** $r + 3 = 4$ **3)** $x - \dfrac{1}{3} = \dfrac{1}{6}$

If more practice is needed, do the Additional Practice Exercises in the margin.

Often it is necessary to simplify one or both sides of an equation before applying the Addition Property of Equality. The following steps are suggested in doing so.

SOLVING LINEAR EQUATIONS USING THE ADDITION PROPERTY OF EQUALITY

1) If necessary, remove any inclusion symbols (parentheses, brackets, braces).

2) If necessary, simplify each side of the equation by adding like terms.

3) If necessary, use the Addition Property of Equality to get the variable on one side and the constant on the other.

4) Check the answer in the original equation.

This list of steps will be modified in Section 3.3 when we discuss the most general types of linear equations.

EXAMPLE 2
Solve the following.

a) $t + 8 - 6 = 11 + 4$ Simplify both sides of the equation.

   $t + 2 = 15$ Since -2 is the additive inverse of 2, add -2 to both sides of the equation.

   $t + 2 + (-2) = 15 + (-2)$ Simplify both sides of the equation.

   $t = 13$ Therefore, 13 is the solution.

   Check:

   $t + 8 - 6 = 11 + 4$ Substitute 13 for t.

   $13 + 8 - 6 = 11 + 4$ Simplify both sides.

   $15 = 15$ Therefore, $t = 13$ is the correct solution.

b) $6.3u + 5.4 - 5.3u = 20.6$ Simplify the left side of the equation.

   $u + 5.4 = 20.6$ Since -5.4 is the additive inverse of 5.4, add -5.4 to both sides of the equation.

   $u + 5.4 - 5.4 = 20.6 - 5.4$ Simplify both sides.

   $u = 15.2$ Therefore, 15.2 is the solution.

   Check:

   $6.3u + 5.4 - 5.3u = 20.6$ Substitute 15.2 for u.

   $6.3(15.2) + 5.4 - 5.3(15.2) = 20.6$ Multiply first.

   $95.96 + 5.4 - 80.56 = 20.6$ Add.

   $20.6 = 20.6$ Therefore, 15.2 is the correct solution.

Note: In Example 2a we added -2 to both sides of the equation. Rather than adding -2 we could have subtracted 2 since subtraction is the same as the addition of the negative. Recall from Chapter 1 that $a - b = a + (-b)$ and $a + (-b) = a - b$. Likewise, in Example 2b we could have subtracted 5.4 from both sides of the equation instead of adding -5.4.

c) $3x - 2(x + 4) = -5$ Apply the distributive property on the left side.

$3x - 2x - 8 = -5$ Simplify the left side.

$x - 8 = -5$ Since 8 is the additive inverse of -8, add 8 to both sides of the equation.

$x - 8 + 8 = -5 + 8$ Simplify both sides.

$x = 3$ Therefore, 3 is the solution.

The check is left as an exercise for the student.

d) $4 = 3(3x - 2) - 4(2x - 1)$ Apply the distributive property on the right.

$4 = 9x - 6 - 8x + 4$ Simplify the right side.

$4 = x - 2$ Since 2 is the additive inverse of -2, add 2 to both sides of the equation.

$4 + 2 = x - 2 + 2$ Simplify both sides of the equation.

$6 = x$ or $x = 6$ Therefore, 6 is the solution.

The check is left as an exercise for the student.

ADDITIONAL PRACTICE

Solve using the addition property.

d) $12 + s - 1 = 5$

e) $5.2x + 7.4 - 4.2x = 11.6$

f) $4(x + 3) - 3x = 7$

g) $6 = 5(x - 2) - 2(2x - 4)$

- **Translating from mathematics to English**

PRACTICE EXERCISES

Solve using the addition property.

4) $3 + t - 6 = 8 - 10$

5) $-5.7a + 4.6 + 6.7a = 16.5 - 9.2$

6) $6(x - 2) - 5x = -10$

7) $3 = 2(4x + 5) - 7(x + 2)$

If you need more practice, do the Additional Practice Exercises in the margin.

In problem solving, we must be able to translate from English to mathematics. We will start by translating from mathematics to English and then translate from English to mathematics. The examples that follow may have few applications in the "real-world", but serve to have you begin to think mathematically. Real-world examples will follow.

EXAMPLE 3

Translate each of the following from mathematics to English. There is more than one translation.

Mathematics	English
a) $t + 8 = 12$	Some number plus eight is twelve, or Eight added to some number is twelve, or The sum of some number and eight is twelve, or Eight more than some number is twelve, or A number increased by eight is twelve.
b) $y - 5 = 11$	Some number minus five is eleven, or Five subtracted from some number is eleven, or The difference of some number and five is eleven, or Five less than some number is eleven, or A number decreased by five equals eleven, or Some number less five is eleven.

Mathematics	English
c) $17 = u + 6$	Seventeen is some number plus six, or Seventeen equals six added to some number, or Seventeen equals a number increased by six, or Seventeen is six more than some number, or Seventeen is the sum of some number and six.
d) $4 = 21 - x$	Four equals twenty-one minus some number, or Four is some number subtracted from twenty-one, or Four is twenty-one decreased by some number, or Four is some number less than twenty-one, or Four equals the difference of twenty-one and some number.

PRACTICE EXERCISES

Translate from mathematical sentences to English sentences. There is more than one possible answer.

8) $r + 7 = 9$ **9)** $t - 4 = 17$ **10)** $15 = x + 6$ **11)** $28 = y - 19$

• **Translating from English into mathematics**

EXAMPLE 4

Translate from English to mathematics and solve. There is only one mathematical translation for each English sentence. Use x as the variable in each exercise.

English	Mathematics
a) A number plus one equals four.	$x + 1 = 4$ $x + 1 - 1 = 4 - 1$ $x = 3$
b) A number minus ten is fifteen.	$x - 10 = 15$ $x - 10 + 10 = 15 + 10$ $x = 25$
c) The sum of twelve and some number is twenty.	$12 + x = 20$ $12 - 12 + x = 20 - 12$ $x = 8$
d) The difference of some number and thirteen is six.	$x - 13 = 6$ $x - 13 + 13 = 6 + 13$ $x = 19$
e) Some number added to eighteen is twenty-nine.	$18 + x = 29$ $18 - 18 + x = 29 - 18$ $x = 11$

PRACTICE EXERCISES

Translate from English into mathematical sentences and solve. Use x as the variable.

12) The sum of a number and eleven is twenty-nine.

13) The difference of a number and six is fourteen.

14) A number increased by four is ten.

15) Eight subtracted from some number is twenty-three.

The following real-world situations may be solved without using algebra if you really think about them. However, later the situations will be far too complicated to solve without the techniques of algebra. Consequently, even though you may be able to do these problems without algebra, use algebraic techniques as practice for the more complicated problems in later sections of this chapter. The procedure we will use in this section is outlined below. As the problems become more complicated we will refine the procedure.

PROCEDURE FOR SOLVING REAL WORLD PROBLEMS WITH ONE VARIABLE

1) Identify the unknown and represent it with a variable.
2) Write a "word" equation relating the known and unknown quantities.
3) Write an algebraic equation using the word equation as a guide.
4) Solve the algebraic equation.
5) Check the solution in the wording of the original problem.

EXAMPLE 5

Let x represent the unknown in each of the following. Write an equation and solve.

a) John scored ten points more on his algebra test than Mike. If John scored 93, what was Mike's score?

Solution:

1) Identify the unknown and represent it with a variable. The unknown is Mike's score, so let x represent Mike's score.

2) Write a "word" equation relating the known and unknown quantities. Since John scored ten points more than Mike,

| Mike's score | plus | ten points | equals | John's score |

3) Write an algebraic equation using the word equation as a guide.

| Mike's score | plus | ten points | equals | John's score |
| x | $+$ | 10 | $=$ | 93 |

4) Solve the equation.

$x + 10 = 93$ Subtract 10 from both sides (add -10).

$x + 10 - 10 = 93 - 10$ Simplify both sides of the equation.

$x + 0 = 83$ $x + 0 = x$.

$x = 83$ Therefore, Mike's score was 83.

5) Check the solution in the wording of the original problem.
Is John's score ten points more than Mike's score? Since 93 is 10 more than 83, the answer is yes and our solution is correct.

Note: Other correct equations are $93 - x = 10$ and $x = 93 - 10$.

b) The limit on the number of crappie you can catch in Florida is 50 per person. Alice has already caught 32. How many more can she catch until she has her limit?

Solution:

1) Identify the unknown and represent it with a variable. The unknown is how many more crappie Alice can catch, so let x represent the number of additional crappie she can catch.

2) Write a "word" equation relating the known and unknown quantities. Since she has already caught 32 and she can catch 50,

| The number Alice has already caught | plus | Additional number she can catch | equals | 50 |

3) Write an algebraic equation using the word equation as a guide.

| The number Alice has already caught | plus | Additional number she can catch | equals | 50 |

$$32 \quad + \quad x \quad = \quad 50$$

4) Solve the equation.

$$32 + x = 50 \qquad \text{Subtract 32 from both sides (add } -32\text{).}$$
$$32 - 32 + x = 50 - 32 \qquad \text{Simplify both sides of the equation.}$$
$$0 + x = 18 \qquad 0 + n = n.$$
$$x = 18 \qquad \text{Therefore, Alice can catch 18 more.}$$

5) Check the solution in the wording of the original problem.
Does the number that she has already caught (32) plus the additional number she can catch (18) equal the limit (50)? Yes. Therefore, the solution is correct.

Note: Other correct equations are $x = 50 - 32$ and $50 - x = 32$.

PRACTICE EXERCISES

Let x represent the unknown in each of the following. Write an equation and solve. There may be more than one correct equation.

16) The cost of a car is $2,500 less than the cost of a minivan. If the cost of the car is $18,350, find the cost of the minivan.

17) Jenny got married 6 years ago. If she is presently 28 years old, how old was she when she got married?

EXERCISE SET 3.1

Solve each of the following.

1) $t + 5 = 13$ **2)** $z + 9 = 12$ **3)** $r + 8 = 2$

4) $s + 9 = 4$ **5)** $a - 3 = 5$ **6)** $b - 6 = 2$

7) $x - 6.3 = 4.2$ **8)** $y - 5.7 = 3.6$

9) $m + 7.3 = 6.1$ **10)** $v + 8.9 = 4.3$

11) $x - \dfrac{2}{3} = \dfrac{4}{3}$ **12)** $x + \dfrac{3}{4} = \dfrac{7}{4}$

13) $x + \dfrac{2}{5} = \dfrac{3}{10}$ **14)** $x - \dfrac{2}{7} = \dfrac{3}{14}$

15) $x - \dfrac{1}{4} = \dfrac{5}{6}$ **16)** $x + \dfrac{1}{3} = \dfrac{3}{4}$

17) $7 + w - 1 = 15$

18) $16 + u - 3 = 23$

19) $18 - 15 + x = 10$

20) $20 - 12 + y = 17$

21) $z - 5 - 4 = -8 + 2$

22) $n - 5 - 3 = -10 + 3$

23) $5t + 7 - 4t = 1 + 5$

24) $3z - 5 - 2z = 1 - 6$

25) $6.1x - 1.3 - 5.1x = 6.8$

26) $4.3x - 3.7 - 3.3x = 8.1$

27) $7 + 5 = 6r + 8 - 5r$

28) $3 + 6 = 7a + 4 - 6a$

29) $-4 - 6 = -8x - 10 + 9x$

30) $-8 - 4 = -3x - 12 + 4x$

31) $4(x + 3) - 3x = 10$

32) $3(x + 5) - 2x = 12$

33) $7x - 3(2x + 4) = -6$

34) $11x - 5(2x + 1) = -2$

35) $16x - 5(3x - 2) - 7 = 5$

36) $17x - 4(4x - 3) - 8 = 6$

37) $3(3.4x - 1.2) - 9.2x = -6.2$

38) $4(2.4x - 1.6) - 8.6x = -3.3$

39) $-2(4.7y - 5.4) + 10.4y = 3.1$

40) $-3(1.6b - 2.4) + 5.8b = -6.2$

41) $6(t + 3) - 5(t + 4) = 10$

42) $8(w - 5) - 7(w - 4) = 15$

43) $3(3x - 5) - 2(4x - 3) = -5$

44) $5(5x - 3) - 6(4x - 2) = -3$

45) $-4(5x - 4) + 7(3x - 2) - 6 = 8 - 2$

46) $-6(4x - 2) + 5(5x - 3) - 9 = 9 - 3$

CHALLENGE EXERCISES (47–48)

47) $3.2(1.5x - 3.4) - 3.8x + 5.32 = -6.33$

48) $4.6(2.5x + 1.4) - 10.5x - 7.34 = -8.21$

Translate the following mathematical sentences into English in at least two ways.

49) $x - 5 = 19$

50) $22 - t = 11$

51) $13 + y = 7$

52) $25 = u + 8$

53) $17 = w - 6$

54) $44 = v + 9$

Translate the following English sentences into mathematical sentences and solve. Use x as the variable.

55) Some number minus two is equal to nine.

56) Twelve increased by some number is nineteen.

57) Some number minus fourteen is eight.

58) Six subtracted from some number is nineteen.

59) The difference of a number and two is fifteen.

60) A number decreased by seven is thirty-two.

61) The sum of five and some number is thirteen.

62) Five more than some number is -2.

63) Eight less than some number is -10.

64) The sum of a number and -3 is -7.

65) The difference of some number and 2 is -8.

Let x represent the unknown in each of the following. Write an equation and solve. A typical "word" equation that results in an equation of the form studied in this section is given.

66) On a trip Francine drove 75 miles more than Hector. If Francine drove 260 miles, how far did Hector drive?

$$\boxed{\begin{array}{c}\text{number of miles}\\ \text{Hector drove}\end{array}} + 75 = \boxed{\begin{array}{c}\text{number of miles}\\ \text{Francine drove}\end{array}}$$

67) Shasta and Charlene share the driving responsibilities on a trip. If the total trip was 550 miles and Shasta drove 275 miles, how far did Charlene drive?

$$\boxed{\begin{array}{c}\text{number of}\\ \text{miles}\\ \text{Shasta drove}\end{array}} + \boxed{\begin{array}{c}\text{number of}\\ \text{miles}\\ \text{Charlene drove}\end{array}} = \boxed{\begin{array}{c}\text{total}\\ \text{miles}\end{array}}$$

68) Horace needs \$465 to buy a new shotgun. If he has already saved \$312, how much more does he need?

$$\boxed{\begin{array}{c}\text{amount}\\ \text{Horace has}\end{array}} + \boxed{\begin{array}{c}\text{amount}\\ \text{Horace needs}\end{array}} = \boxed{\begin{array}{c}\text{cost of}\\ \text{shotgun}\end{array}}$$

69) Jenny and Bryan are going to chip in together to buy their parents an anniversary gift. If the gift costs \$160 and Bryan pays \$75, how much does Jenny pay?

$$\boxed{\begin{array}{c}\text{amount}\\ \text{Jenny pays}\end{array}} + \boxed{\begin{array}{c}\text{amount}\\ \text{Bryan pays}\end{array}} = \boxed{\begin{array}{c}\text{cost of}\\ \text{gift}\end{array}}$$

70) John and Jim shared expenses on a trip from Orlando to Atlanta. John paid \$30 less than Jim. If John paid \$243, how much did Jim pay?

$$\boxed{\begin{array}{c}\text{amount}\\ \text{John pays}\end{array}} + 30 = \boxed{\begin{array}{c}\text{amount}\\ \text{Jim paid}\end{array}}$$

71) In a recent football game the Dolphins scored 8 points less than the Bills. If the Dolphins scored 24 points, how many points did the Bills score?

| number of points Bills scored | − | 8 | = | number of points Dolphins scored |

72) A mixture contains 23 more gallons of water than alcohol. If the mixture contains 47 gallons of water, how many gallons of alcohol does it contain?

| number of gallons of alcohol | + | 23 | = | number of gallons of water |

73) Connie needs a total of 800 points in order to make an A in her English I class. If she already has 632 points, how many more does she need?

| number of points Connie has | + | number of points Connie needs | = | number of points needed for an A |

74) The Orlando Magic scored 13 more points than the New Jersey Nets. If the Nets scored 92 points, how many did the Magic score?

| number of points Nets scored | + | 13 | = | number of points Magic scored |

75) Pamika plans to buy a new T.V. The T.V. costs $46 less at a wholesale store than at an electronics store. If the price of the T.V. is $326 at the electronics store, what is the price at the wholesale store?

| price at wholesale store | + | 46 | = | price at electronics store |

76) The selling price of a computer is $1195. If the markup is $210, find the store's cost.

| store's cost | + | markup | = | selling price |

77) Francine went on a diet and lost 23 pounds. If she weighed 118 pounds after the diet, what did she weigh before the diet?

| weight before diet | − | weight loss | = | weight after diet |

78) The height of Mt. Ranier in Washington is 7,726 ft higher than Mt. Mitchel in North Carolina. If Mt. Mitchel is 6,684 ft high, find the height of Mt. Ranier

| height of Mt. Ranier | − | 7,726 | = | height of Mt. Mitchel |

79) A local department store received $18,000 more from the sale of men's shirts than from the sale of men's pants. If they received $40,000 for the sale of shirts, how much did they receive from the sale of men's pants?

| amount from sale of pants | + | 18,000 | = | amount from sale of shirts |

80) A board is cut into two pieces. The longer piece is 5 feet in length and is 3 ft longer than the shorter piece. Find the length of the shorter piece.

| length of shorter piece | + | 3 | = | length of longer piece |

WRITING EXERCISES

81) What does it mean to say that a number "solves an equation?"

82) Why do we not need a "Subtraction Property of Equality?"

WRITING EXERCISE OR GROUP PROJECT

83) Write two applications problems that require the use of the Addition Property of Equality. Exchange with another group or person and solve.

SECTION 3.2

Multiplication Property of Equality

OBJECTIVES

When you complete this section, you will be able to:

Ⓐ Solve equations using the Multiplication Property of Equality.

Ⓑ Solve equations using the Multiplication Property of Equality which require simplifications.

Ⓒ Solve real-world problems.

The Addition Property of Equality allows us to solve one type of equation. However, it does not allow us to solve equations like $3x = 6$. If we subtracted 3 from both sides we would get $3x - 3 = 3$ which is more complicated than the equation that we began with. We need another technique to be able to solve this type of equation.

In Section 3.1 we found that we could add the same number to both sides of an equation and the equality was preserved. Would the equality be preserved if we multiply both sides of an equation by the same nonzero number?

$6 = 6$	Multiply both sides by 3.
$3(6) \stackrel{?}{=} 3(6)$	Simplify both sides.
$18 = 18$	Therefore, the equality was preserved when both sides were multiplied by 3.
$4 + 3 = 7$	Multiply both sides by 2.
$2(4 + 3) \stackrel{?}{=} 2(7)$	Apply the Distributive Property.
$2(4) + 2(3) \stackrel{?}{=} 2(7)$	Multiply before adding.
$8 + 6 \stackrel{?}{=} 14$	Add.
$14 = 14$	Therefore, the equality was preserved when both sides were multiplied by 2.
$2 \cdot 9 = 18$	Multiply both sides by $\frac{1}{6}$.
$\frac{1}{6} \cdot (2 \cdot 9) = \frac{1}{6} \cdot 18$	$2 \cdot 9 = 18$ and $\frac{1}{6} \cdot 18 = \frac{18}{6}$.
$\frac{18}{6} = \frac{18}{6}$	Divide.
$3 = 3$	Therefore, the equality was preserved when both sides were multiplied by $\frac{1}{6}$.

Note: Multiplying by $\frac{1}{6}$ is the same as dividing by 6.

From the arithmetic equations above it appears that we can multiply (or divide) both sides of an equation by any nonzero number, the equality is preserved. We formally state this as the Multiplication Property of Equality.

MULTIPLICATION PROPERTY OF EQUALITY

If $a = b$, and $c \neq 0$, then $a \cdot c = b \cdot c$. In words, both sides of an equation may be multiplied (or divided) by any nonzero number and the result is still an equation.

As with the Addition Property of Equality, when we apply the Multiplication Property of Equality to an equation with variables, the resulting equation is equivalent to the original.

To solve equations using the Multiplication Property of Equality, we need to recall two properties (Section 2.1). The first is the multiplicative inverse which is also called the reciprocal. Two numbers are multiplicative inverses if their product is 1. For example, the multiplicative inverse (reciprocal) of a is $\frac{1}{a}$, because $\frac{1}{a}(a) = \frac{a}{a} = 1$. The multiplicative inverse of $\frac{a}{b}$ is $\frac{b}{a}$ because $\frac{a}{b} \cdot \frac{b}{a} = \frac{ab}{ba} = 1$. For example, the reciprocal of $\frac{3}{4}$ is $\frac{4}{3}$ because $\frac{3}{4} \cdot \frac{4}{3} = \frac{3 \cdot 4}{4 \cdot 3} = \frac{12}{12} = 1$. The second is the multiplication property of 1 which states that $1 \cdot x = x \cdot 1 = x$.

Suppose we are asked to solve the equation $3u = 15$. As in Section 3.1, the goal in solving this equation is to get the variable with a coefficient of 1 on one side of the equation and the constant on the other. This is done by multiplying both sides of the equation by the reciprocal of the coefficient of the variable.

EXAMPLE 1

Solve the following.

a) $3u = 15$ Since $\frac{1}{3}$ is the reciprocal of 3, multiply both sides by $\frac{1}{3}$.

$\frac{1}{3} \cdot 3u = \frac{1}{3} \cdot 15$ $\frac{1}{3} \cdot 3 = 1$ and $\frac{1}{3} \cdot 15 = 5$.

$1(u) = 5$ $1 \cdot u = u$.

$u = 5$ Therefore, 5 is the solution.

Check:

$3u = 15$ Substitute 5 for u.

$3(5) = 15$ Multiply.

$15 = 15$ Therefore, 5 is the solution.

Note: Division by 3 is the same as multiplication by $\frac{1}{3}$. Consequently, we could have done Example 2a as follows.

$3u = 15$ Divide both sides by 3.

$\frac{3u}{3} = \frac{15}{3}$ $\frac{3}{3} = 1$ and $\frac{15}{3} = 5$.

$u = 5$ Therefore, 5 is the solution.

This is the method that we will use in future examples when appropriate.

b) $-0.6v = 1.8$ Divide both sides by -0.6.

$\frac{-0.6v}{-0.6} = \frac{1.8}{-0.6}$ $\frac{-0.6}{-0.6} = 1$ and $\frac{1.8}{-0.6} = -3$.

$1 \cdot v = -3$ $1 \cdot v = v$.

$v = -3$ Therefore, -3 is the solution.

Check:

$-0.6v = 1.8$ Substitute -3 for v.

$-0.6(-3) = 1.8$ Multiply.

$1.8 = 1.8$ Therefore, -3 is the solution.

c) $\frac{t}{-4} = 7$ Rewrite as a product.

$\frac{1}{-4} t = 7$ Since -4 is the reciprocal of $\frac{1}{-4}$, multiply both sides by -4.

$-4 \cdot \frac{1}{-4} t = -4 \cdot 7$ $-4 \cdot \frac{1}{-4} = 1$ and $-4 \cdot 7 = -28$.

$1 \cdot t = -28$ $1 \cdot t = t$.

$t = -28$ Therefore, -28 is the solution.

Check:

$$\frac{t}{-4} = 7$$ Substitute -28 for t.

$$\frac{-28}{-4} = 7$$ Divide.

$$7 = 7$$ Therefore, -28 is the correct solution.

d) $\frac{3}{2}z = 60$ Since $\frac{2}{3}$ is the reciprocal of $\frac{3}{2}$, multiply both sides by $\frac{2}{3}$.

$$\frac{2}{3} \cdot \frac{3}{2}z = \frac{2}{3} \cdot 60$$ $\frac{2}{3} \cdot \frac{3}{2} = 1$ and $\frac{2}{3} \cdot 60 = \frac{120}{3} = 40$.

$$1 \cdot z = 40$$ $1 \cdot z = z$.

$$z = 40$$ Therefore, 40 is the solution.

Check:

$$\frac{3}{2}z = 60$$ Substitute 40 for z.

$$\frac{3}{2}(40) = 60$$ Multiply.

$$\frac{120}{2} = 60$$ Divide.

$$60 = 60$$ Therefore, 40 is the solution.

e) $-x = 5$ $-x = -1 \cdot x$. Multiply both sides by the reciprocal of -1 which is -1.

$$-1(-x) = -1 \cdot 5$$ Simplify both sides.

$$x = -5$$ Therefore, -5 is the solution.

Check:

$$-x = 5$$ Substitute -5 for x.

$$-(-5) = 5$$ Apply the double negative rule.

$$5 = 5$$ Therefore, -5 is the correct solution.

ADDITIONAL PRACTICE

Solve by use of the multiplication property.

 a) $42 = 6n$

 b) $0.7p = -6.3$

 c) $\frac{x}{-3} = 3$

 d) $\frac{8}{5}s = -32$

 e) $-x = 4$

PRACTICE EXERCISES

Solve the following.

1) $5v = -20$ **2)** $-0.5y = 32.5$

3) $\frac{w}{2} = 6$ **4)** $-\frac{7}{9}u = 21$

5) $3 = -t$

If you need more practice, do the Additional Practice Exercises in the margin.

When applying the Addition Property of Equality in the previous section, it was often necessary to simplify one or both sides of the equation. The same is true when the Multiplication Property of Equality is involved.

EXAMPLE 2

Simplify each of the following. If the equation involves decimals, leave the answer as a decimal. Otherwise, when necessary, leave the answer as a fraction.

a) $8x + 4x = -48$ — Simplify the left side of the equation.

$12x = -48$ — Divide both sides by 12.

$\dfrac{12x}{12} = \dfrac{-48}{12}$ — $\dfrac{12}{12} = 1$ and $\dfrac{-48}{12} = -4$.

$1 \cdot x = -4$ — $1 \cdot x = x$.

$x = -4$ — Therefore, -4 is the solution.

Check:

$8x + 4x = -48$ — Substitute -4 for x.

$8(-4) + 4(-4) = -48$ — Multiply.

$-32 - 16 = -48$ — Add.

$-48 = -48$ — Therefore, -4 is the solution.

b) $5y = 9 + 15$ — Simplify the right side.

$5y = 24$ — Divide both sides by 5.

$\dfrac{5y}{5} = \dfrac{24}{5}$ — $\dfrac{5}{5} = 1$.

$1 \cdot y = \dfrac{24}{5}$ — $1 \cdot y = y$.

$y = \dfrac{24}{5}$ or 4.8 — Therefore, the solution is $\dfrac{24}{5}$.

The check is left as an exercise for the student.

c) $2.4z - 3.6z = 4.8$ — Simplify the left side.

$-1.2z = 4.8$ — Divide both sides by -1.2.

$\dfrac{-1.2z}{-1.2} = \dfrac{4.8}{-1.2}$ — $\dfrac{-1.2}{-1.2} = 1$ and $\dfrac{4.8}{-1.2} = -4$.

$1 \cdot z = -4$ — $1 \cdot z = z$.

$z = -4$ — Therefore, -4 is the solution to nearest tenth.

The check is left as an exercise for the student.

ADDITIONAL PRACTICE

Solve using the multiplication property.

f) $2x + 4x = 24$

g) $2t = -22 + 36$

h) $5.7a - 2.3a = -10.2$

PRACTICE EXERCISES

Solve each of the following. If the equation involves decimals, leave the answer as a decimal. Otherwise, when necessary, leave the answer as a fraction.

6) $16t - 7t = 45$ **7)** $10q = 35 - 14$ **8)** $3.3 = 5.8s - 5.5s$

If you need more practice, do the Additional Practice Exercises in the margin.

As in the previous section, we will translate from mathematics to English and from English to mathematics.

EXAMPLE 3

Translate each of the following from mathematics to English. There may be more than one translation into English.

Mathematics	English
a) $2y = 8$	Twice a number is eight.
	Two times a number equals eight.
	The product of two and a number is eight.
b) $\dfrac{x}{4} = 5$	The quotient of some number and 4 is 5.
	Some number divided by 4 is 5.
c) $32 = -8t$	Thirty-two is negative eight times a number.
	Thirty-two is the product of negative eight and a number.
d) $0.6s = 3$	Six tenths of a number is three.
	The product of six tenths and a number is three.
	Six tenths times some number is 3.

ADDITIONAL PRACTICE

Translate from mathematical sentences to English sentences. There is more than one possible translation.

i) $7y = 28$ **j)** $\dfrac{x}{3} = -4$

k) $15 = -2x$ **l)** $0.4t = 16$

PRACTICE EXERCISES

Translate from mathematical sentences to English sentences. There is more than one possible translation.

9) $9x = 45$ **10)** $\dfrac{v}{8} = 2$

11) $33 = -3w$ **12)** $0.5z = -13$

If more practice is needed, do the Additional Practice Exercises in the margin.

EXAMPLE 4

Translate from English to mathematics and solve. There is only one mathematical translation for each English sentence. Let x be the variable in each case.

English	Mathematics
a) The product of four and a number is twenty.	$4x = 20$
	$\dfrac{4x}{4} = \dfrac{20}{4}$
	$x = 5$
b) The quotient of some number and 3 is -6	$\dfrac{x}{3} = -6$
	$3 \cdot \dfrac{x}{3} = 3(-6)$
	$x = -18$
c) Five and six tenths equals negative ten times a number.	$5.6 = -10x$
	$\dfrac{5.6}{-10} = \dfrac{-10x}{-10}$
	$-0.56 = x$
d) Eight tenths of a number equals negative two and four tenths.	$0.8x = -2.4$
	$\dfrac{0.8x}{0.8} = \dfrac{-2.4}{0.8}$
	$x = -3$

Translate from English to mathematical sentences and solve.

m) The product of six and a number is twelve.

n) The quotient of some number and −3 is 5.

o) Three and five tenths equals seven tenths times a number.

ANSWERS:
Practice 9–12

9) The product of nine and a number is forty-five.

10) The quotient of v and 8 is 2.

11) Thirty-three is negative three times a number.

12) Five tenths of some number is negative thirteen.

Additional Practice i–k

i) Seven times a number is twenty-eight.

j) The quotient of x and 3 is −4.

k) Fifteen is −2 times some number.

l) Four tenths of a number is sixteen.

PRACTICE EXERCISES

Translate from English to mathematical sentences and solve. Use x as the variable.

13) Five times a number is forty.

14) The quotient of some number and −2 is 8.

15) Negative four and two tenths is negative seven times some number.

16) Seven tenths of a number is twenty-one.

If you need more practice, do the Additional Practice Exercises in the margin.

As in the previous section, most of the following real-world problems can be solved without the use of algebra. However, we ask that you continue to follow the same format used in Section 3.1.

EXAMPLE 5
Solve the following.

a) Jude earns $52.00 for working eight hours. What is his hourly wage?

Solution:

1) Identify the unknown and represent it with a variable. The unknown is Jude's hourly wage, so let w represent his hourly wage.

2) Write a "word" equation.

| number of hours Jude worked | times | his hourly wage | equals | total earnings |

3) Write an algebraic equation using the word equation as a guide.

| number of hours Jude worked | times | his hourly wage | equals | total earnings |
| 8 | · | w | = | 52 |

4) Solve the equation.

$$8w = 52 \qquad \text{Divide both sides by 8.}$$
$$\frac{8w}{8} = \frac{52}{8} \qquad \text{Simplify both sides.}$$
$$w = \$6.50 \qquad \text{Therefore, Jude earns \$6.50 per hour.}$$

5) Check the solution in the wording of the original problem.

If Jude worked 8 hours at $6.50 per hour his total earnings would be 8(6.50) which equals $52.00. Therefore, our solution is correct.

b) Sondra works as a salesperson. Her commission is 7% of her total sales. If she earned $245 one week, what were her total sales for that week?

1) Identify the unknown and represent it with a variable. The unknown is Sondra's total sales, so let s represent her sales for the week.

2) Write a "word" equation.

| rate of commission | times | sales for the week | equals | earnings for the week |

3) Write an algebraic equation using the word equation as a guide. Remember, 7% = 0.07.

| rate of commission | times | sales for the week | equals | earnings for the week |
| 0.07 | · | s | = | 245 |

---

4) Solve the equation.

$0.07s = 245$ Divide both sides by 0.07.

$\dfrac{0.07s}{0.07} = \dfrac{245}{0.07}$ Simplify both sides.

$s = 3500$ Therefore, Sondra's sales for the week were $3500.

5) Check the solution in the wording of the original problem. Sondra's earnings are equal to $(.07)(\$3,500)$ which equals $245. Therefore, our solution is correct.

PRACTICE EXERCISES

Solve the following.

17) Hernando drove 230 miles in 5 hours. What is his rate in miles per hour? (rate $\cdot$ time = distance)

18) A bank pays 6% interest per year on savings accounts. If Sally received $1,080 interest on her savings account last year, how much does she have in savings?

EXERCISE SET   **3.2**

Solve the following.

1) $7u = 56$ **2)** $5v = 70$ **3)** $-3n = 36$

4) $-62 = 2m$ **5)** $42 = -14x$ **6)** $-12y = 96$

7) $-64 = -16p$ **8)** $-75 = -15q$ **9)** $7u = 5.6$

10) $6v = 4.8$ **11)** $0.3x = 15$ **12)** $0.8y = 24$

13) $-0.1t = -11$ **14)** $-0.6m = -4.2$

15) $\dfrac{n}{2} = 8$ **16)** $\dfrac{r}{5} = 3$

17) $\dfrac{q}{-3} = 6$ **18)** $\dfrac{w}{-4} = 6$

19) $\dfrac{8}{7}z = -56$ **20)** $\dfrac{8}{9}q = -72$

21) $-\dfrac{2}{3}w = -6$ **22)** $-\dfrac{3}{4}x = -6$

23) $\dfrac{8m}{5} = -120$ **24)** $-91 = \dfrac{7n}{13}$

25) $\dfrac{2}{3}x = \dfrac{1}{5}$ **26)** $\dfrac{3}{5}x = \dfrac{2}{7}$

27) $\dfrac{3}{4}x = -\dfrac{3}{2}$ **28)** $\dfrac{2}{7}x = -\dfrac{4}{5}$

29) $6y = 55 - 13$ **30)** $-4z = -39 + 15$

31) $3x + 5x = 24$ **32)** $4x + 7x = -22$

33) $9x - 3x = -24$ **34)** $11x - 4x = -28$

35) $4x - 7x = 15 - 6$ **36)** $5x - 9x = 12 - 8$

37) $-2.9u + 3.6u = 6.3$ **38)** $-4.6z + 2.2z = 4.8$

39) $3.4x + 5.1x = 30 - 4.5$

Translate the following mathematical sentences to English sentences in at least two ways.

40) $7t = 46$ **41)** $62 = 4z$ **42)** $-14 = 2h$

43) $-39 = 3k$ **44)** $5.6 = 7x$ **45)** $2.7y = 5.6$

Translate from English to mathematical sentences and solve. Use x as the variable.

46) Eight times a number is sixty-four.

47) Five times a number is thirty-five.

48) The product of six and some number is negative forty-eight.

49) The product of four and some number is negative twenty-eight.

50) Six tenths of a number is twelve.

51) Four tenths of a number equals two.

52) The quotient of a number and two is six.

53) The quotient of a number and four is five.

54) Some number divided by five is negative two.

55) Some number divided by negative three is eight.

Solve each of the following.

56) John is buying carpet for his rectangular living room. He arrives at the carpet store and can remember only that the length of the room is 18 feet and the area is 216 square feet. Find the width of the room. (*A* = *LW*)

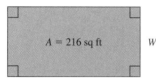

A = 216 sq ft   W

18 ft

57) Leslie has a piece of apple pie a-la-mode. The piece of pie has three times as many calories as the low-calorie ice cream served with it. If the pie had 450 calories, how many calories did the ice cream have?

58) Bill spent $14.50 for four fishing lures all of which had the same price. How much did each cost?

59) Catalina worked twelve hours last week and earned $102. What is her hourly wage?

60) Susan is building a bookcase. How many shelves, each of which is 3.5 feet long, can she cut from a board that is 14 feet long?

61) During a particularly boring speech two-fifths of the audience left. If forty-six people left, how many people were originally in the audience?

62) Two-thirds of a class passed the last exam. If eighteen people passed, how many students are in the class?

63) On her morning walk Frankie passes a tree that is $\frac{1}{6}$ of the total distance that she plans to walk from her house. If she has walked $\frac{2}{3}$ of a mile, how far does she plan to walk?

64) After a stock split Harry will have $1\frac{1}{4}$ times as many shares of stock as he had before the split. If he has

225 shares after the split, how many shares did he have before the split?

65) Three-sevenths of the employees of a company earn at least $28,000 per year. If 123 people earn at least $28,000 per year, how many employees does the company have?

66) Katrina works as a salesperson and is paid a 6% commission on her sales. If she received a $9,000 commission on a sale, what was the amount of the sale?

67) The state sales tax in Florida is 6%. If the sales tax on a car purchased in Florida was $1050, what was the selling price of the car?

68) Teschima is renting a TV for $3.50 per week. If a new TV costs $280, how many weeks will it take for her rental fee to equal the cost of a new TV?

69) A wholesale club charges an annual fee of $80 which entitles its members to a 16% discount. How much would a member have to purchase per year for the savings to equal the amount of the membership fee?

70) A sign is in the shape of an isosceles triangle whose base is 16 inches long. If the area of the sign is 48 square inches, what is the height of the sign? (*A* = $\frac{1}{2}bh$)

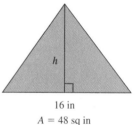

h

16 in
A = 48 sq in

WRITING EXERCISE

71) Why do we not need a "Division Property of Equality"?

WRITING EXERCISE OR GROUP PROJECT

72) Write two applications problems that involve the use of the Multiplication Property of Equality. Exchange with another group or person and solve.

S E C T I O N   3.3

Combining Properties in Solving Linear Equations

OBJECTIVES

When you complete this section, you will be able to:

Ⓐ   Solve linear equations using both the Addition and Multiplication Properties of Equality.

Ⓑ   Solve linear equations that require simplification of one or both sides before solving using the Addition and/or Multiplication Properties of Equality.

In this section we will combine the addition and multiplication properties of equality to solve linear equations in one variable. For example, suppose we were asked to solve $3x + 9 = -6$. In order to solve for x we must get rid of both 3 and 9. Which one do we get rid of first? The procedure for solving such equations is outlined below.

SOLVING LINEAR EQUATIONS

1) If necessary, simplify both sides of the equation as much as possible. This could involve using the distributive, commutative, and associative properties and combining like terms.

2) If necessary, use the Addition Property of Equality to get all the terms with variables on one side of the equation and all the constant terms on the other.

3) If necessary, use the Multiplication Property of Equality to eliminate any coefficient on the variable.

4) Check the answer in the original equation.

Up to this point all the equations that we have solved in this chapter have had all the variables on one side of the equation. If after simplifying both sides of the equation there are terms with variables on both sides, use the Addition Property of Equality to get the terms with the variable on one side of the equation and the constants on the other. We still use the principle of the additive inverse to accomplish this. For example, the additive inverse of $2x$ is $-2x$ since $2x + (-2x) = 0$. In solving equations like $3x + 4 = -5x + 28$, we can eliminate any one of the four terms by adding its additive inverse to both sides of the equation. In Example 1b–e we will first eliminate a term with a variable and then a constant.

EXAMPLE 1

Solve the following.

a) $3x + 9 = -6$ — Subtract 9 from both sides.

$3x + 9 - 9 = -6 - 9$ — Simplify both sides.

$3x = -15$ — Divide both sides by 3.

$\dfrac{3x}{3} = \dfrac{-15}{3}$ — Simplify both sides.

$x = -5$ — Therefore, the solution is -5.

Check:

$3x + 9 = -6$ — Substitute -5 for x.

$3(-5) + 9 = -6$ — Multiply before adding.

$-15 + 9 = -6$ — Add.

$-6 = -6$ — Therefore, -5 is the correct solution.

 b) $3x + 4 = -5x + 28$ — We could eliminate any of the four terms. We will eliminate $-5x$ by adding its inverse $5x$ to both sides of the equation.

$3x + 5x + 4 = -5x + 5x + 28$ — Simplify both sides.

$8x + 4 = 28$ Subtract 4 from both sides.

$8x + 4 - 4 = 28 - 4$ Simplify both sides.

$8x = 24$ Divide both sides by 8.

$\dfrac{8x}{8} = \dfrac{24}{8}$ Simplify both sides.

$x = 3$ Therefore, 3 is the solution.

Check:

$3x + 4 = -5x + 28$ Substitute 3 for x.

$3(3) + 4 = -5(3) + 28$ Multiply before adding.

$9 + 4 = -15 + 28$ Add.

$13 = 13$ Therefore, 3 is the solution.

c) $12z - 3.8 = 5z + 0.4$ We could eliminate any of the four terms. We will eliminate $5z$ by adding its inverse $-5z$ to both sides of the equation.

$12z - 5z - 3.8 = 5z - 5z + 0.4$ Simplify both sides.

$7z - 3.8 = 0.4$ Add 3.8 to both sides.

$7z - 3.8 + 3.8 = 0.4 + 3.8$ Simplify both sides.

$7z = 4.2$ Divide both sides by 7.

$\dfrac{7z}{7} = \dfrac{4.2}{7}$ Simplify both sides.

$z = 0.6$ Therefore, 0.6 is the solution.

Check:

$12z - 3.8 = 5z + 0.4$ Substitute 0.6 for z.

$12(0.6) - 3.8 = 5(0.6) + 0.4$ Multiply before adding.

$7.2 - 3.8 = 3.0 + 0.4$ Add.

$3.4 = 3.4$ Therefore, 0.6 is the solution.

d) $9y - 5 - 8y = 5y + 10 - 3y$ Simplify both sides by adding like terms.

$y - 5 = 2y + 10$ Subtract y from both sides.

$y - y - 5 = 2y - y + 10$ Simplify both sides.

$-5 = y + 10$ Subtract 10 from both sides.

$-5 - 10 = y + 10 - 10$ Simplify both sides.

$-15 = y$ Therefore, -15 is the solution.

The check is left as an exerise for the student.

 e) $7(r - 1) + 10 = 5(2 + r)$ Apply the distributive property.

$7r - 7 + 10 = 10 + 5r$ Simplify both sides by adding like terms.

$7r + 3 = 10 + 5r$ To eliminate $5r$, subtract $5r$ from both sides.

$7r - 5r + 3 = 10 + 5r - 5r$ Simplify both sides.

$2r + 3 = 10$ Subtract 3 from both sides.

$2r + 3 - 3 = 10 - 3$ Simplify both sides.

$2r = 7$ Divide both sides by 2.

$$\frac{2r}{2} = \frac{7}{2}$$ Simplify both sides.

$$r = \frac{7}{2} \text{ or } 3.5$$ Therefore, $\frac{7}{2}$ is the solution.

The check is left as an exerise for the student.

ADDITIONAL PRACTICE

Solve the following.

a) $5x + 2 = 12$

b) $4x - 15 = 5 - 6x$

c) $2.3 - 0.4w = 0.5w - 2.2$

d) $3z + 5 + 5z = 2z - 1$

e) $6(p - 4) - 3p = 18 - 3p$

PRACTICE EXERCISES

Solve each of the following.

1) $4x - 7 = 9$

2) $2r - 4 = 11 - 3r$

3) $0.7 - t = 3t - 2.1$

4) $9 - 6r - 3 - 7r = 2r + 2 - 11r$

5) $4(r + 2) = 3(r + 4) + 6$

If you need more practice, do the Additional Practice Exercises in the margin.

Note: The more complicated equations involving fractions are discussed in Chapter 5 when solving equations with rational expressions.

In solving equations, there are two situations that require special attention. These are **identities** and **contradictions**. An identity is an equation that is true for all values of the variable for which the equation is defined. When solving an equation that is an identity, we arrive at a statement that is obviously true like $5 = 5$. Since any number solves the identity, we will denote the solutions as "all real numbers." A contradiction is an equation that has no solution. When solving an equation that is a contradiction, we arrive at a statement that is obviously false like $3 = 10$. The fact that there is no solution is usually indicated by using the symbol $\varnothing$ which stands for the empty set.

EXAMPLE 2

Determine whether each equation is an identity or a contradiction and indicate the solutions.

a) $3(x + 2) - 4 = x + 2x + 2$ Simplify each side.

$3x + 6 - 4 = 3x + 2$ Add like terms.

$3x + 2 = 3x + 2$ Subtract $3x$ from both sides.

$3x - 3x + 2 = 3x - 3x + 2$ Simplify both sides.

$2 = 2$ This is obviously a true statement. Therefore, this is an identity and the solutions are all real numbers.

b) $4x + 3(x + 5) = 2(2x - 3) + 3(x + 5)$ Simplify both sides.

$4x + 3x + 15 = 4x - 6 + 3x + 15$ Add like terms.

$7x + 15 = 7x + 9$ Subtract $7x$ from both sides.

$7x - 7x + 15 = 7x - 7x + 9$ Add like terms.

$15 = 9$ This is obviously a false statement. Therefore this is a contradiction and there are no solutions. Equivalently, the solution set is $\varnothing$.

PRACTICE EXERCISES

Determine whether each equation is an identity or a contradiction and indicate the solution or solution set.

6) $2(2x + 4) - 11 = 4(x - 2) + 5$ **7)** $4(x - 1) + 2(x - 3) = 4(x - 2) + 2x$

We continue practicing translating mathematical sentences to English and English sentences into mathematics.

EXAMPLE 3

Translate each of the following mathematical sentences to English sentences. There may be more than one translation.

Mathematics	English
a) $2x + 3 = 5$	Three more than twice a number equals five. Twice a number plus three is five. The sum of twice a number and three is five. Twice a number increased by three is five. Three more than the product of two and a number is five.
b) $4u - 12 = 10u$	Four times a number minus twelve equals ten times the number. Four times a number decreased by twelve is ten times the number. The difference between four times a number and twelve equals ten times the number. Twelve less than four times a number is ten times that number.
c) $8(10 - t) = 96$	Eight times the difference of ten and a number is ninety-six. The product of eight and the difference of ten and some number is ninety-six. Eight multiplied by the difference of ten and a number is ninety-six.
d) $1.5 = 0.75v - 0.5v$	One and five tenths equals seventy-five hundredths of a number decreased by five tenths of that number. One and five tenths is the difference of the product of seventy-five hundredths of a number and five tenths of that number. One and five tenths equals seventy-five hundredths of a number minus five tenths of that number.

ADDITIONAL PRACTICE

Translate each of the following mathematical sentences to English sentences. There may be more than one translation.

f) $15 - 5z = 10$

g) $2x - 5 = 4x$

h) $26 = 2(a - 4)$

i) $0.5u - 12 = 0.4$

PRACTICE EXERCISES

Translate each of the following mathematical sentences to English sentences. There is more than one translation.

8) $3w - 7 = 4$ **9)** $5x + 6 = 7x$

10) $12(m - 3) = 48$ **11)** $x + 0.06x = 24$

EXAMPLE 4

Translate the following English sentences to mathematical sentences and solve. Let x be the variable in each case.

English	Mathematics
a) Six more than four times a number equals ten.	$4x + 6 = 10$ $4x + 6 - 6 = 10 - 6$ $4x = 4$ $x = 1$

Note: The numbers are not always written in the same order in mathematics as in English. "Five subtracted from three times a number" is translated "$3x - 5$". Be careful!

b) Five subtracted from three times a number equals four times that number.	$3x - 5 = 4x$ $3x - 3x - 5 = 4x - 3x$ $-5 = x$
c) Twice the sum of some number and five is thirty.	$2(x + 5) = 30$ $2x + 2(5) = 30$ $2x + 10 = 30$ $2x + 10 - 10 = 30 - 10$ $2x = 20$ $x = 10$
d) The total of a number, seven tenths of the number, and negative two tenths of the number equals six tenths.	$x + 0.7x - 0.2x = 0.6$ $x + 0.5x = 0.6$ $1.5x = 0.6$ $x = 0.4$

ADDITIONAL PRACTICE

Translate the following English sentences to mathematical sentences and solve. Use x as the variable.

j) Five more than three times a number is negative four.

k) Six times the difference of a number and one is twenty-four.

l) Six less than eight times a number is fourteen.

m) The sum of six hundredths of a number and the number is seven and forty-two hundredths.

PRACTICE EXERCISES

Translate the following English sentences to mathematical sentences and solve. Use x as the variable.

12) Eight less than three times a number is sixteen.

13) Eighteen decreased by five times a number is equal to four times the number.

14) Four times the difference of a number and five is thirty.

15) Twenty-one equals a number decreased by three tenths of the number.

If you need more practice, do the Additional Practice Exercises in the margin.

Follow the same technique as in the previous two sections to solve the following.

EXAMPLE 5
Solve the following.

 a) An appliance repairman charges a service charge of $25 plus $18 per hour. If the charges for repairing a refrigerator are $79 (excluding parts), how many hours did it take?

Solution:

1) Identify the unknown and represent it with a variable. The unknown is the number of hours it took to repair the refrigerator, so let h represent the number of hours it took to repair the refrigerator.

2) Write the "word" equation.

service charge	plus	charges per hour	times	number of hours	equals	total charges

3) Write the algebraic equation.

service charge	plus	charges per hour	times	number of hours	equals	total charges
25	+	18	·	h	=	79

4) Solve the equation.

$25 + 18h = 79$ Subtract 25 from both sides.

$25 - 25 + 18h = 79 - 25$ Simplify both sides.

$18h = 54$ Divide both sides by 18.

$\dfrac{18h}{18} = \dfrac{54}{18}$ Simplify both sides.

$h = 3$ Therefore, it took 3 hours.

5) Check the answer in the wording of the problem.
   If the repairman works 3 hours, his charges would be $3(18) = \$54$ and $\$54 + \$25 = \$79$. Therefore, the answer is correct.

b) If Pascual had $10 more than he presently has, then twice that amount would be enough to purchase a compact disk player that sells for $280. How much does he presently have?

1) Identify the unknown and represent it with a variable. The unknown is the amount of money Pascual presently has, so let a represent the amount he presently has. Then $a + 10$ represents the amount he would have if he had $10 more.

2) Write a "word" equation.

two	times	$10 more than he presently has	equals	280

3) Write the algebraic equation.

two	times	$10 more than he presently has	equals	280
2	·	$(a + 10)$	=	280

4) Solve the equation.

$2(a + 10) = 280$ Apply the distributive property.

$2a + 20 = 280$ Subtract 20 from both sides.

$2a + 20 - 20 = 280 - 20$ Simplify both sides.

$2a = 260$ Divide both sides by 2.

$$\frac{2a}{2} = \frac{260}{2}$$

$$a = 130$$

Simplify both sides.

Therefore, Pascual presently has $130.

5) Check the answer in the wording of the problem.
If Pascual has $130 then he would have $140 if he had $10 more.
Since 2($140) = $280, the answer is correct.

PRACTICE EXERCISES

Solve the following.

16) A TV repair service charged $22 per hour plus $43 for parts. If the bill for repairing a TV is $76, how many hours did the repair take?

17) Carla earned $390 last week for 40 hours work. Included in the $390 was a $50 bonus. What was her hourly wage?

EXERCISE SET 3.3

Solve the following.

1) $3x + 5 = 11$

2) $5x + 7 = -8$

3) $-7x - 6 = 8$

4) $-3x + 5 = -10$

5) $17 = 4a - 3$

6) $-8 = 6x + 4$

7) $5x - 8 = 4x - 5$

8) $9y + 4 = 8y + 1$

9) $10 - 3u = 12 - 4u$

10) $3r + 5 = 2r + 4$

11) $7 - 9v = v - 3$

12) $4w + 8 = 15 - 6w$

13) $3n - 6 = 2n - 6$

14) $2m + 13 = 4 - 8m$

15) $9p + 3.4 = 8p + 1.4$

16) $0.4q + 0.7 = 0.8 - 0.6q$

17) $3v + 7 + 2v = 18 - 1$

18) $6w + 4 - 8w = -14 + 46$

19) $4k + 5 = 7k - 7 + 9k$

20) $-11b - 5 + 5b = 23 - 10$

21) $3a - 4a + 9 = -12a + 18 - 6a$

22) $17z - 11z - 17 = -4z + 13 - 5z$

23) $2.3y - 1.6 - 0.8y = 4.4$

24) $4.7 - 5.2c - 0.3c = -6.3$

25) $n + 5 = 3(n + 7)$

26) $2t - 9 = 3(t - 2)$

27) $2(5r - 2) = 9r + 1$

28) $2(w - 4) = 5w - 14$

29) $6u - 17 = 4(u + 3) - 3$

30) $2v + 9 = 11 + 7(v - 1)$

31) $14(w - 2) + 13 = 4w + 5$

32) $2x + 2(3x - 4) = -23 + 5x$

33) $2y + 13(y - 1) = 5y + 12$

34) $8(2z - 3) - z = 5z - 34$

35) $0.2(8x - 90) + 0.9x = 0.1x + 6$

36) $1.4u - 0.5(44 - 6u) = 66 + 2.4u$

37) $0.3n + 0.2(4n - 1) = 0.4 + 0.5n$

38) $0.3z + 0.7(2z + 1) = 2.9 - 0.5z$

39) $7(p - 3) = 3(1 - p) - 4$

40) $5(3q - 2) + 20 = 5(q - 4)$

41) $15(n + 2) = 5(n + 4) - 10$

42) $8(m - 5) = 7(m - 5) - 9$

43) $5(u - 2) + 7(3 - u) = 9u$

44) $5(h + 2) - 7(h - 1) = -h$

45) $4(b - 3) + 2(b + 2) = 2(b - 2)$

46) $3(v - 8) + 2(v + 5) = 7(v - 4)$

47) $5(a - 1) - 9(a - 2) = -3(2a + 1) - 2$

48) $4(k + 4) + 13(k - 1) = -3(k - 3) + 14$

49) $9 - 0.2(v - 8) = 0.3(6v - 8)$

50) $0.5(t - 6) + 0.4(1 - t) = -3.4$

Determine whether the following are identities or contradictions and indicate the solutions.

51) $2x + 4 = 2(x + 1) + 2$

52) $3(x + 1) - 2 = 3x + 5$

53) $3(2x - 5) - 4x = 2(x - 3) - 9$

54) $2(4x - 7) + 10 = 2(x - 5) + 2(3x + 3)$

55) $4(2x - 1) - 3(x + 5) = 5(x - 2) + 7$

56) $3(x - 7) - 4(x - 3) = 5 - (x + 3)$

CHALLENGE EXERCISES (57–58)

Solve the following.

57) $0.4(x + 2) = 0.3(x - 5) - 0.5(x - 3) + 2$

58) $1.8(x - 2) + 5(x - 5) + 0.8 = -0.4(8x + 7)$

Translate each of the following mathematical sentences into English sentences. There may be more than one translation.

59) $4w - 3 = 7$ **60)** $5x + 8 = 19$

61) $6v - 9 = 3v$ **62)** $16n - 8 = 6n$

63) $7y + 2.8 = 6.3$ **64)** $q + 0.05q = 45$

65) $12(y - 1) = 9$ **66)** $4(t + 3) = 36$

Translate the following English sentences to mathematical sentences and solve. Use x as the variable.

67) Eight less than three times a number equals thirteen.

68) One is five more than twice a number.

69) The difference of five times a number and six is nine.

70) The sum of three times a number and seven is four.

71) Zero is equal to nine times the sum of a number and three.

72) The product of four and the difference of a number and five is three.

73) Four times the difference of three times a number and two is sixteen.

74) Seven times four more than a number is sixty-three.

Solve the following.

75) A plumber charges $30 for a service call and $45 per hour. Her charge for a repair job was $210 excluding parts. How many hours did the plumber work?

76) An appliance repairman charges $25 dollars for a service call plus his hourly wage. If a repair job costs $109 (excluding parts) and took 3 hours, what is his hourly wage?

77) An auto repair service charges $35 per hour plus parts. If the parts to repair a transmission cost $160 and the total bill was $335, how many hours did the job take?

78) A TV repair shop charges an hourly rate plus the cost of parts. Find the hourly rate for a job that took 1.5 hours if the parts cost $65 and the bill was $113.

79) A sales clerk is paid $25 per day plus 3% commission on all her sales. If Yolanda's total wages for Monday were $70, find the amount of her sales.

80) A performer charges $1000 plus 10% of the gate receipts. She earned $16,000 for her last performance. What were the gate receipts?

81) If Yo Chen had $15 more than she presently has, then with three times that amount she could buy a camera which costs $525. How much does she presently have?

82) Five times $20 less than Bryan presently has is enough to buy a new bike that costs $300. How much does Bryan presently have?

83) Frank earned $395 last week for working 40 hours. Included in the $395 was a bonus of $25. Find his hourly wage.

84) Susie earned $277 for working 28 hours. Included in the $277 was a $60 bonus. Find her hourly wage.

85) For a particular copy machine it costs 15 cents for the first copy and 4 cents for each additional copy. How many copies can be made for $4.95?

86) Joan finds a coupon in the paper for $25 off the cost of repairs at Joe's Garage. If her bill was $147.50 including $60 for parts and 3 hours of labor, how much did Joe charge per hour for labor?

87) A local department store offers $15 off on purchases totaling $100 or more. Ahmed purchases 4 sports coats each costing the same and a silk tie that costs $28. Find the cost of one sports coat if the bill was $355.

88) Francois received a raise of 4% of her present salary. If her new salary is $35,980, what is her present salary?

89) The sales tax in Orlando is 6%. If a college cafeteria in Orlando wants the cost of the lunch special to be $4.24 including tax, what should they charge for the lunch special?

90) Hernando earns $1\frac{1}{4}$ times his hourly wage for all hours worked in excess of 40 hours per week. Last week he worked 46 hours and earned $391.40. What is his hourly wage?

91) Al's company has a limit of $100 per meal for entertaining a client. If there is a 7% sales tax and he leaves a 13% tip, find, to the nearest cent, the maximum amount that he can spend on a meal?

92) A country club offers two types of membership. Plan A charges $500 per year plus $15 per round of golf and plan B charges $675 per year plus $10 per round of golf. How many rounds of golf would have to be played per year for the charges under plans A and B to be equal?

93) Saline solution has 1 part salt to 8 parts water by weight. Find the number of ounces of salt and the number of ounces of water in 108 ounces of the solution.

WRITING EXERCISES

94) When solving an equation of the form $ax + b = c$, why do we eliminate b before eliminating a?

95) How does simplifying the expression $2(x - 3) + 4(x - 6)$ differ from solving the equation $2(x - 3) + 4(x - 6) = 0$?

WRITING EXERCISE OR GROUP PROJECT

If done as a group project, write two applications problems, exchange with another group, and then solve.

96) Write and solve an application problem with equation of the form $ax + b = c$.

S E C T I O N 3.4

Solving Linear Inequalities

OBJECTIVES

When you complete this section, you will be able to:

Ⓐ Solve linear inequalities using the Addition Property of Inequality and graph the solutions.

Ⓑ Solve linear inequalities using the Multiplication Property of Inequality and graph the solutions.

Ⓒ Solve linear inequalities using both the Addition and Multiplication Properties of Inequality and graph the solutions.

INTRODUCTION

In Sections 3.1–3.3 we learned to solve linear equations. In this section we will solve linear inequalities. Before we discuss solving linear inequalities, we will expand on the concept of inequality and how inequalities can be represented on the number line.

The number line represents the integers and all of the numbers between the integers. The numbers between the integers are the rational numbers (to be discussed in Chapter 5) and the irrational numbers (to be discussed in Chapter 8)

Choose a number, say 3. We put a dot on 3 on the number line. Consider some unknown number x. There are three possibilities for the position of x in relation to 3 on the number line.

1) x is to the right of 3 so $x > 3$, or

2) x is right on top of 3 so $x = 3$, or

3) x is to the left of 3 so $x < 3$.

The number line has been divided into three parts, $x > 3$, $x = 3$, and $x < 3$. Expressions like $x > 3$ and $x < 3$ are called inequalities. Those numbers that make an inequality true are called solutions of the inequality. One convenient method of representing the solutions of an inequality is by graphing the solutions on the number line. The expression $x \geq 3$ means $x > 3$ or $x = 3$. The graph of $x = 3$ is a dot on 3 and the graph of $x > 3$ is everything to the right of 3. So the graph of $x \geq 3$ is a dot on 3 and shading on everything to the right. An open dot is used to indicate that the point is not a part of the solution. So the graph of $x > 3$ is an open dot on 3 and shading on everything to the right.

EXAMPLE 1

Graph each of the following on the number line.

a) $x > 5$

Since any number greater than 5 is to the right of 5 on the number line, we put an open dot on 5 (to show that 5 is not a solution) and shade the number line to the right of 5. The shaded region represents all values of x such that $x > 5$ and the arrow indicates that it extends to the right forever.

b) $x < 5$

The numbers less than 5 are to the left of 5 on the number line. We put an open dot on 5 (to show that 5 is not a solution) and then shade to the left of 5 on the number line. The shaded region extends forever as indicated by the arrow.

c) $x \geq -4$

The numbers greater than -4 are to the right of -4 on the number line. We put a solid dot on -4 (to indicate that -4 is a solution) and shade the number line to the right of -4. The arrow indicates the region extends forever.

PRACTICE EXERCISES

Graph the following on the number line.

1) $x \geq -3$

2) $x < -1$

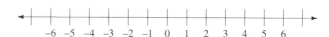

3) $x \leq -3$

4) $x \geq 1$

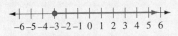

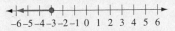
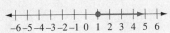
Inequalities of the form $a < x < b$ are called **compound inequalities**. The compound inequality $2 < x < 5$ is read "two is less than x and x is less than 5." On the number line, 2 is to the left of x and x is to the left of 5. Consequently, x is any number between 2 and 5. The graph of this compound inequality is shown below.

The open dots indicate that neither 2 nor 5 are solutions.

EXAMPLE 2

Graph the following on the number line.

 a) $0 \le x < 5$

Put a solid dot on 0 (to indicate that 0 is a solution) and an open dot on 5 (to indicate that 5 is not a solution). Since 0 is to the left of x and x is to the left of 5, x can be any number between 0 and 5. Therefore, shade the region between 0 and 5.

b) $-5 \le x \le 4$

Put a solid dot on both -5 and 4 to indicate they are solutions. Since -5 is to the left of x and x is to the left of 4, x can be any number between -5 and 4. Therefore, shade the region between -5 and 4.

PRACTICE EXERCISES

Graph the following.

5) $-5 < x < 0$

6) $3 < x < 6$

Now we are ready to discuss solving linear inequalities. To solve linear equations we established the Addition and Multiplication Properties of Equality. To solve linear inequalities we will need to establish similar properties for inequalities.

EXAMPLE 3

a) $3 < 8$ Add 5 to both sides.

$3 + 5 \overset{?}{<} 8 + 5$ Simplify both sides.

$8 < 13$ This is a true statement, so adding 5 to both sides of the inequality had no effect on the order symbol.

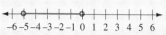
b) $9 > 7$ Subtract 3 from both sides. (Same as adding -3.)

  $9 - 3 \overset{?}{>} 7 - 3$ Simplify both sides.

  $6 > 4$ This is a true statement, so subtracting 3 from both sides of the inequality had no effect on the order symbol.

c) $-12 < -2$ Add 8 to both sides.

  $-12 + 8 \overset{?}{<} -2 + 8$ Simplify both sides.

  $-4 < 6$ This is a true statement, so adding 8 to both sides of the inequality had no effect on the order symbol.

Example 3 leads us to the following statement of the Addition Property of Inequality.

ADDITION PROPERTY OF INEQUALITY

For all real numbers a, b, and c, if $a < b$, then $a + c < b + c$ and if $a > b$, then $a + c > b + c$. In words, any real number may be added to (or subtracted from) both sides of an inequality without affecting the order.

The addition property is also true for the inequalities "$\geq$" and "$\leq$".

As in solving equations with the Addition Property of Equality, we wish to get the variable by itself on one side of the inequality and the constant on the other side. It is often easier to understand the solutions of an inequality if we graph them, so we will graph the solutions of each of the following examples.

EXAMPLE 4
Solve the following and graph the solutions.

a) $x - 6 < 3$ Add 6 to both sides. Adding the same quantity to both sides of an inequality does not change the order symbol.

  $x - 6 + 6 < 3 + 6$ Simplify both sides.

  $x < 9$ Therefore, the solutions are all numbers less than 9.

-2 -1 0 1 2 3 4 5 6 7 8 9 10 11

Unfortunately, there is no easy way to check the solution because there are an infinite number of values of x that make the inequality true. However, we may choose some values of x less than 9 and some greater than 9 and substitute them into the original inequality to get an indication of whether our solution is correct.

Let $x = 8$. From $x - 6 < 3$, we have $8 - 6 < 3$ or $2 < 3$. This is a true statement, so $x = 8$ is a solution and $8 < 9$.

Let $x = 10$. Now we have $10 - 6 < 3$, or $4 < 3$. This not true, so $x = 10$ is not a solution and $10 > 9$.

Therefore, values less than 9 seem to make the inequality true and values greater than 9 seem to make it false. Hence, the solution is reasonable.

b) $2(y - 3) \geq y - 3$ Apply the distributive property.

  $2y - 6 \geq y - 3$ Subtract y from both sides. Subtracting the same quantity from both sides does not change the order symbol.

$2y - y - 6 \geq y - y - 3$ Simplify both sides.

$y - 6 \geq -3$ Add 6 to both sides. Adding the same quantity to both sides of an inequality does not change the order symbol.

$y - 6 + 6 \geq -3 + 6$ Simplify both sides.

$y \geq 3$ Therefore, the solutions are all numbers greater than or equal to 3.

c) $-2(2x - 4) + 5(x + 3) > 11$ Apply the distributive property.

$-4x + 8 + 5x + 15 > 11$ Simplify the left side.

$x + 23 > 11$ Subtract 23 from both sides. Subtracting the same quantity from both sides of an inequality does not change the order symbol.

$x + 23 - 23 > 11 - 23$ Simplify both sides.

$x > -12$ Therefore, the solutions are all numbers greater than -12.

ADDITIONAL PRACTICE

Solve the following inequalities and graph the solutions.

a) $a - 8 + 2 < 0$

b) $6(u + 2) - 5u < 9$

c) $4(4z - 2) - 5(3z - 1) \leq 4$

PRACTICE EXERCISES

Solve the following inequalities and graph the solutions.

7) $t + 8 - 5 > 1$ **8)** $3(s - 2) \leq 2s + 3$

9) $5(2t - 3) - 3(3t - 4) > 2$

If you need more practice, do the Additional Practice Exercises in the margin.

The Addition Property of Inequality cannot be used to solve all linear inequalities just as the Addition Property of Equality alone will not solve all linear equations. In order to solve an inequality like $3x \leq 9$, we need to divide both sides by 3. What effect, if any, will this have on the order symbol? We investigate the effect of multiplication and division in the following examples.

EXAMPLE 5

a) $4 < 9$ Multiply both sides by 2.

$2(4) \overset{?}{<} 2(9)$ Simplify both sides.

$8 < 18$ This is a true statement, so multiplying both sides by 2 had no effect on the order symbol.

b) $5 < 7$ Multiply both sides by -3.

$(-3)5 \overset{?}{<} (-3)7$ Simplify both sides.

$-15 < -21$ This is a false statement. In order to make it true, the $<$ symbol must be changed to a $>$ symbol.

$-15 > -21$ True statement.

7) $t > -2$

8) $s \leq 9$

9) $t > 5$

Additional Practice a–c

a) $a < 6$

b) $u < -3$

c) $z \leq 7$

c) $6 > -1$ Multiply both sides by 4.

   $4(6) \overset{?}{>} 4(-1)$ Simplify both sides.

   $24 > -4$ This is a true statement, so multiplying both sides by 4 had no effect on the order symbol.

d) $8 \geq -9$ Multiply both sides by -5.

   $8(-5) \overset{?}{\geq} -9(-5)$ Simplify both sides.

   $-40 \geq 45$ This is a false statement. In order to make it true, the $\geq$ symbol must be changed to a $\leq$ symbol.

   $-40 \leq 45$ True statement.

      Example 5 suggests the following. If we multiply an inequality by a positive number as in Examples 5a and 5c, the inequality symbol remains unchanged. However, if we multiply an inequality by a negative number as in Examples 5b and 5d, the inequality symbol changes (reverses). In other words, the side that was greater became less and the side that was less became greater. When the order symbol is unchanged we sometimes say the sense of the inequality was not changed and when the order symbol is changed we say the sense of the inequality is reversed. These results are formally stated in the following properties.

MULTIPLICATION PROPERTIES OF INEQUALITY

Positive Factor

For all real numbers a, b, and c, if $a < b$ and $c > 0$, then $a \cdot c < b \cdot c$, and if $a > b$ and $c > 0$, then $a \cdot c > b \cdot c$. In words, if both sides of an inequality are multiplied (or divided) by a positive number, then the order symbol is unchanged. (sense unchanged)

Negative Factor

For all real numbers a, b, and c, if $a < b$ and $c < 0$, then $a \cdot c > b \cdot c$, and if $a > b$ and $c < 0$, then $a \cdot c < b \cdot c$. In words, if both sides of an inequality are multiplied (or divided) by a negative number, then the order symbol must be reversed. (sense reversed)

      The important thing to remember is that the direction of the inequality changes when we multiply (or divide) both sides of the inequality by a negative number. Otherwise, solving linear inequalities is exactly the same as solving linear equations.

      Just as in the addition properties, the multiplication properties are true for the inequalities "$\geq$" and "$\leq$", as well as for "$>$" and "$<$".

EXAMPLE 6

Solve and graph the solution of each of the following.

a) $4x \geq 20$ Divide both sides by 4. Dividing by a positive number does not change the direction of the inequality symbol.

   $\dfrac{4x}{4} \geq \dfrac{20}{4}$ Simplify both sides.

$x \geq 5$ Therefore, the solutions are all numbers greater than or equal to 5.

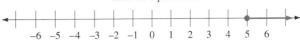

b) $-6y > 18$ Divide both sides by -6. Dividing by a negative number reverses the direction of the order symbol.

$\dfrac{-6y}{-6} < \dfrac{18}{-6}$ Simplify both sides.

$y < -3$ Therefore, the solutions are all numbers less than -3.

c) $\dfrac{x}{-2} \geq 1$ Multiply both sides by -2. Multiplying by a negative number reverses the direction of the order symbol.

$-2 \cdot \dfrac{x}{-2} \leq -2 \cdot 1$ Simplify both sides.

$x \leq -2$ Therefore, the solutions are all numbers less than or equal to -2.

ADDITIONAL PRACTICE

Solve and graph each of the following.

d) $12 \geq 4v$

e) $-3r \geq 27$

f) $\dfrac{x}{-4} \leq -3$

PRACTICE EXERCISES

Solve and graph each of the following.

10) $2z > 4$ **11)** $-5n < 35$ **12)** $\dfrac{u}{-3} > 2$

If more practice is needed, do the Additional Practice Exercises in the margin.

To this point we have discussed linear inequalities that require the use of only one of the properties of inequality. The following examples require both the Addition and Multiplication Properties of Inequality.

EXAMPLE 7

Solve each of the following and graph the solutions.

a) $3x + 5 > x - 3$ Subtract x from both sides. Subtracting from both sides does not affect the order symbol.

$3x - x + 5 > x - x - 3$ Simplify both sides.

$2x + 5 > -3$ Subtract 5 from both sides. Subtracting from both sides does not affect the order symbol.

$2x + 5 - 5 > -3 - 5$ Simplify both sides.

$2x > -8$	Divide both sides by 2. Dividing by a positive number does not affect the order symbol.
$\dfrac{2x}{2} > \dfrac{-8}{2}$	Simplify both sides.
$x > -4$	Therefore, the solutions are all numbers greater than -4.

b) $16x + 8 - 4x > -16 + 10x$ Simplify both sides.

$12x + 8 > -16 + 10x$	Subtract $12x$ from both sides. Subtracting from both sides does not affect the order symbol.
$12x - 12x + 8 > -16 + 10x - 12x$	Simplify both sides.
$8 > -16 - 2x$	Add 16 to both sides. Adding to both sides does not affect the order symbol.
$8 + 16 > -16 + 16 - 2x$	Simplify both sides.
$24 > -2x$	Divide both sides by -2. Dividing by a negative number reverses the order symbol.
$\dfrac{24}{-2} < \dfrac{-2x}{-2}$	Simplify both sides.
$-12 < x$ or $x > -12$	Therefore, the solutions are all numbers greater than -12.

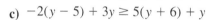

c) $-2(y - 5) + 3y \geq 5(y + 6) + y$ Distribute -2 and 5.

$-2y + 10 + 3y \geq 5y + 30 + y$	Simplify both sides.
$10 + y \geq 6y + 30$	Subtract $6y$ from both sides. Subtracting from both sides does not affect the order symbol.
$10 + y - 6y \geq 6y - 6y + 30$	Simplify both sides.
$10 - 5y \geq 30$	Subtract 10 from both sides. Subtracting from both sides does not affect the order symbol.
$10 - 10 - 5y \geq 30 - 10$	Simplify both sides.
$-5y \geq 20$	Divide both sides by -5. Dividing by a negative number reverses the order symbol.
$\dfrac{-5y}{-5} \leq \dfrac{20}{-5}$	Simplify both sides.
$y \leq -4$	Therefore, the solutions are all numbers less than or equal to -4.

ADDITIONAL PRACTICE

Solve the following.

g) $2t - 6 > 14 - 3t$

h) $6 - 2x \le -4 + 3x$

i) $8z - 3 + 2z \le 10z - 10 - 7z$

j) $7(-8 + u) - 2u \ge -14 + 3(2u - 15)$

PRACTICE EXERCISES

Solve the following and graph the solutions.

13) $5x - 3 > x - 3$

14) $4 - 3x \le -8 + x$

15) $-4t + 7 + 16t > 10t + 12$

16) $20r - 5(6 + 2r) \le 4r + 4(r - 7)$

If you need more practice, do the Additional Practice Exercises in the margin.

Compound inequalities are solved using the same techniques as other linear inequalities. The variable is located between the inequality symbols and our task is to eliminate the constants so the variable will be isolated between the inequality symbols. It is important to remember that whatever you do to the middle part to isolate the variable must be done to all three parts of the inequality.

EXAMPLE 8

Solve each of the following inequalities.

a) $-3 \le x + 2 \le 3$
Subtract 2 from all three expressions. Subtracting does not affect the order symbol.

$-3 - 2 \le x + 2 - 2 \le 3 - 2$
Simplify all three expressions.

$-5 \le x \le 1$
Therefore, the solutions are all numbers greater than or equal to -5 and less than or equal to 1.

b) $1 < 6x + 7 \le 31$
Subtract 7 from all three expressions. Subtracting does not affect the order symbol.

$1 - 7 < 6x + 7 - 7 \le 31 - 7$
Simplify all expressions.

$-6 < 6x \le 24$
Divide each expression by 6. Dividing by a positive number does not affect the order symbols.

$\dfrac{-6}{6} < \dfrac{6x}{6} \le \dfrac{24}{6}$
Simplify all expressions.

$-1 < x \le 4$
Therefore, the solutions are all numbers greater than -1 and less than or equal to 4.

c) $0 \le -1.5y - 3 < 4.5$
Add 3 to all three expressions. Adding does not affect the order symbols.

$0 + 3 \le -1.5y - 3 + 3 < 4.5 + 3$
Simplify all expressions.

$3 \le -1.5y < 7.5$
Divide each expression by -1.5. Dividing by a negative number reverses the order symbols.

ADDITIONAL PRACTICE

Solve the following inequalities.

k) $3 < x + 5 < 7$

l) $-8 < 3v + 4 \leq 19$

m) $-2 < 0.5x - 2.5 < -1$

$$\frac{3}{-1.5} \geq \frac{-1.5y}{-1.5} > \frac{7.5}{-1.5} \qquad \text{Simplify all expressions.}$$

$$-2 \geq y > -5 \text{ or } -5 < y \leq -2 \qquad \begin{array}{l}\text{Therefore, the solutions are all} \\ \text{numbers less than or equal to } -2 \text{ and} \\ \text{greater than } -5.\end{array}$$

PRACTICE EXERCISES

Solve the following inequalities.

17) $-4 \leq x - 3 \leq 2$ **18)** $-13 \leq -4t - 5 < 15$ **19)** $3 < 0.2w + 3.6 < 3.6$

If more practice is needed, do the Additional Practice Exercises in the margin.

Problems are often posed in terms of inequalities as well as equations. Therefore, we need to be able to translate these relationships as well.

EXAMPLE 9

Translate each of the following from mathematics to English. There can be more than one translation.

Mathematics	English
a) $4y > 12$	Four times a number is greater than twelve.
	The product of 4 and a number is more than twelve.
b) $2x - 5 \leq 20$	Twice a number, decreased by five is less than or equal to twenty.
	Twice a number, decreased by five is at most twenty.
	The difference of twice a number and five is less than or equal to twenty.
	Five less than twice a number is at most 20.
c) $3(u + 6) < -9$	Three times the sum of six and a number is less than negative nine.
	The product of three and the sum of a number and six is less than negative nine.
d) $5(2v - 4) + 3 \geq 8$	Three more than five times the difference of twice a number and four is greater than or equal to eight.
	Three more than the product of 5 and the difference of two times a number and 4 is at least eight.

PRACTICE EXERCISES

Translate from mathematical sentences to English sentences.

20) $8x < 4$ **21)** $15 > 3t + 6$

22) $2(x - 4) > 3$ **23)** $4(3x + 2) - 5 \leq 7$

EXAMPLE 10

Translate from English to mathematics and solve. There is only one mathematical translation for each English sentence. Use x as the variable.

English	Mathematics
a) Six times a number is greater than eighteen.	$6x > 18$ $x > 3$
b) Five more than negative three times a number is less than seventeen.	$-3x + 5 < 17$ $-3x + 5 - 5 < 17 - 5$ $-3x < 12$ $\dfrac{-3x}{-3} > \dfrac{12}{-3}$ $x > -4$
c) Seven times the difference of a number and four is at most thirty-five.	$7(x - 4) \le 35$ $7x - 28 \le 35$ $7x - 28 + 28 \le 35 + 28$ $7x \le 63$ $\dfrac{7x}{7} \le \dfrac{63}{7}$ $x \le 9$
d) Five more than three times the difference of a number and four is at least eleven.	$3(x - 4) + 5 \ge 11$ $3x - 12 + 5 \ge 11$ $3x - 7 \ge 11$ $3x - 7 + 7 \ge 11 + 7$ $3x \ge 18$ $\dfrac{3x}{3} \ge \dfrac{18}{3}$ $x \ge 6$

PRACTICE EXERCISES

Translate from English to mathematical sentences and solve. Use x as the variable.

24) The sum of twice a number and nine is less than seventeen.

25) The difference of twelve and four times a number is at most twenty.

26) The difference of seven times a number and fourteen is at least zero.

27) Sixty-three is greater than three times the sum of four times a number and one.

EXERCISE SET 3.4

Graph each of the following.

1) $x > -2$

2) $x < -3$

3) $x > 4$

4) $x > -6$

5) $x \geq 3$

6) $-2 \leq x$

7) $1 < x < 6$

8) $-3 < x < 2$

9) $-5 < x < 3$

10) $-6 < x < -1$

11) $-4 \leq x < -3$

12) $2 < x \leq 6$

Solve by using the addition property of inequality and graph the solutions.

13) $q + 6 < 4$

14) $v - 8 < 5$

15) $6p - 5 - 5p \leq -3$

16) $4w - 2 - 3w \geq 6$

17) $3p - 6 - 2p < -8$

18) $5q + 30 - 4q \leq 14$

19) $3(s + 2) - 2s \geq 7$

20) $4(s - 2) - 3s \leq -3$

21) $4(2x - 5) - 7x + 16 > -2$

22) $2(3y - 5) - 5y + 6 < 2$

23) $8(2u - 3) - 5(3u + 4) > -3$

24) $7(3x - 6) - 4(5x - 6) > -10$

25) $-3(9t - 3) + 4(7t - 2) \leq 3 - 5$

26) $-4(8t + 5) + 3(11t + 5) \leq 4 - 7$

Solve by using the multiplication properties of inequality.

27) $3y < 15$

28) $4t \geq 32$

29) $-14u > 42$

30) $16w \leq -64$

31) $-2a < -14$

32) $-6n \geq -5.4$

33) $\dfrac{x}{-3} < 5$

34) $\dfrac{u}{-2} \geq 4$

35) $\dfrac{r}{-1} > -2$

36) $\dfrac{b}{-6} < -1$

Solve by using the addition and multiplication properties of inequality.

37) $2q - 16 > 8$

38) $3v - 13 < 6$

39) $-4a + 5 \leq -7$

40) $-5r + 2 \geq -8$

41) $8p - 15 \geq 5p$

42) $5v + 4 \leq 2v$

43) $7w + 6 > 4w - 18$

44) $11z - 6 > 4z + 8$

45) $3a - 1.6 \leq 0.8 + a$

46) $4b - 2.4 \geq 9b + 0.6$

47) $12 - 0.3t > -1.1t + 36$

48) $6.8 - 7r \geq 3r - 0.2$

49) $10s - 2 - 12s < 8s + 2 - 6s$

50) $7z + 16 - 4z > 10 - z + 6$

51) $12p - 9p + 12 \geq 7p - 15 + 5p$

52) $15q + 9 - 6q > 4q + 27 + 11q$

53) $7(w - 1) \geq -12$

54) $9(y + 4) \leq 0$

55) $5(2m - 3) > m + 21$

56) $4(3n + 6) < 9n - 3$

57) $x + 6(x - 3) \leq -25$

58) $2(4w - 5) - 3w < 18$

59) $7u - 10(2u + 3) \geq 22$

60) $12(t - 5) - 6t > 3t$

61) $2(1 - 4u) < -3u + 7$

62) $9v - 24 \geq 6(5 - 3v)$

63) $7(n - 2) > 3(n + 4) - 26$

64) $5(m + 4) < 28 - 4(m + 2)$

65) $8(3z - 1) + 15 \leq 11z + 2(5 - 4z)$

66) $10s - 3(4 - 5s) > s + 6(4 + s)$

67) $3(r + 0.9) + 0.2r > 3r + 1.7$

68) $0.7(t - 3) + 0.8 < 1.5 + 0.3t$

69) $-2 < x - 6 < 4$ 70) $6 < y + 7 < 12$

71) $-11 \leq 4x - 3 < 9$ 72) $-3 < 5y - 8 \leq 12$

73) $-3.5 < 2t - 1.5 < 8.5$ 74) $-1 \leq 10s - 3 \leq 5$

CHALLENGE EXERCISES (75-76)

75) $6(3y - 5) + 4(2y - 7) \geq 8y - 13 + 9y$

76) $-5(3 - 2x) + 9(3x - 6) < 7(4x - 1) + 1$

Translate the following mathematical sentences to English sentences in at least two ways.

77) $5x \geq 10$ 78) $9 > 2y$

79) $3s - 5 > 2$ 80) $15 + 4z < 12$

81) $28 \leq 4(t - 2)$ 82) $31 \geq 9(2 - u)$

Translate the following English sentences to mathematical sentences and solve. Use x as the variable.

83) Six times some number is less than eighteen.

84) Thirty-six is at least negative three times a number.

85) Nineteen decreased by four times a number is greater than eighteen.

86) Eight more than five times a number is at most twelve.

87) Three times the sum of twice a number and seven is at least negative three.

88) Seventeen is greater than or equal to twice a number minus five.

89) Four times a number, decreased by six is less than eighteen.

90) Twenty-two is less than or equal to five times a number, plus seven.

WRITING EXERCISES

91) List at least five situations in which we use inequalities, but do not specifically say the inequality. For example, "You must be twenty-one to vote" means your age must be greater than or equal to twenty-one years.

92) Compare solving a linear equation with solving a linear inequality. How are they similar? How do they differ?

93) How do the solutions of linear equations compare with the solutions of linear inequalities?

CRITICAL THINKING

94) Recall that for any two numbers x and a, either $x < a$, $x = a$, or $x > a$. With these in mind, answer the following questions.
a) What does "not greater than" mean?
b) What does "not equal to" mean?
c) What does "not less than or equal to" mean?
d) What does "at least" mean?

SECTION 3.5

Percent Equations

OBJECTIVES

When you complete this section, you will be able to:

Ⓐ Translate mathematical expressions involving percent to English.

Ⓑ Translate English statements into equations involving percents and solve.

Ⓒ Solve real-world percent problems.

INTRODUCTION

The applications of percent are numerous. Percent is used in business, education, economics, medicine, and in many other areas.

We begin this section by translating equations involving percent into verbal descriptions. This will aid us when we translate English phrases into equations which we will then solve.

• **Translation Percent to English**

In translating from mathematics to English, we use the following translations: multiplication symbol "·" translates as "of," the equal sign "=" translates as "is," and the variable translates as "what number," "some number," or "a number."

EXAMPLE 1

Translate each percent equation into two English sentences.

a) $75\% \cdot 80 = x$

Solution:

75% of 80 is what number?

What number is 75% of 80?

The symbol "·" becomes "of," "=" becomes "is," and "x" becomes "what number." The sentence may be turned around and written as:

b) $150\% \cdot x = 18$

Solution:

150% of what number is 18?

18 is 150% of what number?

The symbol "·" becomes "of," "x" becomes "what number," "=" becomes "is." The sentence may be turned around and written as:

c) $x\% \cdot 40 = 14$

Solution:

What percent of 40 is 14?

14 is what percent of 40?

$x\%$ becomes "what percent," "·" becomes "of," and "=" becomes "is." The sentence may be turned around and written as:

ADDITIONAL PRACTICE

Translate each percent equation into two English sentences.

a) $x = 85\% \cdot 400$

b) $150 = 250\% \cdot y$

c) $t\% \cdot 75 = 45$

PRACTICE EXERCISES

Translate the following percent equations into two English sentences.

1) $80\% \cdot 60 = x$ **2)** $75\% \cdot y = 30$ **3)** $60 = t\% \cdot 500$

If more practice is needed, do the Additional Practice Exercises in the margin.

We will now translate from English to mathematics and solve the resulting equation. Remember, in order to perform computations on percents, we must first convert the percent to a decimal. Since percent means hundredths, to convert a percent to a decimal means to write the percent as hundredths without the percent sign. This is done by writing the percent as a fraction whose denominator is one-hundred. If a number is divided by one-hundred, it moves the decimal two places to the left. So, to write a percent as a decimal, move the decimal point two places to the left and drop the percent sign. For example, $49\% = \frac{49}{100} = 0.49$, $150\% = \frac{150}{100} = 1.50$, and $3\% = 0.03$. Conversely, to change a decimal to a percent move the decimal two places to the right and add the percent sign. For example, $0.15 = 15\%$, $3 = 300\%$ and $0.6 = 60\%$.

ANSWERS:
Practice 1–3

1) 80% of 60 is what number?
   What is 80% of 60?

2) 75% of what number is 30?
   30 is 75% of what number?

3) 60 is what percent of 500?
   What percent of 500 is 60?

Additional Practice a–c

a) What number is 85% of 400?
   85% of 400 is what number?

b) 150 is 250% of what number?
   250% of what number is 150?

c) What % of 75 is 45? 45 is
   what % of 75?

EXAMPLE 2

Solve the following.

a) 4% of 1500 is what number?

Solution:

4% of 1500 is what number?	Translate.
$4\% \cdot 1500 = y$.	Rewrite 4% as 0.04.
$0.04(1500) = y$	Multiply.
$60 = y$ or $y = 60$	Hence, 4% of 1500 is 60.

 b) 90% of what number is 63?

Solution:

90% of what number is 63?	Translate.
$90\% \cdot y = 63$	Rewrite 90% as 0.90.
$0.90(y) = 63$	Divide both sides by 0.90.
$\dfrac{0.90y}{0.90} = \dfrac{63}{0.90}$	Simplify.
$y = 70$	Hence, 90% of 70 is 63.

c) What percent of 40 is 8?

Solution:

What percent of 40 is 8?	Translate.
$y\% \cdot 40 = 8$	Divide both sides by 40.
$\dfrac{y\% \cdot 40}{40} = \dfrac{8}{40}$	Simplify both sides.
$y\% = \dfrac{1}{5}$	Rewrite $\frac{1}{5}$ as a decimal (Divide 5 into 1).
$y\% = 0.2$	Rewrite 0.2 as a percent.
$y\% = 20\%$	Drop the percent signs.
$y = 20$	Therefore, 20% of 40 is 8.

ADDITIONAL PRACTICE

Solve the following equations.

d) What number is 20% of 150?

e) 45 is 90% of what number?

f) 54 is what percent of 300?

PRACTICE EXERCISES

Solve the following equations.

4) 10% of 46 is what number?

5) 180% of what number is 45?

6) What percent of 500 is 110?

If more practice is needed, do the Additional Practice Exercises in the margin.

The above percent equations fall into three categories: 1) $a\%$ of b is what number? $(a\% \cdot b = x)$ 2) $a\%$ of what number is b? $(a\% \cdot x = b)$ 3) What percent of a is b? $(x\% \cdot a = b)$ The first step in solving real-world problems involving percents is to first rewrite the problem as one of the above three cases. Then write the corresponding equation and solve.

EXAMPLE 3

Solve the following and round answers to the nearest percent or hundredth where appropriate.

a) Tanya correctly answered 123 questions out of 150 questions on the Arithmetic Final Exam. What percent of the questions did she answer correctly?

Solution:

The question that we are being asked is "What percent of 150 is 123?" Rewrite this as an equation.

$$P\% \cdot 150 = 123$$ Divide both sides by 150.

$$\frac{P\% \cdot 150}{150} = \frac{123}{150}$$ Simplify the left side.

$$P\% = \frac{123}{150}$$ Rewrite $\frac{123}{150}$ as a decimal (Divide 150 into 123).

$$P\% = 0.82$$ Change 0.82 into a percent.

$$P\% = 82\%$$ Drop the percent signs.

$$P = 82$$ Hence, Tanya answered 82% of 150 questions correctly.

b) This year Harold spent $250 on books. If this amount represents 5% of his student loan, how much was the loan?

Solution:

What we are being asked to find is "5% of what is $250?" Rewrite this as an equation.

$$5\% \cdot y = 250$$ Rewrite 5% as 0.05.

$$0.05y = 250$$ Divide both sides by 0.05.

$$\frac{0.05y}{0.05} = \frac{250}{0.05}$$ Divide.

$$y = \$5,000$$ Hence, Harold's student loan is $5,000.

c) A recent study showed that 12.1% of the women giving birth at a certain hospital had taken cocaine within the last 72 hours. If 32,000 women gave birth at this hospital last year, how many of them had taken cocaine within 72 hours of giving birth?

Solution:

What we are being asked is "What is 12.1% of 32,000?" Rewrite this as an equation.

$$12.1\% \cdot 32,000 = z$$ Rewrite 12.1% as 0.121.

$$0.121(32,000) = z$$ Multiply.

$$3872 = z$$ Hence, 3872 mothers had taken cocaine within 72 hours of giving birth.

ADDITIONAL PRACTICE

Solve and round answer to the nearest percent or hundredth.

g) Thirty-five out of 700 light bulbs failed after 1,000 hours of operation. What percent of the bulbs failed?

h) 60% of the students at Edgewater High School scored above 500 on the verbal portion of the SAT. If 330 students scored above 500, how many students took the test?

i) A defense contractor employed 15,000 people before the end of the cold war. Since the cold war ended, they have laid off 43% of their work force. How many were laid off?

PRACTICE EXERCISES

Solve and round answers to the nearest percent or hundredth where appropriate.

7) Out of every 5000 adults in the United States, 4000 do not exercise properly. What percent of the adults do not exercise properly?

8) A score of 70% on a test corresponds to 84 correctly answered test items. How many items were on the test?

9) A family must pay 22% of its income for income tax. How much is the tax if the family earned $65,000 that year?

If more practice is needed, do the Additional Practice Exercise in the margin.

EXERCISE SET 3.5

Translate the following percent equations into two English sentences.

1) $x = 25\% \cdot 36$

2) $50\% \cdot 32 = x$

3) $40\% \cdot 60 = v$

4) $v = 90\% \cdot 150$

5) $20\% \cdot y = 8$

6) $35 = 35\% \cdot y$

7) $52 = 65\% \cdot w$

8) $32\% \cdot w = 16$

9) $t\% \cdot 30 = 21$

10) $32 = t\% \cdot 20$

11) $4.8 = z\% \cdot 80$

12) $z\% \cdot 90 = 7.2$

Solve the following.

13) 5% of 200 is what number?

14) What number is 15% of 80?

15) What number is 80% of 35?

16) 60% of 60 is what number?

17) 60 is 75% of what number?

18) 50% of what number is 45?

19) 0.9% of what number is 54?

20) 18 is 37.5% of what number?

21) 24 is what percent of 80?

22) What percent of 90 is 32.4?

23) What percent of 96 is 12?

24) 30 is what percent of 24?

Solve each of the following and round the answers to the nearest percent or nearest hundredth where appropriate.

25) A recent survey indicated that 405 out of 500 women did not consider physical attraction an essential requirement for a spouse. This represents what percent of the women surveyed?

26) Nicky has read 198 pages of a book which has 660 pages. What percent of the book has she read?

27) Research shows that 63% of all reported traffic accidents involve men. If 1500 traffic accidents are reported, how many involve men?

28) In a University of California study, 58% of 17-year-old girls considered themselves overweight. If 600 girls were studied, how many considered themselves to be overweight? (Actually only 17% were overweight. Data came from *Time*, Special Issue, "Women: the Road Ahead," Fall 1990.)

29) By volume a certain wine is 14% alcohol. If a bottle contains 750 milliliters of wine, how many milliliters of alcohol is in the wine bottle?

30) A solution of salt and water is called a saline solution. One saline solution is 8% salt. How much salt is in 50 milliliters of solution?

31) In mixing salad dressing, we have 3 ounces of water and 2 ounces of olive oil. What percent of the salad dressing is olive oil?

32) If we heat metals to a liquid and combine them, we get a mixture called an alloy. If 18 pounds of copper is mixed with 12 pounds of tin, what percent of the alloy is copper?

33) A family paid $2,210 in real estate taxes last year. If this is a tax rate of 2.6%, how much is their house worth?

34) Out of 650 greeting cards sold at a certain card store, 585 were bought by women. What percent of the cards were bought by women?

35) The body of a male human is 63% water by weight. If a man's body contains 120 pounds of water, how much does he weigh to the nearest pound?

36) At a certain ranch, 2,475 cattle represent 45% of a herd of cattle. How many cattle are in the herd?

37) In 1996 a computer company laid off 1,400 employees which represented 7% of the work force. What was the size of the work force before the lay off?

38) In 1995 an airline laid off 300 employees which represented 10% of the workforce. What was the size of the workforce before the lay off?

CHALLENGE EXERCISES (39–42)

Solve the following.

39) $33\frac{1}{3}$% of 120 is what number?

40) What number is $6\frac{1}{2}$% of 900?

41) Vinegar is 5% acetic acid by volume. How much vinegar would 5 milliliters of acetic acid make?

42) Muriatic acid is a 2% solution of hydrochloric acid. How many gallons of muriatic acid can be made from 1 gallon of hydrochloric acid?

WRITING EXERCISE

43) Write down five household items which are mixtures and list their ingredients by percent.

GROUP PROJECT

44) Write two real-world applications problems of each of the three types studied in this section, exchange with another group, and then solve.

S E C T I O N 3.6

Traditional Applications Problems

OBJECTIVES

When you complete this section, you will be able to:

Ⓐ Solve real-world problems that are general in nature.

Ⓑ Solve real-world problems involving consecutive, consecutive odd, and consecutive even integers.

Ⓒ Solve real-world problems involving distance, time, and rate.

INTRODUCTION

This is the first of two sections of applications problems traditionally found in most algebra books. You will probably never encounter problems of most of these types in your everyday life. Their main purpose is to help you think mathematically and see the power of algebra in problem solving.

In previous sections of this chapter we solved real-world problems that involved only one unknown. In this section we will solve several types of applications problems, many of which involve two or more unknowns. If there are two or more unknowns, we modify our procedure slightly.

If there are two unknowns in a real-world problem, there must be two conditions in the problem: one that gives the relationship between the unknowns and one that gives the information necessary to write the equation.

Often there are special approaches that are used for particular types of real-world problems.

General Problems

EXAMPLE 1

Last week Alena earned $23 more than George at their part-time jobs. Together they earned $145. How much did Alena earn?

Solution:

1) Identify the unknown(s). We are asked to find how much Alena earned. However, we also do not know how much George earned. George earned less than Alena, so: Let n represent the amount that George earned.

2) Represent all other unknowns in terms of the same variable. What condition in the problem gives the relationship between the amounts that George and Alena earned? We are told that Alena earned $23 more than George. Since we let n represent George's total earnings, 23 more than n is $n + 23$, so $n + 23$ represents the amount that Alena earned.

3) Write a "word" equation. What condition gives us the information needed to write the equation? We are told that together they earned $145. Consequently,

| Amount Alena earned | + | Amount George earned | = | Amount earned together |

4) Write an algebraic equation using the "word" equation as a guide.

| Amount Alena earned | + | Amount George earned | = | Amount earned together |
| $(n + 23)$ | + | n | = | 145 |

5) Solve the equation.

$$(n + 23) + n = 145 \qquad \text{Remove the parentheses.}$$
$$n + 23 + n = 145 \qquad \text{Add like terms.}$$
$$2n + 23 = 145 \qquad \text{Subtract 23 from both sides.}$$
$$2n = 122 \qquad \text{Divide both sides by 2.}$$
$$n = 61 \qquad \text{Solution of the equation.}$$

6) Have we answered the question asked? We were asked to find the amount that Alena earned, but n represents the amount that George earned. The amount that Alena earned is $n + 23 = 61 + 23 = \$84$.

7) Check the answer in the wording of the problem. Is the amount that Alena earned \$23 more than George earned? Since \$84 is \$23 more than \$61, yes. Is the total of their earnings \$145? Since \$61 + \$84 = \$145, yes. Therefore, the answers are correct.

PRACTICE EXERCISE

1) Jude purchased a saw and a drill. If the total cost was \$238 and the drill cost \$54 less than the saw, find the cost of the saw.

Consecutive Integer Problems

Consecutive integers are integers that follow in order. For example, 1, 2, 3, 4, and so on, are consecutive integers. To get the next larger consecutive integer, we add 1 to the previous integer. For example, if we begin with 1, then $1 + 1 = 2$, $2 + 1 = 3$, $3 + 1 = 4$, and so on. Therefore, if we let $n =$ some integer, then the second integer is $n + 1$, the third is $(n + 1) + 1 = n + 2$, the fourth is $(n + 2) + 1 = n + 2 + 1 = n + 3$, and so on.

Consecutive odd integers are every other integer starting with an odd integer. For example, 1, 3, 5, 7, and so on, are consecutive odd integers. To get the next larger consecutive odd integer, we add 2 to the previous odd integer. Therefore, if we let $n =$ the first odd integer, then the second odd integer is $n + 2$, the third odd integer is $(n + 2) + 2 = n + 4$, and the fourth odd integer is $(n + 4) + 2 = n + 6$, and so on.

Consecutive even integers are every other integer starting with an even integer. For example, 2, 4, 6, 8, and so on, are consecutive even integers. Again, to get the next larger consecutive even integer, we add 2 to the previous even integer. Hence, if we let $n =$ the first even integer, then $n + 2$ is the second even integer, $n + 4$ is the third even integer, $n + 6$ is the fourth even integer, and so on.

Let us put all three types of integers together for comparison.

Integer	Consecutive	Consecutive Odd	Consecutive Even
First	n	n	n
Second	$n + 1$	$n + 2$	$n + 2$
Third	$n + 2$	$n + 4$	$n + 4$
Fourth	$n + 3$	$n + 6$	$n + 6$
so on	so on	so on	so on

Note: Do not be confused because both consecutive odd and consecutive even integers are represented the same way. The difference is in what you let the first integer be. If the first integer is odd, then that integer plus 2 is also odd. Likewise for even integers.

EXAMPLE 2

a) The sum of three consecutive odd integers is -9. Find the integers.

Solution:

1) Identify the unknown(s). We are asked to find three odd consecutive integers. Let n represent the smallest of the three consecutive odd integers.

2) Represent the other unknown(s) in terms of the same variable. What condition in the problem gives the relationship between the unknowns? The integers are consecutive odd. Consequently: $n + 2$ represents the second and $n + 4$ represents the third.

3) Write a "word" equation. What condition gives the information needed to write the equation? We are told that the sum of the three consecutive odd integers is -9. Thus,

$$\boxed{1^{st}\text{ integer}} \;+\; \boxed{2^{nd}\text{ integer}} \;+\; \boxed{3^{rd}\text{ integer}} \;=\; \boxed{-9}$$

4) Write an algebraic equation using the "word" equation as a guide.

$$\boxed{1^{st}\text{ integer}} \;+\; \boxed{2^{nd}\text{ integer}} \;+\; \boxed{3^{rd}\text{ integer}} \;=\; \boxed{-9}$$
$$\quad\; n \qquad\quad + \qquad (n+2) \qquad + \qquad (n+4) \qquad = \qquad -9$$

5) Solve the equation.

$n + (n + 2) + (n + 4) = -9$	Remove the parentheses.
$n + n + 2 + n + 4 = -9$	Add like terms.
$3n + 6 = -9$	Subtract 6 from both sides.
$3n = -15$	Divide both sides by 3.
$n = -5$	Solution of the equation.

6) Have we answered the question asked? No. We were asked to find three consecutive odd integers. Therefore, the complete solution is $n = -5$, $n + 2 = -3$, and $n + 4 = -1$.

7) Check the answers in the wording of the original problem. Are -1, -3, and -5 consecutive odd integers? Yes. Is the sum of -1, -3, and -5 equal to -9? Yes. Therefore, we have the correct solution.

b) Three times the larger of two consecutive even integers is 10 less than 5 times the smaller. Find the larger integer.

Solution:

1) Identify the unknown(s). We are looking for two consecutive even integers. Let n represent the smaller of the two consecutive even integers.

2) Represent all other unknowns in terms of the same variable. What condition in the problem gives the information needed to represent the other unknown in terms of n? We know the integers are consecutive even. Therefore: $n + 2$ represents the larger of the two consecutive even integers.

3) Write a "word" equation. What condition gives us the information needed to write the equation? We know that three times the larger is 10 less than 5 times the smaller. Thus,

$$\boxed{\text{three times the larger integer}} \;=\; \boxed{\text{10 less than 5 times the smaller integer}}$$

4) Write an algebraic equation using the "word" equation as a guide.

three times the larger integer	=	10 less than 5 times the smaller integer
$3(n + 2)$	=	$5n - 10$

5) Solve the equation.

$3(n + 2) = 5n - 10$	Distribute the 3 on the left side.
$3n + 6 = 5n - 10$	Subtract $3n$ from both sides.
$3n - 3n + 6 = 5n - 3n - 10$	Simplify both sides.
$6 = 2n - 10$	Add 10 to both sides.
$16 = 2n$	Divide both sides by 2.
$8 = n$	Solution of the equation.
or $n = 8$	

6) Have we answered the question asked? No. Since n represents the smaller integer, we have found the smaller integer and the question asked for the larger integer. The larger integer is represented as $n + 2$, so the larger integer is $8 + 2 = 10$.

7) Check the solution in the wording of the problem. Are 8 and 10 consecutive even integers? Yes. Is 3 times the larger ten less than 5 times the smaller? Three times the larger is $3(10) = 30$ and five times the smaller is $5(8) = 40$ and 30 is ten less than 40. Thus we have the correct solution.

PRACTICE EXERCISES

2) Find three consecutive even integers whose sum is 54.

3) Three consecutive odd integers are such that three times the sum of the first and second is seven more than the third. Find the smallest integer.

Distance, Rate, Time Problems

If an object is traveling at a constant rate (speed), the distance traveled depends upon how long it has been traveling (time). This relationship is given by the equation $d = rt$ where d represents the distance traveled, r is the rate (speed), and t is the time. For example, if a truck travels at a constant rate of 50 miles per hour for 5 hours, the distance it has traveled is $d = (50)(5) = 250$ miles.

Even though the problems in this section may be just as easily solved using the procedure previously used, we suggest that you use the procedure below as preparation for problems that follow in Chapter 5.

SOLVING DISTANCE, RATE, TIME PROBLEMS

1) Make a chart with columns for the moving things, distance, rate, and time.

2) Fill in one column with known numerical values.

3) Assign a variable to an unknown and represent any other unknowns in terms of this variable. Fill in another column with these expressions containing variables.

---

4) Fill in the remaining column from the first two using the appropriate relationships between d, r, and t. In this section, $d = rt$.

5) Write the equation using the information in the last column filled in. It is sometimes helpful to draw a diagram.

6) Solve the equation.

7) Be sure that you have answered the question asked.

8) Check the solution in the wording of the original problem.

EXAMPLE 3

a) The distance between Atlanta, Georgia and Charleston, West Virginia is 500 miles. A truck leaves Atlanta traveling toward Charleston at an average rate of 47 miles per hour. At the same time, a bus leaves Charleston traveling toward Atlanta at an average rate of 53 miles per hour. Assuming they are traveling on the same route, how long will it be until they meet?

Solution:

1) Draw a chart and label the columns.

Moving Things	d	r	t
Truck			
Bus			

2) Fill in one column with known numerical values. We know the rate of the truck is 47 miles per hour and the rate of the bus is 53 miles per hour.

Moving Things	d	r	t
Truck		47	
Bus		53	

3) Assign a variable to an unknown and represent any other unknowns in terms of this variable. We are looking for the number of hours it will be until they meet. Since they left at the same time, the truck and car will be traveling for the same number of hours. Let t represent the number of hours for each. Fill in the t column.

Moving Things	d	r	t
Truck		47	t
Bus		53	t

4) Fill in the remaining column using the fact that $d = rt$. Consequently, the distance the truck travels is its rate (47) times its time (t) or $d = 47t$. Likewise for the bus, $d = 53t$.

Moving Things	d	r	t
Truck	$47t$	47	t
Bus	$53t$	53	t

5) Write the equation using the information in the last column filled in. The last column filled in was the distance column which means the equation must involve the distances traveled by the truck and the bus. A diagram might be helpful.

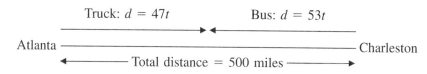

Truck: $d = 47t$ Bus: $d = 53t$

Atlanta ————————————————————————— Charleston

← ————— Total distance = 500 miles ————— →

From the diagram we can see that:

distance the truck traveled	+	distance the bus traveled	=	distance between the cities
47t	+	53t	=	500

6) Solve the equation.

$47t + 53t = 500$ Add like terms.

$100t = 500$ Divide both sides by 100.

$t = 5$ Solution of the equation.

7) Have we answered the question asked? We are looking for the number of hours until they meet and t represents the number of hours that each will be traveling until they meet. Consequently, we have answered the question asked.

8) Check the solution in the wording of the original problem. The total distance traveled by the truck in 5 hours is $5(47) = 235$ miles and the total distance the bus has traveled in 5 hours is $5(53) = 265$ miles. The total miles traveled by both in 5 hours is $235 + 265 = 500$ miles which is the distance between the two cities. Our solution is correct.

b) John leaves home in Lake City, Florida traveling north on I-75 at an average rate of 45 miles per hour. Two hours later his wife Jan leaves home and takes the same route traveling at an average rate of 60 miles per hour. How many hours will it take Jan to catch John?

Solution:

1) Draw a chart and label the columns.

Moving Things	d	r	t
John			
Jan			

2) Fill in one column with known numerical values. We know that John's rate is 45 miles per hour and Jan's rate is 60.

Moving Things	d	r	t
John		45	
Jan		60	

3) Assign a variable to an unknown and represent any other unknowns in terms of this variable. We are looking for the number of hours it will take Jan to catch John, but John left two hours before Jan. If we let t represent John's time, then $t - 2$ will represent Jan's time. Fill in the t column.

Moving Things	d	r	t
John		45	t
Jan		60	$t - 2$

4) Fill in the remaining column using the fact that $d = rt$. Consequently, the distance that John travels is his rate (45) times his time (t) or $d = 45t$. Likewise for Jan, $d = 60(t - 2)$.

Moving Things	d	r	t
John	$45t$	45	t
Jan	$60(t - 2)$	60	$t - 2$

5) Write the equation using the information in the last column filled in. The last column filled in was the distance column which means the equation must involve the distances traveled by John and Jan. A diagram might be helpful.

Lake City ————————————————→
Jan: $d = 60(t - 2)$ ——————→
John: $d = 45t$ ——————→

From the diagram we can see that:

$$\boxed{\text{distance John traveled}} = \boxed{\text{distance Jan traveled}}$$
$$45t = 60(t - 2)$$

6) Solve the equation.

$45t = 60(t - 2)$	Distribute on the right side.
$45t = 60t - 120$	Subtract $60t$ from both sides.
$-15t = -120$	Divide both sides by -15.
$t = 8$	Solution of the equation.

7) Have we answered the question asked? We are looking for the number of hours until Jan catches John and t represents the number of hours that John has been traveling. The number of hours that Jan has been traveling is represented by $t - 2$ which equals $8 - 2 = 6$. Therefore, it will take Jan 6 hours to catch John.

8) Check the solution in the wording of the original problem.
The total distance traveled by John in 8 hours is $8(45) = 360$ miles and the total distance Jan has traveled in 6 hours is $6(60) = 360$ miles. The total miles traveled by each is 360 miles. Thus our solution is correct.

PRACTICE EXERCISES

4) A car traveling at an average rate of 44 miles per hour leaves Jacksonville, Florida traveling toward Jackson, Mississippi. At the same time a bus leaves Jackson traveling toward Jacksonville at an average rate of 36 miles per hour. Assuming they travel the same route and the distance between Jacksonville and Jackson is 600 miles, how long will it be until they meet?

5) A recreational vehicle leaves Tallahassee heading west on I-10 for Houston at an average speed of 35 miles per hour. Four hours later a truck also leaves Tallahassee on I-10 for Houston at an average rate of 55 miles per hour. How many hours will it take the truck to catch the recreational vehicle?

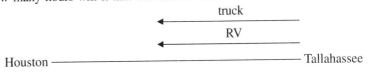

EXERCISE SET   3.6

Solve the following "general" applications problems.

1) A "two by four" (board) which is eight feet long is cut so that one piece is two feet longer than the other. Find the length of each piece.

2) Rick and Hilda were recently married. At the wedding there were 25 more guests for the bride than for the groom. If 157 people attended, how many were guests of the groom?

3) Last year Laura jogged 262 miles more than Pat. If the total number of miles they jogged was 1308, how many miles did Pat jog?

4) The mass transit authority operates two types of buses. A smaller bus, which carries 24 fewer passengers than the larger buses, is used for short distances. If the total number of passengers that can be carried by the two buses is 112, how many passengers can the smaller bus carry?

5) Last week Richard earned $43 less than Patricia. If their combined earnings for the week was $557, how much did each earn?

6) Tara purchased a rocking chair and a bench at a flea market. The total cost for the two items was $111.00. If the bench cost $15 more than the rocking chair, find the cost of each.

7) Marlene is doing her income tax return. She knows that she spent a total of $594 on two business trips last year. If one trip cost $136 more than the other, find the cost of each trip.

8) The Garcia family owns a van and a car. The car gets twice as many miles per gallon of gasoline as the van. If the combined mileage of the two is 54 miles per gallon, how many miles per gallon does the van get?

9) It costs $5 a day extra for after-school care at the nursery school for preschool children. If the total cost for a five-day week of nursery school including after-school care is $85, what is the charge per day for the nursery school without after-school care?

10) It costs $35 per credit hour plus a one time lab fee of $15 to register for courses at Mountain High Community College. If Zane received a $250.00

Solve the following consecutive integer problems.

11) The sum of two consecutive odd integers is 24. Find the two integers.

12) The sum of two consecutive even integers is 98. Find the integers.

13) The sum of three consecutive integers is 75. Find the integers.

14) The sum of three consecutive even integers is 132. Find the integers.

15) Find three consecutive integers such that twice the smallest is 8 less than the largest.

16) Find three consecutive even integers such that three times the smallest is 12 more than the largest.

17) Find three consecutive even integers such that the sum of the first and second is 16 more than the third.

18) Find three consecutive odd integers such that three times the third is 17 more than the sum of the first and second.

19) Find three consecutive even integers such that twice the sum of the first and second is 18 more than the third.

20) Find three consecutive even integers such that 5 times the second equals twice the sum of the first and third.

Solve the following distance, rate, time problems.

21) The distance between Charlotte, North Carolina and Buffalo, New York is 700 miles. Ralph leaves Charlotte toward Buffalo at an average rate of 56 miles per hour. At the same time, Charles leaves Buffalo toward Charlotte at an average rate of 44 miles per hour. Assuming they are traveling on the same route, how long will it be until they meet?

Ralph's distance ──────►◄────── Charles' distance
Charlotte ─────────────── Buffalo
◄──── 700 miles ────►

22) A commercial passenger jet leaves Miami traveling to Reno at an average speed of 560 miles per hour. At the same time a private jet leaves Reno traveling toward Miami at an average rate of 440 miles per hour. How long will it be until they meet if the distance from Miami to Reno is 3,000 miles?

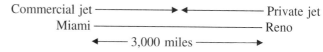

Commercial jet ──────►◄────── Private jet
Miami ─────────────── Reno
◄──── 3,000 miles ────►

23) Frank leaves Memphis traveling toward Albuquerque at an average speed of 60 miles per hour. Two hours later Marsha leaves Albuquerque traveling toward Memphis at an average speed of 50 miles per hour. If the distance from Memphis to Albuquerque is 1000 miles, how long after Marsha leaves will they meet?

24) The distance between Norfolk, Virginia and Buffalo, New York is 595 miles. Grandma Rosie leaves Norfolk traveling toward Buffalo at an average speed of 35 miles per hour. Two hours later Grandpa Julius leaves Buffalo traveling toward Norfolk at an average speed of 40 miles per hour. How long after Grandpa Julius leaves will they meet?

25) Tameka leaves Memphis on I-40 traveling west at an average speed of 30 miles per hour. Three hours later Akiva also leaves Memphis traveling the same route at an average speed of 40 miles per hour. How long will it take Akiva to catch Tameka?

Memphis ————————————————
Tameka's distance ————————————→
Akiva's distance ————————————→

26) Polly leaves Savannah on I-95 traveling north at an average speed of 43 miles per hour. Two hours later Eric also leaves Savannah traveling north on I-95 but at an average speed of 52 miles per hour. How long will it take Eric to catch Polly?

Savannah ————————————————
Polly's distance ————————————→
Eric's distance ————————————→

27) Pia leaves Kansas City traveling west on I-70 at an average speed of 42 miles per hour. At the same time, Alma leaves Kansas City traveling east on I-70 at an average speed of 36 miles per hour. How long will it be until they are 468 miles apart?

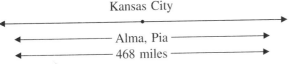

Kansas City
——————— Alma, Pia ———————
——————— 468 miles ———————

28) A truck leaves St. Louis traveling north on I-55 at an average rate of 62 miles per hour. At the same time a bus leaves St. Louis traveling south on I-55 at an

average rate of 55 miles per hour. How long will it be until they are 936 miles apart?

St Louis
←————————————————————————————→
←——————— bus, truck ———————→
←——————— 936 miles ———————→

29) Harold leaves Denver traveling east on I-70 at an average speed of 44 miles per hour. Two hours later Selma leaves Denver traveling west on I-70 at an average rate of 48 miles per hour. How long will it be until they are 364 miles apart?

30) A motorcycle leaves Colorado Springs traveling north on I-25 at an average speed of 47 miles per hour. One hour later a car leaves Colorado Springs traveling south on I-25 at an average speed of 54 miles per hour. How long will it be until they are 552 miles apart?

CHALLENGE EXERCISES (31-32)

31) An architect wishes to build a fountain that has four pipes. If the pipes are arranged in order from the shortest to the longest, each pipe is three feet longer than the previous one. If she has fifty-four feet of pipe, how long should each pipe be?

32) Jim and Amy planned their family so that the two children were born four years apart. If the sum of the ages of the children is 22 years, how old is each child?

WRITING EXERCISE

33) How would you go about teaching a friend how to solve application problems?

WRITING EXERCISES OR GROUP PROJECTS

If done as a group project, have each group write two exercises of each type, exchange with another group and then solve.

34) Write and solve an application problem involving consecutive integers.

35) Write and solve an application problem involving distance, rate, and time.

SECTION 3.7

More Traditional Applications Problems

OBJECTIVES

When you complete this section, you will be able to:

Ⓐ Solve real-world problems involving money.
Ⓑ Solve real-world problems involving investments.
Ⓒ Solve real-world problems involving mixtures.

INTRODUCTION

In this section we continue our discussion of traditional types of applications problems. All the problems in this section can be done using a chart that will be modified slightly (usually the headings) for each different type of problem.

There is an extra step involved in these problems that we did not have to do previously. We have to calculate the value (in the case of money) or the amount of pure substance (in the case of mixtures) before writing the equation.

An important concept often encountered in applications problems is when the total amount is known and parts of the total need to be represented using a variable.

EXAMPLE 1

Represent each of the following.

a) If a board is 12 feet long and a piece x feet long is cut off, represent the length of the remaining piece in terms of x.

Solution:

The length of the other piece is the total length of the board minus the length of the piece cut off. Thus the length of the remaining piece is $12 - x$ feet.

b) A paint contractor buys 36 gallons of paint some of which is acrylic and the remainder oil base. If he buys x gallons of acrylic, represent the number of gallons of oil base paint in terms of x.

Solution:

The number of gallons of oil base paint is the total number of gallons purchased minus the number of gallons of acrylic. Consequently, the number of gallons of oil base paint is $36 - x$ gallons.

PRACTICE EXERCISES

Represent each of the following.

1) Sharla has a piece of ribbon 65 inches long and cuts off x inches to wrap a package. How long is the remaining piece?

2) There is a total of 40 marbles in a jar some of which are black and the remainder white. If there are x black marbles in the jar, represent the number of white marbles in terms of x.

Money Problems

We will discuss two types of money problems. One case will involve purchasing two or more items at different costs. We will be given the total number of items and total cost and we will be asked to find the number of each item purchased. The second type is a special case of the first type involving coins. We suggest the following procedure.

SOLVING MONEY PROBLEMS

1) Draw a chart with four columns and a row for each item. Label the columns Type of Item, Price per Unit, Number of Units, and Total Value. Fill in the Type of Item column.

2) Fill in the Price per Unit column. These values will usually be given, but may involve a variable.

3) Fill in the Number of Units column. This will usually involve a variable. It is usually necessary to assign the variable to the number of units of one item and express the number of units of the other item in terms of this variable.

4) Fill in the Total Value column. The total value is the product of the price per unit and the number of units.

5) Write the "word" equation. This will almost always be the sum of the values of the individual items equals the total value of all the items.

6) Write the algebraic equation using the word equation as a guide.

7) Solve the equation.

8) Be sure you answered the question asked.

9) Check the solution in the wording of the original problem.

EXAMPLE 2

a) A paint contractor paid $372 for 24 gallons of paint for the inside and outside of a house. If the paint for the inside costs $12 per gallon and the paint for the outside costs $18 per gallon, how many gallons of each did he buy?

Solution:

1) Draw a chart and label the columns. Fill in the Type of Item (in this case paint) column.

Type of Paint	Price per Gallon	Number of Gallons	Total Cost
Inside			
Outside			

2) Fill in the Price per Gallon column.

Type of Paint	Price per Gallon	Number of Gallons	Total Cost
Inside	12		
Outside	18		

3) Fill in the Number of Gallons column. Since we are looking for the number of gallons of each type of paint, we must assign a variable. There is no particular advantage (or disadvantage) for letting our variable represent either, so let n represent the number of gallons of inside paint. Since there is a total of 24 gallons, $24 - n$ represents the number of gallons of outside paint.

Type of Paint	Price per Gallon	Number of Gallons	Total Cost
Inside	12	n	
Outside	18	$24 - n$	

4) Fill in the Total Cost column. The total cost is the price per gallon times the number of gallons. Therefore, $12n$ represents the cost of the inside paint in dollars and $18(24 - n)$ represents the cost of the outside paint in dollars.

Type of Paint	Price per Gallon	Number of Gallons	Total Cost
Inside	12	n	$12n$
Outside	18	$24 - n$	$18(24 - n)$

5) Write a "word" equation.

| cost of inside paint | + | cost of outside paint | = | total cost of the paint |

6) Write an algebraic equation using the word equation as a guide.

| cost of inside paint | + | cost of outside paint | = | total cost of the paint |
| $12n$ | + | $18(24 - n)$ | = | 372 |

7) Solve the algebraic equation.

$$12n + 18(24 - n) = 372 \qquad \text{Distribute 18.}$$
$$12n + 432 - 18n = 372 \qquad \text{Add like terms.}$$
$$-6n + 432 = 372 \qquad \text{Subtract 432 from both sides.}$$
$$-6n = -60 \qquad \text{Divide both sides by } -6.$$
$$n = 10 \qquad \text{Solution of the equation.}$$

8) Be sure that you answered the question asked. We were asked to find the number of gallons of each type of paint. Since n represented the number of gallons of inside paint, we still need to find the number of gallons of outside paint. Since $24 - n$ represents the number of gallons of outside paint, we have $24 - 10 = 14$ gallons of outside paint.

9) Check the solutions in the wording of the original problem. Do we have a total of 24 gallons of paint? $10 + 14 = 24$, so yes. Is the total value of the paint $372? Ten gallons of inside paint is worth $10(12) = \$120$ and 14 gallons of outside paint is worth $(14)18 = \$252$. The total value of the paint is $\$120 + \$252 = \$372$, so our solutions are correct.

b) A collection of 23 coins is made up of nickels and dimes and is worth $1.90. How many of each type of coin is there?

Solution:

This is a special case of the money problems involving coins. At first glance, this problem may seem very different from the preceding problem, but they are essentially the same. We will simply change the headings on our columns to Type of Coin, Value of Coin, Number of Coins, Total Value.

1) Draw the chart and label the columns. Fill in the type of coin column.

Type of Coin	Value of Coin	Number of Coins	Total Value
Nickel			
Dime			

2) Fill in the Value of Coin column expressing the value in cents.

Type of Coin	Value of Coin	Number of Coins	Total Value
Nickel	5		
Dime	10		

3) Fill in the Number of Coins column. Since we are looking for the number of nickels and dimes, we need to assign a variable to one of them. Let n represent the number of nickels. Since we have a total of 23 coins, $23 - n$ represents the number of dimes.

Type of Coin	Value of Coin	Number of Coins	Total Value
Nickel	5	n	
Dime	10	$23 - n$	

4) Fill in the Total Value column. The total value of each type of coin is the value of the coin times the number of coins. Therefore, $5n$ represents the value of the nickels in cents and $10(23 - n)$ represents the value of the dimes in cents.

Type of Coin	Value of Coin	Number of Coins	Total Value
Nickel	5	n	$5n$
Dime	10	$23 - n$	$10(23 - n)$

5) Write a "word" equation.

| value of nickels | + | value of the dimes | = | total value of the collection |

6) Write an algebraic equation using the word equation as a guide.

value of nickels	+	value of the dimes	=	total value of the collection
$5n$	+	$10(23 - n)$	=	190

Note: Since the value of the nickels and the value of the dimes were given in cents, it was necessary to change $1.90 into 190¢.

7) Solve the algebraic equation.

$$5n + 10(23 - n) = 190 \qquad \text{Distribute 10.}$$
$$5n + 230 - 10n = 190 \qquad \text{Add like terms.}$$
$$-5n + 230 = 190 \qquad \text{Subtract 230 from both sides.}$$
$$-5n = -40 \qquad \text{Divide both sides by } -5.$$
$$n = 8 \qquad \text{Solution of the equation.}$$

8) Be sure that you answered the question asked. We were asked to find the number of each type of coin. Since n represented the number of nickels, we still need to find the number of dimes. Since $23 - n$ represents the number of dimes, we have $23 - 8 = 15$ dimes.

9) Check the solutions in the wording of the original problem. Do we have a total of 23 coins? $8 + 15 = 23$, so yes. Is the total value of the collection $1.90? Eight nickels are worth $8(0.05) = \$0.40$ and 15 dimes are worth $15(0.10) = \$1.50$. The total value of the collection is $\$.40 + \$1.50 = \$1.90$, so our solutions are correct.

PRACTICE EXERCISES

3) A landscape company pays $7 each for rose bushes and $5 each for azaleas. If they paid $82 for 14 plants, how many of each did they buy?

4) A cashier has a total of 30 nickels and quarters in the cash register. If the total value of the coins is $3.90, how many of each type does he have?

Investment Problems

Investment problems are a special case of money problems. Consequently, the procedure is virtually the same except the column headings are changed to Type of Investment, Interest Rate (expressed as a decimal), Amount Invested, and Interest Earned. The amount of simple interest earned in a year is found by multiplying the annual interest rate expressed as a decimal times the amount invested. For example, the interest earned on $2000 invested at 5% interest for one year is $(0.05)(2000) = \$100$. Also, the total interest earned is the sum of the interests earned on each investment.

EXAMPLE 3

A professional organization has its cash in two investments. Part of the money is in an account paying 8% interest annually and the remainder is in an account paying 14% interest. The amount in the account paying 14% interest annually is $7000 more than the amount in the account paying 8% interest. If the combined interest earned in one year from the two accounts is $4940, find the amount in each account.

Solution:

1) Draw and label a chart.

Type of Investment	Interest Rate	Amount Invested	Interest Earned
8%			
14%			

2) Fill in the Interest Rate column by converting the percents to decimals.

Type of Investment	Interest Rate	Amount Invested	Interest Earned
8%	0.08		
14%	0.14		

3) Fill in the Amount Invested column. We know that there is $7,000 more invested at 14% than at 8%. So if we let x represent the amount invested at 8% then $x + 7000$ represents the amount invested at 14%.

Type of Investment	Interest Rate	Amount Invested	Interest Earned
8%	0.08	x	
14%	0.14	$x + 7000$	

4) Fill in the Interest Earned column. Remember, the interest earned is the interest rate times the amount invested.

Type of Investment	Interest Rate	Amount Invested	Interest Earned
8%	0.08	x	$0.08x$
14%	0.14	$x + 7000$	$0.14(x + 7000)$

5) Write a "word" equation. We know that the total interest earned on the two investments is $4940. Therefore,

| interest earned at 8% | + | interest earned at 14% | = | total interest earned |

6) Write the algebraic equation using the word equation as a guide.

| interest earned at 8% | + | interest earned at 14% | = | total interest earned |

$$0.08x + 0.14(x + 7000) = 4940$$

7) Solve the equation.

$0.08x + 0.14(x + 7000) = 4940$	Distribute the 0.14.
$0.08x + 0.14x + 980 = 4940$	Add like terms.
$0.22x + 980 = 4940$	Subtract 980 from both sides.
$0.22x = 3960$	Divide both sides by 0.22.
$x = 18,000$	Solution of equation.

8) Have we answered the question asked? Partially. Since x represented the amount invested at 8% we have found the amount invested at 8% and still need to find the amount invested at 14%. Since $x + 7000$ represents the amount invested at 14%, we have $18,000 + $7000 = $25,000 invested at 14%.

9) Check the solutions in the wording of the original problem. The interest earned on $18,000 at 8% is $(0.08)(18,000) = 1440 and the interest earned on $25,000 at 14% is $(0.14)(25,000) = 3500. So the total interest on the two investments is $1440 + $3500 = $4940 which is what it is supposed to be. So our solutions are correct.

There are several variations of investment problems.

PRACTICE EXERCISE

5) A college support fund has its money in two investments. The first investment pays 18% interest annually and is $20,000 more than the second investment which pays 12% interest annaully. If the total interest earned from the two investments is $18,600 per year, how much is in the first investment?

Mixture Problems

Mixture problems are very much like money and investment problems. We will consider two types of mixture problems, liquid and dry. We will do a dry mixture problem first. We will again use a chart, but change the headings to Type of Ingredient, Price per Unit, Number of Units, and Total Value. We also need an extra line at the bottom for the mixture. The key to doing dry mixture problems is that the sum of the values of each of the ingredients must equal the value of the mixture.

EXAMPLE 4

How many pounds of peppermint candy worth $1.80 per pound must be mixed with 15 pounds of butterscotch worth $2.30 a pound to get a mixture worth $2.10 per pound?

Solution:

1) Draw the chart, label the columns, and fill in the types of candy.

Type of Candy	Price per Pound	Number of Pounds	Total Value
Peppermint			
Butterscotch			
Mixture			

2) Fill in the Price per Pound column.

Type of Candy	Price per Pound	Number of Pounds	Total Value
Peppermint	1.80		
Butterscotch	2.30		
Mixture	2.10		

3) Since we are looking for the number of pounds of peppermint candy, let x represent the number of pounds of peppermint candy. Also, we know that we have 15 pounds of butterscotch. The number of pounds of candy in the mixture is the number of pounds of peppermint plus the number of

pounds of butterscotch. Consequently, the number of pounds of candy in the mixture is $x + 15$. Fill in the Number of Pounds column.

Type of Candy	Price per Pound	Number of Pounds	Total Value
Peppermint	1.80	x	
Butterscotch	2.30	15	
Mixture	2.10	$x + 15$	

4) The total value of each type of candy is the price per pound times the number of pounds. Use this fact to fill in the Total Value column.

Type of Candy	Price per Pound	Number of Pounds	Total Value
Peppermint	1.80	x	$1.80x$
Butterscotch	2.30	15	$(2.30)(15) = 34.50$
Mixture	2.10	$x + 15$	$2.10(x + 15)$

5) Write the "word" equation.

$$\boxed{\text{Value of peppermint}} + \boxed{\text{Value of butterscotch}} = \boxed{\text{Value of mixture}}$$

6) Write the algebraic equation using the word equation as a guide.

$$\boxed{\text{Value of peppermint}} + \boxed{\text{Value of butterscotch}} = \boxed{\text{Value of mixture}}$$
$$1.80x \qquad + \qquad 34.50 \qquad = \qquad 2.10(x + 15)$$

7) Solve the equation.

$1.80x + 34.50 = 2.10(x + 15)$	Distribute 2.10.
$1.80x + 34.50 = 2.10x + 31.50$	Subtract $1.80x$ from both sides.
$34.50 = 0.3x + 31.50$	Subtract 31.50 from both sides.
$3 = 0.3x$	Divide both sides by 0.3.
$10 = x$	Solution of the equation.

8) Have we answered the question asked? Yes. We were looking for the number of pounds of peppermint candy and x represents the number of pounds of peppermint candy.

9) Check the solution in the wording of the original problem. The value of ten pounds of peppermint candy is $1.80(10) = \$18.00$ and from the chart above we know the value of 15 pounds of butterscotch is $34.50. Consequently, $\$18.00 + \$34.50 = \$52.50$. The mixture sells for $2.10 per pound and we have $10 + 15 = 25$ pounds. Consequently, the value of the mixture is $2.10(25) = \$52.50$ which is the same as the sum of the values of the peppermint and butterscotch. Our solution is correct.

Other variations of this type of problem include giving the number of units and price per unit of each of the ingredients and finding the price per unit of the mixture or giving the price per unit of each of the ingredients, the price per unit and number of units in the mixture, and finding the number of units of each ingredient. The approach is the same in all cases.

Liquid mixture problems are done exactly the same way as dry mixture problems except we are dealing with the amount of pure substance in each ingredient and in the mixture instead of values. Consequently we will not repeat the procedure, but illustrate with an example. The key to doing this type of problem is that the amount of pure substance in a solution (alloy, or similar) is the percent that is pure times the number of units (gallons, pounds, kilograms or other) of the solution. For example, if we have 8 gallons of a solution that is 20% alcohol, the amount of pure alcohol in the solution is $(0.20)(8) = 1.6$ gallons. Remember, to change a percent to a decimal move the decimal two places to the left and drop the % sign.

EXAMPLE 5

How much of a solution that is 30% alcohol must be added to 55 liters of a solution that is 60% alcohol to obtain a solution that is 40% alcohol?

Solution:

1) Draw a chart similar to the one for dry mixtures. Note the headings are different. Fill in the types of solutions.

Type of Solution	Part Pure Alcohol	Volume	Amount Pure Alcohol
30%			
60%			
40% (mixture)			

2) Fill in the Part Pure Alcohol column by converting the percents to decimals.

Type of Solution	Part Pure Alcohol	Volume	Amount Pure Alcohol
30%	0.30		
60%	0.60		
40% (mixture)	0.40		

3) Fill in the Volume column. Since we are looking for the amount (volume) of 30% alcohol, let x represent the volume of 30% alcohol. We know that we have 55 liters of 60% alcohol. Consequently, when the two are mixed we will have $x + 55$ liters of the 40% mixture.

Type of Solution	Part Pure Alcohol	Volume	Amount Pure Alcohol
30%	0.30	x	
60%	0.60	55	
40% (mixture)	0.40	$x + 55$	

4) Fill in the Amount Pure Alcohol column. Remember, the amount of pure alcohol in each solution equals the part pure alcohol times the volume.

Type of Solution	Part Pure Alcohol	Volume	Amount Pure Alcohol
30%	0.30	x	$0.30x$
60%	0.60	55	$(0.60)(55) = 33$
40% (mixture)	0.40	$x + 55$	$0.40(x + 55)$

5) Write the "word" equation.

$$\boxed{\text{amount pure alcohol in 30\% solution}} + \boxed{\text{amount pure alcohol in 60\% solution}} = \boxed{\text{amount of pure alcohol in the 40\% mixture}}$$

6) Write the algebraic equation using the word equation as a guide.

$$\boxed{\text{amount pure alcohol in 30\% solution}} + \boxed{\text{amount pure alcohol in 60\% solution}} = \boxed{\text{amount of pure alcohol in the 40\% mixture}}$$

$$0.30x \quad + \quad 33 \quad = \quad 0.40(x + 55)$$

7) Solve the equation.

$$0.30x + 33 = 0.40(x + 55) \qquad \text{Distribute 0.40 on the right side.}$$
$$0.30x + 33 = 0.40x + 22 \qquad \text{Subtract } 0.30x \text{ from both sides.}$$
$$33 = 0.10x + 22 \qquad \text{Subtract 22 from both sides.}$$
$$11 = 0.10x \qquad \text{Divide both sides by 0.10.}$$
$$110 = x \qquad \text{Solution of the equation.}$$

8) Have we answered the question asked? Yes, since x represented the amount of the 30% solution and that is what we were asked to find.

9) Check the solution in the wording of the original problem. In 110 liters of 30% solution there would be $0.30(110) = 33$ liters of pure alcohol and from the chart above we know that in 55 liters of 60% solution there are 33 liters of pure alcohol. So there are a total of $33 + 33 = 66$ liters of pure alcohol in the two that are mixed. We have a total of $110 + 55 = 165$ liters of the 40% mixture. In this mixture there are $0.40(165) = 66$ liters of pure alcohol which is the same as the amount of pure alcohol in the two that are mixed to get the 40% solution. Thus our solution is correct.

As with dry mixture problems, there are several variations of this type of problem.

PRACTICE EXERCISES

6) If 6 pounds of coffee from the Dominican Republic that sells for $3.00 per pound are mixed with 9 pounds of coffee from Colombia that sells for $4.50 per pound, what should be the selling price of the mixture?

7) How much of a solution that is 20% baking soda must be added to 60 ounces of a solution that is 5% baking soda to make a solution that is 10% baking soda?

EXERCISE SET 3.7

Represent each of the following.

1) Sally is taking a trip of 400 miles. If she has traveled x miles, represent the remainder of the trip in terms of x.

2) Francine walks 5 miles every morning. If she has already walked x miles, represent the distance remaining in terms of x.

3) There is a total of 24 nickels and dimes in a box. If x of these are nickels, represent the number of dimes in terms of x.

4) A seamstress has a box containing 50 buttons, some of which are brass and the remainder glass. If there are x brass buttons, represent the number of glass buttons in terms of x.

5) A nursery has a total of 150 azalea and camellia plants. If there are x azaleas, represent the number of camellias in terms of x.

6) A class has 35 students. If x of the students are girls, represent the number of boys in terms of x.

Solve the following "money" problems.

7) An electrical contractor paid $1040 for 36 light fixtures to be installed in a hotel lobby. If the first type of fixture costs $25 each and the second type of fixture costs $32 each, how many of each kind did she buy?

8) A locksmith installed inside and outside locks in a new classroom building. The locksmith paid $1216 for 80 locks. If inside locks cost $14 each and outside locks cost $26 each, how many of each kind of lock did he buy?

9) The managers of an office building have decided to reduce the cost of air conditioning by installing overhead fans. They paid $5225 for 105 fans, some of which measure 36 inches and the remainder

---

measure 54 inches. If the 36 inch fans cost $45 each and the 54 inch fans cost $65 each, how many of each kind of fan did they buy?

10) An accounting firm purchased a supply of computer disks. They paid $725 for 60 boxes of disks. If the double density disks cost $8 a box and the high density disks cost $15 a box, how many of each kind did they buy?

11) Winston paid $7.46 for 36 stamps. If some of the stamps cost 18¢ each and the remainder cost 25¢ each, how many of each kind did he buy?

12) Gayle paid $17.40 for 42 pens at the campus bookstore. If she bought some pens costing $0.35 and others costing $0.50 each, how many of each kind did she buy?

13) The owners of an exterminator business have replaced their fleet of 27 cars and trucks by purchasing new cars for $12,000 each and new trucks for $9500 each. If they paid a total of $294,000, how many of each type of vehicle did they buy?

14) A building contractor has paid $2088 for 54 doors. If inside doors cost $35 each and outside doors cost $46 each, how many of each kind of door did he buy?

15) At the end of his shift a cashier has only 5 and 20-dollar bills in his drawer. If he has 48 bills worth a total of $900, how many of each kind of bill does he have?

16) A credit union carries only $20 and $50 travelers checks. A customer buys $1,500 worth of travelers checks. If he bought a total of 39 checks, how many $50 checks did he buy?

17) A child goes to a bank to deposit her savings. She has $4.00 worth of nickels and dimes. If she has 5 more nickels than dimes, how many of each type of coin does she have?

18) A certain vending machine accepts only dimes and quarters. If the person who services the machine collects 52 coins with a total value of $8.95, how many of each kind of coin was in the machine?

19) A parking meter will accept dimes and quarters only. At the end of the day, the meter attendants collected 55 coins whose total value was $10.75. How many coins of each type were there?

20) A vending machine will accept nickels and dimes only. The attendant checks the machine and finds a total of 42 coins whose value is $3.40. How many coins of each type were there?

Solve the following investment problems.

21) When he retired, Professor Gomez received a lump sum in cash for his unused sick days. He invested part of the money in bonds that paid 6% interest annually. He invested $16,000 more in certificates of deposit at 10% annual interest than he invested in bonds. The total interest earned per year from the two investments was $3200. How much money did he invest at each rate?

22) A petroleum geologist received a $50,000 bonus for discovering a new oil deposit. She invested part of the money in bonds that paid 15% interest annually and put the rest in a life insurance policy that paid 8% interest annually. How much did she put into the life insurance policy if she expects to earn $5400 from both investments each year?

23) A professional football player received a $200,000 bonus for signing his contract. He invested part of the money in home mortgages that yield 12% interest annually and he put the rest in a business that was yielding a 20% profit annually. How much did he invest in the business if he expects to earn a total of $27,200 per year from both investments?

24) A retired couple sold their home. They invested part of the money in home mortgages at 15% interest annually. They invested $20,000 less than this amount into certificates of deposit at 9% annual interest. How much money did they invest, if they received $15,000 per year interest from the two investments?

Solve the following mixture problems.

25) How many ounces of pipe tobacco that sells for $2.50 per ounce must be mixed with 6 ounces that sells for $3.50 per ounce if the mixture is to sell for $3.10 per ounce.

26) How many pounds of candy that sells for $2.75 per pound must be mixed with 10 pounds that sells for $3.50 per pound to get a mixture that sells for $3.25 per pound?

27) If 8 pounds of almonds that sell for $2.25 per pound are mixed with 4 pounds of cashews that sell for $4.50 per pound, what should be the selling price of the mixture?

28) If 12 ounces of white chocolate that sells for $3.75 per ounce is mixed with 6 ounces of dark chocolate that sells for $4.50 per ounce, what should be the selling price of the mixture?

29) A specialty store owner mixes peanuts worth $1.80 per pound with cashew nuts worth $5.40 per pound. How many pounds of each type of nut is needed to make a 12 pound mixture worth $3.00 a pound?

30) A coffee company wishes to make a special blend of coffee by mixing coffee beans which cost $3.20 a pound with coffee beans which cost $6.20 per pound. How many pounds of each type of coffee bean is needed to make 40 pounds of coffee which costs $3.95 per pound?

31) How many milliliters of a 40% solution of battery acid must be added to 30 milliliters of an 8% solution to make a 20% solution of acid?

32) How many ounces of a 10% baking soda solution must be added to 40 ounces of a 2% baking soda solution to make a 5% baking soda solution?

33) How many tons of an alloy that is 5% nickel must be added to 15 tons of an alloy that is 20% nickel to make an alloy that is 10% nickel?

34) How many cups of a party mix that is 60% peanuts must be added to 5 cups of a party mix that is 30% peanuts to make a party mix that is 45% peanuts?

35) How many liters of pure baking soda must be added to 60 liters of a solution that is 10% baking soda to get a solution that is 15% baking soda?

36) How much pure water must be added to 20 liters of a solution that is 15% salt to make a solution that is 8% salt?

Solve the following exercises. Types are mixed.

37) How many gallons of pure water must be added to 10 gallons of a solution that is 25% soap to make a solution that is 5% soap?

38) How many pounds of an alloy that is 30% tin must be added to 40 pounds of an alloy that is 15% tin to make an alloy that is 25% tin?

39) Professor Matas bought materials for a workshop for elementary school teachers. She bought number stickers for $0.49 each and strings of beads for $1.19 a string. She bought 10 more number stickers than strings of beads. If she paid a total of $30.10, how many of each item did she buy?

40) Jeannette Castillo won $20,000 as part of the state lottery. After her celebration party, she put part of the money into bonds paying 7% interest. The rest, which was $10,000 less than the first part, she invested in a stock that was paying 20% in dividends (like interest). How much did she put in each investment if she earned a total of $2050 per year?

41) How much pure alcohol must be added to 50 gallons of gasohol that is 10% alcohol to make gasohol that is 20% alcohol?

42) Mr. Jacob bought two prescription medications. He paid $0.83 a tablet for the first medication and $1.15 a tablet for the second medication. Together there were 86 tablets for which he paid $79.70. How many of the more expensive tablets did he buy?

CHALLENGE EXERCISES (43–44)

43) A vending machine accepts nickels, dimes, and quarters. At the end of the day there were three more dimes than nickels and four fewer quarters than nickels. If the total value of the coins was $4.10, find the number of each type of coin.

44) A banker has three investments that pay 6%, 5%, and 8% each. She has twice as much invested at 5% as at 6% and $5,000 more invested at 8% than at 6%. If the total interest earned per year from the three investments is $2800, find the amount invested at each rate.

WRITING EXERCISES OR GROUP PROJECTS

If done as a group project, each group should write two exercises of each type, exchange with another group, and solve.

45) Write and solve an application problem involving money.

46) Write and solve an application problem involving mixtures.

47) Write and solve an application problem involving investments.

SECTION **3.8**

Geometric Applications Problems

INTRODUCTION

OBJECTIVE

When you complete this section, you will be able to solve real-world problems involving geometric concepts.

Linear equations are often found in geometric applications. There are essentially two types of problems. The first type are those for which known formulas (Section 1.4) give us a relationship between the unknowns of the problem.

When solving this type of problem, we are either given a value for the variable(s) in the formula or we have to represent the unknowns in the formula in terms of variables. We then substitute into the formula and solve the resulting equation. Problems in Example 1 will be of this type. We will discuss the second type after Example 1. We do not have to look for conditions within the problem to give us the equation since it will be some known formula. We suggest the following.

SOLVING GEOMETRIC APPLICATIONS PROBLEMS

1) Identify the unknown(s) and represent one of them with a variable.

2) Represent all other variables in the problem in terms of the same variable.

3) Write the formula. A sketch might be helpful.

4) Substitute into the formula.

5) Solve the resulting equation.

6) Be sure that you have answered the question asked.

7) Check the solution in the wording of the original problem.

 EXAMPLE 1

a) An urban area has a large park in the shape of a rectangle. The park is 15 city blocks longer than it is wide. If the perimeter is 50 city blocks, find the length and width of the park?

Solution:

1) Identify the unknown(s) and represent one of them with a variable. We are looking for the length and width. Since the width is the smaller,

let x represent the width.

2) Represent all other unknowns in the problem in terms of the same variable. Since the length is 15 more than the width and x represents the width,

$x + 15$ represents the length.

3) Write the formula. The formula needed is the formula for the perimeter of a rectangle. Note that a word equation is not necessary.

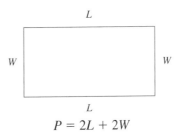

$P = 2L + 2W$

4) Substitute into the formula. We know from the problem that $P = 50$. We have also represented the width with x and the length with $x + 15$. Therefore,

$$50 = 2(x + 15) + 2x$$

5) Solve the equation.

$50 = 2(x + 15) + 2x$	Distribute 2 on the right side.
$50 = 2x + 30 + 2x$	Add like terms.
$50 = 4x + 30$	Subtract 30 from both sides.
$20 = 4x$	Divide both sides by 4.
$5 = x$	Solution of equation.

6) Answered question asked? Partially. We were asked to find both the length and width. Since x represents the width, we have found the width only. The length is represented by $x + 15$, so the length is $5 + 15 = 20$ blocks.

7) Check: Is the length 15 more than the width? Yes, since 20 is 15 more than 5. Is the perimeter 50 blocks? Yes, since $2(20) + 2(5) = 50$? So, our solution is correct.

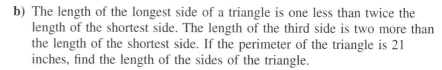

b) The length of the longest side of a triangle is one less than twice the length of the shortest side. The length of the third side is two more than the length of the shortest side. If the perimeter of the triangle is 21 inches, find the length of the sides of the triangle.

Solution:

1) Identify the unknown(s) and represent one of them with a variable. We are looking for the lengths of all three sides of the triangle.

Let x represent the shortest side.

2) Represent all other unknowns in terms of the same variable. We know that the length of the longest side is one less than twice the length of the shortest side, so,

$2x - 1$ represents the length of the longest side.

We also know that the length of the third side is two more than the length of the shortest side, so,

$x + 2$ represents the length of the third side.

3) Write the formula. In this case we need the formula for the perimeter of a triangle.

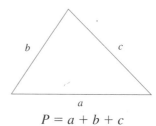

$$P = a + b + c$$

4) Substitute into the formula. We know that $P = 21$ inches and the sides are represented by x, $2x - 1$, and $x + 2$. Therefore,

$$21 = x + (2x - 1) + (x + 2)$$

5) Solve the equation.

$$21 = x + (2x - 1) + (x + 2) \qquad \text{Remove the parentheses.}$$
$$21 = x + 2x - 1 + x + 2 \qquad \text{Add like terms.}$$
$$21 = 4x + 1 \qquad \text{Subtract 1 from both sides.}$$
$$20 = 4x \qquad \text{Divide both sides by 4.}$$
$$5 = x \qquad \text{Solution of the equation.}$$

6) Answered question asked? Partially. We were asked to find the lengths of all three sides. The length of the longest side is $2x - 1 = 2 \cdot 5 - 1 = 10 - 1 = 9$ inches, and the length of the third side is $x + 2 = 5 + 2 = 7$ inches.

7) Check: Is the longest side one less than twice the shortest side? Yes, since $9 = 2 \cdot 5 - 1$. Is the third side two more than the shortest side? Yes, since $7 = 5 + 2$. Is the perimeter 21 inches? Yes, since $5 + 9 + 7 = 21$. Therefore, our solutions are correct.

PRACTICE EXERCISES

1) A rectangular area rug on the living room floor is 4 feet longer than it is wide. If the distance around the rug is 40 feet, find the dimensions (length and width) of the rug.

2) The length of the longest side of a triangle is twice the length of the shortest side. The length of the middle-sized side is 5 feet more than the length of the shortest side. Find the lengths of the sides of the triangle if the perimeter is 33 feet.

The second type of geometric applications problems are those for which we have to represent the unknown(s) in terms of a variable and then write an equation expressing the relationship between the unknowns using some known geometric relationship. The problems in Example 2 will be of this type. These relationships often involve angles. An angle is usually denoted by using the symbol $\angle$. The size of an angle is measured in degrees and is called the measure of the angle. The measure of $\angle A$ is denoted as $m\angle A$. Two angles with the same number of degrees (measure) are called **congruent** angles and the symbol for congruent is $\cong$. These known geometric relationships may come from, but are not limited to, the following.

1) The sum of the measures of all the angles of any triangle is $180°$.

2) Two angles are **supplementary** if the sum of their measures is $180°$. Each angle is the supplement of the other.

3) Two angles are **complementary** if the sum of their measures is $90°$. Each angle is the complement of the other.

4) Angle relationships involving parallel lines. Two or more lines are parallel if they lie in the same plane and do not intersect. Any line intersecting two or more lines is called a **transversal**. In the figure below, L_1 and L_2

are parallel and T is a transversal. The angle relationships are summarized below.

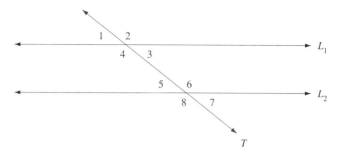

a) Alternate interior angles are congruent. The two pairs of alternate interior angles are $\angle 4$ and $\angle 6$, and $\angle 3$ and $\angle 5$. Consequently, $m\angle 4 = m\angle 6$ and $m\angle 3 = m\angle 5$.

b) Alternate exterior angles are congruent. The two pairs of alternate exterior angles are $\angle 1$ and $\angle 7$, and $\angle 2$ and $\angle 8$. Consequently, $m\angle 1 = m\angle 7$ and $m\angle 2 = m\angle 8$.

c) Corresponding angles are congruent. The four pairs of corresponding angles are $\angle 1$ and $\angle 5$, $\angle 2$ and $\angle 6$, $\angle 3$ and $\angle 7$, and $\angle 4$ and $\angle 8$. Consequently, $m\angle 1 = m\angle 5$, $m\angle 2 = m\angle 6$, $m\angle 3 = m\angle 7$, and $m\angle 4 = m\angle 8$.

d) Interior angles on the same side of the transversal are supplementary. The interior angles on the same side of the transversal are $\angle 3$ and $\angle 6$ and also $\angle 4$ and $\angle 5$. Consequently, $m\angle 3 + m\angle 6 = 180°$ and $m\angle 4 + m\angle 5 = 180°$.

Our procedure for solving will be exactly the same except for writing the equation which, in this case, comes from a known geometric fact rather than a known formula.

EXAMPLE 2

a) In a triangle, the measure of the largest angle is twice the measure of the smallest angle. If the measure of the middle-sized angle is twelve more than the measure of the smallest angle, find the largest angle.

Solution:

1) Identify the unknown(s) and represent one of them with a variable. We are looking for the measure of each of the three angles of the triangle.

   Let t represent the measure of the smallest angle.

2) Represent all other unknowns in the problem in terms of the same variable. We know that the measure of the largest angle is twice the measure of the smallest, so,

   $2t$ represents the measure of the largest angle.

   We also know that the measure of the middle-sized angle is 12 more than the measure of the smallest angle, so,

   $t + 12$ represents the measure of the middle-sized angle.

3) What is the known fact? We know that the sum of the measures of all the angles of a triangle is 180.

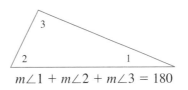

$$m\angle 1 + m\angle 2 + m\angle 3 = 180$$

4) Substitute into the equation. We have represented the angles by t, $2t$, and $t + 12$, so,

$$t + 2t + (t + 12) = 180$$

5) Solve the equation.

$t + 2t + (t + 12) = 180$	Remove parentheses.
$t + 2t + t + 12 = 180$	Simplify the left side.
$4t + 12 = 180$	Subtract 12 from both sides.
$4t + 12 - 12 = 180 - 12$	Simplify both sides.
$4t = 168$	Divide both sides by 4.
$\dfrac{4t}{4} = \dfrac{168}{4}$	Simplify both sides.
$t = 42°$	Solution of equation.

6) Does this answer the question asked? No, the largest angle is $2t = 2(42°) = 84°$.

7) Check: The largest angle is 84°. Is this twice the smallest? The smallest angle is 42° and $84° = 2(42°)$. So, yes. Is the middle angle 12° more than the smallest? The middle angle is $t + 12 = 42° + 12° = 54°$ and 54° is 12° more than 42°. Is the sum of the three angles 180°? Since $42° + 54° + 84° = 180°$, yes. We have the correct solution.

b) Given the figure below with L_1 parallel to L_2, find the value of x and the $m\angle A$ and $m\angle B$.

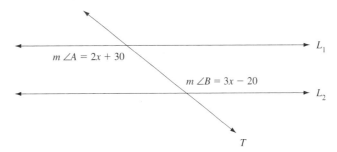

Solution:

1) Identify the unknown(s) and represent one of them with a variable. Since we are given a labeled figure, this is not necessary in this case.

2) Represent all other unknowns in terms of the same variable. Again, this is not necessary.

3) What is the known geometric relationship? $\angle A$ and $\angle B$ are alternate interior angles of two parallel lines. Therefore, their measures are equal.

4) Write the equation.
   $m\angle A = m\angle B$ so,

$$2x + 30 = 3x - 20$$

5) Solve the equation.

$$2x + 30 = 3x - 20 \qquad \text{Subtract } 2x \text{ from both sides.}$$
$$30 = x - 20 \qquad \text{Add 20 to both sides.}$$
$$50 = x \qquad \text{Therefore, } x = 50.$$

6) Does this answer the question asked? Partially. We still need to find $m\angle A$ and $m\angle B$. $m\angle A = 2x + 30 = 2 \cdot 50 + 30 = 100 + 30 = 130°$. $m\angle B = 3x - 20 = 3 \cdot 50 - 20 = 150 - 20 = 130°$.

7) Check: Are the two angles congruent? Yes, since both have measures of 130°.

PRACTICE EXERCISES

3) The measure of the smallest angle of a triangle is 18° less than the measure of the middle angle. If the measure of the largest angle is four times the measure of the smallest angle, find all three angles.

4) The measure of an angle is 30° less than twice the measure of its supplement. Find the measure of each angle.

EXERCISE SET 3.8

1) The width of a rectangular piece of cloth is one foot shorter than the length. If the perimeter is 10 feet, how long is the cloth?

2) A mathematics classroom at Urban College is rectangular in shape and measures 6 meters longer than it is wide. If the perimeter is 44 meters, what are its length and width?

3) A playing field for soccer is rectangular in shape and is 40 yards longer than it is wide. If the perimeter is 220 yards, what are its length and width?

4) A rectangular office building is four times as long as it is wide. If the perimeter is 900 feet, what are the length and width of the building.

5) A rectangular parking space has a length that is two less than three times the width. If the perimeter is 60 feet, what are the dimensions of the parking space?

6) The length of a rectangular sign is 4 inches less than twice the width. The perimeter of the sign is 28 inches. Find the length and the width.

7) The length of the longest side of a triangle is twice the length of the shortest side and the length of the third side is 2 inches more than the shortest side. If the perimeter is 18 inches, find the length of each of the three sides of the triangle.

8) The length of the longest side of a triangle is three times the length of the shortest and the length of the third side is 4 feet more than the length of the shortest. If the perimeter is 19 feet, find the length of each of the three sides of the triangle.

9) Two sides of a triangular flower bed are equal in length and the third side is 4 feet less than twice the length of the equal sides. If the perimeter of the flower bed is 28 feet, find the length of each of the three sides of the flower bed.

10) Two sides of a triangular sign are equal in length and the length of the third side is 3 inches less than twice the length of the equal sides. If the perimeter of the sign is 21 inches, find the length of each of the three sides of the sign.

11) One angle of a triangle measures 90°. One of the remaining two angles is twice as large as the other. Find the measures of the remaining two angles.

12) Two of the angles of a triangle are equal in measure. The measure of the remaining angle is twice the sum of the measures of the two equal angles. Find the measures of the three angles.

13) The measure of the first angle of a triangle is 3 times the measure of the second angle and the measure of the third angle is 5 more than the measure of the first angle. Find the measures of the three angles.

14) The measure of the smallest angle of a triangle is 0.3 of the measure of the largest angle and the measure of the middle angle is 0.5 of the measure of the largest angle. Find all three angles.

15) The measure of the complement of an angle is 60° less than twice the measure of the angle. Find the measures of both angles.

16) The measure of the supplement of an angle is 10° more than four times the measure of the angle. Find the measures of both angles.

17) The measure of the supplement of an angle is 20° more than three times the measure of the angle. Find the measures of both angles.

18) The measure of the complement of an angle is 15° less than four times the measure of the angle. Find the measures of both angles.

19) Given the figure below in which L_1 and L_2 are parallel, find the value of x and the measures of $\angle A$ and $\angle B$.

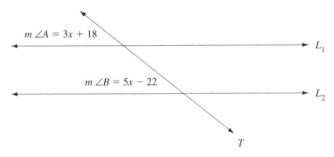

20) Given the figure below in which L_1 and L_2 are parallel, find the value of x and the measures of $\angle A$ and $\angle B$.

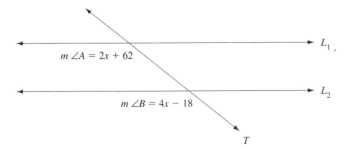

21) Given the figure below in which L_1 and L_2 are parallel, find the value of x and the measures of $\angle A$ and $\angle B$.

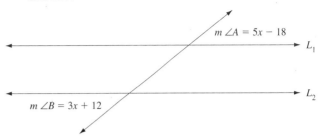

22) Given the figure below in which L_1 and L_2 are parallel, find the value of x and the measures of $\angle A$ and $\angle B$.

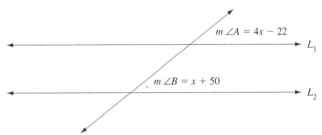

23) Given the figure below in which L_1 and L_2 are parallel, find the value of x and the measures of $\angle A$ and $\angle B$.

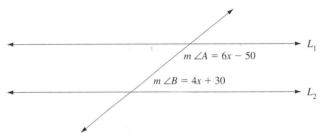

24) Given the figure below in which L_1 and L_2 are parallel, find the value of x and the measures of $\angle A$ and $\angle B$.

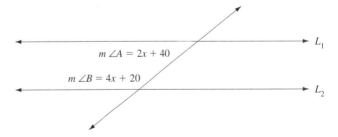

WRITING EXERCISES OR GROUP PROJECT

If done as a group project, each group should write two exercises of each type, exchange with another group, and then solve.

25) Write an application problem involving the perimeter of a triangle.

26) Write an application problem involving angle relationships.

OBJECTIVES

When you complete this section, you will be able to:

Ⓐ   Solve for the unknown when given all of the values in a formula except one.

Ⓑ   Solve a formula for a given variable.

• Solving for the unknown when given all of the values in a formula except one

Formulas are equations that express relationships that apply to a whole class of problems. They are like ready-made equations without specific numbers. We used several geometric formulas in Chapter 2 and in previous sections of this chapter. In Chapter 2 we simply substituted the given values and followed the order of operations since the formula was already solved for the quantity that we were looking for. Often that is not the case and we have to use the techniques for solving linear equations to find the desired value.

In using a formula we must know the values of all of the variables except the one for which we are solving. We substitute the known values for the variables and then solve for the unknown variable using the techniques from the previous sections of this chapter.

EXAMPLE 1

Solve each of the following for the indicated variable.

a) The formula for distance is $d = rt$. If $r = 30$ miles per hour and $t = 5$ hours, find d.

Solution:

$d = rt$	Replace r with 30 and t with 5.
$d = 30 \cdot 5$	Multiply.
$d = 150$ mi	Therefore, the distance is 150 miles.

b) The formula for the perimeter of a rectangle is $P = 2L + 2W$. If $P = 20$ in and $L = 6$ in, find W.

Solution:

$P = 2L + 2W$	Replace P with 20 and L with 6.
$20 = 2(6) + 2W$	Simplify the right side.
$20 = 12 + 2W$	Subtract 12 from both sides.
$8 = 2W$	Divide both sides by 2.
4 in $= W$	Therefore, the width is 4 inches.

c) The formula for finding the circumference (distance around) of a circle is $C = 2\pi r$. If $C = 9.42$ ft, find r. Use $\pi \approx 3.14$.

Solution:

$C = 2\pi r$	Replace C with 9.42 and π with 3.14.
$9.42 = 2(3.14)r$	Simplify the right side.
$9.42 = 6.28r$	Divide both sides by 6.28.
$\dfrac{9.42}{6.28} = \dfrac{6.28r}{6.28}$	Simplify both sides.
1.5 ft $= r$	Therefore, the radius is 1.5 feet.

d) The formula for the area of a triangle is $A = \frac{bh}{2}$. If $A = 60$ in² and $b = 15$ in, find h.

Solution:

$$A = \frac{bh}{2}$$ Replace A with 60 and b with 15.

$$60 = \frac{15h}{2}$$ Multiply both sides by $\frac{2}{15}$.

$$\frac{2}{15}(60) = \frac{2}{15} \cdot \frac{15h}{2}$$ Simplify both sides.

$$8 \text{ in} = 1 \ h$$ Therefore, $h = 8$ inches.

or $h = 8$ in

ADDITIONAL PRACTICE

Solve the following.

a) The formula for interest is $I = prt$. If $I = \$67.50$, $p = \$1,500$, and $t = 0.5$, find r.

b) The formula for the area of a rectangle is $A = LW$. If $W = 3.5$ m and $A = 21$ m², find L.

c) The formula for the area of a circle is $A = \pi r^2$. If $r = 2$ m, find A. Use $\pi \approx 3.14$.

d) The formula for the area of a trapezoid is $A = \frac{h(B + b)}{2}$. If $A = 18$ ft², $h = 4$ ft, and $B = 6$ ft, find b.

PRACTICE EXERCISES

Solve the following.

1) The formula for the volume of a box is $V = LWH$. If $L = 10$ feet, $H = 6$ feet, and $V = 180$ cubic feet, find W.

2) In physics, the formula for work is $W = Fd$ where W represents work, F represents force, and d represents distance. If $F = 40$ pounds and $d = 2$ feet, find W.

3) In electricity, the formula for power is $P = iV$ where P represents power, i represents current, and V represents voltage. If $P = 140$ watts and $i = 10$ amps, find V.

4) The formula for rate is $r = \frac{d}{t}$ where d is the distance and t is the time. Find the distance travelled if the rate is 45 miles per hour and the time is 4 hours.

If more practice is needed, do the Additional Practice Exercises in the margin.

Each variable in a formula may be solved in terms of the other variables of that formula. If we have several problems in which we are asked to find the same variable, we may want to solve the formula for that variable before substituting values for the other variables. This way we do not have to solve several similar equations for the same variable several times.

EXAMPLE 2

Solve the following formulas for the indicated variable.

a) $d = rt$, for t

Solution:

We treat all the other variables as if they were numbers and solve.

$d = rt$ Divide both sides by r.

$\dfrac{d}{r} = \dfrac{rt}{r}$ Simplify the right side.

$\dfrac{d}{r} = t$ Therefore, $t = \dfrac{d}{r}$.

b) $A = \dfrac{1}{2}bh$, for h

$A = \dfrac{1}{2}bh$	Multiply both sides by 2.
$2A = 2\left(\dfrac{1}{2}\right)bh$	Simplify both sides.
$2A = bh$	Divide both sides by b.
$\dfrac{2A}{b} = \dfrac{bh}{b}$	Simplify both sides.
$\dfrac{2A}{b} = h$	Therefore, $h = \dfrac{2A}{b}$.

c) Solve $P = 2L + 2W$ for W.

$P = 2L + 2W$	Subtract $2L$ from both sides.
$P - 2L = 2L - 2L + 2W$	Simplify the right side.
$P - 2L = 2W$	Divide both sides by 2.
$\dfrac{P - 2L}{2} = \dfrac{2W}{2}$	Simplify the right side.
$\dfrac{P - 2L}{2} = W$	Therefore, $W = \dfrac{P - 2L}{2}$.

d) Solve $3x + 4y = 5$ for y.

$3x + 4y = 5$	Subtract $3x$ from both sides.
$3x - 3x + 4y = 5 - 3x$	Simplify the left side.
$4y = 5 - 3x$	Divide both sides by 4.
$\dfrac{4y}{4} = \dfrac{5 - 3x}{4}$	Simplify the left side.
$y = \dfrac{5 - 3x}{4}$	Therefore, $y = \dfrac{5 - 3x}{4}.$

ADDITIONAL PRACTICE

Solve each of the following for the variable indicated.

e) $A = LW$ for L.

f) $P = \dfrac{W}{t}$, for t. (Power)

g) $V = \pi r^2 h$, for h. (Volume of a right circular cylinder)

h) $4x + 2y = 7$ for y.

PRACTICE EXERCISES

Solve each of the following formulas for the given variable.

5) $W = Fd$ for d

6) $E = \dfrac{1}{2}mv^2$ for m

7) Solve $P = 2L + 2W$ for L.

8) $2x + 3y = 9$ for y.

If more practice is needed, do the Additional Practice Exercises in the margin.

EXERCISE SET 3.9

Solve the formula for the indicated variable with the given information.

1) $A = \dfrac{bh}{2}$ for A with b $= 2$ mi and $h = 8$ mi.

2) $A = \dfrac{bh}{2}$ for b with $A = 30$ in^2 and $h = 10$ in.

3) $A = \dfrac{bh}{2}$ for h with $A = 14$ ft^2 and $b = 7$ ft.

4) $d = rt$ for d with $r = 60$ miles per hour and $t = 3$ hours.

5) $d = rt$ for t with $r = 50$ km per hour and $d = 75$ km.

6) $d = rt$ for r with $t = 6$ hours and $d = 720$ km.

7) $C = 2\pi r$ for r with $C = 25.12$ in and $\pi \approx 3.14$.

8) $C = 2\pi r$ for C with $r = 5$ m and $\pi \approx 3.14$.

9) $P = 2W + 2L$ for P with $W = 6$ in and $L = 15$ in.

10) $P = 2W + 2L$ for W with $L = 7$ cm and $P = 24$ cm.

In the following exercises we will omit units since many of them are unfamiliar to us.

Solve each formula for the indicated variable with the given value(s).

11) $A = \dfrac{1}{2}(B + b)h$ for A with $B = 5$, $b = 7$, and $h = 4$.

12) $A = \dfrac{1}{2}(B + b)h$ for h with $B = 6$, $b = 8$, and $A = 35$.

13) $V = iR$ for V with $i = 3$ and $R = 5$.
(Ohm's law in electricity)

14) $V = iR$ for R with $V = 120$ and $i = 3$.

15) $P = iV$ for P with $i = 0.5$ and $V = 120$.
(Electricity: power formula)

16) $P = iV$ for i with $P = 150$ and $V = 120$.

17) $W = mg$ for W with $m = 82$ and $g = 32$.
(Physics: formula for weight)

18) $W = mg$ for m with $W = 170$ and $g = 32$.

19) $P = \dfrac{F}{A}$ for P with $F = 180$ and $A = 15$.
(Physics: pressure on an area)

20) $P = \dfrac{F}{A}$ for F with $P = 50$ and $A = 8$.

Solve each formula for the indicated variable.

21) $K = C + 273$ for C.   **22)** $K - 273 = C$ for K.
(Conversion formulas for Kelvin and Celsius temperatures)

23) $P = kT$ for T.   **24)** $P = kT$ for k.
(Chemistry: gas law)

25) $F = ma$ for m.   **26)** $F = ma$ for a.
(Physics: law of motion)

27) $v = gt$ for t.   **28)** $v = gt$ for g.
(Physics: velocity of falling object)

29) $PV = kT$ for V.   **30)** $PV = kT$ for T.
(Chemistry: universal gas law)

31) $E = mc^2$ for m.   **32)** $E = mc^2$ for c^2.
(Nuclear Physics: conversion from mass to energy)

33) $I = prt$ for r.   **34)** $V = LWH$ for W
(Finance: formula for calculating interest)

35) $a^2 + b^2 = c^2$ for a^2.   **36)** $a^2 + b^2 = c^2$ for b^2.
(Geometry: Pythagorean Theorem for right triangles)

37) $d = \dfrac{1}{2}gt^2$ for g.   **38)** $d = \dfrac{1}{2}gt^2$ for t^2.
(Physics: distance traveled in free fall).

39) $D = \dfrac{M}{V}$ for M.   **40)** $D = \dfrac{M}{V}$ for V.

Solve each of the following for y.

41) $2x + y = 3$ **42)** $3x + y = 5$

43) $3x + 2y = 6$ **44)** $4x + 3y = 8$

45) $6x - 3y = 7$ **46)** $4x - 2y = 7$

CHALLENGE EXERCISES (47–50)

Solve the given formula for the indicated variable.

47) $A = \dfrac{1}{2}(B + b)h$ for B.   **48)** $A = \dfrac{1}{2}(B + b)h$ for b.

49) $F = k\dfrac{q_1 q_2}{r^2}$ for q_1.   **50)** $F = \dfrac{9}{5}C + 32$ for C.

Summary

Definition: [Section 3.1]

Linear Equations: A linear equation is any equation that can be put in the form $ax + b = c$, with only one variable raised to the first power and a, b, and c are constants.

Addition Property of Equality: [Section 3.1]

If $a = b$, then $a + c = b + c$ for any c. In words, we may add any number to both sides of an equation to get an equation with the same solution(s).

Multiplication Property of Equality: [Section 3.2]

If $a = b$, and $c \neq 0$, then $a \cdot c = b \cdot c$. In words, we may multiply both sides of an equation by a non-zero number to get another equation with the same solution(s).

Procedure for Solving Linear Equations: [Section 3.3]

1) If necessary, simplify both sides of the equation as much as possible. This could involve using the distributive, commutative, and associative properties and adding like terms.

2) If necessary, use the Addition Property of Equality to get all the terms with variables on one side of the equation and all the constant terms on the other.

3) If necessary, use the Multiplication Property of Equality to eliminate any coefficient on the variable.

4) Check the answer in the original equation.

Identities and Contradictions: [Section 3.3]

An identity is an equation that is true for all replacements of the variable for which the equation is defined. When solving an identity, we arrive at a statement that is obviously true like $10 = 10$. The solutions of an identity are all real numbers.

A contradiction is an equation that has no solution. When solving a contradiction, we arrive at a statement that is obviously false like $5 = 10$.

Procedure for Graphing Inequalities: [Section 3.4]

Equalities are graphed by placing a dot on the number line at the appropriate point. Inequalities are graphed by:

1) placing an open dot on the number line at the appropriate place and shading to the left for less than ($<$) or shading to the right for greater than ($>$).

2) placing a shaded dot on the number line at the appropriate place and shading to the left for less than

or equal to ($\leq$) or shading to the right for greater than or equal to ($\geq$).

Addition Properties of Inequalities: [Section 3.4]

1) If $a < b$, then $a + c < b + c$ for any c.

2) If $a > b$, then $a + c > b + c$ for any c.

If we add any number to both sides of an inequality we get an inequality with the same solutions.

Multiplication Properties of Inequalities: [Section 3.4]

1) If $a < b$ and $c > 0$, then $a \cdot c < b \cdot c$.

2) If $a > b$ and $c > 0$, then $a \cdot c > b \cdot c$.

If we multiply both sides of the inequality by a positive number, we get another inequality with the same solutions.

1) If $a < b$ and $c < 0$, then $a \cdot c > b \cdot c$.

2) If $a > b$ and $c < 0$, then $a \cdot c < b \cdot c$.

If we multiply both sides of an inequality by a negative number, we must reverse the direction of the inequality symbol.

Percent Equations: [Section 3.5]

Percent equations fall into three categories: 1) $a\%$ of b is what number? ($a\% \cdot b = x$) 2) $a\%$ of what number is b? ($a\% \cdot x = b$) 3) What percent of a is b? ($x\% \cdot a = b$) Translate into the appropriate equation and solve.

Problem Solving Procedures: [Section 3.6]

PROCEDURE FOR SOLVING REAL-WORLD PROBLEMS

1) Identify the unknown(s) and represent one of them with a variable. It is *usually* best to let the variable represent the smaller or smallest quantity.

2) Represent all other unknowns in the problem in terms of the same variable.

3) Write a "word" equation.

4) Write an algebraic equation using the word equation as a guide.

5) Solve the algebraic equation.

6) Be sure that you have answered the question asked.

7) Check the solution in the wording of the original problem.

SOLVING DISTANCE, RATE, TIME PROBLEMS

1) Make a chart with columns for the moving things, distance, rate, and time.

2) Fill in one column with known numerical values.

3) Assign a variable to an unknown and represent any other unknowns in terms of this variable. Fill in another column with these expressions containing variables.

4) Fill in the remaining column from the first two using the appropriate relationships between d, r, and t. In this section, $d = rt$.

5) Write the equation using the information in the last column filled in. It is sometimes helpful to draw a diagram.

6) Solve the equation.

7) Be sure that you have answered the question asked.

8) Check the solution in the wording of the original problem.

Money, Investment, and Mixture Problems [Section 3.7]

PROCEDURE FOR SOLVING MONEY PROBLEMS

1) Draw a chart with four columns and a row for each item. Label the columns Type of Item, Price per Unit, Number of Units, and Total Value. Fill in the Type of Item column.

2) Fill in the Price per Unit column. These values will usually be given, but may involve a variable.

3) Fill in the Number of Units column. This will usually involve a variable. It is usually necessary to assign the variable to the number of units of one item and express the number of units of the other item in terms of this variable.

4) Fill in the Total Value column. The total value is the product of the price per unit and the number of units.

5) Write the "word" equation. This will almost always be the sum of the values of the individual items equals the total value of all the items.

6) Write the algebraic equation using the word equation as a guide.

7) Solve the equation.

8) Be sure you answered the question asked.

9) Check the solution in the wording of the original problem.

For coin problems, change the headings to Type of Coin, Value of Coin, Number of Coins, and Total Value.

For investment problems, change the headings to Type of Investment, Interest Rate (expressed as a decimal), Amount Invested, and Interest Earned.

For mixture problems, change the headings to Type of Ingredient, Price per Unit, Number of Units, and Total Value. Also add an extra row for the mixture.

Geometric Application Problems: [Section 3.8]

Know the formulas for the perimeters, areas, and volumes of common geometric figures.

The sum of all three angles of any triangle is $180°$.

Two angles are supplementary if the sum of their measures is $180°$.

Two angles are complementary if the sum of their measures is $90°$.

If two lines are parallel, then

1) pairs of alternate interior angles are congruent,

2) pairs of alternate exterior angles are congruent,

3) pairs of corresponding angles are congruent, and

4) interior angles on the same side of the transversal are supplementary.

Review Exercises

Solve using the Addition Property of Equality. [Section 3.1]

1) $x - 6 = -2$

2) $x + \dfrac{2}{3} = \dfrac{5}{6}$

3) $6v - 2 = 5v$

4) $10 + z = 7$

5) $5z - 28 - 4z = 12$

6) $6 + 2w + 3 = w$

7) $4u + 13 = 3u + 10$

8) $11 + 15b = 14b + 18$

9) $-4a + 14 + 9a = 5a + 2 - a$

10) $7z - 6 + 2z = 6z - 1 + 2z$

Translate the mathematical sentence to an English sentence in two ways. [Section 3.1]

11) $2x - 5 = 14$

12) $5 + x = -2$

Translate the English sentence into a mathematical sentence and solve. Use x as the variable. [Section 3.1]

13) Twelve more than some number is -2.

14) A number decreased by 4 is 6.

Solve the following.

15) John and Shelley share the cost of a TV. If the TV costs $360 and John pays $220, how much does Shelley pay?

16) Harold bought a suit for $230 after a discount of $50. What was the price of the suit before the discount?

Find the reciprocal of each of the following. [Section 3.2]

17) $\dfrac{-2}{7}$

18) 0.4

Solve using the Multiplication Property of Equality. [Section 3.2]

19) $5x = -40$

20) $35 = -7y$

21) $\dfrac{3}{5}t = 9$

22) $-0.6v = 18$

23) $4.2\,w + 17 = 0.2w$

24) $4.9u - 24 = 2.5u$

Translate the mathematical sentence to an English sentence. [Section 3.2]

25) $14y = -28$

26) $\dfrac{2}{3}x = -6$

Translate the English sentence to a mathematical sentence and solve. Use x as the variable. [Section 3.2]

27) The product of 4 and some number is -12.

28) A number divided by -2 is 3.

Solve the following equations. [Section 3.3]

29) $5w - 13 = 7w + 15$

30) $11 - 4z = 7 + 44$

Solve the following.

31) The cost of a TV is three times the cost of a VCR. If the TV costs $495, find the cost of the VCR.

32) Hans is renting a sofa for $12.75 per month. If a new sofa costs $306, how many months will it take for the rental fees to equal the cost of a new sofa?

33) A high school boasts that $\frac{3}{5}$ of its teachers have a Masters degree. If 45 teachers have a Masters degree, how many teachers does the school have?

34) The value of a piece of property increased by 2% last year which represented an increase of $3600. What was the value of the property before the 2% increase in value?

Simplify, then solve. [Section 3.3]

35) $5t - 11 - 9t + 6 = 19 + 8t$

36) $4 - 5u + 12 + 16u = -6u - 9$

37) $33 - 2(7r + 9) = -9 - 6r$

38) $19 + 24s = 8 - 3(5 - 9s)$

39) $0.6(8 - 5v) + 0.4 = v + 1.2$

40) $3.6 - 4(0.7p - 15) = 2.2p - 1.4$

41) $5(q - 3) - 7(11 - 2q) = 3$

42) $8(10 - 4a) + 15 = -3(a + 7)$

Determine whether the following are identities or contradictions. Give the solutions of each. [Section 3.3]

43) $2(x - 3) + 4x = 6(x - 2)$

44) $3(2x - 1) - 4x = 2(x - 4) + 5$

Translate the mathematical sentence to an English sentence. [Section 3.3]

45) $3x - 2 = 5$

46) $4(2x - 9) = 22$

Translate the English sentence to a mathematical sentence and solve. Use x as the variable. [Section 3.3]

47) The sum of 4 times a number and 3 is 11.

48) Fifty-four is equal to six times the difference of three times a number and nine.

Solve each of the following.

49) The charges for repairing a transmission were $255 including parts. If the mechanic worked for 4 hours and the parts cost $75, how much does the mechanic charge per hour?

50) Rachel paid $37.10 for a blouse which included 6% for taxes. What was the cost of the blouse?

Graph each of the following on the number line.
[Section 3.4]

51) $v \geq 0$

52) $w < 4$

53) $-1 \leq x < 5$

54) $-3 \leq y \leq 2$

Solve each of the following inequalities by using the Addition Property of Inequality. [Section 3.4]

55) $10p - 8 < 9p + 7$

56) $6 - 4q \leq -5q + 18$

57) $1.6x - 4 > 12 - 0.4x$

58) $9.2 - 15y \geq -14y + 7$

59) $14z + 9 - 9z < 6z + 21$

60) $10 - s + 14 > 7 - 2s - 3$

Translate the mathematical sentence to an English sentence in two ways. [Section 3.4]

61) $19 \leq t + 9$

Translate the English sentence to a mathematical sentence. [Section 3.4]

62) Eleven is greater than or equal to some number increased by six.

Solve by using the Multiplication Property of Inequality. [Section 3.4]

63) $6t \leq -42$

64) $-8u < 48$

Solve by using the Addition and Multiplication Properties of Inequalities. [Section 3.4]

65) $9 - 4x > 6x - 11$

66) $33y + 6 \geq 17y + 54$

67) $4z + 32 - 5z < 6z - 13 + 2z$

68) $21 - 8r - 7 > 9r - 1 - 2r$

69) $5(2u - 2) - 6u \leq 14$

70) $7v > 8(3 - v) - 9$

71) $6(2s + 4) - 9 \geq -3(4s + 3)$

72) $3(t - 4) + 6(2t + 3) \leq 6t - 39$

Solve the following. [Section 3.5]

73) What number is 55% of 160?

74) 12% of what number is 17.4?

75) 22.5 is what percent of 150?

76) What percent of 800 is 356?

77) If a saline solution is 25% salt, how many milliliters of salt are in 60 milliliters of the solution?

78) During a slow down of the economy, an airline laid off 8% of its employees. If 6000 employees were laid off, how many employees did the airline have before the lay off?

Solve each of the following. [Section 3.6]

79) In a large lecture class there were 53 more men than women. If there were 329 students in the class, how many women were in the class?

80) Ralph is nine years older than twice his son's age. If the sum of their ages is 60 years, how old is Ralph?

81) This year the legislature increased tuition for college students by 0.1 of last year's amount. If tuition is now $35 per credit hour, what was it last year? (Round to the nearest cent.)

82) The sum of three consecutive integers is 33. Find the integers.

83) Three consecutive odd integers are such that the sum of four times the first and the second is three times the third. Find the integers.

Solve each of the following. [Section 3.7]

84) An office purchased 50 reams of paper for which they paid $260. There were two types of paper. The lighter weight paper costs $4.00 per ream and the heavier paper costs $7.00 per ream. How many reams of each type paper did they purchase?

85) At the end of the day David puts all the nickels and dimes that he has in his pockets in a jar. At the end of the week he has a total of 32 coins whose total value is $2.20. How many of each type coin does he have?

86) Amy matched 5 of the 6 numbers in the Florida Lotto and, after the celebration party, had $6500 left which she decided to invest. She invested part of the money in municipal bonds paying 6% per year and the remainder in certificates of deposit paying 5% per year. How much did she invest in each if her total income from the two investments is $360 per year?

87) A chemist needs 500 ml of a solution that is 40% sulfuric acid but all that she can find in the lab are solutions that are 25% sulfuric acid and 50% sulfuric acid. How much of each should she mix?

88) A supermarket sells a mixture of blueberries and strawberries. How many pounds of blueberries that sell for $2.00 per pound must be mixed with 12 pounds of strawberries that sell for $3.00 per pound if the mixture sells for $2.60 per pound?

Solve the following. [Section 3.8]

89) In a suburban area a residential lot is typically 25 feet longer than it is wide. If the perimeter of the lot is 350 feet, what are the dimensions of the lot?

90) In a triangle, the middle angle is 45° and the largest angle is 15° more than twice the smallest angle. Find the smallest and largest angles.

91) The measure of an angle is 16° more than the measure of its complement. Find the measure of each.

92) Given that L_1 is parallel to L_2 in the figure below, find x.

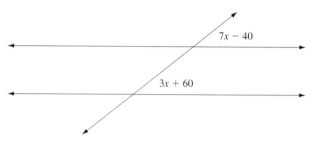

Solve the formula for the indicated variable with the given information. [Section 3.9]

93) $C = 2\pi r$ for C if $r = 3$. Use $\pi \approx 3.14$.

Solve the formula for the indicated variable. [Section 3.9]

94) $ax + b = c$ for x.

Chapter 3 Test

Solve the following equations.

1) $3u + 4 = 2u - 9$

2) $30 - 5v - 19 = 8 - 6v$

3) $-7k = 42$

4) $0.8m = -0.54$

5) $6v - 13 = 8v + 7$

6) $5t + 7 - 9t = 4t - 30$

7) $13 - 0.5(y + 4) = 3.5y - 9$

8) $-3(2a - 6) + 15 = 4(a + 11) - 5a$

Solve the following inequalities.

9) $12p - 3 > 1 + 11p$

10) $14 - 2.3q \leq 6 - 1.3q$

11) $5u - 10 - 3u \geq u + 7$

12) $14x - 7 + 2x > 9x + 42$

13) $10(y - 5) \leq 12y - 38$

14) $23 - 5(4z - 1) < 7(6 - 3z) + 2z$

Translate the mathematical sentences to English sentences.

15) $\frac{x}{3} + 4 = 6$

16) $2(s - 5) = 3s$

Translate the English sentences to mathematical sentences.

17) Five more than seven times a number is equal to three times the number decreased by one.

18) Three times the difference of a number and 6 is one more than the number.

Solve each of the following.

19) 125 is what percent of 200?

20) 100 is 125% of what number?

21) Ericka receives $150 plus a 4% commission on all her sales. If she earned $360 last weekend, what were her sales?

22) Oscar paid $124 for 4 pairs of shorts all of which cost the same. If he had a coupon for $20 off, what was the cost of each pair of shorts?

23) A certain wine is 12% alcohol by volume. If a bottle contains 750 milliliters of wine, how many milliliters of alcohol does the wine contain?

24) The women's basketball team at Central College outscored their opponents by 143 points last season. If the total number of points scored by both Central and opponents when they played each other was 4921 points, how many points did Central score?

25) The sum of three consecutive integers is equal to four times the second integer. Find the integers.

26) A building contractor purchased a total of 25 light bulbs some of which were incandescent and some of which were fluorescent at a total cost of $24. If the incandescent bulbs cost $0.60 each and the fluorescent bulbs cost $1.50 each, how many of each did he buy?

27) A long distance runner left his house running at an average rate of 8 miles per hour. Fifteen minutes (one-fourth of an hour) later his son left his house on his bike traveling on the same route at an average rate of 12 miles per hour. How long will it take the son to catch up with his father?

28) A television picture tube is 9 cm longer than it is wide. If the perimeter of the tube is 142 cm, what are the length and width of the tube?

Solve the following for the indicated variable.

29) $P = 2W + 2L$ for W.   **30)** $A = \frac{1}{2}(B + b)h$ for B.

4

Factoring and Solving Quadratic Equations

4.1 Prime Factorization and Greatest Common Factor

4.2 Factoring Polynomials with Common Factors and by Grouping

4.3 Factoring Binomials

4.4 Factoring Perfect Square Trinomials

4.5 Factoring General Trinomials with Leading Coefficient of One

4.6 Factoring General Trinomials with First Coefficient Other than One

4.7 Mixed Factoring

4.8 Solving Quadratic Equations by Factoring

CHAPTER OVERVIEW

The prerequisite skills for this chapter are in Chapter 2, where you learned the product and quotient laws of exponents and the techniques for multiplying polynomials. As you will see, the procedure for factoring a polynomial is the opposite of finding the product of polynomials. For that reason, you might want to review Chapter 2 prior to beginning Chapter 4.

Knowledge of factoring is crucial for many topics to be discussed in later chapters. In Chapter 5, we must be able to factor in order to reduce rational expressions, to find the products or quotients of rational expressions, and to find the sum or differences of rational expressions.

SECTION 4.1

Prime Factorization and Greatest Common Factor

OBJECTIVES

When you complete this section, you will be able to:

Ⓐ Recognize prime and composite numbers.

Ⓑ Find the prime factorization of composite numbers.

Ⓒ Find the greatest common factor of two or more natural numbers or monomials.

INTRODUCTION

Prime factorization was first discussed in Section 1.2 and we will provide a brief review of the procedure in this section. Recall also that a natural number is prime if it is divisible only by itself and 1. The first few prime numbers are 2, 3, 5, 7, 11, 13, ... A natural number that is not prime is composite.

Following is the procedure for finding the prime factorization of a number first given in Section 1.2.

• Finding prime factorizations

EXAMPLE 1

Find the prime factorization for each of the following.

a) 60   2 is the smallest prime number that divides evenly into 60.

     30   30 is also divisible by 2.
$$2\overline{)60}$$

     15   15 is divisible by 3.
$$2\overline{)30}$$
$$2\overline{)60}$$

      5   5 is prime.
$$3\overline{)15}$$
$$2\overline{)30}$$
$$2\overline{)60}$$

Therefore, $60 = 2 \cdot 2 \cdot 3 \cdot 5 = 2^2 \cdot 3 \cdot 5$, which is the product of all the divisors and the last quotient, is the prime factorization of 60.

b) 36   2 is the smallest prime number that divides evenly into 36.

     18   18 is also divisible by 2.
$$2\overline{)36}$$

      9   3 is the smallest prime that divides into 9 evenly.
$$2\overline{)18}$$
$$2\overline{)36}$$

      3   3 is prime.
$$3\overline{)9}$$
$$2\overline{)18}$$
$$2\overline{)36}$$

Therefore, $36 = 2 \cdot 2 \cdot 3 \cdot 3 = 2^2 \cdot 3^2$, which is the product of all the divisors and the last quotient, is the prime factorization of 36.

We will now use prime factorizations in order to find the **greatest common factor** of two or more natural numbers. First we need to know what is meant by the greatest common factor. The factors of 12 are {1,2,3,4,6,12} and the factors of 18 are {1,2,3,6,9,18}. Therefore, the factors that 12 and 18 have in common are {1,2,3,6}. The largest factor that 12 and 18 have in common is 6. Therefore, 6 is the greatest common factor for 12 and 18. Remember, a number is divisible by all of its factors. So, the greatest common factor for a set of numbers is the largest number that will divide evenly into each number of the set. For 12 and

18, 6 is the greatest common factor, so 6 is the largest number that will divide evenly into both.

Listing the factors and picking out the greatest common factor works for small numbers, but it would certainly be difficult to use this procedure for a pair of large numbers like 108 and 144. We will find the greatest common factor for groups of numbers using the procedure outlined below.

- Finding the greatest common factor of two or more natural numbers

---

FINDING THE GREATEST COMMON FACTOR (GCF)

1) Write each number as the product of its prime factors.

2) Select those factors that are common in each prime factorization.

3) For each factor that is common, select that factor with the smallest power that appears in any of the prime factorizations.

4) The greatest common factor is the product of the factors found in step 3.

We demonstrate this procedure with some examples.

EXAMPLE 2

Find the greatest common factor (GCF) for each of the following.

a) 12 and 18 (We already know the answer is 6 from the introduction.)

$$12 = 2^2 \cdot 3 \qquad \text{Prime factorization.}$$
$$18 = 2 \cdot 3^2 \qquad \text{Prime factorization.}$$

The factors that are common are 2 and 3. The smaller power of 2 that appears in either factorization is 2^1 in $2 \cdot 3^2$ and the smaller power of 3 that appears in either factorization is 3^1 in $2^2 \cdot 3$. Therefore, the GCF is $2^1 \cdot 3^1 = 6$. Consequently, 6 is the largest number that will divide evenly into 12 and 18.

b) 60 and 72

$$60 = 2^2 \cdot 3 \cdot 5 \qquad \text{Prime factorization.}$$
$$72 = 2^3 \cdot 3^2 \qquad \text{Prime factorization.}$$

The factors that are common are 2 and 3 (5 is not a factor of 72). The smaller power of 2 that appears in either factorization is 2^2, in $2^2 \cdot 3 \cdot 5$, and the smaller power of 3 is 3^1, in $2^2 \cdot 3 \cdot 5$. Therefore, the GCF is $2^2 \cdot 3 = 12$. This means that 12 is the largest number that will divide into 60 and 72 evenly.

c) 6 and 35

$$6 = 2 \cdot 3 \qquad \text{Prime factorization.}$$
$$35 = 5 \cdot 7 \qquad \text{Prime factorization.}$$

There are no prime factors in common, so the greatest common factor is 1. Remember, 1 is a factor of every number. In other words, the largest number that will divide evenly into 6 and 35 is 1.

d) 54, 126, and 120

$$54 = 2 \cdot 3^3 \qquad \text{Prime factorization.}$$
$$126 = 2 \cdot 3^2 \cdot 7 \qquad \text{Prime factorization.}$$
$$120 = 2^3 \cdot 3 \cdot 5 \qquad \text{Prime factorization.}$$

The factors that are common are 2 and 3. The smallest power of 2 that appears in any factorization is 2^1 which occurs in $2 \cdot 3^3$ and $2 \cdot 3^2 \cdot 7$. The smallest power of 3 is 3^1 in $2^3 \cdot 3 \cdot 5$. Therefore, the greatest common factor is $2 \cdot 3 = 6$. Consequently, 6 is the largest number that will divide evenly into 54, 126, and 120 evenly.

Why does this procedure work? Since we are looking for the greatest *common* factor, we select only those factors that are *common*. Why the smallest exponent? The exponent indicates how many times the base is used as a *factor*. We are looking for the greatest number of factors common. Hence, the greatest number of times the *factor* is *common* is the smallest power of that *factor*.

ADDITIONAL PRACTICE

Find the greatest common factor for each of the following.

a) 55 and 21

b) 60 and 140

c) 12, 28, and 42

• **Finding the greatest common factor of two or more monomials which contain variables.**

PRACTICE EXERCISES

Find the greatest common factor for each of the following.

1) 24 and 30   **2)** 15 and 77   **3)** 54 and 90   **4)** 56, 112, and 140

If more practice is needed, do the Additional Practice Exercises in the margin.

The procedure for finding the GCF of a monomial containing variables is exactly the same, except usually easier, since we do not have to factor the variables. The same procedure also applies if one or more of the factors is a polynomial other than a monomial.

EXAMPLE 3

Find the GCF for each of the following.

a) a^2b^3 and a^4b^2

The factors a and b are common. The smaller power of a is a^2 and the smaller power of b is b^2. Therefore, the GCF is a^2b^2. This means a^2b^2 is the monomial of greatest degree that divides evenly into a^2b^3 and a^4b^2.

b) a^2b^4 and c^3d^2

There are no factors in common. Therefore, the GCF is 1.

 c) $6c^2d^3$ and $8cd^5$

We need to consider the coefficients and the variables separately. The GCF of 6 and 8 is 2. The variables c and d are common, and the smaller power of c is c^1 and the smaller power of d is d^3. Therefore, the GCF is $2cd^3$. So, $2cd^3$ is the monomial of greatest degree that will divide evenly into $6c^2d^3$ and $8cd^5$.

d) $8x^2yz^4$, $12x^4y^2z^3$, and $14x^2y^3$

The GCF of 8, 12, and 14 is 2. The only variables common are x and y and the smallest powers are x^2 and y^1 respectively. Therefore, the GCF is $2x^2y$.

e) $9x^2(x + 2)$, $6x^4(x + 2)$

The GCF of 9 and 6 is 3. The only variable factor is x and the smallest power on x is 2, but the binomial $(x + 2)$ is also a common factor. Therefore, the GCF is $3x^2(x + 2)$.

Find the GCF of each of the following.

d) p^3q^2 and pq^4

e) $m^3n^2p^3$ and $a^4b^3c^5$

f) $6a^4b^2$ and $9ab^2$

g) $10x^3yz^2$, $15y^3z^4$ and $25x^2y^4z^5$

h) $12a^3(a+5)$, $8a^2(a+5)$

PRACTICE EXERCISES

Find the GCF of each of the following.

5) c^4d^2 and c^3d^5

6) xy^3 and c^2d

7) $18a^3b^4$ and $24a^2b^5$

8) $15x^4y^2z$, $9x^2z^3$, and $18xy^3z^2$

9) $8y^2(y-3)$, $10y(y-3)$

If more practice is needed, do the Additional Practice Exercises in the margin.

ANSWERS:

Practice 1–4

1) 6 **2)** 1 **3)** 18 **4)** 28

Additional Practice a–c

a) 1 **b)** 20 **c)** 2

EXERCISE SET 4.1

Find the prime factorization of each of the following.

1) 100 **2)** 225 **3)** 154 **4)** 232

5) 180 **6)** 126 **7)** 315 **8)** 525

Find the greatest common factor (GCF) of each of the following.

9) 6 and 9 **10)** 8 and 12 **11)** 65 and 35

12) 22 and 55 **13)** 33 and 10 **14)** 65 and 28

15) 12 and 45 **16)** 68 and 76 **17)** 90 and 140

18) 60 and 140 **19)** 180 and 270 **20)** 165 and 175

21) 6, 13 and 21 **22)** 14, 35, and 49

23) 8, 16 and 24 **24)** 12, 30, and 42

25) 16, 32, and 56 **26)** 24, 48 and 60

27) 48, 96 and 120 **28)** 72, 108 and 180

29) mn^3 and m^2n **30)** a^4b and ab^2

31) c^3d^2 and cd^3 **32)** pq^4 and p^3q^4

33) a^3b^5 and a^3b^2 **34)** m^4n^6 and m^2n^6

35) j^3k^4 and a^2b^3 **36)** r^3s^6 and s^3t

37) $a^3b^3c^3$ and a^2bc^2 **38)** $x^4y^3z^2$ and xy^4z^3

39) pq^2r^4 and p^2r^3 **40)** a^4b^2c and a^2c^3

41) $18pq^4$ and $30p^2q$ **42)** $16c^5d$ and $24cd^3$

43) $27s^3t^3$ and $36s^2t^5$

44) $12p^2q^3$ and $24p^5q^4$

45) $9a^2b^4c^2$, $15a^3bc^2$, $12ab^3c^2$

46) $5x^3y^6z$, $10xy^2z$, $20x^2y^2z$

47) $8a^4b^3c^2$, $16a^6b^3c^2$, $4a^2b^5c^5$

48) $12r^5s^4t^2$ $24r^3s^4t^3$, $28r^2s^4t^5$

49) $24p^2q^4$, $33q^2r^3$, $42p^3r^4$

50) $24a^4c^2$, $40a^5b^2$, $56b^4c^2$

51) $14b^3(b+3)$, $21b(b+3)$

52) $15r^4(r-2)$, $12r^6(r-2)$

53) $16c^5d^2(a-3)$, $24cd^3(a-3)$

54) $18a^2b(c+4)$, $24a^3b(c+4)$

Situations similar to the concept of Greatest Common Factor often occur in everyday situations. Answer the following.

55) Juan has three nickels and five dimes and Fred has two nickels and eight dimes. How many of each type of coin do they have in common?

56) Danielle has four dimes and six quarters and Sandy has two dimes and four quarters. How many of each type of coin do they have in common?

57) Frank has six New York Yankees baseball cards and eight Atlanta Braves baseball cards. Bill has four New York Yankees baseball cards and 10 Atlanta Braves baseball cards. How many cards of the same teams do they have in common?

58) Merlene has three yellow blouses and five red blouses. Hazel has six yellow blouses and two red blouses. How many blouses of the same color do they have in common?

59) Bill and Don went fishing. Bill caught five redfish and eight trout while Don caught three redfish and 12 trout. How many fish of the same species do they have in common?

CHALLENGE EXERCISES (60–63)

Find the GCF for the following.

60) 108 and 144

61) 240 and 384

62) $48a^3b^2(x+2)^3, 64a^4b(x+2)^2$

63) $36x^5y^7(a-3)^4, 54x^4y^9(a-3)^7$

WRITING EXERCISES

64) If a and b are both prime numbers with $a \neq b$, what is the GCF of a and b? Explain.

65) Under what circumstances would a number be a common divisor of two other numbers?

66) A friend of yours is having trouble finding the GCF of two large numbers. Describe how you would help him.

GROUP PROJECT

67) Below is an example finding the GCF of 108 and 144 using a different procedure. Study the example and write a paragraph describing the procedure.

2	108	144
2	54	72
3	27	36
3	9	12
	3	4

Therefore, the GCF is $2 \cdot 2 \cdot 3 \cdot 3 = 2^2 \cdot 3^2 = 36$.

SECTION 4.2

Factoring Polynomials with Common Factors and by Grouping

INTRODUCTION

OBJECTIVES

When you complete this section, you will be able to:

Ⓐ Factor a polynomial by removing the greatest common factor.

Ⓑ Factor by grouping.

In section 1.2 we learned how to write a composite number and a monomial as the product of prime factors and reviewed this idea in the last section. This section is the first of six in which we will learn to write composite polynomials (other than monomials) as a product of **prime polynomials**. A prime polynomial, like a prime number, is a polynomial that can be written only as a product of itself and 1. For example, $x + 4$ is prime, but $x^2 - 4$ is not since $x^2 - 4 = (x+2)(x-2)$.

When asked to "factor a polynomial," what are we being asked to do? We know that when two or more numbers are multiplied, each is called a factor of the product. So, factoring a polynomial means to write the polynomial as the product of two or more prime polynomials in the same way factoring a composite number means to write the composite number as the product of two or more prime numbers. Consider the following.

Composite Number	Factors	Composite Polynomial	Factors
$15 = 3 \cdot 5$	3, 5	$x^2 - 4 = (x+2)(x-2)$	$x+2, x-2$
$22 = 2 \cdot 11$	2, 11	$2x^2 + 4x = 2x(x+2)$	$2, x, x+2$

• **Removing the greatest common factor**

The distributive property in the form $ab + ac = a(b + c)$ provides the technique for factoring by removing the greatest common factor. In the binomial $ab + ac$, a is a factor of the first term, ab, and a is also a factor of the second term, ac. Therefore, a is a factor common to both terms. The common factor (a in this case) is written outside the parentheses and the polynomial inside the parentheses is determined by one of two methods.

1) We determine what term must be multiplied by the common factor in order to give the corresponding term of the polynomial being factored. In the above example, $ab + ac = a(\_ + \_)$. The first blank is filled by the term whose product with a is ab. This is b. The second blank is the term whose product with a is ac. This is c. Consequently, $ab + ac = a(b + c)$. This process is sometimes referred to as **inspection**.

2) Find the quotient of the common factor and each of the terms of the polynomial. Using division, divide the first term, ab, by the common factor a, $\frac{ab}{a} = b$. Therefore, the first term inside the parentheses is b. Divide the second term, ac, by the common factor a, $\frac{ac}{a} = c$. Therefore, the second term inside the parentheses is c. So, $ab + ac = a(b + c)$. This division process works because of the definition of division. The common factor is a and the first term of the polynomial to be factored is ab. But $\frac{ab}{a} = b$ (the first term inside the parentheses). By the definition of division,

a (the common factor) $\cdot b$ (the first term inside the parentheses) $= ab$
(the first term of the polynomial being factored).

Any factorization may be verified by finding the product of the factors. If the product is the same as your original polynomial, then your factorization is correct. For example, $2x + 2y = 2(x + y)$ is correct since $2(x + y) = 2x + 2y$. However, we must be careful that the factor that we have removed is the greatest common factor. If the polynomial inside the parentheses still has a common factor, then we have not removed the greatest common factor. Try again!

When adding like terms we were really just removing the greatest common factor and adding the numerical coefficients that appeared inside the parentheses. For example, $4x + 7x$ has a common factor of x. Rewrite $4x + 7x$ as $(4 + 7)x$ by removing the common factor x. Then add 4 and 7 and get $11x$.

In the examples and exercises that follow, you will need to remember the Product Rule for exponents ($a^m \cdot a^n = a^{m+n}$) and the Quotient Rule for exponents ($\frac{a^m}{a^n} = a^{m-n}$).

EXAMPLE 1

Factor each of the following by removing the greatest common factor.

a) $bx + by$

The greatest common factor of bx and by is b. Write b outside the parentheses.

$b(\_ + \_)$

The first blank represents a term whose product with b is bx. The term is x. In expressions which are more complicated, we may need to divide each term of the polynomial by the common factor. If we divide the first term of the polynomial, bx, and by the common factor, b, we get the first term inside the parentheses, x. That is, $\frac{bx}{b} = x$.

$b(x + \_)$

The second blank must be a term whose product with b is by. By inspection or division, the blank represents y. So we now have:

$b(x + y)$

Therefore, $bx + by = b(x + y)$ where the first factor is b and the second factor is $x + y$.

Check:

$b(x + y) = bx + by$. Therefore, the factorization is correct.

b) $4x + 2$ — The GCF of $4x$ and 2 is 2. Write 2 outside the parentheses.

$2(\_ + \_)$ — The first blank represents a term whose product with 2 is $4x$. Since $2 \cdot 2x = 4x$, the blank represents $2x$. Using division, $\frac{4x}{2} = 2x$.

$2(2x + \_)$ — The second blank represents a term whose product with 2 is 2. Since $2 \cdot 1 = 2$, the blank represents 1. Using division, $\frac{2}{2} = 1$.

$2(2x + 1)$ — Therefore, $4x + 2 = 2(2x + 1)$ where the first factor is 2 and the second is $2x + 1$.

Check:

$2(2x + 1) = 4x + 2$. Therefore, the factorization is correct.

c) $8x^2 + 12x^3$ — The greatest common factor of $8x^2$ and $12x^3$ is $4x^2$. Write $4x^2$ outside the parentheses.

$4x^2(\_ + \_)$ — The first blank is a term whose product with $4x^2$ is $8x^2$. Since $4x^2 \cdot 2 = 8x^2$, the blank represents 2. Using division, $\frac{8x^2}{4x^2} = 2$.

$4x^2(2 + \_)$ — The second blank is a term whose product with $4x^2$ is $12x^3$. Since $4x^2 \cdot 3x = 12x^3$, the second blank represents $3x$. Using division, $\frac{12x^3}{4x^2} = 3x$.

$4x^2(2 + 3x)$ — Therefore, $8x^2 + 12x^3 = 4x^2(2 + 3x)$ where the first factor is $4x^2$ and the second is $2 + 3x$.

Check:

$4x^2(2 + 3x) = 8x^2 + 12x^3$. Therefore, the factorization is correct.

d) $12x^2y^3 - 18xy^4$ — The GCF of $12x^2y^3$ and $18xy^4$ is $6xy^3$. Write $6xy^3$ outside the parentheses.

$6xy^3(\_ + \_)$ — The first blank represents $2x$ since $6xy^3 \cdot 2x = 12x^2y^3$ or $\frac{12x^2y^3}{6xy^3} = 2x$.

$6xy^3(2x + \_)$ — The second blank represents $-3y$ since $6xy^3(-3y) = -18xy^4$ or $\frac{-18xy^4}{6xy^3} = -3y$.

$6xy^3(2x - 3y)$ — Therefore, $12x^2y^3 - 18xy^4 = 6xy^3(2x - 3y)$ where the first factor is $6xy^3$ and the second is $2x - 3y$.

Check:

$6xy^3(2x - 3y) = 12x^2y^3 - 18xy^4$. Therefore, the factorization is correct.

e) $10x^2y^3 - 5x^3y - 20x^2y^2$ — The GCF is $5x^2y$. Write $5x^2y$ outside the parentheses.

$5x^2y(\_ + \_ + \_)$ — The first term inside the parentheses is $2y^2$, the second $-x$, and the third $-4y$.

$5x^2y(2y^2 - x - 4y)$ — Therefore, $10x^2y^3 - 5x^3y - 20x^2y^2 = 5x^2y(2y^2 - x - 4y)$.

Check:

$5x^2y(2y^2 - x - 4y) = 10x^2y^3 - 5x^3y - 20x^2y^2$. Therefore, the factorization is correct.

ADDITIONAL PRACTICE

Factor each of the following by removing the greatest common factor.

a) $mn + mp$

b) $8p - 6$

c) $12p^4 + 18p^3$

d) $6c^2d^2 - 9c^3d^4$

e) $8a^2 - 12a + 4$

f) $9x^2y^4 - 15xy^3 - 18x^3y^4$

PRACTICE EXERCISES

Factor each of the following by removing the greatest common factor.

1) $pq + pr$

2) $6a + 9$

3) $8y^3 + 16y^2$

4) $10a^3b^2 - 15a^2b^2$

5) $8a^2b^4 - 16ab^2 - 12a^2b^3$

If more practice is needed, do the Additional Practice Exercises in the margin.

We usually want the coefficient of the first term inside the parentheses to be positive, so it is sometimes necessary to remove a negative common factor as illustrated by the following examples.

EXAMPLE 2

Factor the following by removing a negative common factor.

a) $-2a + 4b$ Since we want the first coefficient inside the parentheses to be positive, the GCF is -2.

$-2(\_ + \_)$ The first term is a since $\frac{-2a}{-2} = a$ and the second term is $-2b$ since $\frac{4b}{-2} = -2b$.

$-2(a - 2b)$ Therefore, $-2a + 4b = -2(a - 2b)$.

Check:
$-2(a - 2b) = -2a + 4b$. Therefore, the factorization is correct.

b) $-3x^3 + 9x^2 - 3x$ The GCF is $-3x$. Write $-3x$ outside the parentheses.

$-3x(\_ + \_ + \_)$ The first term is x^2 since $\frac{-3x^3}{-3x} = x^2$, the second term is $-3x$ since $\frac{9x^2}{-3x} = -3x$, and the third term is 1 since $\frac{-3x}{-3x} = 1$.

$-3x(x^2 - 3x + 1)$ Therefore, $-3x^3 + 9x^2 - 3x = -3x(x^2 - 3x + 1)$.

Check:
$-3x(x^2 - 3x + 1) = -3x^3 + 9x^2 - 3x$. Therefore, the factorization is correct.

c) $-a + b$ We want the first term inside the parentheses to be positive, so factor out -1. We do not write -1, just the "$-$".

$-(\_ + \_)$ The first term is a since $\frac{-a}{-1} = a$ and the second is $-b$ since $\frac{b}{-1} = -b$.

$-(a - b)$ Therefore, $-a + b = -(a - b)$.

Check:
$-(a - b) = -a + b$. Therefore, the factorization is correct.

ADDITIONAL PRACTICE

Factor the following by removing a negative common factor.

g) $-10p + 15q$

h) $-8p^3 + 4p^2 - 16p$

i) $-r + s$

PRACTICE EXERCISES

Factor the following by removing a negative common factor.

6) $-9x + 12y$ **7)** $-4a^4 + 2a^3 - 6a^2$ **8)** $-x + y$

Practice 1–5

1) $p(q + r)$

2) $3(2a + 3)$

3) $8y^2(y + 2)$

4) $5a^2b^2(2a - 3)$

5) $4ab^2(2ab^2 - 4 - 3ab)$

Additional Practice a–f

a) $m(n + p)$

b) $2(4p - 3)$

c) $6p^3(2p + 3)$

d) $3c^2d^2(2 - 3cd^2)$

e) $4(2a^2 - 3a + 1)$

f) $3xy^3(3xy - 5 - 6x^2y)$

Practice 6–8

6) $-3(3x - 4y)$

7) $-2a^2(2a^2 - a + 3)$

8) $-(x - y)$

Additional Practice g–i

g) $-5(2p - 3q)$

h) $-4p(2p^2 - p + 4)$

i) $-(r - s)$

ADDITIONAL PRACTICE

Factor the following by removing the common binomial factor.

j) $r(s + t) + p(s + t)$

k) $z(z + 5) - 6(z + 5)$

• **Factoring by grouping**

If more practice is needed, do the Additional Practice Exercises in the margin.

In the previous examples and practice exercises, the common factor has been a monomial, but that is not always the case. The common factor can also be any polynomial. In the following examples, the common factor is a binomial.

EXAMPLE 3

Factor the following by removing the common binomial factor.

a) $a(c + d) + b(c + d)$ — The factors of the first term are a and $(c + d)$ and the factors of the second term are b and $(c + d)$. So, the common factor is $(c + d)$. Write $(c + d)$ outside the parentheses. To show the entire quantity $(c + d)$ is a factor, put it in parentheses.

$(c + d)(\_ + \_)$ — The first blank is the term whose product with $(c + d)$ is $a(c + d)$ which is a.

$(c + d)(a + \_)$ — The second blank is the term whose product with $(c + d)$ is $b(c + d)$ and is b.

$(c + d)(a + b)$ — Therefore, $a(c + d) + b(c + d) = (c + d)(a + b)$. The check is more difficult in this case. Compare to the distributive property $x(y + z)$ where x is $(c + d)$, y is a, and z is b. Hence, $(c + d)(a + b) = (c + d)a + (c + d)b = a(c + d) + b(c + d)$ by the commutative property of multiplication.

b) $x(x + 2) - 5(x + 2)$ — The common factor is $(x + 2)$.

$(x + 2)(\_ + \_)$ — The first blank is the term whose product with $(x + 2)$ is $x(x + 2)$. This is x.

$(x + 2)(x + \_)$ — The second blank is the term whose product with $(x + 2)$ is $-5(x + 2)$. This is -5.

$(x + 2)(x - 5)$ — Therefore, $x(x + 2) - 5(x + 2) = (x + 2)(x - 5)$.

PRACTICE EXERCISES

Factor the following by removing the common binomial factor.

9) $x(y - z) + w(y - z)$

10) $y(y + 4) - 5(y + 4)$

If more practice is needed, do the Additional Practice Exercises in the margin.

We will now remove both common monomial and binomial factors to factor polynomials with four terms. We will group terms and remove a common factor from each group. The results will have a common binomial factor that we will then remove. Study the following examples carefully.

EXAMPLE 4

Factor the following.

a) $ab + ac + db + dc$ — Group the first two terms together and remove the common factor of a. Group the third and fourth terms together and remove a common factor of d.

$a(b + c) + d(b + c)$ — We now have a common factor of $(b + c)$ which we will remove.

$(b + c)(a + d)$ — Factorization.

Check:
$(b + c)(a + d) = ab + bd + ac + cd$ which is correct, except for the order of the terms. What property states the order of the terms does not matter?

 b) $xy + xz + 3y + 3z$ — Group the first two terms together and remove the common factor of x. Group the third and fourth terms together and remove the common factor of 3.

$x(y + z) + 3(y + z)$ — Now remove the common factor of $y + z$.

$(y + z)(x + 3)$ — Factorization.

Check:
$(y + z)(x + 3) = xy + 3y + xz + 3z$ which is correct.

c) $4x - xy + 20 - 5y$ — Remove the common factor of x from the first two terms and the common factor of 5 from the remaining terms.

$x(4 - y) + 5(4 - y)$ — Now remove the common factof of $4 - y$.

$(4 - y)(x + 5)$ — Factorization.

Check:
$(4 - y)(x + 5) = 4x + 20 - xy - 5y$ which is correct.

Note: When factoring by grouping it is often possible to group the terms in more than one way. Example 1b above could have been done as follows.

d) $xy + xz + 3y + 3z$ — Change the order of the terms to $xy + 3y + xz + 3z$.

$xy + 3y + xz + 3z$ — Group the first two terms together and remove the common factor of y. Group the third and fourth terms together and remove the common factor of z.

$y(x + 3) + z(x + 3)$ — Now remove the common factor of $x + 3$.

$(x + 3)(y + z)$ — Factorization.

Note the answer is the same as above except for the order of the factors.

ADDITIONAL PRACTICE

Factor the following.

l) $mn + mp + qn + qp$

m) $xc - xf + 5c - 5f$

n) $5a + ab + 15 + 3b$

PRACTICE EXERCISES

Factor the following.

11) $ab - ac + xb - xc$ **12)** $xy + xz + 7y + 7z$ **13)** $3y + xy + 12 + 4x$

If more practice is needed, do the Additional Practice Exercises in the margin.

The goal of this type of factoring is to have exactly the same polynomial inside the parentheses after we have grouped the terms and removed the common factors from each group. Consequently, it is sometimes necessary to remove a negative factor in order to make the signs of the binomial factors agree.

EXAMPLE 5
Factor the following.

a) $ax + az - bx - bz$ — Group the first two terms and remove the common factor of a.

$a(x + z) - bx - bz$ — Since we want $x + z$ inside both sets of parentheses, factor $-b$ from the remaining terms. Be careful! This will change the signs of both terms inside the parentheses.

$a(x + z) - b(x + z)$ Now remove the common factor of $x + z$.

$(x + z)(a - b)$ Factorization.

Check:

$(x + z)(a - b) = ax - bx + az - bz$ which is correct.

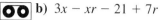 **b)** $3x - xr - 21 + 7r$ Remove the common factor of x from the first two terms.

$x(3 - r) - 21 + 7r$ We want $3 - r$ inside both sets of parentheses, so remove the factor of -7 from the remaining two terms.

$x(3 - r) - 7(3 - r)$ Now remove the common factor of $(3 - r)$.

$(3 - r)(x - 7)$ Factorization.

Check:

$(3 - r)(x - 7) = 3x - 21 - rx + 7r$ which is correct.

c) $3x - xy - 3 + y$ Group the first two terms and remove the common factor of x.

$x(3 - y) - 3 + y$ Notice the two remaining terms are the same as the terms inside the parentheses except the signs are opposite. Factor -1 from the last two terms.

$x(3 - y) - 1(3 - y)$ Now remove the common factor of $3 - y$.

$(3 - y)(x - 1)$ Factorization.

Check:

$(3 - y)(x - 1) = 3x - 3 - xy + y$ which is correct.

ADDITIONAL PRACTICE

Factor the following.

o) $nm + nw - xm - xw$

p) $4x - xr - 20 + 5r$

q) $xe - e - x + 1$

PRACTICE EXERCISES

Factor the following.

14) $xy + hy - xc - hc$ **15)** $4a - ab - 16 + 4b$ **16)** $xy - y - x + 1$

If more practice is needed, do the Additional Practice Exercises in the margin.

EXERCISE SET   4.2

Factor the following by removing the greatest common factor.

1) $cd + cf$

2) $gh + gk$

3) $rs - rt$

4) $xy - xz$

5) $3x + 9$

6) $4x + 8$

7) $8x - 12$

8) $15x - 10$

9) $x^2 + x$

10) $y^2 + y$

11) $r^4 - r^2$

12) $q^3 - q^2$

13) $7a^3 + 14a^2$

14) $5r^4 + 10r^2$

15) $18p^3 - 9p^5$

16) $16r^3 - 8r^4$

17) $15x^3 + 5x$

18) $12d^4 + 6d$

19) $22r^5 - 11r^2$

20) $26t^4 - 13t$

21) $12c^3 + 8c^2$

22) $14b^5 + 8b^3$

23) $18z^2 - 12z^6$

24) $15w^4 - 10w^6$

25) $x^2y^3 + x^2y^2$

26) $r^3s^4 + r^4s^3$

27) $u^4v^2 - u^3v^3$

28) $p^5q^2 - p^2q^4$

29) $c^3d^4 + cd^2$

30) $a^4b^2 + a^3b$

31) $x^2y^2 - x^3y^4$

32) $a^3b^2 - a^5b^5$

33) $18a^2b^4 + 12a^3b^2$

34) $18r^3s^2 + 27r^5s$

35) $9xy^3 - 12x^2y^2$

36) $24fg^4 - 18f^4g^3$

37) $9x^2y^4 + 3xy^3$

38) $8y^3z^2 + 4y^2z$

39) $9x^2 - 12x + 6$

40) $12x^2 - 8x + 12$

41) $15x^4 - 10x^3 - 20x^2$

42) $24y^5 - 16y^4 - 32y^2$

43) $6a^4 + 12a^2 - 24$

44) $9s^5 + 18s^3 - 27$

45) $22x^3 - 33x^2 - 11x$

46) $21a^4 - 28a^2 - 7a$

47) $16c^3d^2 - 24c^2d^4 + 36cd^2$

48) $16r^2s^4 - 8r^3s^4 + 12rs^3$

49) $14x^3y^3 - 21x^2y^4 - 7xy^2$

50) $25y^4z^4 - 30y^3z^2 - 5y^2z^2$

Factor the following by removing a negative common factor.

51) $-3m + 6n$ **52)** $-4a + 8b$

53) $-16c + 8d$ **54)** $-20p + 15q$

55) $-14x + 7$ **56)** $-18a + 9$

57) $-4x^2 + 8x - 16$ **58)** $-6x^2 + 18x - 24$

59) $-10x^3 - 15x^2 + 25x$ **60)** $-8a^3 - 24a^2 + 16a$

Factor the following by removing the common binomial factor.

61) $a(m + n) + b(m + n)$ **62)** $x(p + q) + y(p + q)$

63) $a(c + 2) - b(c + 2)$ **64)** $r(u + 5) - s(u + 5)$

65) $t(t - 3) + 6(t - 3)$ **66)** $y(y - 3) + 3(y - 3)$

67) $a(a - 6) - 7(a - 6)$ **68)** $b(b - 5) - 4(b - 5)$

69) What are the factors in the expression $(2x + 3)(3x - 5)$?

70) What are the factors in the expression $(5x - 2)(2x + 7)$?

CHALLENGE EXERCISES (71–74)

Remove the greatest common factor from each of the following.

71) $360x^6y^5 - 540x^8y^4$ **72)** $756a^9b^6 - 504a^7b^9$

73) $3(a + b)^3 + 9(a + b)^2$ **74)** $4(x + y)^4 - 12(x + y)^2$

Factor each of the following by grouping.

75) $xz + xw + yz + yw$ **76)** $ac + ad + bc + bd$

77) $mn - 4m + 3n - 12$ **78)** $dc - 2d + 5c - 10$

79) $ab + 5a + 15 + 3b$ **80)** $mn + 6n + 24 + 4m$

81) $x^2 - xy + 5x - 5y$ **82)** $a^2 - ab + 3a - 3b$

83) $ab - 3a + b - 3$ **84)** $xy - 5x + y - 5$

85) $xm + xn - my - ny$ **86)** $cd + cf - de - ef$

87) $mn - 6n - 3m + 18$ **88)** $rs - 7s - 4r + 28$

89) $cd - ce - 3d + 3e$ **90)** $pr - pq - 7r + 7q$

91) $2x^2 - xy + 6x - 3y$ **92)** $3r^2 - rs + 12r - 4s$

93) $rs + 4r - s - 4$ **94)** $yz + 5y - z - 5$

95) $cd - 3c - d + 3$ **96)** $ab - 7a - b + 7$

97) $10a + 4ab + 15 + 6b$ **98)** $12x + 9bx + 8 + 6b$

99) $8xz - 4xw - 2yz + yw$

100) $15ac - 5ad - 3bc + bd$

101) $6ac + 4ad - 9bc - 6bd$

102) $6xz + 9xw - 8yz - 12yw$

103) $8x^2 - 2xy + 12x - 3y$

104) $6a^2 - 9ay + 10a - 15y$

WRITING EXERCISES

105) $3a^2b^2(6b + 4a - 8) = 18a^2b^3 + 12a^3b^2 - 24a^2b^2$. However, $3a^2b^2(6b + 4a - 8)$ is not the correct prime factorization of $18a^2b^3 + 12a^3b^2 - 24a^2b^2$. *Why not?*

106) How can we tell if we have removed the greatest common factor? (i.e., What will happen if the factor we remove is a factor of the polynomial but not the greatest common factor?)

107) After factoring a polynomial, how can you determine if your factorization is correct?

SECTION 4.3

Factoring Binomials

OBJECTIVES

When you complete this section, you will be able to:

Ⓐ Factor binomials that are the difference of two squares.

Ⓑ Factor difference of squares in which one or more of the terms is a binomial.

Ⓒ Factor the sum and difference of cubes.

Recall that the product of two binomials that differed only by the signs of the second terms always resulted in a binomial that was the difference of two squares (Section 2.4). For example, $(a + b)(a - b) = a^2 - b^2$. In this case we are multiplying $(a + b)$ and $(a - b)$, so they are factors of $a^2 - b^2$. Consequently, the difference of squares, $a^2 - b^2$, factors into the sum, $(a + b)$, and difference, $(a - b)$, of the expressions being squared. Thus in factored form of $a^2 - b^2 = (a + b)(a - b)$.

FACTORING THE DIFFERENCE OF SQUARES

A binomial that is the difference of squares factors into two binomials. One of the factors is the sum of the expressions being squared and the other factor is the difference of the expressions being squared. Using variables, $a^2 - b^2 = (a + b)(a - b)$. Expressions of the form $a + b$ and $a - b$ are often called conjugate pairs.

To assist you in recognizing perfect squares, refer to Table 1 in the appendix. Each entry in the "n^2" column is the square of the corresponding entry in the "n" column. Consequently, each entry in the n^2 column is a perfect square. You should be able to recognize all perfect squares up to 225 from memory.

EXAMPLE 1

Factor the following difference of squares.

a) $x^2 - y^2 =$ The first term is the square of x and the second term is the square of y. Hence, the correct factorization is the product of the sum and difference of x and y.

$(x + y)(x - y)$ Factorization.

Check:
$(x + y)(x - y) = x^2 - y^2$. Therefore, the factorization is correct.

Note: Due to the commutative property of multiplication, the factors may be written in either order. So, $(x - y)(x + y)$ is also correct.

b) $a^2 - 9 =$ The first term is the square of a and 9 is the square of 3.

$a^2 - 3^2 =$ Therefore, the correct factorization is the product of the sum and difference of a and 3.

$(a + 3)(a - 3)$ Factorization.

Check:
$(a + 3)(a - 3) = a^2 - 9$. Therefore, the factorization is correct.

 c) $y^2 - 4z^2 =$ The first term is the square of y and the second term is the square of $2z$.

$y^2 - (2z)^2 =$ Therefore, the correct factorization is the product of the sum and difference of y and $2z$.

$(y + 2z)(y - 2z)$ Factorization.

Check:
$(y + 2z)(y - 2z) = y^2 - 4z^2$. Therefore, the factorization is correct.

d) $16m^2 - 9n^2 =$

The first term is the square of $4m$ and the second term is the square of $3n$.

$(4m)^2 - (3n)^2 =$

Therefore, the correct factorization is the product of the sum and difference of $4m$ and $3n$.

$(4m + 3n)(4m - 3n)$

Factorization.

Check:
$(4m + 3n)(4m - 3n) = 16m^2 - 9n^2$. Therefore, the factorization is correct.

Be Careful: $x^2 + y^2$ is the sum of squares and cannot be factored with real coefficients. One common error is $x^2 + y^2 = (x + y)^2$. This is incorrect since $(x + y)^2 = x^2 + 2xy + y^2$. There is an extra $2xy$ when it is multiplied out. Another common error is $x^2 + y^2 = (x + y)(x + y)$. This is the same as $(x + y)^2$. Do not confuse $(ab)^n$ with $(a + b)^n$ since $(ab)^n = a^n b^n$, but $(a + b)^n \neq a^n + b^n$ for $n \neq 1$.

ADDITIONAL PRACTICE

Factor the following difference of squares.

a) $p^2 - q^2$

b) $n^2 + 25$

c) $9c^2 - d^2$

d) $4a^2 - 9b^2$

PRACTICE EXERCISES

Factor the following difference of squares.

1) $a^2 - b^2$　　**2)** $c^2 + 16$　　**3)** $x^2 - 9y^2$　　**4)** $25m^2 - 4n^2$

If more practice is needed, do the Additional Practice Exercises in the margin.

Sometimes binomials are the difference of squares, even though they do not appear to be. Also, any factorization is not complete until each of the factors is prime. For example, $12 = 2 \cdot 6$ but the factorization is not complete since 6 is not prime. We will also need to recall that $(a^m)^n = a^{mn}$.

EXAMPLE 2
Factor the following completely.

a) $x^4 - y^4 =$

The first term is the square of x^2 and the second term is the square of y^2.

$(x^2)^2 - (y^2)^2 =$

The factorization is the product of the sum and difference of x^2 and y^2.

$(x^2 + y^2)(x^2 - y^2) =$

$x^2 - y^2$ is still the difference of squares and must be factored again.

$(x^2 + y^2)(x + y)(x - y)$

Complete factorization.

Check:
$(x^2 + y^2)(x + y)(x - y) = (x^2 + y^2)(x^2 - y^2) = x^4 - y^4$. Therefore, the factorization is correct.

b) $a^4 + b^4 = (a^2)^2 + (b^2)^2$

This is the sum of squares and cannot be factored. It is prime.

c) $a^4 - 81 =$

The first term is the square of a^2 and the second term is the square of 9.

$(a^2)^2 - 9^2 =$

The factorization is the product of the sum and difference of a^2 and 9.

$(a^2 + 9)(a^2 - 9) =$ $a^2 - 9$ is also the difference of squares and must be factored.

$(a^2 + 9)(a + 3)(a - 3)$ Complete factorization.

Check:
$(a^2 + 9)(a + 3)(a - 3) = (a^2 + 9)(a^2 - 9) = a^4 - b^4$. Therefore, the factorization is correct.

PRACTICE EXERCISES

Factor the following. If the polynomial cannot be factored, write prime.

5) $p^4 - q^4$ **6)** $r^4 + s^4$ **7)** $x^4 - 16$

In Section 4.2 we found that we could remove a common binomial factor as in $x(x + 2) - 3(x + 2) = (x + 2)(x - 3)$. In factoring the difference of squares, it is also possible for one or more of the expressions being squared to be a binomial. Study the following examples.

EXAMPLE 3
Factor the following.

a) $(a + b)^2 - 4 =$ The first term is the square of $(a + b)$ and 4 is the square of 2.

$(a + b)^2 - 2^2 =$ The factorization is the product of the sum and difference of $(a + b)$ and 2.

$[(a + b) + 2][(a + b) - 2] =$ Remove the parentheses.
$(a + b + 2)(a + b - 2)$ Factorization.

b) $9 - (x - y)^2 =$ The first term is 3^2 and the second term is the square of $(x - y)$.

$3^2 - (x - y)^2 =$ The factorization is the product of the sum and difference of 3 and $(x - y)$.

$[3 + (x - y)][3 - (x - y)] =$ Remove the parentheses.
$(3 + x - y)(3 - x + y)$ Factorization.

ADDITIONAL PRACTICE

Factor the following.

e) $(p - q)^2 - 9$

f) $36 - (c + d)^2$

PRACTICE EXERCISES

Factor the following.

8) $(a + b)^2 - 25$ **9)** $4 - (m - n)^2$

If more practice is needed, do the Additional Practice Exercises in the margin.

In some cases there may be a common factor which must be removed before factoring the difference of squares.

EXAMPLE 4

Factor the following completely.

a) $18x^2 - 8y^2 =$ First remove the common factor of 2.

$2(9x^2 - 4y^2) =$ $9x^2$ is the square of $3x$ and $4y^2$ is the square of $2y$.

$2[(3x)^2 - (2y)^2] =$ Factor the difference of squares as the product of the sum and difference of $3x$ and $2y$.

$2(3x + 2y)(3x - 2y)$ Complete factorization.

b) $25x^2 - 100 =$ First remove the common factor of 25.

$25(x^2 - 4) =$ Factor the difference of squares.

$25(x + 2)(x - 2)$ Complete factorization.

Note: In Example 4b $25x^2 - 100 = (5x)^2 - 10^2$, so it is the difference of squares. If we had factored as $(5x + 10)(5x - 10)$, the factorization would not have been correct since the factors are not prime. Each has a common factor of 5. Remember, always remove the GCF first.

PRACTICE EXERCISES

Factor the following completely.

10) $12x^2 - 27y^2$ **11)** $16a^2 - 64b^2$

Optional

- **Factoring the Sum or Difference of Cubes**

Previously, we learned that products of the form $(a + b)(a^2 - ab + b^2) = a^3 + b^3$ which is the sum of cubes (Section 2.4). We also found that $(a - b)(a^2 + ab + b^2) = a^3 - b^3$ which is the difference of cubes. From these products, we can factor the sum and difference of cubes as follows.

FACTORING THE SUM AND DIFFERENCE OF CUBES

$a^3 + b^3 = (a + b)(a^2 - ab + b^2)$

$a^3 - b^3 = (a - b)(a^2 + ab + b^2)$

Let's make some observations about the factored forms. Notice there is a binomial factor and a trinomial factor. The first term of the binomial factor is the expression whose cube is the first term of the sum or difference of cubes. The second term of the binomial is the expression whose cube is the second term of the sum or difference of cubes. The sign of the binomial is the same as the sign of the sum or difference of cubes. The first term of the trinomial is the square of the first term of the binomial. The second term of the trinomial is the negative of the product of the two terms of the binomial. The third term of the trinomial is the square of the second term of the binomial.

$$\underset{\uparrow}{\textit{Square of first term}} \qquad \underset{\uparrow}{\textit{Square of second term}}$$

$$a^3 + b^3 = (a + b)(a^2 - ab + b^2)$$

$$\underset{\textit{Negative of the product of the two terms}}{\uparrow}$$

The illustration above is for the sum of cubes, but the rule is exactly the same for the difference of cubes.

 EXAMPLE 5

Factor the following.

a) $x^3 - 27 =$ | Rewrite as $x^3 - 3^3$. Therefore, this is the difference of cubes. The first term of the binomial is the expression whose cube is x^3, so it is x. The sign of the binomial is negative since this is a difference of cubes. The second term of the binomial is the expression whose cube is 27. Since $3^3 = 27$, the second term is 3.

$(x - 3)(\_\ \_\ \_) =$ | The first term of the trinomial is the square of the first term of the binomial. Therefore, the first term of the trinomial is x^2.

$(x - 3)(x^2\ \_\ \_) =$ | The second term of the trinomial is the negative of the product of the terms of the binomial which gives $-(-3x) = 3x$.

$(x - 3)(x^2 + 3x\ \_) =$ | The last term of the trinomial is the square of the second term of the binomial. Since $3^2 = 9$, the last term of the trinomial is 9.

$(x - 3)(x^2 + 3x + 9)$ | Factorization. Check by multiplying.

b) $8x^3 + 125 =$ | Rewrite as $(2x)^3 + 5^3$. Therefore, this is the sum of cubes. The first term of the binomial is the expression whose cube is $8x^3$. Therefore, the first term is $2x$. The second term is the expression whose cube is 125. Therefore, the second term is 5. The sign is positive since this is the sum of cubes.

$(2x + 5)(\_\ \_\ \_) =$ | The first term of the trinomial is the square of the first term of the binomial. Since $(2x)^2 = 4x^2$, the first term of the trinomial is $4x^2$.

$(2x + 5)(4x^2\_\ \_) =$ | The second term of the trinomial is the negative of the product of the terms of the binomial. Since $-(2x)(5) = -10x$, the second term of the trinomial is $-10x$.

$(2x + 5)(4x^2 - 10x\ \_) =$ | The last term of the trinomial is the square of the second term of the binomial. Since $5^2 = 25$, the last term of the trinomial is 25.

$(2x + 5)(4x^2 - 10x + 25)$ | Factorization. Check by multiplying.

ADDITIONAL PRACTICE

Factor the following.

g) $y^3 - 1$

h) $8c^3 + 27$

PRACTICE EXERCISES

Factor each of the following.

12) $a^3 - 8$

13) $27x^3 + 64$

If more practice is needed, do the Additional Practice Exercises in the margin.

length and width as binomials. For example, if $A = x^2 - 9$, then $A = (x + 3)(x - 3)$, so L could be $x + 3$ and W could be $x - 3$. See the figure below.

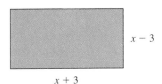

43) If the area of a rectangle is represented by $9x^2 - 49$, how can the length and width be represented?

44) If the area of a rectangle is represented by $16z^2 - 81$, how can the length and width be represented?

CHALLENGE EXERCISES (45–56)

Factor the following completely.

45) $196x^2 - 625$ **46)** $324a^2 - 400$

47) $256c^2 - 361d^2$ **48)** $441r^2 - 576s^2$

49) $338x^2 - 288$ **50)** $720a^2 - 125$

51) $x^8 - y^8$ **52)** $a^6 - b^6$

53) $81(x + 2)^2 - 16y^2$ **54)** $36(3x - 5)^2 - 25y^2$

55) $x^{2n} - y^{2n}$ **56)** $x^{4n} - 81$

Factor the following sum or difference of cubes (optional).

57) $x^3 + y^3$ **58)** $r^3 - s^3$ **59)** $c^3 + 8$

60) $d^3 + 27$ **61)** $n^3 - 64$ **62)** $r^3 - 125$

63) $m^3 + 1$ **64)** $b^3 - 1$ **65)** $8x^3 + y^3$

66) $64a^3 + b^3$ **67)** $27a^3 - 8$ **68)** $125c^3 - 64$

69) $8x^3 + 125y^3$ **70)** $64r^3 - 216s^3$

WRITING EXERCISE

71) What is wrong with the following factorization? $16x^2 - 144 = (4x + 12)(4x - 12)$. What is the correct factorization?

What is wrong with the following factorizations? Refractor each correctly.

72) $8x^3 + 64 = (2x + 4)(4x^2 - 8x + 16)$

73) $x^3 - 64 = (x - 4)(x^2 - 8x + 16)$

E X E R C I S E   S E T   **4.3**

Factor the following difference of squares. If the polynomial cannot be factored, write prime.

1) $m^2 - n^2$ **2)** $r^2 - s^2$ **3)** $x^2 - 4$

4) $p^2 - 1$ **5)** $r^2 - 64$ **6)** $a^2 - 81$

7) $m^2 + n^2$ **8)** $r^2 + s^2$ **9)** $c^2 - 25d^2$

10) $t^2 - 25s^2$ **11)** $a^2 - 100b^2$ **12)** $y^2 - 64z^2$

13) $4x^2 - 25y^2$ **14)** $9a^2 - 16b^2$ **15)** $49p^2 - 81q^2$

16) $25c^2 - 9d^2$ **17)** $9x^2 + 4y^2$ **18)** $36a^2 + 25b^2$

19) $121a^2 - 49b^2$ **20)** $169f^2 - 36g^2$

Factor the following completely.

21) $a^4 - b^4$ **22)** $r^4 - t^4$ **23)** $x^4 - 1$

24) $y^4 - 81$ **25)** $16x^4 - 81$ **26)** $256r^4 - 1$

Factor the following completely.

27) $(x + y)^2 - 9$ **28)** $(p + q)^2 - 25$

29) $36 - (x - y)^2$ **30)** $49 - (r - s)^2$

31) $4(x - y)^2 - 9$ **32)** $16(a - b)^2 - 25$

Factor the following completely.

33) $3x^2 - 75$ **34)** $5y^2 - 20$ **35)** $16x^2 - 36y^2$

36) $50x^2 - 32y^2$ **37)** $27x^2 - 48$ **38)** $20c^2 - 125$

39) $9x^2 - 81$ **40)** $4x^2 - 36$ **41)** $3x^4 - 48$

42) $2y^4 - 162$

Problems 43 and 44 require the use of the formula for the area of a rectangle which is $A = LW$. Represent the

---

S E C T I O N 4.4

Factoring Perfect Square Trinomials

OBJECTIVE

When you complete this section, you will be able to:

Ⓐ Recognize and factor perfect square trinomials.

Natural numbers that result from squaring integers are called **perfect squares**. The first few perfect squares are 1, 4, 9, 16, 25, 36, 49, 64, 81, 100, etc. because $1^2 = 1$, $2^2 = 4$, $3^3 = 9$, etc. In this section we will be factoring perfect square trinomials. A **perfect square trinomial** results from the square of a binomial. For example, $(a + b)^2 = a^2 + 2ab + b^2$. Since $a^2 + 2ab + b^2$ is the result of squaring the binomial $a + b$, it is a perfect square trinomial.

How do we recognize a perfect square trinomial? The answer is in knowing the shortcut method of squaring a binomial (Section 2.4). Recall the procedure for squaring a binomial is:

1) Square the first term of the binomial.

2) Multiply the product of the two terms of the binomial by two.

3) Square the last term of the binomial.

The procedure is diagrammed below for $(a + b)^2$.

First term squared   Last term squared
$$(a + b)^2 = a^2 + 2ab + b^2$$
Two times the product of the terms

For example, $(2x + 3)^2 = (2x)^2 + 2(2x)(3) + 3^2 = 4x^2 + 12x + 9$. Since $4x^2 + 12x + 9$ is the square of $2x + 3$, it is a perfect square trinomial.

Reversing the procedure, we see that a perfect square trinomial must have the following characteristics:

1) The first and last terms of the trinomial must be the squares of other terms.

2) The middle term (ignore the sign for now) must be 2 times the product of the terms whose squares give the first and last terms of the trinomial.

From the example above, we know that $a^2 + 2ab + b^2$ is a perfect square trinomial. The diagram below illustrates the characteristics of a perfect square trinomial using $a^2 + 2ab + b^2$.

The square of a   The square of b
$$a^2 + 2ab + b^2$$
Two times a times b

We illustrate with the following examples.

EXAMPLE 1

Determine if the following are perfect square trinomials.

a) $x^2 - 6x + 9$
$(x)^2 \ 2 \cdot x \cdot 3 \ 3^2$

The first term is the square of x and the last term is the square of 3. The middle term, ignoring the sign, is $2(x)(3)$. Therefore, this is a perfect square trinomial.

b) $4x^2 - 20xy + 25y^2$
$(2x)^2 \ 2 \cdot 2x \cdot 5y \ (5y)^2$

The first term is the square of $2x$ and the last term is the square of $5y$. The middle term, ignoring the sign, is $2(2x)(5y)$. Therefore, this is a perfect square trinomial.

c) $9a^2 - 12ay + 8y^2$ The last term, $8y^2$, is not the square of a term. Therefore, this is not a perfect square trinomial.

d) $9p^2 + 12p + 16$ The first term is the square of $3p$ and the last term is the square of 4. However, the middle term is not equal to $2(3p)(4)$. Therefore, this is not a perfect square trinomial.

PRACTICE EXERCISES

Determine if the following are perfect square trinomials.

1) $y^2 - 8y + 16$ **2)** $16a^2 + 24ab + 9b^2$

3) $25x^2 + 20x + 32$ **4)** $9a^2 - 15a + 25$

Now that we can recognize a perfect square trinomial, how do we find the binomial that when squared results in the perfect square trinomial? Again, the answer lies in the procedure for squaring the binomial. In the introduction we found that $4x^2 + 12x + 9$ resulted from squaring $2x + 3$. Notice the first term of the binomial, $2x$, is the term that when squared gives the first term of the trinomial, $4x^2$. The second term of the binomial, 3, is the term that when squared gives the last term of the trinomial, 9. The sign between the terms of the binomial is the same as the sign of the second term of the perfect square trinomial. Based on the preceding, the procedure for factoring a perfect square trinomial is:

Factoring Perfect Square Trinomial

1) Determine if the trinomial is a perfect square trinomial.

2) If it is, form a binomial with the square of the first term equal to the first term of the trinomial and the square of the second term equal to the last term of the trinomial.

3) The sign between the terms of the binomial is the same as the sign of the middle term of the trinomial.

From previous examples, we know that $a^2 + 2ab + b^2 = (a + b)^2$. So we will illustrate the procedure using $a^2 + 2ab + b^2$.

$$a^2 + 2ab + b^2 = (a$$
$$\uparrow \qquad\qquad\qquad \uparrow$$
$$(a)^2 \qquad \textit{First term is } a$$

$$a^2 + 2ab + b^2 = (a \quad b)$$
$$\uparrow \qquad\quad \uparrow$$
$$(b)^2 \qquad \textit{Second term is } b$$

$$a^2 + 2ab + b^2 = (a + b)$$
$$\uparrow \qquad\qquad \uparrow$$
$$\textit{These two signs are the same}$$

Therefore,
$$a^2 + 2ab + b^2 = (a + b)^2.$$

EXAMPLE 2

Determine whether each of the following is a perfect square trinomial. If it is, factor it as the square of a binomial.

a) $x^2 - 6x + 9 =$ From Example 1a above, this is a perfect square trinomial.

$(x \ \_)^2 =$ x^2 is the square of x, so the first term of the binomial is x.

$(x \quad 3)^2 =$ 9 is the square of 3, so the second term of the binomial is 3.

$(x - 3)^2$ The sign of the middle term of the trinomial is $-$, so the sign of the binomial is $-$.

Check:
$(x - 3)^2 = (x)^2 - 2(x)(3) + 3^2 = x^2 - 6x + 9$. Therefore, the factorization is correct.

b) $x^2 - 10x + 25 =$ The first term is the square of x, the last term is the square of 5, and the middle term is $2(x)(5)$. Therefore, it is a perfect square trinomial.

$(x \ \_)^2 =$ x^2 is the square of x, so the first term of the binomial is x.

$(x \quad 5)^2 =$ 25 is the square of 5, so the second term of the binomial is 5.

$(x - 5)^2$ The sign of the middle term is $-$, so the sign of the binomial is $-$.

Check:
$(x - 5)^2 = (x)^2 + 2(x)(-5) + (-5)^2 = x^2 - 10x + 25.$

c) $4x^2 + 20xy + 25y^2 =$ From Example 1b above, this is a perfect square trinomial.

$(2x \ \_)^2 =$ $4x^2$ is the square of $2x$, so the first term of the binomial is $2x$.

$(2x \quad 5y)^2 =$ $25y^2$ is the square of $5y$, so the second term of the binomial is $5y$.

$(2x + 5y)^2$ The sign of the middle term is $+$, so the sign of the binomial is $+$.

Check:
$(2x + 5y)^2 = (2x)^2 + 2(2x)(5y) + (5y)^2 = 4x^2 + 20xy + 25y^2$. Therefore, the factorization is correct.

d) $9p^2 + 24p + 16 =$ The first term is the square of $3p$, the last term is the square of 4, and the middle term is $2(3p)(4)$. Therefore, this is a perfect square trinomial.

$(3p + 4)^2$ $9p^2$ is the square of $3p$, 16 is the square of 4, and the sign of the middle term is $+$. So, $9p^2 + 24p + 16 = (3p + 4)^2$.

Check:
The check is left as an exercise for the student.

ADDITIONAL PRACTICE

Factor the following if it is a perfect square trinomial.

a) $a^2 - 4a + 4$

b) $9a^2 - 6ab + b^2$

PRACTICE EXERCISES

Determine whether each of the following is a perfect square trinomial. If it is, factor it as the square of a binomial.

5) $y^2 - 8y + 16$

6) $16a^2 + 24ab + 9b^2$

7) $a^2 - 16a + 64$

8) $4x^2 + 20xy + 25y^2$

If more practice is needed, do the Additional Practice Exercises in the margin.

As in the previous section, if there is a GCF, remove it and then see if the resulting polynomial can be factored further.

EXAMPLE 3

Factor the following.

a) $3x^2 - 30x + 75 =$ First remove the GCF of 3.

$3(x^2 - 10x + 25) =$ The resulting trinomial is a perfect square.

$3(x - 5)^2$ The first term is the square of x, the last term is the square of 5, and the sign of the second term is $-$. Check by multiplying.

b) $x^3y + 8x^2y + 16xy =$ Remove the GCF of xy.

$xy(x^2 + 8x + 16) =$ The resulting trinomial is a perfect square.

$xy(x + 4)^2$ x^2 is the square of x, 16 is the square of 4, and the sign of the second term is $+$. Check by multiplying.

ADDITIONAL PRACTICE

Factor the following.

c) $8x^2 - 16x + 8$

d) $x^3 - 18x^2 + 81x$

PRACTICE EXERCISES

Factor the following.

9) $5x^2 - 60x + 180$

10) $x^4y^2 - 14x^3y^2 + 49x^2y^2$

If more practice is needed, do the Additional Practice Exercises in the margin.

E X E R C I S E   S E T   **4.4**

Factor the following, if it is a perfect square. If it is not a perfect square, write prime.

1) $x^2 + 4x + 4$

2) $a^2 - 6a + 9$

3) $x^2 + 2x + 1$

4) $b^2 - 2b + 1$

5) $x^2 + 12x + 36$

6) $c^2 - 14c + 49$

7) $y^2 + 20y + 100$

8) $r^2 + 22r + 121$

9) $25x^2 - 10x + 1$

10) $36x^2 + 12x + 1$

11) $a^2 - 2ab + b^2$

12) $c^2 + 2cd + d^2$

13) $4a^2 + 4ab + b^2$

14) $16a^2 - 8ab + b^2$

15) $16c^2 - 24c + 9$

16) $25x^2 + 20x + 4$

17) $9x^2 + 30x + 25$

18) $4x^2 + 28x + 49$

19) $16x^2 - 8xy + y^2$

20) $49x^2 + 14xy + y^2$

21) $25x^2 + 40xy + 16y^2$

22) $36x^2 - 84xy + 49y^2$

23) $4a^2 - 14ab + 49b^2$

24) $25c^2 + 35cd + 49d^2$

25) $8x^2 - 24x + 18$

26) $27x^2 - 36x + 12$

27) $2x^2 + 28x + 98$

28) $3x^2 - 54x + 243$

29) $4x^2 + 80x + 400$

30) $5x^2 - 60x + 180$

31) $25x^4 - 30x^3 + 9x^2$

32) $4x^3y + 36x^2y + 81xy$

33) $50x^3y + 80x^2y^2 + 32xy^3$

34) $81x^4 - 36x^3y + 4x^2y^2$

35) If $x^2 - 14x + 49$ represents the area of a square, what is the length of each side?

36) If $4a^2 + 36a + 81$ represents the area of a square, what is the length of each side?

CHALLENGE EXERCISES (37–40)

Factor the following.

37) $81x^2 - 90xy + 25y^2$

38) $121x^2 - 286xy + 169y^2$

39) $100a^4 + 300a^2b^2 + 225b^4$

40) $144x^4 + 192x^2y^2 + 64y^4$

Find the value of k so that each of the following will be a perfect square trinomial.

41) $x^2 + 14x + k$ **42)** $4x^2 + kx + 25$

43) $kx^2 + 40x + 16$ **44)** $9a^2 + 30ab + kb^2$

WRITING EXERCISE

45) If the middle term in a perfect square trinomial is negative, why is the middle term of the binomial whose square equals the trinomial also negative?

Problems 35 and 36 require the use of the formula for the area of a square which is $A = s^2$, where s is the length of each side of the square. For example, if $A = x^2 + 2x + 1$, then $A = (x + 1)^2$. This means that each side of the square is $(x + 1)$. See the figure below.

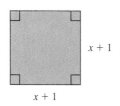

$x + 1$

$x + 1$

SECTION 4.5

Factoring General Trinomials with Leading Coefficient of One

OBJECTIVE

When you complete this section, you will be able to:

Ⓐ Factor a general trinomial whose leading coefficient is one.

INTRODUCTION

When most people think of factoring, it is the factoring of trinomials that usually comes to mind. The factoring techniques discussed in the previous two sections are actually special cases of the more general techniques to be discussed in the next two sections.

We will consider two types of trinomials, those with positive last terms and those with negative last terms. We will first look at the products of binomials that result in trinomials whose last terms are positive.

EXAMPLE 1

Find the following products.

a) $(x + 3)(x + 7) = x^2 + 7x + 3x + 21$ FOIL.
$ = x^2 + 10x + 21$ Add like terms.

Therefore,
$x^2 + 10x + 21 = (x + 3)(x + 7)$ in factored form.

b) $(x - 3)(x - 7) = x^2 - 7x - 3x + 21$ FOIL.

$\qquad\qquad\qquad = x^2 - 10x + 21$ Add like terms.

Therefore,

$x^2 - 10x + 21 = (x - 3)(x - 7)$ in factored form.

From Example 1 we observe that the signs of the last terms of the trinomials are positive. Further observe that the signs of the second terms of both binomials are the same as the sign of the second term of the trinomial.

From Example 1a,

$$x^2 + 10x + 21 = (x + 3)(x + 7)$$

Last sign positive

$$x^2 + 10x + 21 = (x + 3)(x + 7)$$

If the last sign is positive, these three signs are the same.

From Example 1b,

$$x^2 - 10x + 21 = (x - 3)(x - 7)$$

Last sign positive.

$$x^2 - 10x + 21 = (x - 3)(x - 7)$$

If the last sign is positive, these three signs are the same.

When multiplying binomials whose second signs are the same, both the inside and outside products have the same sign. Consequently, we add the absolute values of the outside and inside products in order to find the absolute value of the coefficient of the middle term of the trinomial. This will be helpful when deciding which factors of the last term to try when factoring the trinomial.

$$(x + 3)(x + 7) = x^2 + 7x + 3x + 21 = x^2 + 10x + 21$$

Same sign, so add the absolute values of 7 and 3 to get 10, which is the absolute value of the coefficient of the middle term.

$$(x - 3)(x - 7) = x^2 - 7x - 3x + 21 = x^2 - 10x + 21$$

Same sign, so add the absolute values of 7 and 3 to get 10, which is the absolute value of the coefficient of the middle term.

Consider the following. Remember, we are doing FOIL in reverse and if the last sign of the trinomial is positive, both binomial factors have the same sign as the sign of the middle term of the trinomial.

EXAMPLE 2

Factor the following.

a) $x^2 + 4x + 3$ The sign of the last term is + and the sign of the middle term is +. Therefore, both of the signs of the binomials will be +.

$(\_ + \_)(\_ + \_)$ The first blank in each binomial must be filled by terms with product x^2. These have to be x and x.

$(x + \_)(x + \_)$ The second blank in each binomial must be filled by terms whose product is 3 but which will also give the sum of the outside and inside products of $4x$. The only factors of 3 are $1 \cdot 3$.

$(x + 1)(x + 3)$ Outside product is $3x$ and inside product is x and the sum is $4x$.

$$\begin{array}{c} x \\ 3x \\ \hline 4x \end{array}$$

$(x + 1)(x + 3)$ Factorization.

Check:

$(x + 1)(x + 3) = x^2 + 3x + x + 3 = x^2 + 4x + 3$ which is the original trinomial. Therefore, this factorization is correct.

Note: Since multiplication is commutative, $(x + 3)(x + 1)$ is also correct.

b) $x^2 - 7x + 10$ The sign of the last term of the trinomial is $+$ and the sign of the middle term is $-$. Therefore, both of the signs of the binomials are $-$.

$(\_ - \_)(\_ - \_)$ The first blanks of each binomial must be filled with terms with product x^2. The only reasonable choices are x and x.

$(x - \_)(x - \_)$ The second blanks must be filled by terms whose product is 10 and which give the sum of the outside and inside products of $-7x$. The possible factors of 10 are $1 \cdot 10$ and $2 \cdot 5$. Try 2 and 5.

$(x - 2)(x - 5)$ Outside product is $-5x$ and inside product is $-2x$ and the sum is $-7x$.

$$\begin{array}{c} -2x \\ -5x \\ \hline -7x \end{array}$$

$(x - 2)(x - 5)$ Factorization.

Check:

$(x - 2)(x - 5) = x^2 - 5x - 2x + 10 = x^2 - 7x + 10$ which is the original trinomial. Therefore, this factorization is correct.

c) $a^2 - 6ab + 8b^2$ The sign of the last term is $+$ and the sign of the middle term is $-$. Therefore, both of the signs of the binomials are $-$.

$(\_ - \_)(\_ - \_)$ The first blanks must be filled by terms whose product is a^2. The only choices are a and a.

$(a - \_)(a - \_)$ The second blanks must be filled with terms whose product is $8b^2$ and give a sum of the outside and inside products of $-6ab$. The choices of factors of $8b^2$ are $b \cdot 8b$ or $2b \cdot 4b$. Try $2b$ and $4b$.

$(a - 2b)(a - 4b)$ Outside product is $-4ab$ and inside product is $-2ab$ and the sum is $-6ab$.

$$\begin{array}{c} -2ab \\ -4ab \\ \hline -6ab \end{array}$$

$(a - 2b)(a - 4b)$ Factorization.

Check:

$(a - 2b)(a - 4b) = a^2 - 4ab - 2ab + 8b^2 = a^2 - 6ab + 8b^2$. Therefore, this factorization is correct.

PRACTICE EXERCISES

Factor the following.

1) $a^2 + 8a + 7$ **2)** $b^2 - 9b + 14$

3) $y^2 - 6yz + 5z^2$

If more practice is needed, do the Additional Practice Exercises in the margin.

We will now look at products of binomials which result in trinomials having negative last terms.

EXAMPLE 3

Find the following products.

a) $(x + 3)(x - 7) = x^2 - 7x + 3x - 21$ FOIL.
$\qquad\qquad\qquad = x^2 - 4x - 21$ Add like terms.

Therefore,
$x^2 - 4x - 21 = (x + 3)(x - 7)$ in factored form.

b) $(x - 3)(x + 7) = x^2 + 7x - 3x - 21$ FOIL.
$\qquad\qquad\qquad = x^2 + 4x - 21$ Add like terms.

Therefore,
$x^2 + 4x - 21 = (x - 3)(x + 7)$ in factored form.

Looking at Example 3 we notice the sign of the last term of the trinomial in both examples is negative. In each case, one binomial factor has a + sign and the other a − sign. In other words, the signs are opposites.

From Example 3a,

$$x^2 - 4x - 21 = (x + 3)(x - 7)$$
$$\uparrow$$

Last sign negative.

$$x^2 - 4x - 21 = (x + 3)(x - 7)$$
$$\qquad\qquad\qquad\uparrow\quad\uparrow$$

If the last sign is negative, these signs are opposites.

From Example 3b,

$$x^2 + 4x - 21 = (x - 3)(x + 7)$$
$$\uparrow$$

Last sign negative.

$$x^2 + 4x - 21 = (x - 3)(x + 7)$$
$$\qquad\qquad\qquad\uparrow\quad\uparrow$$

If the last sign is negative, these signs are opposites.

Since the signs of the binomials are opposites, the outside and inside products are opposite in sign. This means the absolute value of the coefficient of the middle term of the trinomial is found by subtracting the absolute values of the coefficients of the outside and inside products of the binomials. This will be helpful when deciding which factors of the last term to try when factoring the trinomial.

From Example 3a,

$$(x + 3)(x - 7) = x^2 - 7x + 3x - 21 = x^2 - 4x - 21$$
$$\qquad\qquad\uparrow\quad\uparrow$$

These are opposite in sign, so subtract the absolute values of 7 and 3 to get 4, which is the absolute value of the coefficient of the middle term of the trinomial.

From Example 3b,

$$(x - 3)(x + 7) = x^2 + 7x - 3x - 21 = x^2 + 4x - 21$$
$$\qquad\qquad\uparrow\quad\uparrow$$

These are opposite in sign, so subtract the absolute values of 7 and 3 to get 4, which is the absolute value of the coefficient of the middle term of the trinomial.

When factoring a trinomial whose last term is negative, we will not be concerned with the signs initially. We will first find binomial factors that give the correct first and last terms of the trinomial and the difference of the outside and inside products of the binomials gives the middle term, disregarding the sign, of the trinomial. When this is accomplished, we will decide which factor has the + sign and which has the − sign. This will become clearer by doing some examples.

EXAMPLE 4

Factor the following.

a) $x^2 - x - 6$ The sign of the last term is −. Therefore, the signs of the binomials are opposites.

$(\underline{\quad}\ \underline{\quad})(\underline{\quad}\ \underline{\quad})$ The first blanks must be filled by terms whose product is x^2. The only choices are x and x. Notice we did not put in any signs since we will determine them last.

$(x\ \underline{\quad})(x\ \underline{\quad})$ The second blanks must be filled in by terms with product of 6 and such that the difference of the outside and inside products is 1. The possible factorizations of 6 are $1 \cdot 6$ and $2 \cdot 3$. Try 1 and 6.

      The inside product is x and the outside product is $6x$. But $6x - x = 5x$ and the middle term of the trinomial is x, disregarding the sign. Therefore, 1 and 6 are not the correct choices. Try 2 and 3.

      The outside product is $3x$ and the inside product is $2x$ and $3x - 2x = x$ which is the middle term of the trinomial, except for the sign.

      Since the middle term of the trinomial is $-x$, we want $-3x$ and $+2x$ as the outside and inside products. Therefore, use -3 and $+2$.

$(x + 2)(x - 3)$ Factorization.

Check:

$(x + 2)(x - 3) = x^2 - 3x + 2x - 6 = x^2 - x - 6$. Therefore, this factorization is correct.

b) $x^2 + 4xy - 12y^2$ The last sign of the trinomial is $-$, so the binomial factors will have opposite signs.

$(\underline{\quad}\ \underline{\quad})(\underline{\quad}\ \underline{\quad})$ The first blanks must be filled by terms whose products are x^2. Again, the only choices are x and x.

$(x\ \underline{\quad})(x\ \underline{\quad})$ The second blanks must be filled with terms whose product is $12y^2$ and the difference of the outside and inside products is $4xy$. The possible factorizations of $12y^2$ are $y \cdot 12y$, $2y \cdot 6y$, and $3y \cdot 4y$. Try y and $12y$.

     The outside product is $12xy$ and the inside product is xy. However, $12xy - xy = 11xy$ and the middle term is $4xy$. Therefore, this is not the correct choice of factors of $12y^2$. Try $2y$ and $6y$.

   The outside product is $6xy$ and the inside product is $2xy$. Also, $6xy - 2xy = 4xy$ which is the correct middle term.

$(x - 2y)(x + 6y)$
$ {-2xy} $
$ 6xy $
$ \overline{4xy} $
    Since the middle term is $+4xy$, we need $+6xy$ and $-2xy$ as the outside and inside products. Therefore, use $+6y$ and $-2y$.

$(x - 2y)(x + 6y)$ Factorization.

Check:
$$(x - 2y)(x + 6y) = x^2 + 6xy - 2xy - 12y^2 = x^2 + 4xy - 12y^2.$$
Therefore, this factorization is correct.

c) $x^2 + 3x - 8$ The last sign is $-$, so the signs of the binomials are opposite.

$(\underline{\quad}\ \underline{\quad})(\underline{\quad}\ \underline{\quad})$ The first blanks must be filled in by terms whose product is x^2. These are, again, x and x.

$(x\ \underline{\quad})(x\ \underline{\quad})$ The second blanks must be filled by terms whose product is 8 and the difference in the outside and inside products is $3x$. The possible factors of 8 are $1 \cdot 8$ and $2 \cdot 4$. Try 1 and 8.

$(x\ \ 1)(x\ \ 8)$ The outside product is $8x$ and the inside product is x, but $8x - x = 7x$ which is not the correct middle term. Try 2 and 4.

$(x\ \ 2)(x\ \ 4)$ The outside product is $4x$ and the inside product is $2x$, but $4x - 2x = 2x$ which is not the correct middle term. Since we have tried all possible combinations of factors of 8 and none work, this trinomial is prime.

ADDITIONAL PRACTICE

Factor the following. If the polynomial cannot be factored, write prime.

d) $x^2 + 5x - 14$

e) $x^2 - 5x - 6$

f) $a^2 + 3a - 12$

PRACTICE EXERCISES

Factor the following. If the polynomial cannot be factored, write prime.

4) $x^2 + 3x - 10$ **5)** $a^2 - 2ab - 15b^2$ **6)** $b^2 + 4b - 10$

If more practice is needed, do the Additional Practice Exercises in the margin.

Following is a summary of the techniques discussed in this section.

SUMMARY

1) If the last sign of the trinomial is positive, both binomial factors will have the same sign as the second sign of the trinomial. Also, the absolute values of the coefficients of the outside and inside products of the binomial factors must have a sum equal to the absolute value of the coefficient of the middle term of the trinomial being factored.

2) If the last term of the trinomial is negative, the signs of the binomial factors will be opposites, Also, the difference of the absolute values of the coefficients of the outside and inside products of the binomial factors must equal the absolute value of the coefficient of the middle term of the trinomial being factored.

We will do one of each type in the following example.

EXAMPLE 5

Factor the following trinomials.

a) $x^2 - 11x + 24$ — The sign of the last term is $+$ and the sign of the second term is $-$. Therefore, the signs of both binomial factors are $-$.

$(\underline{} - \underline{})(\underline{} - \underline{})$ — The first blanks must be filled with terms whose product is x^2. As before, the only choices are x and x.

$(x - \underline{})(x - \underline{})$ — The second blanks must be filled with terms whose product is 24 and which give the sum of the outside and inside products of -11. The possible factors of 24 are $1 \cdot 24$, $2 \cdot 12$, $3 \cdot 8$, and $4 \cdot 6$. Try 3 and 8.

$(x - 3)(x - 8)$ — Outside product is $-8x$ and inside product is $-3x$ and the sum is $-11x$.

$(x - 3)(x - 8)$ — Factorization.

Check:

$(x - 3)(x - 8) = x^2 - 8x - 3x + 24 = x^2 - 11x + 24$ which is the original trinomial. Therefore, this factorization is correct.

b) $b^2 - b - 20$ — The sign of the last term is $-$. Therefore, the signs of the binomials are opposites.

$(\underline{} \ \underline{})(\underline{} \ \underline{})$ — The first blanks must be filled by terms whose product is b^2. The only choices are b and b.

$(b \ \underline{})(b \ \underline{})$ — The second blanks must be filled with terms whose product is 20 and the difference of the outside and inside products is 1. The possible factors of 20 are $1 \cdot 20$, $2 \cdot 10$, and $4 \cdot 5$. Try 2 and 10.

The outside product is $10b$ and the inside product is $2b$. However, $10b - 2b = 8b$ and the middle term of the trinomial is $-b$. Therefore, this is not the correct choice of factors of 20. Try 4 and 5.

$$(b \quad 4)(b \quad 5)$$

The outside product is $5b$ and the inside product is $4b$. Also, $5b - 4b = b$ which is the correct middle term, except for the sign.

$$(b + 4)(b - 5)$$

To get $-b$, we need $-5b$ and $+4b$ as the outside and inside products. Therefore, use -5 and $+4$.

$$(b + 4)(b - 5)$$ Factorization.

Check:

$(b + 4)(b - 5) = b^2 - 5b + 4b - 20 = b^2 - b - 20$ which is the original trinomial. Therefore, this factorization is correct.

ADDITIONAL PRACTICE

Factor the following.

g) $x^2 - x - 42$

h) $c^2 - 12c + 32$

PRACTICE EXERCISES

Factor the following.

7) $a^2 + 13a + 40$

8) $r^2 - 7r - 18$

9) $m^2 + 2m - 24$

10) $x^2 + 16x + 48$

If more practice is needed, do the Additional Practice Exercises in the margin.

As in the previous sections on factoring, if a common factor is present, we must first remove it and try to factor the polynomial inside the parentheses.

EXAMPLE 6

Factor the following completely.

a) $3x^2 + 12x - 63$ Remove the GCF of 3.

$3(x^2 + 4x - 21)$ The last sign of the trinomial is $-$, so the signs of the binomial factors will be opposites.

$3(\_ \ \_)(\_ \ \_)$ The first blanks must be filled with terms whose product is x^2. Again, these are x and x.

$3(x \ \_)(x \ \_)$ The second blanks must be filled with terms whose product is 21 and give a difference of outside and inside products of $4x$. Try 3 and 7.

$3(x \quad 3)(x \quad 7) =$ The outside product is $7x$ and the inside product is $3x$ and $7x - 3x = 4x$ which is the desired middle term. Since we want $+4x$, we need $+7x$ and $-3x$ as the outside and inside products. Therefore, use -3 and $+7$.

$3(x - 3)(x + 7)$ Factorization.

Check:

$3(x - 3)(x + 7) = 3(x^2 + 7x - 3x - 21) = 3(x^2 + 4x - 21) = 3x^2 + 12x - 63$. Therefore, this factorization is correct.

b) $a^3b - 9a^2b^2 + 8ab^3$ Remove the GCF of ab.

$ab(a^2 - 9ab + 8b^2)$ The last sign of the trinomial is $+$ and the middle sign is $-$. Therefore, both signs of the binomials are $-$.

$ab(\_ - \_)(\_ - \_)$ The first blanks are filled by a and a.

$ab(a - \_)(a - \_)$ The second blanks must be filled with terms whose product is $8b^2$ and which give outside and inside products whose sum is $-9ab$. Try b and $8b$.

$ab(a - b)(a - 8b)$ Outside product is $-8ab$ and inside product is $-ab$ and the sum is $-9ab$.

$$-ab$$
$$-8ab$$
$$-9ab$$

$ab(a - b)(a - 8b)$ Factorization.

Check:

$ab(a - b)(a - 8b) = ab(a^2 - 8ab - ab + 8b^2) = ab(a^2 - 9ab + 8b^2) = a^3b - 9a^2b^2 + 8ab^3$. Therefore, this factorization is correct.

ADDITIONAL PRACTICE

Factor the following.

i) $4x^2 - 40x + 36$

j) $m^3n - m^2n^2 - 12mn^3$

PRACTICE EXERCISES

Factor the following.

11) $2x^2 - 18x + 36$

12) $y^3z - 15y^2z^2 - 16yz^3$

If more practice is needed, do the Additional Practice Exercises in the margin.

EXERCISE SET **4.5**

Factor the following if possible. If the polynomial is not factorable, write prime.

1) $x^2 + 3x + 2$ **2)** $x^2 + 6x + 5$ **3)** $x^2 - 4x + 3$

4) $x^2 - 8x + 7$ **5)** $a^2 + 5a + 6$ **6)** $a^2 + 7a + 10$

7) $b^2 - 6b + 8$ **8)** $c^2 - 9c + 14$ **9)** $y^2 - 4y - 5$

10) $z^2 - 2z - 3$ **11)** $a^2 + 2a - 35$

12) $b^2 + 4b - 21$ **13)** $x^2 + 10x + 16$

14) $z^2 + 11z + 18$ **15)** $r^2 + r - 72$

16) $s^2 + s - 20$ **17)** $a^2 - 12a + 27$

18) $c^2 - 13c + 40$ **19)** $x^2 + 4x + 8$

20) $a^2 - 6a + 8$ **21)** $x^2 - 12x + 35$

22) $n^2 - 10n + 24$ **23)** $y^2 + y - 42$

24) $x^2 + x - 72$ **25)** $z^2 + 15z + 36$

26) $b^2 + 11b + 24$ **27)** $a^2 + 4ab + 3b^2$

28) $c^2 + 12cd + 11d^2$ **29)** $r^2 - 8rs + 7s^2$

30) $q^2 - 6qr + 5r^2$ **31)** $c^2 + 7cd - 18d^2$

32) $a^2 + 10ab - 24b^2$ **33)** $r^2 - 7rs - 44s^2$

34) $a^2 - ab - 72b^2$ **35)** $y^2 + 3yz - 40z^2$

36) $m^2 + 11mn - 42n^2$ **37)** $a^2 - 3ab - 54b^2$

38) $x^2 - 4xy - 45y^2$ **39)** $r^2 - 15rs + 36s^2$

40) $a^2 - 6ab - 27b^2$

Factor the following completely.

41) $3a^2 + 15a + 18$ **42)** $4x^2 + 12x + 8$

43) $5z^2 - 25z + 30$ **44)** $6a^2 - 30a + 36$

45) $x^3 - 3x^2 - 28x$ **46)** $y^3 - y^2 - 20y$

47) $y^4 - 20y^3 + 75y^2$ **48)** $c^4 - 10c^3 + 21c^2$

49) $3a^3 - 3a^2 - 12a$ **50)** $4x^3 - 4x^2 - 80x$

51) $x^3y + 2x^2y - 80xy$ **52)** $a^3b + 14a^2b - 51ab$

53) $x^2y^2 + 9xy^2 + 20y^2$ **54)** $y^4z^3 + 13y^3z^3 + 30y^2z^3$

---

55) $4r^2s + 20rs - 56s$　　　56) $3a^2b + 6ab - 45b$

57) $3a^2b - 3ab - 84b$　　　58) $2x^2y - 10xy - 100y$

CHALLENGE EXERCISES (59–61)

Factor the following.

59) $x^2 - 3x - 108$　　　60) $x^2 + 7x - 144$

61) $y^2 + 24y + 128$

Find all integer values of k such that each of the following is factorable over the integers.

62) $x^2 + kx + 6$　　　63) $x^2 + kx + 12$

Recall that the formula for the area of a rectangle is $A = LW$. If the following trinomials represent the areas of rectangles, find binomials that could represent the length and width.

64) $A = x^2 + 19x + 48$　　　65) $A = x^2 + 12x + 32$

66) $A = y^2 + 3y - 54$

The volume of a rectangular solid is $V = LWH$. If each of the following trinomials represents the volume of a rectangular solid, find monomials and/or binomials that could represent the length, width, and height.

67) $V = x^3 + 6x^2 - 40x$

68) $V = 4x^3 - 12x^2 - 112x$

WRITING EXERCISES

69) Describe how $x^2 + 8x + 16$, which is a perfect square trinomial, can be factored using the techniques of this section.

70) Describe how $x^2 - 16$, which is the difference of squares, can be factored using the techniques of this section.

71) If the sign of the last term of a trinomial is positive, why must both signs of the second terms of the binomial factors be the same?

SECTION 4.6

Factoring General Trinomials with First Coefficient Other than One

OBJECTIVE

When you complete this section, you will be able to:

Ⓐ　Factor trinomials whose first coefficient is a natural number other than one.

INTRODUCTION

In the previous section all of the trinomials that we factored had a coefficient of one on the first term. The factorization of these trinomials was made easier since all we needed to do was find the factors of the last term that had either a sum or difference that equaled the coefficient of the middle term. The same trial and error procedure used in the last section applies to trinomials whose first coefficient is not one. However, these are more difficult because the location of the terms in the binomial gives different outside and inside products while using the exact same numbers. In other words, the number of possibilities is greatly increased as illustrated below.

EXAMPLE 1

Factor the following. If the trinomial cannot be factored, write prime.

a) $3x^2 + 5x + 2$　　　The last term is + and the middle term is +, so both signs of the binomial factors are +.

　　$(\_ + \_)(\_ + \_)$　　　The first blanks must be filled by terms whose product is $3x^2$. The only reasonable choice is $3x$ and x.

　　$(3x + \_)(x + \_)$　　　The second blanks must be filled by terms whose product is 2 and which give a sum of the outside and inside products of 5. The only factors of 2 are 2 and 1. Try 1 in the first blank and 2 in the second blank

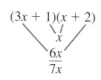

The outside product is 6*x* and the inside product is *x*. However, $6x + x = 7x$ and we needed 5*x*. Therefore, this factorization is incorrect. Try 2 in the first blank and 1 in the second.

The outside product is 3*x* and the inside product is 2*x*, and $3x + 2x = 5x$ which is the correct middle term.

$(3x + 2)(x + 1)$ Factorization.

Check:

$(3x + 2)(x + 1) = 3x^2 + 3x + 2x + 2 = 3x^2 + 5x + 2.$ Therefore, this factorization is correct.

 b) $5x^2 + 8x - 4$

The last sign is $-$. Therefore, the signs of the binomial are different, so we will be looking for the difference of the outside and inside products.

$(\_\ \_)(\_\ \_)$

The first blanks must be filled by terms whose product is $5x^2$. These are 5*x* and *x*.

$(5x\ \_)(x\ \_)$

The second blanks must be filled by terms whose product is 4 and which give a difference of the outside and inside products of 8*x*. The possible factorizations of 4 are $1 \cdot 4$ and $2 \cdot 2$. Try 1 in the first blank and 4 in the second.

The outside product is 20*x* and the inside product is *x*, and $20x - x = 19x$. Since we wanted 8*x*, this is the wrong choice. Try 4 in the first blank and 1 in the second.

The outside product is 5*x* and the inside product is 4*x*, and $5x - 4x = x$. Therefore, this choice is also incorrect. This means that 1 and 4 will not work, so try 2 and 2.

The outside product is 10*x* and the inside product is 2*x*, and $10x - 2x = 8x$ which is what we want.

Since the middle term is $+8x$, we want $+10x$ and $-2x$ for the outside and inside products. Hence, -2 goes in the first blank and 2 goes in the second.

$(5x - 2)(x + 2)$ Factorization.

Check:

$(5x - 2)(x + 2) = 5x^2 + 10x - 2x - 4 = 5x^2 + 8x - 4.$ Therefore, this factorization is correct.

c) $6a^2 - 11a - 10$

The last sign is $-$, so the binomial factors are different in sign.

$(\_\ \_)(\_\ \_)$

The first blanks must be filled by terms whose product is $6a^2$. This could be $6a \cdot a$ or $3a \cdot 2a$. Try $3a \cdot 2a$.

$(3a \_)(2a \_)$	The second blanks must be filled by terms whose product is 10 and such that the difference of the outside and inside products is 11. The possible factors of 10 are $1 \cdot 10$ or $2 \cdot 5$. Try 2 and 5 with 2 in the first blank and 5 in the second.

This gives an outside product of $15a$ and an inside product of $4a$. Also, $15a - 4a = 11a$ which is what we want.

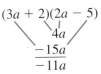

Since the middle term is $-11a$, we need $-15a$ and $+4a$. Put $+$ in the first factor and $-$ in the second.

$(3a + 2)(2a - 5)$ Factorization.

Check:
$$(3a + 2)(2a - 5) = 6a^2 - 15a + 4a - 10 = 6a^2 - 11a - 10.$$ Therefore, this factorization is correct.

Note: In any factorization, the first thing we always look for is a greatest common factor. If there is no greatest common factor (or once it has been removed), no factor may contain a common factor. This fact may greatly reduce the number of trial and error attempts.

d) $4a^2 + 11ab + 6b^2$	The last sign is $+$ and the middle sign is $+$, so both signs of the binomial factors will be $+$. The first terms of the binomial factors must have a product of $4a^2$. The possibilities are $4a \cdot a$ and $2a \cdot 2a$. Let's try $2a$ and $2a$ and then show this choice is impossible.
$(2a + \_)(2a + \_)$	The blanks must be filled by terms whose product is $6b^2$ and such that the sum (since the signs are the same) of the outside and inside products is $11ab$. The possibilities are $6b \cdot b$ and $3b \cdot 2b$. Try $6b$ and b.
$(2a + 6b)(2a + b)$	This cannot work since the first factor has a common factor of 2. It would not help to put $6b$ in the second factor and b in the first since we would then have a common factor of 2 in the second factor. Try $3b$ and $2b$.
$(2a + 3b)(2a + 2b)$	The second factor has a common factor of 2, so this cannot work. The same situation would occur if $2b$ were put in the first factor and $3b$ in the second. Therefore, the first terms cannot be $2a$ and $2a$, so try $4a$ and a.
$(4a + \_)(a + \_)$	Again the possible terms for the second blanks are $6b$ and b or $3b$ and $2b$. Try $6b$ in the first blank and b in the second.
$(4a + 6b)(a + b)$	The first factor has a common factor of 2, so this cannot be correct. Put $6b$ in the second factor and b in the first.
$(4a + b)(a + 6b)$	Neither factor has a common factor, but the outside product is $24ab$ and the inside product is ab. Since $24ab + ab = 25ab$ and the middle term is $11ab$, this is incorrect. Try $3b$ in the first factor and $2b$ in the second.

$$(4a + 3b)(a + 2b)$$

The outside product is $8ab$ and the inside product is $3ab$, and $8ab + 3ab = 11ab$, which is what we need. Notice we could not have put $2b$ in the first factor and $3b$ in the second. Why?

$$(4a + 3b)(a + 2b)$$ Factorization.

Check:

$$(4a + 3b)(a + 2b) = 4a^2 + 8ab + 3ab + 6b^2 = 4a^2 + 11ab + 6b^2.$$
Therefore, this factorization is correct.

Note: After doing a few of these, you should be able to do most, if not all, of the above trial and error steps mentally.

e) $3x^2 + 14x + 10$ The last sign is $+$ and the middle sign is $+$, so both signs of the binomial factors are $+$. The only choice for the first terms are $3x$ and x.

$$(3x + \_)(x + \_)$$ The choices for the second blanks are 1 and 10 or 2 and 5. Try 2 in the first blank and 5 in the second.

$$(3x + 2)(x + 5)$$ Check:
$(3x + 2)(x + 5) = 3x^2 + 15x + 2x + 10 = 3x^2 + 17x + 10$, which is not correct. Try 5 in the first factor and 2 in the second.

$$(3x + 5)(x + 2)$$ Check:
$(3x + 5)(x + 2) = 3x^2 + 6x + 5x + 10 = 3x^2 + 11x + 10$ which is not correct. Since 2 and 5 will not work, try 1 and 10 with 1 in the first factor and 10 in the second.

$$(3x + 1)(x + 10)$$ Check:
$(3x + 1)(x + 10) = 3x^2 + 30x + x + 10 = 3x^2 + 31x + 10$ which is incorrect. Try 10 in the first factor and 1 in the second.

$$(3x + 10)(x + 1)$$ Check:
$(3x + 10)(x + 1) = 3x^2 + 3x + 10x + 10 = 3x^2 + 13x + 10$ which is incorrect. We have tried all possible combinations of the factors of $3x^2$ and 10, and none of them work. Since this exhausts all the possibilities, this polynomial is prime (cannot be factored).

Alternative Method—The "ac" Method

There is another method of factoring trinomials that is often easier than the trial and error method presented earlier called the "ac" method.

Suppose we find the product of $(mx + n)(px + q)$

$$(mx + n)(px + q)$$ FOIL.

$$mpx^2 + mqx + npx + nq$$ Remove the common factor of x from the middle two terms.

$$mpx^2 + (mq + np)x + nq$$

We now make some observations. If we multiply the coefficient of the x^2 term (which is mp) with the constant term (which is nq) we get $mpnq$. If we multiply the terms of the coefficients of the x term (which are mq and np) we get $mqnp$ which, except for the order of the factors, is the same thing we got when multiplying the coefficient of the x^2 term and the constant term. This means that the coefficient of the x term must be the sum of two factors of the product of the coefficient of the x^2 term and the constant term. We summarize the procedure below.

FACTORING TRINOMIALS USING THE "AC" METHOD

Given a trinomial in the form of $ax^2 + bx + c$:

1) Find the product of a and c.

2) Find two factors of ac whose sum is b.

3) Rewrite bx as the sum of two terms whose coefficients are the numbers found in step 2.

4) Factor the resulting polynomial by grouping.

The procedure is essentially the same if the trinomial is of the form $ax^2 + bxy + cy^2$. We illustrate the procedure using the trinomials from Example 1.

EXAMPLE 2

Factor each of the following using the "ac" method of factoring.

a) $3x^2 + 5x + 2 =$

The product of a and c is $3 \cdot 2 = 6$ so we need two factors of 6 whose sum is b which is 5. Six can be factored as $1 \cdot 6$ or $(-1)(-6)$, $2 \cdot 3$ or $(-2)(-3)$. Since $2 + 3 = 5$ (which is b) we rewrite $5x$ and $2x + 3x$.

$3x^2 + 2x + 3x + 2 =$

Factor by grouping. Remove the common factor of x from the first two terms and write $3x + 2$ as $1 \cdot (3x + 2)$.

$x(3x + 2) + 1 \cdot (3x + 2) =$ Remove the common factor of $3x + 2$.

$(3x + 2)(x + 1)$ Factorization.

b) $5x^2 + 8x - 4 =$

$ac = (5)(-4) = -20$ and $b = 8$, so we need two factors of -20 whose sum is 8. We can factor -20 as $(-1)(20)$ or $(1)(-20)$, $(-2)(10)$ or $(2)(-10)$, $(-4)(5)$ or $(4)(-5)$. Since $-2 + 10 = 8$, we rewrite $8x$ as $-2x + 10x$.

$5x^2 - 2x + 10x - 4 =$ Factor by grouping.

$x(5x - 2) + 2(5x - 2) =$ Remove the common factor of $5x - 2$.

$(5x - 2)(x + 2)$ Factorization.

c) $6a^2 - 11a - 10 =$

$ac = 6(-10) = -60$ and $b = -11$ so we need two factors of -60 whose sum is -11. We can factor -60 as $(-1)(60)$ or $(1)(-60)$, $(-2)(30)$ or $(2)(-30)$, $(-3)(20)$ or $(3)(-20)$, $(-4)(15)$ or $(4)(-15)$, $(-5)(12)$ or $(5)(-12)$, $(-6)(10)$ or $(6)(-10)$. Since $4 - 15 = -11$, we write $-11a$ as $4a - 15a$.

$$6a^2 + 4a - 15a - 10 =$$ Factor by grouping.

$$2a(3a + 2) - 5(3a + 2) =$$ Remove the common factor of $3a + 2$.

$$(3a + 2)(2a - 5)$$ Factorization.

d) $4a^2 + 11ab + 6b^2 =$ $\quad$ $ac = 4 \cdot 6 = 24$ and $b = 11$, so we need two factors of 24 whose sum is 11. We can factor 24 as $1 \cdot 24$ or $(-1)(-24)$, $2 \cdot 12$ or $(-2)(-12)$, $3 \cdot 8$ or $(-3)(-8)$, $4 \cdot 6$ or $(-4)(-6)$. Since $3 + 8 = 11$, we rewrite $11ab$ as $3ab + 8ab$.

$$4a^2 + 3ab + 8ab + 6b^2 =$$ Factor by grouping.

$$a(4a + 3b) + 2b(4a + 3b) =$$ Remove the common factor of $4a + 3b$.

$$(4a + 3b)(a + 2b)$$ Factorization.

e) $3x^2 + 14x + 10$ $\quad$ $ac = 3 \cdot 10 = 30$ and $b = 14$, so we need two factors of 30 whose sum is 14. We can factor 30 as $1 \cdot 30$ or $(-1)(-30)$, $2 \cdot 15$ or $(-2)(-15)$, $3 \cdot 10$ or $(-3)(-10)$, $5 \cdot 6$ or $(-5)(-6)$. None of these factors have a sum of 14, so this trinomial is prime.

Each of these methods has its advantages and disadvantages. If you use the "ac" method, you can often save yourself a lot of work by looking at the signs of the original polynomial and use that information as a guide in choosing the factors. For example, in Example 2d, both factors must be positive.

ADDITIONAL PRACTICE

Factor the following. If it will not factor, write prime.

a) $5x^2 - 9x + 4$

b) $7a^2 + 40a - 12$

c) $4b^2 + 5b - 6$

d) $6c^2 - 19cd + 15d^2$

e) $4x^2 - 7x - 3$

PRACTICE EXERCISES

Factor the following. If the trinomial will not factor, write prime.

1) $2x^2 - 5x + 3$ $\qquad$ **2)** $4x^2 + 17x - 15$ $\qquad$ **3)** $6x^2 + x - 12$

4) $6c^2 - 13cd + 6d^2$ $\qquad$ **5)** $2x^2 + 5x + 6$

If more practice is needed, do the Additional Practice Exercises in the margin.

As in all cases of factoring, the first thing we should look for is a common factor and, if there is one, remove it.

EXAMPLE 3

Factor the following. If the expression will not factor, write prime.

a) $18x^2 - 36x + 16$ $\quad$ Remove the common factor of 2.

$\quad 2(9x^2 - 18x + 8)$ $\quad$ Factor the trinomial inside the parentheses.

$\quad 2(3x - 2)(3x - 4)$ $\quad$ Factorization.

Check:
$\quad 2(3x - 2)(3x - 4) = 2(9x^2 - 18x + 8) = 18x^2 - 36x + 16$. Therefore, this factorization is correct.

b) $12a^3 + 27a^2 - 27a$ $\quad$ Remove the common factor of $3a$.

$\quad 3a(4a^2 + 9a - 9)$ $\quad$ Factor the trinomial inside the parentheses.

$\quad 3a(4a - 3)(a + 3)$ $\quad$ Factorization.

Check:
$\quad 3a(4a - 3)(a + 3) = 3a(4a^2 + 9a - 9) = 12a^3 + 27a^2 - 27a$. Therefore, this factorization is correct.

ADDITIONAL PRACTICE

f) $24a^2 + 52a + 24$

g) $16a^2b + 36ab - 10b$

PRACTICE EXERCISES

Factor the following completely. If the expression will not factor, write prime.

6) $9b^2 + 30b - 24$ **7)** $12cd^2 - 38cd + 20c$

If more practice is needed, do the Additional Practice Exercises in the margin.

ANSWERS:
Practice 1–5

1) $(2x - 3)(x - 1)$ **2)** $(4x - 3)(x + 5)$

3) $(3x - 4)(2x + 3)$ **4)** $(3c - 2d)(2c - 3d)$

5) Prime

Additional Practice a–e

a) $(5x - 4)(x - 1)$ **b)** $(7a - 2)(a + 6)$

c) $(4b - 3)(b + 2)$ **d)** $(3c - 5d)(2c - 3d)$

e) Prime

EXERCISE SET   **4.6**

Factor the following completely. If the polynomial will not factor, write prime.

1) $2a^2 + 7a + 5$
2) $3a^2 - 22a + 7$
3) $2a^2 + 5a - 3$
4) $5b^2 + 3b - 2$
5) $7m^2 - 34m - 5$
6) $5y^2 - 2y - 3$
7) $3x^2 - 4x + 7$
8) $5c^2 - 24c + 5$
9) $5c^2 - 2cd + 7d^2$
10) $7x^2 - 20xy - 3y^2$
11) $11x^2 + 16xy + 5y^2$
12) $3c^2 - 10cd + 7d^2$
13) $2x^2 + 13x + 10$
14) $3x^2 - 11x + 6$
15) $5x^2 + 6x - 8$
16) $7a^2 - 9a - 10$
17) $4x^2 + 8x + 3$
18) $6x^2 + 13x + 5$
19) $10x^2 - 33x - 7$
20) $21x^2 + 32x - 5$
21) $2x^2 + 11x + 12$
22) $3x^2 + 26x + 20$
23) $3a^2 - 14a + 16$
24) $2t^2 - 13t + 18$
25) $5c^2 + 11c - 12$
26) $3z^2 - 13z - 30$
27) $2b^2 + 13b - 24$
28) $5r^2 - 36r - 32$
29) $6x^2 + 17x + 5$
30) $6x^2 + 7x + 5$
31) $8x^2 - 22x + 5$
32) $8x^2 - 18x + 7$
33) $24a^2 - 14a - 3$
34) $24c^2 - 13c - 7$
35) $16n^2 - 2n - 5$
36) $18m^2 + 9m - 5$
37) $6x^2 + 19x + 15$
38) $8x^2 + 18x + 9$

39) $12c^2 - 17c + 6$
40) $24n^2 - 42n + 15$
41) $18a^2 + 9a - 35$
42) $16d^2 - 14d - 15$
43) $8w^2 - 6w - 27$
44) $15t^2 + 7t - 30$
45) $20a^2 - 9a + 20$
46) $18y^2 + 3y - 28$
47) $6a^2 + 19a - 15$
48) $12c^2 - 8c - 15$
49) $6a^2 + 29ab + 35b^2$
50) $8a^2 + 34ab + 21b^2$
51) $18c^2 - 9cd - 20d^2$
52) $24r^2 - 25rs - 25s^2$
53) $12a^2 + 7ab - 12b^2$
54) $18x^2 - 9xy - 20y^2$
55) $12f^2 - 32fg + 21g^2$
56) $16a^2 + 48ab + 35b^2$
57) $6a^2 + 3a - 18$
58) $6r^2 - 26r - 20$
59) $4x^2y - 13xy + 3y$
60) $3a^2b^2 - 10a^2b + 8a^2$
61) $6x^3 - 21x^2 + 15x$
62) $2x^3 - 14x^2 + 20x$
63) $18x^3 - 3x^2 - 36x$

CHALLENGE EXERCISES (64–68)

Factor the following. If the polynomial will not factor, write prime.

64) $32x^2 - 134x - 45$
65) $36x^2 + 29x - 20$
66) $24x^2 - 55x - 24$
67) $(x + 2)^2 - (x + 2) - 6$
68) $(x - 3)^2 - 2(x - 3) - 8$

Determine all integer values of k so that the following will be factorable over the integers.

69) $2x^2 + kx + 5$ **70)** $3x^2 + kx + 8$

WRITING EXERCISES

71) How would you explain to a classmate why $x^2 + 9x + 15$ is prime?

72) Explain why $12x^2 - x + 6$ cannot be factored correctly as $(3x + 2)(4x - 3)$.

73) Explain why $6x^2 + 21x + 15$ cannot be factored correctly as $(3x + 3)(2x + 5)$.

74) After factoring a trinomial the results are $(2x - 3)(5x + 7)$. What is the trinomial? Why?

Mixed Factoring

OBJECTIVES

When you complete this section, you will be able to:

Ⓐ Recognize and factor polynomials that have common factors, that are the difference of squares, or that are trinomials.

Ⓑ Recognize and factor polynomials by grouping or (optional) that are the sum or difference of cubes.

INTRODUCTION

In Sections 4.2–4.6 we concentrated on a particular type of factoring in each section. This made matters somewhat easier since we knew which form to look for. When the different types of factoring are mixed, we must first recognize the form in order to know which procedure to use. The following check list is recommended.

ANSWERS:
Practice 6–7

6) $3(3b - 2)(b + 4)$

7) $2c(3d - 2)(2d - 5)$

Additional Practice f–g

f) $4(2a + 3)(3a + 2)$

g) $2b(4a - 1)(2a + 5)$

AIDS TO COMPLETE FACTORING

A) First, remove the greatest common factor if there is one.

B) How many terms does the polynomial have?

   1) If the polynomial has two terms:

      a) Is it the difference of squares? If so, factor using
$a^2 - b^2 = (a + b)(a - b)$.

      b) Is it the sum or difference of cubes? If so, factor using
$a^3 + b^3 = (a + b)(a^2 - ab + b^2)$ or
$a^3 - b^3 = (a - b)(a^2 + ab + b^2)$. (optional in this book)

   2) If the polynomial has three terms:

      a) If necessary, put the trinomial in descending order.

      b) Is it a perfect square trinomial? If so, factor using
$a^2 + 2ab + b^2 = (a + b)^2$ or $a^2 - 2ab + b^2 = (a - b)^2$.

      c) If the trinomial is not a perfect square, factor using trial and error or the "ac" method. Check the factorization using FOIL.

   3) If the polynomial has more than three terms, try factoring by grouping.

C) The factorization is not complete unless each factor is prime. Check each factor and factor any that are not prime.

D) Check your factorization by multiplying and comparing the result with the original polynomial.

We will use the above procedure in the following examples.

EXAMPLE 1

Factor each of the following completely.

a) $x^4 - 10x^3 + 25x^2 =$ Remove the GCF of x^2.

    $x^2(x^2 - 10x + 25) =$ The trinomial is a perfect square.

    $x^2(x - 5)^2$ Factorization is complete.

    Check:

    $x^2(x - 5)^2 = x^2(x^2 - 10x + 25) = x^4 - 10x^3 + 25^2$. Therefore, the factorization is correct.

b) $25x^2 - 81 =$ No common factors.

$(5x)^2 - 9^2 =$ Binomial is the difference of squares.

$(5x + 9)(5x - 9)$ Each factor is prime, so the factorization is complete.

Check:
$(5x + 9)(5x - 9) = 25x^2 - 81.$ Therefore, the factorization is correct.

c) $8x^2 + 2x - 21 =$ No common factors. The trinomial is not a perfect square, so factor using trial and error or the "ac" method.

$(4x + 7)(2x - 3)$ Each factor is prime, so the factorization is complete.

Check:
$(4x + 7)(2x - 3) = 8x^2 - 12x + 14x - 21 = 8x^2 + 2x - 21.$ Therefore, the factorization is correct.

d) $16a^4 - 81 =$ No common factors.

$(4a^2)^2 - 9^2 =$ Binomial is the difference of squares.

$(4a^2 + 9)(4a^2 - 9) =$ $4a^2 + 9$ is prime but $4a^2 - 9$ is the difference of squares.

$(4a^2 + 9)(2a + 3)(2a - 3)$ Each factor is prime, so the factorization is complete.

Check:
$(4a^2 + 9)(2a + 3)(2a - 3) = (4a^2 + 9)(4a^2 - 9) = 16a^4 - 81.$ Therefore, the factorization is correct.

e) $12bc - 9b + 20c - 15 =$ No common factors. There are more than three terms, so factor by grouping. Remove the factor of $3b$ from the first two terms and 5 from the remaining terms.

$3b(4c - 3) + 5(4c - 3) =$ Remove the greatest common factor of $4c - 3$.

$(4c - 3)(3b + 5)$ Each factor is prime, so the factorization is complete.

Check:
$(4c - 3)(3b + 5) = 12bc + 20c - 9b - 15.$ Therefore, the factorization is correct.

f) $12x^2y - 22xy - 70y =$ Remove the greatest common factor of $2y$.

$2y(6x^2 - 11x - 35) =$ The trinomial factor is not a perfect square, so factor by trial and error or the "ac" method.

$2y(2x - 7)(3x + 5)$ Each factor is prime, so the factorization is complete.

Check:
$2y(2x - 7)(3x + 5) = 2y(6x^2 - 11x - 35) = 12x^2y - 22xy - 70y.$ Therefore, the factorization is correct.

g) $6x^2 + 4x + 5$ There is no common factor. The trinomial is not a perfect square. Attempts to factor by trial and error will show the trinomial is prime.

Factor the following completely. If the polynomial will not factor, write prime.

1) $4x^4 + 12x^3 + 9x^2$ **2)** $81a^2 - 49$ **3)** $9x^2 + 6x - 35$

4) $16c^4 - d^4$ **5)** $6xy - 8x - 21y + 28$ **6)** $48x^2y - 20xy - 8y$

7) $4x^2 - x + 6$

The ability to factor quickly and accurately is one of the most important skills in algebra. This skill will be used extensively in Chapter 5 on rational expressions.

EXERCISE SET 4.7

Factor each of the following completely. If the polynomial will not factor, write prime.

1) $16 - 8x^2$ **2)** $12 - 6y^2$

3) $c^2 + 12c + 27$ **4)** $d^2 + 13d + 40$

5) $3a^3 - 9a^2 - 54a$ **6)** $2x^3 + 8x^2 - 64x$

7) $x^2 - 100$ **8)** $a^2 - 144$

9) $c^2 - 18c - 81$ **10)** $r^2 + 22r + 121$

11) $ax + bx + 2a + 2b$ **12)** $bx - 2x + by - 2y$

13) $6a^2 - 48a + 72$ **14)** $5r^2 + 75r + 180$

15) $r^2 + r - 72$ **16)** $b^2 - b - 56$

17) $m^3n^2 + m^2n^4$ **18)** $a^4b^3 + a^2b$

19) $12a^4 - 75a^2$ **20)** $27x^4 - 48x^2$

21) $9a^2 + 30ab + 25b^2$ **22)** $4p^2 - 28pq + 49q^2$

23) $16r^2 + 40r - 96$ **24)** $30a^2 - 87a + 30$

25) $a^2b^2 - 64$ **26)** $p^2q^2 - 81$

27) $5n^2 + 22n + 8$ **28)** $3m^2 + 16m + 16$

29) $8c^2 + 4cd + d^2$ **30)** $10b^2 - 5bc + c^2$

31) $x(y + z) + w(y + z)$ **32)** $r(s - 3) - t(s - 3)$

33) $(a - b)^2 - 36$ **34)** $(r - s)^2 - 169$

35) $10xy - 4y - 15x + 6$ **36)** $10xy + 15y - 8x - 12$

37) $a^4 - 4b^2$ **38)** $25y^4 - 16$

39) $x^2 + 4$ **40)** $a^2 + 25$

41) $16b^4 - 1$ **42)** $n^4 - 625$

43) $6x^2 + 31x + 5$ **44)** $2d^2 + d - 10$

45) $6x^2 + 9xz - 2xy - 3yz$

46) $6a^2 + 4ac - 9ab - 6bc$

47) $ab + 3a - b - 3$

48) $xy - xz - y + z$

49) $16u^3 + 46u^2v - 6uv^2$

50) $18y^3 + 24y^2z - 10yz^2$

51) $2a^2b - 5a^2b^2 + 7ab^2$

52) $x^3y - 3x^2y^2 + 5xy^3$

53) $8c^2 + 33cd + 4d^2$

54) $7r^2 - 50rs + 7s^2$

55) $9ax + 3bx - 9a - 3b$

56) $8ac + 12ad - 6bc - 9bd$

Factor each of the following completely. (optional)

57) $8z^3 + 125$ **58)** $64c^3 + 27$ **59)** $27c^3 - 8d^3$

60) $125t^3 - 64d^3$ **61)** $24c^3 - 3d^3$ **62)** $4t^3 + 500$

CHALLENGE EXERCISES (63–68)

Factor the following completely. (Do not involve optional type of factoring.)

63) $12x^2 - 23x - 24$ **64)** $20x^2 - 39x + 18$

65) $81a^4 - 625b^4$ **66)** $256r^4 - 81s^4$

67) $x^2 + 10x + 25 - y^2$ **68)** $a^2 - 12ab + 36b^2 - 9$

WRITING EXERCISE

69) Why is it important to first remove any GCF when factoring?

GROUP PROJECT

70) Write two polynomials that would be factored by each of the following techniques.
   a) Removing the GCF
   b) Difference of squares
   c) Sum or difference of cubes
   d) Perfect square trinomial
   e) Trial and error using FOIL
   f) Grouping

Solving Quadratic Equations by Factoring

OBJECTIVES

When you complete this section, you will be able to:

Ⓐ   Solve quadratic equations by factoring.

Ⓑ   Solve applications problems using quadratic equations.

INTRODUCTION

We learned how to solve linear equations in Chapter 3 by using the Addition and Multiplication Properties of Equality. In this section we will learn to solve equations of a different type. These equations are called **quadratic equations**. A quadratic equation is any equation of the form $ax^2 + bx + c = 0$ with $a \neq 0$, or any equation which can be put into that form. Solving quadratic equations involves the techniques of solving linear equations and one additional property, the zero product property.

ANSWERS:
Practice 1–7

1) $x^2(2x + 3)^2$

2) $(9a + 7)(9a - 7)$

3) $(3x - 5)(3x + 7)$

4) $(4c^2 + d^2)(2c + d)(2c - d)$

5) $(3y - 4)(2x - 7)$

6) $4y(4x + 1)(3x - 2)$

7) prime

ZERO PRODUCT PROPERTY

If $ab = 0$, then $a = 0$, or $b = 0$, or $a = b = 0$. In words, if the product of two factors is zero, then one or both of the factors is 0.

This property is easily extended to three or more factors. For example, if $abc = 0$, then at least one of the following is true: $a = 0$, $b = 0$, or $c = 0$. When this property is applied to the solutions of equations, it means find the value(s) of the variable that make each factor equal to 0.

EXAMPLE 1

Solve the following using the zero product property.

a) $x(3x + 2) = 0$ Set each factor equal to 0.

     $x = 0$ or $3x + 2 = 0$ Solve the linear equation.

                $3x = -2$

                  $x = -\dfrac{2}{3}$ Therefore, the solutions are 0 and $-\frac{2}{3}$.

b) $(a + 3)(a - 1) = 0$ Set each factor equal to 0.

     $a + 3 = 0$ or $a - 1 = 0$ Solve each linear equation.

     $a = -3$ $a = 1$ Therefore, the solutions are -3 and 1.

c) $(b - 4)(2b + 5)(3b - 4) = 0$ Set each factor equal to 0.

     $b - 4 = 0$ or $2b + 5 = 0$ or $3b - 4 = 0$ Solve each linear equation.

     $b = 4$ $2b = -5$ $3b = 4$

                   $b = -\dfrac{5}{2}$ $b = \dfrac{4}{3}$ The solutions are 4, $-\frac{5}{2}$, and $\frac{4}{3}$.

PRACTICE EXERCISES

Solve the following.

1) $b(3b + 2) = 0$ **2)** $(n + 4)(3n - 5) = 0$

3) $(y - 3)(7y + 2)(5y - 3) = 0$

• Solving quadratic equations

Suppose we were to multiply the factors of Example 1b. We get:

$$(a + 3)(a - 1) = 0$$
$$a^2 + 2a - 3 = 0$$

The result is a quadratic equation for which $(a + 3)(a - 1)$ is the factored form. Consequently, factoring and the zero factor property give us a method for solving some quadratic equations.

SOLVING QUADRATIC EQUATIONS

1) If necessary, rewrite the equation in the form $ax^2 + bx + c = 0$ or $0 = ax^2 + bx + c$.

2) Factor $ax^2 + bx + c$.

3) Set each factor equal to 0. (Zero Product Property)

4) Solve the resulting linear equations.

5) Check the solutions.

EXAMPLE 2

Solve the following.

a) $2x^2 + 5x = 0$ Factor the left side.

   $x(2x + 5) = 0$ Set each factor equal to 0.

   $x = 0, 2x + 5 = 0$ Solve the linear equations.

          $2x = -5$

           $x = -\dfrac{5}{2}$ Therefore, the solutions are 0 and $-\dfrac{5}{2}$.

Check:

$x = 0$ $x = -\dfrac{5}{2}$

Substitute 0 for x.   Substitute $-\dfrac{5}{2}$ for x.

$2(0)^2 + 5(0) = 0$ $2(-\dfrac{5}{2})^2 + 5(-\dfrac{5}{2}) = 0$

$2 \cdot 0 + 0 = 0$ $2(\dfrac{25}{4}) - \dfrac{25}{2} = 0$

$0 + 0 = 0$ $\dfrac{25}{2} - \dfrac{25}{2} = 0$

$0 = 0$ $0 = 0$

Therefore, 0 and $-\dfrac{5}{2}$ are the correct solutions.

b) $x^2 - 2x - 15 = 0$ Factor the left side.

   $(x - 5)(x + 3) = 0$ Set each factor equal to 0.

   $x - 5 = 0, x + 3 = 0$ Solve each linear equation.

   $x = 5, \quad\quad x = -3$ Therefore, the solutions are 5 and -3.

Check:

$x = 5$ $x = -3$

Substitute 5 for x.   Substitute -3 for x.

$5^2 - 2(5) - 15 = 0$ $(-3)^2 - 2(-3) - 15 = 0$

$25 - 10 - 15 = 0$ $9 + 6 - 15 = 0$

$0 = 0$ $0 = 0$

Therefore, 5 and -3 are the correct solutions.

c) $c^2 - 16 = 0$ Factor the left side.

$(c + 4)(c - 4) = 0$ Set each factor equal to 0.

$c + 4 = 0, c - 4 = 0$ Solve each linear equation.

$c = -4, \quad c = 4$ Therefore, the solutions are 4 and -4.

Check:
The check is left as an exercise for the student.

d) $6x^2 - 10 = 11x$ The equation must first be written in the form of $ax^2 + bx + c = 0$. Subtract $11x$ from both sides and write the left side in descending order.

$6x^2 - 11x - 10 = 0$ Factor.

$(2x - 5)(3x + 2) = 0$ Set each factor equal to 0.

$2x - 5 = 0, 3x + 2 = 0$ Solve each linear equation.

$2x = 5, 3x = -2$

$x = \dfrac{5}{2}, \; x = -\dfrac{2}{3}$ Therefore, the solutions are $\dfrac{5}{2}$ and $-\dfrac{2}{3}$.

Check:
The check is left as an exercise for the student.

e) $x(x + 10) = 24$ Simplify the left side.

$x^2 + 10x = 24$ Subtract 24 from both sides.

$x^2 + 10x - 24 = 0$ Factor.

$(x + 12)(x - 2) = 0$ Set each factor equal to 0.

$x + 12 = 0, x - 2 = 0$ Solve each equation.

$x = -12, \quad x = 2$ Therefore, the solutions are -12 and 2.

Check:
The check is left as an exercise for the student.

ADDITIONAL PRACTICE

Solve the following.

a) $4c^2 - 7c = 0$

b) $a^2 + 9a + 14 = 0$

c) $b^2 - 1 = 0$

d) $6x^2 - 3 = 7x$

e) $x(x + 2) = 48$

- **Applications with quadratic equations**

PRACTICE EXERCISES

Solve the following.

4) $2r^2 + 5r = 0$

5) $z^2 - 4z - 21 = 0$

6) $m^2 - 36 = 0$

7) $8x^2 - 15 = 14x$

8) $x(x + 3) = 40$

If more practice is needed, do the Additional Practice Exercises in the margin.

Applications problems often result in quadratic equations. Care must be taken since some solutions to the equations may not be realistic solutions to the original problem. For example, if x represents the length of the side of a rectangle, then x cannot be negative. However, the equation may have a negative solution. We will use the same technique developed in Chapter 3, but will not elaborate as much.

EXAMPLE 3

a) Find two consecutive positive even integers whose product is 80.

Solution:

Let x = the smaller of the two consecutive even integers. Then,
$x + 2$ = the larger of the two consecutive even integers.
Since product means multiply:

Smaller Integer	times	Larger Integer	equals	80
x		$(x + 2)$	$=$	80

$x(x + 2) = 80$	Simplify the left side.
$x^2 + 2x = 80$	Subtract 80 from both sides.
$x^2 + 2x - 80 = 0$	Factor.
$(x + 10)(x - 8) = 0$	Set each factor equal to 0.
$x + 10 = 0,\ x - 8 = 0$	Solve each equation.
$x = -10,\quad x = 8$	Possible solutions are -10 and 8.

Since we were asked for consecutive *positive* even integers, we discard -10. If 8 is the smaller of the two consecutive positive even integers, then $x + 2 = 10$ is the larger. Consequently, the integers are 8 and 10.

Check:

Are 8 and 10 consecutive positive even integers? Yes. Is the product of 8 and 10 equal to 80? Yes. Then 8 and 10 are the correct solutions.

b) The length of a rectangular screen is 3 feet less than twice the width. Find the dimensions of the screen if the area is 54 square feet.

Solution:

Let x = the width. Then,
$2x - 3$ = the length. (The length is 3 less than twice the width.)

Area
54 sq ft x
$2x - 3$

The formula for the area of a rectangle is $A = LW$.

$A = LW$	Substitute for A, L, and W.
$54 = (2x - 3)(x)$	Simplify the right side.
$54 = 2x^2 - 3x$	Subtract 54 from both sides.
$0 = 2x^2 - 3x - 54$	Factor.
$0 = (2x + 9)(x - 6)$	Set each factor equal to 0.
$2x + 9 = 0,\ x - 6 = 0$	Solve each equation.
$2x = -9,\quad x = 6$	
$x = -\dfrac{9}{2}$	

Since the width of a rectangle cannot be negative, we discard $-\frac{9}{2}$. Since we let x represent the width, the width is 6 feet. The length is represented by $2x - 3$, so the length is $2(6) - 3 = 12 - 3 = 9$ feet. The dimensions of the screen are 6 feet wide and 9 feet long.

Check:
    Is the length, which is 9, 3 less than twice the width, which is 6? Yes. Is the area 54 square feet? Since $A = LW = 9 \cdot 6 = 54$ square feet. Yes. Therefore, our solution is correct.

c) If a ball is thrown vertically upward from ground level with an initial velocity of 80 feet per second, the equation giving its height above the ground is $h = 80t - 16t^2$, where h is the height of the ball in feet and t is the number of seconds after the ball was thrown. **1)** Find the number of seconds until the ball returns to the ground. **2)** Find the number of seconds when the ball is 96 feet above the ground.

Solution:
Part 1:
    When the ball returns to ground level its height above the ground is 0, so substitute 0 for h and solve for t.

$0 = 80t - 16t^2$	Remove the common factor of $16t$.
$0 = 16t(5 - t)$	Set each factor equal to 0.
$16t = 0, 5 - t = 0$	Solve each equation.
$t = 0, \quad 5 = t$	$t = 0$ indicates the ball is at ground level before it was thrown. Therefore, the ball returns to ground level after 5 seconds.

Check:
    After 5 seconds, $h = 80(5) - 16(5^2) = 80(5) - 16(25) = 400 - 400 = 0$ which indicates the ball is on the ground.

Part 2:
    If the ball is 96 feet above the ground $h = 96$, so substitute 96 for h and solve for t.

$96 = 80t - 16t^2$	Add $16t^2$ and $-80t$ to both sides.
$16t^2 - 80t + 96 = 0$	Remove the common factor of 16.
$16(t^2 - 5t + 6) = 0$	Factor $t^2 - 5t + 6$.
$16(t - 2)(t - 3) = 0$	Set $t - 2$ and $t - 3$ equal to 0. ($16 \neq 0$)
$t - 2 = 0, t - 3 = 0$	Solve each equation.
$t = 2, \quad t = 3$	Therefore, the ball is 96 feet above the ground after 2 seconds and again after 3 seconds.

Check:
    The check is left as an exercise for the student.

d) The formula for finding the sum of n consecutive positive integers beginning with one is $s = \frac{n(n + 1)}{2}$ where s is the sum and n is the number of positive integers. For example, in the sum $1 + 2 + 3 + 4 + 5 + 6 + 7 + 8 + 9 + 10$, $n = 10$ since we are summing the first 10 positive integers. The sum is $s = \frac{(10)(10 + 1)}{2} = \frac{(10)(11)}{2} = \frac{110}{2} = 55$. How many integers would it take for the sum to be 36?

Solution:
Since the sum is 36, substitute 36 for s and solve for n.

$36 = \dfrac{n(n + 1)}{2}$	Multiply both sides by 2.

$$2(36) = 2 \cdot \frac{n(n+1)}{2}$$ Simplify both sides.

$72 = n^2 + n$ Subtract 72 from both sides.

$0 = n^2 + n - 72$ Factor.

$0 = (n + 9)(n - 8)$ Set each factor equal to 0.

$n + 9 = 0, n - 8 = 0$ Solve each equation.

$n = -9, \quad n = 8$ Since we are looking for the number of positive integers, 8 is the only possible answer.

Check:

Substitute 8 for n and find s. $s = \frac{8(8+1)}{2} = \frac{8(9)}{2} = \frac{72}{2} = 36$ which is what it is supposed to be. Therefore, our answer is correct.

e) A bank has several branches in each large city. If there are n branches in a city and each branch manager is required to visit each of the other branches twice per year, the total number of visits, V, per year is given by $V = n(n - 1)$. How many branches are in a particular city if there is a total of 42 visits in one year?

Since there is a total of 42 visits per year, substitute 42 for V and solve for n.

$42 = n(n - 1)$ Simplify the right side of the equation.

$42 = n^2 - n$ Subtract 42 from both sides.

$0 = n^2 - n - 42$ Factor.

$0 = (n - 7)(n + 6)$ Set each factor equal to 0.

$n - 7 = 0, n + 6 = 0$ Solve each equation.

$n = 7, \quad n = -6$ Since there can not be a negative number of branches, there are 7 branches in this city.

Check:

The check is left as an exercise for the student.

PRACTICE EXERCISES

Solve each of the following.

9) Find two consecutive integers so that twice the square of the larger is 57 more than three times the smaller.

10) The length of a rectangular rug is 5 feet less than twice the width. If the area of the rug is 88 square feet, find the dimensions of the rug.

11) If an object is launched vertically upward with an initial velocity of 128 feet per second, the height of the object above the ground is given by $h = 128t - 16t^2$ where h is the height of the object and t is the number of seconds after it is launched. **a)** Find how long it will take the object to return to the ground. **b)** After how many seconds will the object be 240 feet above the ground?

12) Using the formula from Example 3d, find how many positive integers beginning with 1 it would take for the sum to be 45.

13) Using the formula from Example 3e, find how many branches are in a city if there is a total of 90 visits per year.

E X E R C I S E   S E T   4.8

Solve the following using the zero product property.

1) $(t - 6)(t + 4) = 0$ 2) $(u + 3)(u - 8) = 0$

3) $(2v - 10)(v - 9) = 0$ 4) $(w + 2)(5w + 15) = 0$

5) $r(5r - 2) = 0$ 6) $(3s - 5)s = 0$

7) $6z(7z + 11) = 0$ 8) $(8p - 15)13p = 0$

9) $(a - 3)(a + 2)(a - 5) = 0$

10) $(b + 4)(b - 6)(b - 1) = 0$

11) $(q + 7)(3q - 5)(4q - 1) = 0$

12) $(2x - 9)(6x + 2)(13x - 5) = 0$

13) $(2y - 4)^2 = 0$

14) $(5z + 3)^2 = 0$

Solve the following.

15) $x^2 - x - 6 = 0$ 16) $2d^2 + 7d - 4 = 0$

17) $4m^2 - 23m + 15 = 0$ 18) $n^2 + 10n + 24 = 0$

19) $5s^2 - 10s = 0$ 20) $21k - 7k^2 = 0$

21) $3r^2 - 2r = 0$ 22) $4h - 9h^2 = 0$

23) $v^2 - 4 = 0$ 24) $9 - u^2 = 0$

25) $3w^2 - 75 = 0$ 26) $72 - 2x^2 = 0$

27) $t^2 + 12 = 7t$ 28) $a^2 - 45 = 4a$

29) $3y^2 - 8 = 10y$ 30) $24b^2 - 15 = -2b$

31) $c(c - 12) = -36$ 32) $d(d + 10) = -25$

33) $4y(y + 5) = -25$ 34) $3x(3x - 8) = -16$

Solve each of the following.

35) Find two consecutive positive even integers whose product is 168.

36) Find two consecutive negative odd integers whose product is 143.

37) Find two consecutive negative even integers such that the sum of their squares is 52.

38) Find two consecutive negative odd integers such that the difference of four times the square of the smaller and twice the square of the larger is 34.

39) If the sum of an integer squared and twelve times the integer is -32, find the integer.

40) If the difference of an integer squared and twice the integer is 63, find the integer.

41) If the difference of five times the square of an integer and eight times the integer is 21, find the integer.

42) If the sum of three times the square of an integer and thirteen times the integer is 30, find the integer.

43) An American flag is 4 feet longer than it is wide. Find the dimensions of the flag if the area is 32 square feet. $A = LW$.

44) A rectangular patio is 5 yards shorter than it is long. Find the dimensions of the patio if the area of the patio is 50 square yards. $A = LW$.

45) The length of the cover of a textbook is 3 inches more than its width. Find the dimensions of the cover of the book if the area of the cover is 88 square inches. $A = LW$.

46) The side of a rectangular box is such that the length is 7 centimeters less than twice the width. Find the dimensions of the sides of the box if its area is 130 square centimeters. $A = LW$.

47) A search party is looking for a lost person known to be in a triangular region bounded by three highways. If one of the highways is used as the base of the region, the height to that base is four miles less than the base. If the area of the region is 6 square miles, find the base and height. $A = \frac{bh}{2}$

48) The height of a triangular piece of plywood is two feet more than the base. If the area is 24 square feet, find the base and the height. $A = \frac{bh}{2}$

49) A projectile is launched vertically upward from ground level with an initial velocity of 112 feet per second. The equation giving its height above the ground is $h = 112t - 16t^2$ where h is the height in feet and t is the number of seconds after it was launched. **a)** Find the number of seconds until the projectile returns to the ground. **b)** Find the number of seconds when the projectile is 192 feet above the ground.

50) A rock is thrown vertically upward from ground level with an initial velocity of 96 feet per second. The equation giving its height above the ground is $h = 96t - 16t^2$ where h is the height in feet and t is the number of seconds after it was thrown. **a)** Find the number of seconds until the rock returns to the ground. **b)** Find the number of seconds when the rock is 80 feet above the ground.

51) If a heavy object is dropped from the top of a building that is 256 feet high, the equation giving the height of the object is $h = 256 - 16t^2$ where h is the height above the ground and t is the number of seconds after the object was dropped. How long will it take the object to reach the ground?

52) If a heavy object is dropped from the top of a cliff that is 144 feet high, the equation giving the height of the object is $h = 144 - 16t^2$ where h is the height above the ground and t is the number of seconds after the object was dropped. How long will it take the object to reach the ground?

In Exercises 53–56 use the formula $s = \frac{n(n+1)}{2}$. See Example 3d.

53) How many positive integers, beginning with one, will it take for the sum to be 28?

54) How many positive integers, beginning with one, will it take for the sum to be 90?

55) Logs are stacked in a triangular shaped pile with one log on the top row, two logs on the second row, three logs on the third row, etc. If there is a total of 21 logs in the stack, how many logs are on the bottom row?

56) Fence posts in a lumber yard are stacked in a triangular shaped pile with one post on the top row, two posts on the second row, three posts on the third row, etc. If there is a total of 78 posts in the pile, find the number of posts on the bottom row.

In Exercises 57–58, use the formula $V = n(n-1)$. See Example 3e.

57) How many branches are in a city if there is a total of 72 visits per year.

58) How many branches are in a city if there is a total of 56 visits per year.

CHALLENGE EXERCISES (59–62)

Solve each of the following.

59) $x^4 - 13x^3 + 36x^2 = 0$ **60)** $3a^3 - 147a = 0$

61) $5t^4 + 30t^3 = 0$

62) $32x^3 - 144x^2 + 162x = 0$

WRITING EXERCISES

63) Compare the methods of solving linear equations and quadratic equations.

64) When solving a quadratic equation, why must one side of the equation be equal to 0?

65) Suppose we are solving a quadratic equation by factoring and we arrive at $4(x - 2)(x + 5) = 0$. In order to solve this equation we set the factors $x - 2$ and $x + 5 = 0$. Why do we not also set 4 equal to 0? What happens to the 4?

WRITING EXERCISES OR GROUP PROJECT

If done as a group project, each group should write two exercises of each type, exchange with another group, and then solve.

66) Write and solve an applications problem involving consecutive integers.

67) Write and solve an applications problem involving the area of a rectangular-shaped region.

Summary

Definitions: [Section 4.1]

Factor: The number a is a factor of the number b if b can be written as the product of a and some other number.

Divisible: The number a is divisible by the number b if b divides evenly into a. That is, if b is divided into a, the remainder is 0.

Prime: A natural number is prime if it is greater than one and its only factors are one and itself. Equivalently, a number is prime if it is greater than one and divisible only by one and itself.

Composite: A natural number, other than one, is composite if it is not prime. Equivalently, a number is composite if it has at least one factor other than one and itself.

Prime Factorization: [Section 4.1]

A number is written in prime factorization form if it is written as the product of its prime factors.

Greatest Common Factor: [Section 4.1]

The greatest common factor of two or more numbers is the largest number that is a factor of all the numbers.

The greatest common factor of two or more monomials (other than a number) is the monomial of greatest degree that is a factor of each of the monomials.

Finding the Greatest Common Factor: [Section 4.1]

1) Write each number or monomial as the product of prime factors.

2) Select those factors which are common in each prime factorization.

3) For each factor that is common, select the smallest exponent that occurs in any of the prime factorizations.

4) The greatest common factor is the product of each factor that is common with the smallest power in any of the prime factorizations.

Removing the Greatest Common Factor: [Section 4.2]

Determine the greatest common factor of the terms of the polynomial. Use the distributive property to write the polynomial as the product of the greatest common factor and another polynomial. The terms of the polynomial within the parentheses may be determined either by inspection or by division.

Factoring by Grouping: [Section 4.2]

If the polynomial has more than three terms, factor by grouping. The goal is to get the same binomial as a common factor and then remove it.

Difference of Squares: [Section 4.3]

A binomial, that is the difference of squares, factors into the sum and difference of the expressions being squared.

$$a^2 - b^2 = (a + b)(a - b)$$

Factoring the Sum or Difference of Cubes: [Section 4.3: optional]

The sum or difference of cubes factors into a binomial and a trinomial. The terms of the binomial are the expressions whose cubes give the sum or difference of cubes. The sign of the binomial is the same as the sign of the sum or difference of cubes. The first term of the trinomial is the square of the first term of the binomial. The second term of the trinomial is the negative of the product of the terms of the binomial. The last term of the trinomial is the square of the second term of the binomial.

$$a^3 + b^3 = (a + b)(a^2 - ab + b^2)$$
$$\text{and } a^3 - b^3 = (a - b)(a^2 + ab + b^2)$$

Perfect Square Trinomial: [Section 4.4]

A trinomial is a perfect square trinomial if the first and last terms are squares of expressions and the middle term is two times the product of the expressions being squared. The binomial whose square gives the perfect square trinomial has a first term whose square gives the first term of the trinomial, has a second term whose square gives the last term of the trinomial and has the same middle sign as the second sign of the trinomial.

$$a^2 + 2ab + b^2 = (a + b)^2$$
$$\text{and } a^2 - 2ab + b^2 = (a - b)^2$$

Factoring a General Trinomial: [Sections 4.5 and 4.6]

1) If the last sign of the trinomial is positive, both signs of the binomial factors will be the same as the sign of the middle term of the trinomial. The sum of the outer and inner products of the binomials (ignoring the

signs) will be equal to the second term of the trinomial (ignoring the sign).

2) If the last sign of the trinomial is negative, the signs of the binomial factors will be different. The difference of the outer and inner products (ignoring the signs) will equal the second term of the trinomial (ignoring the sign). Find the terms which give the correct difference, then determine the signs.

3) No binomial factor may contain a common factor.

Factoring a Trinomial Using the "ac" Method

To factor a trinomial of the form $ax^2 + bx + c$:

1) Find the product of a and c.

2) Find two factors of ac whose sum is b.

3) Rewrite bx as the sum of two terms whose coefficients are the numbers found in step 2.

4) Factor the resulting polynomial by grouping.

Factoring a Polynomial: [Section 4.7]

A) First, remove all common factors.

B) How many terms does the polynomial have?
   1) If the polynomial has two terms:
      a) Is it the difference of squares? If so, factor using the form $a^2 - b^2 = (a + b)(a - b)$.
      b) Is it the sum or difference of cubes? If so, factor using the forms

$$a^3 + b^3 = (a + b)(a^2 - ab + b^2) \text{ or}$$
$$a^3 - b^3 = (a - b)(a^2 + ab + b^2).$$

   2) If the polynomial has three terms:
      a) Is it a perfect square trinomial? If so, factor using the forms

$$a^2 + 2ab + b^2 = (a + b)^2 \text{ or}$$
$$a^2 - 2ab + b^2 = (a - b)^2.$$

   b) If the trinomial is not a perfect square, factor using trial and error. Check the factorization using FOIL.

   3) If the polynomial has more than three terms, factor by grouping.

C) The factorization is not complete unless each factor is prime. Check each factor and, if necessary, factor any that are not prime until all factors are prime.

Solving Quadratic Equations: [Section 4.8]

1) The zero factor property: If $ab = 0$, then $a = 0$, $b = 0$, or $a = b = 0$.

2) Using the zero factor property to solve quadratic equations.
   a) If necessary, rewrite the equation in the form $ax^2 + bx + c = 0$.
   b) Factor $ax^2 + bx + c$.
   c) Set each of the factors equal to 0.
   d) Solve the resulting equations.
   e) Check in the original equation.

3) Solving applications problems involving quadratic equations.
   a) Read the problem and assign a variable to an unknown.
   b) Represent any other unknown in terms of the variable.
   c) Write the equation using information from the problem or other known facts.
   d) Solve the equation.
   e) See if all solutions of the equation make sense in the original problem.
   f) Check all solutions which make sense in the original wording of the problem.

Review Exercises

Classify each of the following as prime or composite. [Section 4.1]

1) 87

2) 43

Determine the prime factorization of each of the following. [Section 4.1]

3) 60

4) 308

Find the greatest common factor of each of the following. [Section 4.1]

5) 126 and 75

6) 54 and 140

7) 28, 112, and 294

8) 48, 42, and 180

9) m^2n^4 and m^3n

10) a^3b^4 and b^2c^5

11) $18r^5s^3t^2$ and $24r^2s^2t^3$

12) $15a^3b^2$, $10ab^3$, and $20a^3b^2c$

Factor the following by removing the greatest common factor. [Section 4.2]

13) $ax + ay$

14) $6x^2 + 3$

15) $18x^2y - 24xy$

16) $r^4s^3 + r^6s^4$

17) $18x^2 - 24x + 12$

18) $8a^2b^4 - 12a^3b^5 + 18ab^3$

19) $r(s - 2) - 5(s - 2)$

20) $m(n - p) - q(n - p)$

21) Factor $-15x^3 + 5x^2 - 25x$ by removing a negative common factor.

Completely factor the following by grouping. [Section 4.2]

22) $ax + bx + ay + by$

23) $rs - 8r + 3s - 24$

24) $ab - 6a - 3b + 18$

25) $xy - 7y - 4x + 28$

26) $3ab - 3a - b + 1$

27) $6x^2 + 15x - 4xy - 10y$

Completely factor the following difference of squares. [Section 4.3]

28) $g^2 - h^2$

29) $r^2 - 121$

30) $25x^2 - 16$

31) $49a^2 - 64b^2$

32) $6x^2 - 24$

33) $64c^2 - 16d^2$

34) $(m + n)^2 - 81$

35) $64 - (c - d)^2$

36) $r^2 + s^2$

37) $m^4 - n^4$

Completely factor the following sum or difference of cubes. [Section 4.3: optional]

38) $m^3 + n^3$

39) $r^3 - 1$

40) $a^3 + 125$

41) $64x^3 + 27y^3$

Completely factor the following perfect square trinomials. [Section 4.4]

42) $r^2 + 18r + 81$

43) $x^2 - 24x + 144$

44) $x^2 + 8x + 64$

45) $49a^2 - 14a + 1$

46) $49a^2 + 28ab + 4b^2$

47) $25c^2 + 30cd + 9d^2$

48) $18p^2 + 24p + 32$

49) $4x^2 - 40x + 100$

Completely factor the following trinomials with leading coefficients of one. [Section 4.5]

50) $x^2 + 12x + 11$

51) $x^2 - 12x + 35$

52) $x^2 + 6x - 7$

53) $m^2 - 3m - 10$

54) $x^2 - 11x + 24$

55) $r^2 + rs - 72s^2$

56) $a^2 + 25ab - 54b^2$

57) $m^2 - 19mn - 42n^2$

58) $2x^2 + 24x + 70$

59) $y^3 + 7y^2z - 44yz^2$

Completely factor the following trinomials with leading coefficient other than one. [Section 4.6]

60) $5x^2 - 7x + 2$

61) $3z^2 - 2z - 5$

62) $5x^2 + 18x - 8$

63) $3a^2 + 26a + 16$

64) $8a^2 + 13ab - 7b^2$

65) $18a^2 + a - 5$

66) $35x^2 + 29xy + 6y^2$

67) $20a^2 + 9ab - 18b^2$

68) $20a^2 - 66a - 14$

69) $36c^3 - 51c^2d + 18cd^2$

Completely factor the following polynomials which involve all types of factoring. [Section 4.7]

70) $26a^2 + 13a$

71) $81x^2 - y^2$

72) $2x^2 - 6x - 80$

73) $x^2 - 2x - 63$

74) $22a^2b - 11a^3b^2 + 33a^2b$

75) $9m^2 - 48m + 48$

76) $3xy - 27y - 4x + 36$

77) $4(a + b)^2 - 49$

78) $8x^2 + 28x - 56$

79) $49a^2 - 28ab + 4b^2$

80) $6x^2 + 21x - 45$

81) $6x(y - 5) + 2(y - 5)$

82) $27x^2y - 75y$

83) $121a^2 - 169$

84) $81x^2 - 72xy + 16y^2$

85) $9x^2 + 81$

86) $8x^2 + 24x - 4xy - 12y$

87) $r^2 + 6r - 40$

88) $x^4 + 81$

89) $y^4 - 625$

90) $3x^3y^2 - 12x^2y^3 - 135xy^4$

91) $8x^2 - 15x + 7$

Solve the following equations. [Section 4.8]

92) $x^2 + 5x = 0$ **93)** $x^2 = 6x$

94) $4x^2 - 81 = 0$ **95)** $9x^2 = 64$

96) $x^2 - 2x - 35 = 0$ **97)** $y^2 - 4y - 21 = 0$

98) $2x^2 - 4 = 7x$ **99)** $3x^2 - 8 = -10x$

Solve the following applications problems. [Section 4.8]

100) The length of a rectangle is 5 less than twice the width. If the area of the rectangle is 25 square centimeters, find the dimensions of the rectangle.

101) Find two consecutive odd integers whose product is 63.

102) If 2 is added to the length of each side of a square, the area is increased by 28 square yards. Find the length of a side of the original square.

103) The sum of the square of an integer and 3 times the integer is equal to 28. Find the integer(s).

Chapter 4 Test

1) Classify each of the following as prime or composite.
  a) 111   **b)** 73

2) Find the prime factorization of 252.

Find the greatest common factor for each of the following.

3) 72 and 96   **4)** $48x^3y^5$ and $60x^2yz^3$

Factor each of the following. If the polynomial cannot be factored, write prime.

5) $x^3 + x^2$ **6)** $m^2 - 18m + 81$

7) $m^2 - 3mn - 40n^2$
       8) $49a^2 - 9b^2$

9) $14x^3y - 28x^2y^2 + 35x^4y^3$
       10) $9y^2 - 24y + 16$

11) $16a^2 + 25b^2$
       12) $r(s + 5) - 7(s + 5)$

13) $(x - y)^2 - 100$ **14)** $3q^2 + 6q - 105$

15) $3a^2 - 13a - 30$ **16)** $8a^2 - 34ab + 35b^2$

17) $27x^2 - 12y^2$ **18)** $6ab - 4a + 15b - 10$

19) $5a^2 + 50ab + 125b^2$ **20)** $x^4 - 625$

21) $20a^2 - 7ab - 6b^2$

22) The area of a rectangle is represented by $x^2 - 2x - 24$. How may the length and width be represented as binomials?

Solve the following.

23) $6x^2 + 4x = 0$ **24)** $6x^2 - 19x + 10 = 0$

25) The length of a rectangle is four less than three times the width. Find the dimensions of the rectangle if the area is 32 square inches.

5

Operations with Rational Expressions

5.1 Reducing Rational Expressions to Lowest Terms

5.2 Multiplication of Rational Expressions

5.3 Division of Rational Expressions

5.4 Least Common Multiple, Least Common Denominator, and Equivalent Rational Expressions

5.5 Addition and Subtraction of Rational Expressions

5.6 Complex Fractions

5.7 Solving Equations Containing Rational Numbers and Expressions

5.8 Applications with Rational Expressions

5.9 Ratio and Proportion

I n the last chapter we learned to factor polynomials into prime factors. In this chapter we will use our knowledge of factoring to reduce, multiply, divide, add and subtract rational expressions. In short, we will do all of the things with rational expressions that we normally do with rational numbers.

CHAPTER OVERVIEW

SECTION 5.1

Reducing Rational Expressions to Lowest Terms

OBJECTIVES

When you complete this section, you will be able to:

Ⓐ   Determine the value(s) for which rational expressions are undefined.

Ⓑ   Reduce rational numbers to lowest terms.

Ⓒ   Reduce rational expressions to lowest terms.

INTRODUCTION

A number is rational if it can be written as a fraction with numerator and nonzero denominator that are integers. In algebra, the numerators and the denominators of fractions are often polynomials. Such fractions are called **rational expressions.**

DEFINITION OF RATIONAL EXPRESSION

A rational expression is an algebraic expression of the form $\frac{P}{Q}$ where P and Q are polynomials and $Q \neq 0$.

Identifying values for which
rational expressions are
undefined

Before discussing reducing rational expressions we need to find the value(s) of the variable(s) for which a rational expression is defined. We know that division by 0 is undefined. Consequently, a rational expression is undefined for all value(s) of the variable(s) for which the denominator equals 0. Those values that make the denominator equal to 0 can be found by setting the denominator equal to 0 and solving the resulting equation.

EXAMPLE 1

Find the value(s) of the variable(s) for which the following are defined.

a) $\dfrac{6ab^2}{5a^2b^2}$

Solution:

$\dfrac{6ab^2}{5a^2b^2}$ is defined for all values of a and b that do not make the denominator equal to 0. To find the value(s) that make the denominator equal to 0, set the denominator equal to 0 and solve the equation.

$5a^2b^2 = 0$	The only way $5a^2b^2$ can equal 0 is for $a^2 = 0$ or $b^2 = 0$.
$a^2 = 0,\ b^2 = 0$	The only number whose square is 0 is 0. Therefore,
$a = 0,\ b = 0$	Solutions of the equations.

Since the denominator is equal to 0 for $a = 0$ or $b = 0$, $\dfrac{6ab^2}{5a^2b^2}$ is undefined for $a = 0$ and $b = 0$. Consequently, we say that $a \neq 0$ and $b \neq 0$.

 b) $\dfrac{x}{x + 3}$

Solution:

$\dfrac{x}{x+3}$ is defined for all values of x that do not make the denominator equal to 0. To find the value(s) that make the denominator equal to 0, set the denominator equal to 0 and solve the equation.

$x + 3 = 0$	Subtract 3 from both sides.
$x + 3 - 3 = 0 - 3$	Simplify both sides.
$x = -3$	Solution of the equation.

Since the denominator is equal to 0 for $x = -3$, $\dfrac{x}{x+3}$ is undefined for $x = -3$. Consequently, $x \neq -3$.

 c) $\dfrac{a + b}{a - b}$

Solution:

$\dfrac{a+b}{a-b}$ is defined for all values of a and b that do not that make the denominator equal to 0. To find those values, set the denominator equal to 0 and solve the equation.

$a - b = 0$	Add b to both sides.
$a - b + b = 0 + b$	Simplify both sides.
$a = b$	Solution of the equation.

Since the denominator is equal to 0 when $a = b$, $\dfrac{a+b}{a-b}$ is defined for all values of a and b except for $a = b$. Consequently, $a \neq b$.

d) $\dfrac{x-5}{x^2-2x-24}$

Solution:

$\dfrac{x-5}{x^2-2x-24}$ is defined for all values of x except those that make the denominator equal to 0. To find these values, set the denominator equal to 0 and solve for x.

$x^2 - 2x - 24 = 0$	Factor.
$(x + 4)(x - 6) = 0$	Set each factor equal to 0.
$x + 4 = 0,\ x - 6 = 0$	Solve each equation.
$x = -4,\ x = 6$	Solutions of the equation.

Since the denominator is equal to 0 when $x = -4$ or $x = 6$, $\dfrac{x-5}{x^2-2x-24}$ is defined for all values of x except -4 and 6. Consequently, $x \neq -4$ and $x \neq 6$.

ADDITIONAL PRACTICE

Find the values of the variable(s) for which the following are defined.

a) $\dfrac{4rs^2}{5rs}$ **b)** $\dfrac{x+6}{x+5}$

c) $\dfrac{2a}{a+b}$ **d)** $\dfrac{m-6}{m^2+6m+8}$

PRACTICE EXERCISES

Find the values of the variable(s) for which the following are defined.

1) $\dfrac{3x^2y}{2xy^2}$ **2)** $\dfrac{x-5}{x+4}$

3) $\dfrac{c-d}{2c+d}$ **4)** $\dfrac{c-1}{c^2-4c-21}$

If more practice is needed, do the Additional Practice Exercises in the margin.

In this section we will use our knowledge of factoring to reduce rational expressions to lowest terms in much the same manner that we reduce rational numbers to lowest terms. The technique was first demonstrated in Section 1.2 and may be different than the approach with which you are accustomed. However, this approach works best with rational expressions and we encourage you to use it with rational numbers as well.

Following is a repeat of the definition of lowest terms and the procedure for reducing fractions to lowest terms first given in Section 1.2.

DEFINITION OF LOWEST TERMS

A fraction is reduced to lowest terms if the numerator and the denominator have no common factors other than one.

• **Reducing fractions to lowest terms**

REDUCING A FRACTION TO LOWEST TERMS

1) Write the numerator and the denominator as the product of their prime factors. (It is best not to use exponents in doing so.)

2) Divide the numerator and the denominator by the factors common to both.

3) Multiply the remaining factors.

The same procedure applies whether the fraction is proper or improper. We illustrate this procedure with some examples.

EXAMPLE 2

Reduce the following fractions to lowest terms.

a) $\dfrac{21}{126} =$

Write the numerator and the denominator as the product of prime factors.

$\dfrac{3 \cdot 7}{2 \cdot 3 \cdot 3 \cdot 7} =$

Divide the numerator and the denominator by the factors which are common.

$\dfrac{\cancel{3} \cdot \cancel{7}}{2 \cdot 3 \cdot \cancel{3} \cdot \cancel{7}} =$

We divided by all the factors which were in the numerator. Each time we divide there is a quotient of 1 left. In this case we must write the 1 since it is the only factor left in the numerator.

$\dfrac{1}{6}$

Therefore, $\dfrac{21}{126} = \dfrac{1}{6}$.

b) $\dfrac{60}{198} =$

Write the numerator and the denominator as the product of prime factors.

$\dfrac{2 \cdot 2 \cdot 3 \cdot 5}{2 \cdot 3 \cdot 3 \cdot 11} =$

Divide the numerator and the denominator by the factors that are common.

$\dfrac{\cancel{2} \cdot 2 \cdot \cancel{3} \cdot 5}{\cancel{2} \cdot \cancel{3} \cdot 3 \cdot 11} =$

Multiply the remaining factors.

$\dfrac{10}{33}$

Therefore, $\dfrac{60}{198} = \dfrac{10}{33}$.

Note: Some texts refer to this procedure as "canceling." The author of this text prefers not to use "canceling" since it does not adequately convey the fact that we can divide only by *common factors*. Using canceling can lead to serious problems when reducing rational expressions when parts of terms are often mistakenly "canceled" with parts of other terms.

PRACTICE EXERCISES

Reduce the following fractions to lowest terms.

5) $\dfrac{15}{60}$

6) $\dfrac{180}{378}$

• **Reducing rational expressions to lowest terms**

We will first reduce rational expressions whose numerators and denominators are monomials. We have already discussed this type of rational expression in Chapter 2 where we discussed the division laws of exponents. In this section we will not be using the laws of exponents, but will be reducing these rational expressions using the same procedure we used to reduce rational numbers. Compare the following reduction of a rational number with the reduction of a rational expression where the instructions on the right apply to both exercises.

Rational Numbers	Rational Expressions	
$\dfrac{45}{54}$	$\dfrac{a^3b^4}{a^5b^3}$	Write the numerator and denominator as prime factors.
$\dfrac{3 \cdot 3 \cdot 5}{2 \cdot 3 \cdot 3 \cdot 3} =$	$\dfrac{a \cdot a \cdot a \cdot b \cdot b \cdot b \cdot b}{a \cdot a \cdot a \cdot a \cdot a \cdot b \cdot b \cdot b}$	Divide by the common factors.
$\dfrac{\cancel{3} \cdot \cancel{3} \cdot 5}{2 \cdot \cancel{3} \cdot \cancel{3} \cdot 3} =$	$\dfrac{\cancel{a} \cdot \cancel{a} \cdot \cancel{a} \cdot \cancel{b} \cdot \cancel{b} \cdot \cancel{b} \cdot b}{\cancel{a} \cdot \cancel{a} \cdot \cancel{a} \cdot a \cdot a \cdot \cancel{b} \cdot \cancel{b} \cdot \cancel{b}}$	Multiply the remaining factors.
$\dfrac{5}{6}$	$\dfrac{b}{a^2}$	Reduced forms.

EXAMPLE 3

Reduce the following rational expressions to lowest terms. Assume all denominators have non zero values.

a) $\dfrac{x^2y^4}{x^5y^2} =$

Write the numerator and the denominator in terms of prime factors.

$\dfrac{x \cdot x \cdot y \cdot y \cdot y \cdot y}{x \cdot x \cdot x \cdot x \cdot x \cdot y \cdot y} =$

Divide by factors common to the numerator and the denominator.

$\dfrac{\cancel{x} \cdot \cancel{x} \cdot \cancel{y} \cdot \cancel{y} \cdot y \cdot y}{\cancel{x} \cdot \cancel{x} \cdot x \cdot x \cdot x \cdot \cancel{y} \cdot \cancel{y}} =$

Multiply the remaining factors.

$\dfrac{y^2}{x^3}$

Therefore, $\dfrac{x^2y^4}{x^5y^2} = \dfrac{y^2}{x^3}$.

b) $\dfrac{8x^2y^3}{12x^4y} =$

Write the numerator and the denominator in terms of prime factors.

$\dfrac{2 \cdot 2 \cdot 2 \cdot x \cdot x \cdot y \cdot y \cdot y}{2 \cdot 2 \cdot 3 \cdot x \cdot x \cdot x \cdot x \cdot y} =$

Divide by factors common to the numerator and the denominator.

$\dfrac{\cancel{2} \cdot \cancel{2} \cdot 2 \cdot \cancel{x} \cdot \cancel{x} \cdot \cancel{y} \cdot y \cdot y}{\cancel{2} \cdot \cancel{2} \cdot 3 \cdot \cancel{x} \cdot \cancel{x} \cdot x \cdot x \cdot \cancel{y}} =$

Multiply the remaining factors.

$\dfrac{2y^2}{3x^2}$

Therefore, $\dfrac{8x^2y^3}{12x^4y} = \dfrac{2y^2}{3x^2}$.

Note: Many times reducing these types of rational expressions is easier than reducing arithmetic fractions because the factorization is usually easier. These can also be done using the laws of exponents.

c) $\dfrac{20a^3b^2}{24ab} =$

Write the numerator and the denominator in terms of prime factors.

$\dfrac{2 \cdot 2 \cdot 5 \cdot a \cdot a \cdot a \cdot b \cdot b}{2 \cdot 2 \cdot 2 \cdot 3 \cdot a \cdot b} =$

Divide by factors common to the numerator and the denominator.

$\dfrac{\cancel{2} \cdot \cancel{2} \cdot 5 \cdot \cancel{a} \cdot a \cdot a \cdot \cancel{b} \cdot b}{\cancel{2} \cdot \cancel{2} \cdot 2 \cdot 3 \cdot \cancel{a} \cdot \cancel{b}} =$

Multiply the remaining factors.

$\dfrac{5a^2b}{6}$

Therefore, $\dfrac{20a^3b^2}{24ab} = \dfrac{5a^2b}{6}$.

Reduce the following rational expressions to lowest terms.

e) $\dfrac{a^2b^5}{a^4b}$ **f)** $\dfrac{14cd^3}{28c^3d^2}$

g) $\dfrac{12r^4s^4w^5}{-18r^2s^6w^3}$

PRACTICE EXERCISES

Reduce the following rational expressions to lowest terms.

7) $\dfrac{m^3n^6}{m^4n^2}$ **8)** $\dfrac{18x^4y^2}{21x^2y^3}$ **9)** $\dfrac{-27a^5b^2c^3}{36a^2bc^4}$

If more practice is needed, do the Additional Practice Exercises in the margin.

We know that any number, other than 0, divided by itself is equal to one. For example, $\frac{5}{5} = 1$ and $\frac{-4}{-4} = 1$. We used this fact to reduce rational numbers. We used an extension of this fact to reduce rational expressions with monomial numerators and denominators by indicating that a variable divided by itself also has a value of 1 providing the variable does not equal 0. For example, $\frac{x}{x} = 1$, if $x \neq 0$. We are going to extend this fact again by saying any polynomial divided by itself is equal to one, providing the polynomial is not equal to 0. For example, $\frac{x+3}{x+3} = 1$, if $x + 3 \neq 0$. Since $x + 3$ represents a number for any value of x, we are simply saying that any nonzero number divided by itself is one.

The procedure for reducing rational expressions whose numerators and/or denominators are not monomials is exactly the same as for those whose numerators and denominators are monomials. Compare the following.

Monomials	Nonmonomials	
$\dfrac{8x^2y^3}{12x^4y} =$	$\dfrac{x^2 + 3x + 2}{x^2 - 2x - 3}$	Write the numerator and the denominator in terms of prime factors.
$\dfrac{2 \cdot 2 \cdot 2 \cdot x \cdot x \cdot y \cdot y \cdot y}{2 \cdot 2 \cdot 3 \cdot x \cdot x \cdot x \cdot x \cdot y} =$	$\dfrac{(x+1)(x+2)}{(x+1)(x-3)}$	Divide by factors common to the numerator and the denominator.
$\dfrac{\cancel{2} \cdot \cancel{2} \cdot 2 \cdot \cancel{x} \cdot \cancel{x} \cdot \cancel{y} \cdot y \cdot y}{\cancel{2} \cdot \cancel{2} \cdot 3 \cdot \cancel{x} \cdot \cancel{x} \cdot x \cdot x \cdot \cancel{y}} =$	$\dfrac{\cancel{(x+1)}(x+2)}{\cancel{(x+1)}(x-3)}$	Multiply the remaining factors.
$\dfrac{2y^2}{3x^2}$	$\dfrac{x+2}{x-3}$	Reduced to lowest terms.

EXAMPLE 4

Reduce the following rational expressions. Assume the variables cannot equal any value for which the denominator is equal to 0.

a) $\dfrac{5(x+1)}{8(x+1)} =$ The numerator and the denominator are already factored, so divide the numerator and the denominator by the common factor of $x + 1$.

$\dfrac{5\cancel{(x+1)}}{8\cancel{(x+1)}} =$ $\dfrac{x+1}{x+1} = 1$ if $x \neq -1$.

$\dfrac{5}{8}$ Answer.

b) $\dfrac{4x + 16}{3x + 12} =$ Factor the numerator and the denominator.

$\dfrac{4(x+4)}{3(x+4)} =$ Divide the numerator and the denominator by the common factor of $x + 4$.

$\dfrac{4(x+4)}{3(x+4)} =$ $\dfrac{x+4}{x+4} = 1$ if $x \ne -4$.

$\dfrac{4}{3}$ Answer.

c) $\dfrac{x^2 + x - 6}{x^2 - x - 12} =$ Factor the numerator and the denominator.

$\dfrac{(x+3)(x-2)}{(x+3)(x-4)} =$ Divide the numerator and the denominator by the common factor of $x + 3$.

$\dfrac{(x+3)(x-2)}{(x+3)(x-4)} =$ $\dfrac{x+3}{x+3} = 1$, $x \ne -3$.

$\dfrac{x-2}{x-4}$ Answer.

d) $\dfrac{x+4}{x^2 - 16} =$ Factor the denominator. The numerator is prime.

$\dfrac{x+4}{(x+4)(x-4)} =$ Divide the numerator and the denominator by the common factor of $x + 4$.

$\dfrac{x+4}{(x+4)(x-4)} =$ $\dfrac{x+4}{x+4} = 1$, $x \ne -4$.

$\dfrac{1}{x-4}$ Answer. Remember, each time we divide by a common factor, there is a factor of 1 left. We do not write the 1 unless it is the only factor remaining in the numerator.

Be careful: A common error is to divide by terms instead of factors. Consider the following. $\dfrac{6+3}{3} \ne \dfrac{6+3}{3} \ne 6 + 1 = 7$. If we follow the order of operations, $\dfrac{6+3}{3} = \dfrac{9}{3} = 3$. But $7 \ne 3$. We have two different answers to the same question. Since we followed the order of operations in the second case, we know that 3 is the correct answer. What is wrong with the first case? We have divided by a term and not a factor. Remember, terms are separated by $+$ and $-$ signs and factor implies multiplication. Therefore, before you can reduce a rational expression, you must first write the numerator and the denominator in *factored* form and then you divide by the factors that are common to both. In the following, both the correct and the incorrect methods are shown.

CORRECT

$$\dfrac{x^2 - 6x + 8}{x^2 + 2x - 8} = \dfrac{(x-4)(x-2)}{(x-2)(x+4)} = \dfrac{(x-4)(x-2)}{(x-2)(x+4)} = \dfrac{x-4}{x+4}$$

Since $x - 2$ is a factor common to both the numerator and the denominator, we can divide both the numerator and the denominator by $x - 2$.

x^2 and 8 are terms common to the numerator and the denominator. They are not factors. You cannot divide by terms!

ADDITIONAL PRACTICE

Reduce the following rational expressions to lowest terms.

h) $\dfrac{2(y + 6)}{5(y + 6)}$ **i)** $\dfrac{12x - 4y}{21x - 7y}$

j) $\dfrac{x^2 - 2x - 15}{x^2 + 7x + 12}$ **k)** $\dfrac{x + 1}{x^2 + 4x + 3}$

PRACTICE EXERCISES

Reduce the following rational expressions. Assume the variables cannot equal any value that makes the denominator equal to 0.

10) $\dfrac{4(x + y)}{7(x + y)}$ **11)** $\dfrac{3x + 6}{5x + 10}$

12) $\dfrac{x^2 - 2x - 8}{x^2 + 9x + 14}$ **13)** $\dfrac{x - 3}{x^2 - 7x + 12}$

If more practice is needed, do the Additional Practice Exercises in the margin.

There is one instance in reducing rational expressions that requires special attention. This occurs if the numerator and the denominator contain factors that are exactly the same except for the signs of the terms. For example, $\frac{a - b}{b - a}$. The numerator has $+a$ and $-b$ while the denominator has $-a$ and $+b$, so the signs of the terms are opposites. In order to reduce this type of fraction, we need to recall a technique that was discussed in Section 4.7 where we removed a factor of -1, which we indicated simply as a "$-$", when factoring by grouping.

EXAMPLE 5

Reduce the following rational expressions to lowest terms.

a) $\dfrac{a - b}{b - a} =$

The numerator and the denominator are the same except the signs of the terms are opposites. Factor -1 from the denominator which is indicated simply as "$-$".

$\dfrac{a - b}{-(-b + a)} =$ Apply the commutative property.

$\dfrac{a - b}{-(a - b)} =$ Divide the numerator and the denominator by $a - b$.

$\dfrac{\cancel{a - b}}{-(\cancel{a - b})} =$ Anything (except 0) divided by itself equals 1.

$\dfrac{1}{-1} =$ $\dfrac{1}{-1} = -1$.

-1 Answer.

 b) $\dfrac{x^2 - 1}{1 - x} =$ Factor the numerator.

$\dfrac{(x - 1)(x + 1)}{1 - x} =$ $x - 1$ and $1 - x$ are opposite in signs. Factor -1 from the denominator.

$\dfrac{(x-1)(x+1)}{-(-1+x)} =$ Apply the commutative property.

$\dfrac{(x-1)(x+1)}{-(x-1)} =$ Divide by the common factor of $x-1$.

$\dfrac{\cancel{(x-1)}(x+1)}{-\cancel{(x-1)}} =$ 1 is left in both the numerator and the denominator where the common factors were divided.

$\dfrac{1(x+1)}{-1} =$ $\dfrac{1}{-1} = -1$.

$-(x+1) =$ Distribute the -1.

$-x-1$ Answer.

Note: The answer $-(x+1)$ is usually acceptable and sometimes preferred. ▇

ADDITIONAL PRACTICE

Reduce the following rational expressions to lowest terms.

l) $\dfrac{r-s}{s-r}$ m) $\dfrac{x^2-9}{3-x}$

PRACTICE EXERCISES

Reduce the following rational expressions to lowest terms.

14) $\dfrac{c-d}{d-c}$ 15) $\dfrac{x^2-4}{2-x}$

If more practice is needed, do the Additional Practice Exercises in the margin.

Be careful: Rational expressions of the form $\dfrac{4+x}{4-x}$ are often mistaken for the above type. In this rational expression, not all the terms in the numerator and the denominator are opposites in sign. This expression cannot be reduced because $\dfrac{4+x}{4-x} = \dfrac{4+x}{-(x-4)}$, so there is no common factor.

EXERCISE SET   5.1

Find the value(s) of the variable(s) for which the following are defined.

1) $\dfrac{5}{a}$ 2) $\dfrac{-4}{x}$ 3) $\dfrac{8}{x^2y}$

4) $\dfrac{9}{ab^3}$ 5) $\dfrac{5}{x+5}$ 6) $\dfrac{9}{a-6}$

7) $\dfrac{x+3}{3x+12}$ 8) $\dfrac{y-5}{4y-8}$ 9) $\dfrac{8x}{x+2y}$

10) $\dfrac{7y}{x-3y}$ 11) $\dfrac{5}{(x-7)(x+3)}$

12) $\dfrac{m}{m^2+2m-15}$

Reduce the following fractions to lowest terms.

13) $\dfrac{40}{84}$ 14) $\dfrac{16}{56}$ 15) $\dfrac{36}{54}$ 16) $\dfrac{32}{72}$

17) $\dfrac{72}{108}$ 18) $\dfrac{120}{144}$ 19) $\dfrac{120}{96}$ 20) $\dfrac{72}{124}$

21) $\dfrac{210}{110}$ 22) $\dfrac{300}{108}$ 23) $\dfrac{90}{315}$ 24) $\dfrac{180}{210}$

Reduce the following rational expressions.

25) $\dfrac{a^3b^4}{a^4b^6}$ 26) $\dfrac{u^2v^5}{u^3v^6}$ 27) $\dfrac{12x^5y^3}{3x^2y^2}$

28) $\dfrac{18x^5y^7}{6x^3y^2}$ **29)** $\dfrac{a^5b^3c^6}{a^3b^6c^4}$ **30)** $\dfrac{x^3y^6z^4}{x^2y^4z^6}$

31) $\dfrac{-32m^2n^5p^3}{-18m^2n^3p^4}$ **32)** $\dfrac{-27x^3y^4z^2}{6x^3y^2z^3}$ **33)** $\dfrac{15r^3s^5w}{-25r^4s^6w}$

34) $\dfrac{36qr^4s^7}{-27qr^5s^9}$ **35)** $\dfrac{-32x^4y^3z^2}{16x^2y^2z^2}$ **36)** $\dfrac{42w^3x^4y^3}{-14w^2x^3y}$

Reduce the following rational expressions to lowest terms.

37) $\dfrac{4(x+3)}{9(x+3)}$ **38)** $\dfrac{3(y-2)}{5(y-2)}$ **39)** $\dfrac{4(x-10)}{6(x-10)}$

40) $\dfrac{6(x+5)}{8(x+5)}$ **41)** $\dfrac{2x+12}{3x+18}$ **42)** $\dfrac{3x+6}{5x+10}$

43) $\dfrac{8x+12y}{20x+30y}$ **44)** $\dfrac{24a-16b}{36a-24b}$ **45)** $\dfrac{-3x-15}{4x+20}$

46) $\dfrac{-5x-25}{6x+30}$ **47)** $\dfrac{x^2+3x+2}{x^2-2x-3}$ **48)** $\dfrac{x^2-4x+3}{x^2-5x+6}$

49) $\dfrac{a^2+3a-10}{a^2+8a+15}$ **50)** $\dfrac{c^2+3c-18}{c^2-c-6}$

51) $\dfrac{x-6}{x^2-4x-12}$ **52)** $\dfrac{x-2}{x^2-10x+16}$

53) $\dfrac{x^2-9}{x^2-2x-15}$ **54)** $\dfrac{x^2-25}{x^2-x-30}$

55) $\dfrac{4x^2-9y^2}{6x^2-xy-15y^2}$ **56)** $\dfrac{9x^2-25y^2}{6x^2-5xy-25y^2}$

57) $\dfrac{4a^2-4ab-3b^2}{6a^2-ab-2b^2}$ **58)** $\dfrac{6a^2+13ab+6b^2}{3a^2-7ab-6b^2}$

59) $\dfrac{3(x+7)+5(x+7)}{x+7}$ **60)** $\dfrac{x(y-3)+2(y-3)}{y-3}$

Reduce the following rational expressions to lowest terms.

61) $\dfrac{m-n}{n-m}$ **62)** $\dfrac{q-p}{p-q}$

63) $\dfrac{2b-a}{a-2b}$ **64)** $\dfrac{3m-n}{n-3m}$

65) $\dfrac{a^2-25}{5-a}$ **66)** $\dfrac{c^2-49}{7-c}$

67) $\dfrac{x^2-3x-10}{5-x}$ **68)** $\dfrac{x^2+4x-21}{3-x}$

69) $\dfrac{4-x}{x^2-9x+20}$

WRITING EXERCISES

70) In the rational expression $\dfrac{4}{x-1}$, is $x = 0$ allowed? Why or why not?

71) What is wrong with the following?

$$\frac{3x+4}{2} = \frac{3x+\overset{2}{\cancel{4}}}{\cancel{2}} = 3x+2.$$

72) What is the difference between a term and a factor?

73) Are $a - b$ and $b - a$ always equal? If not, for what values of a and b will $a - b$ and $b - a$ be equal?

74) Why does $\dfrac{b-a}{a-b} = -1$?

75) In the expression $\frac{x+5}{y+5}$ we can not divide the numerator and denominator by 5, but in the expression $\frac{5x+5}{5y+5}$ we can divide the numerator and denominator by 5. Explain.

76) Determine the denominator if $\dfrac{x^2+2x-8}{\text{denominator}} = x + 4$. Explain your reasoning.

77) The expression $\dfrac{x+1}{x^2+1}$ is defined for all values of x. Why?

GROUP PROJECT

78) Make a list of at least five situations from everyday life where a fraction occurs that is not reduced to lowest terms. For example, blood pressure.

S E C T I O N 5.2

Multiplication of Rational Expressions

OBJECTIVE

When you complete this section, you will be able to:

Ⓐ   Multiply two or more rational expressions and leave the product reduced to lowest terms.

INTRODUCTION

From Section 1.2, we recall the procedure for multiplying rational numbers which is given below.

> **MULTIPLICATION OF FRACTIONS**
>
> The product of two fractions is found by calculating the product of the numerators and dividing by the product of the denominators. In symbols, $\frac{a}{b} \cdot \frac{c}{d} = \frac{a \cdot c}{b \cdot d}$. In other words, numerator times numerator divided by denominator times denominator.

The above procedure is easily extended to three or more rational numbers as follows: $\frac{a}{b} \cdot \frac{c}{d} \cdot \frac{e}{f} = \frac{a \cdot c \cdot e}{b \cdot d \cdot f}$ and $\frac{a}{b} \cdot \frac{c}{d} \cdot \frac{e}{f} \cdot \frac{g}{h} = \frac{a \cdot c \cdot e \cdot g}{b \cdot d \cdot f \cdot h}$.

For example, $\frac{3}{5} \cdot \frac{2}{7} = \frac{3 \cdot 2}{5 \cdot 7} = \frac{6}{35}$. Often it is necessary to reduce the product to lowest terms, as in the following.

$\dfrac{3}{15} \cdot \dfrac{10}{9} =$	Put the product of the numerators over the product of the denominators.
$\dfrac{3 \cdot 10}{15 \cdot 9} =$	Write the numerator and the denominator in terms of prime factors.
$\dfrac{3 \cdot 2 \cdot 5}{3 \cdot 5 \cdot 3 \cdot 3} =$	Divide by the common factors of 3 and 5.
$\dfrac{\cancel{3} \cdot 2 \cdot \cancel{5}}{\cancel{3} \cdot \cancel{5} \cdot 3 \cdot 3} =$	Multiply the remaining factors.
$\dfrac{2}{9}$	Product.

Actually, we are more accustomed to dividing by the common factors before multiplying. This is also the method that is most like the way we multiply rational expressions. Compare the following. The instructions on the right apply to both exercises.

Rational Number	Rational Expression	
$\dfrac{3}{4} \cdot \dfrac{10}{21}$	$\dfrac{a^2}{y^2} \cdot \dfrac{y^4}{a^5}$	Write the numerators and the denominators in terms of prime factors.
$\dfrac{3}{2 \cdot 2} \cdot \dfrac{2 \cdot 5}{3 \cdot 7}$	$\dfrac{a \cdot a}{y \cdot y} \cdot \dfrac{y \cdot y \cdot y \cdot y}{a \cdot a \cdot a \cdot a \cdot a}$	Divide by the common factors.
$\dfrac{\cancel{3}}{\cancel{2} \cdot 2} \cdot \dfrac{\cancel{2} \cdot 5}{\cancel{3} \cdot 7}$	$\dfrac{\cancel{a} \cdot \cancel{a}}{\cancel{y} \cdot \cancel{y}} \cdot \dfrac{\cancel{y} \cdot \cancel{y} \cdot y \cdot y}{\cancel{a} \cdot \cancel{a} \cdot a \cdot a \cdot a}$	Multiply the remaining factors.
$\dfrac{5}{14}$	$\dfrac{y^2}{a^3}$	Product.

Thus the procedure for finding the products of rational expressions with monomial numerators and denominators is exactly the same as finding the products of rational numbers and is summarized below.

> **MULTIPLYING RATIONAL EXPRESSIONS**
>
> 1) Factor all numerators and denominators into prime factors.
> 2) Divide by all factors common to the numerators and denominators.
> 3) Multiply the remaining factors.

EXAMPLE 1

Find the following products. Leave answers reduced to lowest terms.

a) $\dfrac{3mn^2}{5m^3n} \cdot \dfrac{10m^2n}{9m^2n^3} =$

Factor the numerators and denominators into prime factors.

$$\dfrac{3 \cdot m \cdot n \cdot n}{5 \cdot m \cdot m \cdot m \cdot n} \cdot \dfrac{2 \cdot 5 \cdot m \cdot m \cdot n}{3 \cdot 3 \cdot m \cdot m \cdot n \cdot n \cdot n} =$$

Divide by common factors.

$$\dfrac{\not{3} \cdot \not{m} \cdot \not{n} \cdot \not{n}}{\not{5} \cdot \not{m} \cdot \not{m} \cdot \not{m} \cdot \not{n}} \cdot \dfrac{2 \cdot \not{5} \cdot \not{m} \cdot \not{m} \cdot \not{n}}{\not{3} \cdot 3 \cdot m \cdot m \cdot \not{n} \cdot \not{n} \cdot n} =$$

Multiply remaining factors.

$$\dfrac{2}{3m^2n}$$

Product.

b) $\dfrac{3xy^2}{4a^2b^2} \cdot \dfrac{2ab}{9x^3y^3} =$

Factor the numerators and denominators into prime factors.

$$\dfrac{3 \cdot x \cdot y \cdot y}{2 \cdot 2 \cdot a \cdot a \cdot b \cdot b} \cdot \dfrac{2 \cdot a \cdot b}{3 \cdot 3 \cdot x \cdot x \cdot x \cdot y \cdot y \cdot y} =$$

Divide by common factors.

$$\dfrac{\not{3} \cdot \not{x} \cdot \not{y} \cdot \not{y}}{\not{2} \cdot 2 \cdot \not{a} \cdot a \cdot \not{b} \cdot b} \cdot \dfrac{\not{2} \cdot \not{a} \cdot \not{b}}{3 \cdot \not{3} \cdot \not{x} \cdot x \cdot x \cdot \not{y} \cdot \not{y} \cdot y} =$$

Multiply the remaining factors. Remember, each time we divide, there is a factor of 1 left which usually is not written. Since 1 is all that is left in the numerator, it must be written.

$$\dfrac{1}{6abx^2y}$$

Product.

Note: The products in Example 1 can also be found by using the laws of exponents developed in Chapter 2. Example 1a can also have been done as follows.

$$\dfrac{3mn^2}{5m^3n} \cdot \dfrac{10m^2n}{9m^2n^3} =$$

Multiply the rational expressions.

$$\dfrac{3mn^2 \cdot 10m^2n}{5m^3n \cdot 9m^2n^3} =$$

Multiply the coefficients and apply $a^m \cdot a^n = a^{m+n}$.

$$\dfrac{30m^3n^3}{45m^5n^4} =$$

Apply $\dfrac{a^m}{a^n} = a^{m-n}$.

$$\dfrac{30}{45}m^{-2}n^{-1} =$$

Reduce the coefficient and apply $a^{-n} = \dfrac{1}{a^n}$.

$$\dfrac{2}{3} \cdot \dfrac{1}{m^2} \cdot \dfrac{1}{n} =$$

Multiply the fractions.

$$\dfrac{2}{3m^2n}$$

Product.

ADDITIONAL PRACTICE

Find the following products. Leave answers reduced to lowest terms.

a) $\dfrac{s^3}{r^2} \cdot \dfrac{r^4}{s^5}$

b) $\dfrac{6r^3s}{5r^2s^2} \cdot \dfrac{15rs^2}{2r^3s^2}$

c) $\dfrac{3rs^2}{14p^2q^3} \cdot \dfrac{7pq^2}{6r^2s^2}$

PRACTICE EXERCISES

Find the following products. Leave answers reduced to lowest terms.

1) $\dfrac{x^3}{y^2} \cdot \dfrac{y^4}{x}$

2) $\dfrac{8a^3b}{7a^3b} \cdot \dfrac{21ab^2}{4a^2b}$

3) $\dfrac{2a^2b}{15xy^3} \cdot \dfrac{5xy}{4a^3b}$

If more practice is needed, do the Additional Practice Exercises in the margin.

The exact same technique is used to multiply rational expressions whose numerators and/or denominators are not integers or monomials. The only difference is the manner in which we factor the numerators and the denominators since they are polynomials instead of integers or monomials. Compare the following.

Monomials	Nonmonomials	
$\dfrac{a^2b^2}{a^3b^3} \cdot \dfrac{a^2b^4}{a^3b}$	$\dfrac{2x-6}{x^2+x-2} \cdot \dfrac{x^2+3x-4}{x^2-9}$	Factor the numerators and denominators.
$\dfrac{a \cdot a \cdot b \cdot b}{a \cdot a \cdot a \cdot b \cdot b \cdot b} \cdot \dfrac{a \cdot a \cdot b \cdot b \cdot b \cdot b}{a \cdot a \cdot a \cdot b}$	$\dfrac{2(x-3)}{(x+2)(x-1)} \cdot \dfrac{(x+4)(x-1)}{(x+3)(x-3)}$	Divide by the common factors.
$\dfrac{\cancel{a} \cdot \cancel{a} \cdot \cancel{b} \cdot \cancel{b}}{\cancel{a} \cdot \cancel{a} \cdot a \cdot \cancel{b} \cdot \cancel{b} \cdot \cancel{b}} \cdot \dfrac{\cancel{a} \cdot \cancel{a} \cdot \cancel{b} \cdot \cancel{b} \cdot b \cdot b}{\cancel{a} \cdot a \cdot a \cdot \cancel{b}}$	$\dfrac{2(x-3)}{(x+2)(x-1)} \cdot \dfrac{(x+4)(x-1)}{(x+3)(x-3)}$	Multiply the remaining factors.
$\dfrac{b^2}{a^2}$	$\dfrac{2(x+4)}{(x+2)(x+3)}$	Product.

EXAMPLE 2

Find the following products. Leave answers reduced to lowest terms.

a) $\dfrac{a+b}{2x^2} \cdot \dfrac{4x}{(a+b)^2} =$ Factor the numerators and the denominators.

$\dfrac{a+b}{2 \cdot x \cdot x} \cdot \dfrac{2 \cdot 2 \cdot x}{(a+b)(a+b)} =$ Divide by the common factors.

$\dfrac{\cancel{a+b}}{2 \cdot \cancel{x} \cdot x} \cdot \dfrac{2 \cdot 2 \cdot \cancel{x}}{(\cancel{a+b})(a+b)} =$ Multiply the remaining factors.

$\dfrac{2}{x(a+b)}$ Product.

b) $\dfrac{2x-6}{3y+12} \cdot \dfrac{2y+8}{5x-15} =$ Factor the numerators and the denominators.

$\dfrac{2(x-3)}{3(y+4)} \cdot \dfrac{2(y+4)}{5(x-3)} =$ Divide by the common factors.

$\dfrac{2(\cancel{x-3})}{3(\cancel{y+4})} \cdot \dfrac{2(\cancel{y+4})}{5(\cancel{x-3})} =$ Multiply the remaining factors.

$\dfrac{2 \cdot 2}{3 \cdot 5} =$ Perform the multiplication.

$\dfrac{4}{15}$ Product.

c) $\dfrac{x^2 - x - 6}{x^2 + 2x - 15} \cdot \dfrac{x^2 + 4x - 12}{x^2 - 4} =$ Factor the numerators and the denominators.

$\dfrac{(x + 2)(x - 3)}{(x + 5)(x - 3)} \cdot \dfrac{(x - 2)(x + 6)}{(x + 2)(x - 2)} =$ Divide by the common factors.

$\dfrac{(x + 2)(x - 3)}{(x + 5)(x - 3)} \cdot \dfrac{(x - 2)(x + 6)}{(x + 2)(x - 2)} =$ Multiply the remaining factors.

$\dfrac{x + 6}{x + 5}$ Product.

ADDITIONAL PRACTICE

Find the following products. Leave answers reduced to lowest terms.

d) $\dfrac{5rs}{(r + s)^2} \cdot \dfrac{2(r + s)}{7r^2s}$

e) $\dfrac{4a - 12}{5b + 20} \cdot \dfrac{2b + 8}{3a - 9}$

f) $\dfrac{a^2 + a - 2}{a^2 - 3a - 10} \cdot \dfrac{a^2 - a - 20}{a^2 + 3a - 4}$

• **Products of rational expressions containing factors of the form $a - b$ and $b - a$**

PRACTICE EXERCISES

Find the following products. Leave answers reduced to lowest terms.

4) $\dfrac{3x^2y}{(w + z)^2} \cdot \dfrac{w + z}{4xy^3}$

5) $\dfrac{3a + 12}{5b - 30} \cdot \dfrac{b - 6}{4a + 16}$

6) $\dfrac{a^2 + 6a + 8}{a^2 + 5a - 6} \cdot \dfrac{a^2 + 3a - 18}{a^2 + a - 12}$

If more practice is needed, do the Additional Practice Exercises in the margin.

In Section 5.1 we reduced rational expressions and sometimes found it necessary to remove a factor of -1 from the numerator or the denominator. This occurred if the terms of a factor in the numerator and the denominator were exactly the same except that the signs were opposites. For example, $\frac{a - b}{b - a}$. The same situation may occur when multiplying rational expressions.

EXAMPLE 3

Find the following products. Reduce answers to lowest terms.

a) $\dfrac{4 - a}{5} \cdot \dfrac{8}{a - 4} =$ Factor -1 from $4 - a$.

$\dfrac{-1(-4 + a)}{5} \cdot \dfrac{8}{a - 4} =$ Commutative property, $-4 + a = a - 4$.

$\dfrac{-1(a - 4)}{5} \cdot \dfrac{8}{a - 4} =$ Divide by the common factors.

$\dfrac{-1(a - 4)}{5} \cdot \dfrac{8}{a - 4} =$ Multiply the remaining factors.

$\dfrac{-8}{5}$ or $-\dfrac{8}{5}$ Product. $\frac{-8}{5} = -\frac{8}{5}$ since a negative divided by a positive is negative.

b) $\dfrac{3 - a}{a^2 + 3a + 2} \cdot \dfrac{a^2 - 4}{a^2 + 2a - 15} =$ Factor everything that is not prime. $3 - a$ and $a - 3$ are the same except the signs are opposites.

$\dfrac{3 - a}{(a + 2)(a + 1)} \cdot \dfrac{(a + 2)(a - 2)}{(a - 3)(a + 5)} =$ Factor -1 from $3 - a$.

$\dfrac{-1(-3 + a)}{(a + 2)(a + 1)} \cdot \dfrac{(a + 2)(a - 2)}{(a - 3)(a + 5)} =$ Rewrite $-3 + a$ as $a - 3$.

$$\dfrac{-1(a-3)}{(a+2)(a+1)} \cdot \dfrac{(a+2)(a-2)}{(a-3)(a+5)} = \qquad \text{Divide by the common factors.}$$

$$\dfrac{-1(a\!\!\!\diagup\!\!\!-3)}{(a\!\!\!\diagup\!\!\!+2)(a+1)} \cdot \dfrac{(a\!\!\!\diagup\!\!\!+2)(a-2)}{(a\!\!\!\diagup\!\!\!-3)(a+5)} = \qquad \text{Multiply the remaining factors.}$$

$$\dfrac{-(a-2)}{(a+1)(a+5)} \qquad \text{Product.}$$

Note: The answer to Example 2b could be written as $-\dfrac{a-2}{(a+1)(a+5)}$ or $\dfrac{2-a}{(a+1)(a+5)}$. Why?

ADDITIONAL PRACTICE

Find the following products. Leave answers reduced to lowest terms.

g) $\dfrac{rs}{b-2} \cdot \dfrac{2-b}{2a}$

h) $\dfrac{m^2-7m+12}{m^2-2m-8} \cdot \dfrac{m-6}{3-m}$

PRACTICE EXERCISES

Find the following products. Leave answers reduced to lowest terms.

7) $\dfrac{7-x}{a} \cdot \dfrac{b}{x-7}$

8) $\dfrac{y^2+3y-10}{y^2-25} \cdot \dfrac{y+4}{2-y}$

If more practice is needed, do the Additional Practice Exercises in the margin.

EXERCISE SET 5.2

Find the following products of rational numbers. Leave answers reduced to lowest terms.

1) $\dfrac{21}{8} \cdot \dfrac{16}{3}$

2) $\dfrac{18}{5} \cdot \dfrac{5}{3}$

3) $-\dfrac{3}{2} \cdot \left(-\dfrac{8}{9}\right)$

4) $-\dfrac{3}{8} \cdot \left(-\dfrac{7}{12}\right)$

5) $\dfrac{5}{8} \cdot \dfrac{32}{15}$

6) $\dfrac{3}{5} \cdot \dfrac{10}{9}$

7) $\dfrac{10}{3} \cdot \dfrac{6}{5}$

8) $\dfrac{8}{7} \cdot \dfrac{6}{5}$

Find the following products of rational expressions with monomial numerators and denominators. Leave answers reduced to lowest terms.

9) $\dfrac{r^3}{s^3} \cdot \dfrac{s^5}{r^6}$

10) $\dfrac{a^5}{b^3} \cdot \dfrac{b^2}{a^3}$

11) $\dfrac{m^2}{n^2} \cdot \dfrac{n^2}{m^4}$

12) $\dfrac{c^3}{d^5} \cdot \dfrac{d^2}{c^3}$

13) $\dfrac{x^2y^2}{a^2b^3} \cdot \dfrac{a^4b}{xy^4}$

14) $\dfrac{r^4s^2}{t^3u^3} \cdot \dfrac{t^5u}{r^6s}$

15) $\dfrac{r^4s}{r^3s^2} \cdot \dfrac{r^2s^3}{r^4s^3}$

16) $\dfrac{a^2b^3}{a^3b^2} \cdot \dfrac{a^2b}{a^3b^3}$

17) $\dfrac{6x^2}{5y^3} \cdot \dfrac{10y^5}{8x}$

18) $\dfrac{12a^4}{7b^5} \cdot \dfrac{14b^2}{16a^2}$

19) $\dfrac{8x^2y^4}{12x^3y^2} \cdot \dfrac{4x^3y^3}{6xy^3}$

20) $\dfrac{10a^2b^2}{18a^2b^3} \cdot \dfrac{12a^3b}{15a^2b^4}$

21) $\dfrac{24a^2bc^3}{16ab^3c^2} \cdot \dfrac{32a^3b^2c^3}{18a^4b^3c^3}$

22) $\dfrac{30x^3y^2z}{42x^2yz^2} \cdot \dfrac{15x^2y^2z^2}{25x^4yz}$

Find the following products of rational expressions whose numerators and denominators are polynomials. Leave answers reduced to lowest terms.

23) $\dfrac{(x+y)^2}{6a} \cdot \dfrac{8a^2}{x+y}$

24) $\dfrac{12x}{(a+b)^2} \cdot \dfrac{a+b}{16x^2}$

25) $\dfrac{3x^2y^4}{10(r+s)} \cdot \dfrac{5(r+s)^2}{12xy^2}$

26) $\dfrac{6(p+q)^2}{5rs} \cdot \dfrac{15r^3s^4}{18(p+q)}$

27) $\dfrac{3x+18}{7x-7} \cdot \dfrac{2x-2}{5x+30}$

28) $\dfrac{2r-8}{3r+6} \cdot \dfrac{5r+10}{3r-12}$

29) $\dfrac{8x+12}{18x-24} \cdot \dfrac{9x-12}{10x+15}$

30) $\dfrac{24x+16}{30x-18} \cdot \dfrac{45x-27}{18x+12}$

49) $\dfrac{b^2 - 16}{b^2 + b - 12} \cdot \dfrac{b^2 + 2b - 15}{4 - b}$

50) $\dfrac{y^2 - y - 20}{y^2 - 2y - 24} \cdot \dfrac{y + 3}{5 - y}$

31) $\dfrac{4a - 16}{3a - 3} \cdot \dfrac{a^2 - 1}{a^2 - 16}$ 32) $\dfrac{x^2 - 9}{x + 3} \cdot \dfrac{x - 5}{x^2 - 25}$

33) $\dfrac{4x^2 - 25}{3x - 4} \cdot \dfrac{9x^2 - 16}{2x + 5}$ 34) $\dfrac{16x^2 - 9}{4x + 3} \cdot \dfrac{25x^2 - 1}{5x + 1}$

35) $\dfrac{a - 6}{7} \cdot \dfrac{9}{6 - a}$ 36) $\dfrac{b - 4}{5} \cdot \dfrac{3}{4 - b}$

37) $\dfrac{3x - 4}{14} \cdot \dfrac{21}{4 - 3x}$ 38) $\dfrac{5 - 2b}{18} \cdot \dfrac{12}{2b - 5}$

39) $\dfrac{4x^2 y^3}{2x - 6} \cdot \dfrac{12 - 4x}{6x^3 y}$ 40) $\dfrac{6b^3 c^3}{70 - 10b} \cdot \dfrac{5b - 35}{8b^2 c^5}$

41) $\dfrac{x^2 - 4}{x^2 - 3x - 10} \cdot \dfrac{x^2 - 8x + 15}{x^2 - 9}$

42) $\dfrac{x^2 + 3x - 10}{x^2 - 16} \cdot \dfrac{x^2 - 9x + 20}{x^2 - 25}$

43) $\dfrac{x^2 - 2x - 15}{x^2 + x - 30} \cdot \dfrac{x^2 + 5x - 6}{x^2 + 7x + 12}$

44) $\dfrac{a^2 - 2a - 24}{a^2 + 9a + 20} \cdot \dfrac{a^2 - 3a - 10}{a^2 - 4a - 12}$

45) $\dfrac{6x^2 + x - 12}{2x^2 - 5x - 12} \cdot \dfrac{3x^2 - 14x + 8}{9x^2 - 18x + 8}$

46) $\dfrac{2x^2 + 3x - 20}{4x^2 + 11x - 3} \cdot \dfrac{4x^2 - 9x + 2}{2x^2 - 9x + 10}$

47) $\dfrac{3x^2 - 5x - 2}{4x^2 - 11x + 6} \cdot \dfrac{4x^2 + 5x - 6}{3x^2 - 8x - 3}$

48) $\dfrac{8x^2 + 10x - 3}{3x^2 + 4x - 4} \cdot \dfrac{x^2 + 6x + 8}{2x^2 + 11x + 12}$

CHALLENGE EXERCISES (51–54)

Find the following products. Reduce answers to lowest terms.

51) $\dfrac{ac + 3a + 2c + 6}{ad + a + 2d + 2} \cdot \dfrac{ad - 5a + 2d - 10}{bc + 3b - 4c - 12}$

52) $\dfrac{mn + 2m + 4n + 8}{np - 3n + 2p - 6} \cdot \dfrac{pq + 5p - 3q - 15}{mq + 4m + 4q + 16}$

53) $\dfrac{a^3 - b^3}{a^2 + ab + b^2} \cdot \dfrac{2a^2 + ab - b^2}{a^2 - b^2}$

54) $\dfrac{3x^2 + 10x - 8}{x^2 - 4} \cdot \dfrac{x^3 - 8}{x^2 + 2x + 4}$

WRITING EXERCISES

55) In finding the product of fractions, why are we allowed to divide by factors common to the numerators and the denominators before multiplying?

56) What is wrong with the following? Rework the problem correctly.

$\dfrac{a^2 + b^2}{a^2 - b^2} \cdot \dfrac{a^2 + 2ab - 3b^2}{a^2 + ab - 6b^2} =$

$\dfrac{(a + b)(a + b)}{(a + b)(a - b)} \cdot \dfrac{(a - b)(a + 3b)}{(a + 3b)(a - 2b)} =$

$\dfrac{\cancel{(a + b)}(a + b)}{\cancel{(a + b)}(a - b)} \cdot \dfrac{\cancel{(a - b)}\cancel{(a + 3b)}}{\cancel{(a + 3b)}(a - 2b)} = \dfrac{a + b}{a - 2b}$

57) What is wrong with the following? Rework the problem correctly.

$\dfrac{a^2 + b^2}{a^2 - b^2} \cdot \dfrac{a^2 + 2ab - 3b^2}{a^2 + ab - 6b^2} =$

$\dfrac{\cancel{a}^2 + \cancel{b}^2}{\cancel{a}^2 - \cancel{b}^2} \cdot \dfrac{\cancel{a} + 2ab - \cancel{3}\cancel{b}^2}{\cancel{a}^2 + ab - \cancel{6}\cancel{b}^2_{\,2}} = \dfrac{2ab}{ab - 2}$

Division of Rational Expressions

OBJECTIVES

When you complete this section, you will be able to:

Ⓐ   Divide rational expressions by factoring.

Ⓑ   Divide rational expressions by long division.

In Section 1.2, we discussed the division of fractions. Recall that two numbers are multiplicative inverses if their product is 1. For example, $\frac{3}{4}$ and $\frac{4}{3}$ are multiplicative inverses (reciprocals) since $\frac{3}{4} \cdot \frac{4}{3} = 1$. In general, for a and $b \neq 0$, the multiplicative inverse of $\frac{a}{b}$ is $\frac{b}{a}$ since $\frac{a}{b} \cdot \frac{b}{a} = 1$. Following is the procedure for the division of fractions from Section 1.2.

DIVISION OF RATIONAL NUMBERS

For all integers a, b, c, and d with b, c, and $d \neq 0$, $\frac{a}{b} \div \frac{c}{d} = \frac{a}{b} \cdot \frac{d}{c}$. In words, to divide by a rational number, multiply by its multiplicative inverse (reciprocal).

This rule is usually stated more loosely as "when dividing rational numbers, invert the one on the right and change the operation to multiplication." The procedure for dividing rational expressions is exactly the same. Consider the following.

Rational Numbers	Rational Expressions	
$\dfrac{4}{15} \div \dfrac{12}{25}$	$\dfrac{a^2b}{c^3d^2} \div \dfrac{ab^3}{c^2d^3}$	Multiply by the reciprocal of the fraction on the right.
$\dfrac{4}{15} \cdot \dfrac{25}{12}$	$\dfrac{a^2b}{c^3d^2} \cdot \dfrac{c^2d^3}{ab^3}$	Factor the numerator and denominator into prime factors.
$\dfrac{2 \cdot 2}{3 \cdot 5} \cdot \dfrac{5 \cdot 5}{2 \cdot 2 \cdot 3}$	$\dfrac{a \cdot a \cdot b}{c \cdot c \cdot c \cdot d \cdot d} \cdot \dfrac{c \cdot c \cdot d \cdot d \cdot d}{a \cdot b \cdot b \cdot b}$	Divide by the common factors.
$\dfrac{\cancel{2} \cdot \cancel{2}}{3 \cdot \cancel{5}} \cdot \dfrac{\cancel{5} \cdot 5}{\cancel{2} \cdot \cancel{2} \cdot 3}$	$\dfrac{\cancel{a} \cdot a \cdot \cancel{b}}{\cancel{c} \cdot \cancel{c} \cdot c \cdot \cancel{d} \cdot \cancel{d}} \cdot \dfrac{\cancel{c} \cdot \cancel{c} \cdot \cancel{d} \cdot \cancel{d} \cdot d}{\cancel{a} \cdot \cancel{b} \cdot b \cdot b}$	Multiply the remaining factors.
$\dfrac{5}{9}$	$\dfrac{ad}{b^2c}$	Product.

Following is the procedure for dividing rational expressions.

DIVISION OF RATIONAL EXPRESSIONS

To find the quotient of two rational expressions:

1) Multiply by the reciprocal of the rational expression on the right.

2) Factor all numerators and denominators into prime factors.

3) Divide by all factors common to the numerators and the denominators.

4) Find the product of all remaining factors.

EXAMPLE 1

Find the following quotients. Leave answers reduced to lowest terms.

a) $\dfrac{21}{8} \div \dfrac{35}{12} =$

Multiply by the reciprocal of $\dfrac{35}{12}$.

$\dfrac{21}{8} \cdot \dfrac{12}{35} =$

Factor the numerator and denominator into prime factors.

$\dfrac{3 \cdot 7}{2 \cdot 2 \cdot 2} \cdot \dfrac{2 \cdot 2 \cdot 3}{5 \cdot 7} =$

Divide by the factors common to the numerator and denominator.

$\dfrac{3 \cdot \cancel{7}}{\cancel{2} \cdot \cancel{2} \cdot 2} \cdot \dfrac{\cancel{2} \cdot \cancel{2} \cdot 3}{5 \cdot \cancel{7}} =$

Multiply the remaining factors.

$\dfrac{9}{10}$

Quotient.

b) $\dfrac{a^2b^3}{xy^2} \div \dfrac{ab^4}{x^2y^3} =$

Multiply by the reciprocal of $\dfrac{ab^4}{x^2y^3}$.

$\dfrac{a^2b^3}{xy^2} \cdot \dfrac{x^2y^3}{ab^4} =$

Write the numerators and the denominators in terms of prime factors.

$\dfrac{a \cdot a \cdot b \cdot b \cdot b}{x \cdot y \cdot y} \cdot \dfrac{x \cdot x \cdot y \cdot y \cdot y}{a \cdot b \cdot b \cdot b \cdot b} =$

Divide by the factors common to the numerator and the denominator.

$\dfrac{\cancel{a} \cdot a \cdot \cancel{b} \cdot \cancel{b} \cdot \cancel{b}}{\cancel{x} \cdot \cancel{y} \cdot \cancel{y}} \cdot \dfrac{\cancel{x} \cdot x \cdot \cancel{y} \cdot \cancel{y} \cdot y}{\cancel{a} \cdot \cancel{b} \cdot \cancel{b} \cdot \cancel{b} \cdot b} =$

Multiply the remaining factors.

$\dfrac{axy}{b}$

Quotient.

c) $\dfrac{4x + 8}{6} \div \dfrac{5x + 10}{9} =$

Multiply by the reciprocal of $\dfrac{5x + 10}{9}$.

$\dfrac{4x + 8}{6} \cdot \dfrac{9}{5x + 10} =$

Factor the numerators and the denominators into prime factors.

$\dfrac{2 \cdot 2(x + 2)}{2 \cdot 3} \cdot \dfrac{3 \cdot 3}{5(x + 2)} =$

Divide by the factors common to the numerator and the denominator.

$\dfrac{2 \cdot 2\cancel{(x + 2)}}{\cancel{2} \cdot \cancel{3}} \cdot \dfrac{\cancel{3} \cdot 3}{5\cancel{(x + 2)}} =$

Multiply the remaining factors.

$\dfrac{6}{5}$

Quotient.

d) $\dfrac{2x^2 + 7x + 3}{x^2 - 9} \div \dfrac{2x^2 + 11x + 5}{x^2 - 3x} =$

Multiply by the reciprocal.

$\dfrac{2x^2 + 7x + 3}{x^2 - 9} \cdot \dfrac{x^2 - 3x}{2x^2 + 11x + 5} =$

Factor the numerators and the denominators.

$\dfrac{(2x + 1)(x + 3)}{(x + 3)(x - 3)} \cdot \dfrac{x(x - 3)}{(2x + 1)(x + 5)} =$

Divide by common factors.

$\dfrac{\cancel{(2x + 1)}\cancel{(x + 3)}}{\cancel{(x + 3)}\cancel{(x - 3)}} \cdot \dfrac{x\cancel{(x - 3)}}{\cancel{(2x + 1)}(x + 5)} =$

Multiply the remaining factors.

$\dfrac{x}{x + 5}$

Quotient.

ADDITIONAL PRACTICE

Find the following quotients. Leave answers reduced to lowest terms.

a) $\dfrac{24}{30} \div \dfrac{18}{15}$

b) $\dfrac{m^4n^2}{a^3b^2} \div \dfrac{m^2n^5}{ab^4}$

c) $\dfrac{10}{5x + 20} \div \dfrac{6}{8x + 32}$

d) $\dfrac{x^2 - x - 6}{x^2 - 16} \div \dfrac{3x^2 - 8x - 3}{x^2 + 4x}$

PRACTICE EXERCISES

Find the following quotients. Leave answers reduced to lowest terms.

1) $\dfrac{10}{9} \div \dfrac{25}{12}$

2) $\dfrac{r^3s^2}{p^2q^2} \div \dfrac{rs^3}{p^4q}$

3) $\dfrac{8x}{2x + 6} \div \dfrac{6}{4x + 12}$

4) $\dfrac{2x^2 + 7x + 6}{x^2 - 4} \div \dfrac{2x^2 - 3x - 9}{x^2 - 2x}$

If more practice is needed, do the Additional Practice Exercises in the margin.

In Section 5.1 we reduced rational expressions by factoring the numerator and denominator and dividing by common factors. This method has some limitations. What if the numerator and the denominator have no common factors or if the numerator and/or denominator are prime? Can we still divide the polynomials? We can and the procedure for dividing polynomials follows almost exactly the same procedure (except often easier!) as dividing whole numbers. We will divide two whole numbers and two polynomials side by side and note the similarities in the procedure.

Divide 1058 by 23.	**Divide** $2x^2 + 5x - 12$ **by** $x + 4$.
$23\overline{\smash{)}1058}$	$x + 4\overline{\smash{)}2x^2 + 5x - 12}$
1) 23 will not go into 10, so divide 23 into 105. We guess 4. Put the 4 above the 5. $$\begin{array}{r} 4 \\ 23\overline{\smash{)}1058} \end{array}$$	**1)** Divide x into $2x^2$. There is no guessing as in long division of whole numbers. $\frac{2x^2}{x} = 2x$. Put the $2x$ above the $2x^2$. $$\begin{array}{r} 2x \\ x + 4\overline{\smash{)}2x^2 + 5x - 12} \end{array}$$
2) Multiply 4 and 23 and put the product beneath 105. $$\begin{array}{r} 4 \\ 23\overline{\smash{)}1058} \\ 92 \end{array}$$	**2)** Multiply $2x$ and $x + 4$ and put the product beneath the like terms of $2x^2 + 5x - 12$. $$\begin{array}{r} 2x \\ x + 4\overline{\smash{)}2x^2 + 5x - 12} \\ 2x^2 + 8x \end{array}$$
3) Subtract 92 from 105. $$\begin{array}{r} 4 \\ 23\overline{\smash{)}1058} \\ -92 \\ \hline 13 \end{array}$$	**3)** Subtract $2x^2 + 8x$ from $2x^2 + 5x$ by changing the signs of the bottom polynomial to $-2x^2 - 8x$ and adding. $$\begin{array}{r} 2x \\ x + 4\overline{\smash{)}2x^2 + 5x - 12} \\ -2x^2 - 8x \\ \hline -3x \end{array}$$
4) Bring down the 8. $$\begin{array}{r} 4 \\ 23\overline{\smash{)}1058} \\ -92 \\ \hline 138 \end{array}$$	**4)** Bring down the -12. $$\begin{array}{r} 2x \\ x + 4\overline{\smash{)}2x^2 + 5x - 12} \\ -2x^2 - 8x \\ \hline -3x - 12 \end{array}$$

Divide 1058 by 23.	Divide $2x^2 + 5x - 12$ by $x + 4$.
5) Divide 23 into 138 and put the quotient over the 8. $\begin{array}{r} 46 \\ 23\overline{)1058} \\ \underline{-92} \\ 138 \end{array}$	**5)** Divide x into $-3x$ and put the quotient (which is -3) over the $5x$. $\begin{array}{r} 2x - 3 \\ x+4\overline{)2x^2 + 5x - 12} \\ \underline{-2x^2 - 8x} \\ -3x - 12 \end{array}$
6) Multiply 6 and 23 and put the product under 138. $\begin{array}{r} 46 \\ 23\overline{)1058} \\ \underline{-92} \\ 138 \\ 138 \end{array}$	**6)** Multiply -3 and $x + 4$ and put the product (which is $-3x - 12$) under $-3x - 12$. $\begin{array}{r} 2x - 3 \\ x+4\overline{)2x^2 + 5x - 12} \\ \underline{-2x^2 - 8x} \\ -3x - 12 \\ -3x - 12 \end{array}$
7) Subtract 138 from 138. $\begin{array}{r} 46 \\ 23\overline{)1058} \\ \underline{-92} \\ 138 \\ \underline{-138} \\ 0 \end{array}$	**7)** Subtract $-3x - 12$ from $-3x - 12$ by changing signs of bottom polynomial to $3x + 12$ and adding. $\begin{array}{r} 2x - 3 \\ x+4\overline{)2x^2 + 5x - 12} \\ \underline{-2x^2 - 8x} \\ -3x - 12 \\ \underline{+3x + 12} \\ 0 \end{array}$
8) Remainder of 0. Therefore, 23 divides evenly into 1058.	**8)** Remainder of 0. Therefore, $x + 4$ divides evenly into $2x^2 + 5x - 12$.

Check:
$(23)(46) = 1058$, so 46 is the correct answer.

Check:
$(x + 4)(2x - 3) = 2x^2 + 5x - 12$, so $2x - 3$ is the correct answer.

As you can see, the procedure is almost exactly the same in both cases. We illustrate the procedure for dividing polynomials with some more examples.

EXAMPLE 2

Find the following quotients using long division.

a) $\dfrac{2x^2 + x - 15}{x + 3}$

Solution:

$\dfrac{2x^2 + x - 15}{x + 3}$ means $(2x^2 + x - 15) \div (x + 3)$ which is written as

$x + 3\overline{)2x^2 + x - 15}$ $\dfrac{2x^2}{x} = 2x$ and put the $2x$ over $2x^2$.

$\begin{array}{r} 2x \\ x+3\overline{)2x^2 + x - 15} \end{array}$ $2x(x + 3) = 2x^2 + 6x$ and put $2x^2 + 6x$ underneath the like terms of $2x^2 + x - 15$.

$$\begin{array}{r} 2x \\ x + 3 \overline{\smash{\big)}\ 2x^2 + x - 15} \\ 2x^2 + 6x \end{array}$$

Subtract $2x^2 + 6x$ by changing all the signs and adding.

$$\begin{array}{r} 2x \\ x + 3 \overline{\smash{\big)}\ 2x^2 + x - 15} \\ \underline{-2x^2 - 6x} \\ -5x \end{array}$$

Bring down -15.

$$\begin{array}{r} 2x \\ x + 3 \overline{\smash{\big)}\ 2x^2 + x - 15} \\ \underline{-2x^2 - 6x} \\ -5x - 15 \end{array}$$

$\dfrac{-5x}{x} = -5$ and put the -5 over the $+x$.

$$\begin{array}{r} 2x - 5 \\ x + 3 \overline{\smash{\big)}\ 2x^2 + x - 15} \\ \underline{-2x^2 - 6x} \\ -5x - 15 \end{array}$$

$-5(x + 3) = -5x - 15$ and put $-5x - 15$ under the other $-5x - 15$.

$$\begin{array}{r} 2x - 5 \\ x + 3 \overline{\smash{\big)}\ 2x^2 + x - 15} \\ \underline{-2x^2 - 6x} \\ -5x - 15 \\ -5x - 15 \end{array}$$

Subtract $-5x - 15$ by changing signs and adding.

$$\begin{array}{r} 2x - 5 \\ x + 3 \overline{\smash{\big)}\ 2x^2 + x - 15} \\ \underline{-2x^2 - 6x} \\ -5x - 15 \\ \underline{+5x + 15} \\ 0 \end{array}$$

The remainder is 0. Therefore, $x + 3$ divides into $2x^2 + x - 15$ evenly.

Check:
$(x + 3)(2x - 5) = 2x^2 + x - 15$, so $2x - 5$ is the correct answer.

  b) $\dfrac{6x^2 + 11x - 10}{3x - 2}$

Solution:
Using long division, this is written as

$$3x - 2 \overline{\smash{\big)}\ 6x^2 + 11x - 10}$$

$\dfrac{6x^2}{3x} = 2x$ and put the $2x$ over $6x^2$.

$$\begin{array}{r} 2x \\ 3x - 2 \overline{\smash{\big)}\ 6x^2 + 11x - 10} \end{array}$$

$2x(3x - 2) = 6x^2 - 4x$.

$$\begin{array}{r} 2x \\ 3x - 2 \overline{\smash{\big)}\ 6x^2 + 11x - 10} \\ 6x^2 - 4x \end{array}$$

Subtract $6x^2 - 4x$ by changing signs and adding.

$$\begin{array}{r} 2x \\ 3x - 2 \overline{\smash{\big)}\ 6x^2 + 11x - 10} \\ \underline{-6x^2 + 4x} \\ 15x \end{array}$$

Bring down the -10.

$$\begin{array}{r} 2x \\ 3x - 2 \overline{\smash{\big)}\ 6x^2 + 11x - 10} \\ \underline{-6x^2 + 4x} \\ 15x - 10 \end{array}$$

$\dfrac{15x}{3x} = 5$ and put the 5 over the $11x$.

$$
\begin{array}{r}
2x + 5 \\
3x - 2\overline{\big)\,6x^2 + 11x - 10} \\
\underline{-6x^2 + 4x} \\
15x - 10
\end{array}
\qquad 5(3x - 2) = 15x - 10.
$$

$$
\begin{array}{r}
2x + 5 \\
3x - 2\overline{\big)\,6x^2 + 11x - 10} \\
\underline{-6x^2 + 4x} \\
15x - 10 \\
15x - 10
\end{array}
\qquad \text{Subtract } 15x - 10 \text{ by changing signs and adding.}
$$

$$
\begin{array}{r}
2x + 5 \\
3x - 2\overline{\big)\,6x^2 + 11x - 10} \\
\underline{-6x^2 + 4x} \\
15x - 10 \\
\underline{-15x + 10} \\
0
\end{array}
\qquad \text{The remainder is 0. Therefore, } 3x - 2 \text{ divides into } 6x^2 + 11x - 10 \text{ evenly.}
$$

Check:
$(3x - 2)(2x + 5) = 6x^2 + 11x - 10$, so $2x + 5$ is the correct answer.

ADDITIONAL PRACTICE

Find the following quotients using long division.

e) $\dfrac{3x^2 + 13x - 30}{x + 6}$

f) $\dfrac{12x^2 + 2x - 2}{3x - 1}$

PRACTICE EXERCISES

Find the following quotients using long division.

5) $\dfrac{x^2 - 2x - 24}{x + 4}$

6) $\dfrac{8x^2 - 6x - 9}{2x - 3}$

If more practice is needed, do the Additional Practice Exercises in the margin.

Actually, all the examples we have done to this point could have been done by factoring the numerator and the denominator and dividing by common factors as though we were reducing rational expressions. In the introductory comments, we posed the question as to whether we could divide two polynomials if there were no common factors. We see that we can by using long division. If the polynomial that is the divisor is not a factor of the dividend, there will be a remainder just as in the division of whole numbers. For example, 3 is not a factor of 11, so $11 \div 3 = 3$ with a remainder of 2. Recall, the remainder is usually written as the numerator with the divisor as the denominator. Consequently, $11 \div 3 = 3\frac{2}{3}$.

Before performing long division, it is necessary to do two things.

1) Write the divisor and the dividend in descending powers of the variable.

2) If there is a missing power of the variable, leave a space for each term with a missing power. This is usually done by inserting 0 times the variable raised to the power that is missing. The 0 serves as a "place holder" much as it does in writing a number like 407.

Note: In the following example, we will not show all the steps, only the actual division. If you get stuck, look back at Example 2.

EXAMPLE 3

Find the following quotients using long division.

a) $\dfrac{5x^2 + x^3 - 12 + 2x}{x + 2}$

Solution:

First we need to rewrite the numerator in descending order, then rewrite as long division.

$$\frac{5x^2 + x^3 - 12 + 2x}{x + 2} = \frac{x^3 + 5x^2 + 2x - 12}{x + 2} = x + 2 \overline{\smash{\big)}x^3 + 5x^2 + 2x - 12}$$

Divide as in Example 1.

$$
\begin{array}{r}
x^2 + 3x - 4 \\
x + 2 \overline{\smash{\big)}x^3 + 5x^2 + 2x - 12} \\
\underline{-x^3 - 2x^2} \\
3x^2 + 2x \\
\underline{-3x^2 - 6x} \\
-4x - 12 \\
\underline{+4x + 8} \\
-4
\end{array}
$$

Since x will not divide into -4, -4 is the remainder. Remember, the remainder is written over the divisor. Therefore,

$$\frac{x^3 + 5x^2 + 2x - 12}{x + 2} = x^2 + 3x - 4 + \frac{-4}{x + 2}$$

is the solution.

Remember, in checking a division problem that has a remainder, we multiply the divisor and the quotient, then add the remainder. The result is equal to the dividend. Earlier we stated that $11 \div 3 = 3$ with a remainder of 2. To check, we have $3 \cdot 3 + 2 = 9 + 2 = 11$. If the above division is correct, then $(x + 2)(x^2 + 3x - 4) + (-4) = x^3 + 5x^2 + 2x - 12$.

Check:

$$
\begin{aligned}
(x + 2)(x^2 + 3x - 4) + (-4) &= \\
x(x^2 + 3x - 4) + 2(x^2 + 3x - 4) - 4 &= \\
x^3 + 3x^2 - 4x + 2x^2 + 6x - 8 - 4 &= \\
x^3 + 5x^2 + 2x - 12.&
\end{aligned}
$$

Therefore, the answer is correct.

b) $\dfrac{8x^3 - 22x + 8}{2x - 3}$

Solution:

The x^2 term is missing, so insert $0x^2$ between $8x^3$ and $-22x$ and divide as before.

$$
\begin{array}{r}
4x^2 + 6x - 2 \\
2x - 3 \overline{\smash{\big)}8x^3 + 0x^2 - 22x + 8} \\
\underline{-8x^3 + 12x^2} \\
12x^2 - 22x \\
\underline{-12x^2 + 18x} \\
-4x + 8 \\
\underline{+4x - 6} \\
2
\end{array}
$$

Therefore, $\dfrac{8x^3 - 22x + 8}{2x - 3} = 4x^2 + 6x - 2 + \dfrac{2}{2x - 3}.$

The check is left as an exercise for the student.

PRACTICE EXERCISES

Use long division to find the following quotients.

7) $\dfrac{16x - 12x^2 + 3x^3 - 12}{x - 2}$ 8) $\dfrac{8x^3 + 12x^2 - 6}{2x - 1}$

If more practice is needed, do the Additional Practice Exercises in the margin.

EXERCISE SET 5.3

Find the following quotients of fractions. Leave answers reduced to lowest terms.

1) $\dfrac{1}{4} \div \dfrac{4}{7}$ 2) $\dfrac{1}{5} \div \dfrac{5}{8}$ 3) $\dfrac{3}{7} \div \left(-\dfrac{5}{7}\right)$

4) $-\dfrac{4}{9} \div \dfrac{7}{9}$ 5) $-\dfrac{18}{25} \div \dfrac{9}{20}$ 6) $\dfrac{28}{45} \div \left(-\dfrac{24}{25}\right)$

7) $\dfrac{8}{9} \div \dfrac{2}{3}$ 8) $\dfrac{14}{15} \div \dfrac{7}{3}$

Find the following quotients of rational expressions. Leave answers reduced to lowest terms.

9) $\dfrac{a}{b^4} \div \dfrac{a^3}{b^2}$ 10) $\dfrac{x^3}{y^3} \div \dfrac{x^5}{y^2}$

11) $\dfrac{8a^2}{9b^3} \div \dfrac{4a^3}{3b^2}$ 12) $\dfrac{12c^2}{7d^3} \div \dfrac{8c^4}{21d^4}$

13) $\dfrac{a^3b^2}{c^4d^3} \div \dfrac{ab^5}{c^2d}$ 14) $\dfrac{m^4n^6}{p^2q^3} \div \dfrac{m^6n^3}{p^4q^4}$

15) $\dfrac{8x^7g^3}{15x^2g^4} \div \dfrac{3xg^2}{4x^2g^3}$ 16) $\dfrac{10c^5g^5}{12c^2g^4} \div \dfrac{5cg^2}{4c^2g^3}$

17) $\dfrac{5x + 15}{8} \div \dfrac{7x + 21}{6}$ 18) $\dfrac{3x - 18}{8} \div \dfrac{4x - 24}{12}$

19) $\dfrac{4a - 8}{15a} \div \dfrac{3a - 6}{5a^2}$ 20) $\dfrac{6b + 24}{7b^2} \div \dfrac{7b + 28}{14b^4}$

21) $\dfrac{4p + 12q}{3p - 6q} \div \dfrac{5p + 15q}{6p - 12q}$ 22) $\dfrac{5r + 10s}{3r - 12s} \div \dfrac{6r + 12s}{4r - 16s}$

23) $\dfrac{10x + 15}{18x - 6} \div \dfrac{20x + 30}{9x - 3}$ 24) $\dfrac{12y + 8}{4y - 10} \div \dfrac{18y + 12}{8y - 20}$

25) $\dfrac{a - b}{6a - 9b} \div \dfrac{b - a}{4a - 6b}$ 26) $\dfrac{x - y}{12x - 3y} \div \dfrac{y - x}{8x - 2y}$

27) $\dfrac{h}{h^2 + 13h + 42} \div \dfrac{4h^2 + 28h}{4h + 24}$

28) $\dfrac{b}{b^2 + 11b + 30} \div \dfrac{6b + 30}{6b^2 + 36b}$

29) $\dfrac{d^2 - d - 12}{2d^2} \div \dfrac{3d^2 + 13d + 12}{d}$

30) $\dfrac{y^2 + y - 20}{7y^2} \div \dfrac{3y^2 + 19y + 20}{y}$

31) $\dfrac{m^2 - 25}{2m - 12} \div \dfrac{3m + 15}{m^2 - 36}$

32) $\dfrac{h^2 - 36}{3h - 21} \div \dfrac{4h + 24}{h^2 - 49}$

33) $\dfrac{x + 8}{x - 8} \div \dfrac{x^2 + 16x + 64}{x^2 - 16x + 64}$

34) $\dfrac{w + 3}{w - 3} \div \dfrac{w^2 + 6w + 9}{w^2 - 6w + 9}$

35) $\dfrac{a^2 - 5a + 6}{a^2 - 9a + 18} \div \dfrac{a^2 - 6a + 8}{a^2 - 9a + 20}$

36) $\dfrac{y^2 + 3y - 40}{y^2 + 3y - 18} \div \dfrac{y^2 + 2y - 48}{y^2 + 2y - 35}$

37) $\dfrac{2x^2 - 7x - 4}{3x^2 - 14x + 8} \div \dfrac{2x^2 + 7x + 3}{3x^2 - 8x + 4}$

38) $\dfrac{8x^2 + 6x - 9}{8x^2 + 14x - 15} \div \dfrac{2x^2 + 11x + 12}{4x^2 + 19x + 12}$

Find the following quotients using long division.

39) $\dfrac{x^2 - 2x - 15}{x + 3}$ 40) $\dfrac{x^2 + 2x - 8}{x + 4}$

41) $\dfrac{x^2 - 4x - 21}{x - 7}$ 42) $\dfrac{x^2 - 2x - 24}{x - 6}$

43) $\dfrac{2x^2 - x - 10}{2x - 5}$ 44) $\dfrac{3x^2 - 22x + 24}{3x - 4}$

45) $\dfrac{6x^2 + 7x - 20}{3x - 4}$ 46) $\dfrac{8x^2 + 6x - 9}{4x - 3}$

47) $\dfrac{4x^2 - 25}{2x + 5}$ 48) $\dfrac{9x^2 - 49}{3x + 7}$

ANSWERS:
Practice 7–8

7) $3x^2 - 6x + 4 + \dfrac{-4}{x - 2}$

8) $4x^2 + 8x + 4 + \dfrac{-2}{2x - 1}$

Additional Practice g–h

g) $3x^2 + x - 10 + \dfrac{-50}{x - 4}$

h) $x^2 + 2x - 3 + \dfrac{-7}{2x - 4}$

49) $\dfrac{2x^2 + 4x - 28}{2x - 6}$

50) $\dfrac{3x^2 - 13x - 25}{3x + 5}$

51) $\dfrac{12x^2 - 5x + 6}{4x - 3}$

52) $\dfrac{12x^2 + 22x + 14}{3x + 7}$

53) $\dfrac{6x^2 + 13x - 16}{2x + 7}$

54) $\dfrac{15x^2 + 29x - 10}{5x - 2}$

55) $\dfrac{2x^3 - 14x^2 + 27x - 12}{x - 4}$

56) $\dfrac{3x^3 + 22x^2 + 33x - 10}{x + 5}$

57) $\dfrac{4x^3 + 6x^2 - 16x + 6}{2x - 1}$

58) $\dfrac{6x^3 - 10x^2 + 25x - 14}{3x - 2}$

59) $\dfrac{6x^3 + 11x^2 - 8x + 9}{2x + 5}$

60) $\dfrac{12x^3 - 17x^2 + 12x - 8}{3x - 2}$

61) $\dfrac{3x^3 - 73x - 10}{x - 5}$

62) $\dfrac{2x^3 - 35x - 12}{x + 4}$

63) $\dfrac{4x^3 - 2x^2 + 18}{2x - 3}$

64) $\dfrac{12x^3 + 22x^2 - 2}{3x + 1}$

65) $\dfrac{16x^3 - x + 10}{4x + 3}$

66) $\dfrac{4x^3 - 42x + 23}{2x - 6}$

67) $\dfrac{x^3 - 8}{x - 2}$

68) $\dfrac{x^3 + 27}{x + 3}$

69) $\dfrac{x^4 - 16}{x - 2}$

70) $\dfrac{x^4 - 81}{x - 3}$

71) $\dfrac{x^3 + 9}{x + 3}$

72) $\dfrac{x^3 - 12}{x - 2}$

CHALLENGE EXERCISES (73–78)

Find the following quotients.

73) $\dfrac{ab + 3a + 2b + 6}{bc + 4b + 3c + 12} \div \dfrac{ac - 3a + 2c - 6}{bc + 4b - 4c - 16}$

74) $\dfrac{xy - 3x + 4y - 12}{xy + 6x - 3y - 18} \div \dfrac{xy + 5x + 4y + 20}{xy + 5x - 3y - 15}$

75) Find the value of k, so $x - 6$ will divide evenly into $2x^2 - 15x + k$.

76) Find the value of k, so $x + 4$ will divide evenly into $3x^2 + 6x + k$.

77) Find the value of k, so $x - 2$ will divide evenly into $4x^2 + kx - 6$.

78) Find the value of k, so $x + 3$ will divide evenly into $3x^2 + kx + 6$.

WRITING EXERCISES

79) When performing the division $\frac{a}{b} \div \frac{c}{d}$, why must b, c, and $d \neq 0$?

80) Is $x + 4$ a factor of $2x^3 + 5x^2 - 11x + 4$? Why or why not?

81) Is $x - 3$ a factor of $3x^3 - 7x^2 - 8x + 6$? Why or why not?

82) When dividing polynomials by long division, how can you check your answer?

83) When dividing polynomials by long division, why is it necessary to insert place holders for missing powers of the variable?

GROUP PROJECTS

84) Find three second degree polynomials that are divisible by $x + 3$.

85) Find three second degree polynomials that leave a remainder of 3 when divided by $x - 2$.

SECTION **5.4**

Least Common Multiple, Least Common Denominator, and Equivalent Rational Expressions

OBJECTIVES

When you complete this section, you will be able to:

Ⓐ Find the least common multiple of two or more integers or polynomials.

Ⓑ Change a rational expression into an equivalent rational expression with a specified denominator.

Ⓒ Find the least common denominator of two or more rational expressions and convert each into an equivalent rational expression with the least common denominator as its denominator.

INTRODUCTION

Previously we discussed the greatest common factor. One interpretation of the greatest common factor is that it is the polynomial of greatest degree that will divide evenly into all the polynomials for which it is the greatest common factor. In this section we discuss the **least common multiple** of two or more polynomials. One interpretation of the least common multiple is that it is the polynomial of smallest degree that is divisible by all the polynomials for which it is the least common multiple.

Before discussing the technique for finding the least common multiple, let us discuss the idea of multiples of integers. To find the multiples of a number, multiply the number by 1, 2, 3, 4, and so on. For example, the multiples of 3 are $3 \cdot 1 = 3$, $3 \cdot 2 = 6$, $3 \cdot 3 = 9$, $3 \cdot 4 = 12$, and so on.

The multiples of 3 are $\{3, 6, 9, \underline{12}, 15, 18, 21, \underline{24},...\}$. The multiples of 4 are $\{4, 8, \underline{12}, 16, 20, \underline{24},...\}$. Looking at the multiples of 3 and 4, we see that 12 and 24 are multiples of both. Consequently, we call 12 and 24 common multiples of 3 and 4. Other common multiples are 36, 48, 60, and so on. Of all the common multiples of 3 and 4, the smallest is 12. Therefore, 12 is the least common multiple of 3 and 4. Notice that each multiple of 3 is divisible by 3 and each multiple of 4 is divisible by 4. Consequently, the common multiples of 3 and 4 are divisible by both 3 and 4.

LEAST COMMON MULTIPLE (LCM)

The integer a is the least common multiple of the integers b and c, if a is the smallest positive integer that is a multiple of both b and c.
Equivalently, a is the least common multiple of b and c, if a is the smallest positive integer that is divisible by both b and c.

The above definition is easily extended to more than two integers. In the case of polynomials, the LCM of the polynomials $P(x)$ and $Q(x)$ is the polynomial of least degree that is a multiple of both $P(x)$ and $Q(x)$.

The technique we use for finding the LCM depends upon finding the prime factorizations of the expressions for which we wish to find the LCM.

• Finding the LCM

FINDING THE LEAST COMMON MULTIPLE (LCM)

Procedure for finding the LCM.

1) Write each expression as the product of its prime factors.
2) Select every factor appearing in any prime factorization. If a factor appears in more than one of the prime factorizations, select the factor with the largest exponent.
3) The LCM is the product of the factors selected in step 2.

This technique for finding the LCM works because any factor of an expression is also a factor of any multiple of that expression. By taking the largest exponent of every factor that appears in any prime factorization, we are assuring ourselves that their product (the LCM) contains every factor of each expression and only those expressions that are factors.

EXAMPLE 1

Find the LCM of the following:

a) 108 and 144

    Solution:

$$108 = 2^2 \cdot 3^3 \qquad \text{Prime factorization.}$$
$$144 = 2^4 \cdot 3^2 \qquad \text{Prime factorization.}$$

    The different factors that appear in either factorization are 2 and 3. The largest exponent on 2 in either factorization is 4 in 2^4, and the largest exponent on 3 in either factorization is 3 in 3^3. Consequently, the LCM is $2^4 \cdot 3^3 = 432$. Note that 432 is divisible by both 108 and 144.

b) 90 and 189

    Solution:

$$90 = 2 \cdot 3^2 \cdot 5 \qquad \text{Prime factorization.}$$
$$189 = 3^3 \cdot 7 \qquad \text{Prime factorization.}$$

    The different factors that appear in either factorization are 2, 3, 5, and 7. The largest exponent on 2 is 1 in 2, the largest exponent on 3 is 3 in 3^3, the largest exponent on 5 is 1 in 5, and the largest exponent on 7 is 1 in 7. Consequently, the LCM is $2 \cdot 3^3 \cdot 5 \cdot 7 = 1890$. Verify that 1890 is divisible by both 90 and 189.

Note: The procedure for finding the LCM of two or more monomials involving variables is exactly the same as finding the LCM from the prime factorizations of integers. Many times it is actually easier since we do not have to find the prime factorizations for the variable factors.

c) x^3y^2z and x^2y^4w

    Solution:

    The different factors that appear are x, y, z, and w. The largest exponent on x is 3 in x^3, the largest exponent on y is 4 in y^4, the largest exponent on z is 1 in z, and the largest exponent on w is 1 in w. Consequently, the LCM is x^3y^4zw. Note that x^3y^4zw is divisible by both x^3y^2z and x^2y^4w.

d) $12r^3s^5$ and $18rs^2$

    Solution:

    We find the prime factorizations of the coefficients and use the same method as before.

$$12 = 2^2 \cdot 3r^3s^5 \qquad \text{Prime factorization of 12.}$$
$$18 = 2 \cdot 3^2rs^2 \qquad \text{Prime factorization of 18.}$$

    The different factors that appear are 2, 3, r, and s. The largest exponent on 2 is 2 in 2^2, the largest exponent on 3 is 2 in 3^2, the largest exponent on r is 3 in r^3, and the largest exponent on s is 5 in s^5. Consequently, the LCM is $36r^3s^5$. Note that $36r^3s^5$ is divisible by $12r^3s^5$ and $18rs^2$.

Note: In order to find the LCM of two or more polynomials that are not monomials, the procedure is still the same. However, we usually leave the LCM in factored form rather than multiplying it out.

e) $x + 3$ and $x - 2$

Solution:

Since $x + 3$ and $x - 2$ are both prime polynomials, the LCM is their product $(x + 3)(x - 2)$. Note that $(x + 3)(x - 2)$ is divisible by both $x + 3$ and $x - 2$.

f) $y - 6$ and $y^2 + 5y$

Solution:

$$y - 6 \qquad\qquad y - 6 \text{ is prime.}$$
$$y^2 + 5y = y(y + 5) \qquad \text{Prime factorization.}$$

The different factors that appear in either prime factorization are y, $y - 6$, and $y + 5$. The largest exponent of each is 1. Therefore, the LCM is $y(y - 6)(y + 5)$. Note that $y(y - 6)(y + 5)$ is divisible by $y - 6$ and $y^2 + 5y$.

 g) $x^2 + 2x - 15$ and $x^2 - 5x + 6$

Solution:

$$x^2 + 2x - 15 = (x - 3)(x + 5) \qquad \text{Prime factorization.}$$
$$x^2 - 5x + 6 = (x - 3)(x - 2) \qquad \text{Prime factorization.}$$

The different factors that appear are $x - 3$, $x + 5$, and $x - 2$. The largest exponent on $x - 3$ is 1 in $x - 3$, the largest exponent on $x + 5$ is 1 in $x + 5$, and the largest exponent on $x - 2$ is 1 in $x - 2$. Consequently, the LCM is $(x - 3)(x + 5)(x - 2)$. Note that $(x - 3)(x + 5)(x - 2)$ is divisible by $x^2 + 2x - 15$ and $x^2 - 5x + 6$.

 h) $x^2 + 5x + 4$ and $x^2 + 2x + 1$

Solution:

$$x^2 + 5x + 4 = (x + 1)(x + 4) \qquad \text{Prime factorization.}$$
$$x^2 + 2x + 1 = (x + 1)^2 \qquad \text{Prime factorization.}$$

The different factors that appear are $x + 1$ and $x + 4$. The largest exponent on $x + 1$ is 2 in $(x + 1)^2$, and the largest exponent on $x + 4$ is 1 in $x + 4$. Consequently, the LCM is $(x + 1)^2(x + 4)$. Note that $(x + 1)^2(x + 4)$ is divisible by $x^2 + 5x + 4$ and $x^2 + 2x + 1$

ADDITIONAL PRACTICE

Find the LCM of the following.

a) 225 and 375

b) 90 and 168

c) $m^2n^4o^2$ and $m^5n^3o^4$

d) $14r^6s^2$ and $20r^2s^2$

e) $b - 5$ and $b + 2$

f) $m - 3$ and $m^2 - 7m + 12$

g) $c^2 + 3c - 18$ and $c^2 + c - 12$

h) $4x^2 - 12x + 9$ and $8x^2 - 18x + 9$

PRACTICE EXERCISES

Find the LCM of the following.

1) 48 and 72

2) 60 and 126

3) r^4st^3 and $r^2s^3u^2$

4) $15a^2b^4$ and $12a^3b^4$

5) $a + 4$ and $a - 3$

6) $b + 3$ and $b^2 + 3b$

7) $y^2 - 7y + 10$ and $y^2 + 3y - 10$

8) $x^2 + 4x + 4$ and $x^2 - 3x - 10$

If more practice is needed, do the Additional Practice Exercises in the margin.

Fractions like $\frac{6}{10}$ and $\frac{3}{5}$ are called equivalent fractions which we discussed previously. Notice that the product of the numerator of the first fraction with the

• **Changing a fraction into an equivalent fraction with a specified denominator**

denominator of the second is equal to the product of the denominator of the first fraction and the numerator of the second. That is, $6 \cdot 5 = 10 \cdot 3$. This observation leads to a more formal definition of equivalent fractions than the one given earlier.

DEFINITION OF EQUIVALENT FRACTIONS

Two or more fractions are equivalent if they represent the same quantity. Equivalently,

$$\frac{a}{b} = \frac{c}{d} \text{ if } a \cdot d = b \cdot c.$$

The above procedure is often referred to as **cross multiplication** because the lines of multiplication "cross" each other.

$$\frac{a}{b} \diagdown\!\!\!\!\diagup \frac{c}{d}$$

$$a \cdot d = b \cdot c$$

In reducing fractions to lowest terms we divide the numerators and the denominators by all common factors to get equivalent fractions reduced to lowest terms. For example, $\frac{6}{10} = \frac{2 \cdot 3}{2 \cdot 5} = \frac{3}{5}$. Now we are going to do the opposite and generate equivalent fractions by multiplying the numerator and the denominator of a fraction by the same number. This is permissible because any nonzero number divided by itself is 1, and any number multiplied by 1 is equivalent to the original number. Since we are multiplying the numerator and the denominator by the same number, we are multiplying the fraction by 1. This is called the fundamental property of fractions.

FUNDAMENTAL PROPERTY OF FRACTIONS

If $\frac{a}{b}$ is a fraction and $c \neq 0$, then $\frac{a}{b} = \frac{a}{b} \cdot \frac{c}{c} = \frac{ac}{bc}$. In words, we can multiply both the numerator and the denominator of a fraction by any number except 0 and get a fraction equivalent to the original fraction.

Note: If we cross multiply $\frac{a}{b} = \frac{ac}{bc}$, we get $abc = bac$. By what properties of multiplication does $abc = bac$?

In the examples that follow, we will use the fundamental property of fractions to create equivalent fractions with a given denominator. We must decide by what we need to multiply the denominator of the first fraction in order to get the denominator of the second fraction. Then we will multiply the numerator and the denominator of the first fraction by this number to get an equivalent fraction with the denominator of the second fraction. Since the purpose of these exercises is to aid you in adding fractions with unlike denominators in a later section, we will limit our discussion to finding missing numerators.

EXAMPLE 2

Find the missing numerator so that the fractions are equal.

a) $\dfrac{3}{4} = \dfrac{?}{12}$ Since $3 \cdot 4 = 12$, multiply $\dfrac{3}{4}$ by $\dfrac{3}{3}$.

$\dfrac{3}{3} \cdot \dfrac{3}{4} = \dfrac{9}{12}$ Therefore, $\dfrac{3}{4} = \dfrac{9}{12}$. Check: $3 \cdot 12 = 4 \cdot 9$, so the answer is correct.

Note: The number that you must multiply the first fraction by to get an equivalent fraction with the given denominator can be found by dividing the denominator of the first fraction into the denominator of the new fraction. In the example above, $12 \div 4 = 3$. Therefore, we multiply $\dfrac{3}{4}$ by $\dfrac{3}{3}$.

b) $\dfrac{7}{12} = \dfrac{?}{108}$ It may not be obvious what we have to multiply 12 by in order to get 108. Since $108 \div 12 = 9$, multiply $\dfrac{7}{12}$ by $\dfrac{9}{9}$.

$\dfrac{9}{9} \cdot \dfrac{7}{12} = \dfrac{63}{108}$ Therefore, $\dfrac{7}{12} = \dfrac{63}{108}$. Check: Does $7 \cdot 108 = 12 \cdot 63$?

Note: The procedure is exactly the same if we have rational expressions instead of rational numbers. In the examples that follow, assume all variables in the denominator have nonzero values.

c) $\dfrac{3}{5x} = \dfrac{?}{15x^3}$ Since $3x^2 \cdot 5x = 15x^3$, multiply $\dfrac{3}{5x}$ by $\dfrac{3x^2}{3x^2}$.

$\dfrac{3x^2}{3x^2} \cdot \dfrac{3}{5x} = \dfrac{9x^2}{15x^3}$ Therefore, $\dfrac{3}{5x} = \dfrac{9x^2}{15x^3}$. Check: Does $3 \cdot 15x^3 = 5x \cdot 9x^2$? Since both sides are equal to $45x^3$, the answer is correct.

d) $\dfrac{12x^2}{7y^2} = \dfrac{?}{28y^4}$ Since $4y^2 \cdot 7y^2 = 28y^4$, multiply $\dfrac{12x^2}{7y^2}$ by $\dfrac{4y^2}{4y^2}$.

$\dfrac{4y^2}{4y^2} \cdot \dfrac{12x^2}{7y^2} = \dfrac{48x^2y^2}{28y^4}$ Therefore, $\dfrac{12x^2}{7y^2} = \dfrac{48x^2y^2}{28y^4}$. Check: Does

$12x^2 \cdot 28y^4 = 7y^2 \cdot 48x^2y^2$?

Note: In Example d it may not be obvious that we need to multiply $7y^2$ by $4y^2$ in order to get $28y^4$. We can use the same procedure we used in Example 1. Since $28y^4 \div 7y^2 = 4y^2$, we need to multiply by $4y^2$.

If the denominators of the rational expressions are polynomials other than monomials, we will need to factor those denominators that are not prime. By doing so, we can determine the expression that we need to multiply the first rational expression by in order to convert it into a rational expression with the same denominator as the second.

e) $\dfrac{5a}{2a + 8} = \dfrac{?}{6a + 24}$ Factor the denominators.

$\dfrac{5a}{2(a + 4)} = \dfrac{?}{6(a + 4)}$ Since $3 \cdot 2(a + 4) = 6(a + 4)$, multiply $\dfrac{5a}{2(a + 4)}$ by $\dfrac{3}{3}$.

$\dfrac{3}{3} \cdot \dfrac{5a}{2(a + 4)} = \dfrac{15a}{6(a + 4)}$ Therefore, $\dfrac{5a}{2a + 8} = \dfrac{15a}{6a + 24}$.

Note: The checks can be done in the usual manner but often may be very difficult. For this reason, equivalent rational expressions are usually not checked.

f)

$$\frac{7}{x^2 - 2x} = \frac{?}{x(x-2)(x+3)}$$

Factor $x^2 - 2x$.

$$\frac{7}{x(x-2)} = \frac{?}{x(x-2)(x+3)}$$

The factor missing is $x + 3$, so we multiply by $\frac{x+3}{x+3}$.

$$\frac{x+3}{x+3} \cdot \frac{7}{x(x-2)} = \frac{7(x+3)}{x(x-2)(x+3)}$$

$7(x+3) = 7x + 21$

$$= \frac{7x + 21}{x(x-2)(x+3)}$$

Therefore, $\dfrac{7}{x^2 - 2x} =$

$\dfrac{7x + 21}{x(x-2)(x+3)}$.

g)

$$\frac{7x}{x^2 + 2x - 8} = \frac{?}{(x-5)(x-2)(x+4)}$$

Factor $x^2 + 2x - 8$.

$$\frac{7x}{(x-2)(x+4)} = \frac{?}{(x-5)(x-2)(x+4)}$$

Since the denominator of the first fraction is missing the factor $x - 5$, we need to multiply by $\frac{x-5}{x-5}$.

$$\frac{x-5}{x-5} \cdot \frac{7x}{x^2 + 2x - 8} = \frac{7x(x-5)}{(x-5)(x-2)(x+4)}$$

$7x(x-5) = 7x^2 - 35x$.

$$= \frac{7x^2 - 35x}{(x-5)(x-2)(x+4)}$$

Therefore, $\dfrac{7x}{x^2 + 2x - 8} =$

$\dfrac{7x^2 - 35x}{(x-5)(x-2)(x+4)}$.

ADDITIONAL PRACTICE

Find the missing numerator so that the rational expressions are equal.

i) $\dfrac{3}{7} = \dfrac{?}{35}$

j) $\dfrac{5}{6} = \dfrac{?}{96}$

k) $\dfrac{3}{8x} = \dfrac{?}{48x^5}$

l) $\dfrac{7y^2}{4z^3} = \dfrac{?}{24z^6}$

m) $\dfrac{9}{4a + 8} = \dfrac{?}{12a + 24}$

n) $\dfrac{4z}{m^2 + 5m} = \dfrac{?}{m(m+4)(m+5)}$

o) $\dfrac{6x}{x^2 + x - 12} = \dfrac{?}{(x+6)(x+4)(x-3)}$

PRACTICE EXERCISES

Find the missing numerator so that the rational expressions are equal.

9) $\dfrac{2}{3} = \dfrac{?}{18}$

10) $\dfrac{5}{6} = \dfrac{?}{144}$

11) $\dfrac{5}{3a} = \dfrac{?}{18a^4}$

12) $\dfrac{6}{11a^2b^3} = \dfrac{?}{55a^3b^3c}$

13) $\dfrac{7x}{3x - 6} = \dfrac{?}{9x - 18}$

14) $\dfrac{2a}{b^2 - 3b} = \dfrac{?}{b(b-3)(b+2)}$

15) $\dfrac{3x}{x^2 - 8x + 15} = \dfrac{?}{(x+4)(x-3)(x-5)}$

If more practice is needed, do the Additional Practice Exercises in the margin.

The least common denominator (LCD) of two or more rational expressions is the least common multiple of the denominators. Hence the LCD of two or more rational expressions is the polynomial of lowest degree that is divisible by all the denominators for which it is the LCD. Finding the least common denominator and writing fractions with the least common denominator as their denominators will be very important in the next section.

• **Finding the LCD of and writing rational expressions with the LCD**

EXAMPLE 3

Find the least common denominator for the following. Then write each as an equivalent rational expression with the LCD as its denominator.

a) $\dfrac{3}{10}$ and $\dfrac{5}{21}$

Solution:

$$10 = 2 \cdot 5 \qquad \text{Prime factorization of the denominators.}$$
$$21 = 3 \cdot 7 \qquad \text{Prime factorization of the denominators.}$$

Therefore, by choosing the highest power of each factor, the LCM of 10 and 21 is $2 \cdot 3 \cdot 5 \cdot 7 = 210$. So the LCD of $\frac{3}{10}$ and $\frac{5}{21}$ is 210. Notice the LCD is the same as $10 \cdot 21$. This is because 10 and 21 have no factors in common. Since $10 \cdot 21 = 210$, we multiply $\frac{3}{10}$ by $\frac{21}{21}$. Since $21 \cdot 10 = 210$, we multiply $\frac{5}{21}$ by $\frac{10}{10}$.

$$\frac{3}{10} = \frac{21}{21} \cdot \frac{3}{10} = \frac{63}{210}.$$
$$\frac{5}{21} = \frac{10}{10} \cdot \frac{5}{21} = \frac{50}{210}.$$

b) $\dfrac{7}{12}$ and $\dfrac{11}{18}$

Solution:

$$12 = 2^2 \cdot 3 \qquad \text{Prime factorization of the denominators.}$$
$$18 = 2 \cdot 3^2 \qquad \text{Prime factorization of the denominators.}$$

Therefore, the LCM of 12 and 18 is $2^2 \cdot 3^2 = 36$. So the LCD of $\frac{7}{12}$ and $\frac{11}{18}$ is also 36. Since $3 \cdot 12 = 36$, we multiply $\frac{7}{12}$ by $\frac{3}{3}$. Since $2 \cdot 18 = 36$, we multiply $\frac{11}{18}$ by $\frac{2}{2}$.

$$\frac{7}{12} = \frac{3}{3} \cdot \frac{7}{12} = \frac{21}{36}.$$
$$\frac{11}{18} = \frac{2}{2} \cdot \frac{11}{18} = \frac{22}{36}.$$

Note: Finding the least common denominators of rational expressions follows the same procedure as finding the least common denominator of fractions.

c) $\dfrac{b}{a^2d}$ and $\dfrac{c}{ad^2}$

Solution:

The LCM of a^2d and ad^2 is a^2d^2. Therefore, the LCD is also a^2d^2. Since $d \cdot a^2d = a^2d^2$, we multiply $\frac{b}{a^2d}$ by $\frac{d}{d}$. Since $a \cdot ad^2 = a^2d^2$, we multiply $\frac{c}{ad^2}$ by $\frac{a}{a}$.

$$\frac{b}{a^2d} = \frac{b}{a^2d} \cdot \frac{d}{d} = \frac{bd}{a^2d^2}.$$
$$\frac{c}{ad^2} = \frac{c}{ad^2} \cdot \frac{a}{a} = \frac{ac}{a^2d^2}.$$

d) $\dfrac{5}{10x^2y^3}$ and $\dfrac{3}{14x^2y^2}$

Solution:

$$10x^2y^3 = 2 \cdot 5x^2y^3 \qquad \text{Prime factorization of the denominators.}$$
$$14x^2y^2 = 2 \cdot 7x^2y^2 \qquad \text{Prime factorization of the denominators.}$$

Therefore, the LCM of $10x^2y^3$ and $14x^2y^2$ is $2 \cdot 5 \cdot 7x^2y^3 = 70x^2y^3$. Consequently, the LCD is $70x^2y^3$. To write $\frac{5}{10x^2y^3}$ as a fraction with a denominator of $70x^2y^3$, we multiply by $\frac{7}{7}$, and to write $\frac{3}{14x^2y^2}$ as a fraction with a denominator of $70x^2y^3$, we multiply by $\frac{5y}{5y}$.

$$\frac{5}{10x^2y^3} = \frac{7}{7} \cdot \frac{5}{10x^2y^3} = \frac{35}{70x^2y^3}.$$

$$\frac{3}{14x^2y^2} = \frac{5y}{5y} \cdot \frac{3}{14x^2y^2} = \frac{15y}{70x^2y^3}.$$

e) $\dfrac{4}{x-3}$ and $\dfrac{7}{x+4}$

Solution:

Since $x - 3$ and $x + 4$ are prime, their LCM is $(x - 3)(x + 4)$. Therefore, the LCD is $(x - 3)(x + 4)$. To write $\frac{4}{x-3}$ as a fraction with a denominator of $(x - 3)(x + 4)$, we multiply it by $\frac{x+4}{x+4}$. To write $\frac{7}{x+4}$ as a fraction with a denominator of $(x - 3)(x + 4)$, we multiply it by $\frac{x-3}{x-3}$.

$$\frac{4}{x-3} = \frac{x+4}{x+4} \cdot \frac{4}{x-3} = \frac{4(x+4)}{(x+4)(x-3)} = \frac{4x+16}{(x-3)(x+4)}.$$

$$\frac{7}{x+4} = \frac{x-3}{x-3} \cdot \frac{7}{x+4} = \frac{7(x-3)}{(x-3)(x+4)} = \frac{7x-21}{(x-3)(x+4)}.$$

Note: It is customary to find the product of the numerators but not the denominators. You will see the reason for this in the next section.

f) $\dfrac{x}{2x+6}$ and $\dfrac{3x}{6x+18}$

Solution:

$$2x + 6 = 2(x + 3) \qquad \text{Prime factorization of the denominators.}$$
$$6x + 18 = 2 \cdot 3(x + 3) \qquad \text{Prime factorization of the denominators.}$$

Therefore, the LCM of $2x + 6$ and $6x + 18$ is $2 \cdot 3(x + 3) = 6(x + 3)$. So the LCD is $6(x + 3)$. To write $\frac{x}{2(x+3)}$ as a fraction with a denominator of $6(x + 3)$, we multiply it by $\frac{3}{3}$ since $3 \cdot 2(x + 3) = 6(x + 3)$. Notice that these rational expressions are equal

$$\frac{3x}{6x+18} = \frac{3x}{6(x+3)}.$$

So the denominator is already the LCD.

$$\frac{x}{2x+6} = \frac{x}{2(x+3)} = \frac{3}{3} \cdot \frac{x}{2(x+3)} = \frac{3x}{6(x+3)}.$$

g) $\dfrac{x+2}{x^2+3x-4}$ and $\dfrac{x-3}{x^2+6x+8}$

Solution:

$x^2 + 3x - 4 = (x+4)(x-1)$ Prime factorization of the denominators.

$x^2 + 6x + 8 = (x+4)(x+2)$ Prime factorization of the denominators.

Therefore, the LCM of $x^2 + 3x - 4$ and $x^2 + 6x + 8$ is $(x+4)(x-1)$ $(x+2)$. So the LCD is $(x+4)(x-1)(x+2)$. Since $\frac{x+2}{x^2+3x-4} = \frac{x+2}{(x+4)(x-1)}$, we need to multiply it by $\frac{x+2}{x+2}$. Since $\frac{x-3}{x^2+6x+8} = \frac{x-3}{(x+4)(x+2)}$, we need to multiply it by $\frac{x-1}{x-1}$.

$$\frac{x+2}{x^2+3x-4} = \frac{x+2}{(x+4)(x-1)} = \frac{x+2}{x+2} \cdot \frac{x+2}{(x+4)(x-1)}$$

$$= \frac{(x+2)(x+2)}{(x+2)(x+4)(x-1)}$$

$$= \frac{x^2+4x+4}{(x+2)(x+4)(x-1)}.$$

$$\frac{x-3}{x^2+6x+8} = \frac{x-3}{(x+4)(x+2)} = \frac{x-1}{x-1} \cdot \frac{x-3}{(x+4)(x+2)}$$

$$= \frac{(x-1)(x-3)}{(x-1)(x+4)(x+2)}$$

$$= \frac{x^2-4x+3}{(x-1)(x+4)(x+2)}.$$

ADDITIONAL PRACTICE

Find the least common denominator of the following. Then write each rational expression as an equivalent rational expression with the LCD.

p) $\dfrac{1}{22}$ and $\dfrac{1}{6}$

q) $\dfrac{7}{36}$ and $\dfrac{11}{24}$

r) $\dfrac{5}{r^4s^2}$ and $\dfrac{2}{r^3s^3}$

s) $\dfrac{3t}{21m^2n^5}$ and $\dfrac{8s}{12mn^2}$

t) $\dfrac{x}{x+1}$ and $\dfrac{y}{x-2}$

u) $\dfrac{x+2}{6x-12}$ and $\dfrac{x-3}{8x-16}$

v) $\dfrac{y+9}{y^2+2y-24}$ and $\dfrac{y-3}{y^2-16}$

PRACTICE EXERCISES

Find the least common denominator of the following. Then write each rational expression as an equal rational expression with the LCD as its denominator.

16) $\dfrac{5}{14}$ and $\dfrac{8}{15}$

17) $\dfrac{9}{20}$ and $\dfrac{7}{50}$

18) $\dfrac{m}{c^3d}$ and $\dfrac{n}{c^2d^2}$

19) $\dfrac{8}{15m^3n^3}$ and $\dfrac{7}{18m^2n}$

20) $\dfrac{a}{a-6}$ and $\dfrac{a}{a+4}$

21) $\dfrac{3y}{4y-20}$ and $\dfrac{6y}{6y-30}$

22) $\dfrac{b-4}{b^2-2b-35}$ and $\dfrac{b+3}{b^2+7b+10}$

If more practice is needed, do the Additional Practice Exercises in the margin.

EXERCISE SET   **5.4**

Find the LCM of the following whole numbers.

1) 9 and 11

2) 13 and 15

3) 70 and 154

4) 30 and 105

5) 28, 63, and 42

6) 36, 60, and 72

7) xy and x^3y^3

8) r^3s^5 and rs

9) a^2b and $a^3b^6c^4$

10) m^8np^2 and m^9p^2

23) $z^2 - 4z - 5$ and $z^2 + 2z + 1$

24) $b^2 - 6b + 9$ and $b^2 - 8b + 15$

25) $t^2 - 9$, $t^2 - 2t - 15$, and $t^2 - 7t + 12$

26) $9y^2 - 16$, $3y^2 + 7y + 4$, $3y^2 - 2y - 8$

Find the missing numerator so that the fractions will be equal.

27) $\dfrac{3}{4} = \dfrac{?}{28}$

28) $\dfrac{6}{7} = \dfrac{?}{35}$

29) $\dfrac{9}{5} = \dfrac{?}{135}$

30) $\dfrac{17}{9} = \dfrac{?}{117}$

31) $\dfrac{c}{e} = \dfrac{?}{ef}$

32) $\dfrac{a}{b} = \dfrac{?}{cb}$

33) $\dfrac{4}{5x} = \dfrac{?}{30x}$

34) $\dfrac{3}{7y^2} = \dfrac{?}{42y^2}$

35) $\dfrac{9r}{11s^2} = \dfrac{?}{121s^5}$

36) $\dfrac{4m}{13n^3} = \dfrac{?}{91n^7}$

37) $\dfrac{12y^2}{7z^4} = \dfrac{?}{63z^9}$

38) $\dfrac{15a^3}{8b^2} = \dfrac{?}{72b^6}$

39) $\dfrac{9t}{14x^2y} = \dfrac{?}{42x^5y^3}$

40) $\dfrac{5r^2}{16ps^4} = \dfrac{?}{64p^4s^7}$

Find the missing numerator so that the rational expressions are equal.

41) $\dfrac{y}{y+3} = \dfrac{?}{5y+15}$

42) $\dfrac{z}{2-z} = \dfrac{?}{8-4z}$

43) $\dfrac{4c}{5c+10} = \dfrac{?}{20c+40}$

44) $\dfrac{6h}{14-7h} = \dfrac{?}{42-21h}$

45) $\dfrac{11}{r-1} = \dfrac{?}{r^2+2r-3}$

46) $\dfrac{2u}{4+u} = \dfrac{?}{12-u-u^2}$

47) $\dfrac{5v}{v^2-5v} = \dfrac{?}{v(v-5)(v+2)}$

48) $\dfrac{7x}{x^2+8x} = \dfrac{?}{x(x-1)(x+8)}$

49) $\dfrac{y-2}{y^2-10y+21} = \dfrac{?}{(y+4)(y-7)(y-3)}$

50) $\dfrac{u+5}{u^2+6u-16} = \dfrac{?}{(u+2)(u-2)(u+8)}$

51) $\dfrac{n-3}{n^2+6n+9} = \dfrac{?}{(n+3)(n-9)(n+3)}$

52) $\dfrac{m+5}{m^2-8m+16} = \dfrac{?}{(m-4)(m+8)(m-4)}$

Find the least common denominator for the following. Then write each as an equivalent rational expression with the LCD as its denominator.

53) $\dfrac{1}{2}$ and $\dfrac{1}{7}$

54) $\dfrac{1}{11}$ and $\dfrac{1}{13}$

55) $\dfrac{2}{5}$ and $\dfrac{7}{15}$

56) $\dfrac{3}{7}$ and $\dfrac{3}{28}$

57) $\dfrac{1}{10}$ and $\dfrac{1}{15}$

58) $\dfrac{5}{24}$ and $\dfrac{6}{25}$

59) $\dfrac{1}{6}, \dfrac{3}{8}$, and $\dfrac{9}{10}$

60) $\dfrac{4}{9}, \dfrac{5}{12}$, and $\dfrac{7}{18}$

ANSWERS:
Practice 16–22

16) LCD = 210
$\dfrac{75}{210}, \dfrac{112}{210}$

17) LCD = 100
$\dfrac{45}{100}, \dfrac{14}{100}$

18) LCD = c^3d^2
$\dfrac{dm}{c^3d^2}, \dfrac{cn}{c^3d^2}$

19) LCD = $90m^3n^3$
$\dfrac{48}{90m^3n^3}, \dfrac{35mn^2}{90m^3n^3}$

20) LCD = $(a-6)(a+4)$
$\dfrac{a^2+4a}{(a-6)(a+4)}$

$\dfrac{a^2-6a}{(a-6)(a+4)}$

21) LCD = $12(y-5)$
$\dfrac{9y}{12(y-5)}$

$\dfrac{12y}{12(y-5)}$

22) LCD = $(b+5)(b-7)(b+2)$
$\dfrac{b^2-2b-8}{(b+5)(b-7)(b+2)}$

$\dfrac{b^2-4b-21}{(b+5)(b-7)(b+2)}$

Additional Practice p–v

p) LCD = 66
$\dfrac{3}{66}, \dfrac{11}{66}$

q) LCD = 72
$\dfrac{14}{72}, \dfrac{33}{72}$

r) LCD = r^4s^3
$\dfrac{5s}{r^4s^3}, \dfrac{2r}{r^4s^3}$

s) LCD = $84m^2n^5$
$\dfrac{12t}{84m^2n^5}, \dfrac{56mn^3s}{84m^2n^5}$

t) LCD = $(x+1)(x-2)$
$\dfrac{x^2-2x}{(x+1)(x-2)}$

$\dfrac{xy+y}{(x+1)(x-2)}$

u) LCD = $24(x-2)$
$\dfrac{4x+8}{24(x-2)}$

$\dfrac{3x-9}{24(x-2)}$

v) LCD = $(y-4)(y+6)(y+4)$
$\dfrac{y^2+13y+36}{(y-4)(y+6)(y+4)}$

$\dfrac{y^2+3y-18}{(y-4)(y+6)(y+4)}$

---

11) $3u^2v$ and $9u^3v^5$

12) $5w^4x^6$ and $15w^8x$

13) $15m^5np^9$ and $45m^3p^7$

14) $18a^3b^3$ and $54ab^4c^3$

15) $9u^3v$, $36u^2v^2$, and $27u^4v^6$

16) $18w^3x^2$, $42w^6x$, and $30w^5x^5$

17) $y+4$ and $y+1$

18) $t-5$ and $t-2$

19) $v-5$ and $v^2-2v-15$

20) $r+9$ and r^2+8r-9

21) x^2-x-12 and $x^2+8x+15$

22) y^2-8y+7 and y^2+y-2

61) $\dfrac{a}{b}$ and $\dfrac{c}{d}$

62) $\dfrac{x}{y}$ and $\dfrac{u}{v}$

63) $\dfrac{r}{stu}$ and $\dfrac{n}{mtu}$

64) $\dfrac{a}{pqr}$ and $\dfrac{b}{rtq}$

65) $\dfrac{x}{wv^2}$ and $\dfrac{y}{sv}$

66) $\dfrac{m}{p^2q}$ and $\dfrac{n}{pz}$

67) $\dfrac{2t}{a^2b^5}$ and $\dfrac{3z}{a^3b^3}$

68) $\dfrac{6u}{h^4k^2}$ and $\dfrac{4v}{h^3k^7}$

69) $\dfrac{7}{6u^2v^4}$ and $\dfrac{3}{8uv^3}$

70) $\dfrac{2}{9a^4b^3}$ and $\dfrac{5}{12ab^2}$

71) $\dfrac{3a}{10w^6z^7}$ and $\dfrac{6b}{25w^2z^6}$

72) $\dfrac{5r}{24ab^6}$ and $\dfrac{3s}{16a^2b^9}$

73) $\dfrac{3}{u+8}$ and $\dfrac{9}{u-10}$

74) $\dfrac{5}{4-v}$ and $\dfrac{6}{2-v}$

75) $\dfrac{t}{3t-9}$ and $\dfrac{5t}{4t-12}$

76) $\dfrac{3z}{5z-10}$ and $\dfrac{z}{7z-14}$

77) $\dfrac{3a}{a^2-1}$ and $\dfrac{a}{a^2+3a-4}$

78) $\dfrac{4b}{b^2-9}$ and $\dfrac{2b}{b^2+b-6}$

79) $\dfrac{v-5}{v^2-2v-3}$ and $\dfrac{3+v}{v^2-5v+6}$

80) $\dfrac{w+6}{w^2+6w+5}$ and $\dfrac{2-w}{w^2+w-20}$

81) $\dfrac{m+1}{m^2+8m+16}$ and $\dfrac{m-4}{m^2+5m+4}$

82) $\dfrac{n-3}{n^2-10n+25}$ and $\dfrac{n-1}{n^2-2n-15}$

83) $\dfrac{x+5}{x^3+x^2-6x}$ and $\dfrac{x-3}{x^4+7x^3+12x^2}$

84) $\dfrac{y-2}{y^5-16y^3}$ and $\dfrac{y-1}{y^3+5y^2+4y}$

CHALLENGE EXERCISES (85–86)

Find the LCM of the following.

85) $105x^2yz^3$, $75x^4y^5z$, and $195x^3y^6z^7$

86) $3y^2 - 75$ and $6y^2 - 60y + 150$

Find the missing numerator so that the rational expressions are equal.

87) $\dfrac{t}{x^2+4x+4} = \dfrac{?}{(x+2)^3}$

88) $\dfrac{4z}{y^2-6y+9} = \dfrac{?}{y^3-6y^2+9y}$

WRITING EXERCISES

89) Why is the LCM divisible by each of the terms for which it is the LCM?

90) In choosing the factors for the LCM, why do we select only factors which appear in the prime factorizations of the terms whose LCM we are finding? What would happen if we selected a factor not found in the prime factorizations?

91) In selecting the factors for the LCM, why is the largest exponent of each factor in the prime factorizations selected?

92) In selecting the factors for the LCM, why must each different factor which appears in any prime factorization be selected?

93) In creating a fraction which is equivalent to a fraction with a given denominator, why is it necessary to multiply both the numerator and the denominator by the *same* expression?

94) How is finding the least common denominator of rational numbers like finding the least common denominator of rational expressions? How is it different?

95) If x and y are prime numbers, what is the LCM of x and y? Why?

96) If x is a multiple of y, what is the LCM of x and y? Why?

Addition and Subtraction of Rational Expressions

OBJECTIVES

When you complete this section, you will be able to:

Ⓐ Add and subtract rational expressions with common denominators.

Ⓑ Add and subtract rational expressions with unlike denominators.

INTRODUCTION

In Section 1.2 we added and subtracted fractions with common denominators using the following procedures.

ADDITION AND SUBTRACTION OF FRACTIONS WITH COMMON DENOMINATORS

To add fractions with common denominators, add the numerators and put the sum over the common denominator. In symbols,

$$\frac{a}{c} + \frac{b}{c} = \frac{a+b}{c}$$

To subtract fractions with common denominators, subtract the numerators and put the difference over the common denominator. In symbols,

$$\frac{a}{c} - \frac{b}{c} = \frac{a-6}{c}$$

This principle is easily extended to three or more fractions: $\frac{a}{d} + \frac{b}{d} + \frac{c}{d} = \frac{a+b+c}{d}$, and so on.

 EXAMPLE 1

Find the following sums or differences. Leave answers reduced to lowest terms.

a) $\dfrac{5}{9} + \dfrac{7}{9} =$ Put the sum of the numerators over the common denominator.

$\dfrac{5+7}{9} =$ $5 + 7 = 12$.

$\dfrac{12}{9} =$ Factor the numerator and the denominator into prime factors.

$\dfrac{2 \cdot 2 \cdot 3}{3 \cdot 3} =$ Divide by the common factor 3.

$\dfrac{2 \cdot 2 \cdot \cancel{3}}{3 \cdot \cancel{3}} =$ Multiply the remaining factors.

$\dfrac{4}{3}$ Sum.

b) $\dfrac{11}{15} - \dfrac{2}{15} =$ Put the difference of the numerators over the common denominator.

$\dfrac{11-2}{15} =$ $11 - 2 = 9$.

$\dfrac{9}{15} =$ Factor the numerator and the denominator into prime factors.

$\dfrac{3 \cdot 3}{3 \cdot 5} =$ Divide by the common factor 3.

$\dfrac{\cancel{3} \cdot 3}{\cancel{3} \cdot 5} =$ Multiply the remaining factors.

$\dfrac{3}{5}$ Difference.

If the denominators contain variables, we use exactly the same procedure. Consider the following.

$\dfrac{5}{18} + \dfrac{7}{18} =$ $\dfrac{5y}{12x} + \dfrac{11y}{12x} =$ Put the sum of the numerators over the common denominator.

$$\frac{5+7}{18} = \qquad \frac{5y+11y}{12x} = \qquad \text{Add the numerators.}$$

$$\frac{12}{18} = \qquad \frac{16y}{12x} = \qquad \text{Reduce to lowest terms.}$$

$$\frac{2}{3} \qquad \frac{4y}{3x} \qquad \text{Sum.}$$

EXAMPLE 2

Find the following sums or differences.

a) $\dfrac{3}{x^2} + \dfrac{y}{x^2} =$ Put the sum of the numerators over the common denominator.

$$\frac{3+y}{x^2} \qquad \text{Sum.}$$

b) $\dfrac{3x}{10y} - \dfrac{7x}{10y} =$ Put the difference of the numerators over the common denominator.

$$\frac{3x-7x}{10y} = \qquad 3x - 7x = -4x.$$

$$\frac{-4x}{10y} = \qquad \text{Reduce to lowest terms.}$$

$$-\frac{2x}{5y} \qquad \text{Difference.}$$

Note: If the common denominators are binomials or trinomials, it is often necessary to factor the numerator of the sum or difference and then reduce.

c) $\dfrac{c}{c+4} - \dfrac{2}{c+4} =$ Put the difference of the numerators over the common denominator.

$$\frac{c-2}{c+4} \qquad \text{Difference.}$$

d) $\dfrac{2y}{y-3} - \dfrac{6}{y-3} =$ Put the difference of the numerators over the common denominator.

$$\frac{2y-6}{y-3} = \qquad \text{Factor the numerator.}$$

$$\frac{2(y-3)}{y-3} = \qquad \text{Divide by the common factor } y-3.$$

$$2 \qquad \text{Difference.}$$

e) $\dfrac{x^2}{x^2+4x-21} + \dfrac{2x-15}{x^2+4x-21} =$ Put the sum of the numerators over the common denominator.

$$\frac{(x^2)+(2x-15)}{x^2+4x-21} = \qquad \text{Remove the parentheses.}$$

$$\frac{x^2+2x-15}{x^2+4x-21} = \qquad \text{Factor the numerator and the denominator.}$$

$$\frac{(x+5)(x-3)}{(x+7)(x-3)} = \qquad \text{Divide by the common factor of } x-3.$$

$$\frac{x+5}{x+7} \qquad \text{Sum.}$$

f) $\dfrac{x^2 + 5x}{x^2 + 7x + 12} - \dfrac{8x + 28}{x^2 + 7x + 12} =$ Put the difference of the numerators over the common denominator.

$\dfrac{(x^2 + 5x) - (8x + 28)}{x^2 + 7x + 12} =$ Remove the parentheses.

$\dfrac{x^2 + 5x - 8x - 28}{x^2 + 7x + 12} =$ Add like terms.

$\dfrac{x^2 - 3x - 28}{x^2 + 7x + 12} =$ Factor the numerator and the denominator.

$\dfrac{(x - 7)(x + 4)}{(x + 4)(x + 3)} =$ Divide by the common factor of $x + 4$.

$\dfrac{x - 7}{x + 3}$ Difference.

Be careful: In exercises like Example 1f above, we are finding the difference of rational expressions. A common error is to forget to change all the signs of the numerator of the second fraction. To avoid making this type of error, you should write the step with the difference of the numerators in parentheses.

ADDITIONAL PRACTICE

Find the following sums or differences of rational expressions. Leave answers reduced to lowest terms.

a) $\dfrac{13}{16} + \dfrac{5}{16}$ **b)** $\dfrac{9}{14} - \dfrac{3}{14}$

c) $\dfrac{v}{u^3} + \dfrac{w}{u^3}$ **d)** $\dfrac{5a}{14b} - \dfrac{11a}{14b}$

e) $\dfrac{y}{3y + 4} + \dfrac{3y}{3y + 4}$

f) $\dfrac{6x}{2x + 5} + \dfrac{15}{2x + 5}$

g) $\dfrac{2x^2}{x^2 - 9} + \dfrac{7x + 3}{x^2 - 9}$

h) $\dfrac{x^2}{x^2 - x - 6} - \dfrac{2x + 8}{x^2 - x - 6}$

PRACTICE EXERCISES

Find the following sums or differences of rational expressions. Leave answers reduced to lowest terms.

1) $\dfrac{19}{24} + \dfrac{7}{24}$ **2)** $\dfrac{11}{18} - \dfrac{7}{18}$ **3)** $\dfrac{m}{n} - \dfrac{6}{n}$

4) $\dfrac{3x}{16z} + \dfrac{35x}{16z}$ **5)** $\dfrac{3a}{a + 2} - \dfrac{a}{a + 2}$ **6)** $\dfrac{x}{x - 4} - \dfrac{4}{x - 4}$

7) $\dfrac{x^2}{x^2 + 3x - 18} + \dfrac{5x - 6}{x^2 + 3x - 18}$ **8)** $\dfrac{x^2 - 3x}{x^2 - 5x + 6} - \dfrac{4x - 10}{x^2 - 5x + 6}$

If more practice is needed, do the Additional Practice Exercises in the margin.

Earlier in this chapter we discussed equivalent forms of negative fractions and we indicated that $-\dfrac{a}{b} = \dfrac{-a}{b}$. It is also true that $\dfrac{a}{-b} = \dfrac{-a}{b}$ since both are equal to $-\dfrac{a}{b}$. For example, $\dfrac{4}{-7} = \dfrac{-4}{7}$ and $\dfrac{5}{-(a - b)} = \dfrac{-5}{a - b}$. In general a fraction has three signs, any two of which may be changed without changing the value of the fraction. For example, $+\dfrac{+2}{3} = +\dfrac{-2}{-3} = -\dfrac{-2}{3} = -\dfrac{2}{-3}$. In general, $-\dfrac{a}{b} = \dfrac{-a}{b} = \dfrac{a}{-b}$.

EXAMPLE 3

Find the following sums or differences. Leave the answers reduced to lowest terms.

a) $\dfrac{2}{5} + \dfrac{3}{-5} =$ Rewrite $\dfrac{3}{-5}$ as $\dfrac{-3}{5}$.

$\dfrac{2}{5} + \dfrac{-3}{5} =$ Find the sum.

$\dfrac{2 + (-3)}{5} =$ $2 + (-3) = -1$.

$\dfrac{-1}{5}$ or $-\dfrac{1}{5}$ Sum.

b)

$$\dfrac{5}{x-2} + \dfrac{3}{2-x} =$$ Rewrite $2-x$ as $-(x-2)$.

$$\dfrac{5}{x-2} + \dfrac{3}{-(x-2)} =$$ Rewrite $\dfrac{3}{-(x-2)}$ as $\dfrac{-3}{x-2}$.

$$\dfrac{5}{x-2} + \dfrac{-3}{x-2} =$$ Find the sum.

$$\dfrac{5+(-3)}{x-2} =$$ $5+(-3)=2$.

$$\dfrac{2}{x-2}$$ Sum.

c)

$$\dfrac{a+b}{a-b} - \dfrac{2a-3b}{b-a} =$$ Rewrite $b-a$ as $-(a-b)$.

$$\dfrac{a+b}{a-b} - \dfrac{2a-3b}{-(a-b)} =$$ Rewrite $\dfrac{2a-3b}{-(a-b)}$ as $\dfrac{-(2a-3b)}{a-b}$.

$$\dfrac{a+b}{a-b} - \dfrac{-(2a-3b)}{a-b} =$$ Remove the parentheses.

$$\dfrac{a+b}{a-b} - \dfrac{-2a+3b}{a-b} =$$ Find the difference.

$$\dfrac{(a+b)-(-2a+3b)}{a-b} =$$ Remove the parentheses.

$$\dfrac{a+b+2a-3b}{a-b} =$$ Add like terms.

$$\dfrac{3a-2b}{a-b}$$ Answer.

ADDITIONAL PRACTICE

Find the following sums or differences. Leave answers reduced to lowest terms.

i) $\dfrac{11}{13} + \dfrac{7}{-13}$

j) $\dfrac{-5}{a-4} + \dfrac{2}{4-a}$

k) $\dfrac{2b-3c}{b-c} - \dfrac{b-2c}{c-b}$

• **Adding fractions with unlike denominators**

PRACTICE EXERCISES

Find the following sums or differences. Leave answers reduced to lowest terms.

9) $\dfrac{5}{7} + \dfrac{2}{-7}$

10) $\dfrac{8}{y-3} + \dfrac{5}{3-y}$

11) $\dfrac{3x+2y}{x-y} - \dfrac{x-5y}{y-x}$

If more practice is needed, do the Additional Practice Exercises in the margin.

Fractions with unlike denominators cannot be added. For example, if Joe eats $\frac{3}{8}$ of a pizza and Sally eats $\frac{2}{5}$ of the pizza, what fractional part of the pizza has been eaten? We cannot add $\frac{3}{8}$ and $\frac{2}{5}$ without first rewriting each as an equivalent fraction with the LCD as its denominator. The procedure first given in Section 1.2 is repeated here.

ADDING FRACTIONS WITH UNLIKE DENOMINATORS

1) Factor all denominators.

2) Find the least common denominator.

3) Change each fraction into an equivalent fraction with the least common denominator as denominator.

4) Add the resulting fractions with common denominators.

Since subtraction of fractions may be expressed in terms of addition, the same procedure applies to subtraction as well.

EXAMPLE 4

Find the following sums or differences. Leave answers reduced to lowest terms.

a) $\dfrac{5}{12} + \dfrac{7}{30} =$ The LCD is 60. Therefore, multiply $\dfrac{5}{12}$ by $\dfrac{5}{5}$ and multiply $\dfrac{7}{30}$ by $\dfrac{2}{2}$.

$\dfrac{5}{5} \cdot \dfrac{5}{12} + \dfrac{2}{2} \cdot \dfrac{7}{30} =$ Multiply the fractions.

$\dfrac{25}{60} + \dfrac{14}{60} =$ Add the fractions.

$\dfrac{39}{60} =$ Reduce to lowest terms.

$\dfrac{13}{20}$ Sum.

b) $\dfrac{3}{10} + \dfrac{7}{15} - \dfrac{9}{25} =$ The LCD is 150. Therefore, multiply $\dfrac{3}{10}$ by $\dfrac{15}{15}$, multiply $\dfrac{7}{15}$ by $\dfrac{10}{10}$, and multiply $\dfrac{9}{25}$ by $\dfrac{6}{6}$.

$\dfrac{15}{15} \cdot \dfrac{3}{10} + \dfrac{10}{10} \cdot \dfrac{7}{15} - \dfrac{6}{6} \cdot \dfrac{9}{25} =$ Multiply the fractions.

$\dfrac{45}{150} + \dfrac{70}{150} - \dfrac{54}{150} =$ Add and subtract the fractions.

$\dfrac{61}{150}$ Answer.

If the denominator contains variables, the procedure is exactly the same. Consider the following.

$\dfrac{5}{6} + \dfrac{3}{8} =$ $\dfrac{5}{6a} + \dfrac{3}{8a} =$ Rewrite each fraction with the least common denominator as its denominator.

$\dfrac{4}{4} \cdot \dfrac{5}{6} + \dfrac{3}{3} \cdot \dfrac{3}{8} =$ $\dfrac{4}{4} \cdot \dfrac{5}{6a} + \dfrac{3}{3} \cdot \dfrac{3}{8a} =$ Multiply.

$\dfrac{20}{24} + \dfrac{9}{24} =$ $\dfrac{20}{24a} + \dfrac{9}{24a} =$ Add the fractions.

$\dfrac{29}{24}$ $\dfrac{29}{24a}$ Sum.

EXAMPLE 5

Find the following sums and/or differences.

a) $\dfrac{3}{4a} - \dfrac{5}{3a} =$

The LCD is $12a$. Therefore, multiply $\frac{3}{4a}$ by $\frac{3}{3}$ and multiply $\frac{5}{3a}$ by $\frac{4}{4}$.

$\dfrac{3}{3} \cdot \dfrac{3}{4a} - \dfrac{4}{4} \cdot \dfrac{5}{3a} =$

Multiply the rational expressions.

$\dfrac{9}{12a} - \dfrac{20}{12a} =$

Subtract the rational expressions.

$-\dfrac{11}{12a}$

Difference.

b) $\dfrac{2x + 4}{4x} + \dfrac{3x - 2}{6x} =$

The LCD is $12x$. Therefore, multiply $\frac{2x + 4}{4x}$ by $\frac{3}{3}$ and $\frac{3x - 2}{6x}$ by $\frac{2}{2}$.

$\dfrac{3}{3} \cdot \dfrac{2x + 4}{4x} + \dfrac{2}{2} \cdot \dfrac{3x - 2}{6x} =$

Multiply the rational expressions.

$\dfrac{6x + 12}{12x} + \dfrac{6x - 4}{12x} =$

Add the rational expressions.

$\dfrac{(6x + 12) + (6x - 4)}{12x} =$

Remove the parentheses.

$\dfrac{6x + 12 + 6x - 4}{12x} =$

Add like terms.

$\dfrac{12x + 8}{12x} =$

Factor.

$\dfrac{4(3x + 2)}{4 \cdot 3x} =$

Divide by the common factor 4.

$\dfrac{3x + 2}{3x}$

Sum.

c) $\dfrac{a - 3}{3a^2} - \dfrac{2a - 3}{2a} =$

The LCD is $6a^2$. Therefore, multiply $\frac{a - 3}{3a^2}$ by $\frac{2}{2}$ and multiply $\frac{2a - 3}{2a}$ by $\frac{3a}{3a}$.

$\dfrac{2}{2} \cdot \dfrac{a - 3}{3a^2} - \dfrac{3a}{3a} \cdot \dfrac{2a - 3}{2a} =$

Multiply the rational expressions.

$\dfrac{2a - 6}{6a^2} - \dfrac{6a^2 - 9a}{6a^2} =$

Subtract the rational expressions.

$\dfrac{(2a - 6) - (6a^2 - 9a)}{6a^2} =$

Remove the parentheses.

$\dfrac{2a - 6 - 6a^2 + 9a}{6a^2} =$

Add like terms.

$\dfrac{-6a^2 + 11a - 6}{6a^2}$

Difference.

Find the following sums or differences. Leave answers reduced to lowest terms.

l) $\dfrac{5}{6} - \dfrac{4}{9}$

m) $\dfrac{2}{3} + \dfrac{3}{5} - \dfrac{7}{12}$

n) $\dfrac{2}{5d} + \dfrac{3}{10d}$

o) $\dfrac{2x-3}{5x} + \dfrac{3x+5}{3x}$

p) $\dfrac{2y+3}{4y^2} - \dfrac{y-6}{2y}$

q) $\dfrac{a+4}{3a^2b} - \dfrac{2a-3}{6ab^2}$

PRACTICE EXERCISES

Find the following sums or differences. Leave answers reduced to lowest terms.

12) $\dfrac{9}{14} - \dfrac{1}{4}$

13) $\dfrac{2}{3} + \dfrac{5}{8} - \dfrac{7}{9}$

14) $\dfrac{7}{6a} - \dfrac{5}{8a}$

15) $\dfrac{b-5}{5b} + \dfrac{b+6}{6b}$

16) $\dfrac{x-2}{2x^2} - \dfrac{x+3}{3x}$

17) $\dfrac{3x+7}{8xy^2} - \dfrac{3x-4}{6x^2y}$

If the denominators are binomials or trinomials, it is often necessary to factor the denominators to find the LCD. If possible, factor the numerator also to see if any of the rational expressions will reduce before performing any operations. Sometimes we have to factor the numerator of the sum or difference, and then reduce the rational expression. For this reason, we usually leave the LCD in factored form. However, the procedure is exactly the same.

EXAMPLE 6

Find the following sums or differences. Leave answers reduced to lowest terms.

a) $\dfrac{5}{x-2} + \dfrac{3}{x+4} =$

The LCD is $(x-2)(x+4)$. Therefore, multiply $\frac{5}{x-2}$ by $\frac{x+4}{x+4}$ and multiply $\frac{3}{x+4}$ by $\frac{x-2}{x-2}$.

$\dfrac{x+4}{x+4} \cdot \dfrac{5}{x-2} + \dfrac{x-2}{x-2} \cdot \dfrac{3}{x+4} =$

Multiply the rational expressions.

$\dfrac{5x+20}{(x+4)(x-2)} + \dfrac{3x-6}{(x+4)(x-2)} =$

Add the rational expressions.

$\dfrac{(5x+20)+(3x-6)}{(x+4)(x-2)} =$

Remove the parentheses.

$\dfrac{5x+20+3x-6}{(x+4)(x-2)} =$

Add like terms.

$\dfrac{8x+14}{(x+4)(x-2)}$

Sum.

Note: In the above example, we could factor the numerator and get $\frac{2(4x+7)}{(x+4)(x-2)}$, but the expression will not reduce. Consequently, we leave the answer as is.

b) $\dfrac{x+2}{4x-8} - \dfrac{x+4}{6x-12} =$

Factor the denominators.

$\dfrac{x+2}{4(x-2)} - \dfrac{x+4}{6(x-2)} =$

The LCD is $12(x-2)$. Therefore, multiply $\frac{x+2}{4(x-2)}$ by $\frac{3}{3}$ and $\frac{x+4}{6(x-2)}$ by $\frac{2}{2}$.

$\dfrac{3}{3} \cdot \dfrac{x+2}{4(x-2)} - \dfrac{2}{2} \cdot \dfrac{x+4}{6(x-2)} =$

Multiply the rational expressions.

12) $\dfrac{11}{28}$ **13)** $\dfrac{37}{72}$

14) $\dfrac{13}{24a}$ **15)** $\dfrac{11}{30}$

16) $\dfrac{-2x^2 - 3x - 6}{6x^2}$

17) $\dfrac{9x^2 + 21x - 12xy + 16y}{24x^2y^2}$

Additional Practice l–q

l) $\dfrac{7}{18}$ **m)** $\dfrac{41}{60}$

n) $\dfrac{7}{10d}$ **o)** $\dfrac{21x + 16}{15x}$

p) $\dfrac{-2y^2 + 14y + 3}{4y^2}$

q) $\dfrac{2ab + 8b - 2a^2 + 3a}{6a^2b^2}$

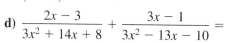

$\dfrac{3x + 6}{12(x - 2)} - \dfrac{2x + 8}{12(x - 2)} =$ Subtract the rational expressions.

$\dfrac{(3x + 6) - (2x + 8)}{12(x - 2)} =$ Remove the parentheses.

$\dfrac{3x + 6 - 2x - 8}{12(x - 2)} =$ Add like terms.

$\dfrac{x - 2}{12(x - 2)} =$ Divide by the common factor $x - 2$.

$\dfrac{1}{12}$ Difference.

c) $\dfrac{4c - 1}{c^2 - 10c + 25} - \dfrac{4}{c - 5} =$ Factor the denominators.

$\dfrac{4c - 1}{(c - 5)^2} - \dfrac{4}{c - 5} =$ The LCD is $(c - 5)^2$. Therefore, multiply $\frac{4}{c - 5}$ by $\frac{c - 5}{c - 5}$.

$\dfrac{4c - 1}{(c - 5)^2} - \dfrac{c - 5}{c - 5} \cdot \dfrac{4}{c - 5} =$ Multiply the rational expressions.

$\dfrac{4c - 1}{(c - 5)^2} - \dfrac{4c - 20}{(c - 5)^2} =$ Subtract the rational expressions.

$\dfrac{(4c - 1) - (4c - 20)}{(c - 5)^2} =$ Remove the parentheses.

$\dfrac{4c - 1 - 4c + 20}{(c - 5)^2} =$ Add like terms.

$\dfrac{19}{(c - 5)^2}$ Difference.

d) $\dfrac{2x - 3}{3x^2 + 14x + 8} + \dfrac{3x - 1}{3x^2 - 13x - 10} =$ Factor the denominators.

$\dfrac{2x - 3}{(3x + 2)(x + 4)} + \dfrac{3x - 1}{(3x + 2)(x - 5)} =$ The LCD is $(3x + 2)(x + 4)(x - 5)$. Multiply $\frac{2x - 3}{(3x + 2)(x + 4)}$ by $\frac{x - 5}{x - 5}$ and $\frac{3x - 1}{(3x + 2)(x - 5)}$ by $\frac{x + 4}{x + 4}$.

$\dfrac{x - 5}{x - 5} \cdot \dfrac{2x - 3}{(3x + 2)(x + 4)} + \dfrac{x + 4}{x + 4} \cdot \dfrac{3x - 1}{(3x + 2)(x - 5)} =$ Multiply.

$\dfrac{2x^2 - 13x + 15}{(3x + 2)(x + 4)(x - 5)} + \dfrac{3x^2 + 11x - 4}{(3x + 2)(x + 4)(x - 5)} =$ Add.

$\dfrac{(2x^2 - 13x + 15) + (3x^2 + 11x - 4)}{(3x + 2)(x + 4)(x - 5)} =$ Remove the parentheses.

$\dfrac{2x^2 - 13x + 15 + 3x^2 + 11x - 4}{(3x + 2)(x + 4)(x - 5)} =$ Add like terms.

$\dfrac{5x^2 - 2x + 11}{(3x + 2)(x + 4)(x - 5)}$ Sum.

Find the following sums or differences. Leave answers reduced to lowest terms.

r) $\dfrac{7}{2x+1} + \dfrac{3}{3x-2}$

s) $\dfrac{x-2}{6x-30} - \dfrac{x-1}{8x-40}$

t) $\dfrac{2y}{y^2-4y+4} - \dfrac{4y+3}{y-2}$

u) $\dfrac{x-4}{x^2-x-20} + \dfrac{2x-1}{x^2+7x+12}$

PRACTICE EXERCISES

Find the following sums or differences. Leave answers reduced to lowest terms.

18) $\dfrac{2x}{x+3} + \dfrac{4}{x-5}$

19) $\dfrac{x+6}{4x+16} - \dfrac{x+7}{6x+24}$

20) $\dfrac{2x+3}{x^2+6x+9} - \dfrac{x+4}{x+3}$

21) $\dfrac{x+2}{2x^2+5x-3} + \dfrac{2x-3}{2x^2-5x+2}$

If more practice is needed, do the Additional Practice Exercises in the margin.

EXERCISE SET 5.5

Find the following sums or differences. Leave your answers reduced to lowest terms.

1) $\dfrac{7}{18} + \dfrac{5}{18}$

2) $\dfrac{4}{25} + \dfrac{9}{25}$

3) $\dfrac{10}{17} - \dfrac{-5}{17}$

4) $\dfrac{9}{22} - \dfrac{-9}{22}$

5) $\dfrac{9}{10} - \dfrac{3}{10}$

6) $\dfrac{17}{25} - \dfrac{12}{25}$

7) $\dfrac{8}{15} + \dfrac{3}{15} - \dfrac{6}{15}$

8) $\dfrac{6}{21} - \dfrac{8}{21} + \dfrac{16}{21}$

Find the sums or differences of the following rational expressions. Leave your answers reduced to lowest terms.

9) $\dfrac{u}{v} - \dfrac{5}{v}$

10) $\dfrac{8}{w} - \dfrac{z}{w}$

11) $\dfrac{11r}{21u} + \dfrac{4m}{21u}$

12) $\dfrac{13p}{30n} + \dfrac{7q}{30n}$

13) $\dfrac{3a}{12x^2} + \dfrac{5a}{12x^2}$

14) $\dfrac{7r}{18a^2} + \dfrac{9r}{18a^2}$

15) $\dfrac{6}{n+5} + \dfrac{3}{n+5}$

16) $\dfrac{4}{2-m} + \dfrac{7}{2-m}$

17) $\dfrac{9}{t-8} - \dfrac{7}{t-8}$

18) $\dfrac{8}{4-z} - \dfrac{12}{4-z}$

19) $\dfrac{5c}{c+2} - \dfrac{2c}{c+2}$

20) $\dfrac{5z}{6-b} - \dfrac{7z}{6-b}$

21) $\dfrac{a+2}{a-5} + \dfrac{a-3}{a-5}$

22) $\dfrac{x-6}{x+4} + \dfrac{x-3}{x+4}$

23) $\dfrac{4r-5}{3r-2} + \dfrac{4r+2}{3r-2}$

24) $\dfrac{6a+1}{4a+3} + \dfrac{2a-5}{4a+3}$

25) $\dfrac{a-6}{a-5} - \dfrac{a+7}{a-5}$

26) $\dfrac{y-6}{y+3} - \dfrac{y+4}{y+3}$

27) $\dfrac{3m+2}{2m+3} - \dfrac{2m-3}{2m+3}$

28) $\dfrac{2n+4}{3n-8} - \dfrac{4n-2}{3n-8}$

29) $\dfrac{x}{x^2-16} - \dfrac{4}{x^2-16}$

30) $\dfrac{x}{x^2-9} + \dfrac{3}{x^2-9}$

31) $\dfrac{a}{a^2+2a-8} + \dfrac{4}{a^2+2a-8}$

32) $\dfrac{m}{m^2-5m+6} - \dfrac{2}{m^2-5m+6}$

33) $\dfrac{x^2}{x^2+x-6} - \dfrac{4}{x^2+x-6}$

34) $\dfrac{x^2}{x^2-x-6} - \dfrac{9}{x^2-x-6}$

35) $\dfrac{x^2}{x^2 + x - 12} + \dfrac{-5x + 6}{x^2 + x - 12}$

36) $\dfrac{y^2}{y^2 - 5y + 4} + \dfrac{11y + 30}{y^2 - 5y + 4}$

37) $\dfrac{t^2 - 4t}{t^2 + 10t + 21} - \dfrac{-6t + 3}{t^2 + 10t + 21}$

38) $\dfrac{v^2 + 6}{v^2 - 3v - 10} - \dfrac{2v + 14}{v^2 - 3v - 10}$

39) $\dfrac{x^2 - 2x}{x^2 - 12x + 36} + \dfrac{-5x + 6}{x^2 - 12x + 36}$

40) $\dfrac{q^2 - 3}{q^2 - 49} - \dfrac{5q + 11}{q^2 - 49}$

Find the sums or differences. Leave the answers reduced to lowest terms.

41) $\dfrac{7}{11} + \dfrac{2}{-11}$

42) $\dfrac{8}{17} + \dfrac{12}{-17}$

43) $\dfrac{3}{t - 4} + \dfrac{11}{4 - t}$

44) $\dfrac{-5}{7 - t} + \dfrac{8}{t - 7}$

45) $\dfrac{2m - n}{m - n} - \dfrac{m - 3}{n - m}$

46) $\dfrac{p - 5q}{p - q} - \dfrac{4p - 6q}{q - p}$

Find the following sums or differences. Leave the answers reduced to lowest terms.

47) $\dfrac{4}{7} + \dfrac{1}{3}$

48) $\dfrac{8}{9} - \dfrac{3}{4}$

49) $\dfrac{3}{10} + \dfrac{2}{5}$

50) $\dfrac{5}{8} + \dfrac{3}{4}$

51) $\dfrac{8}{9} - \dfrac{11}{12}$

52) $\dfrac{14}{15} - \dfrac{7}{10}$

53) $\dfrac{7}{12} + \dfrac{1}{9} - \dfrac{5}{6}$

54) $\dfrac{7}{10} + \dfrac{4}{15} - \dfrac{9}{30}$

Find the following sums or differences. Leave the answers as rational expressions reduced to lowest terms.

55) $\dfrac{2}{a} + \dfrac{a}{b}$

56) $\dfrac{6}{r} + \dfrac{7}{t}$

57) $\dfrac{3z}{10x} + \dfrac{5z}{4x}$

58) $\dfrac{10w}{21y} + \dfrac{7w}{18y}$

59) $\dfrac{x + 4}{5x} + \dfrac{x - 3}{2x}$

60) $\dfrac{a - 4}{3a} + \dfrac{a + 2}{4a}$

61) $\dfrac{m + 8}{2m} - \dfrac{m - 7}{3m}$

62) $\dfrac{n - 9}{7n} - \dfrac{n - 2}{4n}$

63) $\dfrac{b + 3}{2b^2} + \dfrac{b - 2}{7b}$

64) $\dfrac{r + 5}{5r} + \dfrac{r - 6}{3r^2}$

65) $\dfrac{3p + 1}{10p^3q^2} + \dfrac{9p - 2}{6p^2q}$

66) $\dfrac{5t - 8}{16s^5t^2} + \dfrac{2t - 7}{12s^2t^3}$

67) $\dfrac{4}{k} - \dfrac{6}{k + 2}$

68) $\dfrac{5}{j - 5} - \dfrac{9}{j}$

69) $\dfrac{4}{w - 3} + \dfrac{5}{w + 7}$

70) $\dfrac{6}{c + 4} + \dfrac{11}{c - 1}$

71) $\dfrac{x - 4}{3x + 9} - \dfrac{x + 5}{6x + 18}$

72) $\dfrac{y - 3}{2y - 8} - \dfrac{y - 7}{6y - 24}$

73) $\dfrac{2t}{t^2 - 8t + 16} + \dfrac{6}{t - 4}$

74) $\dfrac{4}{s^2 + 14s + 49} + \dfrac{5s}{s + 7}$

75) $\dfrac{u^2 - 3u}{u^2 - 6u + 9} + \dfrac{8}{u - 3}$

76) $\dfrac{v^2 + 4v}{v^2 + 8v + 16} + \dfrac{3}{v + 4}$

77) $\dfrac{2h^2 + 10h}{h^2 - 25} - \dfrac{h}{h + 5}$

78) $\dfrac{21k - k^2}{k^2 - 49} - \dfrac{2k}{k - 7}$

79) $\dfrac{4}{x^2 - x - 12} + \dfrac{2}{x^2 + 5x + 6}$

80) $\dfrac{3}{x^2 - 3x - 10} + \dfrac{5}{x^2 - 2x - 8}$

81) $\dfrac{5}{x^2 + 6x + 8} - \dfrac{3}{x^2 + 3x - 4}$

82) $\dfrac{7}{x^2 + 2x - 15} - \dfrac{4}{x^2 + 3x - 18}$

83) $\dfrac{x + 1}{x^2 - x - 6} + \dfrac{x - 5}{x^2 + 6x + 8}$

84) $\dfrac{y + 3}{y^2 - 3y + 2} + \dfrac{2y + 4}{y^2 + y - 2}$

85) $\dfrac{2v + 5}{v^2 - 16} - \dfrac{v - 9}{v^2 - v - 12}$

86) $\dfrac{4t}{t^2 - 9} + \dfrac{3t - 2}{t^2 - 8t + 15}$

87) $\dfrac{z - 3}{z^2 + 6z + 9} + \dfrac{z + 1}{z^2 + z - 6}$

88) $\dfrac{7w - 2}{w^2 + 2w - 24} + \dfrac{9w - 8}{3w^2 - 16w + 16}$

CHALLENGE EXERCISES (89–96)

89) $\dfrac{x^2 + 2x}{x^3 + 2x^2 - 15x} - \dfrac{15}{x^3 + 2x^2 - 15x}$

90) $\dfrac{y^2 + 9y}{y^3 - 36y} + \dfrac{18}{y^3 - 36y}$

91) $\dfrac{x + 4}{x^3 - 36x} + \dfrac{1}{x^2 - 3x - 18}$

92) $\dfrac{y + 5}{4y^2 + 25y + 25} - \dfrac{y - 1}{3y^2 - 48y + 192}$

93) $\dfrac{x}{x^2 - 2x - 8} + \dfrac{3x}{x^2 + 5x + 6} - \dfrac{2x}{x^2 - x - 12}$

94) $\dfrac{3x}{x^2 + x - 6} - \dfrac{4x}{x^2 + 5x + 6} + \dfrac{x}{x^2 - 4}$

95) $\dfrac{a - 2}{a^2 + 2a - 8} + \dfrac{a + 3}{a^2 + 3a - 4} - \dfrac{a - 1}{a^2 + 3a + 2}$

96) $\dfrac{m + 4}{m^2 - 9} - \dfrac{m - 2}{m^2 + 7m + 12} + \dfrac{m + 1}{m^2 + m - 12}$

WRITING EXERCISES

97) When adding or subtracting rational numbers or rational expressions which have the same denominator, why do we add or subtract the numerators but we do not add or subtract the denominators?

98) Why is it necessary to rewrite each rational number or rational expression with the LCD as its denominator before adding or subtracting?

99) When adding or subtracting rational expressions we need to write each with a common denominator, but when multiplying or dividing we do not need a common denominator. Explain.

100) What is wrong with the following? $\frac{4x-1}{x+6} - \frac{2x-3}{x+6} = \frac{4x-1-2x-3}{x+6} = \frac{2x-4}{x+6}$. What is the correct answer?

SECTION 5.6

Complex Fractions

OBJECTIVES

When you complete this section, you will be able to:

Ⓐ Simplify complex fractions with numerators and/or denominators that are rational numbers.

Ⓑ Simplify complex fractions with numerators and/or denominators that are rational expressions.

INTRODUCTION

A complex fraction is any fraction whose numerator and/or denominator is also a fraction. Examples of complex fractions are

$$\dfrac{\frac{1}{2}}{3}, \quad \dfrac{\frac{3}{4}}{\frac{5}{8}}, \quad \dfrac{\frac{2}{3} + \frac{3}{4}}{\frac{4}{5} + \frac{2}{7}}, \quad \dfrac{\frac{1}{x}}{\frac{x}{y}}, \quad \text{and} \quad \dfrac{\frac{3}{x} + x}{\frac{x^2 - 6}{x}}.$$

One interpretation of a fraction is that it represents division. For example, $\frac{a}{b} = a \div b$. To simplify a complex fraction means to write it as an ordinary fraction. We will first use the division interpretation of a fraction and then show another technique that is sometimes easier. In simplifying complex fractions, using the division interpretation of a fraction, we use the following procedure.

1) If necessary, rewrite the numerator and the denominator as single fractions by performing whatever operations are indicated in either.

2) Rewrite the complex fraction as the division of the numerator by the denominator.

3) Perform the division by inverting the fraction on the right and multiplying.

4) If necessary, reduce to lowest terms.

EXAMPLE 1

Simplify the following complex fractions using division.

a) $\dfrac{\frac{2}{3}}{\frac{5}{6}} =$ Rewrite as division using $\frac{a}{b} = a \div b$.

$\dfrac{2}{3} \div \dfrac{5}{6} =$ Invert $\frac{5}{6}$ and change division to multiplication.

$\dfrac{2}{3} \cdot \dfrac{6}{5} =$ Multiply.

$\dfrac{12}{15} =$ Reduce to lowest terms.

$\dfrac{4}{5}$ Quotient.

Note: If the numerator and/or the denominator is the sum or difference of two or more fractions, we must write the numerator and the denominator as a single fraction before we can simplify.

b) $\dfrac{\frac{2}{3} + \frac{3}{4}}{\frac{3}{8} + 1} =$ The LCD in the numerator is 12 and the LCD of the denominator is 8. Change the fractions in the numerator into equivalent fractions whose denominators are 12, and change the fractions in the denominator into equivalent fractions whose denominators are 8.

$\dfrac{\frac{4}{4} \cdot \frac{2}{3} + \frac{3}{3} \cdot \frac{3}{4}}{\frac{3}{8} + \frac{8}{8} \cdot \frac{1}{1}} =$ Multiply the fractions.

$\dfrac{\frac{8}{12} + \frac{9}{12}}{\frac{3}{8} + \frac{8}{8}} =$ Add the fractions in the numerator and the denominator.

$\dfrac{\frac{17}{12}}{\frac{11}{8}} =$ Rewrite as division.

$$\frac{17}{12} \div \frac{11}{8} =$$ Rewrite as multiplication.

$$\frac{17}{12} \cdot \frac{8}{11} =$$ Write numerators and denominators in terms of prime factors.

$$\frac{17}{2 \cdot 2 \cdot 3} \cdot \frac{2 \cdot 2 \cdot 2}{11} =$$ Divide by the common factors of 2.

$$\frac{17}{\cancel{2} \cdot \cancel{2} \cdot 3} \cdot \frac{\cancel{2} \cdot \cancel{2} \cdot 2}{11} =$$ Multiply the remaining factors.

$$\frac{34}{33}$$ Quotient.

Example 1b is quite involved and has a lot of places where errors may occur. Consequently, a second method in which we multiply the numerator and the denominator of the complex fraction by the LCD of all the fractions that appear in the numerator and/or the denominator is often preferred. The reason we multiply by the LCD of all the fractions is because each denominator divides evenly into the LCD. Therefore, we eliminate all fractions from the numerator and the denominator of the complex fraction. This results in an ordinary fraction. The procedure is summarized below.

SIMPLIFYING COMPLEX FRACTIONS USING THE LCD

1) Find the LCD of all fractions that occur in the numerator and/or the denominator.
2) Multiply both the numerator and the denominator by the LCD.
3) Simplify the results.

We demonstrate this method by reworking the examples from Example 1 plus one other case.

EXAMPLE 2

Simplify the following complex fractions using multiplication.

a) $$\dfrac{\frac{2}{3}}{\frac{5}{6}} =$$ The LCD for $\frac{2}{3}$ and $\frac{5}{6}$ is 6. Therefore, multiply the numerator and the denominator by 6.

$$\dfrac{6\left(\frac{2}{3}\right)}{6\left(\frac{5}{6}\right)} =$$ Perform the multiplications.

$$\frac{4}{5}$$ Simplified form.

b) $\dfrac{\dfrac{2}{3} + \dfrac{3}{4}}{\dfrac{3}{8} + 1} =$

The LCD is 24. Therefore, multiply the numerator and the denominator by 24.

$$\dfrac{24\left(\dfrac{2}{3} + \dfrac{3}{4}\right)}{24\left(\dfrac{3}{8} + 1\right)} =$$

Apply the distributive property.

$$\dfrac{24\left(\dfrac{2}{3}\right) + 24\left(\dfrac{3}{4}\right)}{24\left(\dfrac{3}{8}\right) + 24 \cdot 1} =$$

Perform the multiplications.

$$\dfrac{16 + 18}{9 + 24} =$$

Perform the additions.

$$\dfrac{34}{33}$$

Simplified form.

c) $\dfrac{6}{\dfrac{8}{3}} =$

Since $6 = \frac{6}{1}$, the LCD of 6 and $\frac{8}{3}$ is 3. Therefore, multiply the numerator and the denominator by 3.

$$\dfrac{3\left(\dfrac{6}{1}\right)}{3\left(\dfrac{8}{3}\right)} =$$

Perform the multiplications.

$$\dfrac{18}{8} =$$

Reduce to lowest terms.

$$\dfrac{9}{4}$$

Simplified form.

ADDITIONAL PRACTICE

Simplify the following complex fractions using both methods.

a) $\dfrac{\dfrac{4}{15}}{\dfrac{2}{5}}$

b) $\dfrac{\dfrac{8}{4}}{7}$

c) $\dfrac{3 + \dfrac{2}{3}}{\dfrac{1}{4} + \dfrac{5}{12}}$

d) $\dfrac{\dfrac{7}{10} - 4}{3 - \dfrac{8}{15}}$

PRACTICE EXERCISES

Simplify the following complex fractions using both methods.

1) $\dfrac{\dfrac{3}{8}}{\dfrac{5}{6}}$

2) $\dfrac{\dfrac{8}{9}}{12}$

3) $\dfrac{\dfrac{4}{3} - \dfrac{3}{8}}{1 + \dfrac{7}{6}}$

4) $\dfrac{2 + \dfrac{3}{4}}{5 - \dfrac{5}{12}}$

If more practice is needed, do the Additional Practice Exercises in the margin.

Since the second method in which we use the LCD is always at least as easy as the first using the division interpretation, it is the only method we will use for complex fractions whose numerators and/or denominators contain rational expressions.

EXAMPLE 3

Simplify the following complex fractions.

a) $\dfrac{\dfrac{a}{b}}{\dfrac{c}{d}} =$ The LCD for $\frac{a}{b}$ and $\frac{c}{d}$ is bd. Therefore, multiply the numerator and the denominator by bd.

$\dfrac{bd \cdot \dfrac{a}{b}}{bd \cdot \dfrac{c}{d}} =$ Perform the multiplications.

$\dfrac{ad}{bc}$ Simplified form.

b) $\dfrac{\dfrac{a^2b}{c^2}}{\dfrac{ab^3}{c^3}} =$ The LCD for $\frac{a^2b}{c^2}$ and $\frac{ab^3}{c^3}$ is c^3. Therefore, multiply the numerator and the denominator by c^3.

$\dfrac{c^3\left(\dfrac{a^2b}{c^2}\right)}{c^3\left(\dfrac{ab^3}{c^3}\right)} =$ Perform the multiplications.

$\dfrac{a^2bc}{ab^3} =$ Reduce to lowest terms.

$\dfrac{ac}{b^2}$ Simplified form.

c) $\dfrac{1 - \dfrac{9}{x^2}}{1 + \dfrac{3}{x}} =$ The least common denominator for all the fractions in the numerator and the denominator is x^2. Therefore, multiply the numerator and the denominator by x^2.

$\dfrac{x^2\left(1 - \dfrac{9}{x^2}\right)}{x^2\left(1 + \dfrac{3}{x}\right)} =$ Apply the distributive property.

$\dfrac{x^2 \cdot 1 - x^2\left(\dfrac{9}{x^2}\right)}{x^2 \cdot 1 + x^2\left(\dfrac{3}{x}\right)} =$ Perform the multiplications.

$\dfrac{x^2 - 9}{x^2 + 3x} =$ Factor the numerator and the denominator.

$\dfrac{(x + 3)(x - 3)}{x(x + 3)} =$ Divide by the common factor $(x + 3)$.

$\dfrac{x - 3}{x}$ Simplified form.

$$\frac{1 - \dfrac{2}{x} - \dfrac{15}{x^2}}{1 + \dfrac{7}{x} + \dfrac{12}{x^2}} =$$

d)

The LCD of all the fractions in the numerator and the denominator is x^2. Therefore, multiply the numerator and the denominator by x^2.

$$\frac{x^2\left(1 - \dfrac{2}{x} - \dfrac{15}{x^2}\right)}{x^2\left(1 + \dfrac{7}{x} + \dfrac{12}{x^2}\right)} =$$

Apply the distributive property.

$$\frac{x^2 \cdot 1 - x^2\left(\dfrac{2}{x}\right) - x^2\left(\dfrac{15}{x^2}\right)}{x^2 \cdot 1 + x^2\left(\dfrac{7}{x}\right) + x^2\left(\dfrac{12}{x^2}\right)} =$$

Perform the multiplications.

$$\frac{x^2 - 2x - 15}{x^2 + 7x + 12} =$$

Factor the numerator and the denominator.

$$\frac{(x - 5)(x + 3)}{(x + 4)(x + 3)} =$$

Divide by the common factor $(x + 3)$.

$$\frac{x - 5}{x + 4}$$

Simplified form.

e)

$$\frac{x + \dfrac{9}{x - 6}}{1 + \dfrac{3}{x - 6}} =$$

The LCD of all the fractions in the numerator and the denominator is $x - 6$. Therefore, multiply the numerator and the denominator by $x - 6$.

$$\frac{(x - 6)\left(x + \dfrac{9}{x - 6}\right)}{(x - 6)\left(1 + \dfrac{3}{x - 6}\right)} =$$

Apply the distributive property.

$$\frac{(x - 6)x + (x - 6)\left(\dfrac{9}{x - 6}\right)}{(x - 6) \cdot 1 + (x - 6)\left(\dfrac{3}{x - 6}\right)} =$$

Perform the multiplications.

$$\frac{x^2 - 6x + 9}{x - 6 + 3} =$$

Factor the numerator and combine like terms in the denominator.

$$\frac{(x - 3)^2}{x - 3} =$$

Divide by the common factor of $x - 3$.

$$x - 3$$

Answer.

ADDITIONAL PRACTICE

Simplify the following complex fractions.

e) $\dfrac{\dfrac{r}{s}}{\dfrac{t}{u}}$

f) $\dfrac{\dfrac{4cd^3}{3a}}{\dfrac{2c^4d}{6a^2}}$

g) $\dfrac{1 - \dfrac{25}{a^2}}{1 - \dfrac{5}{a}}$

h) $\dfrac{1 - \dfrac{3}{x} - \dfrac{28}{x^2}}{1 - \dfrac{1}{x} - \dfrac{20}{x^2}}$

i) $\dfrac{4x + \dfrac{9}{x - 3}}{2 + \dfrac{3}{x - 3}}$

PRACTICE EXERCISES

Simplify the following complex fractions.

5) $\dfrac{\dfrac{m}{n}}{\dfrac{p}{q}}$

6) $\dfrac{\dfrac{x^3y^2}{z}}{\dfrac{x^2y^4}{z^3}}$

7) $\dfrac{1 - \dfrac{4}{x^2}}{1 + \dfrac{2}{x}}$

8) $\dfrac{1 - \dfrac{4}{x} - \dfrac{12}{x^2}}{1 + \dfrac{6}{x} + \dfrac{8}{x^2}}$

9) $\dfrac{x + \dfrac{16}{x - 8}}{1 + \dfrac{4}{x - 8}}$

If more practice is needed, do the Additional Practice Exercises in the margin.

EXERCISE SET 5.6

Simplify the following complex fractions.

1) $\dfrac{\frac{2}{3}}{\frac{5}{6}}$ **2)** $\dfrac{\frac{3}{8}}{\frac{5}{4}}$ **3)** $\dfrac{\frac{3}{4}}{\frac{1}{6}}$ **4)** $\dfrac{\frac{5}{9}}{\frac{4}{15}}$

5) $\dfrac{\frac{1}{8}}{2}$ **6)** $\dfrac{3}{\frac{1}{5}}$ **7)** $\dfrac{14}{\frac{7}{2}}$ **8)** $\dfrac{9}{\frac{12}{5}}$

9) $\dfrac{\frac{4}{9}-\frac{1}{3}}{\frac{7}{12}-\frac{5}{18}}$ **10)** $\dfrac{\frac{3}{4}+\frac{1}{12}}{\frac{5}{8}-\frac{7}{20}}$ **11)** $\dfrac{8-\frac{9}{4}}{\frac{7}{8}+\frac{9}{10}}$

12) $\dfrac{\frac{7}{9}+\frac{5}{12}}{3-\frac{7}{3}}$ **13)** $\dfrac{5+\frac{7}{2}}{1+\frac{4}{11}}$ **14)** $\dfrac{9-\frac{5}{4}}{2+\frac{7}{6}}$

15) $\dfrac{\frac{7}{9}+9}{3-\frac{4}{3}}$ **16)** $\dfrac{\frac{12}{5}-6}{8+\frac{7}{10}}$ **17)** $\dfrac{\frac{m}{8}}{\frac{n}{12}}$

18) $\dfrac{\frac{p}{15}}{\frac{q}{10}}$ **19)** $\dfrac{\frac{5}{d}}{\frac{3}{f}}$ **20)** $\dfrac{\frac{7}{g}}{\frac{2}{h}}$

21) $\dfrac{\frac{x}{y}}{\frac{r}{t}}$ **22)** $\dfrac{\frac{u}{v}}{\frac{w}{z}}$ **23)** $\dfrac{\frac{g}{h}}{4}$

24) $\dfrac{\frac{6}{r}}{s}$ **25)** $\dfrac{\frac{a}{b}}{c}$ **26)** $\dfrac{\frac{m}{n}}{p}$

27) $\dfrac{\frac{a}{x^2}}{\frac{b}{x}}$ **28)** $\dfrac{\frac{p}{y}}{\frac{q}{y^3}}$ **29)** $\dfrac{\frac{u}{v^3}}{\frac{w}{v^4}}$

30) $\dfrac{\frac{r}{s^5}}{\frac{t}{s^2}}$ **31)** $\dfrac{\frac{ac}{b^4}}{\frac{ac}{b^3}}$ **32)** $\dfrac{\frac{rs}{t^2}}{\frac{rs}{t^5}}$

33) $\dfrac{\frac{u^7v^2}{w^5}}{\frac{u^3v^4}{w}}$ **34)** $\dfrac{\frac{m^6p^3}{n^7}}{\frac{m^5p^3}{n^4}}$ **35)** $\dfrac{\frac{2}{z}}{1-\frac{z}{2}}$

36) $\dfrac{\frac{s}{3}}{4-\frac{1}{s}}$ **37)** $\dfrac{1+\frac{x}{2}}{1-\frac{x}{2}}$ **38)** $\dfrac{y-\frac{y}{3}}{5+\frac{y}{3}}$

39) $\dfrac{\frac{1}{t}-1}{\frac{1}{t^3}-1}$ **40)** $\dfrac{\frac{6}{s^4}-1}{\frac{2}{s}+1}$

41) $\dfrac{1+\frac{1}{v}-\frac{2}{v^2}}{1-\frac{4}{v}+\frac{3}{v^2}}$ **42)** $\dfrac{2+\frac{13}{w}+\frac{6}{w^2}}{1+\frac{11}{w}+\frac{30}{w^2}}$

43) $\dfrac{\frac{1}{u}-\frac{2}{u^2}}{1+\frac{4}{u}-\frac{12}{u^2}}$ **44)** $\dfrac{\frac{4}{p}+\frac{4}{p^2}}{3-\frac{2}{p}+\frac{5}{p^2}}$

45) $\dfrac{\frac{2}{x}-\frac{7}{x^2}-\frac{30}{x^3}}{2+\frac{11}{x}+\frac{15}{x^2}}$ **46)** $\dfrac{1-\frac{3}{y}-\frac{4}{y^2}}{\frac{3}{y}-\frac{13}{y^2}+\frac{4}{y^3}}$

47) $\dfrac{5+\frac{4}{r+8}}{r+8}$ **48)** $\dfrac{\frac{7}{s-15}-6}{s-15}$

49) $\dfrac{t+\frac{9}{t-7}}{11-\frac{3}{t-7}}$ **50)** $\dfrac{\frac{2}{q+4}-q}{\frac{13}{q+4}+5}$

CHALLENGE EXERCISES (51–52)

51) $\dfrac{\frac{2a}{a+6}+\frac{1}{a}}{9a-\frac{3}{a+6}}$ **52)** $\dfrac{\frac{8}{b}-\frac{7b}{b-1}}{\frac{4}{b-1}+\frac{21}{b}}$

53) In simplifying complex fractions, why do we multiply the numerator and the denominator by the LCD of all the denominators which occur in the numerator and the denominator?

54) Consider the complex fraction $\dfrac{\frac{2}{3}+\frac{5}{6}}{\frac{1}{4}-\frac{1}{3}}$. The LCD of the numerator is 6 and the LCD of the denominator is 12.

Why don't we multiply the numerator by 6 and the denominator by 12? That is, what is wrong with $\dfrac{6\left(\frac{2}{3}+\frac{5}{6}\right)}{12\left(\frac{1}{4}-\frac{1}{3}\right)}$?

55) Explain why $\dfrac{a^{-2}+b^{-2}}{a^{-1}+b^{-1}} \neq \dfrac{a+b}{a^2+b^2}$.

S E C T I O N <u>5.7</u>

Solving Equations Containing Rational Numbers and Expressions

OBJECTIVES

When you complete this section, you will be able to:

Ⓐ Solve equations which contain rational numbers.

Ⓑ Solve equations which contain rational expressions.

INTRODUCTION

The basis for solving equations that contain fractions or rational expressions is the multiplication property of equality introduced in Chapter 3. In words, it states that we may multiply both sides of an equation by any nonzero number and the result is an equation with the same solution(s) as the original equation. In symbols, if $a = b$ and $c \neq 0$, then $ac = bc$.

When solving equations that contain fractions or rational expressions, we multiply both sides of the equation by the LCD of the fractions or rational expressions. Since all the denominators divide evenly into the LCD, multiplying both sides of the equation by the LCD eliminates all fractions. The resulting equation is then solved using the same methods used in Chapter 3. We illustrate with some examples.

- **Solving equations that contain rational numbers**

EXAMPLE 1

Solve the following equations.

a) $\dfrac{x}{3} - \dfrac{x}{5} = 2$

Multiply both sides by the LCD of 15.

$15\left(\dfrac{x}{3} - \dfrac{x}{5}\right) = 15 \cdot 2$

Distribute on the left, multiply on the right.

$15\left(\dfrac{x}{3}\right) - 15\left(\dfrac{x}{5}\right) = 30$

Perform the multiplications.

$5x - 3x = 30$

Add like terms.

$2x = 30$

Divide both sides by 2.

$x = 15$

Solution.

Check:
Substitute 15 for x.

$$\dfrac{15}{3} - \dfrac{15}{5} = 2$$

$$5 - 3 = 2$$

$$2 = 2$$

Therefore, the solution is correct.

---

b) $\dfrac{2}{3}a - \dfrac{1}{6}a = \dfrac{8}{9}$

Multiply both sides by the LCD of 18.

$$18\left(\dfrac{2}{3}a - \dfrac{1}{6}a\right) = 18\left(\dfrac{8}{9}\right)$$

Distribute on the left, multiply on the right.

$$18\left(\dfrac{2}{3}a\right) - 18\left(\dfrac{1}{6}a\right) = 16$$

Perform the multiplications.

$$12a - 3a = 16$$

Combine like terms.

$$9a = 16$$

Divide both sides by 9.

$$a = \dfrac{16}{9}$$

Answer.

Check:

Substitute $\dfrac{16}{9}$ for a.

$$\dfrac{2}{3} \cdot \dfrac{16}{9} - \dfrac{1}{6} \cdot \dfrac{16}{9} = \dfrac{8}{9}$$

Perform the multiplications.

$$\dfrac{32}{27} - \dfrac{16}{54} = \dfrac{8}{9}$$

Reduce $\dfrac{16}{54}$ to lowest terms.

$$\dfrac{32}{27} - \dfrac{8}{27} = \dfrac{8}{9}$$

Add the fractions on the left side.

$$\dfrac{24}{27} = \dfrac{8}{9}$$

Reduce the left side to lowest terms.

$$\dfrac{8}{9} = \dfrac{8}{9}$$

Therefore, the answer is correct.

c) $\dfrac{y + 2}{4} = y + 5$

Multiply both sides by the LCD of 4.

$$4\left(\dfrac{y + 2}{4}\right) = 4(y + 5)$$

Perform the multiplications.

$$y + 2 = 4y + 20$$

Subtract $4y$ from both sides.

$$-3y + 2 = 20$$

Subtract 2 from both sides.

$$-3y = 18$$

Divide both sides by -3.

$$y = -6$$

Solution.

Check:
Substitute -6 for y.

$$\dfrac{-6+2}{4} = -6 + 5$$

Simplify both sides.

$$\dfrac{-4}{4} = -1$$

$$-1 = -1$$

Therefore, the answer is correct.

d) $\dfrac{2x - 3}{5} - \dfrac{3x + 2}{3} = -\dfrac{2}{3}$

Multiply both sides by the LCD of 15.

$$15\left(\dfrac{2x - 3}{5} - \dfrac{3x + 2}{3}\right) = 15\left(-\dfrac{2}{3}\right)$$

Distribute on the left and multiply on the right.

$$15\left(\dfrac{2x - 3}{5}\right) - 15\left(\dfrac{3x + 2}{3}\right) = -10$$

Multiply on the left.

$$3(2x - 3) - 5(3x + 2) = -10 \qquad \text{Distribute on the left.}$$
$$6x - 9 - 15x - 10 = -10 \qquad \text{Combine like terms.}$$
$$-9x - 19 = -10 \qquad \text{Add 19 to both sides.}$$
$$-9x = 9 \qquad \text{Divide both sides by } -9.$$
$$x = -1 \qquad \text{Solution.}$$

Check:

The check is left as an exercise for the student.

ADDITIONAL PRACTICE

Solve the following equations.

a) $\dfrac{a}{9} - \dfrac{a}{4} = \dfrac{5}{6}$

b) $\dfrac{1}{4}x = \dfrac{7}{10}x + \dfrac{1}{2}$

c) $\dfrac{3x + 2}{4} = 2x - 2$

d) $\dfrac{3r + 1}{4} = \dfrac{13}{12} + \dfrac{2r + 5}{6}$

- Solving equations which contain rational expressions

PRACTICE EXERCISES

Solve the following equations and check your answer.

1) $\dfrac{x}{3} - \dfrac{x}{7} = 4$

2) $\dfrac{3}{5}x + \dfrac{3}{10}x = \dfrac{4}{5}$

3) $\dfrac{2x - 5}{5} = x + 2$

4) $\dfrac{2z - 6}{5} - \dfrac{2z + 3}{2} = \dfrac{3}{10}$

If more practice is needed, do the Additional Practice Exercises.

If the denominator(s) contains variables, the procedure is essentially the same with one exception. Since division by 0 is impossible, we must eliminate any value of the variable that makes the denominator equal to zero. It is possible to solve an equation correctly and get a number that does not solve the equation because it makes one or more of the denominators equal to 0. Such numbers are called **apparent** or **extraneous** solutions. For this reason it is mandatory that all answers to equations that contain variables in the denominator be checked in order to determine if they are solutions.

EXAMPLE 2

Solve the following equations. Indicate any value(s) of the variable that make the equation undefined.

a) $\dfrac{5}{x} - \dfrac{3}{4} = \dfrac{1}{2}$ $\qquad x \neq 0.$ Multiply both sides by the LCD, $4x$.

$$4x\left(\dfrac{5}{x} - \dfrac{3}{4}\right) = 4x\left(\dfrac{1}{2}\right) \qquad \text{Distribute on the left, multiply on the right.}$$

$$4x\left(\dfrac{5}{x}\right) - 4x\left(\dfrac{3}{4}\right) = 2x \qquad \text{Perform the multiplications.}$$

$$20 - 3x = 2x \qquad \text{Add } 3x \text{ to both sides.}$$

$$20 = 5x \qquad \text{Divide both sides by 5.}$$

$$4 = x \qquad \text{Possible solution.}$$

Check:

Substitute 4 for x.

$$\dfrac{5}{4} - \dfrac{3}{4} = \dfrac{1}{2}$$

$$\dfrac{2}{4} = \dfrac{1}{2}$$

$$\dfrac{1}{2} = \dfrac{1}{2} \qquad \text{Therefore, 4 is the solution.}$$

b) $\dfrac{3y}{y-4} - \dfrac{12}{y-4} = 6$ $\qquad$ $y \ne 4$. Multiply both sides by the LCD, $y - 4$.

$(y-4)\left(\dfrac{3y}{y-4} - \dfrac{12}{y-4}\right) = (y-4)(6)$ $\qquad$ Distribute on the left, multiply on the right.

$(y-4)\left(\dfrac{3y}{y-4}\right) - (y-4)\left(\dfrac{12}{y-4}\right) = 6y - 24$ $\qquad$ Multiply on the left.

$3y - 12 = 6y - 24$ $\qquad$ Subtract $3y$ from both sides.

$-12 = 3y - 24$ $\qquad$ Add 24 to both sides.

$12 = 3y$ $\qquad$ Divide both sides by 3.

$4 = y$ $\qquad$ Possible solution, but will not work since $y \ne 4$.

Check:
Substitute 4 for y.

$$\dfrac{12}{4-4} - \dfrac{12}{4-4} = 6$$

$$\dfrac{12}{0} - \dfrac{12}{0} = 6$$

Since division by 0 is undefined, there is no solution to this equation. The symbol $\varnothing$ is used to show there is no solution. It is the symbol for the empty set which means there is nothing in the set. Hence, there is no solution.

c) $\dfrac{x}{x+4} = \dfrac{1}{x-2}$

Rather than multiplying both sides of the equation by the LCD, a quicker way is to cross multiply (which is actually the same as multiplying by the LCD). Remember, cross multiplication can be used only when the equation consists of one fraction equal to another fraction.

$\dfrac{x}{x+4} = \dfrac{1}{x-2}$ $\qquad$ $x \ne -4, 2$. Cross multiply.

$x(x-2) = (x+4)(1)$ $\qquad$ Apply the distributive property.

$x^2 - 2x = x + 4$ $\qquad$ Subtract x from both sides.

$x^2 - 3x = 4$ $\qquad$ Subtract 4 from both sides.

$x^2 - 3x - 4 = 0$ $\qquad$ Factor.

$(x-4)(x+1) = 0$ $\qquad$ Set each factor equal to 0.

$x - 4 = 0 \quad x + 1 = 0$ $\qquad$ Solve each equation.

$x = 4 \qquad\quad x = -1$ $\qquad$ Therefore, 4 and -1 are possible solutions.

Check:

$x = 4$ $\qquad\qquad\qquad$ $x = -1$

Replace x with 4. $\qquad\quad$ Replace x with -1.

$\dfrac{4}{4+4} = \dfrac{1}{4-2}$ $\qquad\quad$ $\dfrac{-1}{-1+4} = \dfrac{1}{-1-2}$

$\dfrac{4}{8} = \dfrac{1}{2}$ $\qquad\qquad\quad$ $\dfrac{-1}{3} = \dfrac{1}{-3}$

$\dfrac{1}{2} = \dfrac{1}{2}$ $\qquad\qquad\quad$ $-\dfrac{1}{3} = -\dfrac{1}{3}$ $\qquad$ Therefore, both answers are solutions.

d)

$$\frac{3}{2a + 2} - \frac{4}{3a + 3} = \frac{5}{6}$$

$a \neq -1$. Factor the denominators.

$$\frac{3}{2(a + 1)} - \frac{4}{3(a + 1)} = \frac{5}{6}$$

Multiply both sides by the LCD of $6(a + 1)$.

$$6(a + 1)\left(\frac{3}{2(a + 1)} - \frac{4}{3(a + 1)}\right) = 6(a + 1)\left(\frac{5}{6}\right)$$

Distribute on left, multiply on right.

$$6(a + 1)\left(\frac{3}{2(a + 1)}\right) - 6(a + 1)\left(\frac{4}{3(a + 1)}\right) = 5(a + 1)$$

Multiply.

$$3 \cdot 3 - 2 \cdot 4 = 5a + 5$$

Simplify the left side.

$$9 - 8 = 5a + 5$$

Continue simplifying.

$$1 = 5a + 5$$

Subtract 5 from both sides.

$$-4 = 5a$$

Divide both sides by 5.

$$-\frac{4}{5} = a$$

Possible solution.

Check:

As you can imagine, the check to this example is difficult and is omitted. However, if an equation containing rational expressions is solved correctly, the extraneous solutions always make the denominator equal to 0 as in Example 2b. Often it is advisable to see if the possible solution(s) make a denominator equal to 0 rather than doing a complete check. If so, eliminate that value(s). Since $-\frac{4}{5}$ does not make any denominator equal to 0, it is at least possible if not correct.

e)

$$\frac{4}{x^2 - x - 6} - \frac{2}{x^2 + 3x + 2} = \frac{4}{x^2 - 2x - 3}$$

Factor the denominators.

$$\frac{4}{(x - 3)(x + 2)} - \frac{2}{(x + 2)(x + 1)} = \frac{4}{(x - 3)(x + 1)}$$

$x \neq 3, -2, -1$. Multiply both sides by the LCD of $(x - 3)(x + 2)(x + 1)$.

$$(x - 3)(x + 2)(x + 1)\left(\frac{4}{(x - 3)(x + 2)} - \frac{2}{(x + 2)(x + 1)}\right) =$$
$$(x - 3)(x + 2)(x + 1)\left(\frac{4}{(x - 3)(x + 1)}\right)$$

Distribute on the left side, multiply on the right.

$$(x - 3)(x + 2)(x + 1)\left(\frac{4}{(x - 3)(x + 2)}\right) -$$

Perform the multiplications.

$$(x - 3)(x + 2)(x + 1)\left(\frac{2}{(x + 2)(x + 1)}\right) = (x + 2)4$$

$$(x + 1)4 - (x - 3)2 = 4x + 8$$

Simplify the left side.

$$4x + 4 - 2x + 6 = 4x + 8$$

Add like terms on the left side.

$$2x + 10 = 4x + 8$$

Subtract 2x from both sides.

$$10 = 2x + 8 \qquad \text{Subtract 8 from both sides.}$$

$$2 = 2x \qquad \text{Divide both sides by 2.}$$

$$1 = x \qquad \text{Possible solution.}$$

Check:

Note that 1 does not make any denominator equal to 0 and is therefore likely to be correct. The check is left as an exercise for the student.

ADDITIONAL PRACTICE

Solve the following equations.

e) $\dfrac{3}{x} - \dfrac{3}{2} = \dfrac{6}{x}$

f) $\dfrac{b}{b - 6} - 5 = \dfrac{6}{b - 6}$

g) $\dfrac{x}{x + 4} = \dfrac{4}{x - 2}$

h) $\dfrac{6}{5m - 10} - \dfrac{4}{2m - 4} = \dfrac{1}{5}$

i) $\dfrac{3}{x^2 - 1} + \dfrac{2}{x^2 + 3x + 2} = \dfrac{6}{x^2 + x - 2}$

PRACTICE EXERCISES

Solve the following equations.

5) $\dfrac{5}{3z} - \dfrac{1}{2z} = \dfrac{7}{6}$

6) $\dfrac{x}{x + 5} = 4 - \dfrac{5}{x + 5}$

7) $\dfrac{x}{2x + 1} = \dfrac{-4}{x - 3}$

8) $\dfrac{y}{3y + 6} + \dfrac{3}{4y + 8} = \dfrac{3}{4}$

9) $\dfrac{4}{x^2 - 5x + 6} - \dfrac{6}{x^2 + 2x - 8} = \dfrac{3}{x^2 + x - 12}$

If more practice is needed, do the Additional Practice Exercises.

EXERCISE SET   **5.7**

Solve the following equations.

1) $\dfrac{x}{3} - \dfrac{1}{4} = \dfrac{5}{12}$

2) $\dfrac{y}{6} + \dfrac{5}{2} = 2$

3) $\dfrac{t}{10} + \dfrac{3}{5} = \dfrac{11}{10}$

4) $\dfrac{v}{2} - \dfrac{1}{3} = 14$

5) $\dfrac{1}{3}u - \dfrac{1}{9}u = \dfrac{4}{3}$

6) $\dfrac{1}{7}w + \dfrac{1}{2}w = 3$

7) $\dfrac{2m}{3} - \dfrac{4m}{5} = 1$

8) $\dfrac{3p}{8} + \dfrac{p}{4} = 5$

9) $\dfrac{3}{4}x - \dfrac{1}{3}x = \dfrac{5}{6}$

10) $\dfrac{5}{18}q + \dfrac{2}{9}q = \dfrac{1}{3}$

11) $\dfrac{1}{4}t + \dfrac{1}{3}t + \dfrac{5}{6} = \dfrac{15}{4}$

12) $\dfrac{2}{3}v + \dfrac{1}{6}v - \dfrac{1}{4} = \dfrac{7}{12}$

13) $\dfrac{3x}{4} - \dfrac{2}{5} = \dfrac{3x}{10} + \dfrac{1}{2}$

14) $\dfrac{y}{3} - \dfrac{1}{4} = \dfrac{5y}{12} + \dfrac{1}{6}$

15) $\dfrac{y - 3}{4} = y - 1$

16) $t + 5 = \dfrac{t + 8}{7}$

17) $\dfrac{3x + 6}{2} = x + 2$

18) $2a - 3 = \dfrac{5a + 6}{4}$

19) $\dfrac{4v + 7}{15} = \dfrac{v + 1}{5} + \dfrac{1}{3}$

20) $\dfrac{2y - 1}{5} - 1 = \dfrac{y - 2}{2}$

21) $\dfrac{2q + 1}{4} - \dfrac{3q - 5}{16} = \dfrac{7}{8}$

22) $\dfrac{p - 1}{2} - \dfrac{p + 2}{6} = \dfrac{-7}{12}$

Solve the following equations.

23) $\dfrac{6}{m} + \dfrac{1}{2} = \dfrac{3}{4}$

24) $\dfrac{5}{8} - \dfrac{4}{n} = \dfrac{7}{12}$

25) $\dfrac{5}{r} + \dfrac{3}{r} = 12$

26) $\dfrac{5}{t} - \dfrac{1}{3} = \dfrac{6}{t}$

27) $\dfrac{4}{3u} - \dfrac{1}{2u} = \dfrac{5}{6}$

28) $\dfrac{2}{v} - \dfrac{4}{3v} = \dfrac{3}{5}$

29) $\dfrac{5}{4w} - \dfrac{3}{5w} = \dfrac{-13}{40}$

30) $\dfrac{7}{2z} + \dfrac{11}{6z} = \dfrac{16}{9}$

31) $\dfrac{7}{x - 3} = 8 - \dfrac{1}{x - 3}$

32) $\dfrac{5}{y + 2} + 9 = -\dfrac{4}{y + 2}$

33) $\dfrac{t}{t + 4} = 3 - \dfrac{12}{t + 4}$

34) $\dfrac{8}{z - 5} - 2 = \dfrac{z}{z - 5}$

35) $\dfrac{4a}{a + 6} = \dfrac{10}{a + 6} + 2$

36) $\dfrac{5b}{b - 8} - 4 = \dfrac{3}{b - 8}$

37) $\dfrac{c}{10 - c} = \dfrac{1}{c + 2}$

38) $\dfrac{x}{2 - 3x} = \dfrac{1}{3x + 2}$

39) $\dfrac{3y}{y - 1} = \dfrac{-2}{y + 1}$

40) $\dfrac{2n}{n + 1} = \dfrac{-5}{n + 3}$

41) $\dfrac{m}{m - 1} = \dfrac{3m}{4m - 3}$

42) $\dfrac{5p}{p + 4} = \dfrac{p}{p - 1}$

43) $\dfrac{2y}{3y - 6} + \dfrac{3}{4y - 8} = \dfrac{1}{4}$

44) $\dfrac{2y}{3y + 6} - \dfrac{3}{4y + 8} = \dfrac{1}{4}$

45) $\dfrac{x}{x - 2} - \dfrac{4}{x - 1} = \dfrac{2}{x^2 - 3x + 2}$

46) $\dfrac{20z}{z - 2} + \dfrac{3}{2z - 1} = \dfrac{6}{(z - 2)(2z - 1)}$

47) $\dfrac{6}{t^2 + t - 12} = \dfrac{4}{t^2 - t - 6} + \dfrac{1}{t^2 + 6t + 8}$

48) $\dfrac{7}{v^2 - 6v + 5} - \dfrac{2}{v^2 - 4v - 5} = \dfrac{3}{v^2 - 1}$

49) $\dfrac{5}{w^2 + 2w - 8} + \dfrac{3}{w^2 + w - 6} = \dfrac{4}{w^2 + 7w + 12}$

50) $\dfrac{8}{a^2 + 4a - 12} - \dfrac{5}{a^2 + a - 6} = \dfrac{2}{a^2 + 9a + 18}$

CHALLENGE EXERCISES (51–53)

51) $\dfrac{1}{b^2 - b - 2} = \dfrac{b}{b^2 + 2b - 8} - \dfrac{1}{b^2 + 5b + 4}$

52) $\dfrac{p + 1}{p^2 + 8p + 15} - \dfrac{2}{p^2 + 7p + 10} = \dfrac{1}{p^2 + 5p + 6}$

53) $\dfrac{3}{x^2 + 2x - 24} + \dfrac{x - 5}{x^2 - 16} = \dfrac{x}{x^2 + 10x + 24}$

WRITING EXERCISES

54) In solving equations which involve rational numbers or rational expressions, why do we multiply both sides of the equation by the LCD?

55) Explain the difference between solving an equation involving rational expressions and adding rational expressions. For example, how is solving $\frac{y}{3y + 6} + \frac{3}{4y + 8} = \frac{3}{4}$ different from adding $\frac{y}{3y + 6} + \frac{3}{4y + 8} + \frac{3}{4}$?

56) In Example 2c we solved $\frac{x}{x + 4} = \frac{1}{x - 2}$ by cross multiplying. Show that this is the same as multiplying both sides by the LCD.

SECTION 5.8

Applications with Rational Expressions

OBJECTIVES

When you complete this section, you will be able to:

Ⓐ Solve applications problems involving number properties.

Ⓑ Solve applications problems involving work.

Ⓒ Solve applications problems involving distance, rate, and time.

INTRODUCTION

Applications problems were first discussed in detail in Sections 3.6–3.7. You may wish to review the procedures and suggested approaches before doing this section. Our purpose, in considering applications problems, is to help you to think more mathematically, to better understand the structure of mathematics, to help you translate English into mathematics, to show the power of mathematics in problem solving, and to serve as a foundation for the more realistic problems that some students will encounter in more advanced courses.

In this section we will discuss number problems, work problems, and distance-rate-time problems separately. We discussed some of these types of problems in Chapter 3, but unlike those in Chapter 3, the problems in this section will result in equations containing rational expressions. We will emphasize the approach needed to do each type of problem.

• **Number problems**

Number problems are usually fairly straightforward. You do not have to know special formulas or other facts and all the information needed is contained in the problem. However, there are a few things you need to remember.

a) When working with consecutive integer problems, we let x represent the smallest of the integers, $x + 1$ represents the next largest, $x + 2$ represents the next largest, and so on.

b) When working with consecutive even or consecutive odd integers, let x represent the smallest integer, $x + 2$ represents the next largest, $x + 4$ represents the next largest, and so on.

c) The reciprocal of a number is found by putting 1 over the number. For example, the reciprocal of x is $\frac{1}{x}$, $x \neq 0$.

d) In a fraction, the numerator is the number on top and the denominator is the number on bottom.

We suggest the following procedure.

SOLVING NUMBER PROBLEMS

1) Read the problem and identify the unknown(s).

2) Assign a variable to an unknown.

3) If there is more than one unknown, represent the other unknowns in terms of the variable assigned in step 2.

4) Write the equation using information given in the problem.

5) Solve the equation.

6) Check the solution using the wording of the original problem.

 EXAMPLE 1

a) The denominator of a fraction is 3 more than the numerator. If 4 is added to the numerator and subtracted from the denominator, the resulting fraction is $\frac{9}{4}$. Find the original fraction.

Solution:
Let x be the numerator. Then,
$x + 3$ is the denominator. (The denominator is 3 more than the numerator.) Therefore, the original fraction is $\frac{x}{x+3}$. Since 4 is added to the numerator, the numerator of the new fraction is $x + 4$. Since 4 is also subtracted from the denominator, the denominator of the new fraction is $x + 3 - 4$. Consequently, the new fraction is $\frac{x+4}{x+3-4}$. Since the new fraction is equal to $\frac{9}{4}$, the equation is:

$$\frac{x+4}{x+3-4} = \frac{9}{4}$$ Simplify the denominator on the left.

$$\frac{x+4}{x-1} = \frac{9}{4}$$ Cross multiply.

$$4(x + 4) = (x - 1)9$$ Apply the distributive property.

$$4x + 16 = 9x - 9$$ Subtract $9x$ from both sides.

$$-5x + 16 = -9 \qquad \text{Subtract 16 from both sides.}$$
$$-5x = -25 \qquad \text{Divide both sides by } -5.$$
$$x = 5 \qquad \text{Therefore, the numerator is 5.}$$
$$x + 3 = 5 + 3 = 8 \qquad \text{Therefore, the denominator is 8.}$$

Since the numerator is 5 and the denominator is 8, the original fraction is $\frac{5}{8}$.

Check:

If 4 is added to the numerator of $\frac{5}{8}$, the numerator becomes 9. If 4 is subtracted from the denominator, the denominator becomes 4. The new fraction is $\frac{9}{4}$ which is what it is supposed to be. Therefore, our solution is correct.

b) Two numbers differ by 4. If $\frac{1}{2}$ of the larger number is 3 more than $\frac{1}{3}$ of the smaller, find the numbers.

Solution:

Let x be the larger number.
Then, $x - 4$ is the smaller number. (The numbers differ by 4.)
We represent $\frac{1}{2}$ of the larger number by $\frac{1}{2}x$ and $\frac{1}{3}$ of the smaller number by $\frac{1}{3}(x - 4)$.
Since $\frac{1}{2}$ of the larger number is 3 more than $\frac{1}{3}$ of the smaller, the equation is:

$$\frac{1}{2}x = \frac{1}{3}(x - 4) + 3 \qquad \text{Multiply both sides by 6.}$$

$$6\left(\frac{1}{2}x\right) = 6\left(\frac{1}{3}(x - 4) + 3\right) \qquad \text{Multiply on left, distribute on right.}$$

$$3x = 6 \cdot \frac{1}{3}(x - 4) + 6(3) \qquad \text{Multiply on the right side.}$$

$$3x = 2(x - 4) + 18 \qquad \text{Distribute.}$$
$$3x = 2x - 8 + 18 \qquad \text{Combine like terms.}$$
$$3x = 2x + 10 \qquad \text{Subtract } 2x \text{ from both sides.}$$
$$x = 10 \qquad \text{Therefore, the larger number is 10.}$$
$$x - 4 = 10 - 4 = 6 \qquad \text{Therefore, the smaller number is 6.}$$

Check:

Since 10 and 6 differ by 4 and $\frac{1}{2}$ of 10 is 5 which is 3 more than $\frac{1}{3}$ of 6 which is 2, the solutions are correct.

Note: Since you should now be familiar with the techniques of equation solving, less detail will be given in the explanations to the right of each step. Also, some of the easier steps will be omitted.

c) For two consecutive even integers, the reciprocal of the smaller is added to three times the reciprocal of the larger. If the resulting sum is $\frac{13}{24}$, find the two consecutive even integers.

Solution:

Let x be the smaller of the two consecutive even integers. Then, $x + 2$ is the larger of the two consecutive even integers.

The reciprocal of the smaller is $\frac{1}{x}$ and the reciprocal of the larger is $\frac{1}{x + 2}$. Three times the reciprocal of the larger is $3\left(\frac{1}{x + 2}\right) = \frac{3}{x + 2}$. Since sum means add and the sum is $\frac{13}{24}$, the equation is:

$$\frac{1}{x} + \frac{3}{x+2} = \frac{13}{24}$$

Multiply both sides by the LCD $24x(x+2)$.

$$24x(x+2)\left(\frac{1}{x} + \frac{3}{x+2}\right) = 24x(x+2)\left(\frac{13}{24}\right)$$

Simplify both sides.

$$24x(x+2)\left(\frac{1}{x}\right) + 24x(x+2)\left(\frac{3}{x+2}\right) = 13x(x+2)$$

Perform the multiplications.

$$24(x+2) + 24x(3) = 13x^2 + 26x$$

Simplify the left side.

$$24x + 48 + 72x = 13x^2 + 26x$$

Continue simplifying.

$$96x + 48 = 13x^2 + 26x$$

Subtract $96x$ from both sides.

$$48 = 13x^2 - 70x$$

Subtract 48 from both sides.

$$0 = 13x^2 - 70x - 48$$

Factor the right side.

$$0 = (13x + 8)(x - 6)$$

Set each factor equal to 0.

$$13x + 8 = 0 \text{ or } x - 6 = 0$$

Solve each equation.

$$13x = -8 \qquad x = 6$$

$$x = -\frac{8}{13}$$

The only acceptable answer is $x = 6$ since $-\frac{8}{13}$ is not an integer. If $x = 6$, then $x + 2 = 8$, so the two consecutive even integers are 6 and 8.

Check:
The check is left as an exercise for the student.

PRACTICE EXERCISES

1) The denominator of a fraction is 2 more than the numerator. If 3 is added to the numerator and subtracted from the denominator, the result is equal to $\frac{9}{5}$. Find the original fraction.

2) Two numbers differ by 2. If $\frac{1}{4}$ of the larger number is 4 less than $\frac{1}{5}$ the smaller number, find the numbers.

3) The sum of a fraction and twice its reciprocal is $3\frac{14}{15}$. Find the fraction. (Hint: Let x be the fraction.)

• **Work problems**

The next type of problem to be discussed is the "work" problem. In this type of problem, a task can be performed in a certain amount of time by one individual or thing, and the same task can be performed by another individual or thing in a different amount of time. We are then asked how long it would take for the two working together to perform the same task. There are many variations to this problem. The key to solving this kind of problem is to represent the part of the task that can be done in one unit of time as the rate. For example, if it takes 5 hours to do a job, then the rate is $\frac{1}{5}$, which is the part of the job that can be done in one hour. If it takes x hours to do a job, then the rate is $\frac{1}{x}$, which is the part of the job that can be done in one hour.

The fractional part of the total job done by one individual is the rate times the number of hours worked. For example, if the rate is $\frac{1}{5}$ and the time worked is 3 hours, then that indivdual has done $\frac{1}{5}(3) = \frac{3}{5}$ of the total job. Stated mathematically, the sum of the fractional parts of the job done by each individual is equal to 1, which represents the total job. We suggest using a chart like the one below.

	Time to Do the Job Alone	Rate	Time Worked	Fractional Part of the Total Job Done
Individual A				
Individual B				

Use the following as a guide in filling in the chart.

SOLVING WORK PROBLEMS

1) Assign a variable to the unknown.

2) Find the rate each individual works. This is the part of the job that can be done in one unit of time. (1 divided by the number of hours for the individual to do the job alone.)

3) Find the fractional part of the job that each individual can do by multiplying the rate and the time worked.

4) Write an equation showing the sum of the fractional parts done by each individual equal to one.

5) Solve the equation.

6) See if the results make sense in the words of the original problem.

EXAMPLE 2

a) If Mike can paint his room in 10 hours and Susan can paint the same room in 7 hours, how long would it take them to paint the room working together?

Solution:

Let x be the number of hours to paint the room working together. Consequently, the time worked for each of them is x hours.

	Time to Do the Job Alone	Rate	Time Worked	Fractional Part of the Total Job Done
Mike	10	$\dfrac{1}{10}$	x	$\dfrac{1}{10} \cdot x = \dfrac{x}{10}$
Susan	7	$\dfrac{1}{7}$	x	$\dfrac{1}{7} \cdot x = \dfrac{x}{7}$

Now we write the equation using the following word equation as a guide.

(fractional part Mike paints) + (fractional part Susan paints) = 1 (the total job)

$$\dfrac{x}{10} \qquad + \qquad \dfrac{x}{7} \qquad = 1 \quad \text{Multiply both sides by 70.}$$

$$70\left(\dfrac{x}{10} + \dfrac{x}{7}\right) = 70 \cdot 1 \qquad \text{Distribute on the left, multiply on the right.}$$

$$70\left(\dfrac{x}{10}\right) + 70\left(\dfrac{x}{7}\right) = 70 \qquad \text{Perform the multiplications on the left.}$$

$$7x + 10x = 70 \qquad \text{Combine like terms.}$$

$$17x = 70 \qquad \text{Divide both sides by 17.}$$

$$x = \frac{70}{17} \text{ hours} \qquad \text{Therefore, it takes } \frac{70}{17} \text{ hours working together.}$$

Does the above answer seem reasonable? If both worked at the same rate as Mike, it would take half as long as Mike took which would be 5 hours. If both worked at the same rate as Susan, it would take 3.5 hours. The amount of time working together should be between 3.5 and 5 hours. Is $\frac{70}{17}$ between 3.5 and 5?

b) Sal and Melissa work in an electronics factory. Together they can assemble 30 televisions in 8 hours. If Sal can assemble 30 televisions in 14 hours, how long would it take Melissa to assemble 30 televisions working alone?

Solution:

Let x equal the number of hours for Melissa to assemble 30 televisions working alone.

	Time to Do the Job Alone	Rate	Time Worked	Fractional Part of the Total Job Done
Sal	14	$\frac{1}{14}$	8	$\frac{8}{14}$
Melissa	x	$\frac{1}{x}$	8	$\frac{8}{x}$

Write the equation using the following word equation as a guide.

(part Sal does) + (part Melissa does) = 1

$$\frac{8}{14} \qquad + \qquad \frac{8}{x} \qquad = 1 \qquad \text{Multiply both sides by } 14x.$$

$$14x\left(\frac{8}{14} + \frac{8}{x}\right) = 14x \cdot 1 \qquad \text{Simplify both sides.}$$

$$14x\left(\frac{8}{14}\right) + 14x\left(\frac{8}{x}\right) = 14x \qquad \text{Continue simplifying.}$$

$$8x + 112 = 14x \qquad \text{Subtract } 8x \text{ from both sides.}$$

$$112 = 6x \qquad \text{Divide both sides by 6.}$$

$$\frac{112}{6} = x \qquad \text{Reduce to lowest terms.}$$

$$\frac{56}{3} = x \qquad \text{Therefore, it takes Melissa } 18\frac{2}{3} \text{ hours working alone. Is this answer reasonable?}$$

PRACTICE EXERCISES

Solve the following.

4) If one pipe can fill a tank in 12 hours and a second pipe can fill a tank in 8 hours, how long would it take to fill the tank with both pipes open?

5) John can clean the pool in 6 hours. If he and Johanna work together, they can clean the pool in 4 hours. How long would it take Johanna to clean the pool working alone?

The distance traveled depends upon the rate (speed) and the time traveled according to the formula $d = rt$. For example, if an object moves at the rate of 40 miles per hour for 5 hours, the distance is $d = 40 \cdot 5 = 200$ miles. Equivalent forms of this formula are $r = \frac{d}{t}$ and $t = \frac{d}{r}$. For example, if an object travels 300 miles in 6 hours, then $r = \frac{300}{6} = 50$ mph. If an object travels 240 miles at a rate of 60 miles per hour, then $t = \frac{240}{60} = 4$ hours.

Common kinds of problems that involve distance, rate, and time concern the motion of two different persons or objects. The most efficient way to approach this type of problem is to use a table. The things that are moving are put in a column on the left of the table and the distances, rates, and times are placed in three adjacent columns. We will always know the values of one of distance, rate, or time for each person or object. So one column will be filled in with numbers. We will also know a relationship between two of the remaining quantities. For example, if we know the values of the two distances, then we will also know a relationship between the rates or the times. Suppose we know a relationship between the rates. We let the variable represent one of the rates and express the other rate in terms of the variable. We then use the relationships $d = rt$, $r = \frac{d}{t}$, and $t = \frac{d}{r}$ to fill in the remaining column and write the equation using the information from the last column filled in. This procedure is summarized below.

SOLVING DISTANCE, RATE, TIME PROBLEMS

1) Make a chart with columns for the moving things, distance, rate, and time.

2) Fill in one column with known numerical values.

3) Assign a variable and fill in another column with expressions containing variables.

4) Fill in the remaining column from the first two using the appropriate relationships between d, r, and t.

5) Draw a diagram of the problem.

6) Write the equation using the information in the last column filled in.

7) Solve the equation.

8) Check the solution in the wording of the original problem.

EXAMPLE 3

Solve the following.

a) Tom and John Truong live in cities that are 80 miles apart. Both leave home at 8:00 A.M. traveling toward each other riding bicycles. Tom's speed is two-thirds of John's. Find John's speed if they meet at 10:00 A.M.

Solution:
Consider the chart below.

	d	r	t
Tom			
John			

Since each rider left at 8:00 A.M. and they met at 10:00 A.M., we know that both times are two hours. Put 2 in the time column for each.

	d	r	t
Tom			2
John			2

We also know that Tom's speed is two-thirds of John's. Let John's speed be x, then Tom's speed is $\frac{2}{3}x$. Put these in the table.

	d	r	t
Tom		$\frac{2}{3}x$	2
John		x	2

The rate and time columns have been filled. Since $d = rt$, fill in the d column by finding the products of the expressions in the r and t columns. Therefore, Tom's distance is $\frac{2}{3}x(2) = \frac{4}{3}x$ and John's distance is $2x$.

	d	r	t
Tom	$\frac{4}{3}x$	$\frac{2}{3}x$	2
John	$2x$	x	2

The last column filled in was the distance column, so we write our equation using some known fact(s) about distance. Sometimes it is helpful to draw a diagram.

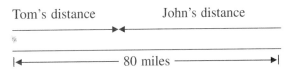

From the diagram, we see that Tom's distance plus John's distance equals 80. Therefore, the equation is:

$\dfrac{4}{3}x + 2x = 80$	Multiply both sides by 3.
$3\left(\dfrac{4}{3}x + 2x\right) = 3 \cdot 80$	Simplify each side.
$3\left(\dfrac{4}{3}x\right) + 3 \cdot 2x = 240$	Continue simplifying.
$4x + 6x = 240$	Add like terms.
$10x = 240$	Divide both sides by 10.
$x = 24$	Therefore, John's rate is 24 mph.

Check:

If John's rate is 24 mph, he will travel 48 miles in two hours. Since Tom's rate is two-thirds of John's, he travels at 16 mph. In two hours, Tom will travel 32 miles. Together they have traveled $48 + 32 = 80$ miles which is the distance between the cities. Therefore, our solution is correct.

b) An airplane flew 450 miles from city A to city B in $1\frac{1}{2}$ hours less than it took to fly 1200 miles from city C to city D. If the airplane flew at the same rate on both trips, find the rate of the airplane.

Solution:

Draw and label the chart.

	d	r	t
A to B			
C to D			

We know the distance from city A to city B is 450 miles and the distance from city C to city D is 1200 miles. We are looking for the rate of the plane. Since it was the same on both trips, let x represent the rate of the plane. Enter these values in the table.

	d	r	t
A to B	450	x	
C to D	1200	x	

The d and r columns are filled in. Since $t = \frac{d}{r}$, fill in the t column by dividing the expressions in the d column by the expressions in the r column. Hence, the time from A to B $= \frac{450}{x}$ and the time from C to D $= \frac{1200}{x}$. Enter these values in the table.

	d	r	t
A to B	450	x	$\frac{450}{x}$
C to D	1200	x	$\frac{1200}{x}$

The last column filled is the time column. Therefore, our equation involves time. We know the trip from A to B took $1\frac{1}{2}$ hours less than the time from C to D. So our equation is of the form (time from A to B) = (time from C to D) $- 1\frac{1}{2}$. Therefore, the equation is:

$$\frac{450}{x} = \frac{1200}{x} - \frac{3}{2}$$
Multiply both sides by $2x$.

$$2x\left(\frac{450}{x}\right) = 2x\left(\frac{1200}{x} - \frac{3}{2}\right)$$
Simplify both sides.

$$2 \cdot 450 = 2x\left(\frac{1200}{x}\right) - 2x\left(\frac{3}{2}\right)$$
Continue simplifying.

$$900 = 2(1200) - 3x$$
Continue simplifying.

$$900 = 2400 - 3x$$
Subtract 2400 from both sides.

$$-1500 = -3x$$
Divide both sides by $-3x$.

$$500 = x$$
Therefore, the rate of the plane is 500 mph.

Check:

If the plane flies at a rate of 500 mph, the trip of 450 miles would take $\frac{450}{500} = \frac{9}{10}$ of an hour and the trip of 1200 miles would take $\frac{1200}{500} = \frac{12}{5}$ of an hour. The difference in times is $\frac{12}{5} - \frac{9}{10} = \frac{24}{10} - \frac{9}{10} = \frac{15}{10} = \frac{3}{2}$ hours. Therefore, our solution is correct.

c) A river has a current of 5 mph. A boat can make a trip of 20 miles upstream in the same time it can make a trip of 30 miles downstream. Find the rate of the boat in still water.

Solution:
Draw and label the chart.

	d	r	t
upstream			
downstream			

We know the distance upstream is 20 miles and the distance downstream is 30 miles. Let x be the speed of the boat in still water. Since the rate of the current is 5 mph, the current will be slowing the rate of the boat upstream by 5 mph. Therefore, the rate of the boat upstream is $x - 5$. When the boat is going downstream, the current is assisting the boat by 5 mph. Therefore, the rate of the boat downstream is $x + 5$. Enter these values in the table.

	d	r	t
upstream	20	$x - 5$	
downstream	30	$x + 5$	

Now we fill in the column for t. Since $t = \frac{d}{r}$, divide the expressions in the d column by the expressions in the r column. Consequently, the time going upstream is $\frac{20}{x-5}$ and the time going downstream is $\frac{30}{x+5}$. Enter these values in the table.

	d	r	t
upstream	20	$x - 5$	$\dfrac{20}{x-5}$
downstream	30	$x + 5$	$\dfrac{30}{x+5}$

The last column filled was the t column. Therefore, our equation involves time. We know the time going upstream is equal to the time going downstream. So our equation is:

$\dfrac{20}{x - 5} = \dfrac{30}{x + 5}$	Cross multiply.
$20(x + 5) = 30(x - 5)$	Simplify both sides.
$20x + 100 = 30x - 150$	Subtract $20x$ from both sides.
$100 = 10x - 150$	Add 150 to both sides.
$250 = 10x$	Divide both sides by 10.
$25 = x$	Therefore, the boat travels at 25 mph in still water.

Check:
If the boat travels at the rate of 25 mph in still water, then its speed against the current is 20 mph. In order to go 20 miles upstream, it would take 1 hour. The speed of the boat with the current is 30 mph. In order to go 30 miles downstream, it would also take 1 hour. The time going upstream equals the time going downstream. Therefore, our solution is correct.

Solve the following.

6) Ellie and Frank live in cities that are 140 miles apart. Both leave home at 7 A.M. traveling by bicycle toward each other. What is Frank's rate if they meet at 11 A.M. and Ellie's rate is three-fourths of Frank's?

7) A plane flew 600 miles in $1\frac{3}{4}$ hours less time than it flew 1125 miles. If the rate is the same for both trips, find the rate of the plane.

8) A plane flew 1500 miles against the wind in the same time it flew 1800 miles with the wind. If the speed of the wind was 50 mph, find the speed of the plane in still air.

9) Shareen makes a trip at an average speed of 60 mph, and Abdul makes the same trip at an average speed of 50 mph. If Abdul's time is $1\frac{1}{2}$ hours more than Shareen's, how much time did it take Shareen to make the trip?

EXERCISE SET   **5.8**

Solve the following using methods shown in Example 1.

1) The numerator and the denominator of a fraction are the same. If 2 is added to the numerator, the new fraction is $\frac{14}{10}$. Find the original numerator.

2) The numerator and the denominator of a fraction are the same. If 4 is subtracted from the denominator, the new fraction equals $\frac{39}{27}$. Find the original fraction.

3) The denominator of a fraction is 3 less than the numerator. If 2 is added to both the numerator and the denominator, the new fraction is $\frac{12}{9}$. Find the original fraction.

4) The numerator of a fraction is 5 more than the denominator. If 3 is subtracted from both the numerator and the denominator, the result is 6. Find the original fraction.

5) One integer is 3 more than another integer. If $\frac{1}{4}$ the larger is equal to $\frac{1}{3}$ the smaller, find the two integers.

6) One integer is 12 less than another integer. If $\frac{1}{7}$ of the larger is equal to $\frac{1}{5}$ of the smaller, find the two integers.

7) Two numbers differ by 6. If $\frac{2}{3}$ of the larger is 5 more than $\frac{1}{2}$ of the smaller, find the two numbers.

8) Two numbers differ by 8. If $\frac{3}{2}$ of the smaller is 6 less than $\frac{3}{4}$ of the larger, find the two numbers.

9) The sum of the reciprocals of two consecutive even integers is $\frac{5}{12}$. Find the two consecutive even integers.

10) Three times the reciprocal of the sum of two consecutive odd integers is $\frac{3}{20}$. Find the two consecutive odd integers.

11) A car costs $\frac{5}{6}$ as much as a van. If together they cost $33,000, how much did the van cost?

12) An electric light bulb uses $\frac{1}{18}$ as much power as an electric toaster. If together they use 1140 watts of power, how many watts does the toaster use?

Solve the following using methods shown in Example 2. Convert fractions to mixed numbers.

13) Jason can wash and wax his car in 4 hours. His younger sister can wash and wax the same car in 6 hours. Working together, in how much time can they wash and wax the car?

14) Jon can clean the house in 7 hours and Linda can clean the house in 5 hours. How long would it take if they worked together?

15) Grandma and Grandpa have volunteered to make quilts for HIV positive babies. If Grandma can make a quilt in 25 days and Grandpa can make a quilt in 35 days, in how many days can they make a quilt working together?

16) It takes Alice 90 minutes to put a futon frame together, and it takes Maya 60 minutes to put the same type of frame together. If they worked together, how long would it take to put a frame together?

17) Working together it takes two roofers 4 hours to put a new roof on a portable classroom. If the first roofer can do the job by himself in 6 hours, how many hours would it take for the second roofer to do the job by himself?

18) Working together, Rita and Tiffany can mow and trim a lawn in 2 hours. If it takes Rita 5 hours to do the lawn by herself, how long would it take Tiffany to do the lawn by herself?

19) With both the cold water and the hot water faucets open, it takes 9 minutes to fill a bathtub. The cold water faucet alone takes 15 minutes to fill the tub. How long would it take to fill the tub with the hot water faucet alone?

20) The cargo hold of a ship has two loading pipes. Used together the two pipes can fill the cargo hold in 6 hours. If the larger pipe alone can fill the hold in 8 hours, how many hours would it take the smaller pipe to fill the hold by itself?

Solve the following using the methods in Example 3.

21) Two buses leave Kansas City traveling in opposite directions, one going east and one going west. The eastbound bus travels at 40 mph and the westbound bus travels at 65 mph. If both buses leave at 10:00 A.M., at what time will they be 525 miles apart?

22) Sailing in opposite directions, an aircraft carrier and a destroyer leave their base in Hawaii at 5:00 A.M. If the destroyer sails at 30 mph and the aircraft carrier sails at 20 mph, at what time will the two ships be 300 miles apart?

23) Rapid City is 360 miles from Sioux Falls. At 6:00 A.M. a freight train leaves Rapid City for Sioux Falls and at the same time a passenger train leaves Sioux Falls for Rapid City. The two trains meet at 9:00 A.M. If the freight train travels $\frac{3}{5}$ as fast as the passenger train, how fast does the passenger train travel?

24) Houston and Calgary are about 2100 miles apart. At 2:00 P.M. an airplane leaves Houston for Calgary flying at 250 mph. At the same time an airplane leaves Calgary for Houston flying at 450 mph. How long will it be before the two airplanes meet?

25) An automobile makes a trip of 270 miles in $2\frac{1}{2}$ hours less than it takes the same auto to make a trip of 420 miles traveling at the same speed. What is the speed of the automobile?

26) A ship leaves port traveling at 15 mph. Two hours later a speed boat leaves the same port traveling at 40 mph. How long will it take for the speed boat to overtake the ship?

27) An airliner flies with the wind from Washington, D.C. to San Francisco in 5 hours. It flies back to Washington, D.C. against the wind in 5.5 hours. If the average speed of the wind is 21 mph, what is the speed of the airliner in still air?

28) A river has a current of 3 mph. A boat goes 40 miles upstream in the same time as it goes 50 miles downstream. Find the speed of the boat in still water.

Solve the following.

29) The denominator of a fraction is twice the numerator. If 3 is added to the numerator and 3 is subtracted from the denominator, the new fraction is $\frac{11}{13}$. Find the original fraction.

30) The same number is added to the numerator and the denominator of the fraction $\frac{4}{5}$. If the new fraction is $\frac{33}{36}$, find the number added.

31) The pitcher of a baseball team has $\frac{2}{3}$ as many hits as the catcher and together they have 85 hits. How many hits does each player have?

32) Two numbers differ by 6. If $\frac{3}{4}$ of the smaller number is $\frac{1}{4}$ less than $\frac{2}{3}$ of the larger, find the two numbers.

33) Hank can load a moving van in 12 hours, and Andy can load a moving van in 15 hours. Working together, how long would it take them to load the moving van?

34) In a bicycle race, the first rider crossed the finish line $2\frac{1}{4}$ hours before the last rider. If the winner had an average speed of 26 mph, and the last rider had an average speed of 16 mph, how long did it take the winner to finish the race?

CHALLENGE EXERCISES (35–37)

35) Find a positive number which is equal to 1 less than 12 times its reciprocal.

36) With both hot and cold water faucets on, it takes 5 minutes to fill a sink. However, the stopper is broken so that water is running out of the sink at the same time that water is running into the sink. If the leak can empty the sink in 15 minutes, how much time will it take the sink to fill to $\frac{1}{2}$ full with both faucets on?

37) Dr. Gilbert spent 4 hours on a bicycle ride in the country. He rode out to the country at 15 mph. His bicycle broke down, and he had to walk back at 3 mph. How far out in the country did he get?

WRITING EXERCISES OR GROUP ACTIVITY

If done as a group activity, each group should write two exercises of each of the following types, exchange with another group, and then solve.

38) Write and solve a "number" word problem.

39) Write and solve a "work" word problem.

40) Write and solve a "distance, time, rate" word problem.

Ratio and Proportion

OBJECTIVES

When you complete this section, you will be able to:

Ⓐ Recognize a proportion.
Ⓑ Simplify proportions.
Ⓒ Solve proportions.
Ⓓ Solve word problems using proportions.

INTRODUCTION

In this section we will discuss a special type of equation that involves rational expressions. First we need some definitions.

DEFINITION OF RATIO

A ratio is a fraction that expresses the comparison of two quantities. A ratio can also be expressed using a colon. In symbols, the ratio of a to b is written as $\frac{a}{b}$ or $a{:}b$.

Since ratios may be written as fractions, they are usually written reduced to lowest terms.

EXAMPLE 1

Write each of the following as ratios in lowest terms.

a) At a large university there are 30,000 students. There are 20,000 men and 10,000 women. What is the ratio of men to women?

Solution:

Since we want the ratio of men to women, the numerator is the number of men and the denominator is the number of women.

$$\frac{\text{number of men}}{\text{number of women}}$$ Substitute for the number of men and women.

$$\frac{20000}{10000}$$ Reduce to lowest terms.

$$\frac{2}{1}$$ Therefore, the ratio of men to women is 2 to 1 or 2:1.

b) A recipe for homemade cereal calls for 6 cups of rolled oats, 1 cup of sunflower seeds, 2 cups of chopped walnuts, and 4 cups of dried fruit. i) What is the ratio of walnuts to rolled oats? ii) What is the ratio of rolled oats to dried fruit?

Solution:

i) Since we want the ratio of walnuts to rolled oats, the numerator is the number of cups of walnuts and the denominator is the number of cups of rolled oats.

$$\frac{\text{number cups walnuts}}{\text{number cups rolled oats}}$$ Substitute.

$$\frac{2}{6}$$ Reduce to lowest terms.

$$\frac{1}{3}$$ Therefore, the ratio of walnuts to rolled oats is 1:3.

ii) Since we want the ratio of rolled oats to dried fruit, the numerator is the number of cups of rolled oats and the denominator is the number of cups of dried fruit.

$$\frac{\text{number cups rolled oats}}{\text{number cups dried fruit}}$$ Substitute.

$$\frac{6}{4}$$ Reduce to lowest terms.

$$\frac{3}{2}$$ Therefore, the ratio of rolled oats to dried fruit is 3:2.

PRACTICE EXERCISES

Write each of the following as ratios in lowest terms.

1) A citrus grove produces 18,000 boxes of oranges and 12,000 boxes of grapefruit. What is the ratio of oranges to grapefruit?

2) A recipe for fruit punch calls for 96 ounces of water, 16 ounces of tea, 12 ounces of frozen lemonade, 64 ounces of cranberry juice, 32 ounces of apple juice, and 16 ounces of orange juice. i) What is the ratio of water to tea? ii) What is the ratio of lemonade to apple juice?

If two ratios are equal, we call the resulting equation **a proportion.**

> **DEFINITION OF PROPORTION**
>
> A proportion is an equation that states that two ratios are equal. In symbols, if $\frac{a}{b}$ and $\frac{c}{d}$ are ratios such that $\frac{a}{b} = \frac{c}{d}$, then $\frac{a}{b} = \frac{c}{d}$ is called a proportion.

Suppose that we simplify a proportion by multiplying both sides by the LCD.

$$\frac{a}{b} = \frac{c}{d}$$ Multiply both sides by the LCD of bd.

$$bd \cdot \frac{a}{b} = bd \cdot \frac{c}{d}$$ Simplify both sides.

$$ad = bc$$

This result is often referred to as cross multiplying and was first discussed in Section 5.4.

$$\frac{a}{b} \times \frac{c}{d}$$

In other words, if two ratios can form a proportion the cross products must be equal.

EXAMPLE 2

Determine if the following fractions can form a proportion.

a) $\dfrac{4}{5}$ and $\dfrac{16}{20}$

$\dfrac{4}{5} \diagup\!\!\!\!\diagdown \dfrac{16}{20}$ Cross multiply.

$4(20) \overset{?}{=} 5(16)$ Multiply.

$80 = 80$ Therefore, the fractions are equal and can form a proportion.

$\dfrac{4}{5} = \dfrac{16}{20}$ Proportion.

b) $\dfrac{2}{3}$ and $\dfrac{9}{12}$

$\dfrac{2}{3} \diagup\!\!\!\!\diagdown \dfrac{9}{12}$ Cross multiply.

$2(12) \overset{?}{=} 3(9)$ Multiply.

$24 \neq 27$ Therefore, the fractions are not equal and cannot form a proportion.

$\dfrac{2}{3} \neq \dfrac{9}{12}$ Not a proportion.

ADDITIONAL PRACTICE

Determine if the following fractions can form a proportion.

a) $\dfrac{3}{4}$ and $\dfrac{15}{30}$

b) $\dfrac{18}{21}$ and $\dfrac{12}{14}$

PRACTICE EXERCISES

Determine if the following fractions can form a proportion.

3) $\dfrac{8}{9}$ and $\dfrac{30}{36}$ 4) $\dfrac{16}{28}$ and $\dfrac{4}{7}$

If more practice is needed, do the Additional Practice Exercises in the margin.

Proportions may be used to express the relationship between both known and unknown quantities. If one of the quantities of a proportion is unknown, we may solve for that quantity by cross multiplying and solving the resulting equation.

PROCEDURE FOR SOLVING PROPORTIONS

To solve a proportion, perform the following steps.

1) If necessary, reduce all fractions to lowest terms.
2) Cross multiply the proportion.
3) Solve the resulting equation.

EXAMPLE 3

Solve the following proportions.

a) $\dfrac{9}{t} = \dfrac{3}{14}$ Neither fraction may be simplified. Cross multiply.

$\quad 9(14) = 3t$ $9(14) = 126.$

$\quad 126 = 3t$ Divide both sides by 3.

$\quad \dfrac{126}{3} = \dfrac{3t}{3}$ Divide.

$\quad 42 = t$ Solution.

b) $\dfrac{x}{20} = \dfrac{9}{12}$ Reduce $\frac{9}{12}$.

$\quad \dfrac{x}{20} = \dfrac{3}{4}$ Cross multiply.

$\quad 4x = 3 \cdot 20$ Multiply.

$\quad 4x = 60$ Divide both sides by 4.

$\quad \dfrac{4x}{4} = \dfrac{60}{4}$ Divide.

$\quad x = 15$ Solution.

c) $\dfrac{2}{3} = \dfrac{8}{x + 6}$ Cross multiply.

$\quad 2(x + 6) = 3 \cdot 8$ Simplify both sides.

$\quad 2x + 12 = 24$ Subtract 12 from both sides.

$\quad 2x + 12 - 12 = 24 - 12$ Simplify both sides.

$\quad 2x = 12$ Divide both sides by 2.

$\quad \dfrac{2x}{2} = \dfrac{12}{2}$ Simplify both sides.

$\quad x = 6$ Solution.

d) $\dfrac{x + 2}{3} = \dfrac{2}{x - 3}$ Cross multiply.

$\quad (x + 2)(x - 3) = 3 \cdot 2$ Simplify both sides.

$\quad x^2 - x - 6 = 6$ Subtract 6 from both sides.

$\quad x^2 - x - 6 - 6 = 6 - 6$ Simplify both sides.

$\quad x^2 - x - 12 = 0$ Factor the left side.

$\quad (x - 4)(x + 3) = 0$ Set each factor equal to 0.

$\quad x - 4 = 0, \; x + 3 = 0$ Solve each equation.

$\quad x = 4, \qquad x = -3$ Solutions.

ADDITIONAL PRACTICE

Solve the following proportions.

c) $\dfrac{6}{y} = \dfrac{12}{8}$

d) $\dfrac{12}{10} = \dfrac{18}{v}$

e) $\dfrac{4}{7} = \dfrac{12}{4x + 1}$

f) $\dfrac{4}{x + 5} = \dfrac{x - 3}{5}$

PRACTICE EXERCISES

Solve the following proportions.

5) $\dfrac{v}{30} = \dfrac{2}{5}$ **6)** $\dfrac{12}{16} = \dfrac{y}{20}$

7) $\dfrac{10}{x + 6} = \dfrac{5}{7}$ **8)** $\dfrac{x - 4}{2} = \dfrac{8}{x + 2}$

If more practice is needed, do the Additional Practice Exercises in the margin.

Proportions appear in many types of applications problems. The following examples illustrate just a few.

EXAMPLE 4

Solve each of the following.

a) An automobile travels 224 miles on 8 gallons of gasoline. How far can it travel on 15 gallons of gasoline?

Solution:

We let x represent the number of miles the car can travel on 15 gallons. The key to the solution is the assumption that the rate of consumption remains constant. We set up the ratios in the form of $\frac{\text{miles}}{\text{gallons}}$.

$$\frac{224 \text{ miles}}{8 \text{ gallons}} = \frac{x \text{ miles}}{15 \text{ gallons}}$$ Drop units.

$$\frac{224}{8} = \frac{x}{15}$$ Reduce $\frac{224}{8}$ to lowest terms.

$$\frac{28}{1} = \frac{x}{15}$$ Cross multiply.

$$28(15) = 1(x)$$ Multiply.

$$420 = x$$ Therefore, the car can travel 420 miles on 15 gallons of gasoline.

Note: There are several other forms of this proportion that will also work. For example, $\frac{8}{224} = \frac{15}{x}$, or $\frac{224}{x} = \frac{8}{15}$, or $\frac{15}{8} = \frac{x}{224}$. Notice that if you cross multiply, before reducing, you will always get $8x = 15(224)$ which is how we can tell that they are all equivalent.

b) The state wildlife department wants to estimate the number of deer in a game preserve. They capture 34 deer, tag them, and release them. Later, they capture 45 deer and find that 17 are tagged. Estimate the number of deer in the preserve.

Solution:

Let y be the number of deer in the preserve. We assume that the ratio of tagged deer to total deer in the sample is the same as the ratio of tagged deer to total deer in the preserve. We set up the ratio in the form $\frac{\text{number tagged}}{\text{total number}}$.

$$\frac{17 \text{ tagged}}{45 \text{ total}} = \frac{34 \text{ tagged}}{y \text{ total}}$$ Drop labels.

$$\frac{17}{45} = \frac{34}{y}$$ Cannot simplify. Cross multiply.

$$17y = 45(34)$$ $45(34) = 1530.$

$$17y = 1530$$ Divide both sides by 17.

$$\frac{17y}{17} = \frac{1530}{17}$$ Simplify both sides.

$$y = 90$$ Therefore, there are about 90 deer in the preserve.

Again, there are several equivalent forms of the proportion that could have been used.

c) The distance between two cities is 150 miles. The scaled distance between the cities on a map is drawn as $1\frac{1}{2}$ inches. If the distance between two other cities is 900 miles, what is the scaled distance between them on the map?

Solution:

Let d be the distance between the two cities on the map. The ratio of miles between the cities to inches on the map remains constant. We set up the ratio in the form $\frac{miles}{inches}$.

$$\frac{150 \text{ miles}}{1\frac{1}{2} \text{ inches}} = \frac{900 \text{ miles}}{d \text{ inches}} \qquad \text{Drop units.}$$

$$\frac{150}{1\frac{1}{2}} = \frac{900}{d} \qquad \text{Convert mixed number to improper fraction.}$$

$$\frac{150}{\frac{3}{2}} = \frac{900}{d} \qquad \text{Cross multiply.}$$

$$150d = \frac{3}{2}(900) \qquad \tfrac{3}{2}(900) = 1350.$$

$$150d = 1350 \qquad \text{Divide both sides by 150.}$$

$$\frac{150d}{150} = \frac{1350}{150} \qquad \text{Simplify both sides.}$$

$$d = 9 \qquad \text{Therefore, the scaled distance between the cities on the map is 9 inches.}$$

PRACTICE EXERCISES

Solve each of the following.

9) A chef knows that he can serve 8 people with 3 pounds of fish fillets. How many pounds of fish fillets does he need to serve 32 people?

10) A manufacturer tested 400 randomly chosen widgets as they came off the production line. If she found 15 defective widgets, how many defective widgets should she expect to find in 2000 widgets?

11) On a map, $1\frac{1}{4}$ inches represents 100 miles on the surface of the earth. What is the distance on the earth between two locations that are 15 inches apart on the map?

EXERCISE SET 5.9

Solve the following using a proportion.

1) One year an auto dealer sold 500 cars and 245 trucks. What was the ratio of cars to trucks?

2) Last month the college bookstore sold 624 T-shirts and 156 sweatshirts. What was the ratio of T-shirts to sweatshirts?

3) In a county referendum, 50,000 people voted for higher taxes and 65,000 voted against higher taxes. What was the ratio of people who voted for the higher taxes to those who voted against higher taxes?

4) On a freight train, there are 42 boxcars and 72 flatcars. What is the ratio of boxcars to flatcars?

Determine if the following ratios can form a proportion.

5) $\dfrac{4}{5}$ and $\dfrac{12}{15}$ 6) $\dfrac{20}{24}$ and $\dfrac{25}{30}$ 7) $\dfrac{2}{3}$ and $\dfrac{8}{12}$

8) $\dfrac{5}{6}$ and $\dfrac{4}{3}$ 9) $\dfrac{4}{9}$ and $\dfrac{20}{45}$ 10) 511 and $\dfrac{7}{9}$

Solve the following proportions.

11) $\dfrac{4}{5} = \dfrac{12}{x}$ 12) $\dfrac{y}{24} = \dfrac{25}{30}$

13) $\dfrac{4}{6} = \dfrac{t}{18}$ 14) $\dfrac{9}{5} = \dfrac{y}{35}$

15) $\dfrac{v}{8} = \dfrac{14}{56}$ 16) $\dfrac{22}{33} = \dfrac{n}{44}$

17) $\dfrac{13}{6} = \dfrac{t}{42}$ 18) $\dfrac{2}{14} = \dfrac{8}{y}$

19) $\dfrac{9}{w} = \dfrac{22.5}{1.5}$ 20) $\dfrac{v}{0.7} = \dfrac{5}{7}$

21) $\dfrac{y}{14} = \dfrac{\frac{3}{7}}{2}$ 22) $\dfrac{\frac{6}{5}}{8} = \dfrac{r}{20}$

23) $\dfrac{2}{3} = \dfrac{8}{x+8}$ 24) $\dfrac{3}{4} = \dfrac{9}{x+5}$

25) $\dfrac{x+5}{6} = \dfrac{7}{3}$ 26) $\dfrac{x+4}{20} = \dfrac{3}{4}$

27) $\dfrac{3}{4} = \dfrac{x+1}{x+3}$ 28) $\dfrac{2}{3} = \dfrac{x+1}{x+6}$

29) $\dfrac{3}{2x} = \dfrac{5}{3x+2}$ 30) $\dfrac{4}{3x-3} = \dfrac{7}{4x+1}$

31) $\dfrac{4x-4}{8} = \dfrac{2x+1}{5}$ 32) $\dfrac{7x-1}{8} = \dfrac{2x+4}{3}$

33) $\dfrac{x-3}{2} = \dfrac{5}{x+6}$ 34) $\dfrac{x+5}{7} = \dfrac{1}{x-1}$

35) $\dfrac{5}{2x+3} = \dfrac{x-5}{3}$ 36) $\dfrac{2}{3x+2} = \dfrac{x-3}{17}$

Solve each of the following.

37) A truck travels 440 miles in 11 hours. How far will the truck travel in 7 hours?

38) An airplane uses 350 pounds of fuel to fly 1400 miles. How much fuel will it use to fly 3,500 miles?

39) In 8 months, Joel can read 5 novels. How many novels can he read in 12 months?

40) A machine can produce 160 toys in 3 hours. How long will it take to produce 400 toys?

41) A baseball player got 6 hits in 14 innings of play. If he continued to hit at the same rate, how many innings must he play to get 33 hits?

42) Martha can do 32 mathematics exercises in 15 minutes. How long will it take her to do 96 exercises?

43) Two rivers that are 270 miles apart are 3 inches apart on a map. Find the distance between two locations on the map that are 675 miles apart on the earth.

44) On a map, $\frac{3}{4}$ inch represents a distance of 75 miles on the surface of the earth. What is the distance on the earth between two cities that are 6 inches apart on the map?

45) The state game commission wants to know how many trout there are in a lake. They release 145 tagged trout into the lake. Six months later they return and capture 160 trout of which 58 are tagged. Approximately how many trout are in the lake?

46) The wildlife commission wants to determine how many alligators are in a swamp. Conservation officers capture, tag, and release 33 alligators. One year later they capture 60 alligators and find that 12 are tagged. Approximately how many alligators are in the swamp?

47) A model airplane is built at a scale of 1:40. If a wing on the model is six inches long, how long is the wing of the airplane in feet?

48) A model car is built at a scale of 1:15. If the bumper of the model car is four inches long, how long is the bumper of the car in feet?

49) If a tile store will install 23 square feet of tile for $92, how much would they charge to install 62 square feet?

50) If a VCR counter registers 660 after playing 22 minutes, what will it register after playing 36 minutes? Assume that the counter measures the distance the tape has travelled.

CHALLENGE EXERCISES (51–55)

Solve the following proportions.

51) $\dfrac{x+1}{4} = \dfrac{6}{4x}$ 52) $\dfrac{2}{x+1} = \dfrac{2x}{6}$

53) One of the professors at Urban University can walk 3 miles in 39 minutes. How long will it take him to walk 5 miles?

54) A physician can see 40 patients in 7 hours. How many can he see in 35 hours (one week). At the rate of $60 per patient, how much can he make in 50 weeks (one year)?

55) A 44-pound bag of fertilizer will cover 5000 square feet of lawn. How many pounds of fertilizer are needed to cover 17,500 square feet of lawn? If the fertilizer costs $13.95 a bag, how much will it cost to fertilize the 17,500 square feet?

WRITING EXERCISES

56) What is the underlying assumption on which all of the above proportion exercises are based?

57) Explain why $\frac{a}{b} = \frac{c}{d}$ and $\frac{a}{c} = \frac{b}{d}$ are the same proportion.

WRITING EXERCISES OR GROUP PROJECTS

58) Interview science teachers or students, nursing teachers or students, and other professionals and make a list of five situations in which proportions are used in their professions.

59) There is a special ratio called the Golden Ratio that often appears in art, architecture, mathematics, and many other fields. Go to the library, research the topic, and write a one-page report on the Golden Ratio.

Summary

Rational Expressions: [Section 5.1]

1) A rational expression is an algebraic expression of the form $\frac{P}{Q}$ where P and Q are polynomials and $Q \neq 0$.

2) To find the value(s) for which a rational expression is undefined, set the denominator equal to 0 and solve the resulting equation.

Lowest Terms: [Section 5.1]

A fraction or rational expression is reduced to lowest terms if the numerator and the denominator have no common factors other than one.

Reducing a Fraction or Rational Expression to Lowest Terms: [Section 5.1]

1) Write the numerator and the denominator as the product of their prime factors.

2) Divide the numerator and the denominator by the common factors.

Multiplication of Rational Numbers or Rational Expressions: [Section 5.2]

1) To multiply two or more rational numbers or expressions, multiply numerator times numerator and denominator times denominator. In symbols, $\frac{a}{b} \cdot \frac{c}{d} = \frac{a \cdot c}{b \cdot d}$.

2) All products should be reduced to lowest terms by using the procedure for reducing rational numbers or expressions to lowest terms. You may divide by factors common to both the numerator and the denominator before multiplying.

Division of Rational Numbers and Expressions: [Section 5.3]

To divide rational numbers or expressions, invert the expression on the right and change the operation to multiplication.

Division of Polynomials (Long Division): [Section 5.3]

Long division of polynomials is done using the same algorithm as division of whole numbers except like terms are aligned, instead of digits with the same place value. The steps are:

1) Divide the first term of the divisor into the first term of the dividend (after the dividend has been put into descending order and missing powers are inserted as 0 times the variable raised to the missing power).

2) Multiply the entire divisor by the first quotient (the result of the first division).

3) Align like terms and subtract the expression found in step 2.

4) Bring down the next term of the dividend.

5) Continue steps 1–4 until all terms of the dividend have been used. The remainder (if any) will have a lower degree than the divisor.

Least Common Multiple: [Section 5.4]

The integer a is the least common multiple of the integers b and c if a is the smallest positive integer which is a multiple of both b and c. Equivalently, a is the least common multiple of b and c if a is the smallest positive integer which is divisible by both b and c.

Finding the Least Common Multiple (LCM): [Section 5.4]

1) Write each number or expression as the product of its prime factors.

2) Select every factor which appears in any prime factorization. If a factor appears in more than one of the prime factorizations, select the factor with the largest exponent.

3) The LCM is the product of the factors selected in step 2.

Equivalent Fractions: [Section 5.4]

Two or more fractions are equivalent if they represent the same quantity. Equivalently, $\frac{a}{b} = \frac{c}{d}$ if $a \cdot d = b \cdot c$.

Fundamental Property of Fractions: [Section 5.4]

If $\frac{a}{b}$ is a fraction and c is any number except 0, then $\frac{a}{b} = \frac{ac}{bc}$. In words, we can multiply both the numerator and the denominator of a fraction by any number except 0 and get a fraction equivalent of the original fraction.

Finding the Missing Numerator to Get Equivalent Fractions: [Section 5.4]

Determine the number or expression that you must multiply the denominator of the given fraction by in order to get the denominator of the new fraction. Multiply both the numerator and the denominator of the given fraction by this number to get an equivalent fraction with the given denominator.

Least Common Denominator (LCD): [Section 5.4]

The least common denominator of two or more rational numbers or rational expressions is the least common multiple of the denominators of the fractions.

Adding Rational Numbers or Rational Expressions with Common Denominators: [Section 5.4]

To add rational numbers or rational expressions with common denominators, add the numerators and put the sum over the common denominator. In symbols, $\frac{a}{b} + \frac{c}{b} = \frac{a+c}{b}$. If necessary, reduce the sum to lowest terms.

Adding Rational Numbers or Rational Expressions with Unlike Denominators: [Section 5.5]

To add rational numbers or rational expressions with unlike denominators, change each rational number or rational expression into an equivalent rational number or rational expression with the least common denominator as its denominator. Then add the fractions using the method of adding rational numbers or rational expressions with common denominators.

Simplifying Complex Fractions Using Division: [Section 5.6]

1) If necessary, rewrite the numerator and/or the denominator as single fractions by performing whatever operations are indicated in either.

2) Rewrite the complex fraction as the division of the numerator by the denominator.

3) Perform the division by inverting the fraction on the right and multiplying.

4) If necessary, reduce to lowest terms.

Simplifying Complex Fractions Using the LCD: [Section 5.6]

1) Find the LCD of all fractions which occur in the numerator and/or the denominator.

2) Multiply both the numerator and the denominator by the LCD.

3) Simplify the results.

Solving Equations Involving Rational Numbers or Rational Expressions: [Section 5.7]

1) Determine the value(s) of the variable, if any, that would make any denominator equal to 0.

2) Multiply both sides of the equation by the LCD of all the rational numbers or rational expressions which appear in the equation.

3) Solve the resulting equation.

4) Check for extraneous solutions.

Solving Number Problems: [Section 5.8]

1) Read the problem and assign a variable to the unknown.

2) If there is more than one unknown, represent the other unknowns in terms of the variable assigned in step 1.

3) Write the equation.

4) Solve the equation.

5) Check the solution(s) using the wording in the original problem.

Solving Work Problems: [Section 5.8]

1) Assign a variable to the unknown.

2) Find the rate each individual works.

3) Find the fractional part of the job that each individual can do.

4) Write an equation showing the sum of the fractional parts done by each individual equal to 1.

5) Solve the equation.

6) See if the results make sense.

Solving Distance, Rate, Time Problems: [Section 5.8]

1) Make a chart with columns for the moving things, distance, rate, and time.

2) Fill in one column with known numerical values.

3) Assign a variable and fill in another column with expressions containing variables.

4) Fill in the remaining column from the first two using the appropriate relationships between d, r, and t.

5) Write the equation using the information in the last column filled in.

6) Solve the equation.

7) Check the solution in the wording of the original problem.

Defining Ratio: [Section 5.9]

A ratio is a fraction that expresses the comparison of two quantities. A ratio can also be expressed using a colon. In symbols, the ratio of a to b is written as $\frac{a}{b}$ or $a:b$.

Defining Proportion: [Section 5.9]

A proportion is an equation which states that two ratios are equal. In symbols, $\frac{a}{b} = \frac{c}{d}$ is a proportion.

Solving a Proportion: [Section 5.9]

1) If necessary, reduce all fractions to lowest terms.

2) Cross multiply the proportion.

3) Solve the resulting equation.

Review Exercises

Find the value(s) of the variable(s) for which the following are defined. [Section 5.1]

1) $\dfrac{x}{x+9}$

2) $\dfrac{x+3}{(x+7)(x+2)}$

3) $\dfrac{x+4}{x-5y}$

4) $\dfrac{3a-4}{a^2+4a-12}$

Reduce the following to lowest terms. [Section 5.1]

5) $\dfrac{36}{42}$

6) $\dfrac{64}{72}$

7) $\dfrac{120}{124}$

8) $\dfrac{84}{96}$

9) $\dfrac{x^3y}{x^2y^3}$

10) $\dfrac{m^2n^4}{m^4n^5}$

11) $\dfrac{r^3s^5}{rs^3}$

12) $\dfrac{32x^2y^5}{8xy^3}$

13) $\dfrac{-42p^6q^2}{27p^3q^4}$

14) $-\dfrac{18m^3n^2}{54m^5n^5}$

15) $\dfrac{7(2x+9)}{5(2x+9)}$

16) $\dfrac{4a-20}{7a-35}$

17) $\dfrac{a^2+4a-21}{a^2+9a+14}$

18) $\dfrac{2x-3}{6x^2-x-12}$

19) $\dfrac{25x^2-9}{10x^2+x-3}$

20) $\dfrac{2c+3d}{8c^2+26cd+21d^2}$

21) $\dfrac{3x-4y}{4y-3x}$

22) $\dfrac{5-2x}{4x^2-25}$

Find the following products. Leave answers reduced to lowest terms. [Section 5.2]

23) $\dfrac{18}{7}\times\dfrac{7}{12}$

24) $\dfrac{21}{10}\times\left(-\dfrac{15}{14}\right)$

25) $\dfrac{x^5}{y^3}\cdot\dfrac{y^2}{x^2}$

26) $\dfrac{28m^2n^3}{15p^2q^6}\cdot\dfrac{25p^4q^3}{14mn}$

27) $\dfrac{m-n}{32m^3n}\cdot\dfrac{36mn^3}{(m-n)^2}$

28) $\dfrac{14x-21}{30x-40}\cdot\dfrac{15x-20}{8x-12}$

29) $\dfrac{25x^2-16}{16x+24}\cdot\dfrac{12x+18}{5x-4}$

30) $\dfrac{2x-7}{12}\cdot\dfrac{10}{7-2x}$

31) $\dfrac{x^2-16}{x^2+6x+8}\cdot\dfrac{x^2-3x-10}{x^2-25}$

32) $\dfrac{8x^2+2x-15}{6x^2+x-12}\cdot\dfrac{3x^2-13x+12}{3x^2-7x-6}$

33) $\dfrac{y^2-9}{y^2-3y-18}\cdot\dfrac{y^2-4y-12}{3-y}$

Find the following quotients. Leave answers reduced to lowest terms. [Section 5.3]

34) $\dfrac{5}{8}\div\dfrac{15}{24}$

35) $\dfrac{32}{25}\div\dfrac{28}{35}$

36) $\dfrac{39a^2b^4}{27x^3y^2}\div\dfrac{26a^3b}{28xy^4}$

37) $\dfrac{8y-20}{9}\div\dfrac{6y-15}{10}$

38) $\dfrac{21a+7b}{16a-24b}\div\dfrac{42a+14b}{24a-36b}$

39) $\dfrac{x-y}{8x-4y}\div\dfrac{y-x}{12x-6y}$

40) $\dfrac{z^2}{z^2-2z-8}\div\dfrac{6z^3+3z^4}{z^2+5z+6}$

41) $\dfrac{4p^2-9}{24p+28}\div\dfrac{10p-15}{36p^2-49}$

42) $\dfrac{16p^2-8pq-3q^2}{8p^2+22pq+5q^2}\div\dfrac{8p^2-10pq+3q^2}{10p^2+pq-3q^2}$

Find the following quotients using long division. [Section 5.3]

43) $\dfrac{12x^2+7x-10}{3x-2}$

44) $\dfrac{15x^2-29x-16}{5x+2}$

45) $\dfrac{6a^3-a^2-9a+7}{3a+4}$

46) $\dfrac{16z^2-16z+12}{4z-3}$

47) $\dfrac{-27x-2x^2+8x^3+18}{2x-3}$

48) $\dfrac{3x^3-24x-7}{x-3}$

Find the LCM of the following. [Section 5.4]

49) 6 and 7

50) 25 and 40

51) 16, 24, 32

52) 18, 27, 36

53) a^2y and ay^4

54) $12t^5v^3$ and $15t^2v^7$

55) $14x^2w^3$ and $21x^5w^4$

56) $z-4$ and $z+3$

57) $a+2$ and a^2+4a+4

58) $y^2+8y+16$, $y+4$, y^2+y-12

Find the missing numerator so that the fractions will be equal. [Section 5.4]

59) $\dfrac{6}{5}=\dfrac{?}{20}$

60) $3=\dfrac{?}{15}$

61) $\dfrac{4r}{7p^3}=\dfrac{?}{35p^5}$

62) $\dfrac{8x^3}{13tz^4}=\dfrac{?}{52t^3z^7}$

63) $\dfrac{4a}{a+5}=\dfrac{?}{2a+10}$

64) $\dfrac{7}{b-3}=\dfrac{?}{b^2-9}$

65) $\dfrac{p - 8}{p^2 + 10p + 25} = \dfrac{?}{(p + 5)(p - 3)(p + 5)}$

66) $\dfrac{2q}{y^2 - 5y} = \dfrac{?}{y(y - 5)(y + 3)}$

Find the least common denominator of the following. Then write each fraction as an equivalent fraction with the LCD as its denominator. [Section 5.4]

67) $\dfrac{4}{7}$ and $\dfrac{5}{2}$

68) $\dfrac{5}{6}$ and $\dfrac{7}{18}$

69) $\dfrac{8}{9}$ and $\dfrac{11}{12}$

70) $\dfrac{6}{15}, \dfrac{3}{5},$ and $\dfrac{13}{20}$

71) $\dfrac{4}{t^2v^5}$ and $\dfrac{6}{tv^3}$

72) $\dfrac{8b}{9p^3q^4}$ and $\dfrac{11c}{12p^2q^5}$

73) $\dfrac{7}{t - 4}$ and $\dfrac{9}{t + 2}$

74) $\dfrac{y}{y - 11}$ and $\dfrac{y}{y - 3}$

75) $\dfrac{4u}{8u + 12}$ and $\dfrac{9u}{14u + 21}$

76) $\dfrac{2w}{w^2 - 16}$ and $\dfrac{6w}{w^2 + 5w + 4}$

77) $\dfrac{3a}{a^2 + 6a + 9}$ and $\dfrac{10a}{a^2 - 2a - 15}$

78) $\dfrac{m + 1}{m^2 - 2m - 8}$ and $\dfrac{m - 1}{m^2 - 3m - 4}$

Find the following sums or differences. Leave the answers reduced to the lowest terms. [Section 5.5]

79) $\dfrac{5}{7} - \dfrac{3}{7}$

80) $\dfrac{3}{8} - \dfrac{-5}{8}$

81) $\dfrac{4}{13} + \dfrac{6}{13} + \dfrac{2}{13}$

82) $\dfrac{9}{11} + \dfrac{5}{11} - \dfrac{7}{11}$

83) $\dfrac{9r}{14x} + \dfrac{2p}{14x}$

84) $\dfrac{4}{y + 2} - \dfrac{3}{y + 2}$

85) $\dfrac{x^2}{a^2 - 3a + 4} - \dfrac{x^2 - 5x}{a^2 - 3a + 4}$

86) $\dfrac{t^3 - 8t}{w^2 - 25} + \dfrac{8t + 1}{w^2 - 25}$

87) $\dfrac{3p^2}{p^2 - 16} + \dfrac{3p - 6}{p^2 - 16}$

88) $\dfrac{x - 3y}{x - y} - \dfrac{5x + 2y}{y - x}$

Find the following sums or differences. Leave the answers reduced to lowest terms. [Section 5.5]

89) $\dfrac{5}{12} + \dfrac{7}{16}$

90) $\dfrac{3}{4} - \dfrac{2}{5}$

91) $\dfrac{7}{m} + \dfrac{b}{n}$

92) $\dfrac{10}{9x} - \dfrac{2}{3x}$

93) $\dfrac{y - 3}{4y} + \dfrac{y + 2}{5y}$

94) $\dfrac{2t - 3}{18t^4u^2} + \dfrac{5t + 1}{12t^3u^5}$

95) $\dfrac{8}{w - 3} - \dfrac{-3}{w}$

96) $\dfrac{6r}{r - 5} + \dfrac{2}{5 - r}$

97) $\dfrac{v + 4}{4v + 8} - \dfrac{v - 2}{2v - 6}$

98) $\dfrac{t^2 - 5t}{t^2 + 8t + 16} + \dfrac{6}{t + 4}$

99) $\dfrac{2w + 5}{w^2 - 25} + \dfrac{6w}{w^2 - 3w - 10}$

100) $\dfrac{z + 9}{4z^2 + 33z + 35} - \dfrac{z - 12}{z^2 + 14z + 49}$

Simplify the following complex fractions using both the division and the LCD methods. [Section 5.6]

101) $\dfrac{\frac{4}{5}}{\frac{3}{10}}$

102) $\dfrac{\frac{5}{18}}{\frac{11}{30}}$

103) $\dfrac{\frac{4}{3} - \frac{8}{9}}{\frac{5}{6} - \frac{4}{12}}$

104) $\dfrac{6 - \frac{10}{7}}{\frac{14}{3} - 2}$

105) $\dfrac{\frac{x}{9}}{\frac{y}{27}}$

106) $\dfrac{\frac{p}{r}}{s}$

107) $\dfrac{\frac{4x}{t^2}}{\frac{20x}{t^3}}$

108) $\dfrac{\frac{u^4v^2}{w}}{\frac{uv^3}{w^2}}$

109) $\dfrac{\frac{x^6y^3}{t^4}}{\frac{x^2y^2}{t}}$

110) $\dfrac{2w - \frac{w}{4}}{6 - \frac{w}{4}}$

111) $\dfrac{1 - \frac{1}{y} - \frac{12}{y}}{1 - \frac{6}{y} + \frac{8}{y}}$

112) $\dfrac{x + \frac{25}{x + 10}}{1 - \frac{5}{x + 10}}$

Solve the following equations. [Section 5.7]

113) $\dfrac{y}{4} - \dfrac{y}{5} = \dfrac{3}{20}$

114) $\dfrac{6p}{7} - \dfrac{5p}{14} = 1$

115) $\dfrac{7}{2}t - \dfrac{4}{9}t = \dfrac{2}{3}$

116) $\dfrac{4w}{5} + \dfrac{1}{6} = \dfrac{7w}{3} - \dfrac{13}{10}$

117) $\dfrac{2v - 5}{4} = 3 - v$

118) $\dfrac{z + 6}{7} - \dfrac{3}{14} = \dfrac{5 - 4z}{21}$

119) $\dfrac{5}{4t} - \dfrac{3}{8t} = \dfrac{1}{2}$

120) $\dfrac{12}{m - 2} = 9 + \dfrac{m}{m - 2}$

121) $\dfrac{1}{q + 4} = \dfrac{q}{3q + 2}$

122) $\dfrac{5}{v^2 + 5v + 6} - \dfrac{2}{v^2 - 2v - 8} = \dfrac{3}{v^2 - 16}$

Solve the following. [Section 5.8]

123) The numerator of a fraction is 3 less than the denominator. If 5 is added to the numerator and 5 is subtracted from the denominator, the result is $\frac{3}{2}$. Find the original fraction.

124) The difference of an integer and ten times its reciprocal is 3. Find the integer.

125) George can write a chapter for a mathematics textbook in 30 days. Leon can write a chapter in 45 days. How long would it take them to write a chapter together?

126) The Gulf Stream is an ocean current off the eastern coast of the United States. A Coast Guard cutter (ship) can sail 200 miles against the current in the same time that it can sail 260 miles with the current. If the current is 3 mph, how fast can the cutter sail in still water?

127) It is 510 road miles from Denver to Salt Lake City. A truck leaves Denver for Salt Lake City at 4:00 A.M. At the same time an automobile leaves Salt Lake City for Denver. If the truck travels at 45 mph and the auto travels at 55 mph, what time would it be when the auto and truck meet? (Give answer to the nearest minute.)

Write each of the following as ratios in lowest terms. [Section 5.9]

128) In a college parking lot, there were 625 compact cars and 350 middle-sized cars. What was the ratio of compact cars to middle-sized cars?

Determine if the following fractions can form a proportion. [Section 5.9]

129) $\frac{12}{15}$ and $\frac{20}{25}$

130) $\frac{0.8}{0.24}$ and $\frac{6}{18}$

Solve the following proportions. [Section 5.9]

131) $\frac{2}{x} = \frac{4}{5}$

132) $\frac{8}{5} = \frac{y}{60}$

133) $\frac{48}{15} = \frac{32}{t}$

134) $\frac{u}{14} = \frac{0}{4}$

135) $\frac{y}{2} = \frac{8.4}{3}$

136) $\frac{11}{3} = \frac{x}{0.6}$

137) $\frac{\frac{2}{3}}{5} = \frac{v}{15}$

138) $\frac{6}{\frac{4}{5}} = \frac{35}{v}$

139) $\frac{4}{7} = \frac{2x + 2}{4x + 1}$

140) $\frac{5}{2x + 3} = \frac{7}{4x - 3}$

Solve each of the following. [Section 5.9]

141) A lawn mower uses 2 quarts of gasoline to mow $\frac{1}{3}$ acre of grass. How many acres can be mowed with 9 quarts of gasoline?

142) A machine can fill 3600 cans in 25 minutes. How many cans can be filled in 70 minutes?

143) On a map, two airports are $4\frac{1}{2}$ inches apart. If 90 miles is represented as $\frac{3}{4}$ inch apart on the map, how many miles apart are the two airports?

144) A naturalist group is interested in how many birds are nesting in a conservation area. They capture 210 birds, tag them, and release them. Later 288 birds are captured and 32 of them are tagged. Approximately how many birds are in the conservation area?

Chapter 5 Test

Simplify the following by reducing to lowest terms.

1) $\dfrac{42a^3b^4}{16ab^6}$

2) $\dfrac{6x^2 + 11x - 10}{4x^2 + 4x - 15}$

Find the following products and quotients.

3) $\dfrac{18}{25} \times \dfrac{35}{24}$

4) $\dfrac{27a^2b^4}{14x^4y} \cdot \dfrac{35xy^3}{18a^5b^2}$

5) $\dfrac{7 - b}{42} \cdot \dfrac{26}{b - 7}$

6) $\dfrac{a^2 + 4a - 21}{a^2 + 3a - 28} \div \dfrac{a^2 + 3a - 18}{a^2 + 8a + 12}$

7) $\dfrac{16y^2 - 25}{6y^2 - 17y - 14} \cdot \dfrac{3y^2 + 2y}{8y^2 - 2y - 15}$

8) Find $\dfrac{12x^2 - 11x - 12}{3x - 5}$ using long division.

9) Find the LCM of $9x^2y^4$ and $15xy^3$.

Find the missing numerator so the fractions are equal.

10) $\dfrac{3t}{4u^3} = \dfrac{?}{20u^5w}$

11) $\dfrac{6p}{q^2 + 4q} = \dfrac{?}{q(q^2 - 16)}$

Find the least common denominator and then write each fraction as an equivalent fraction with the LCD as its denominator.

12) $\dfrac{7a}{a^2 - 3a - 18}$ and $\dfrac{b}{a + 3}$

Find the following sums or differences. Reduce the answers to the lowest terms.

13) $\dfrac{r^3}{r^2 + 16} + \dfrac{r - r^3}{r^2 + 16}$

14) $\dfrac{v + 9}{3v} - \dfrac{v - 6}{9v^2}$

15) $\dfrac{5w}{3w - 12} + \dfrac{7w - 1}{5w - 20}$

16) $\dfrac{t + 3}{t^2 - 10t + 25} - \dfrac{t - 4}{2t^2 - 50}$

Simplify the following complex fractions.

17) $\dfrac{\dfrac{m^2}{3z}}{\dfrac{m}{4z}}$

18) $\dfrac{8 - \dfrac{2u}{3}}{5u + \dfrac{u}{3}}$

Solve the following equations.

19) $\dfrac{5v}{6} - \dfrac{1}{4} = \dfrac{4}{3} - \dfrac{7v}{12}$

20) $\dfrac{p-7}{7} + \dfrac{9}{14} = \dfrac{6-5p}{2}$

21) $\dfrac{11}{t-3} + 6 = \dfrac{7t}{t-3}$

22) $\dfrac{4}{w^2 + 7w + 10} + \dfrac{3}{w^2 - 4} = \dfrac{14}{w^2 + 7w + 10}$

Solve each of the following.

23) Two numbers differ by 4. If $\frac{3}{4}$ of the smaller number is increased by $\frac{1}{2}$ of the larger number, the sum is 2. Find the numbers.

24) Working together it takes Elena and Eduardo 10 days to do an architectural project. If Elena can do the project by herself in 15 days, how long would it take Eduardo to do the project working by himself?

25) An air cargo plane and an airliner take off at the same time from the same airport. They fly in opposite directions. The air cargo plane flies at 420 mph and the airliner flies at 530 mph. How much time will elapse before the two airplanes are 1900 miles apart?

Write as a ratio in lowest terms.

26) At a certain college, 840 faculty teach 84,000 students. What is the ratio of faculty to students?

Solve the following proportions.

27) $\dfrac{15}{27} = \dfrac{y}{18}$

28) $\dfrac{5}{9} = \dfrac{3x-3}{5x-3}$

Solve each of the following.

29) If a washing machine can wash 3 loads of clothes in 75 minutes, how many minutes will it take to wash 7 loads of clothes?

30) In Europe, a forester wants to know how many wild hogs there are in a forest. He captures 12 hogs, tags them, and releases them. A few months later he captures 18 hogs and finds that 4 are tagged. Estimate the number of hogs in the forest.

6

6.1   Ordered Pairs and Solutions of Linear Equations with Two Variables

6.2   Graphing Linear Equations with Two Variables

6.3   Graphing Linear Inequalities with Two Variables

6.4   Slope of a Line

6.5   Writing Equations of Lines

6.6   Relations and Functions

Graphing Linear Equations and Inequalities

I n Chapter 3 we solved linear equations and inequalities that contained only one variable. In Chapter 4 we solved quadratic equations with one variable by factoring. In Chapter 5 we solved equations with one variable that involved rational expressions. In this chapter we will learn to find solutions of linear equations and inequalities that contain two variables.

The solution of a linear equation with one variable is usually a single number, but sometimes there is no solution. The solution of a quadratic equation is usually two numbers, but sometimes one number, or sometimes there was no solution. In order to solve equations with two variables, we will first need to define what is meant by the solution of an equation with two variables. We will also need a method of denoting solutions of equations with two variables which allows us to specify a value of each variable in the equation.

In Chapter 1 we learned to graph whole numbers and integers on the number line. In this chapter we will learn to graph

CHAPTER OVERVIEW

the solutions of equations with two variables. Since a solution will consist of a value for each variable, the solution cannot be graphed on a number line. Hence, graphing a solution of an equation with two variables will require the use of something other than a number line. The graph of the solutions of an equation with two variables will be quite different from the graph of a set of integers.

In Chapter 3 we learned to graph the solutions of linear inequalities. In this chapter we will learn to graph the solutions of linear inequalities with two variables. There will be similarities between the graphs of linear inequalities with two variables and the graphs of linear inequalities with one variable in that both require the shading of regions.

We end the chapter by studying some characteristics of the graphs of linear equations with two variables and by writing equations with two variables that satisfy certain given conditions.

Ordered Pairs and Solutions of Linear Equations with Two Variables

OBJECTIVES

When you complete this section, you will be able to:

Ⓐ Verify solutions of equations with two variables.
Ⓑ Write solutions of linear equations with two variables as ordered pairs.
Ⓒ Determine if a given ordered pair is a solution of a given equation.
Ⓓ Find the missing number in an ordered pair for a given linear equation.
Ⓔ Represent solutions of everyday situations as ordered pairs.

INTRODUCTION

In Chapter 3 we solved linear equations with one variable. The solution of a linear equation with one variable is a single number. The solution can be verified by replacing the variable with the value and simplifying. For example, the solution of $3x + 7 = 19$ is $x = 4$. We can verify that $x = 4$ is a solution of $3x + 7 = 19$ by replacing x with 4 and simplifying. So $3(4) + 7 = 12 + 7 = 19$. Therefore, the solution is correct.

In this chapter we will discuss **linear equations with two variables**. A linear equation with two variables is any equation of the form $ax + by = c$, or any equation that can be put into that form, where a and b are both constants and not equal to 0 at the same time.

Any solution of a linear equation with two variables must have a value for each variable. Hence, any solution has two values. We check a solution of a linear equation with two variables in much the same way that we check a solution of a linear equation with one variable. We replace each variable with its value and simplify.

• **Verifying solutions of linear equations in two variables**

EXAMPLE 1

Verify that the following are solutions of the given equations.

a) $x + 2y = 6$; $x = 4$, $y = 1$ Substitute 4 for x and 1 for y.

    $4 + 2 \cdot 1 = 6$ Multiply 2 and 1.

    $4 + 2 = 6$ Add 4 and 2.

    $6 = 6$ Therefore, the solutions are correct.

b) $x + 2y = 6$; $x = -2$, $y = 4$ Substitute for -2 for x and 4 for y.

    $-2 + 2(4) = 6$ Multiply 2 and 4.

    $-2 + 8 = 6$ Add -2 and 8.

    $6 = 6$ Therefore, the solutions are correct.

Notice two things from Example 1. First, the equation in part a is the same as the equation in part b, but the solutions are different. Consequently, a linear equation in two variables has more than one solution. In fact, it has an infinite number of solutions since we can assign any value to either variable and find the corresponding value for the remaining variable. Second, it is cumbersome to write the solutions in the form given in Example 1. We would have to write "two of the solutions to the linear equation $x + 2y = 6$ are $x = 4$ and $y = 1$, and $x = -2$ and $y = 4$." For this reason we introduce the notion of an **ordered pair**.

An ordered pair consists of two numbers enclosed in parentheses and separated by a comma. They are called ordered pairs because one variable is assigned to the first number of the pair and a different variable is assigned to the second. Hence, the numbers are written in a specific order determined by the variables.

- Writing solutions of linear
  equations with two variables as
  ordered pairs

If a set of ordered pairs are in the form (x,y), the first number in the pair is the x-value and the second is the y-value. So in the ordered pair $(-4,6)$, $x = -4$ and $y = 6$, and in the ordered pair $(2,-3)$, $x = 2$ and $y = -3$. In Example 1 the solution $x = 4$ and $y = 1$ is represented as the ordered pair $(4,1)$, and the solution $x = -2$ and $y = 4$ is represented as $(-2,4)$. This leads us to the following definition.

- Determining if a given ordered
  pair is a solution of a given
  equation

SOLUTIONS OF LINEAR EQUATIONS OF TWO VARIABLES

If the solutions of $Ax + By = C$ are ordered pairs in the form (x,y), the ordered pair (m,n) is a solution if the replacement of x with m and y with n results in a true statement.

EXAMPLE 2

Determine whether the given ordered pair is a solution of the given equation.

a) $3x - 4y = 9$; $(7,3)$ Substitute 7 for x and 3 for y.

    $3(7) - 4(3) = 9$ Multiply before adding.

    $21 - 12 = 9$ Add 21 and -12.

    $9 = 9$ Therefore, $(7,3)$ is a solution.

b) $2x + 3y = 6$; $(-3,0)$ Substitute -3 for x and 0 for y.

    $2(-3) + 3(0) = 6$ Multiply before adding.

    $-6 + 0 = 6$ Add -6 and 0.

    $-6 \neq 6$ Therefore, $(-3,0)$ is not a solution.

c) $4x - y = 4$; $(-1,-8)$ Substitute -1 for x and -8 for y.

    $4(-1) - (-8) = 4$ Multiply before adding and $-(-8) = 8$.

    $-4 + 8 = 4$ Add -4 and 8.

    $4 = 4$ Therefore, $(-1,-8)$ is a solution.

ADDITIONAL PRACTICE

Determine whether the given ordered pair is a solution of the given equation. Assume the ordered pairs are in the form (x,y).

a) $x + 3y = 7$; $(1,2)$

b) $2x - y = 8$; $(-4,-2)$

c) $3x + 5y = 10$; $(0,2)$

PRACTICE EXERCISES

Determine whether the given ordered pair is a solution of the given equation. Assume the ordered pairs are in the form (x,y).

1) $2x - 4y = 10$; $(1,-2)$

2) $3x + y = 8$; $(-3,1)$

3) $3x - y = 7$; $(2,1)$

- Finding the missing member of
  an ordered pair for a given
  equation

If more practice is needed, do the Additional Practice Exercises in the margin.

Suppose we have an equation and know one member of an ordered pair. We can find the other member by substituting the known value for the variable it represents and solving for the other variable.

EXAMPLE 3

Use the given equation to find the missing member of each of the following ordered pairs. Assume the ordered pair is in the form (x,y).

a) $y = 2x - 6$; $(0,\_)$, $(\_,0)$, $(5,\_)$, $(-3,\_)$

Solution:

Since the ordered pairs are in the form of (x,y), the ordered pair $(0,\_)$ means $x = 0$ and we need to find y.

$y = 2(0) - 6$	Substitute 0 for x. $2(0) = 0$.
$y = 0 - 6$	Add 0 and -6.
$y = -6$	$0 - 6 = -6$. Therefore, the ordered pair is $(0,-6)$.

The ordered pair $(\_,0)$ means $y = 0$ and we need to find x.

$0 = 2x - 6$	Substitute 0 for y. Add 6 to both sides.
$0 + 6 = 2x - 6 + 6$	Simplify both sides.
$6 = 2x$	Divide both sides by 2.
$\dfrac{6}{2} = \dfrac{2x}{2}$	Simplify both sides.
$3 = x$	Therefore, the ordered pair is $(3,0)$.

In the ordered pair $(5,\_)$, $x = 5$ and we need to find y. Substitute 5 for x and simplify.

$y = 2(5) - 6$
$y = 10 - 6$
$y = 4$

Therefore, the ordered pair is $(5,4)$.

In the ordered pair $(-3,\_)$, $x = -3$ and we need to find y. Substitute -3 for x and simplify.

$y = 2(-3) - 6$
$y = -6 - 6$
$y = -12$

Therefore, the ordered pair is $(-3,-12)$.

Use the given equation to find the missing member of each of the following ordered pairs. Assume the ordered pair is in the form (x,y).

b) $2x - 3y = 6$; $(0,\_)$, $(\_,0)$, $(6,\_)$, $(\_,-4)$

Solution:

Since the ordered pair is in the form (x,y), the ordered pair $(0,\_)$ means $x = 0$ and we need to find y.

$2(0) - 3y = 6$	Substitute 0 for x. $2(0) = 0$.
$0 - 3y = 6$	$0 - 3y = -3y$.
$-3y = 6$	Divide both sides by -3.
$y = -2$	Therefore, the ordered pair is $(0,-2)$.

The ordered pair $(\_,0)$ means $y = 0$ and we need to find x.

$2x - 3(0) = 6$	Substitute 0 for y. $3(0) = 0$.
$2x - 0 = 6$	$2x - 0 = 2x$.
$2x = 6$	Divide both sides by 2.
$x = 3$	Therefore, the ordered pair is $(3,0)$.

In the ordered pair $(6,\_)$, $x = 6$ and we need to find y. Substitute 6 for x and solve for y.

$2(6) - 3y = 6$
$12 - 3y = 6$
$-3y = -6$
$y = 2$

Therefore, the ordered pair is $(6,2)$.

In the ordered pair $(\_,-4)$, $y = -4$ and we need to find x. Substitute -4 for y and solve for x.

$2x - 3(-4) = 6$
$2x + 12 = 6$
$2x = -6$
$x = -3$

Therefore, the ordered pair is $(-3,-4)$.

Use the given equation to find the missing member of each of the following ordered pairs. Assume the ordered pair is in the form (x,y).

c) $x = 2$; $(\_,0)$, $(\_,-2)$, $(\_,4)$

Solution:

Notice that there is no "y" term in this equation. We can think of the equation as $x + 0y = 2$. For the ordered pair $(\_,0)$, $y = 0$ and we need to find x.

$x + 0(0) = 2$ Substitute 0 for y. $0(0) = 0$.

$x + 0 = 2$ $x + 0 = x$.

$x = 2$ Therefore, the ordered pair is $(2,0)$.

In the ordered pair $(\_,-2)$, $y = -2$ and we need to find x. Substitute -2 for y and simplify.

$x + 0(-2) = 2$
$x + 0 = 2$
$x = 2$

Therefore, the ordered pair is $(2,-2)$.

In the ordered pair $(\_,4)$, $y = 4$ and we need to find x. Substitute 4 for y and simplify.

$x + 0(4) = 2$
$x + 0 = 2$
$x = 2$

Therefore, the ordered pair is $(2,4)$.

Note: In Example 2c, it soon becomes clear that $x = 2$ for any value of y since $0 \cdot y = 0$ for all values of y. When one variable is missing from a linear equation, the remaining variable must always equal the given constant. For example, if $y = -2$, then all ordered pairs solving this equation must have a y-value of -2. The x-value may be anything. Some solutions are $(0,-2)$, $(-3,-2)$, and $(5,-2)$. In general, the solutions are represented by $(x, -2)$ where x can have any value.

ADDITIONAL PRACTICE

Use the given equation to find the missing member of each of the following ordered pairs.

d) $y = -3x + 9$
    $(0,\_)$, $(\_,0)$, $(4,\_)$

e) $x - 3y = 9$
    $(0,\_)$, $(\_,0)$, $(\_,-2)$

f) $x = -4$
    $(\_,0)$, $(\_,-2)$, $(\_,3)$

PRACTICE EXERCISES

Use the given equation to find the missing member of each of the following ordered pairs. Assume the ordered pair is in the form (x,y).

4) $y = 3x + 12$; $(0,\_)$, $(\_,0)$, $(-3,\_)$ **5)** $3x + 4y = 12$; $(0,\_)$, $(\_,0)$, $(-4,\_)$

6) $y = -3$; $(0,\_)$, $(5,\_)$, $(-5,\_)$

If more practice is needed, do the Additional Practice Exercises in the margin.

Ordered pairs can occur in many types of situations. Suppose a car is traveling at a constant rate of 60 miles per hour. At the end of 1 hour the car has traveled $60 \cdot 1 = 60$ miles, at the end of 2 hours the car has traveled $60 \cdot 2 = 120$ miles, and so forth. The distance the car travels is $d = 60t$ where t is the number of hours. If we let the ordered pairs be of the form (t,d), then $(1,60)$ means that after 1 hour the car traveled 60 miles and $(3,180)$ means that after 3 hours the car traveled 180 miles.

EXAMPLE 4

a) The length of a rectangle remains 5 feet while the width is allowed to vary. If W represents the width and A represents the area, represent the area for the given widths using ordered pairs in the form (W,A). Use $W = 2$ft, 4ft, and 9ft. (The formula for the area of a rectangle is $A = LW$.) What does the ordered pair $(8,40)$ represent?

Solution:
Since the length has a constant value of 5 feet, the area is $A = 5W$. Since $W = 2$ft, 4ft, and 9ft, we need to complete the ordered pairs $(2,\_)$, $(4,\_)$, and $(9,)$.

$W = 2$ ft	$W = 4$ ft	$W = 9$ ft
$A = 5 \cdot 2 = 10$ ft^2	$A = 5 \cdot 4 = 20$ ft^2	$A = 5 \cdot 9 = 45$ ft^2
Therefore, the ordered pair is (2,10). This means when the width is 2 ft, the area is 10 ft^2.	Therefore, the ordered pair is (4,20). This means when the width is 4 ft, the area is 20 ft^2.	Therefore, the ordered pair is (9,45). This means when the width is 9 ft, the area is 45 ft^2.

The ordered pair $(8,40)$ means that when the width is 8 ft, the area is 40 ft^2.

b) The distance a free-falling object falls is given by $s = 16t^2$ where s represents the distance in feet and t represents the elapsed time in seconds. Represent the distances for each of the given times using ordered pairs of the form (t,s) for $t = 1$, 4, and 6. What does the ordered pair $(3,144)$ represent?

Solution:
Since $t = 1$, 4, and 6, we need to complete the ordered pairs $(1,\_)$, $(4,\_)$, and $(6,\_)$.

$t = 1$ second	$t = 4$ seconds	$t = 6$ seconds
$s = 16(1)^2 = 16 \cdot 1$ $= 16$ ft	$s = 16(4)^2 = 16 \cdot 16$ $= 256$ ft	$s = 16(6)^2 = 16 \cdot 36$ $= 576$ ft
Therefore, the ordered pair is (1,16). This means the object will travel 16 feet in 1 second.	Therefore, the ordered pair is (4,256). This means the object will travel 256 feet in 4 seconds.	Therefore, the ordered pair is (6,576). This means the object will travel 576 feet in 6 seconds.

The ordered pair $(3,144)$ means that after 3 seconds the object will have fallen 144 feet.

7) In the formula $p = br$, p stands for percentage, b for base, and r for rate. If the rate is fixed at 8% (0.08), the equation is $p = 0.08b$. Represent the percentages for each of the given bases using ordered pairs of the form (b,p) for $b = \$200$, $\$500$, and $\$1000$. What does the ordered pair $(600,48)$ represent?

8) The formula for the area of a circle is $A = \pi r^2$ where A represents the area and r represents the radius. Represent the areas of the following circles using ordered pairs of the form (r,A) for $r = 1$ cm, 3 cm, and 5 cm. Leave A in terms of π. (Do not substitute for π.) What does the ordered pair $(4,16\pi)$ mean?

EXERCISE SET 6.1

Determine whether the given ordered pair is a solution of the given equation.

1) $y = 3x + 2$; $(1,5)$

2) $y = 2x + 4$; $(-1,2)$

3) $y = -4x + 5$; $(-1,1)$

4) $y = -2x - 4$; $(2,0)$

5) $2x + y = 4$; $(0,4)$

6) $x + 3y = 9$; $(9,0)$

7) $x - 4y = 8$; $(0,8)$

8) $3x - y = 12$; $(-4,0)$

9) $2x + 3y = 12$; $(3,2)$

10) $2x + 5y = 10$; $(-5,4)$

11) $3x - 5y = 15$; $(-5,-6)$

12) $4x - 3y = 12$; $(-3,-8)$

13) $4x - 5y = -20$; $(0,-4)$

14) $7x - 3y = -21$; $(-3,0)$

15) $2x + 3y = 6$; $\left(-\dfrac{3}{2},3\right)$

16) $3x + 4y = 8$; $\left(-\dfrac{4}{3},3\right)$

Use the given equation to find the missing member of each of the given ordered pairs. Assume the ordered pairs are in the form (x,y).

17) $y = 2x + 3$; $(0,\_),(\_,0),(3,\_),(\_,7)$

18) $y = 3x - 5$; $(0,\_),(\_,0),(-2,\_),(\_,4)$

19) $y = -2x - 4$; $(0,\_),(\_,0),(-3,\_),(\_,-8)$

20) $y = -3x - 5$; $(0,\_),(\_,0),(-2,\_),(\_,4)$

21) $2x - 3y = 12$; $(0,\_),(\_,0),(-6,\_),(\_,-6)$

22) $3x - 2y = 18$; $(0,\_),(\_,0),(4,\_),(\_,6)$

23) $4x + 5y = 20$; $(0,\_),(\_,0),(-10,\_),(\_,8)$

24) $6x + 5y = 30$; $(0,\_),(\_,0),(10,\_),(\_,12)$

25) $2x + 3y = -6$; $(0,\_),(\_,0),(9,\_),(\_,-4)$

26) $3x + 5y = -30$; $(0,\_),(\_,0),(-10,\_),(\_,9)$

27) $4x - 3y = -12$; $(0,\_),(\_,0),(6,\_),(\_,-12)$

28) $5x - 4y = -20$; $(0,\_),(\_,0),(12,\_),(\_,-20)$

29) $x = 4$; $(\_,0),(\_,3),(\_,-2),(\_,-4)$

30) $y = -3$; $(0,\_),(2,\_),(-2,\_),(-4,\_)$

31) $y = 4$; $(2,\_),(3,\_),(-4,\_),(-1,\_)$

32) $x = -1$; $(\_,2),(\_,3),(\_,-3),(\_,-5)$

Represent each of the following as ordered pairs.

33) A motorcycle is traveling at a constant speed of 40 miles per hour. The formula for the distance traveled at the end of t hours is $d = 40t$. a) Represent the distance traveled after 1, 2, and 5 hours as ordered pairs in the form (t,d). b) What does the ordered pair $(8,320)$ represent?

34) A boat is traveling at a constant speed of 25 miles per hour. The formula for the distance traveled at the end of t hours is $d = 25t$. a) Represent the distance traveled at the end of 1, 3, and 6 hours as ordered pairs in the form (t,d). b) What does the ordered pair $(5,125)$ represent?

35) The width of a rectangle remains constant at 6m and the length is allowed to vary. The formula for the area is $A = L(6) = 6L$. a) Represent the area for $L = 2$m, 4m, and 5m as ordered pairs in the form (L,A). b) What does the ordered pair $(9,54)$ represent?

36) The length of a rectangle remains constant at 5 feet and the width is allowed to vary. The formula for the perimeter is $P = 2(5) + 2W = 10 + 2W$. a) Represent the perimeter for $W = 2$ft, 3ft, and 6ft as ordered pairs in the form (W,P). b) What does the ordered pair (5,20) represent?

37) A particular brand of candy bar costs 60 cents each. If n represents the number of candy bars purchased, the formula for the cost is $c = 60n$ with c in cents. a) Represent the cost of 1, 3, and 7 candy bars, as ordered pairs in the form (n,c). b) What does the ordered pair (4,240) represent?

38) A type of bolt costs 35 cents each. If n represents the number of bolts purchased, the formula for the cost is $c = 35n$ with c in cents. a) Represent the cost of 5, 10, and 15 bolts as ordered pairs in the form (n,c). b) What does the ordered pair (4,140) represent?

39) If v represents the value of a collection of five-dollar bills and n represents the number of bills, then $v = 5n$. a) Represent the value of 2, 5, and 8 five-dollar bills as ordered pairs in the form (n,v). b) What does the ordered pair (10,50) represent?

40) If v represents the value of a collection of ten-dollar bills and n represents the number of bills, then $v = 10n$. a) Represent the value of 3, 5, and 7 ten-dollar bills as ordered pairs in the form (n,v). b) What does the ordered pair (8,80) represent?

41) If an object is thrown upward with an initial velocity of 100 feet per second, then its velocity at the end of t seconds is $v = 100 - 32t$. a) Represent the velocity after 1, 3, and 6 seconds as ordered pairs in the form (t,v). b) What does the ordered pair (2,36) represent?

42) If an object is thrown upward with an initial velocity of 150 feet per second, then its velocity at the end of t seconds is $v = 150 - 32t$. a) Represent the velocity after 2, 3, and 5 seconds as ordered pairs in the form (t,v). b) What does the ordered pair (6,−42) represent?

CHALLENGE EXERCISES (43–48)

43) A particle is moving along a straight line according to the formula $s = 6t^2$ where s represents the distance in inches from the starting point after t seconds. a) Represent the distance after 1, 3, and 5 seconds as ordered pairs in the form (t,s). b) What does the ordered pair (2,24) represent?

44) A particle is moving along a straight line according to the formula $s = 8t^2$ where s represents the distance in feet from the starting point after t seconds. a) Represent the distance after 2, 4, and 6 seconds as ordered pairs in the form (t,s). b) What does the ordered pair (3,72) represent?

45) If an object is thrown upward with an initial velocity of 150 feet per second from a height of 200 feet, the height of the object, s, in feet after t seconds is given by the formula $s = -16t^2 + 150t + 200$. a) Represent the height of the object after 1, 2, and 4 seconds as ordered pairs of the form (t,s). b) What does the ordered pair (3,506) represent?

46) If an object is thrown upward with an initial velocity of 100 feet per second from a height of 125 feet, the height of the object, s, in feet after t seconds is given by the formula $s = -16t^2 + 100t + 125$. a) Represent the height of the object after 2, 4, and 5 seconds as ordered pairs of the form (t,s). b) What does the ordered pair (3,281) represent?

47) A culture of bacteria is growing according to the formula $N = 100(2^t)$ where N is the number present after t hours. a) Represent N after 1, 2, and 3 hours as ordered pairs of the form (t,N). b) What does the ordered pair (4,1600) represent?

48) A culture of bacteria is growing according to the formula $N = 150(3^t)$ where N is the number present after t hours. a) Represent N after 1, 2, and 3 hours as ordered pairs of the form (t,N). b) What does the ordered pair (4,12150) represent?

WRITING EXERCISES

49) A linear equation with two unknowns may be written in the form $ax + by = c$. Solutions of such equations can be represented as ordered pairs of the form (x,y). A linear equation with three unknowns may be written in the form $ax + by + cz = d$. How would you represent solutions to these equations?

50) Write a problem similar to problems 33 through 42 above with solutions that may be represented as ordered pairs. Find three solutions and represent them as ordered pairs.

51) Why does a linear equation with two unknowns have more than one solution? Why does it have an infinite number of solutions?

OBJECTIVES

When you complete this section, you will be able to:

Ⓐ   Plot points on the rectangular coordinate system.

Ⓑ   Graph linear equations with two variables on the rectangular coordinate system.

Ⓒ   Find the *x*- and *y*-intercepts of the graph of a linear equation with two variables.

Ⓓ   Graph vertical and horizontal lines.

INTRODUCTION

In Chapter 3 we solved linear equations with one variable. It is possible to graph the solution of a linear equation on the number line. For example, the solution of $2x + 7 = 5$ is $x = -1$. This solution is graphed on the number line by placing a dot at -1 as shown below.

Since a solution to a linear equation in two variables is an ordered pair, we cannot graph a solution on the number line. Since there are two variables, we need two number lines—one for each variable. Suppose the ordered pairs are in the form of (*x*,*y*). We draw a horizontal number line and label it the *x*-line. At the 0 point of the *x*-line, we draw a vertical number line and put its 0 point at the point of intersection of the two lines. We call this the *y*-line. The *x*-line is called the **x-axis** and the *y*-line is called the **y-axis**. The point of intersection of the two axes is called the **origin**. On the *x*-axis, we label the points to the right of the origin with positive numbers and those to the left with negative numbers. On the *y*-axis, we number the points above the origin with positive numbers and those below with negative numbers. See the figure below.

• **Historical note: The invention of the rectangular coordinate system is attributed to Rene Descartes. One story is that the inspiration for its development came from watching a fly crawling on a ceiling near the corner of the room. Descartes observed that he could give the location of the fly by using only two numbers which gave the perpendicular distance from each of the walls to the fly. Imagine that! A fly is responsible for one of the most important inventions in the history of mathematics!**

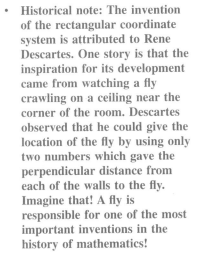

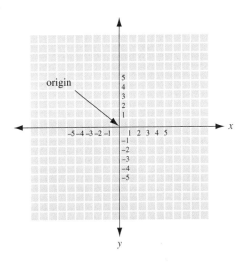

This is known as the rectangular or Cartesian (in honor of its inventor, Rene Descartes) coordinate system. The *x*- and *y*-axes divide the coordinate system into four regions called **quadrants**. The quadrants are numbered counter-clockwise with the first quadrant being the upper right quadrant. See the following figure.

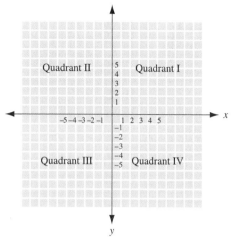

- **Plotting points in the rectangular coordinate system**

Since the *x*-axis is horizontal, the *x*-value of the ordered pair tells us how far and in which direction to go horizontally from the origin. Go to the right if *x* is positive and to the left if *x* is negative. Since the *y*-axis is vertical, the *y*-value tells us how far and in which direction to go vertically. Go up if *y* is positive and down if *y* is negative. For example, to plot the ordered pair (4,6), we begin at the origin and go 4 units to the right and then we go 6 units up as illustrated below.

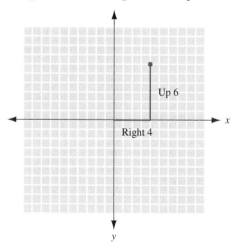

In an ordered pair, the first number of the ordered pair is called the **abscissa** and the second number is called the **ordinate**. Together, they are called the **coordinates** of the point.

EXAMPLE 1
Plot the points represented by the following ordered pairs on the rectangular coordinate system.

a) (3,5), **b)** (−4,3), **c)** (−3,−4), **d)** (3,−5), **e)** (0,2), **f)** (4,0)

Solution:
We first give a description of how the ordered pairs are plotted and then the points are shown on the rectangular coordinate system.

a) To plot the point (3,5), begin at the origin. Go 3 units to the right and 5 units up.

b) To plot the point (−4,3), begin at the origin. Go 4 units to the left and 3 units up.

c) To plot the point (−3,−4), begin at the origin. Go 3 units to the left and 4 units down.

d) To plot the point (3,−5), begin at the origin. Go 3 units to the right and 5 units down.

e) To plot the point (0,2), do not go any units right or left. Go up 2 units from the origin.

f) To plot the point (4,0), begin at the origin. Go 4 units to the right and stay there. Do not go any units up or down.

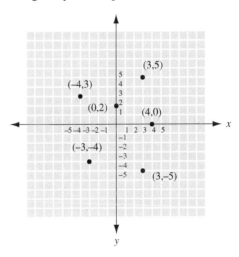

PRACTICE EXERCISES

Plot the points represented by the following ordered pairs on the rectangular coordinate system.

1) (3,4)　　　　　　**2)** (−2,5)　　　　　　**3)** (−4,−1)

4) (4,−3)　　　　　　**5)** (−8,0)　　　　　　**6)** (0,−6)

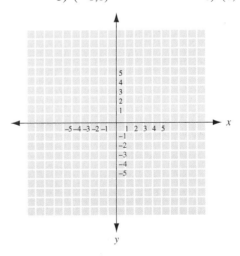

Using the rectangular coordinate system, it is possible to plot solutions of equations with two variables. The plot of the solutions is called the **graph** of the equation. Since there are an infinite number of solutions, it is not possible to plot them all. All the graphs in this section will be straight lines that we will graph using the following procedure.

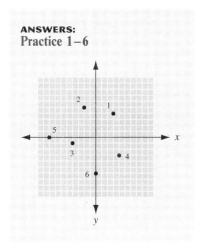

GRAPHING LINEAR EQUATIONS

To graph a linear equation, we perform the following.

1) Determine 3 ordered pairs that solve the equation by letting x have a value and finding y or letting y have a value and finding x.

2) Plot the ordered pairs.

3) Draw the line that contains the three points.

EXAMPLE 2

Draw the graph of each of the following equations.

a) $y = 2x + 1$

Solution:

We need to find some ordered pairs from the solution set. We can let either variable have any value we choose and then solve for the other variable. Since we have to plot the resulting ordered pairs, it is best to use values small in absolute value. This equation is solved for y, so it would be easier to assign values to x and then find y.

Let $x = 0$.	Let $x = 2$.	Let $x = -3$.
$y = 2(0) + 1$	$y = 2(2) + 1$	$y = 2(-3) + 1$
$y = 0 + 1$	$y = 4 + 1$	$y = -6 + 1$
$y = 1$	$y = 5$	$y = -5$
Therefore, the ordered pair is $(0,1)$.	Therefore, the ordered pair is $(2,5)$.	Therefore, the ordered pair is $(-3,-5)$.

Note: One easy way of keeping track of the ordered pairs is to put them in a table. For example, the ordered pairs above could be put in a table as follows.

x	y
0	1
2	5
-3	-5

Now plot the points on the rectangular coordinate system. It appears that the three points all lie on a line, so draw a line through the three points.

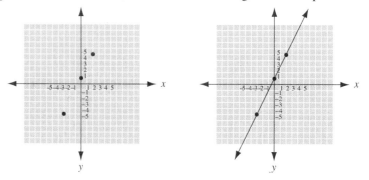

Note: There are a number of things we need to point out at this time. First, it can be shown that the graph of any linear equation with two variables $(ax + by = c)$ is a straight line. That is why such an equation is called *linear*. Second, the coordinates of any point on the line satisfies the equation, and the graph of any ordered pair that satisfies the equation will be on the line. Third, the line continues in both directions indefinitely. This is usually indicated by putting arrow heads on each end of the line.

b) $3x - 2y = 6$

Solution:
We find ordered pairs that satisfy the equation by letting either variable have a value and solving for the other variable.

Let $x = 0$.	Let $y = 0$.	Let $x = -2$.
$3(0) - 2y = 6$	$3x - 2(0) = 6$	$3(-2) - 2y = 6$
$0 - 2y = 6$	$3x - 0 = 6$	$-6 - 2y = 6$
$-2y = 6$	$3x = 6$	$-2y = 12$
$y = -3$	$x = 2$	$y = -6$
Therefore, the ordered pair is $(0, -3)$.	Therefore, the ordered pair is $(2, 0)$.	Therefore, the ordered pair is $(-2, -6)$.

In a table, the ordered pairs would appear as follows.

x	y
0	-3
2	0
-2	-6

Plot the ordered pairs on the rectangular coordinate system and draw a line through them.

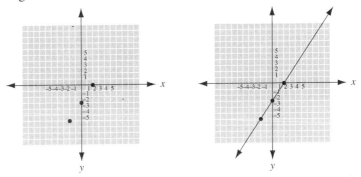

ADDITIONAL PRACTICE

Draw the graph of each of the following equations. Each unit on the coordinate system represents one.

a) $y = 3x - 2$

b) $4x + 2y = 8$

PRACTICE EXERCISES

Draw the graph of each of the following equations. Assume each unit on the coordinate system represents one.

7) $y = x + 3$

8) $2x + y = 4$

If more practice is needed, do the Additional Practice Exercises in the margin.

• Finding the *x*- and *y*-intercepts

ANSWERS:
Practice 7–8

7)

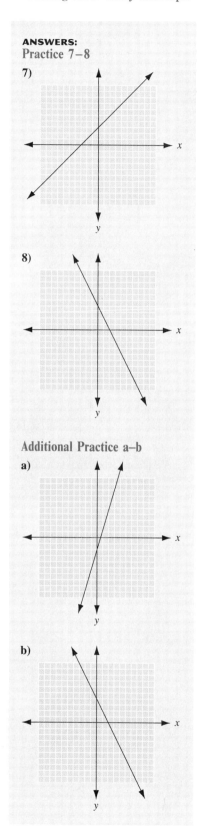

8)

Additional Practice a–b

a)

b)

The point where the graph intersects the *x*-axis is the **x-intercept**, and the point where the graph intersects the *y*-axis is the **y-intercept**. See the figure below.

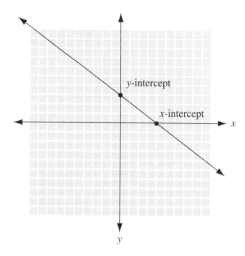

Notice the *y*-value of the *x*-intercept is 0, and the *x*-value of the *y*-intercept is 0. Actually, any point on the *x*-axis has a *y*-value of 0 and any point on the *y*-axis has an *x*-value of 0. Hence, we can find the coordinates of the *x*-intercept by letting $y = 0$ and the coordinates of the *y*-intercept by letting $x = 0$.

Since only one straight line may be drawn through any two distinct points, we need only two ordered pairs from the solution set to draw the graph of a linear equation. The two easiest ordered pairs to find are the *x*- and *y*-intercepts. However, it is a good idea to find a third point as a check since a straight line cannot always be drawn through any three distinct points. The procedure is summarized as follows.

GRAPHING LINEAR EQUATIONS USING THE INTERCEPT METHOD

To graph a linear equation using the intercept method, we perform the following.

1) Let $x = 0$ to find the *y*-intercept and let $y = 0$ to find the *x*-intercept.

2) Find a third point by letting *x* or *y* have any value and solve for the remaining variable.

3) Plot the ordered pairs.

4) Draw a line through the three points.

EXAMPLE 3

Draw the graph of each of the following by finding the x- and y-intercepts. Find a third point as a check. Assume each unit on the coordinate system represents one.

a) $2x + y = 6$

Solution:

To find the
x-intercept, let
$y = 0$.

$2x + 0 = 6$
$2x = 6$
$x = 3$

Therefore, the
x-intercept is (3,0).

To find the
y-intercept, let
$x = 0$.

$2(0) + y = 6$
$0 + y = 6$
$y = 6$

Therefore, the
y-intercept is (0,6).

To find another
point, let $x = -1$.

$2(-1) + y = 6$
$-2 + y = 6$
$y = 8$

Therefore, the
ordered pair is
(−1,8).

In a table the ordered pairs appear as follows.

x	*y*
3	0
0	6
−1	8

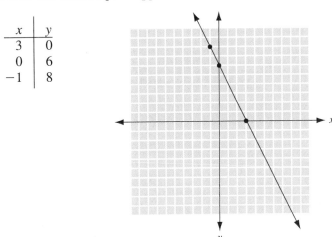

Plot the points on the rectangular coordinate system and draw a line through them.

 b) $2x + 3y = 9$

Solution:

To find the
x-intercept, let
$y = 0$.

$2x + 3(0) = 9$
$2x + 0 = 9$
$2x = 9$
$x = \dfrac{9}{2}$

Therefore, the
x-intercept is $(\frac{9}{2},0)$.

To find the
y-intercept, let
$x = 0$.

$2(0) + 3y = 9$
$0 + 3y = 9$
$3y = 9$
$y = 3$

Therefore, the
y-intercept is (0,3).

To find a third point,
let $x = -3$.

$2(-3) + 3y = 9$
$-6 + 3y = 9$
$3y = 15$
$y = 5$

Therefore, the
ordered pair is
(−3,5).

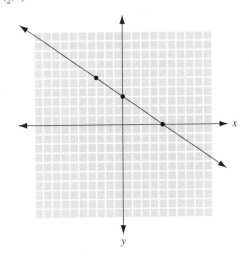

Note: In Example 3b, the x-intercept has a fractional value. In plotting the corresponding point, it was necessary to estimate its location. This can be a real problem if the location of the fraction is difficult to estimate. For example, $(\frac{3}{11}, \frac{12}{5})$ would be difficult to plot accurately. If you can find an x-value that gives an integer value for y, then adding any integer multiple of the coefficient of y to that x-value will give other x-values that results in integer values for y. In other words, if you have an x-value that results in an integer y-value, then (x-value) + (multiple of y-coefficient) = (another x-value that gives an integer y-value).

In Example 3b, the equation was $2x + 3y = 9$ and the y-intercept was $(0,3)$. The coefficient of y in the equation is 3. Therefore, $x = 0 +$ (any integer multiple of 3) will give another x-value that results in an integer value for y. Hence, $x = 3$, 6, 9, -3, -6, -9, etc., will give integer values for y. The opposite applies for y-values. For the equation $2x + 5y = 8$, the x-intercept is $(4,0)$ and the coefficient of x is 2. Hence $y = 0 +$ (any multiple of 2) gives another y-value that results in an integer value for x. Hence $y = 2$, 4, 6, -2, -4, -6, and so on will give integer values for x.

PRACTICE EXERCISES

Draw the graph of each of the following by finding the x- and y-intercepts. Find a third point as a check. Assume each unit on the coordinate system represents one.

9) $x + 4y = 8$ **10)** $2x - 5y = 15$

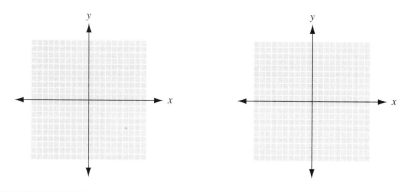

If the equation is of the form $y = ax$, then the x- and y-intercepts are both at $(0,0)$. Hence, we must find two other points in addition to $(0,0)$. Since the equation is solved for y, the easiest way to find these is to assign values to x and find y.

EXAMPLE 4

Graph the following equations. Assume each unit is one.

a) $y = 2x$

Solution:

If we let $x = 0$, then $y = 0$. Hence, the y-intercept is $(0,0)$. If we let $y = 0$, then $x = 0$. Hence, the x-intercept is also $(0,0)$. Find two other points by assigning any value to x and finding y.

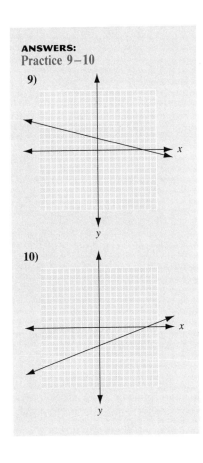
Let $x = 2$.

$y = 2(2)$

$y = 4$

Therefore, the ordered pair is $(2,4)$.

Let $x = -3$.

$y = 2(-3)$

$y = -6$

Therefore, the ordered pair is $(-3,-6)$.

Plot the points on the coordinate system and draw a line through them.

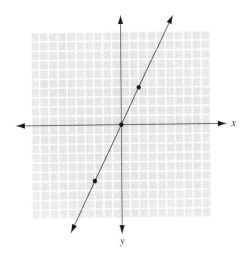

b) $y = -x$

Solution:

If we let $x = 0$, we get $y = 0$. If we let $y = 0$, we get $x = 0$. Hence, both the x- and y-intercepts are $(0,0)$. Find two more points by assigning values to x and finding y.

Let $x = 3$.

$y = -3$

Therefore, the ordered pair is $(3,-3)$.

Let $x = -2$.

$y = -(-2) = 2$

Therefore, the ordered pair is $(-2,2)$.

Plot the points on the coordinate system and draw a line through them.

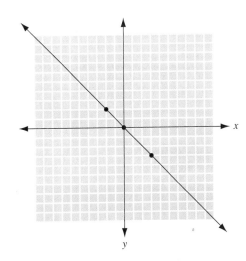

PRACTICE EXERCISES

Graph the following equations. Assume each unit is one.

11) $y = 3x$ **12)** $y = -2x$

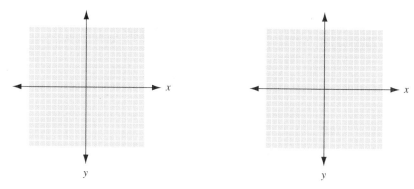

- **Graphing horizontal and vertical lines**

There are two other types of lines that are of particular interest. They are of the form $x = k$ and $y = k$ where k is any constant. We can think of an equation of the form $x = k$ to be $x + 0y = k$. Consequently, regardless of the value for y, $x = k$. Therefore, all solutions to the equation $x = k$ are of the form (k, y) where y can have any value. Similarly, we can think of an equation of the form $y = k$ to be $0x + y = k$. Consequently, $y = k$ for all values of x. Therefore, all solutions of the equation $y = k$ are of the form (x, k) where x can have any value.

EXAMPLE 5

Graph the following equations. Assume each unit is one.

a) $x = 3$

Solution:

Think of $x = 3$ as $x + 0y = 3$. Therefore, $x = 3$ for all values of y. Consequently, solutions of $x = 3$ are of the form $(3, y)$ where y can have any value. Thus, some solutions are $(3, 0)$, $(3, -2)$, and $(3, 4)$.

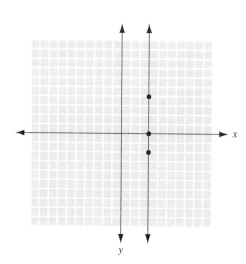

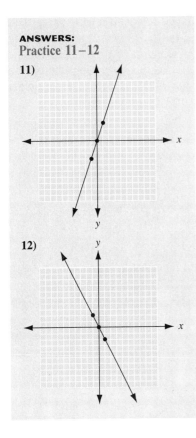
b) $y = -2$

Solution:

Think of $y = -2$ as $0x + y = -2$. Consequently, $y = -2$ for all values of x. Thus, all solutions of $y = -2$ are of the form $(x, -2)$ where x can have any value. Therefore, some solutions are $(0, -2)$, $(3, -2)$ and $(-2, -2)$. Plot the ordered pairs and draw a line through the points.

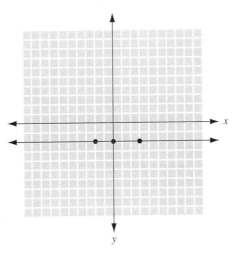

In Example 5, the graph of $x = 3$ is a vertical line and the graph of $y = -2$ is a horizontal line. The graphs of any two ordered pairs with the same x-value will always lie on the same vertical line, and the graphs of any two ordered pairs with the same y-value will always be on the same horizontal line. This leads to the following observation.

VERTICAL AND HORIZONTAL LINES

For any constant k, the graph of an equation of the form $x = k$ is a vertical line intersecting the x-axis at k.

For any constant k, the graph of an equation of the form $y = k$ is a horizontal line intersecting the y-axis at k.

PRACTICE EXERCISES

Graph the following equations. Assume each unit is one.

13) $x = -1$

14) $y = 4$

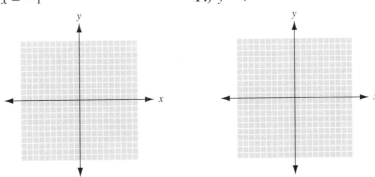

13)

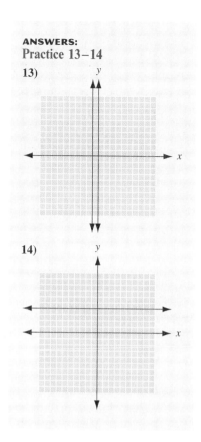

14)

Note: There are a couple of observations we need to make regarding vertical and horizontal lines. If two or more points lie on the same vertical line, they have the same *x*-values. Also, if two or more points have the same *x*-values, they lie on the same vertical line. Likewise, if two or more points lie on the same horizontal line, they have the same *y*-values. Also, if two or more points have the same *y*-value, they lie on the same horizontal line.

15) $4x - y = 2$ 16) $x - 4y = 6$

17) $2x + 3y = 6$ 18) $3x + 2y = 6$

19) $2x - 5y = 10$ 20) $5x - 2y = 10$

21) $2x + 6y = 12$ 22) $6x - 2y = 12$

23) $2x + 5y = 5$ 24) $2x + 7y = 7$

25) $3x - 5y = 10$ 26) $4x - 3y = 8$

27) $4x - 3y = 10$ 28) $3x - 4y = 14$

29) $y = x$ 30) $y = 4x$

31) $y = -3x$ 32) $y = -4x$

33) $2x + y = 0$ 34) $3x - y = 0$

35) $x = 2$ 36) $x = 5$

37) $y = 3$ 38) $y = 7$

39) $x = -4$ 40) $x = -2$

41) $y = -4$ 42) $y = -7$

43) $y = -9$

Write an equation representing each of the following and then graph the equation.

44) The sum of the *x*- and *y*-coordinates is 8.

45) The difference of the *x*- and *y*-coordinates is 4.

46) The sum of three times the *x*-coordinate and 2 times the *y*-coordinate is 12.

47) The difference of 3 times the *x*-coordinate and 4 times the *y*-coordinate is 12.

48) The *y*-coordinate is 5 less than twice the *x*-coordinate.

49) The *y*-coordinate is 2 more than three times the *x*-coordinate.

50) Two times the *y*-coordinate is equal to 6 less than the *x*-coordinate.

E X E R C I S E   S E T   **6.2**

Plot the points represented by the following ordered pairs on the rectangular coordinate system. Assume each unit represents one.

1) $(3,5),(-3,1),(0,5),(2,-4)$

2) $(-4,0),(4,-2),(-3,4),(0,4)$

3) $(3,7),(-2,-4),(0,2),(4,-5)$

4) $(6,-2),(-4,5),(2,-1),(3,4)$

Find three ordered pairs that solve the equation and draw the graph of each of the following. Each unit on the coordinate system represents one.

5) $y = x + 2$ 6) $y = x - 4$

7) $y = 2x + 4$ 8) $y = 3x - 5$

9) $y = -2x + 1$ 10) $y = -3x + 5$

Graph each of the following using the intercept method.

11) $x + y = 3$ 12) $x + y = 6$

13) $x + 3y = 9$ 14) $x + 4y = 8$

Write an equation for each of the following.

51) The Maitland Public Library was having a plant sale. If Katrina bought *x* plants for $2 each and *y* plants for $3 each, write an equation that shows the possible combinations of $2 and $3 plants if the total cost was $24.

52) If *x* represents the length of a rectangle and *y* represents the width, write an equation that shows the possible combinations of lengths and widths if the perimeter is 32 feet.

53) If there are x \$5 bills and y \$10 bills, write an equation that shows the possible combinations of \$5 and \$10 if the total value of the bills is \$450.

54) John receives \$6 per hour for regular time worked and \$9 per hour for overtime. If John works x hours of regular time and y hours of overtime, write an equation that shows the possible combinations of regular hours and overtime hours if John received \$420.

55) Abdul took a trip. For the first part of the trip he averaged 50 miles per hour and for the second part he averaged 60 miles per hour. If Abdul drove x hours at 50 miles per hour and y hours at 60 miles per hour, write an equation that shows the possible combinations of hours if the trip was 500 miles long.

56) In basketball a field goal counts 2 points and a free throw counts 1 point. If Jamal scored x field goals and y free throws, write an equation that shows the possible combinations of field goals and free throws if he scored 27 points.

CHALLENGE EXERCISES (58–60)

57) An object is dropped from the top of a tall building. If y represents the velocity and x represents the time in seconds after the object is dropped, then $y = -32x$.

  a) Find the velocity after 0 seconds, 2 seconds, and 4 seconds. Write the results as ordered pairs.

  b) Draw the graph for $0 \leq x \leq 5$. You will need to change the units on the y-axis.

  c) Use the graph to estimate the number of seconds after the object is thrown when the velocity is -96 feet per second.

58) An object is thrown upward from the top of a tall building with an initial velocity of 150 feet per

second. If y represents the velocity and x the number of seconds after the object is thrown, then $y = -32x + 150$.

  a) Find the velocity after 1 second, 2 seconds, and 3 seconds. Write the results as ordered pairs.

  b) Draw the graph.

  c) Use the graph to estimate the number of seconds after the object is thrown when the velocity is -10 feet per second.

59) If the graph of $ax + by = 15$ has an x-intercept of $(5,0)$ and a y-intercept of $(0,3)$, find a and b.

60) If the graph of $ax + by = -8$ has an x-intercept of $(4,0)$ and a y-intercept of $(0,-2)$, find a and b.

WRITING EXERCISES

61) If the graph of an ordered pair is in the first quadrant, both the x- and y-values of the ordered pair are positive. Describe the x- and y-values of an ordered pair whose graph is in the second quadrant. Why? The third quadrant. Why? The fourth quadrant. Why?

62) What does the graph of an equation represent?

63) Why is the graph of an equation in the form $x = k$ a vertical line?

64) Why is the graph of an equation in the form $y = k$ a horizontal line?

65) Since one and only one line can be drawn through any two distinct points, why is it a good idea to always find three points when graphing a line?

66) When graphing an equation of the form $ax + by = c$, why do we usually find the x- and y-intercepts as two of the points?

67) What is the equation of the x-axis?

SECTION 6.3

Graphing Linear Inequalities with Two Variables

OBJECTIVE

When you complete this section, you will be able to:

Ⓐ Graph linear inequalities with two unknowns.

In Section 6.2 we graphed linear equations. In this section we will graph **linear inequalities**. A linear inequality with two variables is in the form of $ax + by > c$, $ax + by \geq c$, $ax + by < c$, $ax + by \leq c$, or can be put into one of these forms. Just as any ordered pair of numbers that makes a linear equation true is a solution to the equation, any ordered pair that makes a linear inequality true is a solution of the inequality.

EXAMPLE 1

Determine if the given ordered pair is a solution of the given inequality.

a) $y > 3x - 1$; $(2,7)$

Solution:

$y > 3x - 1$	Substitute 2 for x and 7 for y.
$7 > 3(2) - 1$	Simplify the right side.
$7 > 5$	This is true, so $(2,7)$ is a solution.

b) $3x + 2y \geq 6$; $(-4,5)$

Solution:

$3x + 2y \geq 6$; $(-4,5)$	Substitute -4 for x and 5 for y.
$3(-4) + 2(5) \geq 6$	Multiply before adding.
$-12 + 10 \geq 6$	Add.
$-2 \geq 6$	This is a false statement, so $(-4,5)$ is not a solution.

In Chapter 3 we solved and graphed the solutions of linear inequalities with one unknown. For example, the solution of $2x - 3 > 5$ is $x > 4$. To graph $x > 4$, we first graph $x = 4$ with an open dot to indicate that 4 is not a part of the solution. The graph of $x > 4$ is the set of all points of the line on one side of 4 (either to the right or left of 4). Substitute 5 for x and we get $5 \geq 4$ which is a true statement. Since 5 is to the right of 4, the graph of $x > 4$ is all the points to the right of 4. See the following graph.

To graph $x \geq 4$, we would use a closed dot, rather than an open dot, to indicate that 4 is a part of the solution.

The graph of the solutions of a linear inequality with one variable of the form $x > a$ or $x < a$ is all the points of the number line that are on one side of $x = a$ or the other. Similarly, the graph of the solutions of a linear inequality with two variables is all the points of the rectangular coordinate system on one side of the graph of $ax + by = c$ or the other. In other words, all the points on the graph of $ax + by = c$ make the equation true. All other points in the coordinate system result in either $ax + by > c$ or $ax + by < c$ being true. We indicate the solutions by shading all the points on the side of the line that solve the inequality.

To graph the solutions of a linear inequality with two variables, we follow a procedure similar to graphing the solutions of a linear inequality with one variable. The procedure is outlined below.

• Graphing linear inequalities with two unknowns

GRAPHING LINEAR INEQUALITIES WITH TWO VARIABLES

To graph a linear inequality with two variables:

1) Graph the equality $ax + by = c$. If the line is part of the solution ($\leq$ or $\geq$), draw a solid line. If the line is not part of the solution ($<$ or $>$), draw a dashed line.

---

2) Pick a test point not on the line. If the coordinates of the test point solve the inequality, then all points on the same side of the line as the test point solve the inequality. If the coordinates of the test point do not solve the inequality, then all the points on the other side of the line from the test point solve the inequality.

3) Shade the region on the side of the line that contains the solutions of the inequality.

EXAMPLE 2

Graph the following linear inequalities with two variables.

a) $2x + y > 8$

Solution:

Draw the graph of $2x + y = 8$ with a dashed line since the points on the line do not solve the inequality. The x-intercept is $(4,0)$ and the y-intercept is $(0,8)$ and find another point as a check. Remember, to find the x-intercept, let $y = 0$ and the y-intercept, let $x = 0$.

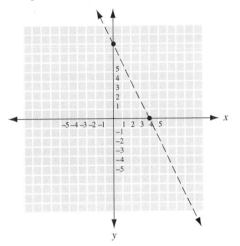

Pick a test point. The origin, $(0,0)$, is often a convenient test point. Substitute $(0,0)$ into the inequality. $2(0) + 0 > 8$, $0 + 0 > 8$, $0 > 8$. This is a false statement, so $(0,0)$ is not a solution of $2x + y > 8$. Therefore, the solutions are on the opposite side of the line from $(0,0)$. Shade the region on the opposite side of the line from $(0,0)$. You may want to test a point on the shaded side just to be sure.

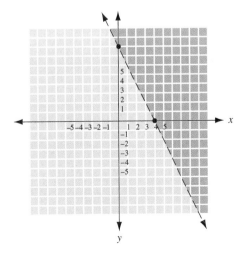

417

b) $3x - y \leq 6$

Solution:

Draw the graph of $3x - y = 6$ as a solid line since the points on the line satisfy the inequality less than or *equal to*. Use the x-intercept $(2,0)$ and the y-intercept $(0,-6)$ and another point as a check.

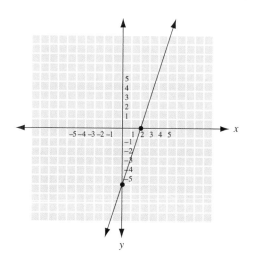

Pick a test point. Again, $(0,0)$ is a convenient choice. Substitute $(0,0)$ into the inequality. $3(0) - 0 \leq 6$, $0 - 0 \leq 6$, $0 \leq 6$. This is a true statement. Therefore, $(0,0)$ solves the inequality. Consequently, all points on the same side of the line as $(0,0)$ solve the inequality. Shade the region on the same side of the line as $(0,0)$.

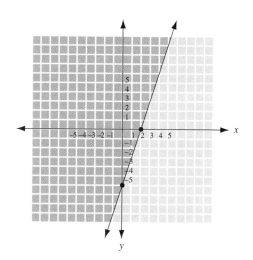

Graph the following linear inequalities.

a) $y \leq 3x - 4$

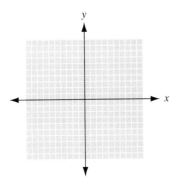

b) $4x + y > 8$

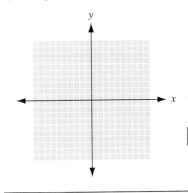

PRACTICE EXERCISES

Graph the following linear inequalities with two variables.

1) $3x - 4y \geq 12$

2) $5x - 2y < 10$

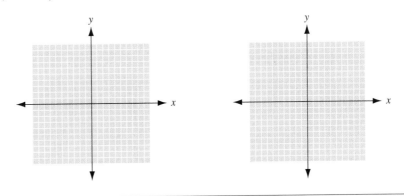

If more practice is needed, do the Additional Practice Exercises in the margin.

If the line passes through the origin, we cannot use the origin as a test point since the test point must lie in a region on one side of the line. In this case, choose any point you wish as long as it is clearly not on the line.

EXAMPLE 3

Graph the following inequality.

 a) $y \leq 3x$

Solution.

Draw the graph of $y = 3x$ as a solid line since the points on the line satisfy the inequality less than or equal to. Since both the x- and y-intercepts are $(0,0)$, we need another point other than the intercepts in order to draw the graph. Choose a value for x and solve for y. Let $x = 3$, then $y = 3(3) = 9$. Therefore, another ordered pair is $(3,9)$.

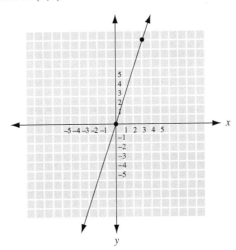

1)

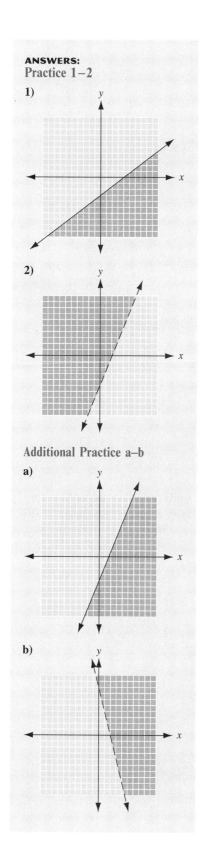

2)

a)

b)

Since (0,0) lies on the line, we cannot use it as the test point. Choose any other point which clearly is not on the line, say (2,0). Substitute (2,0) into the inequality. $0 \leq 3(2)$, $0 \leq 6$. This is true. Therefore, all the solutions are on the same side of the line as (2,0). Shade this region.

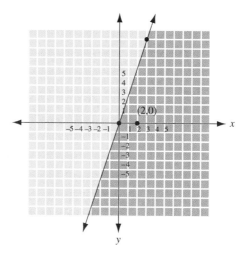

In the case of vertical and horizontal lines, it is not necessary to use a test point. The situation is very much like graphing the solutions to a linear inequality with one variable.

EXAMPLE 4

Graph the following linear inequalities with two variables.

a) $x > -2$

Solution:

Draw the graph of $x = -2$ as a dotted line since the coordinates of the points on the line do not satisfy the inequality. The graph of $x = -2$ is a vertical line intersecting the x-axis at -2.

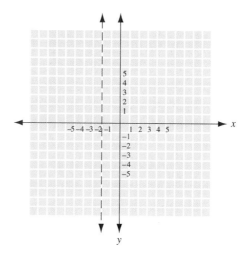

The solutions to $x > -2$ consists of all points whose x-values are greater than -2. Just as on the number line, these points lie to the right of the graph of $x = -2$. Therefore, shade the region to the right of $x = -2$.

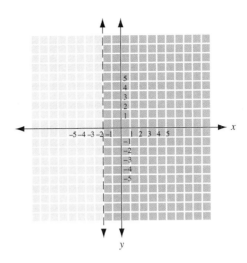

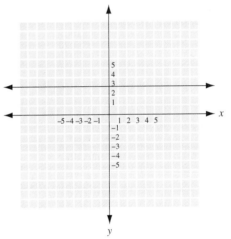

b) $y \leq 3$

Solution:

Draw the graph of $y = 3$ as a solid line since the coordinates of the points on the line satisfy the inequality. The graph of $y = 3$ is a horizontal line that intersects the *y*-axis at 3.

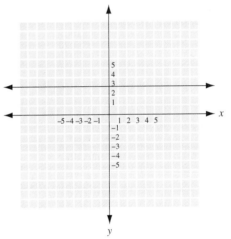

The solutions to $y \leq 3$ consists of all points whose *y*-values are less than 3. These points lie below the line $y = 3$. Therefore, shade the region below the line.

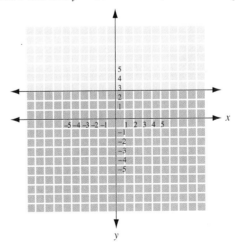

Graph the following inequalities with two variables.

c) $y \geq x$

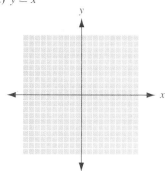

d) $x \geq 2$

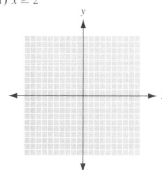

e) $y < -1$

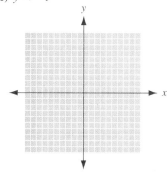

PRACTICE EXERCISES

Graph the following linear inequalities with two variables. Assume each unit on the coordinate system is one.

3) $y < -3x$

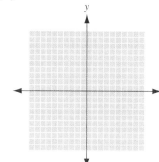

4) $x < -3$

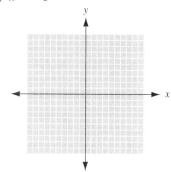

5) $y \geq 2$

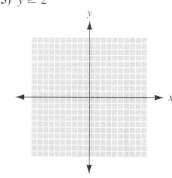

If more practice is needed, do the Additional Practice Exercises in the margin.

Graph the following linear inequalities with two variables. Assume each unit on the coordinate system represents one.

7) $y < x + 2$	8) $y > x - 6$
9) $y \geq 2x + 3$	10) $y \geq 3x - 4$
11) $y < -2x + 5$	12) $y \geq -3x + 2$
13) $y > \frac{1}{3}x + 2$	14) $y \leq \frac{2}{5}x - 3$
15) $2x + y > 6$	16) $3x - y \leq 9$
17) $x + 2y \leq 4$	18) $x + 4y < 8$
19) $2x - 3y > 6$	20) $5x - 2y \leq 15$
21) $3x + 4y \leq 12$	22) $y \leq 2x$
23) $y > x$	24) $y < -4x$
25) $y \geq -x$	26) $y > -2x$
27) $y \leq -4x$	28) $x \geq 2$
29) $x < 4$	30) $y > 3$
31) $y \leq 4$	32) $x \geq -1$
33) $x < -5$	34) $y \leq -2$
35) $y > -6$	36) $y \leq 0$

EXERCISE SET 6.3

Determine whether the given point is a solution of the given inequality.

1) $y > 3x - 2$; (1,0)

2) $y > -2x + 3$; (2,0)

3) $2x + 3y \leq 6$; (-2,5)

4) $3x - 4y \leq 8$; (-2,-3)

5) $3x - 2y \geq 12$; (2,-3)

6) $4x - 3y \leq 8$; (5,4)

3)

c)

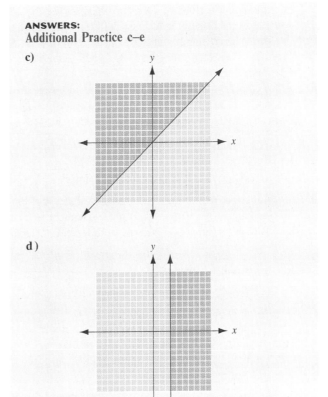

4)

5)

d)

e)

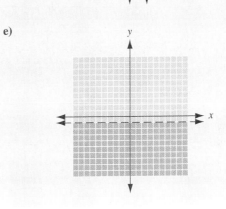

Write an inequality representing each of the following and graph.

37) The sum of twice the *x*-coordinate and four times the *y*-coordinate is less than 12.

38) The sum of three times the *x*-coordinate and four times the *y*-coordinate is greater than or equal to 12.

39) The difference of the *x*-coordinate and two times the *y*-coordinate is greater than or equal to 8.

40) The difference of two times the *x*-coordinate and five times the *y*-coordinate is less than 10.

41) The *y*-coordinate is less than 5 more than three times the *x*-coordinate.

42) The *y*-coordinate is greater than 3 more than four times the *x*-coordinate.

43) Hector is going shopping for some shirts and shorts. The shirts cost $20 each and the shorts cost $25 each. If he buys *x* shirts and *y* shorts, write an inequality that shows the possible combinations of shirts and shorts if Hector can spend no more than $150.

44) Tickets to a high school basketball game cost $3 for students and $5 for adults. If *x* student tickets and *y* adult tickets are sold, write an inequality that shows the possible combinations of student and adult tickets if the receipts from ticket sales must be at least $500.

45) An electronics company assembles TVs and VCRs. It takes 3 hours to assemble a TV and 2 hours to assemble a VCR. If *x* TVs and *y* VCRs are assembled per day, write an inequality that shows the possible combinations of TVs and VCRs if the company must assemble at least 500 units per day to keep up with demand.

CHALLENGE EXERCISES (46–49)

46) A nut company uses x ounces of peanuts in a regular mix and y ounces of peanuts in a special party mix. The company has at most 500 ounces of peanuts in stock.

   a) Write an inequality describing the situation.

   b) Graph the inequality.

   c) Is (100,300) a solution of this inequality?

47) A chemical company needs x liters of sulfuric acid to make chemical A and y liters to make chemical B. The company has at most 200 liters of sulfuric acid in stock.

   a) Write an inequality describing the situation.

   b) Graph the situation.

   c) Is (150,100) a solution of the inequality?

48) A small company is in the business of manufacturing lawn chairs. It takes two hours of labor to make a regular chair and three hours of labor to make a lounge chair. There is a maximum of 84 hours per day of labor available.

   a) Let x represent the number of regular chairs produced and y represent the number of lounge

chairs produced. Write an inequality describing the situation.

   b) Graph the inequality.

   c) Is (20,20) a solution of the inequality?

49) A battery company sells two types of batteries. It makes a profit of $10 on battery A and a profit of $15 on battery B. The company must make a profit of at least $3000 per week to break even.

   a) Let x represent the number of A batteries sold per week and y represent the number of B batteries sold each week. Write an inequality describing the situation.

   b) Graph the inequality.

   c) Is (200,100) a solution of the inequality?

WRITING EXERCISES

50) How is graphing a linear inequality with two variables like graphing a linear inequality with one variable? How does it differ?

51) If your test point is on one side of the boundary line and it fails to solve the inequality, why are the solutions on the other side of the line?

52) In choosing a test point, why can't you choose a point on the line?

SECTION 6.4

Slope of a Line

OBJECTIVES

When you complete this section, you will be able to:

Ⓐ   Find the slope of a line by using the slope formula.
Ⓑ   Graph a line when given a point on the line and the slope of the line.
Ⓒ   Find the slope of a line from the equation of the line.

INTRODUCTION

Thus far, we have been graphing lines by finding at least two points on the line and then drawing a line through the points. In this section we will learn another method. Consider the graphs of the following lines all of which contain the point (3,3).

• Note: You may not be familiar with the use of subscripts. In the notations L_1, L_2, L_3, etc., the L's denote lines. The numbers are called subscripts and are used to denote different lines. Hence, L_1 stands for the first line, L_2 stands for the second line, and so on.

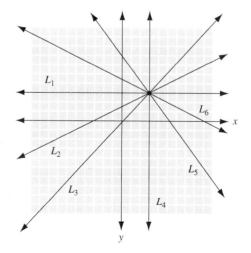

All these lines pass through the same point yet there are substantial differences. L_1 is horizontal, L_4 is vertical, L_5 slants downward (a "falling" line) from left to right, and L_2 slants upward (a "rising" line) from left to right. Also, both L_2 and L_3 slant upward from left to right, but L_3 is steeper than L_2. One way of describing these differences is by using **slope**.

SLOPE OF A LINE

The slope of a line is the ratio of the change in y to the change in x as you go from one point on a line to any other point on the line.

An informal way of thinking of slope is that it is the ratio of the rise of a line to the run of the line between any two points on the line as illustrated by the following figure.

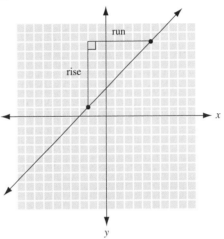

This informal way of thinking of slope is consistent with the formal definition. The rise is a vertical change and, since the y-axis is vertical, any vertical change is a change in y. Also, the run is a horizontal change and, since the x-axis is horizontal, any horizontal change is a change in x.

- **Derivation of the slope formula**

As is often the case in mathematics, the definition gives us no indication of how to actually perform the task. Suppose we want to find the slope of the line that contains the points (2,3) and (7,9). Plot the points and draw the line containing these points. Then go from point (2,3) to (7,9) by first going vertically up from (2,3) and then horizontally to (7,9) as illustrated below.

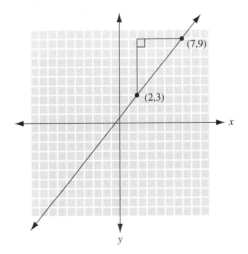

Since any two points on the same vertical line have the same x-value and any two points on the same horizontal line have the same y-value, the point of intersection of the horizontal and vertical lines is (2,9).

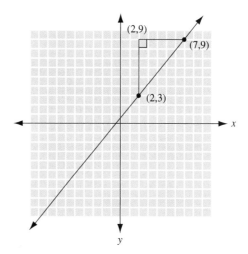

The rise is the change in y as we go from (2,3) to (2,9). The y-value changes from 3 to 9 which is a change of 6. This change in y (rise) can be found by finding the difference of the y-values ($9 - 3 = 6$). The run is the change in x as we go from (2,9) to (7,9). The x-value changes from 2 to 7 which is a change of 5. This change in x (run) can be found by finding the difference of the x-values ($7 - 2 = 5$).

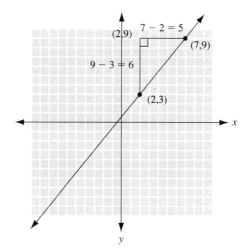

Therefore, the slope is

$$\frac{\text{rise}}{\text{run}} = \frac{\text{change in } y}{\text{change in } x} = \frac{9 - 3}{7 - 2} = \frac{6}{5}.$$

Instead of going through all this each time we want to find the slope of a line, we use any line of the form $ax + by = c$ and any two points on the line. We will denote the points as (x_1, y_1) and (x_2, y_2) and derive a formula for slope. Subscripts are used to indicate that both points have x- and y-values, but these values may be different. The letter **m** is customarily used for slope. Consider the figure below.

- It is not really known why the letter **m** is used to denote slope. One belief is that it comes from the French word "monter" which means "to climb."

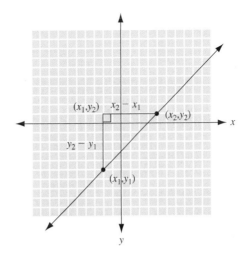

Since any two points on the same vertical line have the same x-value and any two points on the same horizontal line have the same y-value, the coordinates of the point of intersection of the vertical and horizontal line segments from (x_1, y_1) to (x_2, y_2) are (x_1, y_2). As in the previous example, the change in y is the difference in the y-values. Thus, the change in y is $y_2 - y_1$. The change in x is the difference in the x-values. Thus, the change in x is $x_2 - x_1$. Therefore, the formula for the slope of a line containing the points (x_1, y_1) and (x_2, y_2) is $m = \frac{y_2 - y_1}{x_2 - x_1}$.

SLOPE FORMULA

The slope of a line containing the points whose coordinates are (x_1, y_1) and (x_2, y_2) is

$$m = \frac{y_2 - y_1}{x_2 - x_1}.$$

Note: The slope of the line is the difference of the y-values divided by the difference of the x-values. This difference can be found in either order. Consequently, the slope could also be $m = \frac{y_1 - y_2}{x_1 - x_2}$. However, we must be very careful. Whichever y-value is used first in the numerator, the x-value from the same ordered pair must also be used first in the denominator.

- Finding the slope of a line using the slope formula

EXAMPLE 1

Find the slope of the line that contains each of the following pairs of points.

a) (1,3) and (5,9)

Solution:

Let $(x_1, y_1) = (1,3)$ and $(x_2, y_2) = (5,9)$. Hence, $x_1 = 1$, $y_1 = 3$, $x_2 = 5$, and $y_2 = 9$. Substitute these values into the slope formula.

$$m = \frac{y_2 - y_1}{x_2 - x_1} \qquad \text{Substitute for } y_2, y_1, x_2, \text{ and } x_1.$$

$$m = \frac{9 - 3}{5 - 1} \qquad \text{Simplify the numerator and the denominator.}$$

$$m = \frac{6}{4} \qquad \text{Reduce to lowest terms.}$$

$$m = \frac{3}{2} \qquad \text{Therefore, the slope is } \frac{3}{2}.$$

Alternate solution:

If we let $(x_1, y_1) = (5,9)$ and $(x_2, y_2) = (1,3)$, then $m = \frac{3-9}{1-5} = \frac{-6}{-4} = \frac{3}{2}$. Consequently, we get the same result.

b) $(-2,3)$ and $(4,-2)$

Solution:

Let $(x_1, y_1) = (-2,3)$ and $(x_2, y_2) = (4,-2)$. Therefore, $x_1 = -2$, $y_1 = 3$, $x_2 = 4$, and $y_2 = -2$. Substitute these values into the slope formula.

$$m = \frac{y_2 - y_1}{x_2 - x_1} \qquad \text{Substitute for } y_2, y_1, x_2, \text{ and } x_1.$$

$$m = \frac{-2 - 3}{4 - (-2)} \qquad 4 - (-2) = 4 + 2.$$

$$m = \frac{-2 - 3}{4 + 2} \qquad \text{Simplify the numerator and the denominator.}$$

$$m = \frac{-5}{6} = -\frac{5}{6} \qquad \text{Therefore, the slope is } -\frac{5}{6}.$$

c) $(3,2)$ and $(-1,2)$

Solution:

Let $(x_1, y_1) = (3,2)$ and $(x_2, y_2) = (-1,2)$. Therefore, $x_1 = 3$, $y_1 = 2$, $x_2 = -1$, and $y_2 = 2$. Substitute these values into the slope formula.

$$m = \frac{y_2 - y_1}{x_2 - x_1} \qquad \text{Substitute for } y_2, y_1, x_2, \text{ and } x_1.$$

$$m = \frac{2 - 2}{-1 - 3} \qquad \text{Simplify the numerator and denominator.}$$

$$m = \frac{0}{-4} \qquad \text{Simplify.}$$

$$m = 0 \qquad \text{Therefore, the slope is 0.}$$

Note: The graph of the line containing $(3,2)$ and $(-1,2)$ is horizontal.

d) $(3,6)$ and $(3,2)$

Solution:

Let $(x_1, y_1) = (3,6)$ and $(x_2, y_2) = (3,2)$. Therefore, $x_1 = 3$, $y_1 = 6$, $x_2 = 3$, and $y_2 = 2$. Substitute these values into the slope formula.

$$m = \frac{y_2 - y_1}{x_2 - x_1} \qquad \text{Substitute for } y_2, y_1, x_2, \text{ and } x_1.$$

$$m = \frac{2 - 6}{3 - 3}$$ Simplify the numerator and denominator.

$$m = \frac{-4}{0}$$ Division by 0 is undefined. Therefore, the slope of the line through (3,6) and (3,2) is undefined.

Note: The graph containing (3,6) and (3,2) is vertical.

ADDITIONAL PRACTICE

Find the slope of the line which contains each of the following pairs of points.

a) (4,7) and (6,3)

b) (−1,4) and (4,−3)

c) (2,2) and (−4,2)

d) (−5,−2) and (−5,−4)

- **Graphing a line given a point on the line and the slope**

PRACTICE EXERCISES

Find the slope of the line that contains each of the following pairs of points.

1) (2,4), (5,8)   **2)** (−2,3), (5,−2)   **3)** (−3,4), (−1,4)

4) (3,−5), (3,−2)

If more practice is needed, do the Additional Practice Exercises in the margin.

Using slope we now have another method of graphing a line. Remember, the slope represents the change in *y* (vertical change) divided by the change in *x* (horizontal change). If we know a point on the line and the slope, we can find the graph of the line by using the procedure below.

GRAPHING A LINE GIVEN A POINT AND THE SLOPE

1) Plot the given point.

2) From the given point we can find another point by going vertically (up or down) the number of units given by the numerator of the slope and horizontally (right or left) the number of units given by the denominator.

3) Draw a line through these two points.

EXAMPLE 2

Draw the graph of each of the following lines containing the given point and with the given slope.

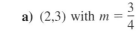

a) (2,3) with $m = \frac{3}{4}$

Solution:

Plot the point with coordinates (2,3). To locate another point on the line, go up 3 units from (2,3) and then go to the right 4 units. Draw the line through these two points.

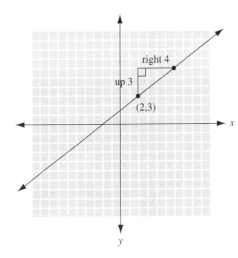

b) $(-2,2)$ and $m = 3$

Solution:

Plot the point with coordinates $(-2,2)$. Think of 3 as $\frac{3}{1}$. Therefore, go 3 units up from $(-2,2)$ and then go 1 unit to the right. Draw the line through these two points.

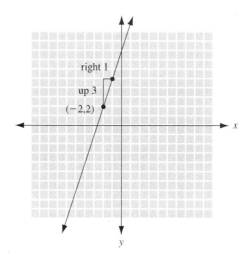

c) $(-3,-1)$ and $m = -\frac{2}{3}$

Solution:

Plot the point with coordinates $(-3,-1)$. The slope $-\frac{2}{3}$ may be thought of as $\frac{-2}{3}$ or $\frac{2}{-3}$. We will use $\frac{-2}{3}$. Therefore, go 2 units down from $(-3,-1)$, then go 3 units to the right. Draw the line through these two points.

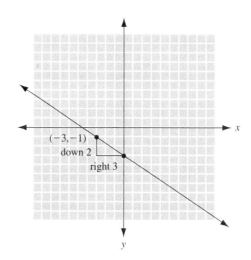

Draw the graph of each of the following lines containing the given point with the given slope.

e) $(-1,4)$, $m = \dfrac{1}{3}$

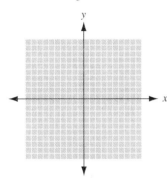

f) $(3,-4)$; $m = 2$

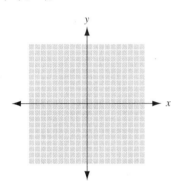

g) $(-3,5)$; $m = -\dfrac{2}{3}$

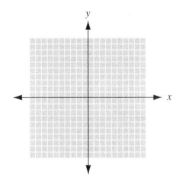

PRACTICE EXERCISES

Draw the graph of each of the following lines containing the given point with the given slope.

5) $(3,4)$, $m = \dfrac{2}{3}$

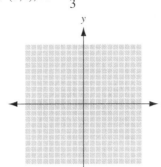

6) $(-4,2)$, $m = -2$

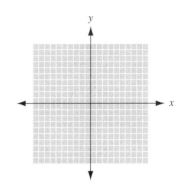

7) $(-5,4)$, $m = -\dfrac{5}{3}$

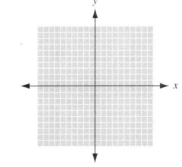

If more practice is needed, do the Additional Practice Exercises in the margin.

We now make two observations about the slope of a line. Since a positive slope means go up from a given point and then go to the right, a line with positive slope rises as we go from left to right. Since a line with a negative slope can mean go down from a given point and then go to the right, a line with negative slope drops as we go from left to right. See the figures below.

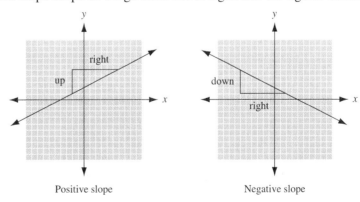

Positive slope   Negative slope

5)

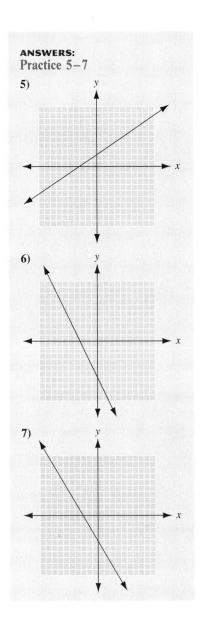

6)

7)

• **Answers to Additional Practice e–g are on page 433.**

The greater the absolute value of the slope, the greater the vertical change is in relation to the horizontal change. Therefore, the greater the absolute value of the slope, the steeper the line. See the figure below.

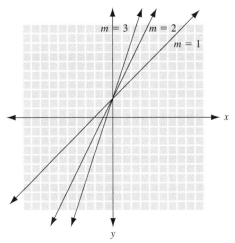

Given the equation of a line, we can find the slope of the line by finding the coordinates of two points on the line. Then substitute the coordinates of the point into the slope formula. The two easiest points to find are the x- and y-intercepts.

EXAMPLE 3

Find the slope of the line represented by each of the following equations.

a) $3x - 2y = 6$

Solution:

Find the x- and y-intercepts by letting $y = 0$ and $x = 0$ respectively.

x-intercept. Let $y = 0$. y-intercept. Let $x = 0$.

$3x - 2(0) = 6$ $3(0) - 2y = 6$
$3x = 6$ $0 - 2y = 6$
$x = 2$ $y = -3$

Therefore, the x-intercept is $(2,0)$.   Therefore, the y-intercept is $(0,-3)$.

Find the slope using the points whose coordinates are $(2,0)$ and $(0,-3)$.

$$m = \frac{y_2 - y_1}{x_2 - x_1} \qquad \text{Substitute for } x_1, y_1, x_2, \text{ and } y_2.$$

$$m = \frac{-3 - 0}{0 - 2} \qquad \text{Simplify.}$$

$$m = \frac{-3}{-2} = \frac{3}{2} \qquad \text{Therefore, the slope is } \frac{3}{2}.$$

b) $3x + 5y = 8$

Solution:

Find the x- and y-intercepts.

x-intercept. Let $y = 0$. y-intercept. Let $x = 0$.

$3x + 5(0) = 8$ $3(0) + 5y = 8$
$3x + 0 = 8$ $0 + 5y = 8$
$3x = 8$ $5y = 8$
$x = \dfrac{8}{3}$ $y = \dfrac{8}{5}$

Therefore, the x-intercept is $(\frac{8}{3},0)$.   Therefore, the y-intercept is $(0,\frac{8}{5})$.

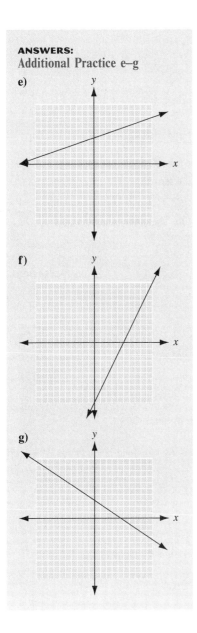

e)

f)

g)

- **Finding the slope of a line from the equation of the line**

Find the slope using the points whose coordinates are $(\frac{8}{3}, 0)$ and $(0, \frac{8}{5})$.

$$m = \frac{y_2 - y_1}{x_2 - x_1}$$ Substitute.

$$m = \frac{\frac{8}{5} - 0}{0 - \frac{8}{3}}$$ Simplify the numerator and denominator.

$$m = \frac{\frac{8}{5}}{-\frac{8}{3}}$$ Rewrite as division.

$$m = \frac{8}{5} \div \left(-\frac{8}{3}\right)$$ Invert and multiply.

$$m = \frac{8}{5} \times \left(-\frac{3}{8}\right)$$ Multiply.

$$m = -\frac{3}{5}$$ Therefore, the slope is $-\frac{3}{5}$.

The x- and y-intercepts in Example 3b involved fractions. This made finding the slope somewhat complicated since it involved simplifying a complex fraction. To show another method of finding the slope of a line when the equation is given, we will solve each of the equations from Example 3 for y and make an observation.

Example 3a

$$3x - 2y = 6$$
$$3x - 3x - 2y = -3x + 6$$
$$-2y = -3x + 6$$
$$\frac{-2y}{-2} = \frac{-3x + 6}{-2}$$
$$y = \frac{3}{2}x - 3$$

Example 3b

$$3x + 5y = 8$$
$$3x - 3x + 5y = -3x + 8$$
$$5y = -3x + 8$$
$$\frac{5y}{5} = \frac{-3x + 8}{5}$$
$$y = -\frac{3}{5}x + \frac{8}{5}$$

In Example 3a, we found the slope of $3x - 2y = 6$ to be $\frac{3}{2}$. When we solved $3x - 2y = 6$ for y, the coefficient of the x-term is also $\frac{3}{2}$. In Example 3b, we found the slope of $3x + 5y = 8$ to be $-\frac{3}{5}$. When we solved $3x + 5y = 8$ for y, we found the coefficient of the x-term is also $-\frac{3}{5}$. This is no coincidence. It can be shown that if a linear equation with two variables is solved for y, the graph of the equation is a line whose slope is the coefficient of x. Further note that the y-intercept of Example 3a is -3 which is the constant when $3x - 2y = 6$ was solved for y giving $y = \frac{3}{2}x - 3$. Likewise for Example 3b the y-intercept is $\frac{8}{5}$ which is the constant when the equation was solved for y giving $y = -\frac{3}{5}x + \frac{8}{5}$. Consequently, when a linear equation is solved for y, we can read both the slope and the y-intercept from the equation.

SLOPE–INTERCEPT FORM OF A LINE

When any linear equation is solved for y, the resulting equation is of the form $y = mx + b$ where m is the slope and b is the y-intercept of the graph of the equation. This is called the slope-intercept form of a line.

EXAMPLE 4

Find the slope and y-intercept of the line represented by each of the following equations.

a) $4x + y = 5$

Solution:
Solve the equation for y.

$4x + y = 5$	Subtract $4x$ from both sides.
$4x - 4x + y = -4x + 5$	Add like terms.
$0 + y = -4x + 5$	$0 + y = y$.
$y = -4x + 5$	Therefore, the slope is -4 and the y-intercept is 5.

 b) $5x - 2y = 7$

Solution:
Solve the equation for y.

$5x - 2y = 7$	Subtract $5x$ from both sides.
$5x - 5x - 2y = -5x + 7$	Add like terms.
$0 - 2y = -5x + 7$	$0 - 2y = -2y$.
$-2y = -5x + 7$	Divide both sides by -2.
$\dfrac{-2y}{-2} = \dfrac{-5x + 7}{-2}$	Simplify both sides.
$y = \dfrac{5}{2}x - \dfrac{7}{2}$	Therefore, the slope is $\frac{5}{2}$ and the y-intercept is $-\frac{7}{2}$.

ADDITIONAL PRACTICE

Find the slope and y-intercept of the line represented by each of the following equations.

h) $5x + y = 9$

i) $7x - 3y = 8$

PRACTICE EXERCISES

Find the slope and y-intercept of the line represented by each of the following equations.

8) $x + 2y = 6$

9) $6x - 5y = 11$

If more practice is needed, do the Additional Practice Exercises in the margin.

The slope-intercept form of a linear equation gives us yet another method of graphing linear equations. The y-intercept is a point of the graph, so knowing a point and the slope we can draw the graph using the technique given earlier in this section.

EXAMPLE 5

Graph the following using the slope and y-intercept.

 a) $y = \dfrac{2}{3}x + 2$

Solution:

From the equation, the slope is $\frac{2}{3}$ and the y-intercept is 2. Plot the y-intercept and, from that point, go up 2 and to the right 3 and plot another point.

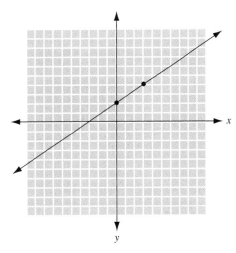

b) $3x + 4y = 8$

Solution:

First we must get the line into slope-intercept form by solving for y.

$$3x + 4y = 8 \qquad \text{Subtract } 3x \text{ from both sides.}$$

$$4y = -3x + 8 \qquad \text{Divide both sides by 4.}$$

$$y = \frac{-3x + 8}{4} \qquad \text{Simplify the right side.}$$

$$y = -\frac{3}{4}x + 2 \qquad \begin{array}{l}\text{Therefore, the slope is } -\frac{3}{4} \text{ and the } y\text{-intercept}\\ \text{is 2.}\end{array}$$

Plot the y-intercept and, from that point, go down 3 and to the right 4.

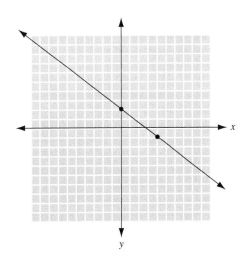

Graph each of the following using the slope-intercept method.

10) $y = \frac{3}{5}x - 1$

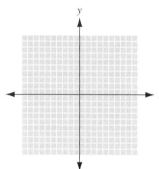

11) $3x + 2y = 6$

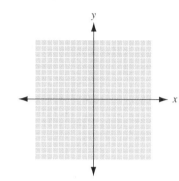

Recall that the graphs of equations of the form $y = k$ are horizontal lines. We can think of $y = k$ as $y = 0x + k$. Consequently, the slope of any horizontal line is 0 and the y-intercept is k. Also recall that the graphs of equations of the form $x = k$ are vertical lines. Since there is no y term, we cannot solve $x = k$ for y. So to find the slope of a vertical line, we need to find two points on the line and use the slope formula.

 EXAMPLE 6

Find the slope of the graph of $x = 3$.

Solution:

All points on the graph of $x = 3$ have an x-value of 3 and y can have any value. So two points on the graph are (3,0) and (3,1). Substitute these values into the slope formula.

$$m = \frac{y_2 - y_1}{x_2 - x_1} \qquad \text{Substitute.}$$

$$m = \frac{1 - 0}{3 - 3} \qquad \text{Simplify.}$$

$$m = \frac{1}{0} \qquad \text{Division by 0 is undefined. Therefore, there is no number which represents the slope of the graph of } x = 4.$$

Note that there is also no y-intercept.

From the above, we make the following generalizations.

SLOPES OF VERTICAL AND HORIZONTAL LINES

1) The graph of any equation of the form $y = b$ is a horizontal line whose slope is 0 and whose y-intercept is b. If the slope of a line is 0, then it is a horizontal line.

2) The graph of any equation of the form $x = a$ is a vertical line intersecting the x-axis at a and whose slope is undefined and there is no y-intercept. If the slope of a line is undefined, then it is a vertical line.

---

10)

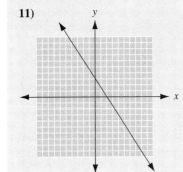

11)

EXAMPLE 7

Find the slope of each of the following.

a) $x = 4$

Solution:
The graph of $x = 4$ is a vertical line whose slope is undefined and there is no y-intercept.

b) $y = -3$

Solution:
The graph of $y = -3$ is a horizontal line whose slope is 0 and the y-intercept is -3.

Draw the graph of each of the following lines containing the given point and with the given slope.

c) $(3, -2)$ with $m = 0$

Solution:
Since the slope is 0, this is a horizontal line through the point $(3, -2)$.

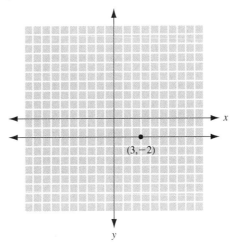

d) $(-5, 2)$ with undefined slope

Solution:
Since the slope is undefined, the graph is a vertical line through $(-5, 2)$.

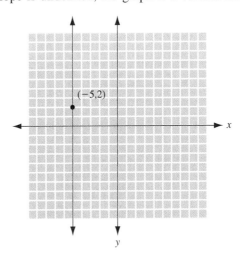

Find the slope of each of the following.

j) $y = -4$ **k)** $x = 5$

Draw the graph of each of the following lines containing the given point and with the given slope.

l) $(4,5)$ with $m = 0$

m) $(-4,2)$ with undefined slope

PRACTICE EXERCISES

Find the slope of each of the following.

12) $x = -2$ **13)** $y = 7$

Draw the graph of each of the following lines containing the given point and with the given slope.

14) $(-2,4)$ with undefined slope

15) $(2,3)$ with $m = 0$

If more practice is needed, do the Additional Practice Exercises in the margin.

SECTION SUMMARY

1) To find the slope of a line containing two given points (x_1,y_1) and (x_2,y_2), use the formula $m = \frac{y_2 - y_1}{x_2 - x_1}$.

2) To graph a line containing a given point and with a given slope, graph the point. From the given point, find another point of the line by going vertically the number of units given in the numerator and horizontally the number of units given in the denominator. Draw the line through these two points.

3) To find the slope and y-intercept of a line, solve the equation for y. The resulting form is $y = mx + b$ where m is the slope and b is the y-intercept.

4) To graph a line in slope-intercept form, plot the y-intercept as the point and use the same technique described in step 2 above.

5) The graph of $x = a$ is a vertical line and the slope is undefined.

6) The graph of $y = b$ is a horizontal line and the slope is 0.

E X E R C I S E   S E T   6.4

Find the slope of the line which contains each of the following pairs of points.

1) $(2,4)$, $(4,5)$ **2)** $(1,5)$, $(3,8)$

3) $(3,1)$, $(1,5)$ **4)** $(3,6)$, $(1,3)$

5) $(-1,4)$, $(2,-2)$ **6)** $(-3,2)$, $(1,-6)$

7) $(3,-2)$, $(-3,2)$ **8)** $(5,-4)$, $(1,-2)$

9) $(-4,-2)$, $(-8,2)$ **10)** $(-5,-2)$, $(-1,-8)$

11) $(0,3)$, $(4,6)$ **12)** $(6,0)$, $(4,-2)$

13) $(-2,-4)$, $(6,0)$ **14)** $(-7,-2)$, $(-2,0)$

15) $(3,5)$, $(-1,5)$ **16)** $(4,8)$, $(-3,8)$

17) $(2,6)$, $(2,3)$ **18)** $(-3,4)$, $(-3,-1)$

19) $(3,6)$, $(-4,6)$ **20)** $(5,-2)$, $(5,4)$

Find the slope of each of the following lines from the graph.

21)   **22)**

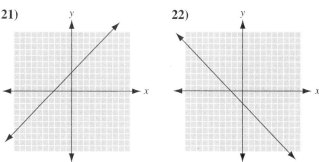

12) Undefined

13) 0

14)

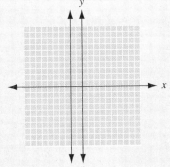

15)

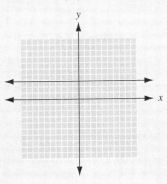

Additional Practice j-m

j) 0

k) Undefined

l)

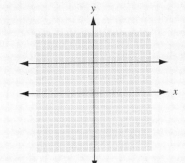

m)

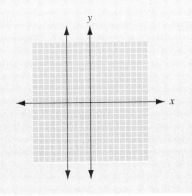

23)

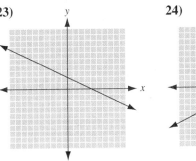

24)

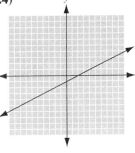

25)

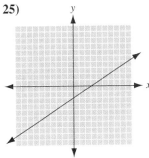

26)

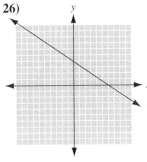

27)

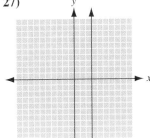

28)

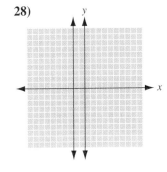

29)

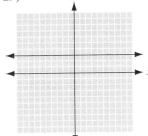

30)

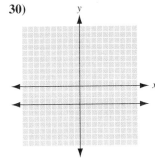

Draw the graph of each of the following lines containing the given point and with the given slope.

31) $(-3,2)$, $m = \dfrac{3}{4}$ **32)** $(4,-3)$, $m = \dfrac{3}{2}$

33) $(-1,-3)$, $m = -\dfrac{5}{3}$ **34)** $(-3,-5)$, $m = -\dfrac{4}{3}$

35) $(0,2)$, $m = 3$ **36)** $(4,0)$, $m = 4$

37) $(-1,3)$, $m = -2$ **38)** $(-4,-1)$, $m = -4$

39) $(2,5)$, undefined slope   **40)** $(-2,4)$, undefined slope

41) $(-2,6)$, $m = 0$ **42)** $(5,-2)$, $m = 0$

Find the slope and the y-intercept of the line represented by each of the following equations. Graph each equation using the slope-intercept method.

43) $y = -2x + 9$ **44)** $y = -3x + 8$

45) $3x - y = 7$ **46)** $2x - y = 11$

47) $2x + 3y = 9$ **48)** $3x + 2y = 12$

49) $5x - 3y = 15$ **50)** $2x - 5y = 10$

51) $2x + y = 0$ **52)** $3x - y = 0$

53) $3x + 2y = 5$ **54)** $5x - 2y = -3$

Find the slope of each of the following.

55) $x = 5$ **56)** $x = 8$ **57)** $x = -1$

58) $x = -6$ **59)** $y = 4$ **60)** $y = 7$

61) $y = -4$ **62)** $y = -8$

Answer the following.

63) A mountain road rises 30 feet vertically over a horizontal distance of 500 feet. a) What is the slope of the road? b) When the slope of a road is written as a percent, it is called the grade. What is the grade?

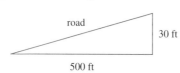

64) The roof of a house rises 6 feet over a span of 20 feet. What is the slope of the roof? (The slope of a roof is often called the pitch.)

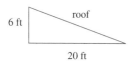

65) A carpenter is working on the outside of a house using a ladder. If the foot of the ladder is 9 feet from the wall and the top of the ladder is at a point 24 feet from the ground, what is the slope of the ladder?

66) The top of a guy wire is attached to a utility pole at a point 30 feet above the ground and the bottom is secured to the ground at a point 8 feet from the bottom of the pole. What is the slope of the guy wire?

CHALLENGE EXERCISES (67–70)

67) A lawn maintenance worker earns $6.00 per hour.

   a) If y represents the worker's earnings for working x hours, write an equation representing the worker's earnings.

   b) Graph the equation.

   c) What is the relationship between the slope of the line and the worker's hourly rate of pay?

68) A local nursery sells only azalea plants. There is an initial start-up cost of $360 per day and a cost of $3.00 per azalea sold per day.

   a) If y represents the cost per day and x the number of azaleas sold per day, write an equation for the daily costs of the nursery.

   b) Graph the equation.

   c) What is the relationship between the slope of the line and the cost per azalea?

69) A salesperson at a paint store earns $100 per week plus a commission of $2.00 on every gallon of paint he/she sells.

   a) If y represents the salesperson's salary for selling x gallons of paint, write an equation that represents the salesperson's salary.

   b) Graph the equation.

   c) What is the relationship between the slope of the line and the commission per gallon of paint sold?

70) A piece of machinery costs $3,000 and depreciates at the rate of $600 per year.

   a) If y represents the value of the machinery after x years, write an equation representing the value of the machinery.

b) Graph the equation.

c) What is the relationship between the slope of the line and the per year depreciation?

WRITING EXERCISES

71) How does the graph of a line whose slope is $\frac{2}{3}$ compare with the graph of a line whose slope is $-\frac{2}{3}$?

72) How does the graph of a line whose slope is $\frac{5}{2}$ compare with the graph of a line whose slope is $\frac{1}{3}$.

73) Why is the slope of a vertical line undefined?

74) Why is the slope of a horizontal line 0?

75) Why does a line with positive slope rise from left to right?

76) Why does a line with negative slope fall from left to right?

77) Give the equation of a line in $ax + by = c$ form that has a) positive slope. b) negative slope.

78) Given the equations $y = -2x - 3$, $y = -\frac{1}{2}x - 3$, $y = -0.75x - 3$ and $y = x - 3$, which equation has the steepest graph? Why?

S E C T I O N 6.5

Writing Equations of Lines

OBJECTIVES

When you complete this section, you will be able to:

Ⓐ Write the equation of a line which contains a given point and has a given slope.

Ⓑ Write the equation of a line which contains two given points.

Ⓒ Determine whether two lines are parallel, perpendicular, or neither parallel nor perpendicular.

Ⓓ Write the equation of a line which contains a given point and is parallel or perpendicular to a line whose equation is given.

INTRODUCTION

If we are given a point on a line and the slope of the line, we can draw a line that contains the given point and that has the given slope. Also, if we are given two points, we can draw the line containing the two points. How do we find the equations of these lines? These are the questions we will answer in this section.

Suppose we want to write the equation of a line that contains (2,3) and has a slope of 2. Let us draw the graph.

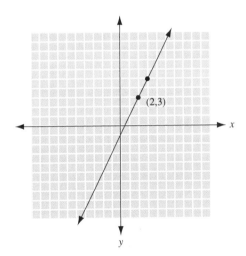

The slope of the line is the ratio of the change in y to the change in x. In this case, this ratio is $\frac{2}{1}$. This does not mean that we must go up two units and to the right one unit in order to get from one point of the line to another. We could go up 4 and to the right 2. We could even go fractional parts of a unit as long as the ratio of the change in y to the change in x is 2. That is, since $\frac{4}{2} = \frac{-10}{-5} = \frac{2}{1}$, we could use any of these ratios to indicate a slope of 2. The ratio is the same for any two points on the line.

If we let (x,y) represent any other point on the line, then the slope from $(2,3)$ to (x,y) is 2. If we go from $(2,3)$ to (x,y) by going vertically and then horizontally, the point of intersection of the vertical and horizontal line segments is $(2,y)$. The change in y is $y - 3$ and the change in x is $x - 2$. See the following figure.

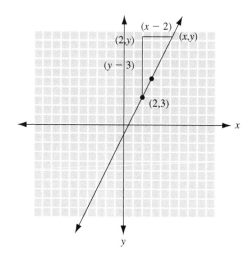

To write the equation of the line, use the slope formula and let $(x_1,y_1) = (2,3)$ and $(x_2,y_2) = (x,y)$.

$$m = \frac{y_2 - y_1}{x_2 - x_1}$$ Substitute for m, x_1, y_1, x_2, and y_2.

$$2 = \frac{y - 3}{x - 2}$$ Think of 2 as $\frac{2}{1}$.

$$\frac{2}{1} = \frac{y - 3}{x - 2}$$ Cross multiply.

$$2(x - 2) = 1(y - 3)$$ Apply the distributive property.

$$2x - 4 = y - 3$$ Add 4 to both sides.

$$2x = y + 1$$ Subtract y from both sides.

$$2x - y = 1$$ Therefore, the equation of the line through $(2,3)$ with slope of 2 is $2x - y = 1$.

- **Writing the equation of a line given a point on the line and the slope of the line**

Now let us use the above procedure to develop a formula for the equation of a line when given a point on the line and the slope of the line. Let (x_1,y_1) be the coordinates of the given point on the line and m be the slope of the line. Let (x,y) be any other point of the line. The slope between (x_1,y_1) and (x,y) is m. Substitute into the slope formula with $(x_2,y_2) = (x,y)$. Consequently,

$$m = \frac{y - y_1}{x - x_1}$$ Think of m as $\frac{m}{1}$.

$$\frac{m}{1} = \frac{y - y_1}{x - x_1}$$ Cross multiply.

$$1(y - y_1) = m(x - x_1) \qquad 1(y - 1) = y - 1.$$
$$y - y_1 = m(x - x_1)$$

Therefore, the equation of a line containing the point (x_1, y_1) and with slope of m is

$$y - y_1 = m(x - x_1).$$

We now state this more formally as the point-slope formula.

POINT-SLOPE FORMULA

The equation of the line containing the given point (x_1, y_1) and with the given slope of m is $y - y_1 = m(x - x_1)$.

Be careful: In using the point slope formula, we substitute the coordinates of the given point for x_1 and y_1 and the value of the slope for m. The variables are x and y, so they are not replaced.

When writing equations of lines, the equations are usually left in standard form, $ax + by = c$, with a, b, and c written as integers and $a > 0$.

EXAMPLE 1

Write the equation of each of the following lines that contain the given point and have the given slope. Leave answer in ax + by = c form.

a) $(3, -4)$ and $m = \dfrac{2}{3}$

Solution:

$(x_1, y_1) = (3, -4)$. Therefore, $x_1 = 3$ and $y_1 = -4$.

$y - y_1 = m(x - x_1)$	Substitute for x_1, y_1, and m.
$y - (-4) = \dfrac{2}{3}(x - 3)$	$y - (-4) = y + 4$ and multiply both sides by 3.
$3(y + 4) = 3 \cdot \dfrac{2}{3}(x - 3)$	Multiply 3 and $\frac{2}{3}$.
$3(y + 4) = 2(x - 3)$	Distribute on both sides.
$3y + 12 = 2x - 6$	Subtract 12 from both sides.
$3y = 2x - 18$	Subtract $2x$ from both sides.
$3y - 2x = 2x - 2x - 18$	Write in $ax + by = c$ form.
$-2x + 3y = -18$	It is preferred that the coefficient of x be positive, so multiply both sides by -1.
$-1(-2x + 3y) = -1(-18)$	Simplify both sides.
$2x - 3y = 18$	Therefore, the equation of the line containing $(3, -4)$ with slope of $\frac{2}{3}$ is $2x - 3y = 18$.

b) $(-2,4)$ and $m = -\dfrac{3}{4}$

Solution:

$(x_1, y_1) = (-2, 4)$ so $x_1 = -2$ and $y_1 = 4$.

$y - y_1 = m(x - x_1)$	Substitute for x_1, y_1, and m.
$y - 4 = -\dfrac{3}{4}(x - (-2))$	$x - (-2) = x + 2$ and multiply both sides by 4.
$4(y - 4) = 4\left(-\dfrac{3}{4}\right)(x + 2)$	Multiply 4 and $-\dfrac{3}{4}$.
$4(y - 4) = -3(x + 2)$	Apply the distributive property.
$4y - 16 = -3x - 6$	Add 16 to both sides.
$4y = -3x + 10$	Add $3x$ to both sides. Write the x-term first.
$3x + 4y = 10$	Therefore, the equation of the line containing $(-2,4)$ and with slope of $-\dfrac{3}{4}$ is $3x + 4y = 10$.

ADDITIONAL PRACTICE

Write the equations of the following lines.

a) $(1, -3)$ and $m = \dfrac{1}{4}$

b) $(-3, 5)$ and $m = -3$

• **Writing the equation of a line containing two given points**

PRACTICE EXERCISES

Write the equation of each of the following lines that contain the given point and have the given slope. Leave answer in ax + by = c form.

1) $(-1, 3)$ and $m = \dfrac{3}{5}$

2) $(4, -2)$ and $m = -\dfrac{3}{2}$

If more practice is needed, do the Additional Practice Exercises in the margin.

The point-slope formula requires that we know a point on the line and the slope of the line. If we know the coordinates of two points on the line, we do not know the slope, but we can find it. Consequently, we can use the point-slope formula to write the equation of a line if we know two points on the line.

EXAMPLE 2

Find the equation of the line that contains the following pair of points. Leave the answers in ax + by = c form.

a) $(2, 4)$ and $(-1, 2)$

Solution:

Using the slope formula with $(x_1, y_1) = (2, 4)$ and $(x_2, y_2) = (-1, 2)$, find the slope of the line.

$m = \dfrac{y_2 - y_1}{x_2 - x_1}$	Substitute for x_1, y_1, x_2, and y_2.
$m = \dfrac{2 - 4}{-1 - 2}$	Simplify.
$m = \dfrac{-2}{-3} = \dfrac{2}{3}$	Therefore, the slope is $\dfrac{2}{3}$.

Since the slope between any two points on the line is the same, we may use either of the given points in the point-slope formula. Use $(2,4)$ as (x_1, y_1).

$y - y_1 = m(x - x_1)$ — Substitute for m, x_1, and y_1.

$y - 4 = \dfrac{2}{3}(x - 2)$ — Multiply both sides by 3.

$3(y - 4) = 3 \cdot \dfrac{2}{3}(x - 2)$ — Multiply 3 and $\dfrac{2}{3}$.

$3(y - 4) = 2(x - 2)$ — Apply the distributive property.

$3y - 12 = 2x - 4$ — Add 12 to both sides.

$3y = 2x + 8$ — Subtract $2x$ from both sides.

$3y - 2x = 8$ — Rewrite in $ax + by = c$ form.

$-2x + 3y = 8$ — Multiply both sides by -1.

$2x - 3y = -8$ — Therefore, the equation of the line containing $(2,4)$ and $(-1,2)$ is $2x - 3y = -8$.

  b) $(-3,-3)$ and $(-1,1)$

Solution:
Find the slope of the line using the slope formula with $(x_1, y_1) = (-3,-3)$ and $(x_2, y_2) = (-1,1)$.

$m = \dfrac{y_2 - y_1}{x_2 - x_1}$ — Substitute for x_1, y_1, x_2, and y_2.

$m = \dfrac{1 - (-3)}{-1 - (-3)}$ — Simplify the numerator and the denominator.

$m = \dfrac{1 + 3}{-1 + 3}$ — Continue simplifying.

$m = \dfrac{4}{2} = 2$ — Therefore, the slope is 2.

Since the slope between any two points on the line is the same, we may use either of the given points in the point-slope formula. Use $(-1,1)$.

$y - y_1 = m(x - x_1)$ — Substitute for m, x_1, and y_1.

$y - 1 = 2(x - (-1))$ — $x - (-1) = x + 1$.

$y - 1 = 2(x + 1)$ — Distribute on the right side.

$y - 1 = 2x + 2$ — Add 1 to both sides.

$y = 2x + 3$ — Subtract $2x$ from both sides.

$-2x + y = 3$ — Multiply both sides by -1.

$2x - y = -3$ — Therefore, the equation of the line containing $(-3,-3)$ and $(-1,1)$ is $2x - y = 1$.

ADDITIONAL PRACTICE

Find the equation of the line that contains the following pairs of points. Leave the answers in $ax + by = c$ form.

c) $(5,2)$ and $(3,8)$

d) $(0,3)$ and $(-2,4)$

PRACTICE EXERCISES

Find the equation of the line that contains the following pairs of points. Leave the answers in $ax + by = c$ form.

3) $(4,3)$ and $(-1,-7)$ **4)** $(-2,5)$ and $(-4,2)$

If more practice is needed, do the Additional Practice Exercises in the margin.

Recall that the graph of a linear equation with two variables of the form $x = k$ is a vertical line. Consequently, all the coordinates of the points on a vertical line have the same x-value. To write the equation of a vertical line, all we need to know is the x-coordinate of one point on the line. The equation is $x =$ (the known x-coordinate). Recall also that the slope of a vertical line is undefined.

Remember that the graph of a linear equation with two variables of the form $y = k$ is a horizontal line. Consequently, all the points on a horizontal line have the same y-coordinate. To write the equation of a horizontal line, all we need to know is the y-coordinate of one point on the line. The equation is $y =$ (the known y-coordinate). Also recall that the slope of a horizontal line is 0.

EXAMPLE 3

Write the equation of each of the following lines.

a) Containing (3,5) and vertical.

Solution:
Since the line is vertical, the equation is of the form $x = k$. We know the x-coordinate of one point on the line is 3. Therefore, the equation is $x = 3$.

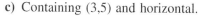

b) Containing $(-2,4)$ with undefined slope.

Solution:
Since the slope is undefined, we know the line is vertical. Therefore, the equation is of the form $x = k$. We know the x-coordinate of one point on the line is -2. Therefore, the equation is $x = -2$.

c) Containing (3,5) and horizontal.

Solution:
Since the line is horizontal, the equation of the line is in the form of $y = k$. We know the y-coordinate of one point on the line is 5. Therefore, the equation is $y = 5$.

d) Containing $(-3,-2)$ and $m = 0$.

Solution:
Since $m = 0$, we know the line is horizontal. Therefore, the equation is of the form $y = k$. We know the y-coordinate of one point on the line is -2. Therefore, the equation of the line is $y = -2$.

ADDITIONAL PRACTICE

Write the equation of each of the following lines.

e) Vertical containing (2,7)

f) Horizontal containing $(-5,7)$

g) Containing $(-3,2)$ with undefined slope.

h) Containing $(5,-3)$ with $m = 0$

PRACTICE EXERCISES

Write the equation of each of the following lines.

5) Containing $(-3,6)$ and vertical.

6) Containing (2,1) with $m = 0$.

7) Containing $(4,-1)$ and horizontal.

8) Containing $(-3,2)$ with undefined slope.

If more practice is needed, do the Additional Practice Exercises in the margin.

• **Parallel and perpendicular lines**

Two lines are **parallel** if they are in the same plane and do not intersect. In the figure below, L_1 is the graph of a line containing $(2,3)$ and with a slope of $\frac{1}{2}$, and L_2 is the graph of a line containing $(2,-2)$ also with a slope of $\frac{1}{2}$.

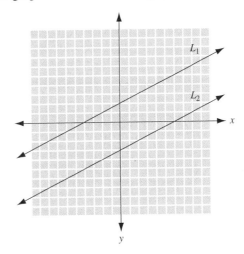

Both lines have slope of $\frac{1}{2}$ and the graphs appear to be parallel lines. It can be shown that two distinct lines are parallel if and only if they have equal slopes. This means that if two distinct lines are parallel, then they have equal slopes. It also means that if two distinct lines have equal slopes, then they are parallel.

Two lines are **perpendicular** if they intersect at right angles. In the figure below, L_1 is the graph of a line containing $(2,3)$ with a slope of $\frac{2}{3}$, and L_2 is the graph of a line containing $(2,3)$ and with a slope of $-\frac{3}{2}$.

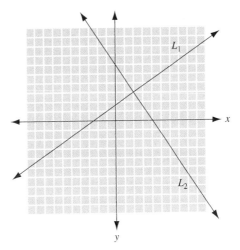

The two lines appear to be perpendicular. That is, they meet at a $90°$ or right angle. Notice the product of their slopes is -1. That is, $\left(\frac{2}{3}\right)\left(-\frac{3}{2}\right) = -1$. It can be shown that two lines are perpendicular if and only if the product of their slopes is -1. This means that if two lines are perpendicular, the product of their slopes is -1. It also means that if the product of the slopes of two lines is -1, the two lines are perpendicular. If m_1 represents the slope of a line and m_2 represents the slope of any line perpendicular to m_1, then $m_1 m_2 = -1$. Solving for m_1 we get $m_1 = -\frac{1}{m_2}$. This means that m_1 is the negative of the reciprocal of m_2. Thus, if a line has a slope of 3, then any line perpendicular to it has a slope of $-\frac{1}{3}$. If a line has a slope of $-\frac{3}{4}$, then any line perpendicular to it has a slope of $\frac{4}{3}$.

PARALLEL AND PERPENDICULAR LINES

Two non-vertical lines are parallel if and only if they have equal slopes.

Two lines are perpendicular if and only if the product of their slopes is -1, provided neither is vertical nor horizontal.

Any two vertical lines are parallel and any vertical and any horizontal line in the same plane are perpendicular to each other.

EXAMPLE 4

Determine whether the lines containing the following pairs of points are parallel, perpendicular, or neither.

a) L_1: $(-3,2)$ and $(1,5)$, L_2: $(2,-1)$ and $(-2,-4)$

Solution:

In order to determine if the lines are parallel or perpendicular, we need the slopes of the lines.

Slope of L_1	Slope of L_2
$m = \dfrac{y_2 - y_1}{x_2 - x_1}$	$m = \dfrac{y_2 - y_1}{x_2 - x_1}$
$m = \dfrac{5 - 2}{1 - (-3)}$	$m = \dfrac{-4 - (-1)}{-2 - 2}$
$m = \dfrac{5 - 2}{1 + 3}$	$m = \dfrac{-4 + 1}{-2 - 2}$
$m = \dfrac{3}{4}$	$m = \dfrac{-3}{-4}$
	$m = \dfrac{3}{4}$

Since the slopes are equal, L_1 and L_2 are parallel.

b) L_1: $(3,-1)$ and $(1,3)$, L_2: $(-1,-3)$ and $(1,-2)$

Solution:

In order to determine if the lines are parallel or perpendicular, we need the slopes of the lines.

Slope of L_1	Slope of L_2
$m = \dfrac{y_2 - y_1}{x_2 - x_1}$	$m = \dfrac{y_2 - y_1}{x_2 - x_1}$
$m = \dfrac{3 - (-1)}{1 - 3}$	$m = \dfrac{-2 - (-3)}{1 - (-1)}$
$m = \dfrac{3 + 1}{1 - 3}$	$m = \dfrac{-2 + 3}{1 + 1}$
$m = \dfrac{4}{-2} = -2$	$m = \dfrac{1}{2}$

Since $(-2)(\frac{1}{2}) = -1$, L_1 and L_2 are perpendicular.

Often all we need from an equation is the slope, not the y-intercept. If we write the general equation of a linear equation ($ax + by = c$) in slope-intercept form we will get a formula for slope.

$$ax + by = c$$ Subtract ax from both sides.

$$by = -ax + c$$ Divide both sides by b.

$$\frac{by}{b} = \frac{-ax + c}{b}$$ Simplify both sides.

$$y = -\frac{a}{b}x + \frac{c}{b}$$ Therefore, the slope is $-\frac{a}{b}$.

FINDING SLOPE FROM AN EQUATION IN STANDARD FORM

If the equation of a line is given in the form $ax + by = c$, then $m = -\dfrac{a}{b}$.

From the preceding we see that if an equation is given in $ax + by = c$ form, the slope is $-\frac{a}{b}$. For example, if $2x + 3y = 5$, then $a = 2$, $b = 3$ and $m = -\frac{2}{3}$ and if $4x - 5y = 7$, then $a = 4$, $b = -5$ and $m = -\frac{4}{-5} = \frac{4}{5}$.

c) L_1: $2x + 5y = 10$ and L_2: $5x - 2y = 4$

Solution:

In order to determine if the lines are parallel or perpendicular, we need to know the slopes.

The slope of $L_1 = -\frac{a}{b} = -\frac{2}{5}$ and the slope of $L_2 = -\frac{a}{b} = -\frac{5}{-2} = \frac{5}{2}$.

Since $(-\frac{2}{5})(\frac{5}{2}) = -1$, L_1 and L_2 are perpendicular.

d) L_1: $3x + 4y = 8$ and L_2: $4x + 3y = 6$

Solution:

In order to determine if the two lines are parallel or perpendicular, we need the slopes. The slope of $L_1 = -\frac{a}{b} = -\frac{3}{4}$ and the slope of $L_2 = -\frac{a}{b} = -\frac{4}{3}$.

Since the slopes are neither equal nor is their product -1, L_1 and L_2 are neither parallel nor perpendicular.

ADDITIONAL PRACTICE

Determine whether the following pairs of lines are parallel, perpendicular or neither.

i) L_1: (1,1) and (2,3)
   L_2: (−3,2) and (−4, 0)

j) L_1: (−2,2) and (1,3)
   L_2: (2,−1) and (3,2)

k) L_1: $2x + y = 4$
   L_2: $2x + y = 6$

l) L_1: $3x - 2y = 4$
   L_2: $2x + 3y = 4$

PRACTICE EXERCISES

Determine whether the lines containing the following pairs of points are parallel, perpendicular or neither.

9) L_1: (4,5) and (−1,2), L_2: (−2,−1) and (3,2)

10) L_1: (5,−2) and (3,1), L_2: (5,6) and (2,4)

Determine whether the graphs of the lines with the given equations are parallel, perpendicular, or neither.

11) L_1: $x + 3y = 6$ and L_2: $3x + y = 4$

12) L_2: $2x - 7y = 5$ and L_2: $2x - 7y = -7$

If more practice is needed, do the Additional Practice Exercises in the margin.

- **Writing the equation of a line containing a given point and parallel or perpendicular to a given line**

ANSWERS:
Practice 9–12

9) parallel

10) perpendicular

11) neither

12) parallel

Additional Practice i–l

i) parallel

j) neither

k) parallel

l) perpendicular

If we know a point on the line and the slope of the line, then we can write the equation of the line. Suppose we are given the equation of a line and the coordinates of a point not on the line. Since parallel lines have equal slopes, we can write the equation of a line containing the given point and parallel to the given line by using the slope of the given line. Similarly, we can write the equation of a line containing the given point and perpendicular to the given line by using the negative of the reciprocal of the slope of the given line.

EXAMPLE 5

Write the equations of the following lines.

 a) Containing $(-2,3)$ and parallel to the graph of $3x + 5y = 8$.

Solution:

To write the equation of a line, we need a point on the line and the slope of the line. We know the point whose coordinates are $(-2,3)$ is on the line. Since parallel lines have equal slope, the slope of the line we are looking for is equal to the slope of the line whose equation is $3x + 5y = 8$. The slope of $3x + 5y = 8$ is $-\frac{3}{5}$.

The slope of any line parallel to this line is also $-\frac{3}{5}$. Therefore, we are looking for the equation of a line containing $(-2,3)$ and with slope of $-\frac{3}{5}$. Since we know a point and the slope, use the point-slope formula.

$y - y_1 = m(x - x_1)$	Substitute for m, x_1, and x_2.
$y - 3 = -\frac{3}{5}(x - (-2))$	$x - (-2) = x + 2$ and multiply both sides by 5.
$5(y - 3) = 5\left(-\frac{3}{5}\right)(x + 2)$	Multiply 5 and $-\frac{3}{5}$.
$5(y - 3) = -3(x + 2)$	Distribute on both sides.
$5y - 15 = -3x - 6$	Add 15 to both sides.
$5y - 15 + 15 = -3x - 6 + 15$	Simplify both sides.
$5y = -3x + 9$	Add $3x$ to both sides.
$3x + 5y = -3x + 3x + 9$	Simplify the right side.
$3x + 5y = 9$	Therefore, the equation of the line containing $(-2,3)$ and parallel to the graph of $3x + 5y = 8$ is $3x + 5y = 9$.

b) Containing $(4,-3)$ and perpendicular to the graph of $2x + 5y = 7$.

Solution:

To write the equation of a line we need a point on the line and the slope of the line. We know the point whose coordinates are $(4,-3)$ is on the line. Since the line we are looking for is perpendicular to the graph of $2x + 5y = 7$, its slope is the negative of the reciprocal of the slope of $2x + 5y = 7$. The slope of $2x + 5y = 7$ is $-\frac{2}{5}$.

The slope of any line perpendicular to the given line is the negative reciprocal of $-\frac{2}{5}$. Therefore, the slope of any line perpendicular to $2x + 5y = 7$ is $\frac{5}{2}$. Hence, we are looking for the equation of a line containing $(4,-3)$ with slope of $\frac{5}{2}$. Use the point-slope formula.

$$y - y_1 = m(x - x_1)$$ Substitute for m, x_1, and x_2.

$$y - (-3) = \frac{5}{2}(x - 4)$$ $y - (-3) = y + 3$ and multiply both sides by 2.

$$2(y + 3) = 2\left(\frac{5}{2}\right)(x - 4)$$ Multiply 2 and $\frac{5}{2}$.

$$2(y + 3) = 5(x - 4)$$ Distribute on both sides.

$$2y + 6 = 5x - 20$$ Subtract $5x$ from both sides.

$$-5x + 2y + 6 = 5x - 5x - 20$$ Simplify both sides.

$$-5x + 2y + 6 = -20$$ Subtract 6 from both sides.

$$-5x + 2y + 6 - 6 = -20 - 6$$ Simplify both sides.

$$-5x + 2y = -26$$ Multiply both sides by -1.

$$-1(-5x + 2y) = -1(-26)$$ Simplify both sides.

$$5x - 2y = 26$$ Therefore, the equation of the line containing $(4, -3)$ and perpendicular to the graph of $2x + 5y = 7$ is $5x - 2y = 26$.

PRACTICE EXERCISES

Write the equations of the following lines. Leave answers in form of $ax + by = c$.

13) Containing $(-1, 2)$ and parallel to the graph of $3x - 6y = 8$.

14) Containing $(-3, -4)$ and perpendicular to $4x + 2y = 7$.

In Section 6.4 we observed that if we solved a linear equation for y we get the slope-intercept form of a line, $y = mx + b$. We can now verify this form of a linear equation. If a given point is the y-intercept, then its coordinates are of the form $(0, b)$. If we substitute $(0, b)$ into the point-slope formula, we get the slope-intercept formula as shown below.

$$y - y_1 = m(x - x_1)$$ Substitute $(0, b)$ for x_1 and y_1.

$$y - b = m(x - 0)$$ Distribute on the right side.

$$y - b = mx$$ Add b to both sides.

$$y = mx + b$$ Slope-intercept form.

We can use this form to write the equation of a line if we know the slope and y-intercept.

EXAMPLE 6

Write the equations of the following lines. Leave the answers in slope-intercept form.

a) Slope of 3 and y-intercept of 4.

Solution:

$$y = mx + b$$ Substitute 3 for m and 4 for b.

$$y = 3x + 4$$ Therefore, the equation of the line with slope of 3 and y-intercept of $(0, 4)$ is $y = 3x + 4$.

b) The slope is $\frac{3}{4}$ and the line contains $(0, \frac{2}{3})$.

Solution:

The point $(0, \frac{2}{3})$ is the y-intercept, so $b = \frac{2}{3}$.

$\quad y = mx + b \quad$ Substitute $\frac{3}{4}$ for m and $\frac{2}{3}$ for b.

$\quad y = \frac{3}{4}x + \frac{2}{3} \quad$ Therefore, the equation of the line with slope of $\frac{3}{4}$ and containing $(0, \frac{2}{3})$ is $y = \frac{3}{4}x + \frac{2}{3}$.

PRACTICE EXERCISES

Write the equations of the following lines. Leave the answers in slope-intercept form.

15) The slope is $\frac{5}{4}$ and whose y-intercept is -3.

16) The slope is $-\frac{3}{5}$ and the line contains $(0, -\frac{3}{4})$.

Frequently real-world situations can be modeled using linear equations with two variables.

EXAMPLE 7

a) A salesperson works for a salary plus commission. If x represents the value of the merchandise sold and y represents the salary earned per month, then $(15,000, 3150)$ indicates that for sales of $15,000 the salary was $3150 and $(10,000, 2150)$ means that on $10,000 worth of sales, the salary was $2150.

1) Write a linear equation that gives the salary, y, in terms of the sales, x.

Solution:

We are being asked to write the equation of a line that contains the points $(15,000, 3150)$ and $(10,000, 2150)$ and leave the answer in the form $y = mx + b$. We will use the point-slope formula, so we need the slope.

$$m = \frac{y_2 - y_1}{x_2 - x_1} \qquad \text{Substitute.}$$

$$m = \frac{3150 - 2150}{15,000 - 10,000} \qquad \text{Simplify.}$$

$$m = \frac{1000}{5000} \qquad \text{Reduce.}$$

$$m = \frac{1}{5}$$

Now use either of the points and the point-slope formula. We will use $(10,000, 2150)$.

$$y - y_1 = m(x - x_1) \qquad \text{Substitute.}$$

$$y - 2150 = \frac{1}{5}(x - 10,000) \qquad \text{Distribute } \frac{1}{5}.$$

$$y - 2150 = \frac{1}{5}x - 2000 \qquad \text{Add 2150 to both sides.}$$

$$y = \frac{1}{5}x + 150 \qquad \text{Therefore, the salary, } y, \text{ in terms of the sales, } x, \text{ is } y = \frac{1}{5}x + 150.$$

2) Use the equation found in part 1 to find the salary for sales of $18,000.

Solution:

Since x represents the sales, we replace x with 18,000 and find y, the salary.

$$y = \frac{1}{5}x + 150 \qquad \text{Substitute 18,000 for } x.$$

$$y = \frac{1}{5}(18,000) + 150 \qquad \text{Multiply } \frac{1}{5} \text{ and 18,000.}$$

$$y = 3600 + 150 \qquad \text{Add.}$$

$$y = 3750 \qquad \text{Therefore, the salesperson earns \$3750 for sales of \$18,000.}$$

3) Use the equation found in part 1 to find the total value of the sales in a month when the salary was $1150.

Solution:

Since y represents the salary, we will replace y with $1150 and find x, the sales.

$$y = \frac{1}{5}x + 150 \qquad \text{Substitute 1150 for } y.$$

$$1150 = \frac{1}{5}x + 150 \qquad \text{Subtract 150 from both sides.}$$

$$1000 = \frac{1}{5}x \qquad \text{Multiply both sides by 5.}$$

$$5000 = x \qquad \text{Therefore, the salesperson's sales were \$5000 in order to earn \$1150 in a month.}$$

b) A college buys a mower for $10,000. After 6 years it has depreciated to a value of $4000. Assuming the mower depreciates the same each year (called linear or "straight-line" depreciation) answer the following.

1) Write a linear equation giving the value of the mower, y, in terms of the number of years after it was purchased, x.

Solution:

Since x represents the number of years after it was purchased and y represents the value of the mower, the fact that the mower cost $10,000 can be represented as the ordered pair (0, 10,000) and the fact that after 6 years the mower was worth $4000 can be written as (6, 4000). We need the equation of a line containing these two points and leave the answer in $y = mx + b$ form. First find the slope.

$$m = \frac{y_2 - y_1}{x_2 - x_1} \qquad \text{Substitute.}$$

$$m = \frac{4000 - 10,000}{6 - 0} \qquad \text{Simplify.}$$

$$m = \frac{-6000}{6} \qquad \text{Divide.}$$

$$m = -1000$$

Now use the point-slope formula and either of the points. We will use (0, 10,000).

$$y - y_1 = m(x - x_1) \qquad \text{Substitute.}$$
$$y - 10,000 = -1000(x - 0) \qquad \text{Simplify.}$$
$$y - 10,000 = -1000x \qquad \text{Add 10,000 to both sides of the equation.}$$
$$y = -1000x + 10,000 \qquad \text{Therefore, the value of the mower, } y, \text{ in terms of the number of years after it was purchased, } x, \text{ is } y = -1000x + 10,000.$$

2) Use the equation found in part 1 to find the value of the mower 8 years after it was purchased.

Solution:

Since x represents the number of years after it was purchased, replace x with 8 and find y.

$$y = -1000x + 10,000 \qquad \text{Substitute 8 for } x.$$
$$y = -1000(8) + 10,000 \qquad \text{Multiply } -1000 \text{ and } 8.$$
$$y = -8000 + 10,000 \qquad \text{Add.}$$
$$y = 2000 \qquad \text{Therefore, 8 years after it was purchased, the mower will be worth \$2000.}$$

3) How many years after it was purchased will the mower have no value?

Solution:

Since y represents the value of the mower, we are being asked to find x when $y = 0$.

$$y = -1000x + 10,000 \qquad \text{Substitute 0 for } y.$$
$$0 = -1000x + 10,000 \qquad \text{Add } 1000x \text{ to both sides of the equation.}$$
$$1000x = 10,000 \qquad \text{Divide both sides of the equation by 1000.}$$
$$x = 10 \qquad \text{Therefore, the mower will have no value after 10 years.}$$

PRACTICE EXERCISES

17) John is a book salesperson and earns a salary of \$100 per week plus \$7.50 for each book sold.

  a) Write a linear equation that gives his weekly earnings, y, in terms of the number of books sold, x.

  b) Use the equation found in part a to find John's earnings during a week in which he sold 15 books.

  c) How many books would he have to sell in order to make \$400 in one week?

18) The Jones family purchased a lot for \$20,000. After 5 years the value of the lot had increased to \$30,000. Assuming the lot increases by the same amount each year, answer the following.

  a) Write a linear equation that gives the value of the lot, y, in terms of the number of years after it was purchased, x.

  b) Use the equation found in part a to find the value of the lot 9 years after it was purchased.

  c) How many years after it was purchased will the lot have a value of \$50,000?

E X E R C I S E   S E T   **6.5**

Write the equation of each of the following lines which contain the given point and have the given slope. Leave the answer in $ax + by = c$ form.

1) $(3,1)$ and $m = 2$

2) $(5,2)$ and $m = 3$

3) $(-2,4)$ and $m = -2$

4) $(4,-2)$ and $m = -3$

5) $(-3,5)$ and $m = \dfrac{1}{4}$

6) $(4,-5)$ and $m = \dfrac{1}{2}$

7) $(2,-4)$ and $m = -\dfrac{1}{3}$

8) $(-2,-1)$ and $m = -\dfrac{1}{5}$

9) $(5,1)$ and $m = \dfrac{4}{3}$

10) $(6,-2)$ and $m = \dfrac{5}{2}$

11) $(-2,-5)$ and $m = -\dfrac{2}{3}$

12) $(4,-1)$ and $m = -\dfrac{5}{3}$

13) $(3, 4)$ and vertical

14) $(-2,5)$ and vertical

15) $(2,3)$ and horizontal

16) $(-1,-7)$ and horizontal

17) $(5,2)$ and undefined slope

18) $(-6,2)$ and undefined slope

19) $(5,4)$ and $m = 0$

20) $(-6,-1)$ and $m = 0$

Write the equation of the line which contains the following pair of points. Leave the answers in $ax + by = c$ form.

21) $(4,-3)$ and $(-1,7)$

22) $(2,-5)$ and $(4,-2)$

23) $(1,5)$ and $(-3,1)$

24) $(4,-3)$ and $(-2,-1)$

25) $(-1,-5)$ and $(-4,1)$

26) $(5,-2)$ and $(-3,4)$

27) $(6,0)$ and $(-3,6)$

28) $(0,4)$ and $(3,5)$

29) $(5,-3)$ and $(5,1)$

30) $(-2,4)$ and $(-2,-1)$

31) $(5,-4)$ and $(-1,-4)$

32) $(7,2)$ and $(-3,2)$

Determine whether the following lines are parallel, perpendicular, or neither.

33) L_1: $(2,-4)$ and $(4,2)$,
   L_2: $(2,-5)$ and $(6,-1)$

34) L_1: $(1,-3)$ and $(4,0)$,
   L_2: $(-1,4)$ and $(1,10)$

35) L_1: $(4,-3)$ and $(-1,-1)$,
   L_2: $(-2,2)$ and $(-4,-3)$

36) L_1: $(-4,5)$ and $(2,3)$,
   L_2: $(-1,-2)$ and $(0,1)$

37) L_1: $(-6,2)$ and $(-2,0)$,
   L_2: $(8,-3)$ and $(10,1)$

38) L_1: $(-5,-1)$ and $(7,7)$,
   L_2: $(1,5)$ and $(-2,3)$

39) L_1: $(3,-7)$ and $(-1,-1)$,
   L_2: $(8,-2)$ and $(2,2)$

40) L_1: $(4,-5)$ and $(5,-1)$,
   L_2: $(3,-4)$ and $(7,-3)$

41) L_1: $(2,4)$ and $(2,-2)$,
   L_2: $(-4,1)$ and $(-4,6)$

42) L_1: $(3,5)$ and $(-2,5)$,
   L_2: $(-1,2)$ and $(4,2)$

43) L_1: $(-2,3)$ and $(4,3)$,
   L_2: $(6,1)$ and $(6,7)$

44) L_1: $(-4,1)$ and $(-4,6)$,
   L_2: $(-3,5)$ and $(3,5)$

Determine whether the graphs of the lines with the given equations are parallel, perpendicular, or neither.

45) L_1: $4x + y = 7$,
   L_2: $4x + y = -3$

46) L_1: $3x - y = 5$,
   L_2: $3x - y = 9$

47) L_1: $4x + 3y = 6$,
   L_2: $3x - 4y = 12$

48) L_1: $5x - 3y = 11$,
   L_2: $3x + 5y = 8$

49) L_1: $4x + 6y = 5$,
   L_2: $8x + 12y = 9$

50) L_1: $5x + 10y = 15$,
   L_2: $4x - 2y = 5$

51) L_1: $x = 4$,
   L_2: $x = -1$

52) L_1: $y = 5$,
   L_2: $y = -5$

53) L_1: $x = 6$,
   L_2: $y = 2$

54) L_1: $y = -5$,
   L_2: $x = 3$

Write the equations of the following lines.

55) Containing $(3,-1)$ and parallel to the graph of $2x + 4y = 10$.

56) Containing $(-4,6)$ and parallel to the graph of $3x + 5y = 4$.

57) Containing $(-3,-4)$ and parallel to the graph of $4x + 8y = 9$.

58) Containing $(-4,2)$ and parallel to the graph of $3x + 9y = 10$.

59) Containing $(4,-1)$ and perpendicular to the graph of $x + 4y = 7$.

60) Containing $(-5,0)$ and perpendicular to the graph of $3x - y = 6$.

61) Containing $(-5,3)$ and perpendicular to the graph of $2x + 6y = 7$.

62) Containing $(4,-1)$ and perpendicular to the graph of $8x + 4y = 15$.

63) Containing $(6,-3)$ and parallel to the graph of $x = -5$.

64) Containing $(7,-3)$ and parallel to the graph of $y = 2$.

65) Containing $(4,-4)$ and perpendicular to the graph of $x = 6$.

66) Containing $(-2,3)$ and perpendicular to the graph of $y = 3$.

Write the equations of the following lines. Leave the answers in $y = mx + b$ form.

67) Slope is 2 and y-intercept is 4

68) Slope is -5 and y-intercept is -2

69) Slope is $\frac{3}{5}$ and y-intercept is -1

70) Slope is $\frac{5}{2}$ and y-intercept is 4

71) Slope is $-\frac{2}{3}$ and y-intercept is $\frac{4}{5}$

72) Slope is $-\frac{6}{5}$ and y-intercept is $-\frac{5}{4}$

73) The cost to drive a rental car depends on the number of miles driven. The ordered pair $(100,40)$ indicates that it costs $40 to drive 100 miles and the ordered pair $(200,55)$ indicates that it costs $55 to drive 200 miles.

a) Write a linear equation giving the cost, y, in terms of the number of miles, x.

b) Using the equation found in part a, find how much it costs to drive 500 miles.

c) For how many miles driven would the cost be $62.50?

74) A salesperson works for a fixed salary plus a commission on sales. The ordered pair $(1000,250)$ indicates earnings of $250 for $1000 in sales and the ordered pair $(3000,450)$ indicates earnings of $450 for $3000 in sales.

a) Write a linear equation giving the salary, y, in terms of the value of merchandise sold, x.

b) Use the equation in part a to find the salary for sales of $4,000.

c) Find the total sales for a week in which the salesperson earned $750.

75) The cost to ride a taxi depends on the number of miles ridden. The ordered pair $(2,2.80)$ means that a ride of 2 miles costs $2.80 and $(6,6)$ means that a ride of 6 miles costs $6.

a) Write a linear equation that gives the cost, y, in terms of the number of miles, x.

b) Use the equation found in part a to find the cost of a taxi ride of 8 miles.

c) Find the length of a taxi ride that costs $5.20.

76) The price of an item depends on the number produced. The ordered pair $(2,80)$ indicates that when 2 items are produced the price per item is $80 and the ordered pair $(3,70)$ indicates that when 3 items are produced the price per item is $70 for $x \le 6$.

a) Write a linear equation giving the cost per item, y, in terms of the number of items produced, x.

b) Use the equation in part a to find the price per item if 5 items are produced.

c) Use the equation in part a to find the number of items produced if the price per item is $40.

77) A farmer purchases a tractor for $50,000. After 5 years the tractor has depreciated to a value of $25,000. Assuming linear depreciation, answer the following.

a) Write a linear equation that gives the value of the tractor, y, in terms of the number of years after it was purchased, x.

b) Use the equation found in part a to find the value of the tractor after 8 years.

c) Use the equation found in part a to find the number of years until the tractor has no value.

78) A basketball trading card cost $1.50 new and after 4 years has a value of $2.70. Assuming the card increases in value by the same amount each year, answer the following.

a) Write a linear equation that gives the value of the card, y, in terms of the number of years after it was purchased, x.

b) Use the equation found in part a to find the value of the card 10 years after it was purchased.

c) Use the equation found in part a to find the number of years until the card is worth $9.00.

79) Sally is a salesperson at a model home center and her salary is $200 per month plus a 5% commission on her sales.

a) Write a linear equation that gives her salary, y, in terms of her sales, x.

b) Use the equation found in part a to find how much she will earn during a month in which her sales are $250,000.

c) Use the equation found in part a to find her sales for a month in which she earned $8950.

80) The cost of renting a car is $30 per day plus $0.15 per mile.

a) Write a linear equation giving the cost, y, in terms of the number of miles driven, x, for a one-day rental.

b) Use the equation found in part a to find the cost if the car is driven 300 miles in one day.

c) Use the equation found in part a to find how many miles the car was driven in one day if the cost was $52.50.

WRITING EXERCISES

81) Why must parallel lines have equal slopes?

82) Are the lines with equations $y = 2x + 3$ and $y = -2x + 3$ perpendicular? Why or why not?

83) Can two perpendicular lines both have negative slopes? Why or why not?

84) How can you tell if two lines are parallel without graphing them?

SECTION 6.6

Relations and Functions

OBJECTIVES

When you complete this section, you will be able to:

Ⓐ Recognize functions when given as a set of ordered pairs, as a graph or as an equation.

Ⓑ Determine the domain and range of relations and functions.

Ⓒ Use functional notation.

INTRODUCTION

In Section 1.1 a set was defined as any collection of objects and the objects that make up the set are called the elements of the set. In algebra, these elements are often numbers. Recall that sets are indicated by using braces, { }. In this section we will discuss sets whose elements are ordered pairs of numbers. In an ordered pair, the first number is often called the first component and the second number the second component. For example, in the ordered pair $(2, -3)$ the first component is 2 and the second is -3. With this in mind we have the following definitions.

DEFINITION OF RELATION

A relation is any set of ordered pairs. The set of all first components of the ordered pairs is the **domain** of the relation and the set of all second components is called the **range**.

If the ordered pairs are given in the form (x, y), the domain is the set of all x-values and the range is the set of all y-values.

 EXAMPLE 1

Give the domain and range of each of the following relations.

a) $\{(1,2), (-3,4), (6,3), (-4,-5)\}$

Solution:

The domain is the set of all first components which is $\{1, -3, 6, -4\}$.
The range is the set of all second components which is $\{2, 4, 3, -5\}$.

b) $\{(-3,2), (-1,-4), (2,-4), (2,6)\}$

Solution:
The domain is the set of all first components which is $\{-3, -1, 2\}$.
The range is the set of all second components which is $\{2, -4, 6\}$.

Note: In Example 1b two ordered pairs have the first component 2 and two ordered pairs have the same second component -4, but these are listed only once in the domain and range respectively.

A function is a special type of relation.

DEFINITION OF FUNCTION

A function is a relation in which every first component is paired with exactly one second component.

• **Recognizing functions as sets of ordered pairs.**

Another way of thinking of a function is that no two ordered pairs can have the same first components and different second components. More specifically, if the ordered pairs are in the form (x,y), then a function is a correspondence that assigns exactly one value of y to each value of x. As such y is said to be a function of x. The variable x is the **independent variable** and y is the **dependent variable**. Since a function is a relation, the domain of the function is the set of all first components and the range is the set of all second components.

EXAMPLE 2
Determine which of the following relations are functions and give the domain and range of each.

a) $\{(-3,4), (-1,-3), (0,2), (3,5)\}$

Solution:
Since no two ordered pairs have the same first component and different second components, this is a function. The domain is $\{-3, -1, 0, 3\}$ and the range is $\{4, -3, 2, 5\}$.

b) $\{(-4,2), (-2,4), (1,-7), (3,4)\}$

Solution:
Since no two ordered pairs have the same first component and different second components, this is a function. Notice that the ordered pairs $(-2,4)$ and $(3,4)$ have the same second component paired with different first components. This does not violate the definition of a function. The domain is $\{-4, -2, 1, 3\}$ and the range is $\{2, 4, -7\}$.

c) $\{(-5,3), (3,6), (-4,-2), (3,1), (5,3)\}$

Solution:
This is not a function since the first component 3 is paired with two different second components, 6 and 1. The domain is $\{-5, 3, -4, 5\}$ and the range is $\{3, 6, -2, 1\}$.

PRACTICE EXERCISES

Determine which of the following relations are functions and give the domain and range of each.

1) $\{(-5,-3), (-2,-1), (3,5), (4,-6)\}$

2) $\{(3,-2), (-4,6), (3,2), (5,7), (-3,3)\}$

3) $\{(-5,2), (6,-4), (-4,5), (5,2)\}$

• **Recognizing graphs of functions**

The relation in Example 2c was not a function. Let's take a look at the graph of this relation.

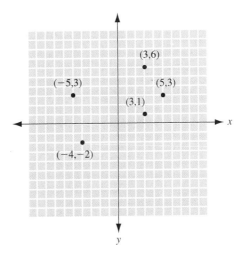

The two ordered pairs that prevented this relation from being a function were (3,6) and (3,1) because the first component 3 was paired with two second components, 6 and 1. We know from Section 6.2 that the graph of any linear equation of the form $x = k$ is a vertical line. Consequently, if two ordered pairs have the same x-value, then their graph must be on the same vertical line. You will note that the graphs of (3,6) and (3,1) are on the same vertical line as demonstrated below.

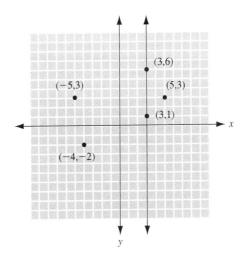

This observation leads to the following graphical test for a function.

VERTICAL LINE TEST

If any vertical line intersects the graph of a relation in more than one point, then that relation is not a function.

An equivalent way of stating the above test is that no vertical line may intersect the graph of a function in more than one point.

Although we have not graphed anything except points and straight lines in this book, this test can be applied to any graph.

EXAMPLE 3

Determine whether each of the following is the graph of a function.

a)

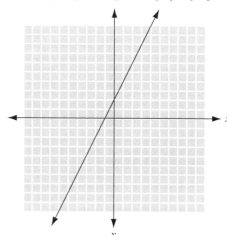

b)

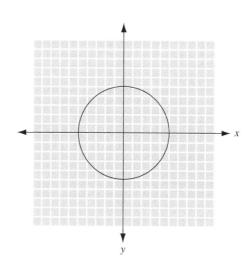

c)

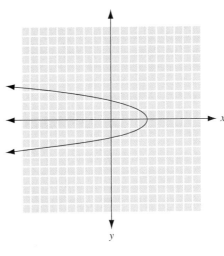

 d)

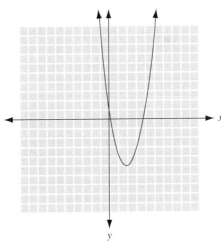

Solutions:

The graphs a and d are the graphs of functions, but the graphs b and c are not since each may be intersected by a vertical line in more than one point as illustrated below.

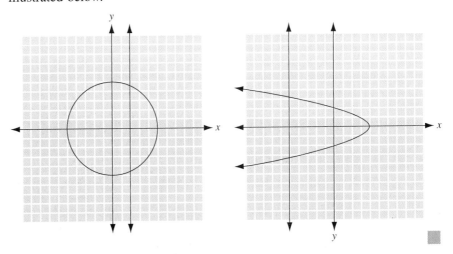

Determine whether each of the following is the graph of a function.

4)

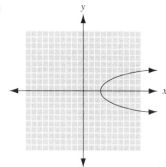

5)

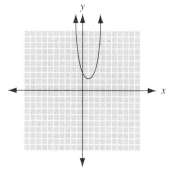

6)

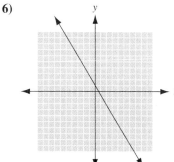

7)

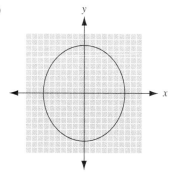

• **Recognizing functions as equations**

Rather than giving functions as sets of ordered pairs or graphs, functions are usually given in the form of equations. If the equation defines y as a function of x, then for each value assigned to x the equation must give exactly one value for y.

▎ **EXAMPLE 4**

Determine whether the following equations define y as a function of x.

a) $y = 3x + 5$

Solution:

For each value of x we find the corresponding value of y by multiplying the value of x by 3 and adding 5. This results in exactly one value for y. Therefore, y is a function of x. For example, if $x = 2$ then $y = 3 \cdot 2 + 5 = 6 + 5 = 11$ and only 11. This corresponds to the ordered pair $(2,11)$ and this is the only ordered pair with an x-value of 2.

b) $y^2 = x - 3$

Solution:

Suppose x has a value of 4, then $y^2 = 4 - 3 = 1$. Then $y = 1$ or -1 since $1^2 = 1$ and $(-1)^2 = 1$. This corresponds to the ordered pairs $(4,1)$ and $(4,-1)$ which are two ordered pairs with the same x-value and different y-values. Therefore, this does not define y as a function of x. ▎

ADDITIONAL PRACTICE

*Determine whether the following
equations define y as a function of x.*

a) $y = 2x - 3$

b) $y^2 - 4 = x$

ANSWERS:
Practice 4–7

4) no   **5)** yes

6) yes   **7)** no

• **Using functional notation**

PRACTICE EXERCISES

Determine whether the following equations define y as a function of x.

8) $y = x^2$ **9)** $x = y^2 - 2$

If more practice is needed, do the Additional Practice Exercises in the margin.

If y is a function of x it is convenient to name the function using a special notation called **functional notation** in which y is replaced with a symbol of the form $f(x)$ which is read "the value of f at x" or more commonly "f of x." For example, if $y = 2x + 3$ then y is a function of x. Replacing y with $f(x)$ we have $f(x) = 2x + 3$. Function names are usually lower case letters so any lower case letter may be used as a function name. So $g(x) = 2x + 3$ is the same function.

---

f(x) NOTATION

When using notation of the form $f(x)$:

1) f is the name of the function. (Frequently other symbols, especially other lower case letters, are used instead of f.)

2) x is a value from the domain.

3) $f(x)$ is the value of the function in the range that corresponds with the value of x from the domain. It is a y-value, so the results can be written as the ordered pair $(x, f(x))$.

When finding the value of $f(x)$ for a specific value of x we are **evaluating the function**. This is accomplished by substituting the value for x and evaluating the expression using the order of operations. For example, if $f(x) = 2x + 3$ and we wish to find the value of the function when $x = 3$, then we are finding $f(3)$. To do this, replace x with 3 in the equation of the function and evaluate. So, $f(3) = 2 \cdot 3 + 3 = 6 + 3 = 9$. This can be written as the ordered pair $(3,9)$.

Actually, any letter or symbol can be used for both the name of the function and the independent variable. For example, $f(x) = 3x + 5$, $g(t) = 3t + 5$, and $h(r) = 3r + 5$ all represent the same function since they all represent the same set of ordered pairs.

EXAMPLE 5

Given $f(x) = 3x - 4$, find each of the following. Represent each result as an ordered pair.

a) $f(2)$

b) $f(-3)$

c) $f(0.5)$

d) $f(a)$

Solution:

a) To find $f(2)$ replace x with 2 and simplify.

$$f(2) = 3 \cdot 2 - 4 \qquad \text{Multiply.}$$
$$= 6 - 4 \qquad \text{Add.}$$
$$= 2 \qquad \text{Therefore } f(2) = 2.$$

As an ordered pair, this is (2,2).

b) To find $f(-3)$ replace x with -3 and simplify.

$$f(-3) = 3(-3) - 4 \qquad \text{Multiply.}$$
$$= -9 - 4 \qquad \text{Add.}$$
$$= -13 \qquad \text{Therefore } f(-3) = -13.$$

As an ordered pair, this is $(-3, -13)$.

c) To find $f(0.5)$ replace x with 0.5 and simplify.

$$f(0.5) = 3(0.5) - 4 \qquad \text{Multiply.}$$
$$= 1.5 - 4 \qquad \text{Add.}$$
$$= -2.5 \qquad \text{Therefore } f(0.5) = -2.5.$$

As an ordered pair, this is $(0.5, -2.5)$.

d) To find $f(a)$ replace x with a.

$$f(a) = 3a - 4 \qquad \text{Since this can not be simplified, } f(a) = 3a - 4.$$

As an ordered pair, this is $(a, 3a - 4)$.

ADDITIONAL PRACTICE

Given $h(t) = 2t + 5$, find the following.

c) $h(4)$ **d)** $h(-1)$

e) $h(1.3)$ **f)** $h(b)$

PRACTICE EXERCISES

Given $g(x) = 4x - 2$, find the following. Represent each result as an ordered pair.

10) $g(2)$ **11)** $g(-4)$ **12)** $g\left(\dfrac{1}{2}\right)$ **13)** $g(c)$

If more practice is needed, do the Additional Practice Exercises in the margin.

EXERCISE SET   **6.6**

Determine which of the following relations are functions and give the domain and range of each.

1) $\{(-6,2), (-3,-2), (5,-7), (7,9)\}$

2) $\{(-8,-5), (-3,6), (0,7), (2,3)\}$

3) $\{(-4,2), (-1,3), (3,2), (5,1)\}$

4) $\{(7,-3), (4,7), (-5,2), (3,-3)\}$

5) $\{(-8,2), (-6,3), (-8,-3), (5,1)\}$

6) $\{(-5,1), (3,-2), (6,2), (-5,4)\}$

13) 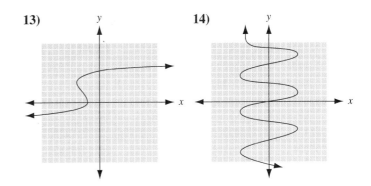   **14)**

Determine whether each of the following is the graph of a function.

7) 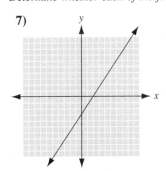   **8)**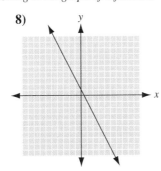

9) 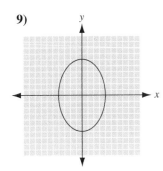   **10)**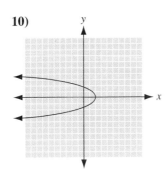

11) 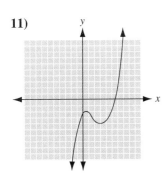   **12)**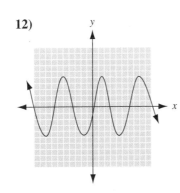

Determine whether each of the following define y as a function of x.

15) $y = -2x + 3$ **16)** $y = 6x - 5$
17) $y = 2x^2$ **18)** $y = -3x^2 + 2$
19) $y^2 = 2x - 3$ **20)** $y^2 = -x^2 + 5$
21) $2x + y^2 = 6$ **22)** $y^2 - 5x = 2$
23) $x^2 + y^2 = 16$ **24)** $2x^2 + y^2 = 16$
25) $y = x^3$ **26)** $y = x^3 - 1$
27) $y = |2x + 3|$ **28)** $y = |x + 7|$
29) $y < x - 6$ **30)** $y \geq 2x + 3$
31) $y = (x - 4)^2$ **32)** $y = (x + 5)^2$

Given $f(x) = 3x + 5$, find the following. Write each result as an ordered pair.

33) $f(0)$ **34)** $f(4)$ **35)** $f(a)$

Given $g(z) = z^2 - 2$, find the following. Write each result as an ordered pair.

36) $g(3)$ **37)** $g(-1)$ **38)** $g(a)$

Given $h(t) = 16t^2$, find the following. Write each result as an ordered pair.

39) $h(1)$ **40)** $h(-2)$ **41)** $h(a)$

Given $r(a) = a^3 + 2$, find the following. Write each result as an ordered pair.

42) $r(0)$ **43)** $r(-2)$ **44)** $r(b)$

Given $f(x) = |2x - 3|$, find the following. Write each result as an ordered pair.

45) $f(-2)$ **46)** $f(4)$ **47)** $f(z)$

48) The cost, $C(x)$, as a function of the number of items produced, x, is given by $C(x) = 20x + 150$. a) Find the cost of producing 10 items. b) Find the cost of producing 15 items. c) What does $C(30) = 750$ mean?

49) For a rental car, the cost, $C(x)$, as a function of the number of miles driven, x, is given by $C(x) = 0.12x + 22$. a) Find the cost of driving 100 miles. b) Find the cost of driving 350 miles. c) What does $C(250) = 52$ mean?

50) The value, $V(x)$, of a car as a function of the number of years after it was purchased, x, is given by $V(x) = -1500x + 18{,}000$. a) Find the value of the car after 3 years. b) Find the value of the car after 6 years. c) What does $V(10) = 3000$ mean?

51) The monthly salary, $S(x)$, of a salesperson as a function of her sales, x, is given by

$S(x) = 0.10x + 175$. a) Find her salary for sales of $15,000. b) Find her salary for sales of $10,000. c) What does $S(20{,}000) = 2175$ mean?

CHALLENGE EXERCISES (52–55)

Determine whether the following define y as a function of x.

52) $y = |x + 3|$

53) $|y| = 2x - 1$

54) $x = |y| + 3$

55) $|x| = y + 7$

Summary

Linear Equations with Two Variables: [Section 6.1]

A linear equation with two variables is any equation of the form $ax + by = c$, or any equation which can be put into that form, with a and b both not equal to zero at the same time.

Ordered Pairs: [Section 6.1]

An ordered pair is a pair of numbers enclosed in parentheses, separated by a comma, and with a variable assigned to each number. The first number is the abscissa and the second is the ordinate.

Solutions of Linear Equations with Two Variables: [Section 6.1]

If the solutions of $ax + by = c$ are ordered pairs in the form (x,y), the ordered pair (m,n) is a solution if the replacement of x with m and y with n into the given equation results in a true statement.

Finding a Missing Member of an Ordered Pair: [Section 6.1]

If an equation and one member of an ordered pair are given, find the other member of the ordered pair by substituting the given member for the variable it represents and solving the equation for the other variable.

Rectangular (Cartesian) Coordinate System: [Section 6.2]

The rectangular coordinate system is made up of a horizontal and a vertical number line which intersect at the zero point of each. The point of intersection is called the origin. The horizontal number line is the x-axis and the vertical number line is the y-axis. The four regions which the x- and y-axis divide the coordinate plane are called quadrants. The quadrants are numbered counterclockwise beginning with the top right.

Plotting Points on the Rectangular Coordinate System: [Section 6.2]

When plotting points, ordered pairs are assumed to be in the form (x,y). To plot a point, begin at the origin and go to the left or right the number of units given by the first number of the ordered pair. From that point, go up or down the number of units given by the second number of the ordered pair.

Graphing Linear Equations with Two Variables: [Section 6.2]

To graph a linear equation with two variables, find at least two ordered pairs (three is recommended) which are solutions of the equation. Plot the points on the rectangular coordinate system and draw a straight line through them.

x- and y-Intercepts: [Section 6.2]

The point(s) where a graph crosses the x-axis is (are) called the x-intercept(s). To find the x-intercept(s), let $y = 0$ and solve for x.

The point(s) where a graph crosses the y-axis is (are) called the y-intercept(s). To find the y-intercept(s), let $x = 0$ and solve for y.

Graphing Lines which Contain the Origin: [Section 6.2]

The graph of any equation of the form $y = ax$ contains the origin. Since both intercepts are zero, find at least one other point by assigning values to x and solving for y.

Vertical and Horizontal Lines: [Section 6.2]

The graph of any equation of the form $x = a$, where a is a constant, is a vertical line.

The x-values of all points on the vertical line are all a.

The graph of any equation of the form $y = b$, where b is a constant, is a horizontal line.

The y-values of all points on the horizontal line are all b.

Graphing Linear Inequalities with Two Variables: [Section 6.3]

To graph a linear inequality with two variables:

1) Graph the equality $ax + by = c$. If the line is part of the solution ($\leq$ or $\geq$), draw a solid line. If the line is not part of the solution ($<$ or $>$), draw a dashed line.

2) Pick a test point. If the coordinates of the test point satisfy the inequality, then all points on the same side of the line as the test point solve the inequality. If the coordinates of the test point do not satisfy the inequality, then all the points on the other side of the line from the test point solve the inequality.

3) Shade the region on the side of the line which contains the solutions of the inequality.

Definition of Slope: [Section 6.4]

The slope of a line is the ratio of the change in y to the change in x between any two points on the line. Alternately, the slope is the ratio of the rise over the run between any two points on the line.

Slope Formula: [Section 6.4]

The slope of the line containing the two points (x_1, y_1) and (x_2, y_2) is $m = \frac{y_2 - y_1}{x_2 - x_1}$.

Graphing a Line Given a Point and the Slope: [Section 6.4]

1) Plot the given point.

2) From the given point we can find another point by going vertically (up or down) the number of units given by the numerator and horizontally (right or left) the number of units given by the denominator.

3) Draw a line through these two points.

Observations about Slope: [Section 6.4]

1) A line with positive slope rises from left to right.

2) A line with negative slope falls from left to right.

3) The greater the absolute value of the slope, the steeper the line.

Slopes of Vertical and Horizontal Lines: [Section 6.4]

The slope of a vertical line is undefined. The slope of a horizontal line is 0.

Determining Slope and y-Intercept from the Equation: [Section 6.4]

To find the slope of the graph of a line from its equation, solve the equation for y. The resulting equation is in the form $y = mx + b$ where m is the slope and b is the y-intercept.

Parallel and Perpendicular Lines: [Section 6.5]

Two lines are parallel if and only if their slopes are equal. Two lines are perpendicular if and only if the product of their slopes is -1.

Writing Equations of Lines: [Section 6.5]

1) To write the equation of a line given a point (x_1, y_1) on the line and the slope m of the line, use the point-slope formula $y - y_1 = m(x - x_1)$. Substitute for m, x_1, and y_1. Never substitute for x and y.

2) To write the equation of a line which contains two given points: a) Find the slope of the line. b) Substitute into the point-slope formula using the

slope which you found and either of the two given points as (x_1, y_1).

3) To write the equation of a vertical line, set x equal to the x-value of any point on the line.

4) To write the equation of a horizontal line, set y equal to the y-value of any point on the line.

5) To write the equation of a line containing a given point and parallel to a given line, determine the slope of the line. The slope of the line you are looking for is equal to the slope of the given line. Substitute the slope and the coordinates of the given point into the point-slope formula.

6) To write the equation of a line containing a given point and perpendicular to a given line, determine the slope of the line. The slope of the line you are looking for is equal to the negative reciprocal of the slope of the given line. Substitute the slope and the coordinates of the given point into the point-slope formula.

7) To write the equation of a line when given the slope and y-intercept, substitute for the slope and the y-intercept into the slope-intercept formula, $y = mx + b$.

Relations: [Section 6.6]

A relation is any set of ordered pairs. In algebra, relations are usually given in the form of equations.

Functions: [Section 6.6]

A function is a relation in which each first component of the ordered pair is paired with exactly one second component. Equivalently, no two ordered pairs can have the same first components and different second components.

Vertical Line Test: [Section 6.6]

If any vertical line intersects the graph of a relation in more than one point, the graph does not represent a function.

Determining if an Equation Represents a Function: [Section 6.6]

An equation represents a function if each value of the independent variable (usually x) results in exactly one value of the dependent variable (usually y).

$f(x)$ Notation: [Section 6.6]

When using $f(x)$ notation:

a) f is the name of the function.

b) x is a value from the domain.

c) $f(x)$ is the y-value from the range that is paired with x from the domain.

Review Exercises

Determine whether the given ordered pair is a solution of the given equation. Assume the ordered pairs are in the form of (x,y). [Section 6.1]

1) $y = 4x + 5$; $(-2,-3)$

2) $y = -3x + 5$; $(2,1)$

3) $4x + 5y = 20$; $(10,-4)$

4) $7x - 2y = 14$; $(4,7)$

5) $x = 3$; $(3,5)$

6) $y = -5$; $(-5,5)$

Use the given equation to find the missing member of each of the following ordered pairs. Assume the ordered pairs are in the form (x,y). [Section 6.1]

7) $y = -3x + 3$; $(0,\_)$, $(\_,0)$, $(3,\_)$, $(\_,-3)$

8) $y = 4x + 2$; $(0,\_)$, $(\_,0)$, $(-5,\_)$, $(\_,-6)$

9) $3x + 2y = 12$; $(0,\_)$, $(\_,0)$, $(-4,\_)$, $(\_,3)$

10) $5x - 3y = 15$; $(0,\_)$, $(\_,0)$, $(6,\_)$, $(\_,5)$

11) A motorcycle is traveling at a constant speed of 60 miles per hour. The formula for the distance traveled after t hours is $d = 60t$. Represent the distance traveled after 2, 4, and 7 hours as ordered pairs in the form (t,d). What does the ordered pair $(3,180)$ represent?

12) An accountant is paid $25 per hour. If p represents her pay and n represents the number of hours worked, represent the accountant's pay for 2, 5, and 7 hours work as ordered pairs in the form of (n,p). What does the ordered pair $(4,100)$ represent?

Plot the points represented by the following ordered pairs on the rectangular coordinate system. Assume each unit represents one. [Section 6.2]

13) $(0,2)$, $(-3,4)$, $(0,1)$

14) $(5,-2)$, $(4,-2)$, $(-6,7)$

Draw the graph of each of the following equations. Assume each unit represents one. [Section 6.2]

15) $y = 3x - 6$

16) $y = -2x - 1$

17) $3x + y = 9$

18) $4x - y = 8$

19) $2x + 3y = -6$

20) $2x - 6y = 12$

21) $y = -5x$

22) $y = 3x$

23) $x = 7$

24) $y = -6$

25) Write an equation for the following. The sum of twice the x-coordinate and four times the y-coordinate is 16.

26) Francine works two part-time jobs. She receives $5 per hour when working at the supermarket and $7 per hour when working at the drug store. If x represents the number of hours that she worked at the supermarket and y the number of hours that she works at the drug store, write an equation showing the possible combinations of hours if she earned $145 last week.

Graph the following linear inequalities with two variables. Assume each unit represents one. [Section 6.3]

27) $y < x + 5$

28) $y \geq \frac{5}{2}x - 3$

29) $2x - 5y \geq 10$

30) $y < -5x$

31) $x \geq -6$

32) $y \leq 8$

33) Write an inequality for the following. The difference of three times the y-coordinate and two times the x-coordinate is at least 8.

34) Sharon goes shopping for some slacks and blouses. The slacks cost $18 each and the blouses cost $15 each. If x represents the number of pairs of slacks and y the number of blouses, write an inequality that shows the possible combinations of slacks and blouses if Sharon can spend no more than $180.

Find the slope of the line which contains each of the following pairs of points. [Section 6.4]

35) $(4,2)$ and $(5,4)$

36) $(6,3)$ and $(3,1)$

37) $(4,-1)$ and $(-2,2)$

38) $(-2,-7)$ and $(0,-2)$

39) $(4,2)$ and $(4,-2)$

40) $(2,-5)$ and $(-4,-5)$

Draw the graph of each of the following lines containing the given point and with the given slope. Assume each unit represents one. [Section 6.4]

41) $(2,-3)$, $m = \frac{4}{3}$

42) $(-3,4)$, $m = -\frac{2}{3}$

43) $(3,-1)$, $m = 3$

44) $(-1,-4)$, $m = -2$

45) $(5,-2)$, undefined slope

46) $(-2,4)$, $m = 0$

Find the slope and y-intercept of the line represented by each of the following equations. [Section 6.4]

47) $x + 2y = 8$ **48)** $5x - 2y = 15$

49) $x = 5$ **50)** $y = -3$

51) The operator of a crane lowers an object onto a loading dock at a point 30 feet from the crane. If the tip of the crane is 20 feet directly above the dock, what is the slope of the crane?

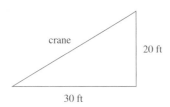

52) A road drops 8 feet vertically for every 100 feet horizontally. What is the slope of the road?

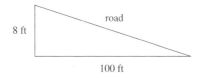

Write the equation of each of the following lines which contain the given point and have the given slope. Leave answers in ax + by = c form. [Section 6.5]

53) $(1,3)$ and $m = 3$

54) $(-2,5)$ and $m = -2$

55) $(4,-1)$ and $m = -\dfrac{5}{2}$

56) $(-5,4)$ and $m = \dfrac{4}{3}$

57) $(-7,-1)$ and undefined slope

58) $(3,-5)$ and $m = 0$

59) $(6,2)$ and horizontal **60)** $(-8,5)$ and vertical

Write the equation of the line which contains the following pairs of points. Leave answers in ax + by = c form. [Section 6.5]

61) $(-3,4)$ and $(7,-1)$ **62)** $(5,1)$ and $(1,-3)$

63) $(0,6)$ and $(6,-3)$ **64)** $(4,0)$ and $(5,3)$

65) $(-3,5)$ and $(2,5)$ **66)** $(2,6)$ and $(2,-3)$

Determine whether the following lines are parallel, perpendicular, or neither. [Section 6.5]

67) L_1: $(4,-1)$ and $(10,1)$,
L_2: $(-4,2)$ and $(2,4)$

68) L_1: $(2,-2)$ and $(-3,-4)$,
L_2: $(-1,-1)$ and $(-3,4)$

69) L_1: $(-7,3)$ and $(-1,-1)$,
L_2: $(-2,8)$ and $(2,2)$

70) L_1: $3x + 4y = 8$,
L_2: $4x - 3y = 9$

71) L_1: $4x + 6y = 5$,
L_2: $2x + 3y = 6$

72) L_1: $5x + 4y = 12$,
L_2: $4x + 5y = 15$

73) L_1: $x = 3$,
L_2: $x = -\dfrac{1}{3}$

74) L_1: $y = 4$,
L_2: $x = 5$

Write the equation of each of the following lines. Leave answers in ax + by = c form. [Section 6.5]

75) Contains $(6,-4)$ and perpendicular to the graph of $5x - 3y = 15$.

76) Contains $(-4,3)$ and parallel to the graph of $9x + 3y = 11$.

77) Contains $(-5,4)$ and parallel to the graph of $4x + 6y = 18$.

78) Contains $(-1,-1)$ and perpendicular to the graph of $4x - y = 5$.

79) Contains $(2,-5)$ and parallel to the graph of $x = 6$.

80) Contains $(4,-1)$ and parallel to the graph of $y = 3$.

81) Contains $(5,3)$ and perpendicular to the graph of $x = -1$.

Write the equation of each of the following lines. Leave the answer in slope-intercept form. [Section 6.5]

82) $m = -3$ and y-intercept is $(0,\frac{5}{6})$

83) $m = \frac{6}{5}$ and y-intercept is $(0,4)$

84) Horizontal line with y-intercept $(0,-3)$

Answer the following. [Section 6.5]

85) The fixed cost of operating a sandwich shop is $250 per day. In addition, each sandwich they sell costs the shop an average of $1.50.

a) Write a linear equation that gives the daily operating costs, y, in terms of the number of sandwiches sold, x.

b) Use the equation found in part a to find the cost for a day when 300 sandwiches were sold.

c) Using the equation found in part a, find the number of sandwiches sold on a day when the costs were $587.50.

Determine whether the following relations are functions and give the domain and range of each.

86) {(−3,2), (−1,4), (0,7), (−1,5)}

87) {(−4,2), (−1,4), (3,2), (6,5)}

Determine whether each of the following is the graph of a function.

88)

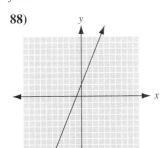

89)

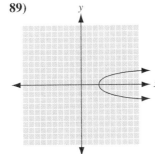

90)

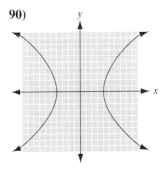

91)

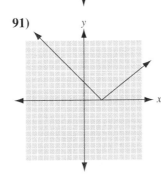

Determine whether each of the following define y as a function of x.

92) $y = -2x + 1$

93) $y = |2x + 6|$

94) $y^2 = x + 6$

95) $x^2 = y + 5$

96) Given $f(x) = 2x^2 + 3$, find each of the following. Represent each result as an ordered pair.

   a) $f(2)$

   b) $f(-3)$

   c) $f(a)$

97) Given $g(x) = x^3 - 3x^2 + 2$, find each of the following. Represent each result as an ordered pair.

   a) $g(-2)$

   b) $g(0)$

   c) $g(b)$

98) A company purchases a piece of machinery for $30,000. The value, $V(x)$, of the machinery as a function of the number of years after it was purchased, x, is given by $-1800x + 30,000$.

   a) Find the value of the machinery after 4 years.

   b) Find the value of the machinery after 10 years.

   c) What does $V(8) = 15,600$ mean?

Chapter 6 Test

1) If the solutions of $5x - 3y = 9$ are in the form of (x,y), is the ordered pair $(-3,-8)$ a solution?

2) Use the given equation to find the missing member of each of the given ordered pairs. Assume the ordered pairs are in the form (x,y).
$3x + 4y = -24$; $(0,\_)$, $(\_,0)$, $(-4,\_)$, $(\_,3)$

3) A particular type of pipe costs $2.00 per foot. The cost, C, of L feet of pipe is represented as $C = 2L$. Represent the cost of 1 ft, 2 ft, and 4 ft of this pipe as ordered pairs in the form of (L,C) where L is the length of the piece of pipe and C is the cost. What does the ordered pair $(10,20)$ represent?

Graph each of the following. Assume each unit on the coordinate system represents one.

4) $y = 2x - 6$

5) $3x + 5y = -15$

6) $3x + 4y = -6$

7) $x = -5$

8) A hardware store sells hammers for $9 each and shovels for $20 each. If x represents the number of hammers and y the number of shovels sold, write an equation showing all possible combinations of hammers and shovels if the hardware store received $480 for the sale of hammers and shovels.

Graph the following.

9) $y \geq -\dfrac{2}{3}x + 4$

10) $4x - 5y > 20$

11) $y \leq -6$

12) Containing $(-1,4)$ with slope $= \dfrac{3}{5}$.

13) Containing $(0,4)$ with undefined slope.

Find the slope of the lines which contain the following pairs of points.

14) $(3,-2)$ and $(-4,5)$

15) $(4,-2)$ and $(-2,-2)$

Find the slope of the lines with the given equations.

16) $x - 4y = 7$

17) $5x - 2y = 3$

18) $x = -3$

Determine whether the following pairs of lines are parallel, perpendicular, or neither parallel nor perpendicular.

19) L_1: contains $(4,-5)$ and $(-1,-3)$; L_2 contains $(-7,2)$ and $(-3,12)$.

20) L_1 is the graph of $4x - 7y = 5$ and L_2 is the graph of $8x - 14y = 11$.

Write the equation of each of the following lines. Leave answers in ax + by = c form.

21) Contains $(4,-2)$ with $m = -\dfrac{5}{2}$.

22) Contains $(6,-1)$ and $(2,-3)$.

23) Contains $(-2,3)$ and parallel to the graph of $x - 3y = 6$.

24) Contains $(5,-6)$ and is horizontal.

25) How does the graph of a line with slope -2 compare with the graph of a line with slope 2?

26) Does the following set of ordered pairs represent a function? Give the domain and range.
$\{(-4,7), (8,-2), (6,-4), (8,3)\}$

27) Does the following graph represent the graph of a function?

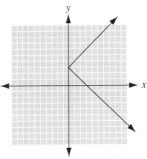

28) Does $y = 2x^2 + 1$ define y as a function of x?

29) Given $f(x) = x^2 - 3x + 2$, find each of the following. Write the results as ordered pairs.
   a) $f(2)$ b) $f(-1)$

30) The cost, $C(x)$, of operating a small business as a function of the number of days it operates, x, is given by $C(x) = 500x + 2,000$.

   a) What is the cost of operating 5 days?

   b) What does $C(10) = 7000$ mean?

7.1 Defining Linear Systems and Solving by Graphing

7.2 Solving Systems of Linear Equations Using Elimination by Addition

7.3 Solving Systems of Linear Equations Using Substitution

7.4 Applications Using Systems of Linear Equations

7.5 Systems of Linear Inequalities

Systems of Linear Equations and Inequalities

CHAPTER OVERVIEW

In this chapter we continue our study of linear equations with two variables. We know from Chapter 6 that a linear equation with two variables has an infinite number of solutions that are written as ordered pairs. In this chapter we will be finding the single ordered pair (if there is one) that solves two linear equations with two variables. We will learn three techniques for finding this point, if it exists. These techniques also provide us with new methods for solving applications problems.

SECTION 7.1

Defining Linear Systems and Solving by Graphing

OBJECTIVES

When you complete this section, you will be able to:

Ⓐ Determine whether a given ordered pair is the solution of a linear system with two variables.

Ⓑ Solve a linear system with two variables by graphing.

Ⓒ Determine whether a linear system with two variables is consistent, inconsistent, or dependent.

INTRODUCTION

As mentioned in the introduction, in the last chapter we found that a linear equation with two variables has an infinite number of solutions, each of which is an ordered pair. In this section we will be looking for one ordered pair that is the solution of two linear equations. For example, some solutions of $x + y = 4$ are $(0,4)$, $(4,0)$, $(1,3)$, $(2,2)$, $(5,-1)$, $(-1,5)$, and $(-2,6)$. Some solutions of $x - y = 6$ are $(0,-6)$, $(6,0)$, $(1,-5)$, $(3,-3)$, $(5,-1)$, $(-2,-8)$, and $(-3,-9)$. By examining the ordered pairs that are solutions of each equation, we see that the ordered pair $(5, -1)$ is a solution of both equations. The equations $x + y = 4$ and $x - y = 6$ are an example of a **system of linear equations**. Therefore, $(5, -1)$ is a solution of this system of equations. Systems of equations are usually written with one equation beneath the other as below.

$$\begin{cases} x + y = 4 \\ x - y = 6 \end{cases}$$

SYSTEM OF LINEAR EQUATIONS

A system of linear equations is two or more linear equations with the same variables.

In this text we will consider only systems of two linear equations with two variables.

We observed that the ordered pair $(5, -1)$ solved each of the equations $x + y = 4$ and $x - y = 6$. Any ordered pair that solves all the equations in a system of equations is a solution of the system.

SOLUTION(S) OF A SYSTEM OF LINEAR EQUATIONS

The solution(s) of a system of linear equations with two variables consists of all ordered pairs that solve all the equations of the system.

To determine if an ordered pair is a solution of a system of linear equations, we must show that the ordered pair solves each equation of the system.

EXAMPLE 1

Determine if the given ordered pair is a solution of the system of linear equations. The ordered pair is in the form (x,y).

a) $(3, -1)$
$$\begin{cases} 2x + y = 5 \\ 3x - y = 10 \end{cases}$$

Solution:

To determine whether the given ordered pair is a solution of the system, we must determine if it solves each equation in the system. Substitute 3 for x and -1 for y into each equation and simplify.

$2x + y = 5$	$3x - y = 10$
$2(3) + (-1) = 5$	$3(3) - (-1) = 10$
$6 - 1 = 5$	$9 + 1 = 10$
$5 = 5$	$10 = 10$

Since $(3, -1)$ solves both equations of the system, $(3, -1)$ is a solution of the system.

b) $(2,-3)$
$$\begin{cases} 4x + 3y = -1 \\ 2x - 5y = -11 \end{cases}$$

Solution:
Substitute 2 for x and -3 for y into each equation and simplify.

$$\begin{array}{ll} 4x + 3y = -1 & 2x - 5y = -11 \\ 4(2) + 3(-3) = -1 & 2(2) - 5(-3) = -11 \\ 8 - 9 = -1 & 4 + 15 = -11 \\ -1 = -1 & 19 \neq -11 \end{array}$$

Since $(2,-3)$ does not solve both of the equations in the system, $(2,-3)$ is not a solution of the system. ■

ADDITIONAL PRACTICE

Determine whether the given ordered pair is a solution of the system of linear equations. The ordered pair is in the form (x, y).

a) $(2,-4)$
$$x - 2y = -10$$
$$x + 4y = -14$$

b) $(2,-1)$
$$5x + 3y = 7$$
$$4x - y = 9$$

PRACTICE EXERCISES

Determine whether the given ordered pair is a solution of the system of linear equations. The ordered pair is in the form (x,y).

1) $(4,1)$
$$\begin{cases} 3x + y = 13 \\ 2x + y = 9 \end{cases}$$

2) $(-3,2)$
$$\begin{cases} 2x - 3y = -12 \\ 4x + 5y = 2 \end{cases}$$

If more practice is needed, do the Additional Practice Exercises in the margin.

In the previous examples and practice exercises, the ordered pair was given and we were asked to determine whether the ordered pair was a solution of the system. Now we will find the ordered pair that is the solution of a system of linear equations. Recall from Chapter 6 that if a point lies on the graph of a line, then the coordinates of the point must solve the equation of the line. Consequently, if the graphs of two lines intersect at a point, then the coordinates of the point of intersection must solve both equations. This leads to the following method of solving a system of two linear equations with two variables.

SOLVING A SYSTEM OF LINEAR EQUATIONS BY GRAPHING

To solve a system of two linear equations with two variables, graph each equation of the system. If the graphs of the equations intersect, the coordinates of the point(s) of intersection are the solutions of the system.

EXAMPLE 2
Find the solution(s) of the following systems of linear equations by graphing.

a) $\begin{cases} x + y = 6 \\ x - y = -2 \end{cases}$

Solution:

To find the solution(s) of the system, we graph both equations on the same coordinate system and find the point(s) of intersection of the graphs. Remember, the quickest way to graph the equations is by using the intercepts.

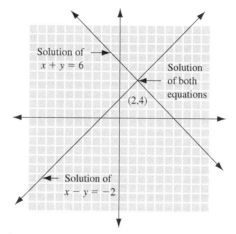

From the graph we see that the two lines seem to intersect at the point whose coordinates are (2,4). Therefore, (2,4) is a possible solution of the system.

Check:

If (2,4) is the solution of the system, then (2,4) must solve both equations. Substitute 2 for x and 4 for y into both equations and simplify.

$$
\begin{array}{ll}
x + y = 6 & x - y = -2 \\
2 + 4 = 6 & 2 - 4 = -2 \\
6 = 6 & -2 = -2
\end{array}
$$

Since (2,4) solves both equations, (2,4) is the solution of the system.

b) $\begin{cases} 3x + 2y = 4 \\ 2x + 3y = 1 \end{cases}$

Solution:

To find the solution(s) of the system, we graph both equations on the same coordinate system and find the point(s) of intersection of the graphs.

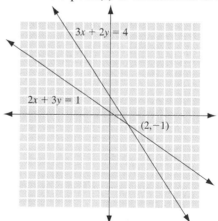

From the graph, we see that the two lines seem to intersect at the point whose coordinates are (2,−1). Therefore, (2,−1) is a possible solution of the system.

Check:

If (2,−1) is the solution of the system, then (2,−1) must solve both equations in the system. Substitute 2 for x and −1 for y into both equations and simplify.

$$3x + 2y = 4 \qquad 2x + 3y = 1$$
$$3(2) + 2(-1) = 4 \qquad 2(2) + 3(-1) = 1$$
$$6 - 2 = 4 \qquad 4 - 3 = 1$$
$$4 = 4 \qquad 1 = 1$$

Since $(2, -1)$ solves both equations, $(2, -1)$ is the solution of the system.

ADDITIONAL PRACTICE

Find the solution(s) of the following linear systems with two variables by graphing. Assume each unit on the coordinate system represents one.

c) $x - y = -6$
$\quad x + 2y = 0$

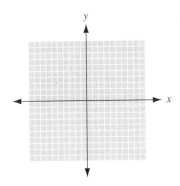

d) $3x - 2y = -12$
$\quad x - 2y = -8$

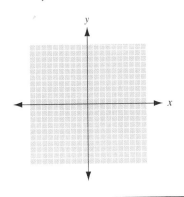

PRACTICE EXERCISES

Find the solution(s) of the following linear systems by graphing. Assume each unit on the coordinate system represents one.

3) $\begin{cases} x + y = 1 \\ x - y = -5 \end{cases}$

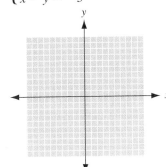

4) $\begin{cases} 2x + 3y = 12 \\ 2x + y = 8 \end{cases}$

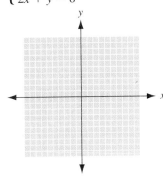

If more practice is needed, do the Additional Practice Exercises in the margin.

One of the problems with solving systems of equations graphically is that the two lines do not always intersect at a point whose coordinates are integers. In this case it may not be possible to determine the exact solution of the system by the graphing method, as illustrated in the example below.

EXAMPLE 3

Find the solution(s) of the following linear system by graphing. Assume each unit on the coordinate system represents one.

$$\begin{cases} 6x + 6y = 5 \\ 2x - 3y = -10 \end{cases}$$

Solution:

To find the solution(s) of the system, we graph both equations on the same coordinate system and find the point(s) of intersection of the graphs.

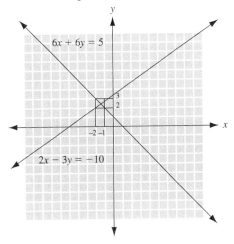

From the graph we see that the x-value of the point of intersection is between -1 and -2 and the y-value is between 2 and 3. The only way to find the exact coordinates of the point of intersection using the graphical method is to approximate the coordinates and then substitute them into both equations. This can be very time consuming and there is no guarantee that the exact solution will ever be found. For that reason, in this section we will limit ourselves to systems whose graphs intersect at points whose coordinates are integers. We will learn two other methods that are more efficient for solving systems of linear equations in Sections 7.2 and 7.3.

If a system of two linear equations with two variables is graphed on one coordinate system, there are three possible outcomes.

1) The lines can intersect in exactly one point as in Examples 2 and 3. If this occurs, the system is said to be **consistent** and the system has one solution which is the coordinates of the point of intersection.

2) The lines can be parallel. If this occurs, the system is said to be **inconsistent** and the system has no solution, since parallel lines do not intersect.

3) The lines can coincide. If this occurs, the system is said to be **dependent** and there are an infinite number of solutions since the lines intersect at an infinite number of points. Any ordered pair which solves one equation of a dependent system also solves the other.

Examples of inconsistent and dependent systems are given below.

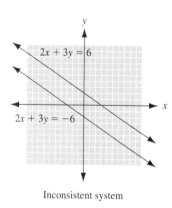

Inconsistent system

No solution

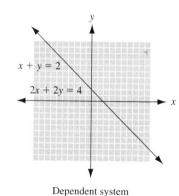

Dependent system

Infinite number of solutions

EXAMPLE 4

Determine whether the following systems are consistent, inconsistent, or dependent. If the system is consistent, find the solution of the system.

a) $\begin{cases} 2x + y = 4 \\ 4x + 2y = -7 \end{cases}$

Solution:

To determine whether the system is consistent, inconsistent, or dependent, we can draw the graph of the system on one coordinate system.

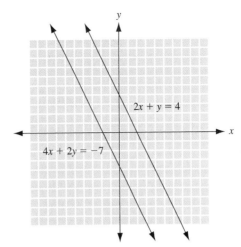

The two lines appear to be parallel. To confirm that they are, we need to find the slope of each line. Remember that parallel lines have equal slopes. To find the slope of each line, write each equation in slope-intercept form.

$$2x + y = 4$$
$$2x - 2x + y = -2x + 4$$
$$y = -2x + 4$$
Therefore, the slope is -2.

$$4x + 2y = -7$$
$$4x - 4x + 2y = -4x - 7$$
$$2y = -4x - 7$$
$$\frac{2y}{2} = \frac{-4x - 7}{2}$$
$$y = -2x - \frac{7}{2}$$
Therefore, the slope is -2.

Since both lines have the same slope, the lines are parallel. Since the lines have different y-intercepts, they are not the same line. Consequently, the system is inconsistent.

Note: It is possible to determine if a system is inconsistent without graphing by writing each equation in slope-intercept form. The lines of an inconsistent system are parallel so they must have the same slope. In order to be distinct lines, rather than the same line, they must have different y-intercepts.

b) $\begin{cases} 4x + 6y = 4 \\ 6x + 9y = 6 \end{cases}$

To determine whether the system is consistent, inconsistent, or dependent, we can draw the graph of the system on one coordinate system.

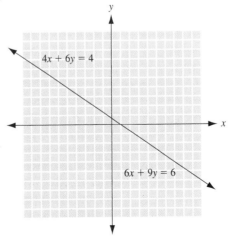

The graphs of the two lines appear to be the same line. To confirm that they are, write each equation in slope-intercept form.

$$4x + 6y = 4$$
$$4x - 4x + 6y = -4x + 4$$
$$6y = -4x + 4$$
$$\frac{6y}{6} = \frac{-4x + 4}{6}$$
$$y = -\frac{4}{6}x + \frac{4}{6}$$
$$y = -\frac{2}{3}x + \frac{2}{3}$$

Therefore, the slope is $-\frac{2}{3}$ and the y-intercept is $\frac{2}{3}$.

$$6x + 9y = 6$$
$$6x - 6x + 9y = -6x + 6$$
$$9y = -6x + 6$$
$$\frac{9y}{9} = \frac{-6x + 6}{9}$$
$$y = -\frac{6}{9}x + \frac{6}{9}$$
$$y = -\frac{2}{3}x + \frac{2}{3}$$

Therefore, the slope is $-\frac{2}{3}$ and the y-intercept is $\frac{2}{3}$.

Since both lines have the same slope and the same y-intercept, they are the same line. Consequently, the system is dependent. The solutions of a dependent system are usually written as the set of ordered pairs which solve either of the equations. The solutions of this system could be written as $\{(x,y): 4x + 6y = 4\}$. This is read "the set of all ordered pairs such that $4x + 6y = 4$."

Note: It is possible to determine whether a system is dependent by writing both equations in slope-intercept form. If both lines have the same slope and the same y-intercept, they are the same line. Consequently, the system is dependent.

c) $\begin{cases} 2x - y = 5 \\ 3x + 4y = 24 \end{cases}$

To determine whether the system is consistent, inconsistent, or dependent, we draw the graph of the system on one coordinate system.

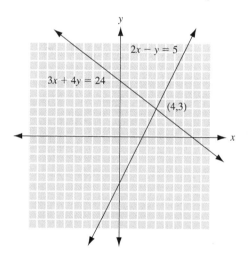

Since the lines intersect at a single point, the system is consistent. The coordinates of the point of intersection are (4,3). Therefore, (4,3) is a possible solution of the system. The check is left as an exercise for the student.

Note: It is possible to determine if a system is consistent without graphing by writing the equations of the system in slope-intercept form. Since any two lines in the same plane which are not parallel must intersect, the system is consistent if the slopes of the lines are not the same.

ADDITIONAL PRACTICE

Determine whether the following systems are consistent, inconsistent, or dependent. If the system is consistent, find the solution of the system.

e) $x + 2y = 6$
$2x = -4y - 7$

f) $4x + y = 7$
$2y = -8x + 14$

g) $x + 2y = 3$
$y = -2x + 12$

PRACTICE EXERCISES

Determine whether the following systems are consistent, inconsistent, or dependent. If the system is consistent, find the solution of the system.

5) $\begin{cases} 3x + 2y = 12 \\ 6x + 4y = -12 \end{cases}$

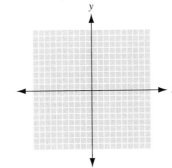

6) $\begin{cases} x + 2y = 6 \\ 2x = 12 - 4y \end{cases}$

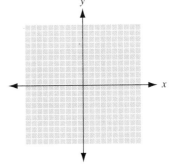

7) $\begin{cases} 5x + 3y = 30 \\ x + y = 8 \end{cases}$

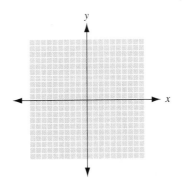

If more practice is needed, do the Additional Practice Exercises in the margin.

EXERCISE SET 7.1

Determine whether the given ordered pair is a solution of the system of linear equations. The ordered pair is in the form (x, y).

1) (3,4)
$\begin{cases} x + 3y = 15 \\ x + 2y = 11 \end{cases}$

2) (−4,1)
$\begin{cases} 2x + y = -7 \\ 3x + y = -11 \end{cases}$

3) (−2,−3)
$\begin{cases} 4x + 2y = 5 \\ x + 3y = 1 \end{cases}$

4) (1,6)
$\begin{cases} x - y = -1 \\ 2x + y = 7 \end{cases}$

5) (−1,2)
$\begin{cases} x - 3y = -7 \\ 6x + 2y = -2 \end{cases}$

6) (5,1)
$\begin{cases} 4x - 5y = 13 \\ x + y = 17 \end{cases}$

7) (−2,0)
$\begin{cases} 2x + 3y = 4 \\ x + 2y = 4 \end{cases}$

8) (0,3)
$\begin{cases} 5x + 3y = 9 \\ -4x + 7y = 21 \end{cases}$

Find the solution(s) of the following systems of linear equations by graphing. Assume each unit on the coordinate system represents one.

9) $\begin{cases} x + y = 1 \\ x - y = -1 \end{cases}$

10) $\begin{cases} x + y = -1 \\ x - y = 3 \end{cases}$

11) $\begin{cases} 3x - 2y = -12 \\ 3x + y = -3 \end{cases}$

12) $\begin{cases} x - 2y = 2 \\ x + 4y = 8 \end{cases}$

13) $\begin{cases} x - y = -7 \\ x = -3 \end{cases}$

14) $\begin{cases} y = 5 \\ x + y = 4 \end{cases}$

15) $\begin{cases} 2x - y = -4 \\ 2x - 3y = 0 \end{cases}$

16) $\begin{cases} x - 4y = 8 \\ x + 2y = 2 \end{cases}$

17) $\begin{cases} x - y = 4 \\ x + 5y = 10 \end{cases}$

18) $\begin{cases} x - 2y = -12 \\ x + 4y = 6 \end{cases}$

ANSWERS:
Practice 5–7

5) Inconsistent   6) Dependent

7) Consistent; (3,5)

Additional Practice e–g

e) Inconsistent   f) Dependent

g) Consistent; (7,−2)

---

29) $\begin{cases} x + y = -5 \\ 6x - y = 12 \end{cases}$
30) $\begin{cases} 3x - 2y = 12 \\ x + y = -1 \end{cases}$

CHALLENGE EXERCISES (31–32)

Determine whether the given ordered pair is a solution of the system of linear equations. The ordered pair is in the form (x, y).

31) $\left(\dfrac{1}{4}, 2\right)$
$\begin{cases} 4x + y = 3 \\ 8x + 3y = 8 \end{cases}$

32) $(-24, 12)$
$\begin{cases} 2x - 3y = -84 \\ x + 5y = 36 \end{cases}$

19) $\begin{cases} y = 3 \\ 2x - y = 3 \end{cases}$
20) $\begin{cases} x = -5 \\ 4x - 5y = -40 \end{cases}$

21) $\begin{cases} 6x - 3y = 0 \\ 3x - 2y = 0 \end{cases}$
22) $\begin{cases} 4x - 2y = 4 \\ 2x + 3y = -6 \end{cases}$

Determine whether the following systems are consistent, inconsistent, or dependent. If the system is consistent, find the solution of the system.

23) $\begin{cases} x - 3y = -12 \\ 4x - 12y = -36 \end{cases}$
24) $\begin{cases} 3x + 2y = 6 \\ 6x + 4y = 24 \end{cases}$

25) $\begin{cases} 2x + 3y = 9 \\ 4x + 6y = 12 \end{cases}$
26) $\begin{cases} 3x + y = 6 \\ 6x + 2y = 18 \end{cases}$

27) $\begin{cases} 3x + y = -6 \\ 9x + 3y = 9 \end{cases}$
28) $\begin{cases} -2x - y = 4 \\ 6x + 3y = 18 \end{cases}$

WRITING EXERCISES

33) If the slopes of two lines are not equal, what type of system is composed of these lines? Why?

34) If the slopes of two lines are equal, what type of system(s) is/are composed of these lines? Why?

35) Why are the coordinates of the point of intersection of two lines the solution of the system?

CRITICAL THINKING

36) Find a system of linear equations with solution (2,3), draw the graph of the system, and explain how you found the equations of the system.

37) Find a system of equations with no solution and describe how you found the equations of the system.

SECTION 7.2

Solving Systems of Linear Equations Using Elimination by Addition

OBJECTIVES

When you complete this section, you will be able to:

Ⓐ   Solve consistent systems of linear equations with two variables by eliminating a variable using the addition method.

Ⓑ   Recognize inconsistent and dependent systems of equations when using the addition method.

INTRODUCTION

We found in Section 7.1 that one of the difficulties with solving systems of equations using the graphing method was that the lines did not always intersect at a point with integer coordinates. In this section we will develop another method of solving systems of linear equations with two variables without resorting to graphing. It can be shown that if two lines intersect at a point, then the sum of the equations of the lines, or any multiples of these equations, is the equation of another line that intersects the two given lines at the same point.

In Chapter 3 we introduced the addition property of equality which stated that if $a = b$, then $a + c = b + c$. We will need an extension of this property that is given as follows.

ADDITION OF EQUALS

If $a = b$ and $c = d$, then $a + c = b + d$. In words, equals added to equals results in equals.

Below we repeat Example 2 from Section 7.1 along with a new method of solution.

Find the solution(s) of the following system of linear equations by graphing.

a) $\begin{cases} x + y = 6 \\ x - y = -2 \end{cases}$

Solution:

To find the solution of the system, we graph both equations on the same coordinate system and find the point of intersection of the graphs.

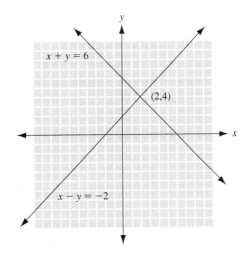

From the graph we see that the two lines seem to intersect at the point with coordinates (2,4), Therefore, (2,4) is a possible solution of the system. Notice that if we add the two equations, y drops out.

$x + y =$	6	
$x - y = -2$		Add the equations.
$2x \quad = 4$		Divide both sides by 2.
$x \quad = 2$		Note: This is the x-value of the point of intersection of the two lines because it represents the equation of a vertical line that passes through the point of intersection. To find the y-value, substitute 2 for x in either of the original equations. Use $x + y = 6$.
$x + y = 6$		Substitute 2 for x.
$2 + y = 6$		Subtract 2 from both sides.
$y = 4$		Note: This is the y-value of the point of intersection. Consequently, we have solved the system without graphing it.

Summarizing, the Addition of Equals Property and the fact that the sums of multiples of the equations of a system gives another equation with a graph that passes through the same point of intersection as the original system, leads to another method of solving systems of linear equations. Following is an outline of the procedure. It will be clearer after some examples.

We illustrate the procedure above with some examples.

EXAMPLE 1

Solve the following systems of equations using elimination by addition.

a) $\begin{cases} x + y = 3 \\ x - y = 9 \end{cases}$

Solution:

It is easier to eliminate y than x because we can eliminate y simply by adding the two equations.

$$\begin{array}{rcl} x + y = & 3 \\ \underline{x - y = & 9} \end{array}$$ Add the equations.

$2x \quad = 12$ Divide both sides by 2.

$x \quad = 6$ Substitute $x = 6$ in either equation to find y.
    Use $x + y = 3$.

$x + y = \quad 3$ Substitute 6 for x.

$6 + y = \quad 3$ Subtract 6 from both sides.

$y = -3$ Therefore, the solution of the system is $(6, -3)$.

Check:

Substitute 6 for x and -3 for y in both equations.

$$\begin{array}{ll} x + y = 3 & x - y = 9 \\ 6 + (-3) = 3 & 6 - (-3) = 9 \\ 3 = 3 & 6 + 3 = 9 \\ & 9 = 9 \end{array}$$

Since $(6, -3)$ solves both equations, it is the correct solution.

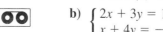 b) $\begin{cases} 2x + 3y = 1 \\ x + 4y = -7 \end{cases}$

Solution:

Notice that neither variable will be eliminated if we add the equations just as they are. In order for a variable to be eliminated, the coefficients of that variable must be additive inverses. It appears easier to make the coefficients of x additive inverses since we would have to multiply only one equation by a constant. Consequently, we will eliminate x by multiplying $x + 4y = -7$ by -2 and then adding the equations.

$$2x + 3y = 1 \qquad \qquad \text{Leave } 2x + 3y = 1 \text{ unchanged.}$$
$$x + 4y = -7 \qquad \qquad \text{Multiply } x + 4y = -7 \text{ by } -2.$$

$$2x + 3y = 1 \qquad \qquad \text{Leave } 2x + 3y = 1 \text{ unchanged.}$$
$$-2(x + 4y) = -2(-7) \qquad \text{Simplify both sides.}$$

$$\begin{aligned} 2x + 3y &= 1 \\ -2x - 8y &= 14 \\ \hline -5y &= 15 \end{aligned} \qquad \text{Add the equations.}$$

$\qquad\qquad\qquad\qquad\quad$ Divide both sides by -5.

$$y = -3 \qquad \qquad \text{Substitute } -3 \text{ for } y \text{ in either equation to find } x.$$
$\qquad\qquad\qquad\qquad\qquad\qquad$ Use $x + 4y = -7$.

$$x + 4y = -7 \qquad \qquad \text{Substitute } -3 \text{ for } y.$$
$$x + 4(-3) = -7 \qquad \qquad \text{Multiply } 4 \text{ and } -3.$$
$$x - 12 = -7 \qquad \qquad \text{Add } 12 \text{ to both sides.}$$
$$x = 5 \qquad \qquad \text{Therefore, the solution is } (5, -3).$$

Check:

Substitute 5 for x and -3 for y in both equations.

$$\begin{array}{ll} 2x + 3y = 1 & x + 4y = -7 \\ 2(5) + 3(-3) = 1 & 5 + 4(-3) = -7 \\ 10 - 9 = 1 & 5 - 12 = -7 \\ 1 = 1 & -7 = -7 \end{array}$$

Since $(5, -3)$ solves both equations, it is the correct solution.

c) $\begin{cases} 3x - 2y = 7 \\ 4x - 3y = 10 \end{cases}$

Solution:

If we add the equations as they are, neither variable would be eliminated. Also, we can not multiply just one equation by an integer to make the coefficients of either variable additive inverses. Consequently, we must multiply both equations. If we choose to eliminate x, we must make the coefficients of x the least common multiple of 3 and 4 but with opposite signs. The LCM of 3 and 4 is 12. If we choose to eliminate y, we must make the coefficients of y the LCM of 2 and 3 but with opposite signs. The LCM of 2 and 3 is 6. Since 6 is smaller than 12 (It's as good of a reason as any!), we choose to eliminate y by making one coefficient 6 and the other -6.

$$3x - 2y = 7 \qquad \qquad \text{Multiply } 3x - 2y = 7 \text{ by } 3.$$
$$4x - 3y = 10 \qquad \qquad \text{Multiply } 4x - 3y = 10 \text{ by } -2.$$

$$3(3x - 2y) = 3(7) \qquad \qquad \text{Simplify both sides.}$$
$$-2(4x - 3y) = -2(10) \qquad \qquad \text{Simplify both sides.}$$

$$9x - 6y = 21$$
$$\underline{-8x + 6y = -20}$$
$$x = 1$$

Add the equations.

Substitute 1 for x in either equation and solve for y.
Use $3x - 2y = 7$.

$$3x - 2y = 7$$

Substitute 1 for x.

$$3(1) - 2y = 7$$

Multiply 3 and 1.

$$3 - 2y = 7$$

Subtract 3 from both sides.

$$-2y = 4$$

Divide both sides by -2.

$$\frac{-2y}{-2} = \frac{4}{-2}$$

Simplify both sides.

$$y = -2$$

Therefore, the solution is $(1, -2)$.

Check:

The check is left as an exercise for the student.

Note: Example 1d below exhibits the advantage of the addition method when the solutions are fractions instead of integers.

 d) $\begin{cases} x + 3y = -1 \\ 3x + 6y = -1 \end{cases}$

Solution:

Let us eliminate x by multiplying $x + 3y = -1$ by -3 and adding the equations.

$$x + 3y = -1$$

Multiply $x + 3y = -1$ by -3.

$$3x + 6y = -1$$

Leave $3x + 6y = -1$ unchanged.

$$-3(x + 3y) = -3(-1)$$

Simplify both sides.

$$3x + 6y = -1$$

Leave unchanged.

$$-3x - 9y = 3$$
$$\underline{3x + 6y = -1}$$

Add the equations.

$$-3y = 2$$

Divide both sides by -3.

$$y = -\frac{2}{3}$$

Substitute $-\frac{2}{3}$ for y in either equation and solve for x. Use $x + 3y = -1$.

$$x + 3y = -1$$

Substitute $-\frac{2}{3}$ for y.

$$x + 3\left(-\frac{2}{3}\right) = -1$$

Multiply 3 and $-\frac{2}{3}$.

$$x - 2 = -1$$

Add 2 to both sides.

$$x = 1$$

Therefore, $\left(1, -\frac{2}{3}\right)$ is the solution.

Check:

Substitute 1 for x and $-\frac{2}{3}$ for y in both equations.

$$x + 3y = -1 \qquad\qquad 3x + 6y = -1$$
$$1 + 3\left(-\frac{2}{3}\right) = -1 \qquad 3(1) + 6\left(-\frac{2}{3}\right) = -1$$
$$1 - 2 = -1 \qquad\qquad 3 - 4 = -1$$
$$-1 = -1 \qquad\qquad\quad -1 = -1$$

Since $\left(1, -\frac{2}{3}\right)$ solves both equations, $\left(1, -\frac{2}{3}\right)$ is the correct solution.

*Solve the following systems of
equations.*

a) $x + 2y = 10$
$\ x - 2y = -6$

b) $x + 3y = -3$
$\ 5x + 2y = 24$

c) $4x + 5y = 2$
$\ 3x - 2y = -10$

d) $4x - 4y = -1$
$\ 12x - 8y = 3$

PRACTICE EXERCISES

Solve the following systems of equations using elimination by addition.

1) $\begin{cases} x - y = 5 \\ x + y = 1 \end{cases}$ **2)** $\begin{cases} 3x + 4y = 6 \\ x - 3y = -11 \end{cases}$

3) $\begin{cases} 4x - 3y = -23 \\ 5x - 4y = -30 \end{cases}$ **4)** $\begin{cases} 12x + 18y = 17 \\ 6x + 10y = 9 \end{cases}$

If more practice is needed, do the Additional Practice Exercises in the margin.

When one of the variables has a fractional value and this value is substituted back into one of the equations to find the other variable, it is often necessary to combine fractions. Sometimes it is easier to repeat the elimination procedure to find the remaining variable rather than substituting a fraction and solving the resulting equation with fractions.

EXAMPLE 2

Solve the following system of linear equations.

$$\begin{cases} 3x - 4y = 15 \\ 4x + 2y = 7 \end{cases}$$

Solution:

Since we can eliminate y by multiplying $4x + 2y = 7$ by 2 and then adding, we choose to eliminate y.

$3x - 4y = 15$	Leave $3x - 4y = 15$ unchanged.
$4x + 2y = 7$	Multiply $4x + 2y = 7$ by 2.
$3x - 4y = 15$	Leave unchanged.
$2(4x + 2y) = 2(7)$	Simplify both sides.
$3x - 4y = 15$	Add the equations.
$\underline{8x + 4y = 14}$	
$11x = 29$	Divide both sides by 11.
$x = \dfrac{29}{11}$	Therefore, the x-value of the solution is $\dfrac{29}{11}$.

We could find y by substituting $\frac{29}{11}$ for x in either equation and solving for y, but that does not look like it would be a whole lot of fun. Instead, let us repeat the elimination procedure and eliminate x this time. Since the LCM of 3 and 4 is 12, we need to make one coefficient of x equal to 12 and the other -12.

$3x - 4y = 15$	Multiply $3x - 4y = 15$ by 4.
$4x + 2y = 7$	Multiply $4x + 2y = 7$ by -3.
$4(3x - 4y) = 4(15)$	Simplify both sides.
$-3(4x + 2y) = -3(7)$	Simplify both sides.
$12x - 16y = 60$	Add the equations.
$\underline{-12x - 6y = -21}$	
$-22y = 39$	Divide both sides by -22.
$y = -\dfrac{39}{22}$	Therefore, the y-value of the solution is $-\dfrac{39}{22}$.

Consequently, the solution of the system is $\left(\dfrac{29}{11}, -\dfrac{39}{22}\right)$.

Check:

Substitute $\frac{29}{11}$ for x and $-\frac{39}{22}$ for y in both equations.

$$3x - 4y = 15 \qquad\qquad 4x + 2y = 7$$

$$3\left(\frac{29}{11}\right) - 4\left(-\frac{39}{22}\right) = 15 \qquad 4\left(\frac{29}{11}\right) + 2\left(-\frac{39}{22}\right) = 7$$

$$\frac{87}{11} + \frac{156}{22} = 15 \qquad\qquad \frac{116}{11} - \frac{78}{22} = 7$$

$$\frac{174}{22} + \frac{156}{22} = 15 \qquad\qquad \frac{232}{22} - \frac{78}{22} = 7$$

$$\frac{330}{22} = 15 \qquad\qquad\qquad \frac{154}{22} = 7$$

$$15 = 15 \qquad\qquad\qquad\quad 7 = 7$$

Since $\left(\frac{29}{11}, -\frac{39}{22}\right)$ solves both equations, it is the correct solution.

PRACTICE EXERCISES

Solve the following systems of equations.

5) $\begin{cases} 3x + 2y = 9 \\ 2x - 5y = 12 \end{cases}$ **6)** $\begin{cases} 4x + y = 6 \\ 3x + 2y = 9 \end{cases}$

Before the elimination by addition method can be used, both equations should be written in the form of $ax + by = c$ where a, b, and c are integers. If necessary, rewrite the equations of the system in this form before attempting to solve.

EXAMPLE 3

Solve the following systems of equations using elimination by addition.

a) $\begin{cases} x = 3y + 22 \\ 5y = -2x - 22 \end{cases}$

Solution:

We first have to rewrite each equation in $ax + by = c$ form.

$$\begin{array}{ll} x = 3y + 22 & 5y = -2x - 22 \\ x - 3y = 3y - 3y + 22 & 2x + 5y = -2x + 2x - 22 \\ x - 3y = 22 & 2x + 5y = -22 \end{array}$$

Now solve the system $x - 3y = 22$ and $2x + 5y = -22$ by eliminating x.

$x - 3y = 22$	Multiply $x - 3y = 22$ by -2.
$2x + 5y = -22$	Leave $2x + 5y = -22$ unchanged.
$-2(x - 3y) = -2(22)$	Simplify both sides.
$2x + 5y = -22$	Leave unchanged.
$-2x + 6y = -44$	Add the equations.
$\underline{2x + 5y = -22}$	
$\qquad\quad 11y = -66$	Divide both sides by 11.

$y = -6$	Substitute $y = -6$ in either of the original equations and find x. Use $x = 3y + 22$.
$x = 3y + 22$	Substitute -6 for y.
$x = 3(-6) + 22$	Multiply 3 and -6.
$x = -18 + 22$	Add -18 and 22.
$x = 4$	Therefore, the solution of the system is $(4, -6)$.

Check:

Substitute 4 for x and -6 for y in both of the original equations. The check is left as an exercise for the student.

b) $\begin{cases} \dfrac{1}{5}x + \dfrac{1}{2}y = \dfrac{1}{5} \\ \dfrac{1}{2}x + \dfrac{1}{3}y = -\dfrac{4}{3} \end{cases}$

Solution:

The coefficients of x and y should be integers since it is difficult to work with fractions. Multiply each equation by its LCD to eliminate fractions.

$\dfrac{1}{5}x + \dfrac{1}{2}y = \dfrac{1}{5}$	Multiply this equation by the LCD 10.
$\dfrac{1}{2}x + \dfrac{1}{3}y = -\dfrac{4}{3}$	Multiply this equation by the LCD 6.
$10\left(\dfrac{1}{5}x + \dfrac{1}{2}y\right) = 10\left(\dfrac{1}{5}\right)$	Simplify both sides.
$6\left(\dfrac{1}{2}x + \dfrac{1}{3}y\right) = 6\left(-\dfrac{4}{3}\right)$	Simplify both sides.
$2x + 5y = 2$	To eliminate x, multiply this equation by -3.
$3x + 2y = -8$	To eliminate x, multiply this equation by 2.
$-3(2x + 5y) = -3(2)$	Simplify both sides.
$2(3x + 2y) = 2(-8)$	Simplify both sides.

$$\begin{array}{rl} -6x - 15y = & -6 \\ \underline{6x + 4y = -16} & \quad \text{Add the equations.} \\ -11y = -22 & \quad \text{Divide both sides by } -11. \\ y = 2 & \end{array}$$

Substitute $y = 2$ into either of the original equations to find x. Use $\dfrac{1}{5}x + \dfrac{1}{2}y = \dfrac{1}{5}$.

$\dfrac{1}{5}x + \dfrac{1}{2}y = \dfrac{1}{5}$	Substitute 2 for y.
$\dfrac{1}{5}x + \dfrac{1}{2}(2) = \dfrac{1}{5}$	Multiply $\dfrac{1}{2}$ and 2.
$\dfrac{1}{5}x + 1 = \dfrac{1}{5}$	Subtract 1 from both sides.
$\dfrac{1}{5}x + 1 - 1 = \dfrac{1}{5} - 1$	Add. Write 1 as $\dfrac{5}{5}$.
$\dfrac{1}{5}x = \dfrac{1}{5} - \dfrac{5}{5}$	Add.

$$\frac{1}{5}x = -\frac{4}{5}$$ Multiply both sides by 5.

$$5\left(\frac{1}{5}x\right) = 5\left(-\frac{4}{5}\right)$$ Simplify both sides.

$$x = -4$$ Therefore, the solution of the system is $(-4, 2)$.

Check:

The check is left as an exercise for the student.

ADDITIONAL PRACTICE

Solve the following systems of equations.

e) $y = 2x + 2$
    $x = -3y + 13$

f) $\frac{1}{3}x + \frac{1}{5}y = \frac{1}{3}$
    $x + \frac{1}{2}y = \frac{1}{2}$

PRACTICE EXERCISES

Solve the following systems of equations.

7) $\begin{cases} 5y = 5 - 3x \\ x = -2y + 1 \end{cases}$

8) $\begin{cases} \dfrac{1}{4}x - \dfrac{1}{2}y = -\dfrac{1}{4} \\ \dfrac{1}{3}x + \dfrac{1}{6}y = \dfrac{4}{3} \end{cases}$

If more practice is needed, do the Additional Practice Exercises in the margin.

Remember, if a system is inconsistent, it has no solutions. If a system is dependent, it has an infinite number of solutions. How can we tell if a system is inconsistent or dependent if we are using the addition method and do not have the graphs? Study the following examples.

EXAMPLE 4

Solve the following systems of equations.

a) $\begin{cases} 4x - 2y = 7 \\ 2x - y = 4 \end{cases}$

Solution:

Let us eliminate y. Make one coefficient -2 and the other 2.

$4x - 2y = 7$ Leave $4x - 2y = 7$ unchanged.
$2x - y = 4$ Multiply $2x - y = 4$ by -2.

$4x - 2y = 7$ Leave $4x - 2y = 7$ unchanged.
$-2(2x - y) = -2(4)$ Simplify both sides.

$$\begin{array}{r} 4x - 2y = 7 \\ \underline{-4x + 2y = -8} \\ 0 = -1 \end{array}$$ Add the equations.

This is a false statement which indicates there is no solution. This is denoted by the symbol $\varnothing$. The lines are parallel and the system is inconsistent.

b) $\begin{cases} 4x + 2y = 6 \\ 6x + 3y = 9 \end{cases}$

Solution:

Since the coefficients of y are smaller than the coefficients of x, we will eliminate y. Since the LCM of 2 and 3 is 6, we will make one coefficient of y equal to 6 and the other equal to -6.

$4x + 2y = 6$ Multiply $4x + 2y = 6$ by 3.

$6x + 3y = 9$ Multiply $6x + 3y = 9$ by -2.

$3(4x + 2y) = 3(6)$ Simplify both sides.

$-2(6x + 3y) = -2(9)$ Simplify both sides.

$$\begin{aligned} 12x + 6y &= 18 \\ -12x - 6y &= -18 \\ \hline 0 &= 0 \end{aligned}$$

Add the equations.

This is a true statement for all values of x and y. Hence, there are an infinite number of solutions which means the system is dependent. We indicate the solutions as $\{(x,y): 4x + 2y = 6\}$.

ADDITIONAL PRACTICE

Solve the following systems of equations. Indicate whether the system is consistent, inconsistent, or dependent.

g) $6x + 3y = 5$
    $4x + 2y = 7$

h) $x = -2y + 6$
    $\dfrac{1}{2}x + y = 3$

PRACTICE EXERCISES

Solve the following systems of equations. Indicate whether the system is consistent, inconsistent, or dependent.

9) $\begin{cases} 6x + 2y = 6 \\ y = -3x + 3 \end{cases}$ **10)** $\begin{cases} 12x + 6y = 9 \\ 4x + 2y = 5 \end{cases}$

If more practice is needed, do the Additional Practice Exercises in the margin.

EXERCISE SET   7.2

Solve the following systems of equations using elimination by addition and verify the solution from the graph.

1) $\begin{cases} x - y = 2 \\ x + y = 2 \end{cases}$ **2)** $\begin{cases} y + x = 2 \\ y - x = 8 \end{cases}$

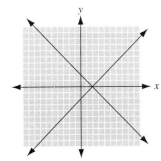   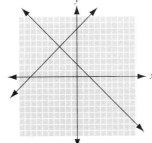

3) $\begin{cases} x + y = 3 \\ 3x - y = 1 \end{cases}$ **4)** $\begin{cases} x + 3y = -9 \\ -x + 2y = -11 \end{cases}$

5) $\begin{cases} 5x + y = -17 \\ 4x - y = -10 \end{cases}$ **6)** $\begin{cases} -x + 6y = -30 \\ x - 4y = 20 \end{cases}$

7) $\begin{cases} 5x + y = -1 \\ x - 3y = -13 \end{cases}$ **8)** $\begin{cases} x - 2y = -4 \\ -2x + 5y = 9 \end{cases}$

9) $\begin{cases} 4x + 5y = 24 \\ -x + 3y = -6 \end{cases}$ **10)** $\begin{cases} 3x - 5y = -17 \\ 4x + y = -15 \end{cases}$

11) $\begin{cases} 7x - 4y = 4 \\ 5x + y = 26 \end{cases}$ **12)** $\begin{cases} 5x - 2y = 0 \\ -x + 4y = 0 \end{cases}$

13) $\begin{cases} 3x + 7y = 8 \\ 2x - 4y = 14 \end{cases}$ **14)** $\begin{cases} 12x - 2y = -54 \\ 13x + 4y = -40 \end{cases}$

15) $\begin{cases} 4x + 5y = -4 \\ 3x + 8y = -20 \end{cases}$ **16)** $\begin{cases} 10x - 3y = 14 \\ -3x + 9y = 12 \end{cases}$

17) $\begin{cases} 3x - 4y = -14 \\ 5x + 7y = 45 \end{cases}$ **18)** $\begin{cases} -5x + 4y = -31 \\ 13x - 6y = 85 \end{cases}$

19) $\begin{cases} 3x + 2y = 8 \\ x + 4y = 1 \end{cases}$ **20)** $\begin{cases} 4x + y = -2 \\ 8x - y = 11 \end{cases}$

21) $\begin{cases} 3x - y = 6 \\ 9x + 2y = -2 \end{cases}$ **22)** $\begin{cases} 14x + 7y = 6 \\ 7x + 6y = 8 \end{cases}$

23) $\begin{cases} 5x + 6y = 2 \\ 10x + 3y = -2 \end{cases}$ **24)** $\begin{cases} 7x - 6y = 1 \\ 8x - 12y = 6 \end{cases}$

25) $\begin{cases} 8x + 3y = -1 \\ 4x - 6y = -3 \end{cases}$ **26)** $\begin{cases} 8x - 18y = -7 \\ 16x + 9y = 11 \end{cases}$

27) $\begin{cases} 13x - 22y = -1 \\ 26x + 33y = 12 \end{cases}$ **28)** $\begin{cases} 30x + 12y = -3 \\ 15x - 24y = 11 \end{cases}$

---

29) $\begin{cases} 14x + 14y = 1 \\ 28x - 7y = 23 \end{cases}$

30) $\begin{cases} 20x + 8y = 4 \\ 10x - 12y = 22 \end{cases}$

31) $\begin{cases} x + 2y = 2 \\ 4x = 3y + 36 \end{cases}$

32) $\begin{cases} 5x = -2y - 1 \\ 3x - y = 6 \end{cases}$

33) $\begin{cases} x + 4y + 16 = 0 \\ 3y = 9 - 6x \end{cases}$

34) $\begin{cases} 4x = 8 - 2y \\ 3x = 6y + 21 \end{cases}$

35) $\begin{cases} \dfrac{1}{2}x - \dfrac{1}{4}y = \dfrac{1}{4} \\ \dfrac{2}{3}x + \dfrac{1}{5}y = \dfrac{7}{15} \end{cases}$

36) $\begin{cases} \dfrac{3}{4}x - \dfrac{2}{7}y = -5 \\ \dfrac{5}{8}x + \dfrac{1}{4}y = -\dfrac{3}{4} \end{cases}$

37) $\begin{cases} \dfrac{3}{2}x + \dfrac{1}{4}y = 1 \\ \dfrac{2}{5}x + \dfrac{1}{3}y = \dfrac{4}{3} \end{cases}$

38) $\begin{cases} \dfrac{1}{2}x - \dfrac{8}{3}y = \dfrac{7}{3} \\ \dfrac{3}{7}x + \dfrac{4}{9}y = \dfrac{88}{63} \end{cases}$

39) $\begin{cases} 0.4x + 0.3y = 4.2 \\ 1.2x + 0.6y = 12 \end{cases}$

40) $\begin{cases} 2.4x + 3y = 7.2 \\ 0.8x + 2.4y = 8 \end{cases}$

41) $\begin{cases} 0.3x - 1.8y = 2.7 \\ 3.6x + 4y = 6.8 \end{cases}$

42) $\begin{cases} 6.7x + 3.2y = 2.6 \\ 0.08x + 0.5y = 2.34 \end{cases}$

Solve the following systems of equations. Indicate whether the system is consistent, inconsistent, or dependent.

43) $\begin{cases} 4x - 2y = 3 \\ -8x + 4y = -6 \end{cases}$

44) $\begin{cases} 3x + y = 1 \\ -4x - 2y = -6 \end{cases}$

45) $\begin{cases} -2x + 5y = 8 \\ x + 4y = 22 \end{cases}$

46) $\begin{cases} x = 2y + 6 \\ 3x = 6y + 9 \end{cases}$

47) $\begin{cases} 2x - 5y = 4 \\ -8x + 20y = -20 \end{cases}$

48) $\begin{cases} 5x + y = 4 \\ 3y = 12 - 15x \end{cases}$

CHALLENGE EXERCISES (49–54)

Solve the following systems of equations.

49) $\begin{cases} 6x + 9y = -3 \\ 5x + 37y = -32 \end{cases}$

50) $\begin{cases} 5x - 7y = -73 \\ -12x + 6y = 132 \end{cases}$

51) $\begin{cases} 3x - 8y = 11 \\ 12x + 10y = 23 \end{cases}$

52) $\begin{cases} 85 = 24x + 24y \\ 55 = 18x + 16y \end{cases}$

53) $\begin{cases} \dfrac{5}{6} = \dfrac{1}{4}x - \dfrac{1}{6}y \\ 3\dfrac{1}{5} = 6x + \dfrac{1}{5}y \end{cases}$

54) $\begin{cases} \dfrac{1}{2}x - 3y = -0.21 \\ -\dfrac{1}{3}x + \dfrac{1}{2}y = 0.11 \end{cases}$

WRITING EXERCISES

55) If the results of applying the addition method is the equation $-2 \neq 0$, what does this mean?

56) If the result of applying the addition method is the equation $4 = 4$, what does this mean?

S E C T I O N   7.3

Solving Systems of Linear Equations Using Substitution

OBJECTIVE

When you complete this section, you will be able to:

Ⓐ   Solve systems of linear equations with two variables using substitution.

INTRODUCTION

In this section we will learn another method for solving linear systems with two variables. This method is based on the principle of substitution which allows us to replace a quantity with any other quantity that is equal to it.

To solve a system of linear equations with two variables using substitution, we will solve one equation for one of its variables, if necessary. This expression will then be substituted for that variable in the other equation because at the point of intersection, these two expressions are equal. This results in an equation with one variable that we can solve using techniques learned previously. We will illustrate the process with some examples and then formalize the technique.

EXAMPLE 1

Solve the following system of equations using substitution.

$$\begin{cases} 2x + y = 7 \\ 3x + 2y = 12 \end{cases}$$

Solution:

Solve $2x + y = 7$ for y since it is the easiest variable to solve for in either equation. Substitute the resulting expression for y in the other equation $3x + 2y = 12$.

$2x + y = 7$	Subtract $2x$ from both sides.
$2x - 2x + y = 7 - 2x$	Add like terms.
$y = 7 - 2x$	Substitute $7 - 2x$ for y in $3x + 2y = 12$.
$3x + 2(7 - 2x) = 12$	Distribute 2.
$3x + 14 - 4x = 12$	Add like terms.
$-x + 14 = 12$	Subtract 14 from both sides.
$-x = -2$	Multiply both sides by -1.
$-1(-x) = -1(-2)$	Perform the multiplications.
$x = 2$	Substitute 2 for x in either of the original equations and solve for y. Use $2x + y = 7$.
$2x + y = 7$	Substitute 2 for x and solve for y.
$2(2) + y = 7$	Perform the multiplication.
$4 + y = 7$	Subtract 4 from both sides.
$y = 3$	Therefore, (2,3) is the solution of the system.

Check:

Substitute 2 for x and 3 for y in both of the original equations.

$2x + y = 7$	$3x + 2y = 12$
$2(2) + 3 = 7$	$3(2) + 2(3) = 12$
$4 + 3 = 7$	$6 + 6 = 12$
$7 = 7$	$12 = 12$

Since (2,3) solves both equations, it is the correct solution.
We summarize the procedure below.

SOLVING SYSTEMS OF LINEAR EQUATIONS USING SUBSTITUTION

1) If necessary, solve one of the equations for one of its variables.

2) Substitute the expression found in step 1 into the other equation for the variable solved for in step 1.

3) Solve the resulting equation.

4) Substitute the value found in step 3 into either of the original equations of the system and solve for the other variable.

5) Check the solution of the system in both of the original equations.

The substitution method is most practical if the coefficient of one of the variables is 1 or -1 since solving for that variable will avoid fractions. If an equation has a variable with a coefficient of 1 or -1, we will solve that equation for that variable. If neither of the variables has a coefficient of one in either equation, then solving for either variable will result in fractions that are not easy to work with.

The substitution method gives us another method of solving systems of linear equations, and this technique is useful in later courses when the systems may not be linear.

EXAMPLE 2

Solve the following systems of equations using substitution.

a) $\begin{cases} 4x + 3y = 10 \\ y = -3x \end{cases}$

Solution:

Since $y = -3x$ is already solved for y, we do not need to solve one of the equations for a variable.

$y = -3x$	Substitute $-3x$ for y in $4x + 3y = 10$.
$4x + 3(-3x) = 10$	Multiply 3 and $-3x$.
$4x - 9x = 10$	Add like terms.
$-5x = 10$	Divide both sides by -5.
$x = -2$	Substitute -2 for x in either of the original equations and solve for y. Use $y = -3x$.
$y = -3x$	Substitute -2 for x.
$y = -3(-2)$	Multiply.
$y = 6$	Therefore, the solution to the system is $(-2,6)$.

Check:

Since the check is done in the same manner as solving using the graphical and elimination by addition methods, the checks will be omitted.

b) $\begin{cases} x + y = -2 \\ 2x + 7y = -29 \end{cases}$

Solution:

We must solve one of the equations for one of its variables. If possible, we would like to avoid fractions. Since the coefficients of both x and y are one in the equation $x + y = -2$, solving for either x or y would avoid fractions.

$x + y = -2$	Solve for x. Subtract y from both sides.
$x + y - y = -2 - y$	Add like terms.
$x = -2 - y$	Substitute $-2 - y$ for x in $2x + 7y = -29$.
$2(-2 - y) + 7y = -29$	Distribute 2.
$-4 - 2y + 7y = -29$	Add like terms.
$-4 + 5y = -29$	Add 4 to both sides.
$5y = -25$	Divide both sides by 5.
$y = -5$	Substitute -5 for y in either of the original equations and solve for x. Use $x + y = -2$.
$x + y = -2$	Substitute -5 for y.
$x + (-5) = -2$	Add 5 to both sides.
$x = 3$	Therefore, the solution is $(3, -5)$.

c) $\begin{cases} 2x + 5y = -4 \\ 3x + y = 20 \end{cases}$

Solution:

The only variable with a coefficient of one is y in $3x + y = 20$. Therefore, we will solve $3x + y = 20$ for y.

$3x + y = 20$	Subtract $3x$ from both sides.
$3x - 3x + y = 20 - 3x$	Add like terms.
$y = 20 - 3x$	Substitute $20 - 3x$ for y in $2x + 5y = -4$.
$2x + 5(20 - 3x) = -4$	Distribute 5.
$2x + 100 - 15x = -4$	Add like terms.
$-13x + 100 = -4$	Subtract 100 from both sides.
$-13x = -104$	Divide both sides by -13.
$x = 8$	Substitute 8 for x in either of the original equations and solve for y. Use $3x + y = 20$.
$3x + y = 20$	Substitute 8 for x.
$3(8) + y = 20$	Multiply 3 and 8.
$24 + y = 20$	Subtract 24 from both sides.
$y = -4$	Therefore, the solution is $(8, -4)$.

ADDITIONAL PRACTICE

Solve the following systems of equations using substitution.

a) $x + 4y = 12$
    $x = 2y$

b) $x - 3y = -26$
    $x + y = 2$

c) $2x + 3y = 1$
    $x + 4y = -7$

PRACTICE EXERCISES

Solve the following systems of equations using substitution.

1) $\begin{cases} 2x + 3y = 8 \\ y = -2x \end{cases}$ **2)** $\begin{cases} 3x + 4y = 13 \\ x + y = 5 \end{cases}$ **3)** $\begin{cases} x - 3y = -5 \\ 4x + 5y = -3 \end{cases}$

If more practice is needed, do the Additional Practice Exercises in the margin.

If neither variable has a coefficient of one in either equation, the substitution method is made more difficult by the introduction of fractions. Examine both equations carefully and try to determine which variable would be best to solve for. As before, substitute the resulting expression into the other equation for that variable. Actually, solving linear equations of this type is usually easier using the addition method of Section 7.2. Substitution is included here because it is often the only method of solving nonlinear systems of equations found in more advanced algebra courses.

EXAMPLE 3

Solve the following systems of equations using substitution.

a) $\begin{cases} 3x + 2y = 4 \\ 2x + 3y = 1 \end{cases}$

Solution:

Neither variable has a coefficient of one in either equation. There seems to be no particular advantage, or disadvantage, in solving either equation for either variable. We have to make a choice, so let us solve $3x + 2y = 4$ for y.

$3x + 2y = 4$	Subtract $3x$ from both sides.
$3x - 3x + 2y = 4 - 3x$	Add like terms.
$2y = 4 - 3x$	Divide both sides by 2.
$\dfrac{2y}{2} = \dfrac{4 - 3x}{2}$	Apply $\dfrac{a + b}{c} = \dfrac{a}{c} + \dfrac{b}{c}$.
$y = 2 - \dfrac{3}{2}x$	Substitute $2 - \dfrac{3}{2}x$ for y in $2x + 3y = 1$.
$2x + 3\left(2 - \dfrac{3}{2}x\right) = 1$	Distribute 3.
$2x + 6 - \dfrac{9}{2}x = 1$	Multiply both sides by the LCD of 2.
$2\left(2x + 6 - \dfrac{9}{2}x\right) = 2(1)$	Distribute on the left. Multiply on right.
$4x + 12 - 9x = 2$	Add like terms.
$-5x + 12 = 2$	Subtract 12 from both sides.
$-5x = -10$	Divide both sides by -5.
$x = 2$	Substitute 2 for x in either of the original equations and solve for y. Use $3x + 2y = 4$.
$3x + 2y = 4$	Substitute 2 for x and solve for y.
$3(2) + 2y = 4$	Multiply 3 and 2.
$6 + 2y = 4$	Subtract 6 from both sides.
$2y = -2$	Divide both sides by 2.
$y = -1$	Therefore, the solution is $(2,-1)$.

b) $\begin{cases} 15x + 6y = 14 \\ 6x + 14y = 23 \end{cases}$

Solution:

Neither variable has a coefficient of one in either equation. Since the coefficient of y in $15x + 6y = 14$ is the smallest coefficient in either equation, we will solve $15x + 6y = 14$ for y. It is as good a reason as any!

$15x + 6y = 14$	Subtract $15x$ from both sides.
$15x - 15x + 6y = 14 - 15x$	Simplify the left side.
$6y = 14 - 15x$	Divide both sides by 6.
$\dfrac{6y}{6} = \dfrac{14 - 15x}{6}$	Apply $\dfrac{a + b}{c} = \dfrac{a}{c} + \dfrac{b}{c}$.
$y = \dfrac{14}{6} - \dfrac{15}{6}x$	Reduce to lowest terms.
$y = \dfrac{7}{3} - \dfrac{5}{2}x$	Substitute $\frac{7}{3} - \frac{5}{2}x$ for y in $6x + 14y = 23$ and solve for x.
$6x + 14\left(\dfrac{7}{3} - \dfrac{5}{2}x\right) = 23$	Distribute 14.
$6x + \dfrac{98}{3} - 35x = 23$	Add like terms.
$-29x + \dfrac{98}{3} = 23$	Multiply the equation by the LCD 3.

$$3\left(-29x + \frac{98}{3}\right) = 3(23)$$ Distribute on the left. Multiply on right.

$$-87x + 98 = 69$$ Subtract 98 from both sides.

$$-87x = -29$$ Divide both sides by -87.

$$x = \frac{29}{87} = \frac{1}{3}$$ Substitute $\frac{1}{3}$ for x in either of the original equations and solve for y. Use $15x + 6y = 14$.

$$15x + 6y = 14$$ Substitute $\frac{1}{3}$ for x.

$$15\left(\frac{1}{3}\right) + 6y = 14$$ Multiply 15 and $\frac{1}{3}$.

$$5 + 6y = 14$$ Subtract 5 from both sides.

$$6y = 9$$ Divide both sides by 6.

$$y = \frac{9}{6} = \frac{3}{2}$$ Therefore, the solution is $\left(\frac{1}{3}, \frac{3}{2}\right)$.

ADDITIONAL PRACTICE

Solve the following systems using the substitution method.

d) $7x - 3y = 34$
    $6x + 5y = 14$

e) $12x + 3y = 8$
    $6x + 6y = 1$

PRACTICE EXERCISES

Solve the following systems using the substitution method.

4) $\begin{cases} 2x + 5y = 19 \\ 4x - 3y = -27 \end{cases}$ **5)** $\begin{cases} 3x + 5y = 8 \\ 3x - 10y = -10 \end{cases}$

If more practice is needed, do the Additional Practice Exercises in the margin.

We recognize inconsistent and dependent systems the same way when using the substitution method as when using the addition method. That is, if we arrive at a false statement, we know the system is inconsistent. If we arrive at a true statement, we know the system is dependent.

EXAMPLE 4

If possible, solve the following systems of equations. If the system is not consistent, indicate whether it is inconsistent or dependent.

a) $\begin{cases} 2x - y = 7 \\ 4x - 2y = -3 \end{cases}$

Solution:

Since the coefficient of y in $2x - y = 7$ is -1, solve $2x - y = 7$ for y.

$$2x - y = 7$$ Subtract $2x$ from both sides.

$$2x - 2x - y = 7 - 2x$$ Add like terms.

$$-y = 7 - 2x$$ Multiply both sides by -1.

$$-1(-y) = -1(7 - 2x)$$ Multiply on the left. Distribute on the right.

$$y = -7 + 2x$$ Substitute $-7 + 2x$ for y in $4x - 2y = -3$.

$$4x - 2(-7 + 2x) = -3$$ Distribute -2.

$$4x + 14 - 4x = -3$$ Add like terms.

$$14 = -3$$ Since this is a false statement, the system is inconsistent and has no solution. Therefore, the solution set is $\varnothing$.

b) $\begin{cases} 3x - y = 4 \\ 2y - 6x = -8 \end{cases}$

Solution:
Since the coefficient of y in $3x - y = 4$ is -1, solve $3x - y = 4$ for y.

$3x - y = 4$	Subtract $3x$ from both sides.
$3x - 3x - y = 4 - 3x$	Add like terms.
$-y = 4 - 3x$	Multiply both sides by -1.
$-1(-y) = -1(4 - 3x)$	Multiply on the left. Distribute on the right.
$y = -4 + 3x$	Substitute $-4 + 3x$ for y in $2y - 6x = -8$.
$2(-4 + 3x) - 6x = -8$	Distribute 2.
$-8 + 6x - 6x = -8$	Add like terms.
$-8 = -8$	Since this is a true statement, the system is dependent and all solutions are common. Therefore, the solutions may be written as $\{(x,y): 3x - y = 4\}$.

PRACTICE EXERCISES

If possible, solve the following systems of equations. If the system is not consistent, indicate whether it is inconsistent or dependent.

6) $\begin{cases} x + 3y = 7 \\ 6y + 2x = -9 \end{cases}$ **7)** $\begin{cases} 3x + 2y = 6 \\ \dfrac{3}{8}x + \dfrac{1}{4}y = \dfrac{3}{4} \end{cases}$

EXERCISE SET   **7.3**

Solve the following system of equations using substitution. Verify your solution using the graph.

1) $\begin{cases} x = 2 \\ 4x + 5y = 3 \end{cases}$ **2)** $\begin{cases} 3x + y = 5 \\ y = 2 \end{cases}$

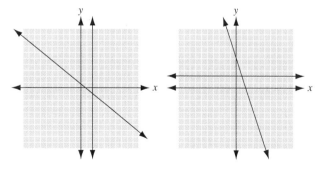

3) $\begin{cases} 3x - y = -2 \\ 5x = y \end{cases}$ **4)** $\begin{cases} 2x - y = 9 \\ x = 2y \end{cases}$

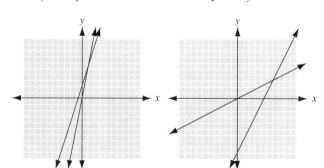

21) $\begin{cases} 5x - 6y = -3 \\ 3x - 5y = 1 \end{cases}$ **22)** $\begin{cases} 2x - y = -4 \\ 4x - 5y = -51 \end{cases}$

If possible, solve the following systems of equations. If the system is not consistent, indicate whether it is inconsistent or dependent.

23) $\begin{cases} x - 2y = 6 \\ -2x + 4y = -12 \end{cases}$ **24)** $\begin{cases} x - 3y = -4 \\ -5x + 15y = 6 \end{cases}$

25) $\begin{cases} 8x + 5y = 11 \\ 2x + y = 5 \end{cases}$ **26)** $\begin{cases} 6x + 2y = 8 \\ 9x + 3y = 12 \end{cases}$

27) $\begin{cases} 3x - 2y = 6 \\ -6x + 4y = 15 \end{cases}$ **28)** $\begin{cases} 2x + 5y = 8 \\ 3x + 4y = -2 \end{cases}$

29) $\begin{cases} 5x + y = 3 \\ 6x - 7y = -62 \end{cases}$ **30)** $\begin{cases} -5x + 3y = 15 \\ 15x - 9y = 30 \end{cases}$

5) $\begin{cases} 4x + y = 5 \\ 3x = -2y \end{cases}$ **6)** $\begin{cases} 3y = -8x \\ 3x - 2y = -25 \end{cases}$

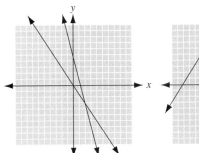   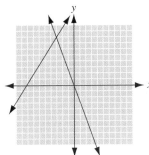

Solve the following system of equations using the substitution method.

7) $\begin{cases} x + y = -5 \\ x - 2y = -2 \end{cases}$ **8)** $\begin{cases} x + 3y = 1 \\ x + y = -3 \end{cases}$

9) $\begin{cases} 2x - 3y = -6 \\ x - y = -1 \end{cases}$ **10)** $\begin{cases} 2x + 5y = -7 \\ x - y = -7 \end{cases}$

11) $\begin{cases} x + y = 6 \\ 2x - y = 3 \end{cases}$ **12)** $\begin{cases} x + y = -10 \\ 4x - 7y = 15 \end{cases}$

13) $\begin{cases} -x + 2y = -12 \\ 2x - 3y = 20 \end{cases}$ **14)** $\begin{cases} -x + 2y = -9 \\ 4x + 5y = -3 \end{cases}$

15) $\begin{cases} x - 2y = -9 \\ 2x + 3y = -4 \end{cases}$ **16)** $\begin{cases} x - 3y = 10 \\ 4x + 5y = 6 \end{cases}$

17) $\begin{cases} 2x - y = -4 \\ -x + 3y = -3 \end{cases}$ **18)** $\begin{cases} x - 3y = -19 \\ 2x + y = -3 \end{cases}$

19) $\begin{cases} 8x - 3y = 10 \\ 4x + 3y = 14 \end{cases}$ **20)** $\begin{cases} 2x + 3y = 1 \\ 5x + 3y = -7 \end{cases}$

CHALLENGE EXERCISES (31–36)

31) $\begin{cases} \dfrac{1}{2}x + y = 1 \\ \dfrac{1}{4}x + y = 0 \end{cases}$ **32)** $\begin{cases} \dfrac{2}{3}x + y = 1 \\ \dfrac{5}{6}x + \dfrac{2}{5}y = -3 \end{cases}$

33) $\begin{cases} 2.1x + 0.5y = 16.6 \\ 4x - 5y = -16 \end{cases}$ **34)** $\begin{cases} 6x - 7y = 14 \\ 4.8x + 3.2y = -6.4 \end{cases}$

35) $\begin{cases} \dfrac{1}{2}x + \dfrac{2}{9}y = 2 \\ 0.8x + 0.03y = 3.2 \end{cases}$ **36)** $\begin{cases} \dfrac{3}{4}x - \dfrac{2}{5}y = 6 \\ 15x - 8y = 120 \end{cases}$

WRITING EXERCISES

37) When solving a system of linear equations by substitution, how do you determine which equation to solve for which variable?

38) If the results of solving a system of linear equations by substitution is a statement like $8 = 8$, which type of system is it?

39) If the results of solving a system of linear equations by substitution is a statement like $0 \neq 8$, which type of system is it?

S E C T I O N **7.4**

Applications Using Systems of Linear Equations

OBJECTIVES

When you complete this section, you will be able to solve the following types of applications problems using systems of linear equations.

A Numbers   **D** Distance, rate, and time
B Geometry formulas   **E** Percent
C Money   **F** Mixture

INTRODUCTION

Previously, applications problems had only one variable. One of the most difficult things in doing applications problems that have more than one unknown is representing all the unknowns in terms of the one variable. Using systems of equations, we can assign a different variable to each unknown and eliminate this problem. Some of the examples and exercises in this section are the same problems as in Chapter 3, except the approach is different. Consequently, when you complete this section you will have a choice of methods to use when solving some types of applications problems.

An important fact to remember is that there must be the same number of equations as there are unknowns. Since the systems we have been solving involve two variables, the following applications problems will have two unknowns and will require two equations. Below is a general procedure.

SOLVING APPLICATIONS PROBLEMS USING SYSTEMS OF EQUATIONS

1) Represent both of the unknowns in the problem with a variable.

2) From the information given in the problem, write two equations.

3) Solve the system of equations using either the addition or the substitution method.

4) Check the solutions in the wording of the original problem.

EXAMPLE 1

One number is twice the other. If the sum of the two numbers is 51, find the numbers.

Solution:

The unknowns in the problem are the two numbers.
Let x represent the smaller number, and
Let y represent the larger number. Then,

$y = 2x$	The larger number is twice the smaller.
$x + y = 51$	The sum of the two numbers is 51.

Since $y = 2x$ is already solved for y, we will use the substitution method.

$y = 2x$	Substitute $2x$ for y in $x + y = 51$.
$x + (2x) = 51$	Add like terms.
$3x = 51$	Divide both sides by 3.
$x = 17$	Therefore, the smaller number is 17. To find the larger number, substitute 17 for x in either of the original equations. Use $y = 2x$.
$y = 2x$	Substitute 17 for x.
$y = 2(17)$	Multiply 2 and 17.
$y = 34$	Therefore, the larger number is 34.

Check:

Is one number twice the other? Yes, since 34 is twice 17. Is the sum of the numbers 51? Yes, since $17 + 34 = 51$. Therefore, our solutions are correct.

1) Find two numbers such that the sum of twice the larger and the smaller is 23 and the difference of the larger and three times the smaller is -6.

EXAMPLE 2

A paint contractor paid $372 for 24 gallons of paint to paint the inside and outside of a house. If the paint for the inside costs $12 per gallon and the paint for the outside costs $18 per gallon, how many gallons of each did he buy?

Solution:

We will use a chart as in Chapter 3 except use two unknowns instead of one.

The unknowns are the number of gallons that he bought for the inside and the number of gallons that he bought for the outside.

Let x represent the number of gallons for the inside, and

Let y represent the number of gallons for the outside.

Type of Paint	Price per Gallon	Number of Gallons	Total Cost
Inside	12	x	$12x$
Outside	18	y	$18y$
Total		24	372

From the table we see that $x + y = 24$ (the number of gallons of inside paint plus the number of gallons of outside paint equals the total number of gallons of paint purchased) and $12x + 18y = 372$ (the cost of the inside paint plus the cost of the outside paint equals the total cost of the paint).

Therefore, we need to solve the system. Let's eliminate x.

$x + y = 24$	Multiply $x + y = 24$ by -12.
$12x + 18y = 372$	Leave unchanged.
$-12(x + y) = -12(24)$	Distribute on the left. Multiply on the right.
$12x + 18y = 372$	Leave unchanged.

$$\begin{aligned} -12x - 12y &= -288 \\ \underline{12x + 18y} &= \underline{372} \\ 6y &= 84 \end{aligned}$$

Add the equations.

Divide both sides by 6.

$$y = 14$$

Therefore, he bought 14 gallons of paint for the outside. Substitute 14 for y in either of the original equations and solve for x. Use $x + y = 24$.

$x + y = 24$	Substitute 14 for y.
$x + 14 = 24$	Subtract 14 from both sides.
$x = 10$	Therefore, he bought 10 gallons of paint for the inside.

Check:

Did he buy a total of 24 gallons? Yes, since $14 + 10 = 24$. Does 10 gallons of inside paint at $12 per gallon and 14 gallons of outside paint at $18 per gallon cost $372? $10(12) + 14(18) = 120 + 252 = 372$. Yes, so our solutions are correct.

PRACTICE EXERCISE

2) A landscape company pays $7 each for rose bushes and $5 each for azaleas. If they paid $82 for 14 plants, how many of each did they buy?

EXAMPLE 3

A collection of 23 coins, made up of nickels and dimes, is worth $1.90. How many of each type of coin are there?

Solution:

Again we use a table as in Chapter 3.
The two unknowns are the number of nickels and the number of dimes.
Let x represent the number of nickels, and
Let y represent the number of dimes.

Type of Coin	Value of Coin	Number of Coins	Total Value
Nickel	5	x	$5x$
Dime	10	y	$10y$
Total		23	190

From the table we see that $x + y = 23$ (the number of nickels + the number of dimes = 23) and $5x + 10y = 190$ (value of the nickels + value of the dimes = $1.90).

We need to solve the system. Let's eliminate x.

$x + y = 23$	Multiply $x + y = 23$ by -5.
$5x + 10y = 190$	Leave this equation unchanged.
$-5(x + y) = -5(23)$	Simplify both sides.
$5x + 10y = 190$	Leave unchanged.
$-5x - 5y = -115$	
$\underline{5x + 10y = 190}$	Add the equations.
$5y = 75$	Divide both sides by 5.
$y = 15$	Therefore, there are 15 dimes. Substitute 15 for y in either of the original equations and solve for x. Use $x + y = 23$.
$x + y = 23$	Substitute 15 for y.
$x + 15 = 23$	Subtract 25 from both sides.
$x = 8$	Therefore, there are 8 nickels.

Check:

Is there a total of 23 coins? Yes, since $15 + 8 = 23$. Is the value of 8 nickels and 15 dimes equal to $1.90? $8(5) + 15(10) = 40 + 150 = 190¢ = \$1.90$. Yes, so our solutions are correct.

PRACTICE EXERCISE

3) A cashier has a total of 30 nickels and quarters in the cash register. If the coins are worth $3.90, how many of each does he have?

EXAMPLE 4

In 4 hours, a boat can go 152 kilometers downstream or 104 kilometers upstream. Find the speed of the boat in still water and the speed of the current.

Solution:

The unknowns are the speed of the boat and the speed of the current.
Let x represent the speed of the boat in still water, and
Let y represent the speed of the current.
Recall in Section 6.8, we used a chart to do distance, time, and rate problems.

	d	r	t
downstream	152	$x + y$	4
upstream	104	$x - y$	4

Since the rate of the boat is x and the rate of the current is y, the rate downstream is $x + y$ and the rate upstream is $x - y$.

Since $d = rt$,

$152 = (x + y)4$ (distance downstream) = (rate downstream)(time downstream) and,

$104 = (x - y)4$ (distance upstream) = (rate upstream)(time upstream)

Simplify the equations and solve the system.

$152 = 4x + 4y$ Add the equations.
$\underline{104 = 4x - 4y}$

$256 = 8x$ Divide both sides by 8.

$32 = x$ Therefore, the rate of the boat in still water is 32 kilometers per hour. Substitute 32 for x in either original equation and solve for y. Use $152 = (x + y)4$.

$152 = (x + y)4$ Substitute 32 for x.

$152 = (32 + y)4$ Distribute on the right side.

$152 = 128 + 4y$ Subtract 128 from both sides.

$24 = 4y$ Divide both sides by 4.

$6 = y$ Therefore, the speed of the current is 6 kilometers per hour.

Check:

Will the boat go 152 kilometers downstream in 4 hours? The rate downstream is the rate of the boat plus the rate of the current which is $32 + 6 = 38$ kilometers per hour. In 4 hours the distance downstream is $4(38) = 152$ kilometers. Will the boat go 104 kilometers upstream in 4 hours? The rate upstream is the rate of the boat minus the rate of the current which is $32 - 6 = 26$ kilometers per hour. In 4 hours the distance upstream is $4(26) = 104$ kilometers. Therefore, our solutions are correct.

PRACTICE EXERCISE

4) A car is traveling at an average rate which is 20 miles per hour more than the rate of a bus. They leave the same town and travel in opposite directions. Find the rate of each if the distance between them is 300 miles after 3 hours.

EXAMPLE 5

Roxanne received an inheritance of $10,000. She invested part of it at 8% and the remainder at 11%. How much did she invest at each rate if the total interest from both investments was $1010 per year?

Solution:

We use a table as before.

The two unknowns are the amount invested at 8% and the amount invested at 11%.

Let x represent the amount invested at 8%, and
Let y represent the amount invested at 11%.

Type of Investment	Interest Rate	Amount Invested	Interest Earned
8%	0.08	x	$0.08x$
11%	0.11	y	$0.11y$
Total		10,000	1010

From the table we see that $x + y = 10{,}000$ (the total amount invested was $10,000) and $0.08x + 0.11y = 1010$ (the interest earned at 8% + the interest earned at 11% = $1010).

Solve the system.

$x + y = 10{,}000$	Leave unchanged.
$0.08x + 0.11y = 1010$	Multiply by 100 to eliminate the decimals.
$x + y = 100$	Leave unchanged.
$100(0.08x + 0.11y) = 100(1010)$	Simplify both sides.
$x + y = 10{,}000$	Multiply $x + y = 10{,}000$ by -8.
$8x + 11y = 101{,}000$	Leave unchanged.
$-8(x + y) = -8(10{,}000)$	Simplify both sides.
$8x + 11y = 101{,}000$	Leave unchanged.

$$
\begin{array}{rcl}
-8x - 8y &=& -80{,}000 \\
\underline{8x + 11y} &=& \underline{101{,}000} \\
3y &=& 21{,}000 \\
y &=& 7000
\end{array}
$$

Add the equations.

Divide both sides by 3.

Therefore, there was $7000 invested at 11%. Substitute 7000 for y in either of the original equations. Use $x + y = 10{,}000$.

$x + y = 10{,}000$	Substitute 7000 for y.
$x + 7000 = 10{,}000$	Subtract 7000 from both sides.
$x = 3000$	Therefore, there was $3000 invested at 7%.

Check:

Is there a total of $10,000 invested? Yes, since $7000 + $3000 = $10,000. Is the total interest earned $1010? The interest earned on $3000 at 8% = 0.08(3000) = $240. The interest earned on $7000 at 11% is 0.11(7000) = $770. Since $240 + $770 = $1010, our solutions are correct.

5) Jerome received $50,000 from the sale of some property. He invested part of it in a certificate of deposit at 9% and the remainder in bonds at 12%. Find the amount invested in each if the total interest received from the two investments is $5400 per year.

EXAMPLE 6

A candy store owner mixes some candy worth $2.50 per pound with some worth $3.50 per pound. How many pounds of each would he need if he wants 10 pounds of the mixture worth $2.90 per pound?

Solution:

The two unknowns are the number of pounds of candy worth $2.50 per pound and the number of pounds of candy worth $3.50 per pound.

Let x represent the number of pounds of candy worth $2.50 per pound, and

Let y represent the number of pounds of candy worth $3.50 per pound.

Type of Candy	Price per Pound	Number of Pounds	Total Value
$2.50 per pound	2.50	x	$2.50x$
$3.50 per pound	3.50	y	$3.50y$
Mixture	2.90	10	$2.90(10) = 29$

From the table see that $x + y = 10$ (The number of pounds of $2.50 candy + the number of pounds of $3.50 candy = the number of pounds in the mixture.) and $2.50x + 3.50y = 29$ (the value of the $2.50 per pound candy) + (the value of the $3.50 per pound candy) = (the value of the mixture).

Sometimes a diagram of this type of problem, referred to as mixture problems, is helpful.

$2.50 per lb		$3.50 per lb		$2.90 per lb
x lb	+	y lb	=	10 lb
value of x lb	+	value of y lb	=	value of mixture
$2.50x$	+	$3.50y$	=	$2.90(10)$

Solve the system.

$x + y = 10$ Leave unchanged.

$2.5x + 3.5y = 29$ Multiply both sides by 10 to eliminate the decimals.

$x + y = 10$ Leave unchanged.

$10(2.5x + 3.5y) = 10(29)$ Simplify both sides.

$x + y = 10$

$25x + 35y = 290$

ANSWER:
Practice 5

5) 20,000 at 9%
   30,000 at 12%

Let us use the substitution method. So solve $x + y = 10$ for y.

$x + y = 10$	Subtract x from both sides.
$x - x + y = 10 - x$	Add like terms.
$y = 10 - x$	Substitute $10 - x$ for y in $25x + 35y = 290$.
$25x + 35(10 - x) = 290$	Distribute 35.
$25x + 350 - 35x = 290$	Add like terms.
$-10x + 350 = 290$	Subtract 350 from both sides.
$-10x = -60$	Divide both sides by -10.
$x = 6$	Therefore, he would need 6 pounds of candy worth $2.50 per pound. Substitute 6 for x in either of the original equations and find y. Use $x + y = 10$.
$6 + y = 10$	Subtract 6 from both sides.
$y = 4$	Therefore, he needs 4 pounds of candy worth $3.50 per pound.

Check:

Does the number of pounds of $2.50 per pound candy plus the number of pounds of $3.50 per pound candy add up to the number of pounds of candy in the mixture? Yes, since $6 + 4 = 10$. Does the value of the $2.50 per pound candy + the value of the $3.50 per pound candy = the value of the mixture? The value of the $2.50 per pound candy is $6(\$2.50) = \$15.00$. The value of the $3.50 per pound candy is $4(\$3.50) = \$14.00$. The value of the mixture is $10(\$2.90) = \$29$. Since $\$15.00 + \$14.00 = \$29.00$, the answer is yes. Therefore, our solutions are correct.

PRACTICE EXERCISE

6) The Smokehouse restaurant specializes in a special barbecue that is a mixture of pork and beef. They need 20 pounds of this mixture for a picnic they are catering. If pork barbecue sells for $3.50 per pound and beef barbecue sells for $5.00 per pound, how many pounds of each will they need if the mixture sells for $4.10 per pound?

EXAMPLE 7

How much of a solution that is 40% alcohol must be added to a solution that is 60% alcohol to obtain 50 liters of a solution that is 55% alcohol?

Solution:

We use a chart as before.

The unknowns are the number of liters of 40% alcohol and the number of liters of 60% alcohol.

Let x represent the number of liters of 40% alcohol and let y represent the number of liters of 60% alcohol.

Type of Solution	Part Pure Alcohol	Volume	Amount Pure Alcohol
40%	0.40	x	$0.40x$
60%	0.60	y	$0.60y$
55% (mixture)	0.55	50	$0.55(50) = 27.5$

From the table we see that $x + y = 50$ (the number of liters of 40% alcohol plus the number of liters of 60% alcohol is equal to the number of liters in the mixture) and $0.40x + 0.60y = 27.5$ (the amount of pure alcohol in the 40% solution plus the amount of pure alcohol in the 60% solution is equal to the amount of pure alcohol in the mixture).

Solve the system

$x + y = 50$	Solve for x.
$0.40x + 0.60y = 27.5$	Leave unchanged.
$x = 50 - y$	Substitute for x in $0.40x + 0.60y = 27.5$.
$0.40(50 - y) + 0.60y = 27.5$	Distribute 0.40.
$20 - 0.40y + 0.60y = 27.5$	Add like terms.
$20 + 0.20y = 27.5$	Subtract 20 from both sides.
$0.20y = 7.5$	Divide both sides by 0.20.
$y = 37.5$	Therefore, 37.5 liters of the 60% solution are needed. To find x, substitute 37.5 for y in the equation $x + y = 50$.
$x + 37.5 = 50$	Subtract 37.5 from both sides.
$x = 12.5$	Therefore, 12.5 liters of the 40% solution are needed.

Check:
The check is left as an exercise for the student.

PRACTICE EXERCISE

7) How many pounds of an alloy that is 30% tin is to be mixed with an alloy that is 70% tin in order to get 100 pounds of an alloy that is 60% tin?

EXERCISE SET 7.4

Solve each of the following applied problems using a system of linear equations.

1) The difference of two numbers is 8. If 5 times the smaller number is 3 times the larger, find the numbers.

2) One positive number is 3 times another positive number. If the difference of the two numbers is 16, find the numbers.

3) The sum of two numbers is 15. If the difference of twice one of the numbers and the other number is 18, find the numbers.

4) The difference of two numbers is 8. If 5 times the first number is 3 times the second number, find the numbers.

5) An electrical contractor paid $1040 for 36 light fixtures to be installed in a hotel lobby. If the first type of fixture costs $25 each and the second type of fixture costs $32 each, how many of each kind did she buy?

6) A locksmith installed inside and outside locks in a new classroom building. The locksmith paid $1216 for 80 locks. If inside locks cost $14 each and outside locks cost $26 each, how many of each kind of lock did he buy?

7) The managers of an office building have decided to reduce the cost of air conditioning by installing overhead fans. They paid $5225 for 105 fans some of which measure 36 inches and the remainder measure 54 inches. If the 36-inch fans cost $45 each and the 54-inch fans cost $65 each, how many of each kind of fan did they buy?

8) An accounting firm purchased a supply of computer disks. They paid $725 for 60 boxes of disks. If the double density disks cost $8 a box and the high density disks cost $15 a box, how many of each kind did they buy?

9) A certain vending machine accepts only dimes and quarters. If the person who services the machine

collects 52 coins with a total value of $8.95, how many of each kind of coin was in the machine?

10) At the end of his shift a cashier has only 5- and 20-dollar bills in his drawer. If he has 48 bills worth a total of $900, how many of each kind of bill does he have?

11) Winston paid $7.46 for 36 stamps. If he bought 8 fewer 25-cent stamps than 18-cent stamps, how many of each kind did he buy?

12) Gayle paid $17.40 for 42 pens at the campus bookstore. If she bought some pens costing $0.35 each and others costing $0.50 each, how many of each kind did she buy?

13) A freight train and a passenger train leave Memphis at the same time traveling in opposite directions. The freight train is traveling 15 miles per hour slower than the passenger train. Find the rate of each train if the distance between them is 525 miles after 5 hours.

14) At 8:00 A.M. two airplanes leave cities which are 2250 miles apart flying toward each other. The rate of one airplane is 150 miles per hour less than twice the rate of the other. If the planes meet at 11:00 A.M., how fast is each plane flying?

15) In 7 hours, a Coast Guard cutter can sail 126 miles with an ocean current or it can sail 84 miles against the current. Find the speed of the cutter in still water and find the speed of the current.

16) An airplane flies 3000 miles coast to coast in 6 hours with a tail wind pushing it. When it flies back against the wind, it takes $7\frac{1}{2}$ hours. What is the speed of the airplane in still air and what is the speed of the wind?

17) Rochelle received a $20,000 bonus for selling over $1,000,000 worth of life insurance. She put part of the money into a savings account paying 6% interest and the rest of the money into municipal bonds paying 8.5% interest. How much did she invest at each rate if the total interest from both investments was $1650 per year?

18) The manager of a supermarket received a $5000 bonus for exceeding the sales quota for her store. She put part of the money into a savings account paying 5% interest and the rest into a certificate of deposit paying 9% interest. How much did she invest at each rate if the total interest from both investments was $370 per year?

19) A college development fund has managed to raise $250,000. They invested part of the money in home mortgages with an expected return of 22% per year and the rest was invested in common stock with an expected return of 28% per year. If the fund earns $61,000 in one year, how much was put into each type of investment?

20) A certain college invests its operating funds in two types of U.S. government securities. The average amount invested is $750,000 which is split between U.S. government guaranteed mortgages at 12% interest and U.S. government treasury bills at 9% interest. If a total of $79,500 is earned per year, how much is invested in each type of security?

21) A specialty store owner mixes peanuts worth $1.80 per pound with cashew nuts worth $5.40 per pound. How many pounds of each type of nut is needed to make a 12-pound mixture worth $3.00 a pound?

22) A coffee company wishes to make a special blend of coffee by mixing coffee beans which cost $3.20 per pound with coffee beans which cost $6.20 per pound. How many pounds of each type of coffee bean are needed to make 40 pounds of coffee that costs $3.95 per pound?

23) A 12% solution of salt is mixed with a 4% solution. How many liters of each is needed to obtain 30 liters of an 8% solution of salt?

24) A 70% solution of alcohol is mixed with a 30% solution. How many gallons of each are needed to obtain 25 gallons of a 60% solution of alcohol?

25) The sum of two numbers is 25. If the difference of these numbers is 5, find the two numbers.

26) One half of a number is equal to three times another number. If the difference of the two numbers is 20, find the numbers.

27) The owners of an extermination business have replaced their fleet of 27 cars and trucks by purchasing new cars for $12,000 each and new trucks for $9500 each. If they paid a total of $294,000, how many of each type of vehicle did they buy?

28) A building contractor has paid $2088 for 54 doors. If inside doors cost $35 each and outside doors cost $46 each, how many of each kind of door did he buy?

29) The metro rail (rapid transit system) accepts only quarters and silver dollars. If at the end of the day 900 coins worth $375 are collected, how many of each type of coin were in the machines?

30) Mr. and Mrs. Wong had a birthday party at the movie theater for their 8-year-old daughter. They paid $89 for 22 people to attend the movie. If a child's ticket is $3.50 and an adult ticket is $5.50, how many children and how many adults attended the party?

31) In 4 hours, a speed boat can go 180 kilometers downstream or 140 kilometers upstream. Find the speed of the boat in still water and the speed of the current.

32) Two ships leave port at 5:00 A.M., both traveling in the same direction. At noon, the ships are 56 miles apart and the sum of the distances they have traveled is 224 miles. How fast is each ship traveling?

33) A book editor received $3000 for signing 6 authors last year. She invested part of the money in a mutual fund paying 18% interest and the remainder in a certificate of deposit paying 9% interest. If she earns $414 per year interest from her investments, how much did she put in each investment?

34) A well-known author received a $160,000 check for his new novel. He decides to invest part of the money in bonds at 8% and part of the money in stocks at a 24% expected return. If at the end of the year he expects to earn $25,600, how much did he put in each investment?

35) Pure rice (100%) and a mixture which is 50% rice and 50% beans are mixed. How many cups of each are needed to get 20 cups of a mixture that is 70% rice and 30% beans?

36) A solution that is 15% baking soda is mixed with a solution that is 5% baking soda. How many liters of each are needed to obtain 12 liters of a 10% solution of baking soda?

37) The length of a rectangular sheet of plywood is 4 feet more than the width. If the perimeter is 32 feet, find the length and width.

38) The length of a rectangular picture is two inches less than twice the width. If the perimeter is 26 inches, find the length and width of the picture.

39) One side of a triangular sign is 8 cm. The length of the smaller of the remaining two sides is 2 cm less than the length of the longer side. If the perimeter is 20 cm, find the lengths of the remaining two sides.

40) The parallel sides of a trapezoid are 6 m and 9 m in length. The length of the longer of the two non-parallel sides is 1 less than twice the length of the shorter side. If the perimeter is 23 m, find the lengths of the non-parallel sides.

41) Recall from Section 1.5 that two angles are supplementary if the sum of their measures is 180. If two angles are supplementary and the measure of the larger is 30 more than the measure of the smaller, find the measure of each angle.

42) Recall from Section 1.5 that two angles are complementary if the sum of their measures is 90. If two angles are complementary and the measure of the smaller is 14 less than the measure of the larger, find the measure of each angle.

WRITING EXERCISE

43) Which do you think is easier—solving applications problems using a system of two variables or solving applications problems using one variable as we did in Chapters 3 and 7? Why?

WRITING EXERCISES OR GROUP PROJECT

If done as a group project, each group should write two exercises of each of the following types, exchange with another group, and then solve.

44) Write and solve a number problem using a system of linear equations.

45) Write and solve a rate, time, distance problem using a system of linear equations.

46) Write and solve a mixture problem using a system of linear equations.

SECTION 7.5

Systems of Linear Inequalities

OBJECTIVE

When you complete this section, you will be able to:

Ⓐ   Graphically represent the solutions of a system of linear inequalities.

INTRODUCTION

In Section 6.3 we learned to graph linear inequalities with two variables. The procedure is given below.

GRAPHING LINEAR INEQUALITIES WITH TWO VARIABLES

1) Graph the equality $ax + by = c$. If the line is part of the solution ($\leq$ or $\geq$), draw a solid line. If the line is not part of the solution ($<$ or $>$), draw a dashed line.

2) Pick a test point. If the coordinates of the test point satisfy the inequality, then all points on the same side of the line as the test point solve the inequality. If the coordinates of the test point do not satisfy the inequality, then all points on the other side of the line from the test point solve the inequality.

3) Shade the region on the side of the line which contains the solutions of the inequality.

Example 1b from Section 6.3 is repeated below to remind you of this procedure.

EXAMPLE 1
Graph $3x - y \leq 6$.

Solution:

Draw the graph of $3x - y = 6$ as a solid line since the points on the line solve the inequality less than or *equal to*. Use the x-intercept $(2,0)$ and the y-intercept $(0,-6)$.

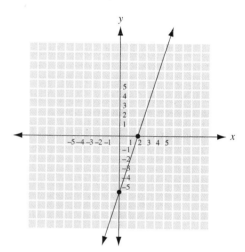

Pick a test point. The origin, $(0,0)$, is a convenient choice. Substitute $(0,0)$ into the inequality. $3(0) - 0 \leq 6$, $0 - 0 \leq 6$, $0 \leq 6$. This is a true statement. Therefore, $(0,0)$ solves the inequality. Consequently, the coordinates of all points on the same side of the line as $(0,0)$ solve the inequality. Therefore, shade the region on the same side of the line as $(0,0)$.

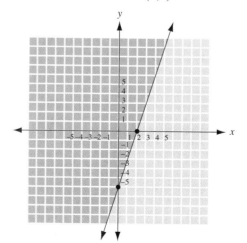

A system of linear inequalities with two variables is much like a system of linear equations with two variables.

> **SYSTEM OF LINEAR INEQUALITIES**
>
> A system of linear inequalities is two or more linear inequalities with the same variables.

To solve a system of linear inequalities, we graph each inequality. The solutions of the system are the coordinates of all points that solve all the inequalities. Consequently, the region where the solutions overlap (intersect) represents the solutions of the system.

EXAMPLE 2

Graph the solutions of the following systems of linear inequalities.

a) $\begin{cases} x + y > 6 \\ x - y < -2 \end{cases}$

Solution:

The solutions of the system are represented by the region where the solutions of $x + y > 6$ and $x - y < -2$ overlap. Therefore, we draw the graphs of $x + y > 6$ and $x - y < -2$ on the same set of axes. First draw the graph of $x + y > 6$.

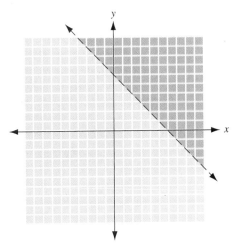

Now draw the graph of $x - y < -2$ on the same set of axes as the graph of $x + y > 6$.

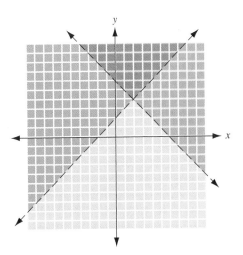

The heavily shaded region where the solutions of $x + y > 6$ and $x - y < -2$ overlap represents the points with coordinates that are the solutions of the system. Since the lines are dotted, the coordinates of the points on the lines are not solutions of the system.

b) $\begin{cases} 2x + 3y \le 12 \\ 2x + y \le 8 \end{cases}$

Solution:

The solutions of the system are represented by the region where the solutions of $2x + 3y \le 12$ and $2x + y \le 8$ overlap. Therefore, draw the graphs of $2x + 3y \le 12$ and $2x + y \le 8$ on the same set of axes.

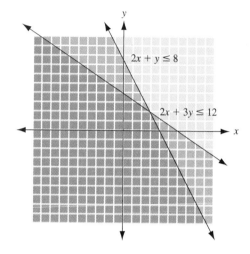

The heavily shaded region where the solutions of $2x + 3y \le 12$ and $2x + y \le 8$ overlap (the intersection of the two sets) represents the points whose coordinates are the solutions of the system. Since the lines are solid, the coordinates of the points on the lines (within the shaded region) are also solutions of the system.

c) $\begin{cases} x + 4y > 8 \\ x \ge 5 \end{cases}$

Solution:

The solutions of the system are represented by the region where the solutions of $x + 4y > 8$ and $x \ge 5$ overlap. Therefore, draw the graphs of $x + 4y > 8$ and $x \ge 5$ on the same set of axes.

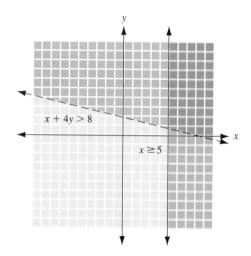

The heavily shaded region where the solutions of $x + 4y > 8$ and $x \ge 5$ overlap represents the points with coordinates that are the solutions of the system. Within the shaded region, the coordinates of the points on the line $x + 4y = 8$ are not solutions of the system but the coordinates of the points on the line $x = 5$ are.

Graph the solutions of the following systems of linear inequalities.

a) $x + y < 3$
$x - y > 5$

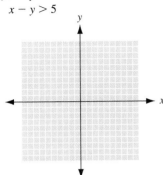

b) $x + 3y \leq 6$
$x - 2y \geq 2$

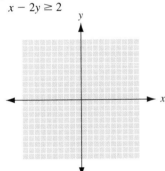

c) $3x - 2y \geq 6$
$x < 6$

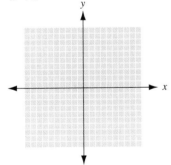

d) $x > 4$
$y < -3$

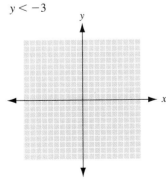

d) $\begin{cases} x \leq -4 \\ y > 3 \end{cases}$

Solution:

The solutions of the system are represented by the region where the solutions of $x \leq -4$ and $y > 3$ overlap. Therefore, draw the graphs of $x \leq -4$ and $y > 3$ on the same set of axes.

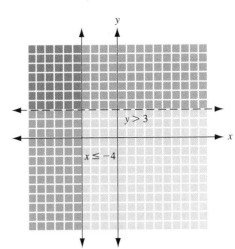

The heavily shaded region where the solutions of $x \leq -4$ and $y > 3$ overlap represents the points with coordinates that are the solutions of the system. Within the shaded region, the coordinates of the points on the line $x = -4$ are solutions of the system but the coordinates of the points on the line $y = 3$ are not.

PRACTICE EXERCISES

Graph the solution of the following systems of linear inequalities.

1) $\begin{cases} x + y > 4 \\ x - y < 6 \end{cases}$

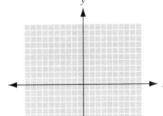

2) $\begin{cases} 2x - y \leq 6 \\ x + 2y \geq 8 \end{cases}$

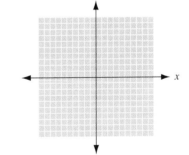

3) $\begin{cases} 3x - 2y \leq 6 \\ y > 5 \end{cases}$

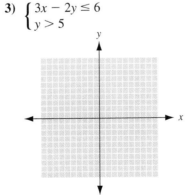

4) $\begin{cases} x \geq 4 \\ y < -3 \end{cases}$

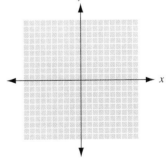

If more practice is needed, do the Additional Practice Exercises in the margin.

1)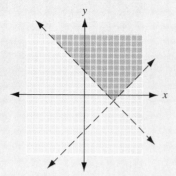

2)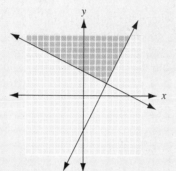

3)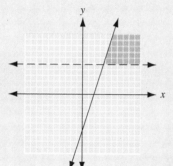

4)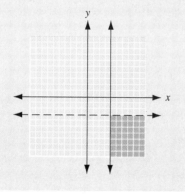

a)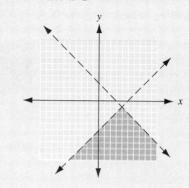

b)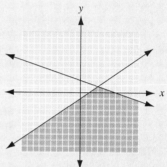

c)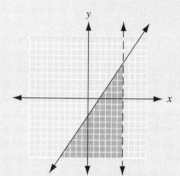

d)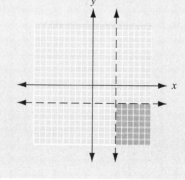

Graph the solutions of the following systems of linear inequalities.

1) $\begin{cases} x - y > 2 \\ x + y > 4 \end{cases}$ **2)** $\begin{cases} x - y < -5 \\ 2x - y < -7 \end{cases}$

3) $\begin{cases} x + y < -2 \\ x - y > 6 \end{cases}$ **4)** $\begin{cases} y - x > 5 \\ x + y < 3 \end{cases}$

5) $\begin{cases} 2x + y \geq -1 \\ x - y \leq 5 \end{cases}$ **6)** $\begin{cases} x + 3y \leq 11 \\ x - 2y \leq 1 \end{cases}$

7) $\begin{cases} 2x + 3y < 1 \\ x - 4y \geq 3 \end{cases}$ **8)** $\begin{cases} 4x - y > -14 \\ 5x + 2y \leq 2 \end{cases}$

9) $\begin{cases} 2x + 5y \leq 7 \\ 3x + y > -9 \end{cases}$ **10)** $\begin{cases} x - 2y > 3 \\ 2x - 4y \leq 20 \end{cases}$

11) $\begin{cases} x + y \geq 4 \\ y \geq 2 \end{cases}$ **12)** $\begin{cases} x - 2y > 0 \\ x < 0 \end{cases}$

13) $\begin{cases} 2x + y \geq 0 \\ x < 3 \end{cases}$ **14)** $\begin{cases} 5x - 6y > -12 \\ y > 2 \end{cases}$

15) $\begin{cases} 5x + 3y < -4 \\ 2y \geq 4 \end{cases}$ **16)** $\begin{cases} 3x - y \geq -13 \\ 4x < 12 \end{cases}$

17) $\begin{cases} x \geq -2 \\ y < 4 \end{cases}$ **18)** $\begin{cases} y \geq -4 \\ x < 3 \end{cases}$

19) $\begin{cases} x < 1 \\ y \geq 0 \end{cases}$ **20)** $\begin{cases} x \geq -2 \\ y \geq 1 \end{cases}$

CHALLENGE EXERCISES (21–22)

21) $\begin{cases} 2x + y \geq 1 \\ x - y \geq -1 \\ x > 2 \end{cases}$ **22)** $\begin{cases} x + y \geq 1 \\ y - x \geq -5 \\ y > -2 \end{cases}$

WRITING EXERCISE

23) Describe the regions that are the solutions of the following systems of inequalities.
   a) $x > 0$ and $y > 0$.   b) $x > 0$ and $y < 0$.

Summary

Definition of a System of Linear Equations:
[Section 7.1]

A system of linear equations is two or more linear equations with the same variables.

Solutions of a System of Linear Equations: [Section 7.1]

The solution(s) of a system of linear equations with two variables consists of all ordered pairs which solve both the equations of the system.

Solving a System of Linear Equations by Graphing:
[Section 7.1]

To solve a system of linear equations with two variables graphically, graph each equation of the system. The coordinates of the point(s) of intersection are the solutions of the system.

Types of Systems of Equations: [Section 7.1]

A system is consistent if the graphs of the equations of the system intersect at a single point. The coordinates of the point of intersection is the solution of the system.

A system is inconsistent if the graphs of the equations of the system are parallel lines. Consequently, the system has no solution. This is usually indicated by the symbol $\varnothing$.

A system is dependent if the graphs of the equations of the system are the same line. There are an infinite number of solutions which are represented by $\{(x,y): ax + by = c\}$ where $ax + by = c$ is either of the equations.

Solving Linear Systems Using Elimination by Addition: [Section 7.2]

1) Choose the variable to be eliminated.

2) If necessary, multiply one or both of the equations of the system by the appropriate constant(s) so the coefficients of the chosen variable will be additive inverses.

3) Add the equations.

4) If necessary, solve the equation resulting from step 3 for the remaining variable.

5) Substitute the value found for the variable in step 4 into either of the equations of the system and solve for the other variable.

6) Check the solution in both equations.

Recognizing Types of Equations When Using Addition:
[Section 7.2]

If the system is consistent, the solution of the system is found.

If the system is inconsistent, a false statement will result when the equations are added.

If the system is dependent, a true statement will result when the equations are added.

Solving Linear Systems Using Substitution:
[Section 7.3]

1) If necessary, solve one of the equations for a chosen variable.

2) Substitute the expression found in step 1 for the chosen variable in the other equation.

3) Solve the resulting equation.

4) Substitute the value found in step 3 into either of the original equations and solve for the remaining variable.

5) Check the solution of the system in both of the original equations.

Recognizing Types of Equations When Using Substitution: [Section 7.3]

Same as when using the addition method.

Solving Applications Problems Using Systems of Equations: [Section 7.4]

1) Represent each of the unknowns in the problem with a variable.

2) From the information given in the problem, write two equations.

3) Solve the system of equations using either the addition or the substitution method.

4) Check the solutions in the wording of the original problem.

Systems of Linear Inequalities: [Section 7.5]

To graphically represent the solutions of a system of linear inequalities, graph all the inequalities of the system on the same coordinate axes. The region where the individual solutions overlap represents all points with coordinates that solve the system.

Review Exercises

Determine whether the given ordered pair is a solution of the system of linear equations. The ordered pair is in the form (x,y). [Section 7.1]

1) $(5,-5)$
$$\begin{cases} 7x - 3y = 36 \\ 4x + 2y = 2 \end{cases}$$

2) $(\frac{2}{3}, -2)$
$$\begin{cases} 3x + 4y = -6 \\ 3x - y = 4 \end{cases}$$

3) $(2,1)$
$$\begin{cases} \frac{1}{2}x + 7y = 8 \\ \frac{1}{4}x + 3y = \frac{7}{2} \end{cases}$$

4) $(-3,4)$
$$\begin{cases} 0.3x + 0.7y = -1.9 \\ x - 0.5y = 5 \end{cases}$$

Find the solution(s) of the following systems of linear equations by graphing. Assume that each unit on the coordinate system represents one. [Section 7.1]

5) $\begin{cases} x = -1 \\ 4x + y = -2 \end{cases}$

6) $\begin{cases} x + y = 1 \\ 2x - 3y = 12 \end{cases}$

7) $\begin{cases} x - y = -1 \\ 2x - 3y = -9 \end{cases}$

8) $\begin{cases} 8x + 5y = 0 \\ 2x + y = -2 \end{cases}$

Determine whether the following systems are consistent, inconsistent, or dependent. If the system is consistent, find the solution of the system. Use the method of your choice. [Sections 7.1–7.3]

9) $\begin{cases} 3x - 7y = 4 \\ 12x - 28y = 16 \end{cases}$

10) $\begin{cases} 3x - 5y = 42 \\ x + y = -2 \end{cases}$

11) $\begin{cases} 6x - 5y = 52 \\ -6x + 2y = -10 \end{cases}$

12) $\begin{cases} -6x + 24 = -9 \\ 2x - 8y = 3 \end{cases}$

Solve the following systems of equations using the elimination by addition method. If the system is not consistent, indicate whether it is inconsistent or dependent. [Section 7.2]

13) $\begin{cases} x - y = 12 \\ x + y = 14 \end{cases}$

14) $\begin{cases} 2x - y = -2 \\ 2x - 3y = -30 \end{cases}$

15) $\begin{cases} 4x + 2y = -4 \\ 5x - 3y = -38 \end{cases}$

16) $\begin{cases} 3x - 7y = -10 \\ 8x - 3y = -11 \end{cases}$

17) $\begin{cases} x - 8y = -5 \\ 3x + 12y = -6 \end{cases}$

18) $\begin{cases} 6x + 6y = 7 \\ 9x - 10y = 20 \end{cases}$

19) $\begin{cases} 12x + 11y = 12 \\ 24x - 22y = 4 \end{cases}$

20) $\begin{cases} x = 1 - 7y \\ 2y = 3x + 20 \end{cases}$

21) $\begin{cases} \frac{1}{2}x - \frac{3}{4}y = -8 \\ \frac{1}{4}x - \frac{3}{8}y = 2 \end{cases}$

22) $\begin{cases} 0.5x + 0.3y = 3 \\ 0.4x - 0.7y = 2.3 \end{cases}$

23) $\begin{cases} 3x - 4y = 2 \\ -15x + 20y = -10 \end{cases}$

24) $\begin{cases} 5x - 6y = 4 \\ 30x - 36y = 12 \end{cases}$

Solve the following systems of equations using the substitution method. If the system is not consistent, indicate whether it is inconsistent or dependent. [Section 7.3]

25) $\begin{cases} x = -2y \\ 6x - 5y = -17 \end{cases}$

26) $\begin{cases} x - y = -2 \\ x - 2y = -7 \end{cases}$

27) $\begin{cases} 2x - y = 2 \\ 8x - 7y = -10 \end{cases}$

28) $\begin{cases} 9x + 4y = 18 \\ 5x - 2y = -28 \end{cases}$

29) $\begin{cases} 4x - 3y = 7 \\ 8x - 3y = -13 \end{cases}$

30) $\begin{cases} 5x - y = 3 \\ 10x - 2y = 5 \end{cases}$

Solve the following systems of linear equations by the most appropriate method. If the system is not consistent, indicate whether it is inconsistent or dependent. [Sections 7.1–7.3]

31) $\begin{cases} x + y = -5 \\ 3x - y = -7 \end{cases}$

32) $\begin{cases} 2x = 1 + y \\ 5x - 4y = -8 \end{cases}$

33) $\begin{cases} 6x - 15y = 9 \\ -2x + 5y = -3 \end{cases}$

34) $\begin{cases} 5x - 2y = 0 \\ 3x + 3y = -21 \end{cases}$

35) $\begin{cases} y = -4x \\ 11x + 5y = 9 \end{cases}$

36) $\begin{cases} 8x - 11y = 7 \\ 16x - 22y = 10 \end{cases}$

37) $\begin{cases} 4x + y = 4 \\ x - 0.5y = -0.5 \end{cases}$

38) $\begin{cases} 8x - 5y = 4 \\ \frac{1}{2}x - y = -\frac{1}{40} \end{cases}$

Solve each of the following using a system of linear equations. [Section 7.4]

39) The difference of two numbers is 7. If the sum of the numbers is 25, find the numbers.

40) A university in the western part of the country changes to a new phone system. They pay a total of $6273 for 425 basic phone sets and 17 executive phone sets. If the cost of a basic phone and an executive phone together is $33, how much did each type of phone set cost?

41) A 9-year-old boy finds that his piggy bank has $5.25 in nickels and dimes. If there are 65 coins in the piggy bank, how many nickels and how many dimes does he have?

42) A motorcyclist and a bicyclist leave Buffalo going in opposite directions. The motorcyclist travels 4 times as fast as the bicyclist. If they are 225 miles apart after 3 hours, how fast is each traveling?

43) A professional athlete received a $50,000 bonus for signing a contract to skate in an ice show. She invested part of the money in a certificate of deposit which pays 12% interest and part of the money in tax-free bonds paying an average of 7.5% interest. If she earns $4200 interest, how much did she put in each investment?

44) Pure antifreeze (100%) and a solution which is 40% antifreeze are mixed. How many quarts of each are needed to obtain 8 quarts of 70% antifreeze solution?

Graph the solutions of the following systems of linear inequalities. [Section 7.5]

45) $\begin{cases} x + y > 5 \\ x - y < -1 \end{cases}$

46) $\begin{cases} x + y \geq 2 \\ x + 2y \leq 6 \end{cases}$

47) $\begin{cases} 2x - y \leq 3 \\ 2x > 6 \end{cases}$

48) $\begin{cases} 3x < 15 \\ y \leq -4 \end{cases}$

49) $\begin{cases} 2x + 3y \leq 4 \\ y + x > 0 \end{cases}$

50) $\begin{cases} y - x \leq 0 \\ x \geq 1 \end{cases}$

Chapter 7 Test

Determine whether the given ordered pair is a solution of the system of linear equations. The ordered pair is in the form (x,y).

1) $(-3,-1)$
$$\begin{cases} 2x + 5y = 1 \\ \frac{1}{3}x - y = 2 \end{cases}$$

Find the solution(s) of the following systems of linear equations by graphing.

2) $\begin{cases} x - y = 8 \\ 2x - y = 12 \end{cases}$

3) $\begin{cases} 4x - 3y = -18 \\ 2x - 3y = -12 \end{cases}$

Determine whether the following system is consistent, inconsistent, or dependent. If the system is consistent, find the solution of the system.

4) $\begin{cases} 4x - 6y = 12 \\ 2x - 3y = 6 \end{cases}$

Solve the following systems of equations using the elimination by addition method. If the system is not consistent, indicate whether it is inconsistent or dependent.

5) $\begin{cases} 2x + 3y = 24 \\ 5x + 4y = 46 \end{cases}$

6) $\begin{cases} 0.9x - 0.2y = 3.1 \\ 11x - 8y = -1 \end{cases}$

7) $\begin{cases} \frac{2}{3}x + \frac{5}{8}y = 3 \\ \frac{5}{6}x - \frac{1}{8}y = -\frac{7}{2} \end{cases}$

Solve the following systems of equations using the substitution method. If the system is not consistent, indicate whether it is inconsistent or dependent.

8) $\begin{cases} x + 2y = -6 \\ 3x + 4y = -10 \end{cases}$

9) $\begin{cases} 4x + y = 17 \\ 3x + 2y = 14 \end{cases}$

10) $\begin{cases} 3x + 4y = 14 \\ 2x - 3y = -19 \end{cases}$

Solve the following systems of linear equations by the most appropriate method. If the system is not consistent, indicate whether it is inconsistent or dependent.

11) $\begin{cases} 3x = y - 5 \\ 4x = 3y + 5 \end{cases}$

12) $\begin{cases} 10x + 5y = -2 \\ 5x + 2y = 0 \end{cases}$

13) $\begin{cases} y - x = 1 \\ 0.8x - 0.7y = -0.4 \end{cases}$

14) $\begin{cases} x = \frac{1}{2}y - 6 \\ \frac{2x}{7} - 3y = 4 \end{cases}$

Solve each of the following using a system of linear equations.

15) The sum of 3 times a number and 8 is equal to another number. Four times the first number equals the difference of the second number and 6. Find the numbers.

16) The machines at a coin laundry take only dimes and quarters. At the end of the day, one washer has 90 coins with a total value of $13.50. How many of each kind of coin were in the washer?

17) Two cities are 600 miles apart. A car and a bus leave the two cities at the same time, traveling toward each other. The car is traveling 10 miles an hour faster than the bus. If they meet after 6 hours, how fast was each vehicle traveling?

18) Christopher has just won $16,000 in a video game contest. His parents put part of the money into a mutual fund paying an expected return of 22% and part of the money into tax-free bonds paying 7% interest. If he earns $2920 in a year in interest, how much was put into each investment?

Graph the solutions of the following systems of linear inequalities.

19) $\begin{cases} x - y \geq 5 \\ 2x + 3y > -3 \end{cases}$

20) $\begin{cases} y - x < 3 \\ x \geq 2 \end{cases}$

8

Roots and Radicals

8.1   Defining and Finding Roots

8.2   Simplifying Radicals

8.3   Products and Quotients of Radicals

8.4   Addition, Subtraction, and Mixed Operations with Radicals

8.5   Rationalizing the Denominator

8.6   Solving Equations with Radicals

8.7   Pythagorean Theorem

CHAPTER OVERVIEW

In Section 1.2 we learned to raise numbers to powers. For example, $2^2 = 4$ and $3^3 = 27$. In this chapter we will learn the "reverse" procedure called finding a root. For example, we will be asked questions like, "The square of what number is 4?" or, "The cube of what number is 27?"

The result of finding a root is not always a rational number. For example, there is no rational number whose square is 3 and there is no rational number whose cube is 4. Thus, we will introduce a new set of numbers, called **irrational numbers**, that will be used for solving equations in Chapter 9.

We know how to perform the operations of addition, subtraction, multiplication, division, and raising to powers on rational numbers. In this chapter we will learn how to perform these same operations on radicals. In addition, we will define the simplest form for a radical and learn to write radicals in this form.

We also know how to solve equations involving rational numbers. In Section 8.5 we will learn to solve equations involving radicals and irrational numbers.

We end the chapter with a very important relationship from geometry involving the sides of a right triangle. Finding one side of a right triangle when given the other two sides often results in an irrational number.

Defining and Finding Roots

OBJECTIVES

When you complete this section, you will be able to:

Ⓐ Find square roots and higher order roots of numbers.
Ⓑ Determine whether a root is rational or irrational.
Ⓒ Find decimal approximations of square and cube roots of irrational numbers using a calculator or a table.

INTRODUCTION

• **Defining roots of numbers**

The operations of addition and subtraction are inverse operations since one "undoes" the other. Begin with a number, say 3. Add any number to 3. From this sum subtract the same number you previously added. The result will be 3. Multiplication and division (except by 0) are also inverse operations.

In Section 1.2 we learned to raise numbers to positive integral powers. In this section we will learn the inverse operation that, with some restrictions, is finding the roots of numbers.

The inverse of squaring a number is finding the square root. A square root of a number is a number that we must square to get the given number. Since $4^2 = 16$, a square root of 16 is 4. Another square root of 16 is -4 since $(-4)^2 = 16$.

The inverse of cubing a number is finding the cube root. We know that $2^3 = 8$. Consequently, the cube root of 8 is 2. Note that 2 is the only real number whose cube is 8, so 2 is the only real cube root of 8.

The inverse of raising a number to the fourth power is finding the fourth root. We know that $3^4 = 81$. So, a fourth root of 81 is 3. Since $(-3)^4 = 81$, -3 is also a fourth root of 81.

Based on these examples, we make the following definitions.

DEFINITION OF ROOTS OF A NUMBER

If the number b is a square root of the number a, then $b^2 = a$.

If the number b is a cube root of the number a, then $b^3 = a$.

If the number b is a fourth root of the number a, then $b^4 = a$.

If the number b is an nth root of the number a, then $b^n = a$.

In words, the square root of a given number is a number whose square equals the given number. The cube root of a given number is a number whose cube equals the given number. The fourth root of a given number is a number whose fourth power equals the given number. In general, the nth root of a given number is a number whose n^{th} power equals the given number.

To indicate the root of a number we use the symbol, $\sqrt{}$, called a **radical sign**. In the expression $\sqrt[n]{a}$, read "the nth root of a," n is called the **root index** and indicates which root we are to find. If no root index is given, we assume it to be 2 which means find the square root. The number a is called the **radicand**. The entire expression is called a **radical** and any expression containing a radical is called a **radical expression**.

Using radicals, we restate the definition given above.

If $\sqrt{a} = b$, then $b^2 = a$. That is, $\sqrt{a}$ is the number whose square is a.

If $\sqrt[3]{a} = b$, then $b^3 = a$. That is, $\sqrt[3]{a}$ is the number whose cube is a.

If $\sqrt[4]{a} = b$, then $b^4 = a$. That is, $\sqrt[4]{a}$ is the number whose fourth power is a.

In general, if $\sqrt[n]{a} = b$, then $b^n = a$. That is, $\sqrt[n]{a}$ is the number whose n^{th} power is a.

Based on the above, if $\sqrt{9} = 3$ then $3^2 = 9$. Also, since $5^2 = 25$, $\sqrt{25} = 5$. If $\sqrt[3]{27} = 3$ then $3^3 = 27$ and since $4^3 = 64$ then $\sqrt[3]{64} = 4$. Similar statements can be made for other powers and roots.

Using radicals, $\sqrt{4}$ is read "the square root of 4," $\sqrt[3]{8}$ is read "the cube root of 8," $\sqrt[4]{16}$ is read "the fourth root of 16," and $\sqrt[5]{32}$ is read "the fifth root of 32."

Previously, we noted that 16 has two square roots, 4 and -4. This is usually written as ± 4. Also, 81 has two fourth roots, ± 3. Consequently, without some agreement, this can lead to confusion as to what the symbols represent. If the root index is even, then $\sqrt[n]{a}$ denotes the non-negative root only and is called the **principal root**. We write the negative even root of a number as $-\sqrt[n]{a}$. We write both the positive and negative even roots as $\pm\sqrt[n]{a}$. Consider the following symbols and their meanings for even values of n.

Symbol	Meaning
$\sqrt{16}$	4. (The positive square root of 16.)
$-\sqrt{16}$	-4. (The negative square root of 16.)
$\pm\sqrt{16}$	± 4. (The positive and negative square roots of 16.)
$\sqrt[4]{81}$	3. (The positive fourth root of 81.)
$-\sqrt[4]{81}$	-3. (The negative fourth root of 81.)
$\pm\sqrt[4]{81}$	± 3. (The positive and negative fourth roots of 81.)
$\sqrt[n]{a}$	The positive n^{th} root of a.
$-\sqrt[n]{a}$	The negative n^{th} root of a.
$\pm\sqrt[n]{a}$	The positive and negative n^{th} roots of a.

It should also be noted that if the root index is even, the radicand must be positive. For example, $\sqrt{-4}$ does not exist as a real number since there is no real number whose square is -4. Also, $\sqrt[4]{-81}$ does not exist as a real number since there is no real number whose fourth power is -81. Numbers representing even roots of negative numbers are called **imaginary numbers** and are usually studied in more advanced math courses.

If the root index is odd, there is no problem with finding roots of any real number. For example, $\sqrt[3]{27} = 3$ and only 3 since 3 is the only real number whose cube is 27. Also, $\sqrt[3]{-27} = -3$ and only -3 since -3 is the only real number whose cube is -27. We summarize as follows.

EXISTENCE OF ROOTS

a) If n is even and $a > 0$, then $\sqrt[n]{a}$ exists as a real number.

b) If n is even and $a < 0$, then $\sqrt[n]{a}$ does not exist as a real number.

c) If n is odd, then $\sqrt[n]{a}$ exists as a real number for all values of a.

EXAMPLE 1

Find the following roots if they exist. If the root does not exist, write "this root does not exist as a real number."

a) $\sqrt{9} = 3$ since 3 is positive and $3^2 = 9$.

b) $-\sqrt{9} = -3$ since -3 is negative and $(-3)^2 = 9$.

c) $\sqrt{-9}$ does not exist as a real number since there is no real number whose square is -9.

d) $\pm\sqrt{9} = \pm 3$ since we want both roots and $(\pm 3)^2 = 9$.

e) $\sqrt[3]{8} = 2$ since $2^3 = 8$.

f) $\sqrt[3]{-8} = -2$ since $(-2)^3 = -8$.

g) $\sqrt[4]{16} = 2$ since 2 is positive and $2^4 = 16$.

ADDITIONAL PRACTICE

Find the following roots if they exist.

a) $\sqrt{64}$ **b)** $-\sqrt{64}$

c) $\sqrt{-64}$ **d)** $\pm\sqrt{64}$

e) $\sqrt[5]{32}$ **f)** $-\sqrt[5]{32}$

g) $\sqrt[5]{-32}$ **h)** $-\sqrt[3]{125}$

• Determining whether a root is rational or irrational

PRACTICE EXERCISES

Find the following roots if they exist. If the root does not exist, write "this root does not exist as a real number."

1) $\sqrt{36}$ 2) $-\sqrt{36}$ 3) $\sqrt{-36}$ 4) $\pm\sqrt{36}$

5) $\sqrt[3]{64}$ 6) $-\sqrt[3]{64}$ 7) $\sqrt[3]{-64}$ 8) $-\sqrt[4]{81}$

If more practice is needed, do the Additional Practice Exercises in the margin.

Only certain numbers have roots that are rational numbers. Remember, a rational number is any number that can be written as the ratio of two integers. As a decimal, a rational number either terminates or repeats.

The numbers 1, 4, 9, 16, 25, 36, . . . which have rational square roots are called **perfect squares**. The numbers ± 1, ± 8, ± 27, ± 64, . . . which have rational cube roots are called **perfect cubes**. Similarly, we have numbers that have rational fourth roots, fifth roots, and beyond. Table 1 in the appendix gives the squares and cubes of the integers from 1 to 100. Consequently, Table 1 gives the perfect square integers from 1 to 10,000 and the perfect cube integers from 1 to 1,000,000.

If a number is not a perfect square, then it is impossible to represent the square root exactly as a terminating or repeating decimal. For example, we cannot represent $\sqrt{3}$ exactly as a decimal since there is no decimal whose square is 3. However, we may approximate $\sqrt{3}$ as closely as we please by using a sufficient number of decimal places. To the nearest tenth, $\sqrt{3} = 1.7$ and $(1.7)^2 = 2.89$. To the nearest hundredth, $\sqrt{3} = 1.73$ and $(1.73)^2 = 2.9929$ which is closer to 3 than $(1.7)^2$, so 1.73 is a better approximation of $\sqrt{3}$ than

• **Finding decimal approximations of irrational square and cube roots**

1.7. To the nearest thousandth, $\sqrt{3} = 1.732$ and $(1.732)^2 = 2.999824$ which is closer to 3 than $(1.73)^2$, so 1.732 is a better approximation of $\sqrt{3}$ than 1.73.

Numbers like $\sqrt{3}$ are called irrational numbers. Irrational numbers cannot be represented by terminating or repeating decimals since terminating and repeating decimals are rational numbers. Hence, irrational numbers are non-repeating, non-terminating decimals. In general, any root of a number that is not rational is irrational. Decimal approximations of irrational roots may be found using a calculator or tables. Table 1 in the appendix gives decimal approximations of square and cube roots of numbers from 1 to 100 and square roots of 10 times any number from 1 to 100. Calculators afford an easy method for approximating irrational roots.

Using Table 1

The headings at the top of Table 1 appear as below.

n	n^2	n^3	$\sqrt{n}$	$\sqrt[3]{n}$	$\sqrt{10n}$

Under the n column are the integers 1–100. The other columns are self-explanatory. For example, if we go down to 8 under the n column, we find the row,

8	64	512	2.828	2.000	8.944

From this row we see that $8^2 = 64$, $8^3 = 512$, $\sqrt{8} = 2.828$, $\sqrt[3]{8} = 2.000$, and $\sqrt{80} = 8.944$ to the nearest thousandth.

We may also use the table in another way. Since 64 is in the n^2 column, $8^2 = 64$ which means $\sqrt{64} = 8$. Likewise, since 512 is in the n^3 column, $8^3 = 512$ which means $\sqrt[3]{512} = 8$. In general, any number in the n column is the square root of the corresponding number in the n^2 column and the cube root of any number in the n^3 column.

EXAMPLE 2

Identify each of the following as rational or irrational. If the number is rational, find the root exactly. If the number is irrational, approximate the number to the nearest thousandth by using a calculator or Table 1. The symbol "≈" means "approximately equal to."

a) $\sqrt{12}$ Irrational. $\sqrt{12} \approx 3.464$.

b) $\sqrt[3]{30}$ Irrational. $\sqrt[3]{30} \approx 3.107$.

c) $\sqrt{81}$ Rational. $\sqrt{81} = 9$.

d) $\sqrt{5184}$ Rational. $5184 = 72^2$, so $\sqrt{5184} = 72$.

e) $\sqrt{9.61}$ Rational. $9.61 = 3.1^2$, so $\sqrt{9.61} = 3.1$.

f) $\sqrt[3]{753571}$ Rational. $753571 = 91^3$, so $\sqrt[3]{753571} = 91$.

g) $\sqrt[3]{68}$ Irrational. $\sqrt[3]{68} \approx 4.082$.

h) $\sqrt{670}$ Irrational. $\sqrt{670} \approx 25.884$. (Look under $\sqrt{10n}$ column where $n = 67$.)

i) $\sqrt{24}$ **j)** $\sqrt[3]{49}$

k) $\sqrt{49}$ **l)** $\sqrt{1156}$

m) $\sqrt{8.41}$ **n)** $\sqrt[3]{2744}$

o) $\sqrt[3]{21}$ **p)** $\sqrt{450}$

PRACTICE EXERCISES

Identify each of the following as rational or irrational. If the number is rational, find the root exactly. If the number is irrational, approximate the number to the nearest thousandth by using a calculator or Table 1.

9) $\sqrt{19}$ **10)** $\sqrt[3]{63}$ **11)** $\sqrt{121}$ **12)** $\sqrt{3364}$

13) $\sqrt{12.25}$ **14)** $\sqrt[3]{262144}$ **15)** $\sqrt[3]{86}$ **16)** $\sqrt{440}$

If more practice is needed, do the Additional Practice Exercises in the margin.

The definition of a root can also be used to find roots with variable radicands. Since even roots of negative radicands do not exist, all variables are assumed to represent non-negative numbers. Recall that $(a^m)^n = a^{mn}$.

EXAMPLE 3

Find the following roots. Assume all variables represent non-negative numbers.

a) $\sqrt{x^2} = (\quad)$ By definition of square root, $(\quad)^2 = x^2$. Since $(x)^2 = x^2$, the $(\quad)$ represents x. Therefore,

$\sqrt{x^2} = x$ Remember, x must be non-negative.

b) $\sqrt{a^4} = (\quad)$ By definition of square root, $(\quad)^2 = a^4$. Since $(a^2)^2 = a^4$, the $(\quad)$ represents a^2. Therefore,

$\sqrt{a^4} = a^2$

c) $\sqrt[3]{y^6} = (\quad)$ By definition of cube root, $(\quad)^3 = y^6$. Since $(y^2)^3 = y^6$, the $(\quad)$ represents y^2. Therefore,

$\sqrt[3]{y^6} = y^2$

Note: You may have noticed that to find the root of a variable radicand, the exponent on the variable must be divisible by the root index. In fact, the exponent on the root is the exponent on the radicand divided by the index. For example, to find $\sqrt[4]{y^8}$, divide 8 by 4. Since $8 \div 4 = 2$, the exponent on the root is 2. Therefore, $\sqrt[4]{y^8} = y^2$. This works because of the definition of multiplication. If $\sqrt[4]{y^8} = y^n$ then $(y^n)^4 = y^8$ which gives $y^{4n} = y^8$. Consequently $4n = 8$ or $n = \frac{8 \text{ (the exponent)}}{4 \text{ (the index)}}$.

q) $\sqrt{r^4}$ **r)** $\sqrt{p^8}$

s) $\sqrt[5]{x^{15}}$

PRACTICE EXERCISES

Find the following roots. Assume all expressions have non-negative values.

17) $\sqrt{z^2}$ **18)** $\sqrt{c^6}$ **19)** $\sqrt[4]{w^{12}}$

If more practice is needed, do the Additional Practice Exercises in the margin.

Often the radicands of radical expressions are not monomials. However, the same procedures apply.

EXAMPLE 4

Find the following roots. Assume all expressions have non-negative values.

a) $\sqrt{(a-2)^2} =$ We need an expression with a square of $(a-2)^2$, which is $a-2$.

$a-2$ Therefore, $\sqrt{(a-2)^2} = a-2$.

b) $\sqrt{(2x-3y)^2} =$ We need an expression with a square of $(2x-3y)^2$, which is $2x-3y$.

$2x-3y$ Therefore, $\sqrt{(2x-3y)^2} = 2x-3y$.

c) $\sqrt{x^2 + 6x + 9} =$ Rewrite $x^2 + 6x + 9$ as $(x+3)^2$.

$\sqrt{(x+3)^2} =$ We need the expression with a square of $(x+3)^2$, which is $x+3$.

$x+3$ Therefore, $\sqrt{x^2 + 6x + 9} = x+3$.

PRACTICE EXERCISES

Find the following roots. Assume all radicands have non-negative values.

20) $\sqrt{(a-2)^2}$ **21)** $\sqrt{(4x-5y)^2}$ **22)** $\sqrt{a^2 + 4a + 4}$

28) $\sqrt{400}$ **29)** $\sqrt[3]{50}$ **30)** $\sqrt[3]{21}$

31) $\sqrt[3]{64}$ **32)** $\sqrt[3]{27}$ **33)** $\sqrt{1.69}$

34) $\sqrt{0.64}$ **35)** $\sqrt{810}$ **36)** $\sqrt{640}$

37) $\sqrt[3]{.125}$ **38)** $\sqrt[3]{.216}$ **39)** $\sqrt[3]{100}$

40) $\sqrt[3]{80}$ **41)** $\sqrt{2116}$ **42)** $\sqrt{3844}$

43) $\sqrt[3]{9261}$ **44)** $\sqrt[3]{24389}$

Find the following roots. Assume all variables represent non-negative numbers.

45) $\sqrt{b^2}$ **46)** $\sqrt{c^2}$ **47)** $\sqrt{n^4}$ **48)** $\sqrt{m^4}$

49) $\sqrt{r^8}$ **50)** $\sqrt{a^{10}}$ **51)** $\sqrt[3]{n^3}$ **52)** $\sqrt[3]{m^3}$

53) $\sqrt[3]{r^6}$ **54)** $\sqrt[3]{a^{12}}$ **55)** $\sqrt[4]{c^4}$ **56)** $\sqrt[4]{d^8}$

57) $\sqrt[5]{s^{10}}$ **58)** $\sqrt[5]{y^{15}}$ **59)** $\sqrt[6]{a^6}$ **60)** $\sqrt[6]{d^{12}}$

Find the following roots. Assume all expressions have non-negative values.

61) $\sqrt{(x-5)^2}$ **62)** $\sqrt{(r+1)^2}$

63) $\sqrt{(2a+b)^2}$ **64)** $\sqrt{(3x-2y)^2}$

65) $\sqrt{x^2 + 10x + 25}$ **66)** $\sqrt{a^2 - 8a + 16}$

67) $\sqrt{4a^2 - 12ab + 9b^2}$ **68)** $\sqrt{9a^2 - 30ab + 15b^2}$

WRITING EXERCISES

69) Why is $\sqrt{-16}$ not a real number?

70) In order for $\sqrt[n]{x}$ to exist as a real number when n is even, x must be non-negative. However, if n is odd, then $\sqrt[n]{x}$ exists for all values of x whether positive or negative. Explain.

EXERCISE SET **8.1**

Find the following roots if they exist. If the root does not exist, write "this root does not exist as a real number."

1) $\sqrt{25}$ **2)** $\sqrt{-25}$ **3)** $-\sqrt{25}$

4) $\pm\sqrt{25}$ **5)** $-\sqrt{100}$ **6)** $\sqrt{100}$

7) $\sqrt{-100}$ **8)** $\pm\sqrt{100}$ **9)** $\sqrt{625}$

10) $-\sqrt{289}$ **11)** $\sqrt{169}$ **12)** $\pm\sqrt{256}$

13) $\sqrt[3]{27}$ **14)** $-\sqrt[3]{27}$ **15)** $\sqrt[3]{-27}$

16) $-\sqrt[3]{216}$ **17)** $\sqrt[3]{-216}$ **18)** $\sqrt[3]{216}$

19) $-\sqrt[3]{-216}$ **20)** $\sqrt[4]{625}$ **21)** $\sqrt[4]{-625}$

22) $\sqrt[5]{243}$ **23)** $-\sqrt[5]{243}$ **24)** $\sqrt[5]{-243}$

Identify each of the following as rational or irrational. If the number is rational, find the root exactly. If the number is irrational, approximate the number to the nearest thousandth by using a calculator or Table 1.

25) $\sqrt{24}$ **26)** $\sqrt{32}$ **27)** $\sqrt{144}$

71) Interview some science instructors, engineering instructors or others who use mathematics in their work. Then make a list of a total of five situations in which radical expressions occur in areas other than mathematics.

S E C T I O N 8.2

Simplifying Radicals

OBJECTIVES

When you complete this section, you will be able to:

Ⓐ   Simplify square roots using the product rule.

Ⓑ   Simplify higher order roots using the product rule.

Ⓒ   Use the alternative rule to simplify radicals.

INTRODUCTION

Fractions are not considered to be in simplest form unless they are reduced to lowest terms. Likewise, radical expressions are not considered to be in simplest form unless they satisfy certain conditions. In order to simplify radicals we need to develop some properties of radicals.

CALCULATOR EXPLORATION ACTIVITY—OPTIONAL

Use your calculator to evaluate items in columns A and B below. Round your answer to the nearest ten thousandth. Compare the results obtained in each line of the columns.

Column A	Column B
a) $\sqrt{2} \cdot \sqrt{3}$ = _____	$\sqrt{6}$ = _____
b) $\sqrt{3} \cdot \sqrt{6}$ = _____	$\sqrt{18}$ = _____
c) $\sqrt{5} \cdot \sqrt{6}$ = _____	$\sqrt{30}$ = _____
d) $\sqrt{20} \cdot \sqrt{30}$ = _____	$\sqrt{600}$ = _____

1) Based on your answers to the corresponding lines in Columns A and B, $\sqrt{a} \cdot \sqrt{b} =$ _____. We justify these results below.

Consider the following. If we find each root and then multiply, we have $\sqrt{9} \cdot \sqrt{16} = 3 \cdot 4 = 12$. If we multiply the radicands and then find the root, we have

$$\sqrt{9} \cdot \sqrt{16} = \sqrt{9 \cdot 16} = \sqrt{144} = 12.$$

Note that we arrived at the same answer using either method. This suggests the following rule for multiplying radicals.

• Multiplying radicals

PRODUCTS OF RADICALS

If $\sqrt[n]{a}$ and $\sqrt[n]{b}$ are both real numbers, then $\sqrt[n]{a} \cdot \sqrt[n]{b} = \sqrt[n]{a \cdot b}$.

In this section, we will be using this rule in the form $\sqrt[n]{ab} = \sqrt[n]{a} \cdot \sqrt[n]{b}$. For example, $\sqrt{4 \cdot 3} = \sqrt{4} \cdot \sqrt{3} = 2\sqrt{3}$.

• **Simplifying radicals**

A radical is in **simplified form** if the radicand contains no factor that can be written to a power greater than or equal to the index. In other words, the radicand of a square root cannot contain a factor that is a perfect square, the radicand of a cube root cannot contain a factor that is a perfect cube, and so on. Consequently, the first step in simplifying a *square* root is to write the radicand as the product of the largest possible perfect square and a number that has no perfect square factors.

Recall from Section 4.1 that any number is divisible by all of its factors. Consequently, if a radicand has a perfect square factor it must be divisible by an integer that is a perfect square. If so, by the definition of division, the number of times that the perfect square divides into the radicand is the other factor of the radicand. Remember, the integers that are perfect squares are 1, 4, 9, 16, 25, 36, ... Since 1 is a factor of every number, we ignore 1 when finding perfect square factors of a radicand.

EXAMPLE 1

Write each of the following as the product of the largest possible perfect square and another number that has no perfect square factors.

a) 27

Solution:

See if 27 is divisible by a perfect square integer. Is 27 divisible by 4? No. Is 27 divisible by 9? Yes, $27 \div 9 = 3$. Since 3 has no perfect square factors, 9 is the largest perfect square factor of 27. Therefore, $27 = 9 \cdot 3$.

b) 150

Solution:

See if 150 is divisible by a perfect square integer. Is 150 divisible by 4? No. Is 150 divisible by 9? No. Is 150 divisible by 16? No. Is 150 divisible by 25? Yes, $150 \div 25 = 6$. Since 6 has no perfect square factors, 25 is the largest perfect square factor of 150. Therefore, $150 = 25 \cdot 6$.

c) 80

Solution:

See if 80 is divisible by a perfect square integer. Is 80 divisible by 4? Yes, $80 \div 4 = 20$. Does 20 have a perfect square factor? Yes, $20 \div 4 = 5$. Consequently, 4 is not the largest perfect square factor of 80. Keep going. Is 80 divisible by 9? No. Is 80 divisible by 16? Yes, $80 \div 16 = 5$. Since 5 has no perfect square factors, 16 is the largest perfect square factor of 80. Therefore, $80 = 16 \cdot 5$.

Note: We can determine the greatest perfect square factor of a number by using a procedure similar to the one we used to write a number in terms of its prime factors. The procedure is as follows.

DETERMINING THE GREATEST PERFECT SQUARE FACTOR

1) Divide the number by any perfect square factor.

2) If possible, divide the resulting quotient by any perfect square factor.

3) Continue dividing the quotients by perfect square factors until you arrive at a quotient with no perfect square factors.

4) The greatest perfect square factor is the product of all the perfect square divisors and the last quotient is the other factor.

Use of the procedure on Example 1c would proceed as follows. Four is a perfect square factor of 80, so divide 4 into 80.

$$\begin{array}{r} 20 \\ 4\overline{\smash)80} \end{array}$$ 20 has a perfect square factor of 4, so divide 4 into 20.

$$\begin{array}{r} 5 \\ 4\overline{\smash)20} \\ 4\overline{\smash)80} \end{array}$$ 5 has no perfect square factors.

Therefore, $80 = 4 \cdot 4 \cdot 5 = 16 \cdot 5$ where 16 is the greatest perfect square factor of 80.

d) 72

Solution:

We will use the procedure above. Four is a perfect square factor of 72, so divide 4 into 72.

$$\begin{array}{r} 18 \\ 4\overline{\smash)72} \end{array}$$ 9 is a perfect square factor of 18, so divide 9 into 18.

$$\begin{array}{r} 2 \\ 9\overline{\smash)18} \\ 4\overline{\smash)72} \end{array}$$ 2 has no perfect square factors.

Therefore, $72 = 4 \cdot 9 \cdot 2 = 36 \cdot 2$ where 36 is the greatest perfect square factor of 72.

PRACTICE EXERCISES

Write each of the following as the product of the largest possible perfect square and another number that has no perfect square factors.

1) 40   **2)** 125   **3)** 96   **4)** 108

The procedure for simplifying *square roots* is outlined below.

SIMPLIFYING SQUARE ROOTS

1) Write the radicand as the product of the largest possible perfect square and a number that has no perfect square factors.

2) Apply the product rule in the form $\sqrt{a \cdot b} = \sqrt{a} \cdot \sqrt{b}$ with a as a perfect square.

3) Find the square root of the perfect square radicand.

EXAMPLE 2

Simplify the following radical expressions.

a) $\sqrt{18} =$ The largest perfect square factor of 18 is 9. Rewrite 18 as $9 \cdot 2$.

$\sqrt{9 \cdot 2} =$ Apply $\sqrt{ab} = \sqrt{a} \cdot \sqrt{b}$.

$\sqrt{9} \cdot \sqrt{2} =$ $\sqrt{9} = 3$.

$3\sqrt{2}$ Therefore, $\sqrt{18} = 3\sqrt{2}$.

b) $\sqrt{72} =$ The largest perfect square factor of 72 is 36. Rewrite 72 as $36 \cdot 2$.

$\sqrt{36 \cdot 2} =$ Apply $\sqrt{ab} = \sqrt{a} \cdot \sqrt{b}$.

$\sqrt{36} \cdot \sqrt{2} =$ $\sqrt{36} = 6$.

$6\sqrt{2}$ Therefore, $\sqrt{72} = 6\sqrt{2}$.

Note: Even though 4 is a perfect square factor of 72, we should not write $\sqrt{72} = \sqrt{4 \cdot 18}$ because 18 has 9 as a perfect square factor. However, we could proceed as follows:

$$\sqrt{72} = \sqrt{4 \cdot 18}$$
$$= \sqrt{4} \cdot \sqrt{18}$$
$$= 2\sqrt{18}$$
$$= 2\sqrt{9 \cdot 2}$$
$$= 2\sqrt{9} \cdot \sqrt{2}$$
$$= 2 \cdot 3\sqrt{2}$$
$$= 6\sqrt{2},$$

and arrive at the same answer, but with more steps. That is why it is preferable to find the largest perfect square factor.

 c) $2\sqrt{80} =$ The largest perfect square factor of 80 is 16. Rewrite 80 as $16 \cdot 5$.

$2\sqrt{16 \cdot 5} =$ Apply $\sqrt{ab} = \sqrt{a} \cdot \sqrt{b}$.

$2\sqrt{16} \cdot \sqrt{5} =$ $\sqrt{16} = 4$.

$2 \cdot 4\sqrt{5} =$ Multiply 2 and 4.

$8\sqrt{5}$ Therefore, $2\sqrt{80} = 8\sqrt{5}$.

Alternative Method (optional)

Sometimes finding the greatest perfect square factor of the radicand, especially if the radicand is a large number can be troublesome. Another method, using the prime factorization of the radicand, is often useful under these circumstances. A perfect square integer can be thought of as an integer that can be expressed as the product of two identical factors. For example, $49 = 7 \cdot 7$ and $121 = 11 \cdot 11$. The square root of a perfect square is one of these two identical factors. For example, $\sqrt{49} = \sqrt{7 \cdot 7} = 7$ which is one of the two identical factors of 49. We can use this idea to simplify a radical by writing the radicand as the product of its prime factors without using exponents. Each factor that appears twice will have a square root equal to one of the two factors. The remaining factors stay underneath the radical sign. We illustrate with some examples.

EXAMPLE 3

Simplify the following radicals using prime factorizations.

a) $\sqrt{72} =$ Write 72 as the product of its prime factors.

$\sqrt{2 \cdot 2 \cdot 2 \cdot 3 \cdot 3} =$ $\sqrt{2 \cdot 2} = 2$ and $\sqrt{3 \cdot 3} = 3$.

$2 \cdot 3\sqrt{2} =$ $2 \cdot 3 = 6$.

$6\sqrt{2}$ Therefore, $\sqrt{72} = 6\sqrt{2}$.

b) $\sqrt{375} =$ Write 375 as the product of its prime factors.

$\sqrt{\underline{5 \cdot 5} \cdot 5 \cdot 3} =$ $\sqrt{5 \cdot 5} = 5$.

$5\sqrt{5 \cdot 3} =$ $3 \cdot 5 = 15$.

$5\sqrt{15}$ Therefore, $\sqrt{375} = 5\sqrt{15}$.

ADDITIONAL PRACTICE

Simplify the following square roots.

a) $\sqrt{20}$ **b)** $3\sqrt{45}$

PRACTICE EXERCISES

Simplify the following square roots.

5) $\sqrt{8}$ **6)** $\sqrt{50}$ **7)** $2\sqrt{54}$ **8)** $5\sqrt{108}$

If more practice is needed, do the Additional Practice Exercises in the margin.

We can apply either procedure for simplifying radicals with radicands that contain variables. Remember, to find the root of a variable expression, the exponent on the variable must be divisible by the root index. Consequently, variable factors that are perfect squares must have even exponents. Therefore, to find the square root, we rewrite the radicand as the product of the variable with the largest possible even exponent and the variable to the first power.

EXAMPLE 4

Simplify the following roots. Assume all variables have non-negative values.

a) $\sqrt{x^3} =$ The largest perfect square factor of x^3 is x^2. Rewrite x^3 as $x^2 \cdot x$.

$\sqrt{x^2 \cdot x} =$ Apply $\sqrt{ab} = \sqrt{a} \cdot \sqrt{b}$.

$\sqrt{x^2} \cdot \sqrt{x} =$ $\sqrt{x^2} = x$.

$x\sqrt{x}$ Therefore, $\sqrt{x^3} = x\sqrt{x}$.

b) $a^2\sqrt{a^7}$ The largest perfect square factor of a^7 is a^6. Rewrite a^7 as $a^6 \cdot a$.

$a^2\sqrt{a^6 \cdot a} =$ Apply $\sqrt{ab} = \sqrt{a} \cdot \sqrt{b}$.

$a^2\sqrt{a^6} \cdot \sqrt{a} =$ $\sqrt{a^6} = a^3$.

$a^2 \cdot a^3\sqrt{a} =$ Multiply $a^2 \cdot a^3$.

$a^5\sqrt{a}$ Therefore, $a^2\sqrt{a^7} = a^5\sqrt{a}$.

c) $\sqrt{c^5 d^3} =$ The largest perfect square factor of c^5 is c^4, and the largest perfect square factor of d^3 is d^2. Rewrite $c^5 d^3$ as $c^4 d^2 \cdot cd$.

$\sqrt{c^4 d^2 \cdot cd} =$ Apply $\sqrt{ab} = \sqrt{a} \cdot \sqrt{b}$.

$\sqrt{c^4 d^2} \cdot \sqrt{cd} =$ Apply $\sqrt{ab} = \sqrt{a} \cdot \sqrt{b}$ to $\sqrt{c^4 d^2}$.

$\sqrt{c^4} \cdot \sqrt{d^2} \cdot \sqrt{cd} =$ $\sqrt{c^4} = c^2$ and $\sqrt{d^2} = d$.

$c^2 d\sqrt{cd}$ Therefore, $\sqrt{c^5 d^3} = c^2 d\sqrt{cd}$.

Note: If we used the alternative method of simplifying square roots, Example 4c would appear as follows.

$\sqrt{c^5 d^3} =$ Write $c^5 d^3$ as the product of its prime factors.

$\sqrt{\underline{c \cdot c} \cdot \underline{c \cdot c} \cdot c \cdot \underline{d \cdot d} \cdot d}$ $\sqrt{c \cdot c} = c$ and $\sqrt{d \cdot d} = d$.

$$c \cdot c \cdot d\sqrt{c \cdot d} =$$ $c \cdot c = c^2.$
$$c^2 d\sqrt{cd} =$$ Therefore, $\sqrt{c^5 d^3} = c^2 d\sqrt{cd}.$

 d) $3n\sqrt{28n^9}$ The largest perfect square factor of 28 is 4 and the largest perfect square factor of n^9 is n^8. Rewrite $28n^9$ and $4n^8 \cdot 7n$.

$$3n\sqrt{4n^8 \cdot 7n} =$$ Apply $\sqrt{ab} = \sqrt{a} \cdot \sqrt{b}.$
$$3n\sqrt{4n^8} \cdot \sqrt{7n} =$$ Apply $\sqrt{ab} = \sqrt{a} \cdot \sqrt{b}$ to $\sqrt{4n^8}.$
$$3n\sqrt{4} \cdot \sqrt{n^8} \cdot \sqrt{7n} =$$ $\sqrt{4} = 2$ and $\sqrt{n^8} = n^4.$
$$3n \cdot 2n^4\sqrt{7n} =$$ Multiply $3n \cdot 2n^4.$
$$6n^5\sqrt{7n}$$ Therefore, $3n\sqrt{28n^9} = 6n^5\sqrt{7n}.$

PRACTICE EXERCISES

Simplify the following square roots.

9) $\sqrt{a^5}$ **10)** $c^2\sqrt{c^{11}}$ **11)** $\sqrt{r^7 s^3}$ **12)** $4b\sqrt{27b^7}$

• **Simplifying higher order roots**

To simplify cube roots, we rewrite the radicand as the product of a perfect cube and some number that has no perfect cube factors. Recall that the integers that are perfect cubes are 1, 8, 27, 64, 125, ... To determine if the radicand has a perfect cube factor, see if the radicand is divisible by any integer that is a perfect cube. For example, 32 has the perfect cube 8 as a factor since $32 \div 8 = 4$. Consequently, $32 = 8 \cdot 4$.

Remember, in finding cube roots of variables, the exponent on the radicand must be divisible by the index 3. So variable factors must be written as the product of the variable to the largest possible power divisible by 3 and the variable to a power smaller than 3.

EXAMPLE 5

Simplify the following radical expressions.

a) $\sqrt[3]{24} =$ The largest perfect cube factor of 24 is 8. Rewrite 24 as $8 \cdot 3.$

$$\sqrt[3]{8 \cdot 3} =$$ Apply the product rule.
$$\sqrt[3]{8} \cdot \sqrt[3]{3} =$$ $\sqrt[3]{8} = 2.$
$$2\sqrt[3]{3}$$ Therefore, $\sqrt[3]{24} = 2\sqrt[3]{3}.$

 b) $5\sqrt[3]{108} =$ The largest perfect cube factor of 108 is 27. Rewrite 108 as $27 \cdot 4.$

$$5\sqrt[3]{27 \cdot 4} =$$ Apply the product rule.
$$5\sqrt[3]{27} \cdot \sqrt[3]{4} =$$ $\sqrt[3]{27} = 3.$
$$5 \cdot 3\sqrt[3]{4} =$$ Multiply 5 and 3.
$$15\sqrt[3]{4}$$ Therefore, $5\sqrt[3]{108} = 15\sqrt[3]{4}.$

c) $\sqrt[3]{x^5} =$ The largest perfect cube factor of x^5 is x^3. Rewrite x^5 as $x^3 \cdot x^2.$

$$\sqrt[3]{x^3 \cdot x^2} =$$ Apply the product rule.
$$\sqrt[3]{x^3} \cdot \sqrt[3]{x^2} =$$ $\sqrt[3]{x^3} = x.$
$$x\sqrt[3]{x^2}$$ Therefore, $\sqrt[3]{x^5} = x\sqrt[3]{x^2}.$

d) $\sqrt[3]{40x^{10}} =$ The largest perfect cube factor of 40 is 8 and the largest perfect cube factor of x^{10} is x^9. Rewrite $40x^{10}$ as $8x^9 \cdot 5x$.

$\sqrt[3]{8x^9 \cdot 5x} =$ Apply the product rule.

$\sqrt[3]{8x^9} \cdot \sqrt[3]{5x} =$ Apply the product rule to $\sqrt[3]{8x^9}$.

$\sqrt[3]{8} \cdot \sqrt[3]{x^9} \cdot \sqrt[3]{5x} =$ $\sqrt[3]{8} = 2$ and $\sqrt[3]{x^9} = x^3$.

$2x^3\sqrt[3]{5x}$ Therefore, $\sqrt[3]{40x^{10}} = 2x^3\sqrt[3]{5x}$.

Alternative Method (Optional)

The alternative method of simplifying radicands is easily extended to roots other than square roots. For example, a perfect cube can be thought of as a number that can be written as the product of three identical factors. For example, $8 = 2 \cdot 2 \cdot 2$. Therefore, the cube root of a perfect cube is one of its three identical factors. For example, $\sqrt[3]{8} = \sqrt[3]{2 \cdot 2 \cdot 2} = 2$ which is one of the three identical factors of 8. Using the alternative method, Example d above would appear as follows.

$\sqrt[3]{40x^{10}}$ Write $40x^{10}$ as the product of its prime factors.

$\sqrt[3]{2 \cdot 2 \cdot 2 \cdot 5 \cdot x \cdot x \cdot x \cdot x \cdot x \cdot x \cdot x \cdot x \cdot x \cdot x}$ $\sqrt[3]{2 \cdot 2 \cdot 2} = 2$ and $\sqrt[3]{x \cdot x \cdot x} = x$.

$2 \cdot x \cdot x \cdot x\sqrt[3]{5x}$ $x \cdot x \cdot x = x^3$.

$2x^3\sqrt[3]{5x}$ Therefore, $\sqrt[3]{40x^{10}} = 2x^3\sqrt[3]{5x}$.

ADDITIONAL PRACTICE

Simplify the following roots.

c) $\sqrt[3]{72}$ d) $4\sqrt[3]{48}$

e) $\sqrt[3]{r^4}$ f) $\sqrt[3]{250m^{14}}$

PRACTICE EXERCISES

Simplify the following radical expressions.

13) $\sqrt[3]{56}$ 14) $6\sqrt[3]{128}$ 15) $\sqrt[3]{a^8}$ 16) $\sqrt[3]{81c^{13}}$

If more practice is needed, do the Additional Practice Exercises in the margin.

EXERCISE SET   8.2

Simplify the following.

1) $\sqrt{12}$ 2) $\sqrt{28}$ 3) $\sqrt{98}$

4) $\sqrt{48}$ 5) $\sqrt{128}$ 6) $\sqrt{180}$

7) $5\sqrt{54}$ 8) $3\sqrt{63}$ 9) $6\sqrt{80}$

10) $4\sqrt{50}$ 11) $\sqrt{a^3}$ 12) $\sqrt{d^5}$

13) $\sqrt{x^2y^4}$ 14) $\sqrt{a^6b^2}$ 15) $\sqrt{x^6y^8z^{10}}$

16) $\sqrt{p^4q^8r^8}$ 17) $c\sqrt{c^3d^6}$ 18) $x\sqrt{x^7y^4}$

19) $rs^2\sqrt{r^9s^5}$ 20) $a^2b^3\sqrt{a^{11}b^3}$ 21) $\sqrt{20x^6}$

22) $\sqrt{45y^4}$ 23) $3\sqrt{72x^5}$ 24) $6\sqrt{75c^5d^3}$

25) $\sqrt[3]{32}$ 26) $\sqrt[3]{54}$ 27) $4\sqrt[3]{40}$

28) $7\sqrt[3]{81}$ 29) $\sqrt[3]{x^7}$ 30) $\sqrt[3]{b^{11}}$

31) $\sqrt[3]{x^6y^5}$ 32) $\sqrt[3]{m^{13}n^9}$ 33) $\sqrt[3]{128z^8}$

34) $\sqrt[3]{48h^{14}}$ 35) $2\sqrt[3]{24}$ 36) $4\sqrt[3]{250}$

CHALLENGE EXERCISES (37–43)

Simplify the following.

37) $\sqrt[4]{162}$ 38) $\sqrt[4]{x^9}$ 39) $\sqrt[4]{a^6b^7}$

40) $\sqrt[5]{64}$ 41) $\sqrt[5]{972}$ 42) $\sqrt[5]{x^{12}}$

43) $\sqrt[5]{a^9b^{16}}$

WRITING EXERCISES

44) When simplifying square root radicals, we wrote the radicand as the product of a perfect square and another number that had no perfect square factors. When simplifying cube roots, we wrote the radicand as the product of a perfect cube and another number with no perfect cube factors. Using this same line of reasoning, how would you rewrite the radicand of a fourth root before simplifying? a fifth root?

45) To find the root of a variable expression, the exponent must be divisible by the root index. If the root index is 4, what exponents would you want on the factor of the radicand whose fourth root you will find? fifth root?

CRITICAL THINKING

46) Write a procedure similar to that for finding the largest perfect square factor following Example 1c for finding the largest perfect cube factor.

47) Use $a = 6$ and $b = 8$ to answer the following.

a) Does $\sqrt{a^2 + b^2} = \sqrt{a^2} + \sqrt{b^2} = a + b$?

b) Does $\sqrt{(a^2 + b^2)} = a + b$?

SECTION 8.3

Products and Quotients of Radicals

OBJECTIVES

When you complete this section, you will be able to:

Ⓐ Find the products of radicals and simplify the results.

Ⓑ Find the quotients of radicals and simplify the results.

Ⓒ Simplify radical expressions using both the product and quotient rules.

INTRODUCTION

We know how to perform the operations of addition, subtraction, multiplication, and division on rational numbers, but how do you perform these operations on radical expressions which are irrational numbers? In this section we will multiply and divide radicals. In Section 8.4 we will add and subtract radical expressions.

We learned how to multiply radicals in Section 8.2. For reference, the product rule for radicals is repeated below.

• Multiplying radicals

PRODUCTS OF RADICALS

If $\sqrt[n]{a}$ and $\sqrt[n]{b}$ are both real numbers, then $\sqrt[n]{a} \cdot \sqrt[n]{b} = \sqrt[n]{a \cdot b}$.

In this section we will be using this rule in the form $\sqrt[n]{a} \cdot \sqrt[n]{b} = \sqrt[n]{ab}$.

EXAMPLE 1

Find the following products of radicals. Assume all variables have non-negative values.

a) $\sqrt{3} \cdot \sqrt{5}$ Apply $\sqrt{a} \cdot \sqrt{b} = \sqrt{ab}$.

    $\sqrt{3 \cdot 5}$ Multiply 3 and 5.

    $\sqrt{15}$ Therefore, $\sqrt{3} \cdot \sqrt{5} = \sqrt{15}$.

b) $\sqrt{2} \cdot \sqrt{8} =$ Apply $\sqrt{a} \cdot \sqrt{b} = \sqrt{ab}$.

     $\sqrt{2 \cdot 8} =$ Multiply 2 and 8.

     $\sqrt{16} =$ Find the $\sqrt{16}$.

     4 Therefore, $\sqrt{2} \cdot \sqrt{8} = 4$.

c) $\sqrt{7} \cdot \sqrt{x} =$ Apply $\sqrt{a} \cdot \sqrt{b} = \sqrt{ab}$.

     $\sqrt{7x}$ Therefore, $\sqrt{7} \cdot \sqrt{x} = \sqrt{7x}$.

d) $\sqrt{x} \cdot \sqrt{x} =$ Apply $a \cdot b = ab$.

     $\sqrt{x \cdot x} =$ Multiply x and x.

     $\sqrt{x^2} =$ Find the $\sqrt{x^2}$.

     x Therefore, $\sqrt{x} \cdot \sqrt{x} = x$.

ADDITIONAL PRACTICE

Find the following roots. Assume all variables have non-negative values.

a) $\sqrt{6} \cdot \sqrt{5}$

b) $\sqrt{2} \cdot \sqrt{32}$

c) $\sqrt{7} \cdot \sqrt{p}$

d) $\sqrt{q} \cdot \sqrt{q}$

PRACTICE EXERCISES

Find the following products of radicals. Assume all variables have non-negative values.

1) $\sqrt{3} \cdot \sqrt{7}$ **2)** $\sqrt{3} \cdot \sqrt{27}$ **3)** $\sqrt{5} \cdot \sqrt{a}$ **4)** $\sqrt{r} \cdot \sqrt{r}$

If more practice is needed, do the Additional Practice Exercises in the margin.

Often it is necessary to simplify products of radicals. If the radical expressions have coefficients, multiply the coefficients and then multiply the radicals in much the same manner that terms containing variables are multiplied.

EXAMPLE 2

Find the following products. Leave answers in simplified form. Remember, you may use the alternative method to simplify the radicals.

a) $\sqrt{2} \cdot \sqrt{6} =$ Apply the product rule in the form $a \cdot b = ab$.

     $\sqrt{2 \cdot 6} =$ Multiply.

     $\sqrt{12} =$ 12 has a perfect square factor of 4. Rewrite 12 as $4 \cdot 3$.

     $\sqrt{4 \cdot 3} =$ Apply the product rule in the form $\sqrt{ab} = \sqrt{a} \cdot \sqrt{b}$.

     $\sqrt{4} \cdot \sqrt{3} =$ $\sqrt{4} = 2$.

     $2\sqrt{3}$ Therefore, $\sqrt{2} \cdot \sqrt{6} = 2\sqrt{3}$.

b) $5\sqrt{3} \cdot 3\sqrt{15} =$ Regroup and multiply 5 and 3, and $\sqrt{3}$ and $\sqrt{15}$.

     $(5 \cdot 3)(\sqrt{3} \cdot \sqrt{15}) =$ Multiply.

     $15\sqrt{45} =$ 45 has a perfect square factor of 9. Rewrite 45 as $9 \cdot 5$.

     $15\sqrt{9 \cdot 5} =$ Apply the product rule in the form $\sqrt{ab} = \sqrt{a} \cdot \sqrt{b}$.

     $15\sqrt{9} \cdot \sqrt{5} =$ $\sqrt{9} = 3$.

     $15 \cdot 3\sqrt{5} =$ Multiply 15 and 3.

     $45\sqrt{5}$ Therefore, $5\sqrt{3} \cdot 3\sqrt{15} = 45\sqrt{5}$.

c) $\sqrt{x^3} \cdot \sqrt{x^5} =$ Apply the product rule in the form $\sqrt{a} \cdot \sqrt{b} = \sqrt{ab}$.

     $\sqrt{x^3 \cdot x^5} =$ $x^3 \cdot x^5 = x^{3+5} = x^8$.

     $\sqrt{x^8} =$ Find the root.

     x^4 Therefore, $\sqrt{x^3} \cdot \sqrt{x^5} = x^4$.

d) $2\sqrt{5a^2} \cdot 3\sqrt{10a^5} =$

$(2 \cdot 3)(\sqrt{5a^2} \cdot \sqrt{10a^5}) =$

$2 \cdot 3\sqrt{5a^2 \cdot 10a^5} =$

$6\sqrt{50a^7}$

Regroup and multiply 2 and 3 and $\sqrt{5a^2}$ and $\sqrt{10a^5}$.

Apply $\sqrt{a} \cdot \sqrt{b} = \sqrt{ab}$

Multiply.

The largest perfect square factor of 50 is 25, and the largest perfect square factor of a^7 is a^6. Rewrite $50a^7$ as $25a^6 \cdot 2a$.

$6\sqrt{25a^6 \cdot 2a} =$

$6\sqrt{25a^6} \cdot \sqrt{2a} =$

$6 \cdot 5a^3\sqrt{2a} =$

$30a^3\sqrt{2a}$

Apply the product rule in the form $\sqrt{ab} = \sqrt{a} \cdot \sqrt{b}$.

Using the product rule, $\sqrt{25a^6} = 5a^3$.

Multiply 6 and 5.

Therefore, $2\sqrt{5a^2} \cdot 3\sqrt{10a^5} = 30a^3\sqrt{2a}$.

ADDITIONAL PRACTICE

Find the following products of radicals. Leave answers in simplified form.

e) $\sqrt{6} \cdot \sqrt{15}$

f) $4\sqrt{14} \cdot 2\sqrt{6}$

g) $\sqrt{a} \cdot \sqrt{a^3}$

h) $5\sqrt{10m^2} \cdot \sqrt{15m^5}$

PRACTICE EXERCISES

Find the following products. Leave answers in simplified form. Assume all variables have non-negative values.

5) $\sqrt{3} \cdot \sqrt{21}$ 6) $4\sqrt{6} \cdot 7\sqrt{15}$

7) $\sqrt{a^5} \cdot \sqrt{a^5}$ 8) $4\sqrt{10c} \cdot 2\sqrt{6c^4}$

If more practice is needed, do the Additional Practice Exercises in the margin.

The product rule for radicals can be used to develop a very important property of radicals. In Example 1d, we found that the product $\sqrt{x} \cdot \sqrt{x} = x$. By the definition of an exponent, $\sqrt{x} \cdot \sqrt{x} = (\sqrt{x})^2$. Consequently, $(\sqrt{x})^2 = x$. Also, $\sqrt[3]{x} \cdot \sqrt[3]{x} \cdot \sqrt[3]{x} = \sqrt[3]{x^3} = x$. But $\sqrt[3]{x} \cdot \sqrt[3]{x} \cdot \sqrt[3]{x} = (\sqrt[3]{x})^3$ also. Consequently, $(\sqrt[3]{x})^3 = x$. This leads to the following generalization.

POWERS OF ROOTS

If $\sqrt[n]{a}$ is a real number, then $(\sqrt[n]{a})^n = a$.

EXAMPLE 3

Find the following. Assume all variables have non-negative values.

a) $(\sqrt{2})^2 = 2$

b) $(\sqrt[3]{a})^3 = a$

c) $(\sqrt[4]{x})^4 = x$

d) $(\sqrt[5]{9})^5 = 9$

PRACTICE EXERCISES

Find the following. Assume all variables have non-negative values.

9) $(\sqrt{5})^2$ 10) $(\sqrt[4]{y})^4$

A rule similar to the product rule can be established for quotients.

CALCULATOR ACTIVITY—OPTIONAL

Use your calculator to evaluate columns A and B below and round your answer off to the nearest ten-thousandth. Compare the results obtained in each line.

Column A   Column B

a) $\dfrac{\sqrt{10}}{\sqrt{2}} = $ _____ $\sqrt{5} = $ _____

b) $\dfrac{\sqrt{18}}{\sqrt{6}} = $ _____ $\sqrt{3} = $ _____

c) $\dfrac{\sqrt{48}}{\sqrt{8}} = $ _____ $\sqrt{6} = $ _____

d) $\dfrac{\sqrt{42}}{\sqrt{6}} = $ _____ $\sqrt{7} = $ _____

1) Based on your answers to the corresponding lines of Columns A and B, $\dfrac{\sqrt{a}}{\sqrt{b}} = $ _____. We justify this result below.

• Dividing radicals

Consider the following. Simplify $\dfrac{\sqrt{36}}{\sqrt{4}}$ by first evaluating the numerator and the denominator: $\dfrac{\sqrt{36}}{\sqrt{4}} = \dfrac{6}{2} = 3$. If we divide the radicands, we get $\dfrac{\sqrt{36}}{\sqrt{4}} = \sqrt{\dfrac{36}{4}} = \sqrt{9} = 3$. Note that we arrived at the same answer. Consequently, $\dfrac{\sqrt{36}}{\sqrt{4}} = \sqrt{\dfrac{36}{4}}$. This suggests the following rule for quotients of radicals.

QUOTIENTS OF RADICALS

If $\sqrt[n]{a}$ and $\sqrt[n]{b}$ both exist and $b \neq 0$, then $\dfrac{\sqrt[n]{a}}{\sqrt[n]{b}} = \sqrt[n]{\dfrac{a}{b}}$.

Like the product rule, this rule may be applied in either order as $\dfrac{\sqrt[n]{a}}{\sqrt[n]{b}} = \sqrt[n]{\dfrac{a}{b}}$ or as $\sqrt[n]{\dfrac{a}{b}} = \dfrac{\sqrt[n]{a}}{\sqrt[n]{b}}$.

EXAMPLE 4

Find the following quotients. Assume all variables have positive values.

a) $\dfrac{\sqrt{28}}{\sqrt{7}} = $ Apply $\dfrac{\sqrt{a}}{\sqrt{b}} = \sqrt{\dfrac{a}{b}}$.

$\sqrt{\dfrac{28}{7}} = $ Divide.

$\sqrt{4} = $ Simplify.

2 Therefore, $\dfrac{\sqrt{28}}{\sqrt{7}} = 2$.

---

b) $\sqrt{\dfrac{4}{9}} =$ Apply $\sqrt{\dfrac{a}{b}} = \dfrac{\sqrt{a}}{\sqrt{b}}$.

$\dfrac{\sqrt{4}}{\sqrt{9}} =$ Simplify the numerator and the denominator.

$\dfrac{2}{3}$ Therefore, $\sqrt{\dfrac{4}{9}} = \dfrac{2}{3}$.

c) $\dfrac{\sqrt{32x^5}}{\sqrt{2x}} =$ Apply $\dfrac{\sqrt{a}}{\sqrt{b}} = \sqrt{\dfrac{a}{b}}$.

$\sqrt{\dfrac{32x^5}{2x}} =$ Divide.

$\sqrt{16x^4} =$ Simplify.

$4x^2$ Therefore, $\dfrac{\sqrt{32x^5}}{\sqrt{2x}} = 4x^2$.

d) $\dfrac{\sqrt{81x^4y^3}}{\sqrt{3xy^2}} =$ Apply $\dfrac{\sqrt{a}}{\sqrt{b}} = \sqrt{\dfrac{a}{b}}$.

$\sqrt{\dfrac{81x^4y^3}{3xy^2}} =$ Divide.

$\sqrt{27x^3y}$ The largest perfect square factor of 27 is 9, and the largest perfect square factor of x^3 is x^2. Rewrite $27x^3y$ as $9x^2 \cdot 3xy$.

$\sqrt{9x^2 \cdot 3xy} =$ Apply $\sqrt{ab} = \sqrt{a} \cdot \sqrt{b}$.

$\sqrt{9x^2} \cdot \sqrt{3xy} =$ $\sqrt{9x^2} = 3x$.

$3x\sqrt{3xy}$ Therefore, $\dfrac{\sqrt{81x^4y^3}}{\sqrt{3xy^2}} = 3x\sqrt{3xy}$.

ADDITIONAL PRACTICE

Find the following quotients.

i) $\dfrac{\sqrt{45}}{\sqrt{5}}$ **j)** $\sqrt{\dfrac{64}{25}}$

k) $\dfrac{\sqrt{54x^7}}{\sqrt{6x^3}}$ **l)** $\dfrac{\sqrt{32m^4n^3}}{\sqrt{4m^2n^2}}$

- **Simplifying expressions using both the product and quotient rules**

PRACTICE EXERCISES

Find the following quotients. Assume all variables have positive values.

11) $\dfrac{\sqrt{75}}{\sqrt{3}}$ **12)** $\sqrt{\dfrac{16}{25}}$ **13)** $\dfrac{\sqrt{48x^3}}{\sqrt{12x}}$ **14)** $\dfrac{\sqrt{16a^5b^4}}{\sqrt{2a^2b^2}}$

If more practice is needed, do the Additional Practice Exercises in the margin.

We can combine the product and quotient rules to simplify radical expressions involving both multiplication and division.

EXAMPLE 5

Simplify the following.

a) $\sqrt{\dfrac{2}{3}} \cdot \sqrt{\dfrac{10}{3}} =$ Apply the product rule.

$\sqrt{\dfrac{2}{3} \cdot \dfrac{10}{3}} =$ Multiply the fractions.

$$\sqrt{\dfrac{20}{9}} =$$ Apply the quotient rule.

$$\dfrac{\sqrt{20}}{\sqrt{9}} =$$ Rewrite 20 as $4 \cdot 5$ and $\sqrt{9} = 3$.

$$\dfrac{\sqrt{4 \cdot 5}}{3} =$$ Apply the product rule.

$$\dfrac{\sqrt{4} \cdot \sqrt{5}}{3} =$$ $\sqrt{4} = 2.$

$$\dfrac{2\sqrt{5}}{3}$$ Therefore, $\sqrt{\dfrac{2}{3}} \cdot \sqrt{\dfrac{10}{3}} = \dfrac{2\sqrt{5}}{3}.$

Note: We could also have applied the quotient rule first and then the product rule and done Exercise 5a as follows:

$$\sqrt{\dfrac{2}{3}} \cdot \sqrt{\dfrac{10}{3}} = \dfrac{\sqrt{2}}{\sqrt{3}} \cdot \dfrac{\sqrt{10}}{\sqrt{3}}$$

$$= \dfrac{\sqrt{2 \cdot 10}}{\sqrt{3 \cdot 3}}$$

$$= \dfrac{\sqrt{20}}{\sqrt{9}}$$

$$= \dfrac{\sqrt{4 \cdot 5}}{3}$$

$$= \dfrac{\sqrt{4} \cdot \sqrt{5}}{3}$$

$$= \dfrac{2\sqrt{5}}{3}.$$

b) $\sqrt{\dfrac{x}{5}} \cdot \sqrt{\dfrac{x^3}{125}} =$ Apply the product rule.

$$\sqrt{\dfrac{x}{5} \cdot \dfrac{x^3}{125}} =$$ Multiply the rational expressions.

$$\sqrt{\dfrac{x^4}{625}} =$$ Apply the quotient rule.

$$\dfrac{\sqrt{x^4}}{\sqrt{625}} =$$ $\sqrt{x^4} = x^2$ and $\sqrt{625} = 25.$

$$\dfrac{x^2}{25}$$ Therefore, $\sqrt{\dfrac{x}{5}} \cdot \sqrt{\dfrac{x^3}{125}} = \dfrac{x^2}{25}.$

PRACTICE EXERCISES

Simplify the following.

15) $\sqrt{\dfrac{3}{5}} \cdot \sqrt{\dfrac{6}{5}}$ 16) $\sqrt{\dfrac{2}{a}} \cdot \sqrt{\dfrac{8}{a}}$

EXERCISE SET **8.3**

Find the following products or powers of roots. Leave all answers in simplified form. Assume all variables have non-negative values.

1) $\sqrt{6} \cdot \sqrt{5}$
2) $\sqrt{6} \cdot \sqrt{7}$
3) $\sqrt{7} \cdot \sqrt{5}$
4) $\sqrt{3} \cdot \sqrt{13}$
5) $\sqrt{2} \cdot \sqrt{18}$
6) $\sqrt{2} \cdot \sqrt{50}$
7) $\sqrt{3} \cdot \sqrt{75}$
8) $\sqrt{27} \cdot \sqrt{3}$
9) $\sqrt{5} \cdot \sqrt{x}$
10) $\sqrt{11} \cdot \sqrt{c}$
11) $\sqrt{15} \cdot \sqrt{y}$
12) $\sqrt{13} \cdot \sqrt{k}$
13) $\sqrt{x} \cdot \sqrt{y}$
14) $\sqrt{a} \cdot \sqrt{b}$
15) $\sqrt{k} \cdot \sqrt{k}$
16) $\sqrt{m} \cdot \sqrt{m}$
17) $\sqrt{x} \cdot \sqrt{x^3}$
18) $\sqrt{s^3} \cdot \sqrt{s}$
19) $(\sqrt{w})^2$
20) $(\sqrt{z})^2$
21) $(\sqrt{x^3})^2$
22) $(\sqrt{b^5})^2$
23) $(\sqrt[3]{w})^3$
24) $\sqrt{3} \cdot \sqrt{6}$
25) $\sqrt{6} \cdot \sqrt{8}$
26) $\sqrt{3} \cdot \sqrt{21}$
27) $\sqrt{5} \cdot \sqrt{15}$
28) $2\sqrt{7} \cdot 3\sqrt{14}$
29) $4\sqrt{6} \cdot 2\sqrt{15}$
30) $5\sqrt{10} \cdot 3\sqrt{14}$
31) $2\sqrt{6} \cdot 5\sqrt{21}$
32) $4\sqrt{10} \cdot \sqrt{20}$
33) $\sqrt{a^3} \cdot \sqrt{a}$
34) $\sqrt{r^3} \cdot \sqrt{r^5}$
35) $\sqrt{y^3} \cdot \sqrt{y^2}$
36) $\sqrt{n^5} \cdot \sqrt{n^2}$
37) $\sqrt{m^7} \cdot \sqrt{m^4}$
38) $\sqrt{x^2 y^3} \cdot \sqrt{x^4 y^3}$
39) $x\sqrt{x^5 y^2} \cdot y\sqrt{xy^3}$
40) $c\sqrt{c^5 d^3} \cdot c\sqrt{c^5 d^2}$
41) $\sqrt{a^5 b} \cdot \sqrt{a^2 b^2}$
42) $2\sqrt{5a} \cdot 3\sqrt{10a^3}$
43) $4\sqrt{6c^3} \cdot 3\sqrt{10c^5}$
44) $6\sqrt{15c^2} \cdot 2\sqrt{10c^5}$

Find the following quotients. Assume all variables have positive values. Leave answers in simplest form.

45) $\sqrt{\dfrac{1}{4}}$
46) $\sqrt{\dfrac{9}{25}}$
47) $\sqrt{\dfrac{49}{64}}$
48) $\sqrt{\dfrac{81}{64}}$

49) $\dfrac{\sqrt{32}}{\sqrt{8}}$
50) $\dfrac{\sqrt{125}}{\sqrt{5}}$
51) $\dfrac{\sqrt{98}}{\sqrt{2}}$
52) $\dfrac{\sqrt{27}}{\sqrt{3}}$
53) $\dfrac{6\sqrt{48}}{2\sqrt{6}}$
54) $\dfrac{8\sqrt{54}}{4\sqrt{3}}$
55) $\dfrac{9\sqrt{160}}{3\sqrt{8}}$
56) $\dfrac{10\sqrt{280}}{2\sqrt{10}}$
57) $\dfrac{\sqrt{x^3}}{\sqrt{x}}$
58) $\dfrac{\sqrt{a^5}}{\sqrt{a^3}}$
59) $\dfrac{\sqrt{c^5 d^4}}{\sqrt{cd^3}}$
60) $\dfrac{\sqrt{m^6 n^5}}{\sqrt{m^3 n^3}}$
61) $\dfrac{\sqrt{18x^3}}{\sqrt{2x}}$
62) $\dfrac{\sqrt{20r^5}}{\sqrt{5r^3}}$
63) $\dfrac{8\sqrt{45a^5}}{2\sqrt{5a}}$
64) $\dfrac{6\sqrt{48n^7}}{3\sqrt{3n^3}}$
65) $\dfrac{12\sqrt{72c^5}}{4\sqrt{6c^2}}$
66) $\dfrac{15\sqrt{48a^7}}{5\sqrt{2a^2}}$
67) $\dfrac{36\sqrt{96x^6 y^{11}}}{4\sqrt{3x^2 y^4}}$
68) $\dfrac{54\sqrt{240r^9 s^{10}}}{9\sqrt{5r^6 s^4}}$

Simplify the following.

69) $\sqrt{\dfrac{3}{7}} \cdot \sqrt{\dfrac{8}{7}}$
70) $\sqrt{\dfrac{8}{5}} \cdot \sqrt{\dfrac{6}{5}}$
71) $\sqrt{\dfrac{a^3}{2}} \cdot \sqrt{\dfrac{a^5}{2}}$
72) $\sqrt{\dfrac{c^2}{6}} \cdot \sqrt{\dfrac{c^5}{6}}$
73) $\sqrt{\dfrac{3x^5}{2}} \cdot \sqrt{\dfrac{15x^5}{8}}$
74) $\sqrt{\dfrac{5y}{3}} \cdot \sqrt{\dfrac{10}{27}}$

WRITING EXERCISE

75) Compare simplifying $(3x)(2x)$ with simplifying $(3\sqrt{2})(2\sqrt{2})$.

Addition, Subtraction, and Mixed Operations with Radicals

OBJECTIVES

When you complete this section, you will be able to:

Ⓐ Add and subtract like radicals.
Ⓑ Simplify radicals and then add and/or subtract like radicals.
Ⓒ Simplify radical expressions which contain mixed operations.

INTRODUCTION

We have found that $\sqrt[n]{a} \cdot \sqrt[n]{b} = \sqrt[n]{ab}$ and $\frac{\sqrt[n]{a}}{\sqrt[n]{b}} = \sqrt[n]{\frac{a}{b}}$ provided that $\sqrt[n]{a}$ and $\sqrt[n]{b}$ are real numbers. Does $\sqrt[n]{a} + \sqrt[n]{b} = \sqrt[n]{a+b}$?

CALCULATOR EXPLORATION ACTIVITY–OPTIONAL

Use your calculator to evaluate columns A and B below and round your answers off to the nearest ten-thousandth. Compare the results of the corresponding lines.

Column A	Column B
a) $\sqrt{2} + \sqrt{3}$ = _____	$\sqrt{5}$ = _____
b) $\sqrt{5} + \sqrt{12}$ = _____	$\sqrt{17}$ = _____
c) $\sqrt{23} + \sqrt{19}$ = _____	$\sqrt{42}$ = _____
d) $\sqrt{31} + \sqrt{43}$ = _____	$\sqrt{74}$ = _____

1) Based on your answers to the corresponding lines in columns A and B, does $\sqrt{a} + \sqrt{b} = \sqrt{a+b}$?

Addition and subtraction of radicals is done in much the same manner as addition and subtraction of variables. The distributive property that was used to add like terms will be used to add like radicals. Remember that like terms have the same variables and the same exponents on the variables. Similarly, **like radical expressions** have the same radicand and the same root indices. Study the following comparisons of addition of terms and addition of radical expressions.

$3x + 5x$

These are like terms so we can apply the distributive property and add.
$3x + 5x = (3 + 5)x = 8x$

$3\sqrt{2} + 5\sqrt{2}$

These are like radical expressions so we can apply the distributive property and add.
$3\sqrt{2} + 5\sqrt{2} = (3 + 5)\sqrt{2} = 8\sqrt{2}$

$3x + 5y$

These are not like terms because the variables are different. Consequently, we cannot add these terms.

$3\sqrt{2} + 5\sqrt{3}$

These are not like radical expressions because the radicands are different. Consequently, we cannot add these radicals.

$3x^2 + 5x^3$

These are not like terms because the exponents on the variables are different. Consequently, we cannot add these terms.

$3\sqrt{2} + 5\sqrt[3]{2}$

These are not like radical expressions because the root indices are different. Consequently, we cannot add these radicals.

Instead of applying the distributive property each time we added like terms, we observed that you add the coefficients and leave the variable portion unchanged. Similarly, we have the following rule for adding like radical expressions.

ADDING LIKE RADICAL EXPRESSIONS

To add like radicals, add the coefficients and leave the radical portion unchanged.

Note: Since $a - b = a + (-b)$, we do not need a separate rule for differences.

EXAMPLE 1
Find the following sums of radicals.

a) $7\sqrt{5} + 2\sqrt{5} =$
   $9\sqrt{5}$
   Add the coefficients and leave the radicals unchanged.
   Therefore, $7\sqrt{5} + 2\sqrt{5} = 9\sqrt{5}$.

b) $4\sqrt{x} - 8\sqrt{x} =$
   $-4\sqrt{x}$
   Add the coefficients and leave the radicals unchanged.
   Therefore, $4\sqrt{x} - 8\sqrt{x} = -4\sqrt{x}$.

c) $6\sqrt[3]{4} + \sqrt[3]{4} =$
   $7\sqrt[3]{4}$
   Add the coefficients and leave the radicals unchanged.
   Therefore, $6\sqrt[3]{4} + \sqrt[3]{4} = 7\sqrt[3]{4}$.

d) $6x\sqrt[4]{3} - 2x\sqrt[4]{3} =$
   $4x\sqrt[4]{3}$
   Add the coefficients and leave the radicals unchanged.
   Therefore, $6x\sqrt[4]{3} - 2x\sqrt[4]{3} = 4x\sqrt[4]{3}$.

e) $3a^2\sqrt{2a} - 7a^2\sqrt{2a} =$
   $-4a^2\sqrt{2a}$
   Add the coefficients and leave the radicals unchanged.
   Therefore, $3a^2\sqrt{2a} - 7a^2\sqrt{2a} = -4a^2\sqrt{2a}$.

Be Careful: A very common error is to add radicands when finding the sums of radicals. The following illustrates that this cannot be done.

$$\sqrt{9} + \sqrt{16} \neq \sqrt{9 + 16}$$
$$3 + 4 \neq \sqrt{25}$$
$$7 \neq 5$$

Never add radicands!

ADDITIONAL PRACTICE

Find the following sums of radicals.

a) $4\sqrt{5} + 6\sqrt{5}$

b) $3\sqrt{y} - 7\sqrt{y}$

c) $9\sqrt[3]{3x} - \sqrt[3]{3x}$

d) $z\sqrt{5} + 8z\sqrt{5}$

e) $5a^2\sqrt{7a} + 3a^2\sqrt{7a}$

PRACTICE EXERCISES

Find the following sums of radicals.

1) $5\sqrt{6} + 2\sqrt{6}$

2) $3\sqrt{c} - 9\sqrt{c}$

3) $5\sqrt[4]{5} + \sqrt[4]{5}$

4) $7y\sqrt[3]{7} - 12y\sqrt[3]{7}$

5) $7y^3\sqrt{6y} + 8y^3\sqrt{6y}$

If more practice is needed, do the Additional Practice Exercises in the margin.

It is often necessary to simplify radicands before they can be added.

EXAMPLE 2

a) $\sqrt{12} + \sqrt{75} =$ Rewrite 12 as $4 \cdot 3$ and 75 as $25 \cdot 3$.

  $\sqrt{4 \cdot 3} + \sqrt{25 \cdot 3} =$ Apply the product rule.

  $\sqrt{4} \cdot \sqrt{3} + \sqrt{25} \cdot \sqrt{3} =$ Find $\sqrt{4}$ and $\sqrt{25}$.

  $2\sqrt{3} + 5\sqrt{3} =$ Add like radicals.

  $7\sqrt{3}$ Therefore, $\sqrt{12} + \sqrt{75} = 7\sqrt{3}$.

b) $3\sqrt{80} - 2\sqrt{245} =$ Rewrite 80 as $16 \cdot 5$ and 245 as $49 \cdot 5$.

  $3\sqrt{16 \cdot 5} - 2\sqrt{49 \cdot 5} =$ Apply the product rule.

  $3\sqrt{16} \cdot \sqrt{5} - 2\sqrt{49} \cdot \sqrt{5} =$ Find 16= and $\sqrt{49}$.

  $3 \cdot 4\sqrt{5} - 2 \cdot 7\sqrt{5} =$ Multiply.

  $12\sqrt{5} - 14\sqrt{5} =$ Add like radicals.

  $- 2\sqrt{5}$ Therefore, $3\sqrt{80} - 2\sqrt{245} = -2\sqrt{5}$.

c) $\sqrt{48x^3} + \sqrt{12x^3} =$ Rewrite $48x^3$ as $16x^2 \cdot 3x$ and $12x^3$ as $4x^2 \cdot 3x$.

  $\sqrt{16x^2 \cdot 3x} + \sqrt{4x^2 \cdot 3x} =$ Apply the product rule.

  $\sqrt{16x^2} \cdot \sqrt{3x} + \sqrt{4x^2} \cdot \sqrt{3x} =$ Find $\sqrt{16x^2}$ and $\sqrt{4x^2}$.

  $4x\sqrt{3x} + 2x\sqrt{3x} =$ Add like radicals.

  $6x\sqrt{3x}$ Therefore, $\sqrt{48x^3} + \sqrt{12x^3} = 6x\sqrt{3x}$.

ADDITIONAL PRACTICE

Find the following sums of radicals.

f) $\sqrt{72} + \sqrt{32}$

g) $5\sqrt{48} - 2\sqrt{75}$

h) $\sqrt{28a^5} - \sqrt{63a^5}$

• **Mixed operations with radicals**

PRACTICE EXERCISES

Find the following sums.

6) $\sqrt{45} + \sqrt{20}$ 7) $4\sqrt{24} - 6\sqrt{54}$ 8) $\sqrt{50x^5} - \sqrt{18x^5}$

If more practice is needed, do the Additional Practice Exercises in the margin.

Now that we know how to perform the operations of addition, subtraction, multiplication, and division with radicals, it is possible to simplify radical expressions that contain more than one operation. The order of operations applies to radical expressions just as with any other numbers. All answers should be left in simplest form as described below.

SIMPLEST FORM OF RADICALS

A radical expression is in simplest form if:

1) All rational roots have been found.

2) No factors are left in a radicand that can be removed. For example, no square root may have a radicand with a perfect square factor.

3) All possible products, quotients, sums, and differences have been found.

4) There are no radicals in the denominator of a fraction. (Discussed in Section 8.5.)

EXAMPLE 3

Perform the following operations.

a) $\sqrt{2}\cdot\sqrt{10} + \sqrt{3}\cdot\sqrt{15} =$ Apply the product rule.

$\sqrt{2\cdot10} + \sqrt{3\cdot15} =$ Find the products.

$\sqrt{20} + \sqrt{45} =$ Rewrite 20 as $4\cdot5$ and 45 as $9\cdot5$.

$\sqrt{4\cdot5} + \sqrt{9\cdot5} =$ Apply the product rule.

$\sqrt{4}\cdot\sqrt{5} + \sqrt{9}\cdot\sqrt{5} =$ Find 9= and $\sqrt{4}$.

$2\sqrt{5} + 3\sqrt{5} =$ Add like radicals.

$5\sqrt{5}$ Therefore, $\sqrt{2}\cdot\sqrt{10} + \sqrt{3}\cdot\sqrt{15} = 5\sqrt{5}$.

b) $\dfrac{\sqrt{54}}{\sqrt{3}} + \sqrt{32} =$ Apply the quotient rule and rewrite 32 as $16\cdot2$.

$\sqrt{\dfrac{54}{3}} + \sqrt{16\cdot2} =$ Divide 54 by 3. Apply the product rule to $\sqrt{16\cdot2}$.

$\sqrt{18} + \sqrt{16}\cdot\sqrt{2} =$ Rewrite 18 as $9\cdot2$ and find $\sqrt{16}$.

$\sqrt{9\cdot2} + 4\sqrt{2} =$ Apply the product rule to$9\cdot2=$.

$\sqrt{9}\cdot\sqrt{2} + 4\sqrt{2} =$ Find $\sqrt{9}$.

$3\sqrt{2} + 4\sqrt{2} =$ Add like radicals.

$7\sqrt{2}$ Therefore, $\dfrac{\sqrt{54}}{\sqrt{3}} + \sqrt{32} = 7\sqrt{2}$.

Note: It is possible to perform operations on radical expressions using calculators and find decimal approximations for the results. For example, the decimal approximation for the results of Example 3b above, accurate to the nearest thousandth, is 9.899.

ADDITIONAL PRACTICE

Perform the following operations with radicals.

i) $\sqrt{5}\cdot\sqrt{15} + 2\sqrt{7}\cdot\sqrt{21}$

j) $\dfrac{\sqrt{72}}{\sqrt{6}} + \sqrt{75}$

PRACTICE EXERCISES

Perform the following operations.

9) $2\sqrt{3}\cdot\sqrt{6} + 4\sqrt{7}\cdot\sqrt{14}$ **10)** $\sqrt{63} + \dfrac{\sqrt{140}}{\sqrt{5}}$

If more practice is needed, do the Additional Practice Exercises in the margin.

Products involving sums and differences of radicals are found in much the same way as products of polynomials.

• **Products involving sums**

EXAMPLE 4

Find the following products. Assume all variables have non-negative values.

a) $\sqrt{3}(\sqrt{3} + \sqrt{15}) =$ Apply the distributive property.

$\sqrt{3}\cdot\sqrt{3} + \sqrt{3}\cdot\sqrt{15} =$ Apply the product rule.

$\sqrt{3\cdot3} + \sqrt{3\cdot15} =$ Find the products.

$\sqrt{9} + \sqrt{45} =$ Find $\sqrt{9}$ and simplify $\sqrt{45}$.

$3 + 3\sqrt{5}$ Therefore, $\sqrt{3}(\sqrt{3} + \sqrt{15}) = 3 + 3\sqrt{5}$.

b) $2\sqrt{6}(3 + 5\sqrt{5}) =$ Apply the distributive property.

$2\sqrt{6} \cdot 3 + 2\sqrt{6} \cdot 5\sqrt{5} =$ Apply the product rule.

$2 \cdot 3\sqrt{6} + 2 \cdot 5\sqrt{6 \cdot 5} =$ Find the products.

$6\sqrt{6} + 10\sqrt{30}$ Therefore, $2\sqrt{6}(3 + 5\sqrt{5}) = 6\sqrt{6} + 10\sqrt{30}$.

c) $(2 + \sqrt{3})(\sqrt{5} - \sqrt{6}) =$ Multiply using FOIL.

$2\sqrt{5} - 2\sqrt{6} + \sqrt{3} \cdot \sqrt{5} - \sqrt{3} \cdot \sqrt{6} =$ Apply the product rule.

$2\sqrt{5} - 2\sqrt{6} + \sqrt{15} - \sqrt{18} =$ Simplify $\sqrt{18}$.

$2\sqrt{5} - 2\sqrt{6} + \sqrt{15} - 3\sqrt{2}$ Therefore, $(2 + \sqrt{3})(\sqrt{5} - \sqrt{6}) = 2\sqrt{5} - 2\sqrt{6} + \sqrt{15} - 3\sqrt{2}$.

d) $(3\sqrt{x} + \sqrt{y})(2\sqrt{x} - 5\sqrt{y}) =$ Multiply using FOIL.

$3\sqrt{x} \cdot 2\sqrt{x} - 3\sqrt{x} \cdot 5\sqrt{y} + \sqrt{y} \cdot 2\sqrt{x} - \sqrt{y} \cdot 5\sqrt{y} =$ Apply the product rule.

$6x - 15\sqrt{xy} + 2\sqrt{xy} - 5y =$ Add like radicals.

$6x - 13\sqrt{xy} - 5y$ Therefore, $(3\sqrt{x} + \sqrt{y})(2\sqrt{x} - 5\sqrt{y}) = 6x - 13\sqrt{xy} - 5y$.

e) $(5 + \sqrt{3})^2 =$ Apply $(a + b)^2 = a^2 + 2ab + b^2$.

$5^2 + 2 \cdot 5\sqrt{3} + (\sqrt{3})^2 =$ Simplify.

$25 + 10\sqrt{3} + 3 =$ Add 25 and 3.

$28 + 10\sqrt{3}$ Therefore, $(5 + \sqrt{3})^2 = 28 + 10\sqrt{3}$.

ADDITIONAL PRACTICE

Find the following products.

k) $\sqrt{5}(\sqrt{6} - \sqrt{2})$

l) $7\sqrt{13}(3\sqrt{2} - 4)$

m) $(\sqrt{3} - 5)(\sqrt{7} - \sqrt{5})$

n) $(3\sqrt{n} + 2\sqrt{m})(5\sqrt{n} - 3\sqrt{m})$

o) $(4 - \sqrt{3})^2$

PRACTICE EXERCISES

Find the following products.

11) $\sqrt{7}(\sqrt{3} + \sqrt{7})$ **12)** $2\sqrt{11}(3 - 3\sqrt{6})$

13) $(4 - \sqrt{3})(\sqrt{2} - \sqrt{11})$ **14)** $(2\sqrt{a} + 3\sqrt{b})(\sqrt{a} - 3\sqrt{b})$

15) $(3 + \sqrt{5})^2$

If more practice is needed, do the Additional Practice Exercises in the margin.

One form of products of sums and differences of radicals is of particular interest. Recall that product of binomials of the form $(a + b)(a - b) = a^2 + ab - ab - b^2 = a^2 - b^2$. For example, $(x + 3)(x - 3) = x^2 - 3^2 = x^2 - 9$. If square roots are involved, expressions of the form $a + b$ and $a - b$ are called **conjugates**. For example, $3 + \sqrt{2}$ and $3 - \sqrt{2}$ are conjugates and their product is $(3 + \sqrt{2})(3 - \sqrt{2}) = 3^2 - (\sqrt{2})^2 = 9 - 2 = 7$. (Remember that $(\sqrt{x})^2 = x$ if $x \geq 0$.)

ANSWERS:
Practice 11–15

11) $\sqrt{21} + 7$

12) $6\sqrt{11} - 6\sqrt{66}$

13) $4\sqrt{2} - 4\sqrt{11} - \sqrt{6} + \sqrt{33}$

14) $2a - 3\sqrt{ab} - 9b$

15) $14 + 6\sqrt{5}$

Additional Practice k–o

k) $\sqrt{30} - \sqrt{10}$

l) $21\sqrt{26} - 28\sqrt{13}$

m) $\sqrt{21} - \sqrt{15} - 5\sqrt{7} + 5\sqrt{5}$

n) $15n + \sqrt{mn} - 6m$

o) $19 - 8\sqrt{3}$

EXAMPLE 5

Find the following products using the fact that $(a + b)(a - b) = a^2 - b^2$.

a) $(2 + \sqrt{3})(2 - \sqrt{3}) =$ Apply $(a + b)(a - b) = a^2 - b^2$.

$2^2 - (\sqrt{3})^2 =$ $2^2 = 4$ and $(\sqrt{3})^2 = 3$.

$4 - 3 =$ Add 4 and -3.

1 Therefore, $(2 + \sqrt{3})(2 - \sqrt{3}) = 1$.

b) $(\sqrt{5} - 3\sqrt{3})(\sqrt{5} + 3\sqrt{3}) =$ Apply $(a + b)(a - b) = a^2 - b^2$.

$(\sqrt{5})^2 - (3\sqrt{3})^3 =$ $(\sqrt{5})^2 = 5$ and $(3\sqrt{3})^2 = 3^2(\sqrt{3})^2 = 9 \cdot 3$.

$5 - 9 \cdot 3 =$ $9 \cdot 3 = 27$.

$5 - 27 =$ Add 5 and -27.

-22 Therefore, $(\sqrt{5} - 3\sqrt{3})(\sqrt{5} + 3\sqrt{3}) = -22$.

Note: The product of conjugates always results in a rational number. This will be useful in the next section on rationalizing the denominator.

ADDITIONAL PRACTICE

Find the following products.

p) $(\sqrt{7} - 3)(\sqrt{7} + 3)$

q) $(5\sqrt{7} - \sqrt{3})(5\sqrt{7} + \sqrt{3})$

PRACTICE EXERCISES

Find the following products.

16) $(4 + \sqrt{10})(4 - \sqrt{10})$ 17) $(3\sqrt{5} + 2\sqrt{6})(3\sqrt{5} - 2\sqrt{6})$

If more practice is needed, do the Additional Practice Exercises in the margin.

In Section 9.3 it will be necessary to simplify and reduce expressions of the form $\frac{a + b\sqrt{c}}{d}$ when solving quadratic equations. It will be helpful if we practice this procedure at this time. Remember, when reducing fractions you must divide the numerator and the denominator by the greatest factor common to both the numerator and the denominator.

EXAMPLE 6

Simplify the following.

a) $\dfrac{4 + 2\sqrt{5}}{6} =$ Remove the common factor of 2 from the numerator.

$\dfrac{2(2 + \sqrt{5})}{6} =$ Divide the numerator and the denominator by 2.

$\dfrac{2 + \sqrt{5}}{3}$ Therefore, $\dfrac{4 + 2\sqrt{5}}{6} = \dfrac{2 + \sqrt{5}}{3}$.

b) $\dfrac{6 + 4\sqrt{18}}{6} =$ Simplify $\sqrt{18}$.

$\dfrac{6 + 4 \cdot 3\sqrt{2}}{6} =$ Multiply 4 and 3.

$\dfrac{6 + 12\sqrt{2}}{6} =$ Remove the common factor of 6 from the numerator.

$\dfrac{6(1 + 2\sqrt{2})}{6} =$ Divide the numerator and the denominator by 6.

$1 + 2\sqrt{2}$ Therefore, $\dfrac{6 + 4\sqrt{18}}{6} = 1 + 2\sqrt{2}$.

Simplify the following.

r) $\dfrac{12 + 6\sqrt{3}}{3}$

s) $\dfrac{4 - 3\sqrt{8}}{4}$

Simplify the following.

18) $\dfrac{8 - 4\sqrt{2}}{2}$

19) $\dfrac{10 + 3\sqrt{75}}{5}$

If more practice is needed, do the Additional Practice Exercises in the margin

ANSWERS:
Practice 16–17

16) 6   **17)** 21

Additional Practice p–q

p) -2 **q)** 172

E X E R C I S E   S E T   **8.4**

Find the following sums of radicals. Assume all variables have non-negative values.

1) $2\sqrt{7} + 5\sqrt{7}$

2) $3\sqrt{3} + 8\sqrt{3}$

3) $9\sqrt{6} - 15\sqrt{6}$

4) $2\sqrt{10} - 13\sqrt{10}$

5) $7\sqrt{a} + 2\sqrt{a}$

6) $5\sqrt{y} + 7\sqrt{y}$

7) $-5\sqrt{n} + 2\sqrt{n}$

8) $-8\sqrt{m} + 4\sqrt{m}$

9) $6x\sqrt[3]{9} - 3x\sqrt[3]{9}$

10) $4y\sqrt[3]{3} - y\sqrt[3]{3}$

11) $6x^2\sqrt[4]{5x} + 12x^2\sqrt[4]{5x}$

12) $3y^3\sqrt[4]{8y} - 9y^3\sqrt[4]{8y}$

Simplify the following radicals and then find the sums. Assume all variables have non-negative values.

13) $\sqrt{27} + \sqrt{12}$

14) $\sqrt{8} + \sqrt{32}$

15) $\sqrt{48} - \sqrt{75}$

16) $\sqrt{80} - \sqrt{20}$

17) $\sqrt{27x} + \sqrt{75x}$

18) $\sqrt{80y} - \sqrt{125y}$

19) $\sqrt{48x^2} + \sqrt{75x^2}$

20) $\sqrt{24n^2} - \sqrt{96n^2}$

21) $\sqrt{18y^5} + \sqrt{72y^5}$

22) $\sqrt{48a^7} + \sqrt{108a^7}$

23) $12\sqrt{5} + \sqrt{45}$

24) $14\sqrt{6} - \sqrt{24}$

25) $\sqrt{80} - 4\sqrt{45}$

26) $\sqrt{20} - 2\sqrt{180}$

27) $3\sqrt{96} - 2\sqrt{54}$

28) $5\sqrt{63} + 2\sqrt{28}$

29) $6\sqrt{48a^3} - 2\sqrt{75a^3}$

30) $4\sqrt{98y^5} - 7\sqrt{128y^5}$

31) $\sqrt{150} - \sqrt{54} + \sqrt{24}$

32) $\sqrt{20} + \sqrt{125} - \sqrt{80}$

33) $2\sqrt{8} - 3\sqrt{48} + 2\sqrt{98} - \sqrt{75}$

34) $3\sqrt{216} - \sqrt{147} - 4\sqrt{96} - \sqrt{108}$

35) $4\sqrt{72x^2y} - 2x\sqrt{128y} + 5\sqrt{32x^2y}$

36) $3\sqrt{24a^3b^2} - 2a\sqrt{150ab^2} + 4ab\sqrt{96a}$

Simplify the following.

37) $\sqrt{3} \cdot \sqrt{15} + \sqrt{8} \cdot \sqrt{10}$

38) $\sqrt{6} \cdot \sqrt{8} + \sqrt{5} \cdot \sqrt{15}$

39) $3\sqrt{3} \cdot \sqrt{18} - 4\sqrt{18} \cdot \sqrt{12}$

40) $6\sqrt{8} \cdot \sqrt{10} - 5\sqrt{12} \cdot \sqrt{15}$

41) $\dfrac{\sqrt{40}}{\sqrt{5}} + \sqrt{50}$

42) $\dfrac{\sqrt{60}}{\sqrt{5}} + \sqrt{48}$

43) $\dfrac{\sqrt{540}}{\sqrt{3}} - 4\sqrt{125}$

44) $\dfrac{\sqrt{288}}{\sqrt{6}} - 6\sqrt{108}$

Simplify the following products. Assume all variables have non-negative values.

45) $\sqrt{2}(3 + \sqrt{2})$

46) $\sqrt{5}(4 - \sqrt{5})$

47) $\sqrt{3}(\sqrt{3} - \sqrt{15})$

48) $\sqrt{6}(\sqrt{6} + \sqrt{2})$

49) $\sqrt{5}(\sqrt{3} + 2\sqrt{15})$

50) $\sqrt{7}(\sqrt{5} - 3\sqrt{14})$

51) $4\sqrt{3}(2\sqrt{3} - 4\sqrt{6})$

52) $6\sqrt{2}(3\sqrt{2} + 2\sqrt{10})$

53) $(3 + \sqrt{5})(4 - \sqrt{2})$

54) $(3 + \sqrt{7})(7 - \sqrt{3})$

55) $(3 + \sqrt{x})(2 + \sqrt{x})$

56) $(5 - \sqrt{a})(2 + \sqrt{a})$

57) $(4 - 2\sqrt{6})(2 - 5\sqrt{6})$

58) $(9 - 2\sqrt{2})(1 + 4\sqrt{2})$

59) $(2 + 3\sqrt{3})(3 + 5\sqrt{2})$

60) $(7 - 3\sqrt{5})(2 - 2\sqrt{10})$

61) $(\sqrt{2} + \sqrt{3})(\sqrt{3} + \sqrt{5})$

62) $(\sqrt{5} + \sqrt{2})(\sqrt{2} + \sqrt{7})$

63) $(\sqrt{x} + \sqrt{y})(\sqrt{x} - 2\sqrt{y})$

64) $(2\sqrt{a} + \sqrt{b})(\sqrt{a} + \sqrt{b})$

65) $(\sqrt{3} + 2\sqrt{2})(\sqrt{3} - 4\sqrt{2})$

66) $(\sqrt{5} - 4\sqrt{6})(\sqrt{5} + 2\sqrt{6})$

67) $(5\sqrt{2} - 3\sqrt{5})(2\sqrt{2} + 6\sqrt{5})$

68) $(2\sqrt{2} + 3\sqrt{3})(3\sqrt{2} - 4\sqrt{3})$

69) $(4\sqrt{2} + 2\sqrt{5})(3\sqrt{7} - 3\sqrt{3})$

70) $(8\sqrt{2} - 2\sqrt{3})(2\sqrt{5} + 3\sqrt{10})$

71) $(2\sqrt{a} + 3\sqrt{b})(4\sqrt{a} - \sqrt{b})$

72) $(\sqrt{m} - 4\sqrt{n})(2\sqrt{m} - 3\sqrt{n})$

73) $(4 + \sqrt{6})^2$ 74) $(1 - \sqrt{2})^2$ 75) $(6 + \sqrt{3})^2$

76) $(5 + \sqrt{2})^2$ 77) $(2 + 2\sqrt{3})^2$ 78) $(3 + 2\sqrt{5})^2$

Find the following products of conjugates. Assume all variables have non-negative values.

79) $(2 + \sqrt{3})(2 - \sqrt{3})$

80) $(3 + \sqrt{5})(3 - \sqrt{5})$

81) $(4 + 2\sqrt{3})(4 - 2\sqrt{3})$

82) $(5 + 3\sqrt{3})(5 - 3\sqrt{3})$

83) $(3 + \sqrt{x})(3 - \sqrt{x})$

84) $(5 + \sqrt{y})(5 - \sqrt{y})$

85) $(\sqrt{2} + 4)(\sqrt{2} - 4)$

86) $(\sqrt{7} - 6)(\sqrt{7} + 6)$

87) $(5\sqrt{2} + 5)(5\sqrt{2} - 5)$

88) $(6\sqrt{3} - 2)(6\sqrt{3} + 2)$

89) $(\sqrt{3} + \sqrt{2})(\sqrt{3} - \sqrt{2})$

90) $(\sqrt{5} - \sqrt{3})(\sqrt{5} + \sqrt{3})$

91) $(\sqrt{x} + \sqrt{y})(\sqrt{x} - \sqrt{y})$

92) $(\sqrt{a} - \sqrt{b})(\sqrt{a} + \sqrt{b})$

93) $(3\sqrt{7} + \sqrt{13})(3\sqrt{7} - \sqrt{13})$

94) $(4\sqrt{5} - 3\sqrt{2})(4\sqrt{5} + 3\sqrt{2})$

95) $(3\sqrt{5} - 2\sqrt{6})(3\sqrt{5} + 2\sqrt{6})$

96) $(5\sqrt{2} - 2\sqrt{7})(5\sqrt{2} + 2\sqrt{7})$

Simplify the following.

97) $\dfrac{6 + 3\sqrt{2}}{3}$ 98) $\dfrac{8 + 4\sqrt{5}}{4}$

99) $\dfrac{10 + 20\sqrt{6}}{5}$ 100) $\dfrac{12 - 18\sqrt{2}}{3}$

101) $\dfrac{3 + 2\sqrt{45}}{3}$ 102) $\dfrac{12 - 3\sqrt{20}}{6}$

103) $\dfrac{9 - 2\sqrt{54}}{3}$ 104) $\dfrac{10 - 3\sqrt{50}}{5}$

105) $\dfrac{4 + 6\sqrt{3}}{8}$ 106) $\dfrac{6 - 3\sqrt{5}}{9}$

107) $\dfrac{8 - 4\sqrt{6}}{6}$ 108) $\dfrac{12 + 6\sqrt{7}}{9}$

CHALLENGE EXERCISES (109–116)

Perform the following operations on radical expressions whose indices are greater than 2.

109) $\sqrt[3]{16} + \sqrt[3]{54}$

110) $\sqrt[3]{24} + \sqrt[3]{81}$

111) $\sqrt[3]{250x^4y^5} + \sqrt[3]{128x^4y^5}$

112) $\sqrt[3]{108a^6b^4} + \sqrt[3]{32a^6b^4}$

113) $\sqrt[4]{48} + \sqrt[4]{243}$

114) $\sqrt[4]{64} - \sqrt[4]{324}$

115) $(\sqrt[3]{2} + \sqrt[3]{4})(2\sqrt[3]{2} - 3\sqrt[3]{4})$

116) $(\sqrt[3]{9} - 2\sqrt[3]{3})(2\sqrt[3]{9} + 3\sqrt[3]{3})$

WRITING EXERCISES

117) Compare finding the product $(2x + 3y)(3x - 4y)$ with finding the product $(2\sqrt{2} + 3\sqrt{5})(3\sqrt{2} - 4\sqrt{5})$. How are they alike and how are they different?

118) Compare adding $3x + 2x$ with adding $3\sqrt{2} + 2\sqrt{2}$.

119) Why is the product of conjugates always a rational number instead of an irrational number?

120) Why is $\sqrt{a} + \sqrt{b} \neq \sqrt{a+b}$? Use examples if necessary.

Rationalizing the Denominator

OBJECTIVES

When you complete this section, you will be able to:

Ⓐ Rationalize the denominator when the denominator contains a single radical which is a square root or a cube root.

Ⓑ Rationalize the denominator when the denominator contains an expression of the form $a \pm b$ where a and/or b is a square root.

INTRODUCTION

In Section 8.4 we discussed certain conditions that must be satisfied in order for a radical to be in simplest form. We did not discuss the condition that the denominator of a fraction cannot contain a radical which is an irrational number. The process of making the denominator a rational number is called **rationalizing the denominator**. One reason to rationalize denominators is to make finding decimal approximations easier. Consider the following two methods of approximating $\dfrac{2}{\sqrt{3}}$.

Method 1		Method 2	
$\dfrac{2}{\sqrt{3}}$	Replace $\sqrt{3}$ with 1.732.	$\dfrac{2}{\sqrt{3}}$	Multiply by $\dfrac{\sqrt{3}}{\sqrt{3}}$.
$\dfrac{2}{1.732}$	Divide.	$\dfrac{2}{\sqrt{3}} \cdot \dfrac{\sqrt{3}}{\sqrt{3}}$	Simplify.
1.155	Therefore, $\dfrac{2}{\sqrt{3}} \approx 1.155$.	$\dfrac{2\sqrt{3}}{\sqrt{9}}$	Replace $\sqrt{3}$ with 1.732 and $\sqrt{9}$ with 3.
		$\dfrac{2(1.732)}{3}$	Simplify.
		1.155	Therefore, $\dfrac{2}{\sqrt{3}} \approx 1.155$.

Method 2 is a little longer, but the arithmetic is much easier. With the use of calculators, ease in approximating radical expressions is no longer an important reason for rationalizing the denominator. Still, skill in rationalizing denominators (and numerators) is important for other reasons in more advanced math courses.

• **Rationalizing denominators of the form $\sqrt{a}$**

The technique used in rationalizing the denominator is to multiply the fraction by an appropriate radical expression divided by itself (hence, with value equal to 1), so that the denominator of the fraction becomes a radical expression with rational value. This means that if the denominator of the original fraction is a square root, we will multiply this fraction by a square root divided by itself (thus, with value equal to 1), so that the radicand in the denominator is a perfect square. If the radicand of the original fraction contains variables, we must multiply the fraction by a radical expression divided by itself that will make the exponent(s) of the variable(s) in the radicand of the denominator divisible by two. Cube roots will be discussed later.

EXAMPLE 1

Rationalize the denominators of the following fractions. Assume all variables have positive values.

a) $\dfrac{3}{\sqrt{5}} =$ Since $\sqrt{5} \cdot \sqrt{5} = \sqrt{25} = 5$, multiply by $\dfrac{\sqrt{5}}{\sqrt{5}} \left(\dfrac{\sqrt{5}}{\sqrt{5}} = 1 \right)$.

$\dfrac{3}{\sqrt{5}} \cdot \dfrac{\sqrt{5}}{\sqrt{5}} =$ Multiply.

$\dfrac{3\sqrt{5}}{\sqrt{25}} =$ $\sqrt{25} = 5$.

$\dfrac{3\sqrt{5}}{5}$ Therefore, $\dfrac{3}{\sqrt{5}} = \dfrac{3\sqrt{5}}{5}$.

b) $\dfrac{8}{\sqrt{6}} =$ Since $\sqrt{6} \cdot \sqrt{6} = \sqrt{36} = 6$, multiply by $\dfrac{\sqrt{6}}{\sqrt{6}} \left(\dfrac{\sqrt{6}}{\sqrt{6}} = 1 \right)$.

$\dfrac{8}{\sqrt{6}} \cdot \dfrac{\sqrt{6}}{\sqrt{6}} =$ Multiply.

$\dfrac{8\sqrt{6}}{\sqrt{36}} =$ $\sqrt{36} = 6$.

$\dfrac{8\sqrt{6}}{6} =$ Reduce to lowest terms.

$\dfrac{4\sqrt{6}}{3}$ Therefore, $\dfrac{8}{\sqrt{6}} = \dfrac{4\sqrt{6}}{3}$.

c) $\dfrac{\sqrt{3}}{\sqrt{8}} =$ Since $\sqrt{8} \cdot \sqrt{2} = \sqrt{16} = 4$, multiply by $\dfrac{\sqrt{2}}{\sqrt{2}} \left(\dfrac{\sqrt{2}}{\sqrt{2}} = 1 \right)$.

$\dfrac{\sqrt{3}}{\sqrt{8}} \cdot \dfrac{\sqrt{2}}{\sqrt{2}} =$ Multiply.

$\dfrac{\sqrt{6}}{\sqrt{16}} =$ $\sqrt{16} = 4$.

$\dfrac{\sqrt{6}}{4}$ Therefore, $\dfrac{\sqrt{3}}{\sqrt{8}} = \dfrac{\sqrt{6}}{4}$.

Note: One method of determining the radicand of the radical expression by which you multiply the original fraction is to find the smallest perfect square that is divisible by the radicand of the original fraction. For example, in Example 1c the smallest perfect square that is divisible by 8 is 16 and $16 \div 8 = 2$. Therefore, the radicand of the expression that we multiply by is 2. As another example, given $\dfrac{1}{\sqrt{12}}$, the smallest perfect square divisible by 12 is 36 and $36 \div 12 = 3$. Therefore, we would multiply $\dfrac{1}{\sqrt{12}}$ by $\dfrac{\sqrt{3}}{\sqrt{3}}$.

d) $\dfrac{3}{\sqrt{x}} =$ Since $\sqrt{x} \cdot \sqrt{x} = \sqrt{x^2} = x$, multiply by $\dfrac{\sqrt{x}}{\sqrt{x}}\left(\dfrac{\sqrt{x}}{\sqrt{x}} = 1\right)$.

$\dfrac{3}{\sqrt{x}} \cdot \dfrac{\sqrt{x}}{\sqrt{x}} =$ Multiply.

$\dfrac{3\sqrt{x}}{\sqrt{x^2}} =$ $\sqrt{x^2} = x$.

$\dfrac{3\sqrt{x}}{x}$ Therefore, $\dfrac{\sqrt{x}}{\sqrt{x}} = \dfrac{3\sqrt{x}}{x}$.

Alternative Method (Optional)

In Section 8.2 we gave an alternative method of simplifying radicals in which we wrote the radicand in terms of prime factors. Each time a factor appeared two times, the square root was one of the two equal factors. For example, $\sqrt{2 \cdot 2} = 2$, $\sqrt{2 \cdot 2 \cdot 2 \cdot 2} = 2 \cdot 2 = 4$, and $\sqrt{x \cdot x} = x$. We illustrate this technique by reworking Examples 1a and 1c.

EXAMPLE 2

Rationalize the following denominators using prime factorization. Assume all variables have positive values.

a) $\dfrac{3}{\sqrt{5}} =$ Since $\sqrt{5 \cdot 5} = 5$, we need another factor of 5 in the radicand of the denominator. So multiply by $\dfrac{\sqrt{5}}{\sqrt{5}}$.

$\dfrac{3}{\sqrt{5}} \cdot \dfrac{\sqrt{5}}{\sqrt{5}} =$ Multiply.

$\dfrac{3\sqrt{5}}{\sqrt{5 \cdot 5}} =$ $\sqrt{5 \cdot 5} = 5$.

$\dfrac{3\sqrt{5}}{5}$ Therefore, $\dfrac{3}{\sqrt{5}} = \dfrac{3\sqrt{5}}{5}$.

b) $\dfrac{\sqrt{3}}{\sqrt{8}} =$ Write 8 in terms of prime factors.

$\dfrac{\sqrt{3}}{\sqrt{2 \cdot 2 \cdot 2}} =$ Since we need an even number of factors of 2 in the radicand of the denominator, multiply by $\dfrac{\sqrt{2}}{\sqrt{2}}$.

$\dfrac{\sqrt{3}}{\sqrt{2 \cdot 2 \cdot 2}} \cdot \dfrac{\sqrt{2}}{\sqrt{2}} =$ Multiply.

$\dfrac{\sqrt{6}}{\sqrt{2 \cdot 2 \cdot 2 \cdot 2}} =$ $\sqrt{2 \cdot 2 \cdot 2 \cdot 2} = 2 \cdot 2 = 4$.

$\dfrac{\sqrt{6}}{4}$ Therefore, $\dfrac{\sqrt{3}}{\sqrt{8}} = \dfrac{\sqrt{6}}{4}$.

a) $\dfrac{7}{\sqrt{5}}$ **b)** $\dfrac{10}{\sqrt{18}}$

c) $\dfrac{\sqrt{3}}{\sqrt{7}}$ **d)** $\dfrac{c}{\sqrt{d}}$

PRACTICE EXERCISES

Rationalize the denominators of the following fractions. Assume all variables have positive values.

1) $\dfrac{5}{\sqrt{2}}$ **2)** $\dfrac{8}{\sqrt{10}}$ **3)** $\dfrac{\sqrt{5}}{\sqrt{12}}$ **4)** $\dfrac{b}{\sqrt{a}}$

If more practice is needed, do the Additional Practice Exercises in the margin.

If the radicand is a fraction, apply the quotient rule for radicals. This results in a fraction with a radical in the denominator that is rationalized using the technique of Example 1.

EXAMPLE 3

Simplify the following. Assume all variables have positive values.

a) $\sqrt{\dfrac{7}{2}} =$ Rewrite using the quotient rule.

$\dfrac{\sqrt{7}}{\sqrt{2}} =$ Since $\sqrt{2} \cdot \sqrt{2} = \sqrt{4} = 2$, multiply by $\dfrac{\sqrt{2}}{\sqrt{2}}$.

$\dfrac{\sqrt{7}}{\sqrt{2}} \cdot \dfrac{\sqrt{2}}{\sqrt{2}} =$ Multiply.

$\dfrac{\sqrt{14}}{\sqrt{4}} =$ $\sqrt{4} = 2$.

$\dfrac{\sqrt{14}}{2}$ Therefore, $\sqrt{\dfrac{7}{2}} = \dfrac{\sqrt{14}}{2}$.

b) $\sqrt{\dfrac{a}{b^3}} =$ Rewrite using the quotient rule.

$\dfrac{\sqrt{a}}{\sqrt{b^3}} =$ Since $\sqrt{b^3} \cdot \sqrt{b} = \sqrt{b^4} = b^2$, multiply by $\dfrac{\sqrt{b}}{\sqrt{b}}$.

$\dfrac{\sqrt{a}}{\sqrt{b^3}} \cdot \dfrac{\sqrt{b}}{\sqrt{b}} =$ Multiply.

$\dfrac{\sqrt{ab}}{\sqrt{b^4}} =$ $\sqrt{b^4} = b^2$.

$\dfrac{\sqrt{ab}}{b^2}$ Therefore, $\sqrt{\dfrac{a}{b^3}} = \dfrac{\sqrt{ab}}{b^2}$.

c) $\sqrt{\dfrac{2}{3x}} =$ Rewrite using the quotient rule.

$\dfrac{\sqrt{2}}{\sqrt{3x}} =$ Since $\sqrt{3x} \cdot \sqrt{3x} = \sqrt{9x^2} = 3x$, multiply by $\dfrac{\sqrt{3x}}{\sqrt{3x}}$.

$$\dfrac{\sqrt{2}}{\sqrt{3x}} \cdot \dfrac{\sqrt{3x}}{\sqrt{3x}} = \qquad \text{Multiply.}$$

$$\dfrac{\sqrt{6x}}{\sqrt{9x^2}} = \qquad \sqrt{9x^2} = 3x.$$

$$\dfrac{\sqrt{6x}}{3x} \qquad \text{Therefore, } \sqrt{\dfrac{2}{3x}} = \dfrac{\sqrt{6x}}{3x}.$$

PRACTICE EXERCISES

Simplify the following. Assume all variables have positive values.

5) $\sqrt{\dfrac{5}{3}}$

6) $\sqrt{\dfrac{c}{d^3}}$

7) $\sqrt{\dfrac{5}{2y}}$

• **Rationalizing denominators that are cube roots**

If the denominator contains a cube root instead of a square root, the technique is essentially the same. We multiply the original fraction by the cube root of a quantity divided by itself so that the resulting radicand in the denominator is a perfect cube.

EXAMPLE 4

Rationalize the denominators of the following.

a) $\dfrac{3}{\sqrt[3]{2}} = \qquad$ Since $\sqrt[3]{2} \cdot \sqrt[3]{4} = \sqrt[3]{8} = 2$, multiply by $\dfrac{\sqrt[3]{4}}{\sqrt[3]{4}}$.

$$\dfrac{3}{\sqrt[3]{2}} \cdot \dfrac{\sqrt[3]{4}}{\sqrt[3]{4}} = \qquad \text{Multiply.}$$

$$\dfrac{3\sqrt[3]{4}}{\sqrt[3]{8}} = \qquad \sqrt[3]{8} = 2.$$

$$\dfrac{3\sqrt[3]{4}}{2} \qquad \text{Therefore, } \dfrac{3}{\sqrt[3]{2}} = \dfrac{3\sqrt[3]{4}}{2}.$$

Note: We can also determine the radicand of the radical expression by which we multiply the original fraction by finding the smallest perfect cube divisible by the radicand. For example, in Example 4a the smallest perfect cube that is divisible by 2 is 8 and $8 \div 2 = 4$. Therefore, the radicand of the radical expression that we multiply by is 4.

b) $\dfrac{\sqrt[3]{a}}{\sqrt[3]{b}} = \qquad$ Since $\sqrt[3]{b} \cdot \sqrt[3]{b^2} = \sqrt[3]{b^3} = b$, multiply by $\dfrac{\sqrt[3]{b^2}}{\sqrt[3]{b^2}}$.

$$\dfrac{\sqrt[3]{a}}{\sqrt[3]{b}} \cdot \dfrac{\sqrt[3]{b^2}}{\sqrt[3]{b^2}} = \qquad \text{Multiply.}$$

$$\dfrac{\sqrt[3]{ab^2}}{\sqrt[3]{b^3}} = \qquad \sqrt[3]{b^3} = b.$$

$$\dfrac{\sqrt[3]{ab^2}}{b} \qquad \text{Therefore, } \dfrac{\sqrt[3]{a}}{\sqrt[3]{b}} = \dfrac{\sqrt[3]{ab^2}}{b}.$$

c)

$$\sqrt[3]{\dfrac{5}{9a^2}} =$$

Rewrite $\sqrt[3]{\dfrac{5}{9a^2}}$ as $\dfrac{\sqrt[3]{5}}{\sqrt[3]{9a^2}}$.

$$\dfrac{\sqrt[3]{5}}{\sqrt[3]{9a^2}} =$$

Since $\sqrt[3]{9a^2} \cdot \sqrt[3]{3a} = \sqrt[3]{27a^3} = 3a$,

multiply by $\dfrac{\sqrt[3]{3a}}{\sqrt[3]{3a}}$.

$$\dfrac{\sqrt[3]{5}}{\sqrt[3]{9a^2}} \cdot \dfrac{\sqrt[3]{3a}}{\sqrt[3]{3a}} =$$

Multiply.

$$\dfrac{\sqrt[3]{15a}}{\sqrt[3]{27a^3}} =$$

$\sqrt[3]{27a^3} = 3a$.

$$\dfrac{\sqrt[3]{15a}}{3a}$$

Therefore, $\sqrt[3]{\dfrac{5}{9a^2}} = \dfrac{\sqrt[3]{15a}}{3a}$.

Alternative Method (Optional)

The method of rationalization using prime factorization also works very well for radicands containing cube roots. Each time a factor appeared three times, the cube root is one of the three equal factors. For example, $\sqrt[3]{a \cdot a \cdot a} = a$, and $\sqrt[3]{2 \cdot 2 \cdot 2 \cdot 2 \cdot y \cdot y \cdot y \cdot y \cdot y} = 2y\sqrt[3]{2y^2}$. We illustrate this technique by reworking Examples 4a and 4c.

EXAMPLE 5

Rationalize the following denominators using prime factorization. Assume all variables have positive values.

a)

$$\dfrac{3}{\sqrt[3]{2}} =$$

Since $\sqrt[3]{2 \cdot 2 \cdot 2} = 2$, we need two more factors of 2 in the radicand of the denominator. So multiply by $\dfrac{\sqrt[3]{2 \cdot 2}}{\sqrt[3]{2 \cdot 2}}$.

$$\dfrac{3}{\sqrt[3]{2}} \cdot \dfrac{\sqrt[3]{2 \cdot 2}}{\sqrt[3]{2 \cdot 2}} =$$

Multiply.

$$\dfrac{3\sqrt[3]{2 \cdot 2}}{\sqrt[3]{2 \cdot 2 \cdot 2}} =$$

$\sqrt[3]{2 \cdot 2 \cdot 2} = 2$.

$$\dfrac{3\sqrt[3]{4}}{2}$$

Therefore, $\dfrac{3}{\sqrt[3]{2}} = \dfrac{3\sqrt[3]{4}}{2}$.

b)

$$\sqrt[3]{\dfrac{5}{9a^2}} =$$

Rewrite $\sqrt[3]{\dfrac{5}{9a^2}}$ as $\dfrac{\sqrt[3]{5}}{\sqrt[3]{9a^2}}$.

$$\dfrac{\sqrt[3]{5}}{\sqrt[3]{9a^2}} =$$

Write $9a^2$ in terms of prime factors.

$$\dfrac{\sqrt[3]{5}}{\sqrt[3]{3 \cdot 3 \cdot a \cdot a}} =$$

Since $\sqrt[3]{3 \cdot 3 \cdot 3} = 3$ and $\sqrt[3]{a \cdot a \cdot a} = a$, we need another factor of 3 and a in the radicand of the denominator. So multiply by $\dfrac{\sqrt[3]{3a}}{\sqrt[3]{3a}}$.

$$\dfrac{\sqrt[3]{5}}{\sqrt[3]{3 \cdot 3 \cdot a \cdot a}} \cdot \dfrac{\sqrt[3]{3a}}{\sqrt[3]{3a}} =$$

Multiply.

$$\dfrac{\sqrt[3]{15a}}{\sqrt[3]{3 \cdot 3 \cdot 3 \cdot a \cdot a \cdot a}} =$$

$\sqrt[3]{3 \cdot 3 \cdot 3 \cdot a \cdot a \cdot a} = 3a$.

$$\dfrac{\sqrt[3]{15a}}{3a}$$

Therefore, $\sqrt[3]{\dfrac{5}{9a^2}} = \dfrac{\sqrt[3]{15a}}{3a}$.

Rationalize the denominators of the following.

e) $\dfrac{5}{\sqrt[3]{9}}$ **f)** $\sqrt[3]{\dfrac{3}{x^2}}$

g) $\dfrac{8}{\sqrt[3]{25c}}$

- **Rationalizing the denominators of the form $a \pm b$ where a and/or b are square roots**

PRACTICE EXERCISES

Rationalize the denominators of the following.

8) $\dfrac{6}{\sqrt[3]{3}}$ **9)** $\sqrt[3]{\dfrac{m}{n}}$ **10)** $\dfrac{4}{\sqrt[3]{4y^2}}$

If more practice is needed, do the Additional Practice Exercises in the margin.

Recall from Section 8.4 that the product of conjugates always results in a rational number. Consequently, if the denominator contains an expression of the form $a + b$ where a and/or b is a square root, multiply the original fraction by the conjugate of the denominator divided by itself.

EXAMPLE 6

Rationalize the denominators of the following.

a) $\dfrac{3}{4 + \sqrt{5}} =$ Since the conjugate of $4 + \sqrt{5}$ is $4 - \sqrt{5}$, multiply by $\dfrac{4 - \sqrt{5}}{4 - \sqrt{5}}$.

$\dfrac{3}{4 + \sqrt{5}} \cdot \dfrac{4 - \sqrt{5}}{4 - \sqrt{5}} =$ Multiply the fractions.

$\dfrac{3(4 - \sqrt{5})}{(4 + \sqrt{5})(4 - \sqrt{5})} =$ Perform the multiplications.

$\dfrac{12 - 3\sqrt{5}}{4^2 - (\sqrt{5})^2} =$ Simplify the denominator.

$\dfrac{12 - 3\sqrt{5}}{16 - 5} =$ Add 16 and -5.

$\dfrac{12 - 3\sqrt{5}}{11}$ Therefore, $\dfrac{3}{4 + \sqrt{5}} = \dfrac{12 - 3\sqrt{5}}{11}$.

b) $\dfrac{2 + \sqrt{2}}{4 - \sqrt{2}} =$ Since the conjugate of $4 - \sqrt{2}$ is $4 + \sqrt{2}$, multiply by $\dfrac{4 + \sqrt{2}}{4 + \sqrt{2}}$.

$\dfrac{2 + \sqrt{2}}{4 - \sqrt{2}} \cdot \dfrac{4 + \sqrt{2}}{4 + \sqrt{2}} =$ Multiply the fractions.

$\dfrac{(2 + \sqrt{2})(4 + \sqrt{2})}{(4 - \sqrt{2})(4 + \sqrt{2})} =$ Perform the multiplications.

$\dfrac{8 + 2\sqrt{2} + 4\sqrt{2} + \sqrt{4}}{4^2 - (\sqrt{2})^2} =$ Simplify.

$\dfrac{8 + 6\sqrt{2} + 2}{16 - 2} =$ Continue simplifying.

$\dfrac{10 + 6\sqrt{2}}{14} =$ Factor the numerator.

$\dfrac{2(5 + 3\sqrt{2})}{14} =$ Reduce to lowest terms.

$\dfrac{5 + 3\sqrt{2}}{7}$ Therefore, $\dfrac{2 + \sqrt{2}}{4 - \sqrt{2}} = \dfrac{5 + 3\sqrt{2}}{7}$.

c) $\dfrac{1 + \sqrt{y}}{\sqrt{x} - \sqrt{y}} =$

$\dfrac{1 + \sqrt{y}}{\sqrt{x} - \sqrt{y}} \cdot \dfrac{\sqrt{x} + \sqrt{y}}{\sqrt{x} + \sqrt{y}} =$

$\dfrac{(1 + \sqrt{y})(\sqrt{x} + \sqrt{y})}{(\sqrt{x} - \sqrt{y})(\sqrt{x} + \sqrt{y})} =$

$\dfrac{\sqrt{x} + \sqrt{y} + \sqrt{xy} + \sqrt{y^2}}{(\sqrt{x})^2 - (\sqrt{y})^2} =$

$\dfrac{\sqrt{x} + \sqrt{y} + \sqrt{xy} + y}{x - y}$

Since the conjugate of $\sqrt{x} - \sqrt{y}$ is $\sqrt{x} + \sqrt{y}$, multiply by $\dfrac{\sqrt{x} + \sqrt{y}}{\sqrt{x} + \sqrt{y}}$.

Multiply the fractions.

Perform the multiplications.

Simplify.

Therefore, $\dfrac{1 + \sqrt{y}}{\sqrt{x} - \sqrt{y}} = \dfrac{\sqrt{x} + \sqrt{y} + \sqrt{xy} + y}{x - y}$.

ADDITIONAL PRACTICE

Rationalize the denominators of the following.

h) $\dfrac{7}{3 + \sqrt{7}}$

i) $\dfrac{3 + \sqrt{3}}{5 - \sqrt{3}}$

j) $\dfrac{\sqrt{5} + \sqrt{2}}{\sqrt{5} - \sqrt{2}}$

PRACTICE EXERCISES

Rationalize the denominators of the following. Assume all variables have non-negative values.

11) $\dfrac{7}{6 + \sqrt{6}}$

12) $\dfrac{4 - \sqrt{2}}{3 - \sqrt{2}}$

13) $\dfrac{\sqrt{3} + \sqrt{5}}{\sqrt{3} - \sqrt{5}}$

If more practice is needed, do the Additional Practice Exercises in the margin.

EXERCISE SET 8.5

Rationalize the following denominators. Assume all variables have positive values.

1) $\dfrac{3}{\sqrt{7}}$ 2) $\dfrac{2}{\sqrt{3}}$ 3) $\dfrac{5}{\sqrt{5}}$

4) $\dfrac{6}{\sqrt{6}}$ 5) $\dfrac{6}{\sqrt{2}}$ 6) $\dfrac{12}{\sqrt{3}}$

7) $\dfrac{20}{\sqrt{10}}$ 8) $\dfrac{30}{\sqrt{15}}$ 9) $\dfrac{4}{\sqrt{x}}$

10) $\dfrac{6}{\sqrt{y}}$ 11) $\dfrac{m}{\sqrt{n}}$ 12) $\dfrac{r}{\sqrt{s}}$

13) $\dfrac{\sqrt{5}}{\sqrt{2}}$ 14) $\dfrac{\sqrt{7}}{\sqrt{5}}$ 15) $\dfrac{\sqrt{3}}{\sqrt{8}}$

16) $\dfrac{\sqrt{11}}{\sqrt{18}}$ 17) $\dfrac{\sqrt{5}}{\sqrt{32}}$ 18) $\dfrac{\sqrt{7}}{\sqrt{27}}$

19) $\dfrac{\sqrt{a}}{\sqrt{b}}$ 20) $\dfrac{\sqrt{c}}{\sqrt{d}}$ 21) $\dfrac{\sqrt{d}}{\sqrt{e}}$

Simplify the following. Assume all variables have positive values.

22) $\sqrt{\dfrac{3}{2}}$ 23) $\sqrt{\dfrac{5}{3}}$ 24) $\sqrt{\dfrac{7}{8}}$ 25) $\sqrt{\dfrac{7}{12}}$

26) $\sqrt{\dfrac{u}{v}}$ 27) $\sqrt{\dfrac{p}{q}}$ 28) $\sqrt{\dfrac{3}{y^3}}$ 29) $\sqrt{\dfrac{7}{x^5}}$

30) $\sqrt{\dfrac{3}{2x}}$ 31) $\sqrt{\dfrac{7}{5y}}$ 32) $\sqrt{\dfrac{4}{3b}}$ 33) $\sqrt{\dfrac{9}{7a}}$

Rationalize the denominators of the following. Assume all variables have nonzero values.

34) $\dfrac{5}{\sqrt[3]{4}}$ 35) $\dfrac{2}{\sqrt[3]{9}}$ 36) $\dfrac{8}{\sqrt[3]{2}}$ 37) $\dfrac{10}{\sqrt[3]{25}}$

38) $\dfrac{\sqrt[3]{x}}{\sqrt[3]{y}}$ 39) $\dfrac{\sqrt[3]{r}}{\sqrt[3]{s}}$ 40) $\dfrac{\sqrt[3]{a^2}}{\sqrt[3]{b^2}}$ 41) $\dfrac{\sqrt[3]{x}}{\sqrt[3]{y^2}}$

42) $\sqrt[3]{\dfrac{2}{3a}}$ 43) $\sqrt[3]{\dfrac{5}{4b}}$ 44) $\sqrt[3]{\dfrac{5}{9x^2}}$ 45) $\sqrt[3]{\dfrac{7}{2d^2}}$

11) $\dfrac{42 - 7\sqrt{6}}{30}$

12) $\dfrac{10 + \sqrt{2}}{7}$

13) $-4 - \sqrt{15}$

Additional Practice h–j

h) $\dfrac{21 - 7\sqrt{7}}{2}$

i) $\dfrac{9 + 4\sqrt{3}}{11}$

j) $\dfrac{7 + 2\sqrt{10}}{3}$

61) $\dfrac{\sqrt{5} - \sqrt{3}}{\sqrt{5} + \sqrt{3}}$

62) $\dfrac{\sqrt{a} + \sqrt{b}}{\sqrt{a} - \sqrt{b}}$

63) $\dfrac{\sqrt{x} + \sqrt{y}}{\sqrt{x} - \sqrt{y}}$

64) $\dfrac{4 + \sqrt{2}}{\sqrt{3}}$

65) $\dfrac{\sqrt{5} - \sqrt{6}}{\sqrt{2}}$

CHALLENGE EXERCISES (66–73)

Rationalize the denominators of the following.

66) $\dfrac{4 - 2\sqrt{3}}{3\sqrt{2} - 2\sqrt{5}}$

67) $\dfrac{2\sqrt{5} + 3\sqrt{6}}{3\sqrt{2} - 3\sqrt{7}}$

68) $\dfrac{2}{\sqrt[4]{2}}$

69) $\dfrac{3}{\sqrt[4]{9}}$

70) $\dfrac{x^2}{\sqrt[4]{x^2}}$

71) $\dfrac{y^2}{\sqrt[3]{y^3}}$

72) $\dfrac{2}{\sqrt[5]{2}}$

73) $\dfrac{6}{\sqrt[5]{8}}$

WRITING EXERCISES

74) In your own words, describe the procedures used in rationalizing the denominator.

75) What is wrong with the following? $\dfrac{9}{16} = \dfrac{3}{4}$

76) If we were to rationalize the denominator of $\dfrac{1}{\sqrt[3]{a} + \sqrt[3]{b}}$, we would not multiply by $\dfrac{\sqrt[3]{a} - \sqrt[3]{b}}{\sqrt[3]{a} - \sqrt[3]{b}}$. Why not?

Rationalize the denominators of the following.

46) $\dfrac{23}{5 + \sqrt{2}}$

47) $\dfrac{-1}{2 + \sqrt{5}}$

48) $\dfrac{19}{5 - \sqrt{6}}$

49) $\dfrac{7}{3 - \sqrt{3}}$

50) $\dfrac{x}{5 - \sqrt{x}}$

51) $\dfrac{a}{7 + \sqrt{a}}$

52) $\dfrac{3 + \sqrt{2}}{2 + \sqrt{2}}$

53) $\dfrac{4 + \sqrt{3}}{3 - \sqrt{3}}$

54) $\dfrac{\sqrt{6} - 5}{\sqrt{6} + 3}$

55) $\dfrac{\sqrt{5} - 6}{\sqrt{5} + 4}$

56) $\dfrac{3 + \sqrt{2}}{2 + \sqrt{3}}$

57) $\dfrac{5 + \sqrt{3}}{4 - \sqrt{5}}$

58) $\dfrac{2 - \sqrt{b}}{3 + \sqrt{b}}$

59) $\dfrac{7 + \sqrt{a}}{2 - \sqrt{a}}$

60) $\dfrac{\sqrt{3} + \sqrt{2}}{\sqrt{3} - \sqrt{2}}$

S E C T I O N   **8.6**

Solving Equations with Radicals

OBJECTIVES

When you complete this section, you will be able to:

Ⓐ Solve equations that contain one radical (square root).

Ⓑ Solve equations that have two radicals (square roots).

INTRODUCTION

Solving equations that contain radicals that are square roots depends upon two concepts. The first was presented in Section 8.1 and is $(\sqrt{x})^2 = x$. For example, $(\sqrt{3})^2 = 3$, $(\sqrt{x + 2})^2 = x + 2$, etc. The second was presented in Section 2.4 and is $(a + b)^2 = a^2 + 2ab + b^2$. For example, $(x + 5)^2 = x^2 + 2 \cdot 5x + 5^2 = x^2 + 10x + 25$, and $(2 + \sqrt{x})^2 = 2^2 + 2 \cdot 2\sqrt{x} + (\sqrt{x})^2 = 4 + 4\sqrt{x} + x$.

The key to solving equations that contain one radical expression which is a square root is to isolate the radical on one side of the equation. After this is accomplished, square both sides of the equation. However, there may be a problem in doing so. When both sides of an equation are raised to a power, the resulting equation may have solutions that do not solve the original equation. Consider the following:

• Solving equations with one radical

Begin with the equation $x = 4$.

$x = 4$	The solution is obviously 4. Square both sides.
$x^2 = 4^2$	$4^2 = 16$.
$x^2 = 16$	Subtract 16 from both sides.
$x^2 - 16 = 0$	Factor $x^2 - 16$.
$(x + 4)(x - 4) = 0$	Set each factor equal to 0.
$x + 4 = 0, x - 4 = 0$	Solve each equation.
$x = -4, \quad x = 4$	Therefore, the solutions of $x^2 = 4^2$ are 4 and -4.

You will note that the equation that resulted from squaring both sides of the original equation not only has the solution of the original equation (4), but also has another solution (-4). This leads to the following observation.

OBSERVATION

If both sides of an equation are raised to a power, the resulting equation contains all the solutions of the original equation and perhaps contains some solutions that are not solutions of the original equation. Any solution of the resulting equation that does not solve the original equation is called an **extraneous** solution and is discarded.

This means that when we solve an equation by raising both sides to a power, we will get all the solutions of the original equation, but we may get some solutions that will not solve the original equation. Consequently, we must check all solutions of the resulting equation in the original equation. This difficulty results because the symbol, $\sqrt{}$, denotes the positive square root only. Following is the procedure for solving equations that contain one square root.

SOLVING EQUATIONS CONTAINING ONE SQUARE ROOT

If an equation contains one square root, it can be solved by:

1) Isolating the square root on one side of the equation.

2) Squaring both sides of the equation.

3) Solving the resulting equation.

4) Checking all solutions in the original equation.

EXAMPLE 1

Solve the following equations.

a)

$\sqrt{x} = 4$	Since $(\sqrt{x})^2 = x$, square both sides.
$(\sqrt{x})^2 = 4^2$	Simplify both sides.
$x = 16$	Therefore, 16 is a possible solution.

Check:

$\sqrt{x} = 4$	Substitute 16 for x.
$\sqrt{16} = 4$	Remember, the radical sign denotes the positive root only.
$4 = 4$	Therefore, 16 is the correct solution.

---

b) $\sqrt{2x + 3} - 5 = 0$ — Isolate the radical by adding 5 to both sides.

$\sqrt{2x + 3} - 5 + 5 = 0 + 5$ — Simplify both sides.

$\sqrt{2x + 3} = 5$ — Square both sides.

$(\sqrt{2x + 3})^2 = 5^2$ — Simplify both sides.

$2x + 3 = 25$ — Subtract 3 from both sides.

$2x = 22$ — Divide both sides by 2.

$x = 11$ — Therefore, 11 is a possible solution.

Check:

$\sqrt{2x + 3} - 5 = 0$ — Substitute 11 for x.

$\sqrt{2(11) + 3} - 5 = 0$ — Simplify.

$\sqrt{22 + 3} - 5 = 0$ — Continue simplifying.

$\sqrt{25} - 5 = 0$ — $\sqrt{25} = 5$.

$5 - 5 = 0$ — Simplify.

$0 = 0$ — Therefore, 11 is the correct solution.

c) $\sqrt{3x - 3} + 6 = 0$ — Isolate the radical by subtracting 6 from both sides.

$\sqrt{3x - 3} = -6$ — Square both sides.

$(\sqrt{3x - 3})^2 = (-6)^2$ — Simplify both sides.

$3x - 3 = 36$ — Add 3 to both sides.

$3x = 39$ — Divide both sides by 3.

$x = 13$ — Therefore, 13 is a possible solution.

Check:

$\sqrt{3x - 3} + 6 = 0$ — Substitute 13 for x.

$\sqrt{3(13) - 3} + 6 = 0$ — Simplify.

$\sqrt{39 - 3} + 6 = 0$ — Continue simplifying.

$\sqrt{36} + 6 = 0$ — $\sqrt{36} = 6$.

$6 + 6 = 0$ — Simplify.

$12 \neq 0$ — Therefore, 13 is not a solution.

Since 13 is the only possible solution, $\sqrt{3x - 3} + 6 = 0$ has no solution. Remember, this is usually denoted by $\varnothing$.

d) $\sqrt{3x + 10} = x + 2$ — Square both sides.

$(\sqrt{3x + 10})^2 = (x + 2)^2$ — Simplify both sides.

$3x + 10 = x^2 + 2 \cdot 2x + 4$ — Simplify the right side.

$3x + 10 = x^2 + 4x + 4$ — Subtract $3x$ and 10 from both sides.

$3x - 3x + 10 - 10 = x^2 + 4x - 3x + 4 - 10$ — Simplify both sides.

$0 = x^2 + x - 6$ — Factor the right side.

$0 = (x + 3)(x - 2)$ — Set each factor equal to 0.

$x + 3 = 0, \; x - 2 = 0$ — Solve each equation.

$x = -3, \quad x = 2$ — Therefore, the possible solutions are -3 and 2.

Check:

$x = -3$

$\sqrt{3x + 10} = x + 2$ Substitute -3 for x.

$\sqrt{3(-3) + 10} = -3 + 2$ Simplify.

$\sqrt{-9 + 10} = -1$ Continue simplifying.

$\sqrt{1} = -1$ $\sqrt{1} = 1$.

$1 \neq -1$ Therefore, -3 is not a solution.

$x = 2$

$\sqrt{3x + 10} = x + 2$ Substitute 2 for x.

$\sqrt{3(2) + 10} = 2 + 2$ Simplify.

$\sqrt{6 + 10} = 4$ Continue simplifying.

$\sqrt{16} = 4$ $\sqrt{16} = 4$.

$4 = 4$ Therefore, 2 is a solution.

Since -3 did not check, 2 is the only solution of the equation. ◼

ADDITIONAL PRACTICE

Solve the following equations.

a) $\sqrt{b} - 5 = 0$

b) $\sqrt{3x + 6} = 6$

c) $\sqrt{4 - 2x} - 8 = 0$

d) $\sqrt{3x + 3} = 2x - 1$

• **Solving equations with two radicals**

PRACTICE EXERCISES

Solve the following equations.

1) $\sqrt{a} - 6 = 0$ **2)** $\sqrt{4x - 8} - 4 = 0$

3) $\sqrt{2x - 5} + 7 = 0$ **4)** $\sqrt{x - 1} = x - 3$

If more practice is needed, do the Additional Practice Exercises in the margin.

If the equation contains more than one radical, the procedure for solving is more complicated and is outlined below.

SOLVING EQUATIONS CONTAINING TWO SQUARE ROOTS

If an equation contains two square roots, it can be solved by:

1) Isolating one of the radicals on one side of the equation.

2) Squaring both sides of the equation and, if necessary, simplifying each side.

3) Isolating the radical and squaring both sides of the equation again if the equation still contains a radical.

4) Checking all solutions in the original equation.

EXAMPLE 2

Solve the following equations.

a) $\sqrt{4x - 3} - \sqrt{2x + 7} = 0$ Isolate $\sqrt{4x - 3}$ by adding $\sqrt{2x + 7}$ to both sides.

$\sqrt{4x - 3} - \sqrt{2x + 7} + \sqrt{2x + 7} = 0 + \sqrt{2x + 7}$ Simplify.

$\sqrt{4x - 3} = \sqrt{2x + 7}$ Square both sides.

$(\sqrt{4x - 3})^2 = (\sqrt{2x + 7})^2$ Simplify.

$4x - 3 = 2x + 7$ Subtract $2x$ from both sides.

$2x - 3 = 7$ Add 3 to both sides.

$2x = 10$ Divide both sides by 2.

$x = 5$ Therefore, 5 is a possible solution.

Check:

$\sqrt{4x - 3} - \sqrt{2x + 7} = 0$ Substitute 5 for x.

$\sqrt{4(5) - 3} - \sqrt{2(5) + 7} = 0$ Simplify.

$\sqrt{20 - 3} - \sqrt{10 + 7} = 0$ Continue simplifying.

$\sqrt{17} - \sqrt{17} = 0$ Add like radicals.

$0 = 0$ Therefore, 5 is the solution.

 b) $\sqrt{x + 4} = 3 - \sqrt{x - 2}$ Since one radical is isolated, square both sides.

$(\sqrt{x + 4})^2 = (3 - \sqrt{x - 2})^2$ Simplify both sides.

$x + 4 = 3^2 - 2 \cdot 3\sqrt{x - 2} + (\sqrt{x - 2})^2$ Simplify the right side.

$x + 4 = 9 - 6\sqrt{x - 2} + x - 2$ Continue simplifying the right side.

$x + 4 = x + 7 - 6\sqrt{x - 2}$ Subtract x from both sides.

$4 = 7 - 6\sqrt{x - 2}$ Subtract 7 from both sides.

$-3 = -6\sqrt{x - 2}$ Divide both sides by -3.

$1 = 2\sqrt{x - 2}$ Square both sides.

$1^2 = (2\sqrt{x - 2})^2$ Simplify both sides.

$1 = 4(x - 2)$ Simplify the right side.

$1 = 4x - 8$ Add 8 to both sides.

$9 = 4x$ Divide both sides by 4.

$\dfrac{9}{4} = x$ Therefore, $\frac{9}{4}$ is a possible solution.

Check:

$\sqrt{x + 4} = 3 - \sqrt{x - 2}$ Substitute $\frac{9}{4}$ for x.

$\sqrt{\dfrac{9}{4} + 4} = 3 - \sqrt{\dfrac{9}{4} - 2}$ Get common denominators.

$\sqrt{\dfrac{9}{4} + \dfrac{16}{4}} = 3 - \sqrt{\dfrac{9}{4} - \dfrac{8}{4}}$ Add the fractions.

$\sqrt{\dfrac{25}{4}} = 3 - \sqrt{\dfrac{1}{4}}$ Find the square roots.

$\dfrac{5}{2} = 3 - \dfrac{1}{2}$ Common denominator on the right.

$\dfrac{5}{2} = \dfrac{6}{2} - \dfrac{1}{2}$ Add the fractions.

$\dfrac{5}{2} = \dfrac{5}{2}$ Therefore, $\frac{9}{4}$ is the correct solution.

PRACTICE EXERCISES

Solve the following equations.

5) $\sqrt{3x + 2} - \sqrt{x + 4} = 0$ **6)** $\sqrt{12 + x} = 6 - \sqrt{x}$

If more practice is needed, do the Additional Practice Exercises in the margin.

Radical equations frequently occur in real-world situations.

EXAMPLE 3

a) The distance a car will skid when the brakes are applied is approximated by the formula $s = k\sqrt{d}$ where s is the speed of the car in miles per hour, d is the distance the car will skid in feet, and k is a constant determined by road conditions, type of tires, and so on. If $k = 4.21$, find the following:

1) If the length of the skid mark is 196 feet, find the speed of the car.

Solution:

$s = k\sqrt{d}$	Substitute 4.21 for k and 196 for d.
$s = 4.21\sqrt{196}$	$\sqrt{196} = 14$.
$s = 4.21(14)$	Multiply.
$s = 58.94$	Therefore, the speed of the car was 58.94 miles per hour.

2) If the car is traveling at a speed of 63.15 miles per hour, how far will it take it to skid to a stop?

Solution:

$s = k\sqrt{d}$	Substitute 63.15 for s and 4.21 for k.
$63.15 = 4.21\sqrt{d}$	Divide both sides by 4.21.
$15 = \sqrt{d}$	Square both sides of the equation.
$225 = d$	Therefore, it would take the car 225 feet to stop.

b) Under certain conditions, the distance one can see to the horizon is given by $s = \sqrt{1.75h}$ where s is the distance in miles and h is the elevation in feet.

1) How far can a person see to the horizon if their elevation is 1372 feet?

Solution:

$s = \sqrt{1.75h}$	Substitute 1,372 for h.
$s = \sqrt{(1.75)(1,372)}$	Multiply.
$s = \sqrt{2401}$	$\sqrt{2401} = 49$
$s = 49$	Therefore, you would be able to see 49 miles to the horizon.

2) What would the elevation have to be in order to see 70 miles to the horizon?

Solution:

$$s = \sqrt{1.75h}$$ Substitute 70 for s.

$$70 = \sqrt{1.75h}$$ Square both sides.

$$4{,}900 = 1.75h$$ Divide both sides by 1.75.

$$2{,}800 = h$$ Therefore, the elevation would have to be 2,800 feet.

PRACTICE EXERCISES

Answer the following.

7) The number of amperes of current flowing in an electrical system is calculated by the formula amperes $= \sqrt{\frac{\text{watts}}{\text{ohms}}}$.

   a) How many amperes of current flow if there are 8,000 watts of power used through a 20 ohm resistance?

   b) How many watts of power are used by an appliance that uses 8 amperes when the resistance is 35 ohms?

8) After being dropped, the amount of time it takes an object to fall s feet is given by the formula $t = \frac{\sqrt{s}}{4}$ where t is the time in seconds and s is distance in feet, given that air resistance is negligible.

   a) How long will it take an object to reach the ground if it is dropped from a height of 256 feet?

   b) If an object is dropped from the top of a tall building and it takes 6 seconds for it to hit the ground, how tall is the building?

EXERCISE SET 8.6

Solve the following equations.

1) $\sqrt{a} = 2$

2) $\sqrt{c} = 9$

3) $\sqrt{z} - 7 = 0$

4) $\sqrt{n} - 5 = 0$

5) $\sqrt{a} + 4 = 0$

6) $\sqrt{b} + 2 = 0$

7) $2\sqrt{x} = 8$

8) $3\sqrt{y} = 9$

9) $\sqrt{y} - 4 = 4$

10) $\sqrt{x} + 2 = 6$

11) $\sqrt{x} + 3 = 8$

12) $\sqrt{y} - 4 = 2$

13) $\sqrt{5h - 21} = 3$

14) $\sqrt{2d + 39} = 7$

15) $\sqrt{7r - 26} - 4 = 0$

16) $\sqrt{7d - 24} - 2 = 0$

17) $\sqrt{8r - 15} = r$

18) $\sqrt{4d + 32} = d$

19) $\sqrt{2m + 24} - m = 0$

20) $\sqrt{6h - 8} - h = 0$

21) $\sqrt{w - 3} = w - 5$

22) $\sqrt{2t + 9} = t - 3$

23) $\sqrt{4k + 25} = k - 5$

24) $\sqrt{4f + 25} = f - 5$

25) $\sqrt{w - 4} - w + 6 = 0$

26) $\sqrt{x - 2} - x + 4 = 0$

27) $\sqrt{y - 4} - y + 4 = 0$

28) $\sqrt{3n + 16} - n + 4 = 0$

29) $\sqrt{4f - 20} = \sqrt{f - 2}$

30) $\sqrt{10n - 3} = \sqrt{4n + 3}$

31) $\sqrt{5n - 21} = \sqrt{n - 1}$

32) $\sqrt{6f - 4} = \sqrt{4f + 8}$

33) $\sqrt{3a - 7} - \sqrt{a - 1} = 0$

34) $\sqrt{6r - 26} - \sqrt{r - 1} = 0$

35) $\sqrt{4 - x} = 2 - \sqrt{x}$

36) $\sqrt{9 - 11x} = 3 - \sqrt{x}$

37) $\sqrt{a - 4} = 4 - \sqrt{a + 4}$

38) $\sqrt{b + 8} = 6 - \sqrt{b - 4}$

39) $\sqrt{c + 10} + \sqrt{c + 2} = 2$

40) $\sqrt{n + 7} + \sqrt{n - 4} = 1$

41) $\sqrt{x + 5} + \sqrt{x - 3} = 4$

42) $\sqrt{m - 4} + \sqrt{m + 5} = 3$

Answer the following.

43) Using the formula $s = k\sqrt{d}$ from Example 3a with $k = 3.7$, find the following.

    a) Find how fast a car was traveling if it leaves a skid mark of 324 feet while stopping.

    b) Find the length of the skid mark for a car stopping from a speed of 74 miles per hour.

44) The distance a person can see to the horizon on a particular day is given by $s = \sqrt{1.8h}$ where s is the distance in miles and h is the elevation of the person in feet. Answer the following.

    a) How far can a person see to the horizon from a height of 200 feet?

    b) How high would a person have to be in order to see a distance of 5 miles?

45) Using the formula amperes $= \sqrt{\frac{watts}{ohms}}$ given in Practice Exercise 7, find the following.

    a) Find the number of amperes of current flowing if the number of watts of power used is 4000 and the number of ohms of resistance is 10.

    b) Find the number of watts used by an electrical system that draws 15 amperes of current when the resistance is 120 ohms.

46) Using the formula $t = \frac{\sqrt{s}}{4}$ given in Practice Exercise 8, find the following.

    a) Find how long it takes an object to hit the ground if it is dropped from a height of 3600 feet.

    b) Find the height of a cliff if it takes 16 seconds for a rock dropped from the top to reach the ground below.

47) The equation for the length of time it takes a pendulum to make one swing is $t = 2\pi\sqrt{\frac{L}{32}}$ where t is the time in seconds and L is the length of the pendulum in feet. Find the following.

    a) Find how much time it takes a pendulum that is 2 feet long to make one swing.

    b) If a pendulum makes one swing every 1.5 seconds, find the length of the pendulum to the nearest one-hundredth of a foot.

48) If air resistance is neglected, the velocity of an object, v, in feet per second, after it has fallen from rest a distance of d feet, is given by $v = 8\sqrt{d}$. Find the following.

    a) Find the velocity of an object that has fallen 289 feet.

    b) If an object hits the ground with a velocity of 120 feet per second, how many feet has it fallen?

CHALLENGE EXERCISES (49–58)

Solve the following equations.

49) $\sqrt{2x - 3} - \sqrt{2x - 11} = 2$

50) $\sqrt{2x + 20} + \sqrt{2x + 5} = 3$

51) $\sqrt[3]{x} = 2$ **52)** $\sqrt[3]{x} = -3$

53) $\sqrt[3]{x + 3} = -2$ **54)** $\sqrt[3]{x - 5} = 1$

55) $\sqrt[4]{x} = 2$ **56)** $\sqrt[4]{x + 2} = 1$

57) $\sqrt[4]{2x - 1} = 3$ **58)** $\sqrt[4]{2x + 4} = 2$

WRITING EXERCISE

59) What is wrong with the following "solution" of the equation?

$$\sqrt{3x - 9} = 5 + \sqrt{2x + 6}$$
$$(\sqrt{3x - 9})^2 = (5 + \sqrt{2x + 6})^2$$
$$3x - 9 = 25 + 2x + 6$$
$$3x - 9 = 2x + 31$$
$$x - 9 = 31$$
$$x = 40$$

Pythagorean Theorem

When you complete this section, you will be able to:

Ⓐ Find the unknown side of a right triangle.

Ⓑ Find the length of a diagonal of a rectangle or square.

Ⓒ Find the length of an unknown side of a rectangle when given the length of a diagonal and the length of one side.

Ⓓ Find the length of a diagonal of a square.

Ⓔ Find the length of each side of a square when given the length of a diagonal.

Ⓕ Solve applications problems involving right triangles.

INTRODUCTION

- There is some doubt as to whether Pythagoras himself actually discovered the relationship between the sides of a right triangle or whether it was one of his followers.

- Conclusions drawn from observations are never accepted as fact until they are proven mathematically. Many proofs of the Pythagorean Theorem exist but none are included in this text. Once a fact has been proven, it is called a theorem.

Radicals occur in real-world situations when it is necessary to find an unknown side of a right triangle. The procedure for finding an unknown side of a right triangle is attributed to the Greek mathematician and philosopher, Pythagoras, and for that reason is called the Pythagorean Theorem.

A triangle in which one of the angles is a right (90°) angle is a **right triangle**. The sides that form the right angle are called the **legs** of the right triangle and the side opposite the right angle is called the **hypotenuse**. See the figure below.

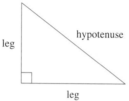

The Pythagorean Theorem probably came from observing that if squares were constructed on each of the sides of a right triangle, the sum of the areas of the squares on the legs was equal to the area of the square on the hypotenuse. See the figure below in which the length of the legs are a and b and the length of the hypotenuse is c.

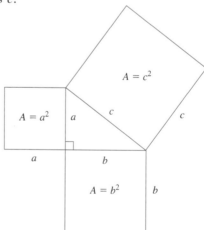

Thus, we have $a^2 + b^2 = c^2$. The Pythagorean Theorem is no longer stated in terms of areas of squares, but is stated as follows.

PYTHAGOREAN THEOREM

If the legs of a right triangle have lengths of a and b and the hypotenuse has length of c, then $a^2 + b^2 = c^2$.

In words, the sum of the squares of the lengths of the legs of a right triangle is equal to the square of the length of the hypotenuse.

Before finding an unknown side of a right triangle, we need one other concept which will be discussed in detail in Section 9.1. Consider an equation like $x^2 = 16$. This asks the question "the square of what number is 16?" This is the definition of square root. Consequently, the number(s) with square(s) equal to 16 are the square roots of 16 which are 4 and -4. Likewise, if $x^2 = 3$, then $x = \sqrt{3}$ and $-\sqrt{3}$. Since the lengths of sides of triangles are positive, we will be limited to taking the *positive square roots only*. Since a and b represent the legs, we may substitute the lengths of a leg for either a or b interchangeably.

EXAMPLE 1

Find the length of the unknown side of each of the following right triangles.

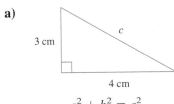

a)

Solution:

The lengths of the legs are 3 cm and 4 cm. We need to find the length of the hypotenuse. Use the Pythagorean Theorem.

$a^2 + b^2 = c^2$	Substitute for a and b.
$3^2 + 4^2 = c^2$	Square 3 and 4.
$9 + 16 = c^2$	Add 9 and 16.
$25 = c^2$	Find $\sqrt{25}$.
$5 = c$	Therefore, the length of the hypotenuse is 5 cm long.

b)

Solution:

We know the length of the hypotenuse and one of the legs. Use the Pythagorean Theorem to find the length of the other leg.

$a^2 + b^2 = c^2$	Substitute for a and c.
$7^2 + b^2 = 8^2$	Square 7 and 8.
$49 + b^2 = 64$	Subtract 49 from both sides.
$b^2 = 15$	Take $\sqrt{15}$.
$b = \sqrt{15}$	Therefore, the length of the other leg is $\sqrt{15}$ inches long.

c)

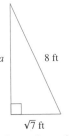

Solution:

We know the length of the hypotenuse and one leg. Use the Pythagorean Theorem to find the length of the other leg.

$a^2 + b^2 = c^2$	Substitute for b and c.
$a^2 + (\sqrt{7})^2 = 8^2$	Square $\sqrt{7}$ and 8.
$a^2 + 7 = 64$	Subtract 7 from both sides.
$a^2 = 57$	Take $\sqrt{57}$.
$a = \sqrt{57}$	Therefore, the length of the other leg is $\sqrt{57}$ feet long.

d) Find the length of the other leg of a right triangle if one leg is 8 feet long and the hypotenuse is 12 feet long.

Solution:
Use the Pythagorean Theorem. Substitute for a or b and c.

$a^2 + b^2 = c^2$	Substitute for b and c.
$a^2 + 8^2 = 12^2$	Square 8 and 12.
$a^2 + 64 = 144$	Subtract 64 from both sides.
$a^2 = 80$	Take $\sqrt{80}$.
$a = \sqrt{80}$	Simplify $\sqrt{80}$.
$a = 4\sqrt{5}$	Therefore, the unknown leg is $4\sqrt{5}$ feet.

ADDITIONAL PRACTICE

Find the length of the unknown side of each of the following right triangles.

a)

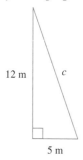

b)

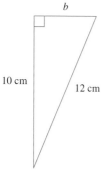

c)

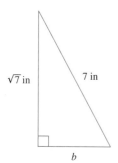

d) The hypotenuse of a right triangle is 9 inches long and one leg is 5 inches long. Find the length of the other leg.

PRACTICE EXERCISES

Find the length of the unknown side of each of the following right triangles.

1)

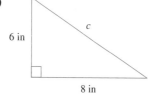

2)

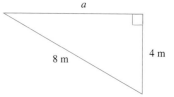

3)

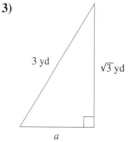

4) Find the length of the other leg of a right triangle if one leg is 7 inches long and the hypotenuse is 13 inches long.

If more practice is needed, do the Additional Practice Exercises in the margin.

A diagonal of a square or a rectangle divides the square or rectangle into two right triangles. See the figure below.

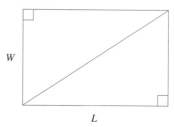

The sides of the rectangle become the legs of the right triangle and the diagonal of the rectangle becomes the hypotenuse of the right triangle. Consequently, if we know the lengths of the sides of the rectangle, we can find the length of the diagonal. Also, if we know the length of the diagonal and the length of one of the sides of the rectangle, we can find the length of the other side.

EXAMPLE 2

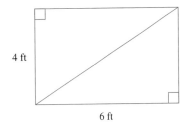 **a)** Find the length of the diagonal of the rectangle below.

4 ft

6 ft

Solution:
The lengths of the legs of the right triangle are 6 feet and 4 feet. The diagonal is the hypotenuse of a right triangle.

$a^2 + b^2 = c^2$	Substitute for a and b.
$6^2 + 4^2 = c^2$	Square 6 and 4.
$36 + 16 = c^2$	Add 36 and 16.
$52 = c^2$	Take $\sqrt{52}$.
$\sqrt{52} = c$	Simplify $\sqrt{52}$.
$2\sqrt{13} = c$	Therefore, the diagonal of the rectangle is $2\sqrt{13}$ feet.

 b) The length of a diagonal of a rectangle is $\sqrt{26}$ meters and the width of the rectangle is $2\sqrt{2}$ meters. Find the length.

Solution.
Draw and label a figure.

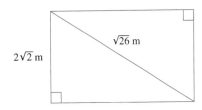

$2\sqrt{2}$ m

$\sqrt{26}$ m

The hypotenuse is $\sqrt{26}$ meters long and one leg is $2\sqrt{2}$ meters long.

$a^2 + b^2 = c^2$	Substitute for a (or b) and c.
$(2\sqrt{2})^2 + b^2 = (\sqrt{26})^2$	Square $2\sqrt{2}$ and $\sqrt{26}$.
$2^2(\sqrt{2})^2 + b^2 = 26$	Square 2 and $\sqrt{2}$.
$4 \cdot 2 + b^2 = 26$	Multiply 4 and 2.
$8 + b^2 = 26$	Subtract 8 from both sides.
$b^2 = 18$	Take $\sqrt{18}$.
$b = \sqrt{18} = 3\sqrt{2}$	Therefore, the length is $3\sqrt{2}$ meters.

c) The diagonal of a square is $4\sqrt{2}$ inches long. Find the length of the sides of the square.

Solution:

Draw and label a figure. Remember, all four sides of a square are equal in length. Therefore, let each side have length of x.

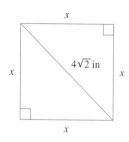

The legs are each x inches long and the hypotenuse is $4\sqrt{2}$ inches long.

$a^2 + b^2 = c^2$ Substitute x for a and b and $4\sqrt{2}$ for c.

$x^2 + x^2 = (4\sqrt{2})^2$ Add x^2 and x^2 and square $4\sqrt{2}$.

$2x^2 = 4^2(\sqrt{2})^2$ Square 4 and $\sqrt{2}$.

$2x^2 = 16 \cdot 2$ Multiply 16 and 2.

$2x^2 = 32$ Divide both sides by 2.

$x^2 = 16$ Take $\sqrt{16}$.

$x = \sqrt{16} = 4$ Therefore, each side of the square is 4 inches.

ADDITIONAL PRACTICE

e) Find the length of the diagonal of the rectangle below.

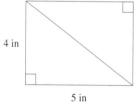

4 in

5 in

f) The diagonal of a rectangle is $\sqrt{30}$ feet long and the width is $\sqrt{14}$ feet long. Find the length.

g) The diagonal of a square is $\sqrt{48}$ meters long. Find the length of each side of the square.

PRACTICE EXERCISES

5) Find the length of the diagonal of the rectangle below.

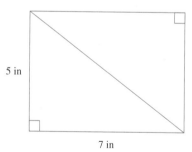

5 in

7 in

6) The diagonal of a rectangle is $2\sqrt{11}$ centimeters long and the length is $2\sqrt{7}$ centimeters. Find the width.

7) The diagonal of a square is $3\sqrt{2}$ yards long. Find the length of the sides of the square.

If more practice is needed, do the Additional Practice Exercises in the margin.

Applications of right triangles occur in many real-world situations. Knowledge of how to find the unknown side of a right triangle is important for solution of many kinds of problems.

EXAMPLE 3

Solve the following.

a) A building contractor wants to be sure the walls of a building meet at right angles. If he marks a point 6 feet from the corner of one wall and a point 8 feet from the corner of the other wall, what should be the distance between these two points?

Solution:

If the walls meet at right angles, the distances measured along each wall will be the legs of a right triangle and the distance between the points will be the hypotenuse. See the figure below.

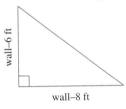

Use the Pythagorean Theorem to find the distance between the points.

$a^2 + b^2 = c^2$ Substitute for a and b.

$6^2 + 8^2 = c^2$ Square 6 and 8.

$36 + 64 = c^2$ Add 36 and 64.

$100 = c^2$ Find $\sqrt{100}$.

$10 = c$ If the walls meet at right angles, it will be ten feet between the two points.

b) A house is 48 feet wide. The roof is supported by rafters that are 7 feet above the level of the walls at the center of the house. Find the length of each rafter. See the figure below.

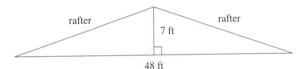

Solution:

The length of each rafter is the length of the hypotenuse of a right triangle. Since the rafters meet at the center of the house, the point below where the rafters meet is 24 feet from each wall. See the figure below.

Thus, we have two right triangles with legs of lengths 24 feet and 7 feet. Use the Pythagorean Theorem to find the hypotenuse.

$a^2 + b^2 = c^2$ Substitute for a and b.

$24^2 + 7^2 = c^2$ Square 24 and 7.

$576 + 49 = c^2$ Add 576 and 49.

$625 = c^2$ Find $\sqrt{625}$.

$25 = c$ Therefore, each rafter is 25 feet long.

PRACTICE EXERCISES

Solve the following.

8) Television screens are measured diagonally. For example, if a television is advertised as having a 17-inch screen, that means the screen measures 17 inches diagonally. If a television is advertised as having a 25-inch rectangular screen and the height of the screen is 20 inches, how wide is the screen?

9) The range for a particular brand of marine radio is 20 miles. Bill and Don are in separate boats which are equipped with these radios. Their boats are traveling on courses that are at right angles to each other. If Bill is 8 miles from port and Don is 15 miles from the same port, can they contact each other by radio? Why or why not?

EXERCISE SET   **8.7**

Find the value of x in each of the following right triangles.

1)

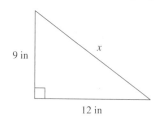

2)

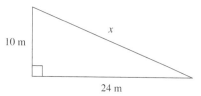

3)

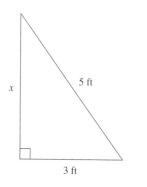

4)

5)

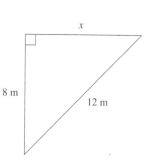

6)

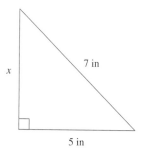

13) One leg is 4 meters long and the hypotenuse is 6 meters long.

14) One leg is 10 centimeters long and the hypotenuse is 12 centimeters long.

15) The lengths of the legs are $\sqrt{38}$ inches and $\sqrt{14}$ inches.

16) The lengths of the legs are $\sqrt{10}$ yards and $\sqrt{22}$ yards.

17) One leg is $2\sqrt{6}$ centimeters long and the hypotenuse is $4\sqrt{2}$ centimeters long.

18) One leg is $2\sqrt{7}$ inches long and the hypotenuse is $4\sqrt{3}$ inches long.

Find the lengths of the diagonals of the following rectangles or squares.

7)

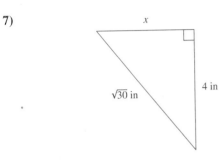

8)

9)

10)

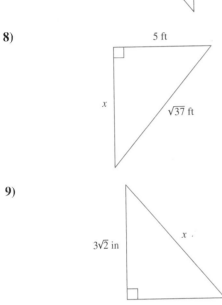

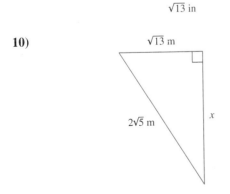

Find the length of the unknown side of each of the following right triangles.

11) The lengths of the legs are 6 feet and 9 feet.

12) The lengths of the legs are 7 inches and 12 inches.

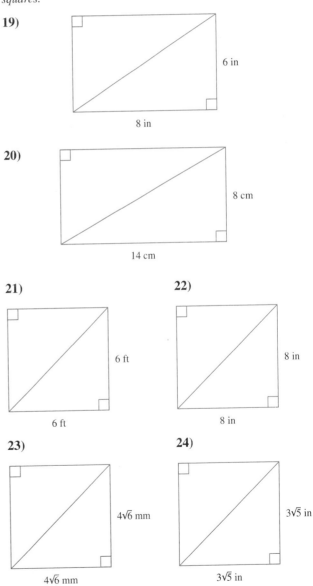

19) 6 in / 8 in

20) 8 cm / 14 cm

21) 6 ft / 6 ft

22) 8 in / 8 in

23) $4\sqrt{6}$ mm / $4\sqrt{6}$ mm

24) $3\sqrt{5}$ in / $3\sqrt{5}$ in

Find the length of the unknown side(s) of each of the following rectangles.

25)　　　　　　　　　26)

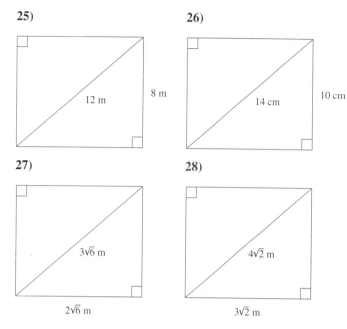

27)　　　　　　　　　28)

Solve the following.

29) The length of a rectangle is 5 centimeters and the width is 8 centimeters. Find the length of a diagonal.

30) The length of a rectangle is 9 meters and the width is 6 meters. Find the length of a diagonal.

31) The length of a diagonal of a rectangle is 12 inches and the length of one side is $4\sqrt{5}$ inches. Find the width.

32) The length of a diagonal of a rectangle is 14 yards and the length of one side is $4\sqrt{6}$ yards. Find the width.

33) The length of each side of a square is 7 feet. Find the length of a diagonal.

34) The length of each side of a square is 5 feet. Find the length of a diagonal.

35) The length of each side of a square is $5\sqrt{2}$ centimeters. Find the length of a diagonal.

36) The length of each side of a square is $2\sqrt{7}$ feet. Find the length of a diagonal.

37) The length of a diagonal of a square is 8 inches. Find the length of each side of the square.

38) The length of a diagonal of a square is 10 yards. Find the length of each side of the square.

Solve the following.

39) In Example 3a, a building contractor wanted to be sure the walls of a building met at right angles. Suppose he marks a point on one wall 8 feet from the corner and a point 15 feet from the corner on the other wall. If the walls are perpendicular, how far is it between the two points?

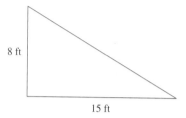

40) In Example 3b, we found the length of some rafters. How long would the rafters be for a house 48 feet wide if the rafters meet at the center of the house at a point 10 feet above the level of the walls?

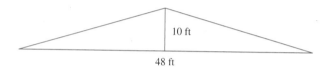

41) Practice Exercise 8 dealt with the measurement of television screens. If a television has a 13 inch screen that is 10 inches high, how wide is it?

42) Practice Exercise 9 dealt with a particular brand of marine radio. Suppose a marine radio has a range of 25 miles. Bill and Don are traveling in separate boats equipped with these radios on courses that are at right angles to each other as in Practice Exercise 9. If Bill is 12 miles from port and Don is 16 miles from the same port, can they contact each other by radio? Why or why not?

43) When hurricanes threaten, a common practice is to tape windows to keep the glass from shattering. If a rectangular window is 40 inches long and 30 inches wide, how long is a piece of tape that goes diagonally across the window?

44) Suppose a plate glass window 72 inches long and 60 inches wide is taped as in Exercise 43. How long is a piece of tape that goes diagonally across this window?

45) A newly transplanted tree is supported by three ropes that are attached to the tree at a point 24 feet above

the ground. These ropes are tied to pegs that are driven into the ground 7 feet from the base of the tree. What is the total length of the three pieces of rope?

46) A guy wire is attached to a telephone pole at a point 30 feet above the ground. How long is the wire if the other end is attached at a point on the ground 10 feet from the base of the pole?

47) An advertising sign is in the shape of a rectangle. A stripe is painted diagonally across the sign. If the length of the stripe is 8 feet and the sign is 6 feet long, find the width of the sign.

48) Two pieces of wood are nailed together at right angles. To give added support, a third piece of wood is nailed to each of these at points 3 feet from the corner. Find the length of this piece of wood.

WRITING EXERCISE

49) Write and solve an application problem involving the Pythagorean Theorem.

GROUP PROJECT

50) Make a list of five situations in which right triangles occur in "everyday" situations.

Summary

Root of a Number: [Section 8.1]

A number b is an nth root of a number a if $b^n = a$. Written symbolically, $\sqrt[n]{a} = b$ if $b^n = a$. If n is even, then a must be non-negative. If n is odd, then a may have any real number value.

Radical: [Section 8.1]

A radical is an algebraic expression of the form $\sqrt[n]{a}$. The symbol, $\sqrt{}$, is the radical sign, n is the root index, and a the radicand.

Perfect n^{th} Powers: [Section 8.1]

The perfect squares are $\{1, 4, 9, 16, 25, ...\}$.

The perfect cubes are $\{1, 8, 27, 64, 125, ...\}$.

The perfect fourth powers are $\{1, 16, 81, 256, ...\}$.

And so on ... .

Numbers that are perfect n^{th} powers have rational n^{th} roots.

Irrational Numbers: [Section 8.1]

An irrational number is any real number that is not rational. Written as a decimal, irrational numbers neither terminate nor repeat. Examples are $\sqrt{2}$, $\sqrt[3]{5}$, and $\sqrt[5]{8}$.

Products of Radicals: [Section 8.2]

If $\sqrt[n]{a}$ and $\sqrt[n]{b}$ both exist, then $\sqrt[n]{a} \cdot \sqrt[n]{b} = \sqrt[n]{ab}$. This rule is also used in the form of $\sqrt[n]{ab} = \sqrt[n]{a} \cdot \sqrt[n]{b}$.

Radicals to Powers: [Section 8.2]

In general, if $\sqrt[n]{a}$ exists, then $(\sqrt[n]{a})^n = a$.

Quotients of Radicals: [Section 8.3]

If $\sqrt[n]{a}$ and $\sqrt[n]{b}$ both exist and $b \neq 0$, then $\dfrac{\sqrt[n]{a}}{\sqrt[n]{b}} = \sqrt[n]{\dfrac{a}{b}}$. This rule is also used in the form $\sqrt[n]{\dfrac{a}{b}} = \dfrac{\sqrt[n]{a}}{\sqrt[n]{b}}$.

Like Radicals: [Section 8.3]

Like radicals must have the same root index and the same radicand.

Addition of Like Radical Expressions: [Section 8.3]

To add like radicals, add the coefficients and leave the radical portion unchanged. Never add the radicands. Often it is necessary to simplify the radical expressions before they can be added.

Simplest Form of a Radical: [Sections 8.3 and 8.4]

1) Find all rational roots.

2) Remove all possible factors from the radicand. For example, no square root may have a radicand with a perfect square factor.

3) Find all possible products, quotients, sums, and differences.

4) Rationalize all the denominators. This also means no radicand may contain a fraction.

Rationalizing the Denominator: [Section 8.4]

1) If the denominator is in the form of $\sqrt{a}$, multiply the numerator and the denominator of the original fraction by the same appropriate radical so that the denominator of the original fraction becomes a radical expression with rational value.

2) If the denominator is of the form $a + b$ where a and/or b is a square root, multiply the numerator and the denominator of the fraction by the conjugate of the denominator. The conjugate is found by changing the middle sign of the expression $a + b$ where a and/or b is a square root.

Roots of Equations Containing Radicals: [Section 8.5]

If both sides of an equation are raised to a power, the resulting equation contains all the solutions of the original equation and perhaps contains some solutions which do not solve the original equation.

Extraneous Solutions: [Section 8.5]

Any solution of an equation, resulting from raising both sides of an equation to a power, which is not a solution of the original equation is called an extraneous solution and is discarded.

Solving Equations Containing Square Roots: [Section 8.5]

1) If the equation contains only one square root, isolate the radical on one side of the equation. Then square both sides of the equation and solve the

resulting equation. Check all solutions in the original equation.

2) If the equation contains two square roots, isolate one radical on one side of the equation. Then square both sides of the equation and simplify the resulting equation. Isolate the remaining radical on one side of the equation and square both sides again. Solve the

resulting equation and check all solutions in the original equation.

Pythagorean Theorem: [Section 8.6]

If the lengths of the legs of a right triangle are a and b and the length of the hypotenuse is c, then $a^2 + b^2 = c^2$.

Review Exercises

Find the following roots if they exist. If the root does not exist, write "this root does not exist as a real number." Assume all variables have non-negative values. [Sections 8.1 and 8.2]

1) $\sqrt{16}$ 2) $\sqrt{-16}$ 3) $-\sqrt{16}$

4) $\pm\sqrt{16}$ 5) $\sqrt[3]{125}$ 6) $\sqrt[3]{-125}$

7) $-\sqrt[3]{125}$ 8) $-\sqrt[3]{-125}$ 9) $\sqrt[4]{81}$

10) $\sqrt[4]{-81}$ 11) $-\sqrt[4]{81}$ 12) $\sqrt{b^4}$

13) $\sqrt{n^5}$ 14) $\sqrt[3]{m^5}$ 15) $\sqrt[5]{z^8}$

16) $\sqrt{m^5 n^6}$ 17) $\sqrt[3]{a^4 b^6}$ 18) $\sqrt[4]{x^6 y^7}$

Identify each of the following as rational or irrational. If the number is rational, find the root exactly. If the number is irrational, approximate the number to the nearest thousandth by using a calculator or Table 1. [Section 8.1]

19) $\sqrt{32}$ 20) $\sqrt{169}$ 21) $\sqrt[3]{63}$

22) $\sqrt[3]{1331}$ 23) $\sqrt{1.44}$ 24) $\sqrt[3]{.027}$

Simplify the following radicals. [Section 8.2]

25) $\sqrt{63}$ 26) $\sqrt{44}$ 27) $4\sqrt{125}$

28) $5\sqrt{108}$ 29) $\sqrt[3]{54}$ 30) $6\sqrt[3]{32}$

31) $\sqrt{162x^7}$ 32) $\sqrt{200a^8}$ 33) $3\sqrt{40z^5 y^3}$

34) $\sqrt[3]{128a^4}$ 35) $5\sqrt[3]{108n^3 m^7}$ 36) $\sqrt[4]{162z^5}$

Find the following products or powers of roots. Leave answers in simplified form. Assume all variables have non-negative values. [Section 8.3]

37) $(\sqrt{r})^2$ 38) $(\sqrt{x^3})^2$ 39) $(\sqrt{5})^2$

40) $\sqrt{5}\cdot\sqrt{7}$ 41) $\sqrt{6}\cdot\sqrt{r}$ 42) $\sqrt{m}\cdot\sqrt{n}$

43) $\sqrt{7}\cdot\sqrt{7}$ 44) $\sqrt{5}\cdot\sqrt{35}$ 45) $\sqrt{45}\cdot\sqrt{5}$

46) $4\sqrt{6}\cdot 2\sqrt{24}$ 47) $5\sqrt{75}\cdot 3\sqrt{6}$

48) $7\sqrt{18}\cdot 3\sqrt{6}$ 49) $\sqrt{c^3}\cdot\sqrt{c}$

50) $\sqrt{s^5}\cdot\sqrt{s^3}$ 51) $\sqrt{v}\cdot\sqrt{v^4}$

52) $2\sqrt{30n^3}\cdot 3\sqrt{10n^3}$ 53) $6\sqrt{5v^4}\cdot 2\sqrt{15v^3}$

Find the following quotients. Leave all answers in simplified form. Assume all variables have positive values. [Section 8.3]

54) $\sqrt{\dfrac{25}{9}}$ 55) $\sqrt{\dfrac{81x^2}{4y^4}}$ 56) $\sqrt{\dfrac{13}{25}}$

57) $\dfrac{\sqrt{48}}{\sqrt{3}}$ 58) $\dfrac{6\sqrt{72}}{3\sqrt{8}}$ 59) $\dfrac{\sqrt{x^7}}{\sqrt{x^3}}$

60) $\dfrac{\sqrt{c^5 d^4}}{\sqrt{c^2 d^2}}$ 61) $\dfrac{10\sqrt{63a^5}}{5\sqrt{7a}}$ 62) $\dfrac{6a^3\sqrt{140n^5}}{2a\sqrt{5n^2}}$

Simplify the following. Leave answers in simplified form. Assume all variables have positive values. [Section 8.3]

63) $\sqrt{\dfrac{3}{2}}\cdot\sqrt{\dfrac{5}{8}}$ 64) $\sqrt{\dfrac{n^2}{m^3}}\cdot\sqrt{\dfrac{n^5}{m}}$

65) $\sqrt{\dfrac{7a}{18b}}\cdot\sqrt{\dfrac{3a}{2b}}$

Find the following sums. Leave answers in simplified form. Assume all variables have non-negative values. [Section 8.4]

66) $7\sqrt{5} - 9\sqrt{5}$ 67) $6\sqrt{a} + 12\sqrt{a}$

68) $4a\sqrt[3]{x} - 6a\sqrt[3]{x}$ 69) $\sqrt{3} + \sqrt{75}$

70) $\sqrt{48} - \sqrt{3}$ 71) $4\sqrt{27} - 5\sqrt{12}$

72) $-4\sqrt{28} - 5\sqrt{63}$ 73) $2a\sqrt{20} - 5a\sqrt{45}$

74) $4\sqrt{63a^3} + 6\sqrt{64a^3}$

Simplify the following. [Section 8.4]

75) $\sqrt{6}\cdot\sqrt{2} + \sqrt{6}\cdot\sqrt{8}$

76) $\dfrac{\sqrt{300}}{\sqrt{2}} - \sqrt{96}$

77) $\sqrt{5}(3 - \sqrt{30})$

78) $\sqrt{3}(\sqrt{3} + 2\sqrt{6})$

79) $(4 + \sqrt{5})(5 - \sqrt{5})$

80) $(2\sqrt{3} + \sqrt{5})(\sqrt{3} - 4\sqrt{5})$

81) $(3\sqrt{3} + \sqrt{6})(3\sqrt{3} - \sqrt{6})$

82) $(2 - \sqrt{7})^2$

Reduce the following to lowest terms. [Section 8.4]

83) $\dfrac{6 - 2\sqrt{2}}{12}$ 84) $\dfrac{8 + 4\sqrt{12}}{4}$

Rationalize the following denominators. Leave answers in simplest form. Assume all variables have positive values. [Section 8.5]

85) $\dfrac{6}{\sqrt{7}}$ 86) $\dfrac{8}{\sqrt{18}}$ 87) $\sqrt{\dfrac{5}{3}}$

88) $\dfrac{a}{\sqrt{b}}$ 89) $\sqrt{\dfrac{m}{n}}$ 90) $\dfrac{5}{\sqrt{5n}}$

91) $\dfrac{4}{\sqrt[3]{9}}$ 92) $\dfrac{n}{\sqrt[3]{n^2}}$ 93) $\dfrac{4a}{\sqrt[3]{2a}}$

94) $\dfrac{4}{2 - \sqrt{5}}$ 95) $\dfrac{\sqrt{7}}{\sqrt{7} + 2}$ 96) $\dfrac{6}{\sqrt{3} - \sqrt{5}}$

97) $\dfrac{\sqrt{12}}{3 - \sqrt{6}}$ 98) $\dfrac{2 + \sqrt{5}}{2 - \sqrt{5}}$ 99) $\dfrac{5 + \sqrt{3}}{\sqrt{5} - \sqrt{3}}$

Solve the following equations. [Section 8.6]

100) $\sqrt{a} = 9$

101) $3\sqrt{n} - 18 = 0$

102) $\sqrt{x + 4} = 5$

103) $2\sqrt{2x + 4} + 6 = 2$

104) $\sqrt{3n - 5} = \sqrt{5n + 9}$

105) $4\sqrt{s} = \sqrt{15s - 5}$

106) $\sqrt{7x - 10} = x$

107) $\sqrt{x - 3} = x - 3$

108) $\sqrt{n + 2} + \sqrt{n - 6} = 4$

Find the value of x in each of the following. [Section 8.7]

109)

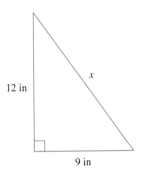

110)

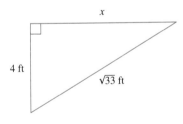

111)

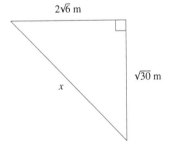

112)

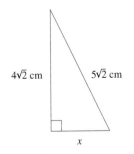

113)

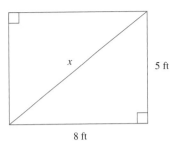

114)

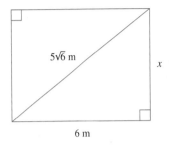

115) The lengths of the legs of a right triangle are 9 inches and 10 inches. Find the length of the hypotenuse.

116) The length of the hypotenuse of a right triangle is 12 meters and the length of one leg is 4 meters. Find the length of the other leg.

117) A rectangle has a length of $6\sqrt{3}$ inches and a width of $2\sqrt{5}$ inches. Find the length of a diagonal.

118) A rectangle has a diagonal length of 9 yards and a width of $3\sqrt{2}$ yards. Find the length.

119) Each side of a square is 10 inches long. Find the length of a diagonal.

120) The length of a diagonal of a square is $3\sqrt{10}$ feet. Find the length of each side of the square.

121) A television antenna is supported by two guy wires. The wires are attached to the antenna at a point 15 feet above the ground. The other ends of the wires are anchored to the ground at a point 5 feet from the base of the antenna. Find the length of each of the wires.

122) A tree was broken over during a storm with the broken portion still resting on the stump forming a right triangle. The point where the tree broke is 15 feet above the ground. The top of the tree is on the ground at a point 36 feet from the base of the tree. How high was the tree before it broke?

Chapter 8 Test

Find the following roots if they exist. If the root does not exist, write "the root does not exist as a real number."

1) $\sqrt{81}$

2) $\sqrt[3]{-64}$

3) $\sqrt{x^8}$

4) $\sqrt[3]{c^9 d^6}$

Simplify the following.

5) $\sqrt{96}$

6) $3\sqrt{98x^5}$

7) $4\sqrt[3]{40}$

8) $3\sqrt[3]{64y^7}$

Find the following products and/or quotients. Leave answers in simplified form. Assume all variables have positive values.

9) $2\sqrt{35} \cdot 3\sqrt{5}$

10) $3\sqrt{15a^3} \cdot 4\sqrt{10a^4}$

11) $\dfrac{\sqrt{96}}{\sqrt{6}}$

12) $\dfrac{8\sqrt{56c^7}}{\sqrt{7c^2}}$

13) $\sqrt{\dfrac{21c^3}{5}} \cdot \sqrt{\dfrac{3c}{5}}$

Find the following sums.

14) $6\sqrt{6} - 12\sqrt{6}$

15) $3\sqrt{80} + 5\sqrt{20}$

16) $\sqrt{180a^3} + 4\sqrt{45a^3}$

Simplify the following. Leave answers in simplified form. Assume all variables have non-negative values.

17) $\sqrt{3}(\sqrt{6} + 4\sqrt{18})$

18) $(\sqrt{5} + 2\sqrt{6})(3\sqrt{5} - \sqrt{6})$

19) $(4 + 2\sqrt{7})(4 - 2\sqrt{7})$

20) $(5 - \sqrt{x})^2$

Rationalize the denominators of the following. Assume all variables have positive values.

21) $\dfrac{8}{\sqrt{12}}$

22) $\dfrac{8}{\sqrt[3]{16}}$

23) $\sqrt{\dfrac{m}{n}}$

24) $\dfrac{12}{7 - \sqrt{5}}$

29) Find the value of x in the right triangle below.

25) $\dfrac{5 + \sqrt{6}}{2 - \sqrt{6}}$

Solve the following equations. If there is no solution, write $\varnothing$.

26) $2\sqrt{b} - 10 = 0$

27) $\sqrt{7z + 58} - 11 = 0$

30) Two cars approach the same intersection at right angles to each other. If one car is 45 feet from the intersection and the other is 60 feet from the intersection, find the distance between the two cars.

28) $\sqrt{x + 27} - \sqrt{x + 6} = 3$

9

9.1   Solving Incomplete Quadratic Equations
9.2   Solving Quadratic Equations by Completing the Square
9.3   Solving Quadratic Equations by the Quadratic Formula
9.4   Solving Mixed Equations
9.5   Applications Involving Quadratic Equations

Solving Quadratic Equations

CHAPTER OVERVIEW

Quadratic equations were introduced in Section 4.8. Recall that a quadratic equation is any equation that may be put in the form of $ax^2 + bx + c = 0$ with $a \neq 0$. We solved quadratic equations by factoring and the solutions were rational numbers.

However, if asked to solve the equation $x^2 + x - 5 = 0$ by factoring, we could not since $x^2 + x - 5$ cannot be factored. In this chapter we will learn two methods of solving this type of equation. One method is called completing the square and involves creating a perfect square trinomial, which is the square of a binomial. The procedure for squaring a binomial was discussed in Section 2.4. You may want to review this procedure prior to doing Section 9.2. The second method involves use of the quadratic formula. The quadratic formula is derived by completing the square on the general equation $ax^2 + bx + c = 0$. The solutions of quadratic equations are often irrational numbers that were discussed in Chapter 8. Section 9.4 contains types of equations that we have solved previously (linear, quadratic, radical, rational), except now many of the radical and rational equations will result in quadratic equations after simplification.

Quadratic equations also allow us to solve a wide variety of applications problems and we end the chapter with a section on applications problems that can be solved only by the use of quadratic equations.

SECTION 9.1

Solving Incomplete Quadratic Equations

OBJECTIVES

When you complete this section, you will be able to:

Ⓐ Solve quadratic equations of the form $ax^2 + bx = 0$.
Ⓑ Solve quadratic equations of the form $x^2 = k$ by extraction of roots.
Ⓒ Solve quadratic equations of the form $(ax + b)^2 = c$ by extraction of roots.

INTRODUCTION

As mentioned in the chapter introduction, quadratic equations were first introduced in Section 4.8, and a quadratic equation is any equation that can be put in the form $ax^2 + bx + c = 0$ with $a \neq 0$. In Chapter 4 we solved quadratic equations by factoring. Recall that after the equation was written in $ax^2 + bx + c = 0$ form, we factored $ax^2 + bx + c$ and set each factor equal to 0. We then solved the resulting linear equations. For example, solve $x^2 + 2x - 15 = 0$.

$$x^2 + 2x - 15 = 0 \qquad \text{Factor } x^2 + 2x - 15.$$
$$(x + 5)(x - 3) = 0 \qquad \text{Set each factor equal to 0.}$$
$$x + 5 = 0,\ x - 3 = 0 \qquad \text{Solve each equation.}$$
$$x = -5, \quad x = 3 \qquad \text{Therefore, the solutions are } -5 \text{ and } 3.$$

However, this method cannot be used to solve all quadratic equations since not all quadratic polynomials are factorable. For example, $x^2 + x + 3 = 0$ cannot be solved by factoring since $x^2 + x + 3$ cannot be factored. In this section we will learn a technique that will be used in Section 9.2 to develop a procedure that will allow us to solve any quadratic equation.

A quadratic equation in the form of $ax^2 + bx + c = 0$ in which either b or $c = 0$ is called an **incomplete quadratic equation**. If $c = 0$, the equation is of the form $ax^2 + bx = 0$. This type of equation can be solved by factoring and was discussed in Section 4.8. We will do a couple of examples to refresh your memory.

• **Solving quadratic equations of the form $ax^2 + bx = 0$**

EXAMPLE 1

Solve the following equations.

a) $2x^2 + 3x = 0$ Remove the common factor x.

$\quad x(2x + 3) = 0$ Set each factor equal to 0.

$\quad x = 0 \text{ or } 2x + 3 = 0$ $x = 0$ is one solution. Solve $2x + 3 = 0$.

$\qquad\qquad\qquad 2x = -3$ Therefore, the solutions are 0 and $-\frac{3}{2}$.

$$x = -\frac{3}{2}$$

Check:

$$x = 0 \qquad\qquad\qquad x = -\frac{3}{2}$$

Substitute 0 for x.   Substitute $-\frac{3}{2}$ for x.

$$2x^2 + 3x = 0 \qquad\qquad 2x^2 + 3x = 0$$

$$2(0)^2 + 3(0) = 0 \qquad 2\left(-\frac{3}{2}\right)^2 + 3\left(-\frac{3}{2}\right) = 0$$

$$2(0) + 0 = 0 \qquad\qquad 2\left(\frac{9}{4}\right) - \frac{9}{2} = 0$$

$$0 + 0 = 0 \qquad\qquad\qquad \frac{9}{2} - \frac{9}{2} = 0$$

$$0 = 0 \qquad\qquad\qquad\qquad 0 = 0$$

Therefore, $x = 0$ and $x = -\frac{3}{2}$ are the correct solutions.

b) $4x^2 = 8x$ Subtract $8x$ from both sides.

$\ 4x^2 - 8x = 0$ Remove the common factor of $4x$.

$\ 4x(x - 2) = 0$ Set each factor equal to 0.

$\ 4x = 0 \text{ or } x - 2 = 0$ Solve each equation.

$\ x = 0 \qquad x = 2$ Therefore, the solutions are 0 and 2.

Check:

The check is left as an exercise for the student.   ▪

PRACTICE EXERCISES

Solve the following equations.

1) $3x^2 + 9x = 0$ **2)** $5x^2 = 2x$

- **Solving quadratic equations that can be put in the form $x^2 = c$**

Incomplete quadratic equations with $b = 0$ have the form $ax^2 + c = 0$. We also solved some special cases of this type of equation in Section 4.8 by factoring. For example, $x^2 - 9 = 0$ can be solved by factoring. However, $x^2 - 3 = 0$ cannot be solved by factoring. The technique we will use in this section to solve equations of the form $ax^2 + c = 0$ is known as **extraction of roots** and was used in Section 8.7 on the Pythagorean Theorem.

We begin with the equation $x^2 - 9 = 0$. This equation can be solved by factoring as follows:

$$x^2 - 9 = 0$$
$$(x + 3)(x - 3) = 0$$
$$x + 3 = 0, x - 3 = 0$$
$$x = -3, \quad x = 3$$

Therefore, the solutions are -3 and 3.

However, $x^2 - 9 = 0$ can be written as $x^2 = 9$. This equation is asking the question, "The square of what number is equal to 9?" This is precisely the definition of square root. Thus, we can solve the equation $x^2 = 9$ by extracting the two square roots of 9. Since the two square roots of 9 are 3 and -3, the solutions of $x^2 = 9$ are 3 and -3. This procedure is summarized below.

SOLVING QUADRATIC EQUATIONS BY EXTRACTION OF ROOTS

To solve quadratic equations of the form $ax^2 + c = 0$,

1) Solve the equation for x^2. This results in an equation of the form $x^2 = k$ where k is some constant.

2) The solutions of $x^2 = k$ are both the positive and negative square roots of k written as $x = \pm\sqrt{k}$ (read "x equals the positive and negative square roots of k"). That is, $x = \sqrt{k}$ or $x = -\sqrt{k}$.

3) Write the answers in simplified form.

Note: If k is a negative number, then $\sqrt{k}$ is not a real number. Therefore, the equation has no real number solutions. As mentioned in Section 8.1, the square roots of negative numbers result in imaginary numbers and are usually discussed in Intermediate Algebra. Consequently, when we encounter such situations in this text, we will write "there are no real number solutions" and we will indicate the solution set as $\varnothing$.

EXAMPLE 2

Solve the following equations.

a) $x^2 = 16$ Take the positive and negative square roots of 16.

   $x = \pm 4$ Therefore, the solutions are 4 and -4.

Check:

$x = 4$	$x = -4$
Substitute 4 for x.	Substitute -4 for x.
$x^2 = 16$	$x^2 = 16$
$4^2 = 16$	$(-4)^2 = 16$
$16 = 16$	$16 = 16$

Therefore, ± 4 are the correct solutions.

b) $x^2 - 27 = 0$ Add 27 to both sides of the equation.

   $x^2 = 27$ Take the positive and negative square roots of 27.

   $x = \pm\sqrt{27}$ Rewrite 27 as $9 \cdot 3$.

   $x = \pm\sqrt{9 \cdot 3}$ Apply the product rule for radicals.

   $x = \pm\sqrt{9} \cdot \sqrt{3}$ $\sqrt{9} = 3$.

   $x = \pm 3\sqrt{3}$ Therefore, the solutions are $3\sqrt{3}$ and $-3\sqrt{3}$.

Check:

$x = 3\sqrt{3}$	$x = -3\sqrt{3}$
Substitute $3\sqrt{3}$ for x.	Substitute $-3\sqrt{3}$ for x.
$x^2 - 27 = 0$	$x^2 - 27 = 0$
$(3\sqrt{3})^2 - 27 = 0$	$(-3\sqrt{3})^2 - 27 = 0$
$3^2(\sqrt{3})^2 - 27 = 0$	$(-3)^2(\sqrt{3})^2 - 27 = 0$
$9 \cdot 3 - 27 = 0$	$9 \cdot 3 - 27 = 0$
$27 - 27 = 0$	$27 - 27 = 0$
$0 = 0$	$0 = 0$

Therefore, $\pm 3\sqrt{3}$ are the correct solutions.

c) $2x^2 - 6 = 0$ Add 6 to both sides of the equation.

   $2x^2 = 6$ Divide both sides of the equation by 2.

   $x^2 = 3$ Take the positive and negative square roots of 3.

   $x = \pm\sqrt{3}$ Therefore, the solutions are $\sqrt{3}$ and $-\sqrt{3}$.

Check:

The check is left as an exercise for the student.

d) $3x^2 - 7 = 0$ Add 7 to both sides of the equation.

$3x^2 = 7$ Divide both sides by 3.

$x^2 = \dfrac{7}{3}$ Take the positive and negative square roots of $\frac{7}{3}$.

$x = \pm\sqrt{\dfrac{7}{3}}$ Apply the quotient rule for radicals.

$x = \pm\dfrac{\sqrt{7}}{\sqrt{3}}$ Rationalize the denominator.

$x = \pm\dfrac{\sqrt{7}}{\sqrt{3}} \cdot \dfrac{\sqrt{3}}{\sqrt{3}}$ Apply the product rule for radicals.

$x = \pm\dfrac{\sqrt{21}}{\sqrt{9}}$ $\sqrt{9} = 3$.

$x = \pm\dfrac{\sqrt{21}}{3}$ Therefore, the solutions are $\frac{\sqrt{21}}{3}$ and $-\frac{\sqrt{21}}{3}$.

Check:
The check is left as an exercise for the student.

e) $x^2 + 4 = 0$ Subtract 4 from both sides.

$x^2 = -4$ Take the positive and negative square roots of -4.

$x = \pm\sqrt{-4}$ Since $\sqrt{-4}$ does not exist as a real number, this equation has no real number solutions. So the solution set is $\varnothing$.

Be careful: Do not confuse solving equations like $x^2 = 9$ and finding $\sqrt{9}$. Remember, the symbol $\sqrt{}$ means find the positive square root only. Hence, $\sqrt{9} = 3$ only. *There is no radical sign in $x^2 = 9$ so we are not limited to just the positive square root. We are asked to find the number(s) with square equal to 9.* Consequently, we find both the positive and negative square roots of 9.

ADDITIONAL PRACTICE

Solve the following equations.

a) $x^2 = 81$

b) $x^2 - 18 = 0$

c) $6x^2 - 42 = 0$

d) $7x^2 - 5 = 0$

e) $2x^2 + 18 = 0$

- **Solving quadratic equations of the form** $(ax + b)^2 = c$

PRACTICE EXERCISES

Solve the following equations.

3) $x^2 = 49$ **4)** $x^2 - 8 = 0$ **5)** $4x^2 - 24 = 0$

6) $5x^2 - 2 = 0$ **7)** $x^2 + 16 = 0$

If more practice is needed, do the Additional Practice Exercises in the margin.

The method of extraction of roots can be extended to solve equations like $(2x + 3)^2 = 9$. We illustrate with some examples.

EXAMPLE 3

Solve the following equations. If there are no real number solutions, indicate the solution set as $\varnothing$.

a) $(x + 3)^2 = 16$ Take the positive and negative square roots of 16.

$x + 3 = \pm 4$ Rewrite as two equations.

$x + 3 = 4, \; x + 3 = -4$ Solve each equation.

$x = 1, \qquad x = -7$ Therefore, the solutions are 1 and -7.

Check:

$x = 1$	$x = -7$
Substitute 1 for x.	Substitute -7 for x.
$(x + 3)^2 = 16$	$(x + 3)^2 = 16$
$(1 + 3)^2 = 16$	$(-7 + 3)^2 = 16$
$4^2 = 16$	$(-4)^2 = 16$
$16 = 16$	$16 = 16$

Therefore, 1 and -7 are the correct solutions.

b) $(x + 3)^2 = 5$ Take the positive and negative square roots of 5.

$x + 3 = \pm\sqrt{5}$ Rewrite as two equations.

$x + 3 = \sqrt{5}, \, x + 3 = -\sqrt{5}$ Solve each equation.

$x = -3 + \sqrt{5}, \, x = -3 - \sqrt{5}$ Therefore, the solutions are $-3 + \sqrt{5}$ and $-3 - \sqrt{5}$.

Note: The solutions of Example 3b are usually written as $-3 \pm \sqrt{5}$.

Check:

$x = -3 + \sqrt{5}$	$x = -3 - \sqrt{5}$
Substitute $-3 + \sqrt{3}$ for x.	Substitute $-3 - \sqrt{3}$ for x.
$(x + 3)^2 = 5$	$(x + 3)^2 = 5$
$(-3 + \sqrt{5} + 3)^2 = 5$	$(-3 - \sqrt{5} + 3)^2 = 5$
$(\sqrt{5})^2 = 5$	$(-\sqrt{5})^2 = 5$
$5 = 5$	$5 = 5$

Therefore, $-3 + \sqrt{5}$ and $-3 - \sqrt{5}$ are the correct solutions.

c) $(2x + 4)^2 = 25$ Take the positive and negative square roots of 25.

$2x + 4 = \pm 5$ Rewrite as two equations.

$2x + 4 = 5, \, 2x + 4 = -5$ Solve each equation.

$2x = 1, \qquad 2x = -9$

$x = \dfrac{1}{2}, \qquad x = -\dfrac{9}{2}$ Therefore, the solutions are $\frac{1}{2}$ and $-\frac{9}{2}$.

Check:

The check is left as an exercise for the student.

d) $(3x - 1)^2 = 7$ — Take the positive and negative square roots of 7.

$$3x - 1 = \pm\sqrt{7}$$ — Rewrite as two equations.

$3x - 1 = \sqrt{7}, \; 3x - 1 = -\sqrt{7}$ — Solve each equation.

$3x = 1 + \sqrt{7}, \; 3x = 1 - \sqrt{7}$

$x = \dfrac{1 + \sqrt{7}}{3}, \; x = \dfrac{1 - \sqrt{7}}{3}$ — Therefore, the solutions are $\dfrac{1 + \sqrt{7}}{3}$ and $\dfrac{1 - \sqrt{7}}{3}$.

Note: The solutions to Example 3d are often written as $\dfrac{1 \pm \sqrt{7}}{3}$.

Check:

$x = \dfrac{1 + \sqrt{7}}{3}$

Substitute $\dfrac{1 + \sqrt{7}}{3}$ for x.

$(3x - 1)^2 = 7$

$\left(3\left(\dfrac{1 + \sqrt{7}}{3}\right) - 1\right)^2 = 7$

$(1 + \sqrt{7} - 1)^2 = 7$

$(\sqrt{7})^2 = 7$

$7 = 7$

$x = \dfrac{1 - \sqrt{7}}{3}$

Substitute $\dfrac{1 - \sqrt{7}}{3}$ for x.

$(3x - 1)^2 = 7$

$\left(3\left(\dfrac{1 - \sqrt{7}}{3}\right) - 1\right)^2 = 7$

$(1 - \sqrt{7} - 1)^2 = 7$

$(-\sqrt{7})^2 = 7$

$7 = 7$

Therefore, $\dfrac{1 \pm \sqrt{7}}{3}$ are the correct solutions.

e) $(2a - 3)^2 + 12 = 0$ — Subtract 12 from both sides.

$(2a - 3)^2 = -12$ — Take the positive and negative square roots of -12.

$2a - 3 = \pm\sqrt{-12}$ — Since $\sqrt{-12}$ does not exist as a real number, this equation has no real number solutions. So the solution set is $\varnothing$.

ADDITIONAL PRACTICE

Solve the following equations.

f) $(x - 6)^2 = 64$

g) $(x + 4)^2 = 2$

h) $(4x + 3)^2 = 9$

i) $(5x - 1)^2 = 11$

j) $(x + 5)^2 + 25 = 0$

PRACTICE EXERCISES

Solve the following equations. If there are no real number solutions, indicate the solution set as $\varnothing$.

8) $(x + 5)^2 = 36$ **9)** $(x - 4)^2 = 8$ **10)** $(3x - 2)^2 = 4$

11) $(5x - 3)^2 = 10$ **12)** $(3x - 1)^2 + 16 = 0$

If more practice is needed, do the Additional Practice Exercises in the margin.

Quadratic equations are also used in the solution of applications problems.

EXAMPLE 4

a) The square of three less than a number is 36. Find the number(s).

Solution:

Let x represent the number. Then,
$x - 3$ represents three less than the number. Therefore, the equation is:

$(x - 3)^2 = 36$	Take the positive and negative square roots of 36.
$x - 3 = \pm 6$	Rewrite as two equations.
$x - 3 = 6, \ x - 3 = -6$	Solve each equation.
$x = 9, \qquad x = -3$	Therefore, the numbers are 9 and -3.

Check:

Three less than 9 is $9 - 3 = 6$ and $6^2 = 36$. Therefore, 9 is a solution. Three less than -3 is $-3 - 3 = -6$ and $(-6)^2 = 36$. Therefore, -3 is also a solution.

b) Find the length of each side of a square with area of 49 square inches.

Solution:

The formula for the area of a square is $A = s^2$, where s is the length of a side of the square.

$A = s^2$	Substitute 49 for A.
$49 = s^2$	Take the positive and negative square roots of 49.
$\pm 7 = s$	Since the length of a side of a square must be positive, the only solution is 7 inches.
$7 = s$	

PRACTICE EXERCISES

13) The square of four more than a number is 64. Find the number(s).

14) Find the length of each side of a square with area of 169 square centimeters.

EXERCISE SET 9.1

Solve each of the following equations. If there are no real number solutions, indicate the solution set as $\varnothing$.

1) $4x^2 + 8x = 0$

2) $2x^2 + 10x = 0$

3) $7a^2 = -21a$

4) $5r^2 = -45r$

5) $9z^2 = 6z$

6) $12t^2 = 30t$

7) $x(x + 2) = 4x$

8) $n(n - 4) = 2n$

9) $r^2 = 25$

10) $y^2 = 100$

11) $a^2 = -36$

12) $p^2 = -64$

13) $x^2 - 121 = 0$

14) $x^2 - 144 = 0$

15) $x^2 - 7 = 0$

16) $x^2 - 11 = 0$

17) $x^2 + 5 = 0$

18) $t^2 + 7 = 0$

19) $x^2 - 48 = 0$

20) $x^2 - 32 = 0$

21) $w^2 - 28 = 0$

22) $m^2 - 45 = 0$

23) $2x^2 - 8 = 0$

24) $3x^2 - 27 = 0$

25) $4x^2 - 20 = 0$

26) $5x^2 - 30 = 0$

27) $2m^2 - 40 = 0$

28) $3r^2 - 54 = 0$

29) $5q^2 - 40 = 0$ **30)** $7b^2 - 56 = 0$

31) $7x^2 + 28 = 0$ **32)** $3a^2 + 30 = 0$

33) $2x^2 - 7 = 0$ **34)** $3x^2 - 5 = 0$

35) $5a^2 - 11 = 0$ **36)** $7r^2 - 5 = 0$

37) $3z^2 - 4 = 0$ **38)** $6a^2 - 25 = 0$

39) $(x + 4)^2 = 25$ **40)** $(a - 2)^2 = 100$

41) $(x + 5)^2 = 44$ **42)** $(n - 7)^2 = 54$

43) $(3c + 4)^2 = 49$ **44)** $(2a - 5)^2 = 64$

45) $(2r - 3)^2 = 121$ **46)** $(3s - 6)^2 = 81$

47) $(x - 3)^2 = 7$ **48)** $(x + 4)^2 = 3$

49) $(m + 5)^2 = 11$ **50)** $(n - 9)^2 = 13$

51) $(2a + 4)^2 = 2$ **52)** $(3s - 2)^2 = 5$

53) $(s - 2)^2 = 98$ **54)** $(n - 5)^2 = 75$

55) $(3x + 2)^2 = 52$ **56)** $(2a - 6)^2 = 72$

Solve the following applications problems. Remember from Section 4.9 that if the solution of an application problem requires a quadratic equation, not all solutions of the equation will always satisfy the conditions of the problem. Always check the solution(s) in the wording of the problem.

57) The square of three more than a number is 25. Find the number(s).

58) The square of two less than three times a number is 16. Find the number(s).

Exercises 59 through 62 require the use of the formula for the area of a square. The formula for the area of a square is $A = s^2$ where s represents the length of each side.

59) Find the length of each side of a square if the area is 64 square inches.

60) Find the length of each side of a square if the area is 121 square meters.

61) If the length of each side of a square is doubled and then increased by four, the area would be 100 square centimeters. Find the length of each side of the square.

62) If the length of each side of a square is tripled and then decreased by three, the area would be 36 square feet. Find the length of each side of the square.

CHALLENGE EXERCISES (63–66)

Solve the following equations. If there are no real number solutions, indicate the solution set as $\varnothing$.

63) $3(x + 5)^2 = 7$ **64)** $2(a - 3)^2 = 5$

65) $5(2x + 3)^2 - 11 = 0$ **66)** $7(4x - 1)^2 - 15 = 0$

WRITING EXERCISES

67) Compare and contrast the meanings of $\sqrt{49}$ and $x^2 = 49$. How are they similar and how are they different?

68) Explain in detail why the equation $(x + 2)^2 = -9$ has no real number solution.

SECTION **9.2**

Solving Quadratic Equations by Completing the Square

OBJECTIVES

When you complete this section, you will be able to:

Ⓐ   Determine the number to be added to a polynomial of the form $x^2 + bx$ in order to form a perfect square trinomial.

Ⓑ   Solve quadratic equations by completing the square.

INTRODUCTION

Recall from Section 4.4 that a perfect square trinomial is a trinomial that is the square of a binomial. For example, $(x + 3)^2 = (x)^2 + 2(x)(3) + 3^2 = x^2 + 6x + 9$. Since $x^2 + 6x + 9 = (x + 3)^2$, it is a perfect square trinomial. By examining $x^2 + 6x + 9 = (x + 3)^2$ and the procedure for squaring a binomial, we make the following observations about a binomial that when squared gives a perfect square trinomial.

1) The first term of the binomial is the square root of the first term of the trinomial.

2) The second term of the binomial is the square root of the last term of the trinomial.

3) The sign between the terms of the binomial is the same as the sign of the middle term of the trinomial.

Using the above example, we have:

$$x^2 + 6x + 9$$
$$(\sqrt{x^2} + \sqrt{9})$$

• **Completing the square**

That is, $x^2 + 6x + 9 = (\sqrt{x^2} + \sqrt{9})^2 = (x + 3)^2$.

Suppose we know the first two terms of a trinomial whose squared term has a coefficient of 1 and we want to find the constant term necessary for the trinomial to be a perfect square trinomial. For example, what constant term is necessary for $x^2 + 6x + \_$ to be a perfect square trinomial, and what is the binomial whose square gives this perfect square trinomial?

If we represent the last term of the above trinomial with a^2, we have $x^2 + 6x + a^2$. We need to find a. The first term of the binomial whose square gives $x^2 + 6x + a^2$ is $\sqrt{x^2} = x$. The second term of the binomial is $\sqrt{a^2} = a$. Since the sign of the second term of the trinomial is $+$, we have:

$$x^2 + 6x + a^2 = (x + a)^2$$

Expand $(x + a)^2$ and we have:

$$x^2 + 6x + a^2 = x^2 + 2ax + a^2$$

Since two polynomials are equal if and only if their corresponding terms are equal, we conclude that:

$$6 = 2a$$

$$\frac{6}{2} = a$$

$$3 = a$$

Since the last term of the perfect square trinomial is a^2 and $a = 3$, the number needed in order for $x^2 + 6x + \_$ to be a perfect square trinomial is $3^2 = 9$. Hence, $x^2 + 6x + 9$ is a perfect square trinomial and $x^2 + 6x + 9 = (x + 3)^2$.

The procedure for finding the number that will make a polynomial of the form $x^2 + bx$ a perfect square trinomial is called **completing the square**. We do not want to go through all these steps each time we need to complete the square. The key to the procedure lies in the steps which we repeat below.

$$6 = 2a$$

$$\frac{6}{2} = a$$

$$3 = a$$

We then squared 3 to get the last term of the perfect square trinomial. Since 6 is the coefficient of the first degree term of the trinomial, we found the last term of the perfect square trinomial by taking half the coefficient of the first degree term and squaring it. We generalize this discussion to the following procedure for completing the square.

COMPLETING THE SQUARE

To find the number needed to make a polynomial of the form $x^2 + bx$ a perfect square trinomial, multiply the coefficient of the first degree term by $\frac{1}{2}$ and square the result. Since the square of a negative number is positive, we ignore the sign of the first degree term.

EXAMPLE 1

Find the number needed to make each of the following a perfect square trinomial. Then write each trinomial as the square of a binomial.

a) $x^2 + 8x +$ __   Take $\frac{1}{2}$ of 8 and square it. $\frac{1}{2}(8) = 4$ and $4^2 = 16$. Therefore, the number needed is 16.

     $x^2 + 8x + 16$ Write as the square of a binomial.

     $(x + 4)^2$

b) $a^2 - 10a +$ __   Take $\frac{1}{2}$ of 10 and square it. $\frac{1}{2} \cdot 10 = 5$ and $5^2 = 25$. Therefore, the number needed is 25.

     $a^2 - 10a + 25$ Write as the square of a binomial.

     $(a - 5)^2$

c) $b^2 + 3b +$ __   Take $\frac{1}{2}$ of 3 and square it. $\frac{1}{2}(3) = \frac{3}{2}$ and $\left(\frac{3}{2}\right)^2 = \frac{9}{4}$. Therefore, the number needed is $\frac{9}{4}$.

     $b^2 + 3b + \dfrac{9}{4}$ Write as the square of a binomial.

     $\left(b + \dfrac{3}{2}\right)^2$

d) $r^2 - \dfrac{7}{3}r +$ __   Take $\frac{1}{2}$ of $\frac{7}{3}$ and square it. $\frac{1}{2}\left(\frac{7}{3}\right) = \frac{7}{6}$ and $\left(\frac{7}{6}\right)^2 = \frac{49}{36}$.

     Therefore, the number needed is $\frac{49}{36}$.

     $r^2 - \dfrac{7}{3}r + \dfrac{49}{36}$ Write as the square of a binomial.

     $\left(r - \dfrac{7}{6}\right)^2$

ADDITIONAL PRACTICE

Find the number needed to make each of the following a perfect square trinomial. Then write each trinomial as the square of a binomial.

a) $x^2 + 12x +$ __

b) $r^2 - 4r +$ __

c) $c^2 + 9c +$ __

d) $r^2 - \dfrac{3}{5}r +$ __

PRACTICE EXERCISES

Find the number needed to make each of the following a perfect square trinomial. Then write each trinomial as the square of a binomial.

1) $x^2 + 2x +$ __   **2)** $n^2 - 14n +$ __   **3)** $m^2 + 7m +$ __   **4)** $y^2 - \dfrac{5}{2}y +$ __

If more practice is needed, do the Additional Practice Exercises in the margin.

By combining the techniques of completing the square and extracting of roots, we can solve any quadratic equation using the procedure outlined below.

SOLVING QUADRATIC EQUATIONS BY COMPLETING THE SQUARE

To solve a quadratic equation by completing the square:

1) If the coefficient of the x^2 term is not 1, divide by the coefficient.

2) Write the equation in the form $x^2 + bx = c$.

3) Add the constant needed to make $x^2 + bx$ a perfect square to both sides of the equation.

4) Write the perfect square trinomial as the square of a binomial.

5) Solve by extraction of roots.

6) Check all solutions in the original equation.

• Solving quadratic equations by completing the square

Although the equations in Example 2 could be solved by factoring we will solve them by completing the square for illustrative purposes.

EXAMPLE 2

Solve the following equations by completing the square.

a) $x^2 - 2x - 8 = 0$ Add 8 to both sides of the equation.

$x^2 - 2x = 8$ Find the number that must be added to $x^2 - 2x$ to make it a perfect square trinomial and add it to both sides of the equation.

$x^2 - 2x + \_ = 8 + \_$ Take $\frac{1}{2}$ of 2 and square it. $\frac{1}{2}(2) = 1$ and $1^2 = 1$. Therefore, add 1 to both sides of the equation.

$x^2 - 2x + 1 = 8 + 1$ Write $x^2 - 2x + 1$ as the square of a binomial.

$(x - 1)^2 = 9$ Solve using extraction of roots.

$x - 1 = \pm 3$ Write as two equations.

$x - 1 = 3, \ x - 1 = -3$ Solve each equation.

$x = 4, \qquad x = -2$ Therefore, the solutions are 4 and -2.

Check:

$x = 4$	$x = -2$
Substitute 4 for x.	Substitute -2 for x.
$x^2 - 2x - 8 = 0$	$x^2 - 2x - 8 = 0$
$4^2 - 2(4) - 8 = 0$	$(-2)^2 - 2(-2) - 8 = 0$
$16 - 8 - 8 = 0$	$4 + 4 - 8 = 0$
$0 = 0$	$0 = 0$

Therefore, 4 and -2 are the correct solutions.

b) $a^2 + 5a + 4 = 0$
           Subtract 4 from both sides of the equation.

$a^2 + 5a = -4$
           Find the number that must be added to $a^2 + 5a$ to make it a perfect square trinomial and add it to both sides of the equation.

$a^2 + 5a + \_ = -4 + \_$
           Take $\frac{1}{2}$ of 5 and square it. $\frac{1}{2}(5) = \frac{5}{2}$ and $\left(\frac{5}{2}\right)^2 = \frac{25}{4}$.

           Therefore, add $\frac{25}{4}$ to both sides of the equation.

$a^2 + 5a + \dfrac{25}{4} = -4 + \dfrac{25}{4}$
           Write the left side of the equation as the square of a binomial and write the right side with the LCD of 4.

$\left(a + \dfrac{5}{2}\right)^2 = -\dfrac{16}{4} + \dfrac{25}{4}$
           Simplify the right side.

$\left(a + \dfrac{5}{2}\right)^2 = \dfrac{9}{4}$
           Solve by using extraction of roots.

$a + \dfrac{5}{2} = \pm\dfrac{3}{2}$
           Write as two equations.

$a + \dfrac{5}{2} = \dfrac{3}{2}, a + \dfrac{5}{2} = -\dfrac{3}{2}$
           Solve each equation.

$a = \dfrac{3}{2} - \dfrac{5}{2}, a = -\dfrac{3}{2} - \dfrac{5}{2}$

$a = -\dfrac{2}{2} = -1, a = -\dfrac{8}{2} = -4$
           Therefore, the solutions are -1 and -4.

Check:

To check, substitute -1 and -4 in the original equation. This is left as an exercise for the student.

ADDITIONAL PRACTICE

Solve the following equations by completing the square.

e) $x^2 - 6x + 8 = 0$

f) $b^2 + 3b + 2 = 0$

PRACTICE EXERCISES

Solve the following equations by completing the square.

5) $x^2 - 8x + 12 = 0$ **6)** $r^2 + 3r - 18 = 0$

If more practice is needed, do the Additional Practice Exercises in the margin.

As indicated above, the equations in Example 2 could have been solved by factoring. The real advantage gained from completing the square concerns equations that cannot be solved by factoring.

EXAMPLE 3

Solve the following equations by completing the square.

a) $x^2 - 4x + 2 = 0$ Subtract 2 from both sides of the equation.

$x^2 - 4x = -2$ Find the number that must be added to $x^2 - 4x$ to make it a perfect square trinomial and add it to both sides of the equation.

$x^2 - 4x + \_ = -2 + \_$ Take $\frac{1}{2}$ of 4 and square it. $\frac{1}{2}(4) = 2$ and $2^2 = 4$. Therefore, add 4 to both sides.

$x^2 - 4x + 4 = -2 + 4$ Write the left side as the square of a binomial.

$(x - 2)^2 = 2$ Solve using extraction of roots.

$x - 2 = \pm\sqrt{2}$ Write as two equations.

$x - 2 = \sqrt{2}, x - 2 = -\sqrt{2}$ Solve each equation.

$x = 2 + \sqrt{2}, x = 2 - \sqrt{2}$ Therefore, the solutions are $2 \pm \sqrt{2}$.

Check:

$x = 2 + \sqrt{2}$
Substitute $2 + \sqrt{2}$ for x.
$x^2 - 4x + 2 = 0$
$(2 + \sqrt{2})^2 - 4(2 + \sqrt{2}) + 2 = 0$
$4 + 4\sqrt{2} + 2 - 8 - 4\sqrt{2} + 2 = 0$
$0 = 0$

$x = 2 - \sqrt{2}$
Substitute $2 - \sqrt{2}$ for x.
$x^2 - 4x + 2 = 0$
$(2 - \sqrt{2})^2 - 4(2 - \sqrt{2}) + 2 = 0$
$4 - 4\sqrt{2} + 2 - 8 + 4\sqrt{2} + 2 = 0$
$0 = 0$

Therefore, the correct solutions are $2 \pm \sqrt{2}$.

b) $a^2 = -7a - 1$ We need the equation to be in $a^2 + ba = c$ form, so add $7a$ to both sides of the equation.

$a^2 + 7a = -1$ Find the number that must be added to $a^2 + 7a$ to make it a perfect square trinomial and add it to both sides of the equation.

$a^2 + 7a + \_ = -1 + \_$ Take $\frac{1}{2}$ of 7 and square it. $\frac{1}{2}(7) = \frac{7}{2}$ and $\left(\frac{7}{2}\right)^2 = \frac{49}{4}$. Therefore, add $\frac{49}{4}$ to both sides.

$a^2 + 7a + \dfrac{49}{4} = -1 + \dfrac{49}{4}$ Write the left side as the square of a binomial and write the right side with the LCD of 4.

$\left(a + \dfrac{7}{2}\right)^2 = -\dfrac{4}{4} + \dfrac{49}{4}$ Simplify the right side.

$\left(a + \dfrac{7}{2}\right)^2 = \dfrac{45}{4}$ Solve using extraction of roots.

$$a + \frac{7}{2} = \pm\sqrt{\frac{45}{4}}$$

Apply the quotient rule and write 45 as $9 \cdot 5$.

$$a + \frac{7}{2} = \pm\frac{\sqrt{9 \cdot 5}}{\sqrt{4}}$$

Apply the product rule to $\sqrt{9 \cdot 5}$ and $\sqrt{4} = 2$.

$$a + \frac{7}{2} = \pm\frac{\sqrt{9} \cdot \sqrt{5}}{2}$$

$\sqrt{9} = 3$.

$$a + \frac{7}{2} = \pm\frac{3\sqrt{5}}{2}$$

Write as two equations.

$$a + \frac{7}{2} = \frac{3\sqrt{5}}{2}, a + \frac{7}{2} = -\frac{3\sqrt{5}}{2}$$

Solve each equation.

$$a = -\frac{7}{2} + \frac{3\sqrt{5}}{2}, a = -\frac{7}{2} - \frac{3\sqrt{5}}{2}$$

Add the fractions.

$$a = \frac{-7 + 3\sqrt{5}}{2}, a = \frac{-7 - 3\sqrt{5}}{2}$$

Therefore, the solutions are $x = \frac{-7 \pm 3\sqrt{5}}{2}$.

Check:

The check is complicated and is left as a challenge exercise for the student.

c) $x^2 + 8x + 20 = 0$

Subtract 20 from both sides.

$x^2 + 8x + \_ = -20 + \_$

Take $\frac{1}{2}$ of 8 and square it. $\frac{1}{2}(8) = 4$ and $4^2 = 16$. Therefore, add 16 to both sides.

$x^2 + 8x + 16 = -20 + 16$

Write the left side as the square of a binomial. Simplify the right sides.

$(x + 4)^2 = -4$

Solve using extraction of roots.

$x + 4 = \pm\sqrt{-4}$

Since $\sqrt{-4}$ is not a real number, this equation has no real number solutions. Hence the solution set is $\varnothing$.

ADDITIONAL PRACTICE

Solve the following equations by completing the square.

g) $r^2 = -6r + 4$

h) $x^2 - 4 = -5x$

i) $a^2 - 6a + 13 = 0$

PRACTICE EXERCISES

Solve the following equations by completing the square. If there is no real solution, indicate the solution set as $\varnothing$.

7) $x^2 - 4x - 6 = 0$ **8)** $b^2 - 5 = -5b$ **9)** $x^2 - 10x + 34 = 0$

If more practice is needed, do the Additional Practice Exercises in the margin.

The procedure for completing the square works only if the coefficient of the second degree term is one. If the coefficient of the second degree term is not one, we must divide both sides of the equation by that coefficient in order to make it one. We illustrate with some examples.

ANSWERS:
Practice 7–9

7) $x = 2 \pm \sqrt{10}$

8) $b = \dfrac{-5 \pm 3\sqrt{5}}{2}$

9) $\varnothing$

Additional Practice g–i

g) $-3 \pm \sqrt{13}$

h) $x = \dfrac{-5 \pm \sqrt{41}}{2}$

i) $\varnothing$

EXAMPLE 4

Solve the following equations by completing the square.

a) $2x^2 - 12 = 5x$ — Rewrite in the form $ax^2 + bx = c$.

$2x^2 - 5x = 12$ — Divide both sides of the equation by 2 to get x^2.

$\dfrac{2x^2 - 5x}{2} = \dfrac{12}{2}$ — Simplify both sides.

$x^2 - \dfrac{5}{2}x = 6$ — Find the number that must be added to $x^2 - \frac{5}{2}x$ to make it a perfect square trinomial and add it to both sides of the equation.

$x^2 - \dfrac{5}{2}x + \_\_ = 6 + \_\_$ — Take $\frac{1}{2}$ of $\frac{5}{2}$ and square it. $\frac{1}{2} \cdot \frac{5}{2} = \frac{5}{4}$ and $\left(\frac{5}{4}\right)^2 = \frac{25}{16}$. Add $\frac{25}{16}$ to both sides.

$x^2 - \dfrac{5}{2}x + \dfrac{25}{16} = 6 + \dfrac{25}{16}$ — Write the left side as the square of a binomial and write the right side with the LCD of 16.

$\left(x - \dfrac{5}{4}\right)^2 = \dfrac{96}{16} + \dfrac{25}{16}$ — Simplify the right side.

$\left(x - \dfrac{5}{4}\right)^2 = \dfrac{121}{16}$ — Solve using extraction of roots.

$x - \dfrac{5}{4} = \pm\dfrac{11}{4}$ — Write as two equations.

$x - \dfrac{5}{4} = \dfrac{11}{4}, x - \dfrac{5}{4} = -\dfrac{11}{4}$ — Solve each equation.

$x = \dfrac{11}{4} + \dfrac{5}{4}, x = -\dfrac{11}{4} + \dfrac{5}{4}$ — Add the fractions.

$x = \dfrac{16}{4} = 4, x = -\dfrac{6}{4} = -\dfrac{3}{2}$ — Therefore, the solutions are 4 and $-\frac{3}{2}$.

Check:

The check is left as an exercise for the student.

b) $3x^2 + 6x + 1 = 0$ — Subtract 1 from both sides.

$3x^2 + 6x = -1$ — Divide both sides by 3 to get x^2.

$\dfrac{3x^2 + 6x}{3} = \dfrac{-1}{3}$ — Simplify the left side.

$x^2 + 2x = -\dfrac{1}{3}$ — Find the number that must be added to $x^2 + 2x$ to make it a perfect square trinomial and add it to both sides.

$x^2 + 2x + \_\_ = -\dfrac{1}{3} + \_\_$ — Take $\frac{1}{2}$ of 2 and square it. $\frac{1}{2}(2) = 1$ and $1^2 = 1$. Add 1 to both sides.

$x^2 + 2x + 1 = -\dfrac{1}{3} + 1$ — Write the left side as a binomial squared and write the right side with the LCD of 3.

$(x + 1)^2 = -\dfrac{1}{3} + \dfrac{3}{3}$ — Simplify the right side.

$(x + 1)^2 = \dfrac{2}{3}$ — Solve using extraction of roots.

$$x + 1 = \pm\sqrt{\frac{2}{3}}$$ Apply the quotient rule for radicals.

$$x + 1 = \pm\frac{\sqrt{2}}{\sqrt{3}}$$ Rationalize the denominator.

$$x + 1 = \pm\frac{\sqrt{2}}{\sqrt{3}} \cdot \frac{\sqrt{3}}{\sqrt{3}}$$ Simplify the right side.

$$x + 1 = \pm\frac{\sqrt{6}}{3}$$ Rewrite as two equations.

$$x + 1 = \frac{\sqrt{6}}{3}, x + 1 = -\frac{\sqrt{6}}{3}$$ Solve each equation.

$$x = -1 + \frac{\sqrt{6}}{3}, x = -1 - \frac{\sqrt{6}}{3}$$ Write the right side with the LCD of 3.

$$x = -\frac{3}{3} + \frac{\sqrt{6}}{3}, x = -\frac{3}{3} - \frac{\sqrt{6}}{3}$$ Add the fractions.

$$x = \frac{-3 + \sqrt{6}}{3}, x = \frac{-3 - \sqrt{6}}{3}$$ Therefore, the solutions are $\frac{-3 \pm \sqrt{6}}{3}$.

Check:
The check is complicated and is left as a challenge exercise for the student.

ADDITIONAL PRACTICE

Solve the following equations by completing the square.

j) $-3x + 5 = 2x^2$

k) $3x^2 + 12x = 5$

PRACTICE EXERCISES

Solve the following equations by completing the square. If there is no real solution, indicate the solution set as $\varnothing$.

10) $3x^2 + 2 = 5x$ **11)** $2x^2 + 6x = -3$

If more practice is needed, do the Additional Practice Exercises in the margin.

As you have no doubt noticed, solving quadratic equations by completing the square can be very tedious and sometimes difficult. In the next section we will use completing the square to develop another method that usually is easier.

EXERCISE SET 9.2

Find the number needed to make each of the following a perfect square trinomial. Then write each trinomial as the square of a binomial.

1) $a^2 + 6a + \_$

2) $r^2 + 14r + \_$

3) $x^2 - 18x + \_$

4) $a^2 - 20x + \_$

5) $n^2 + 5n + \_$

6) $m^2 + 11m + \_$

7) $b^2 - 7b + \_$

8) $r^2 - 9r + \_$

9) $x^2 + \frac{5}{2}x + \_$

10) $c^2 + \frac{5}{3}c + \_$

11) $a^2 - \frac{3}{7}a + \_$

12) $n^2 - \frac{1}{2}n + \_$

Solve the following equations by completing the square. Most of the exercises 13 through 32 have rational solutions or no solutions. If there is no real solution, indicate the solution set as $\varnothing$.

13) $r^2 - 8r + 15 = 0$ **14)** $x^2 - 6x + 5 = 0$

15) $x^2 = -8x - 12$ **16)** $x^2 = 6x - 8$

17) $z^2 - 4z = 12$ **18)** $b^2 - 2b = 24$

19) $x^2 - 10 = -3x$ **20)** $a^2 + 6 = -9a$

21) $n^2 + 9n = -20$ **22)** $n^2 - 5n = 14$

23) $x^2 - \frac{7}{2}x + 3 = 0$ **24)** $x^2 - \frac{13}{2}x + 10 = 0$

25) $x^2 - \dfrac{7}{3}x = 2$

26) $y^2 - \dfrac{24}{5}y = 1$

27) $2x^2 = 3x + 20$

28) $2y^2 = 15y - 18$

29) $3x^2 + 6 = 11x$

30) $3x^2 - 24 = 14x$

31) $a^2 - 4a + 6 = 0$

32) $n^2 + 6n + 10 = 0$

Solve the following by completing the square. Most of the exercises 33 through 52 have irrational solutions or no real solutions. If there is no real solution, indicate the solution set as $\varnothing$.

33) $x^2 + 4x - 1 = 0$

34) $a^2 + 6a - 4 = 0$

35) $r^2 - 8r = -5$

36) $z^2 - 10z = -6$

37) $a^2 - 2 = 14a$

38) $t^2 - 4 = 3t$

39) $x^2 + 8x + 3 = 0$

40) $y^2 + 6y + 2 = 0$

41) $x^2 - 10x = 5$

42) $z^2 - 2z = 6$

43) $t^2 - 3 = -5t$

44) $z^2 - 1 = -7z$

45) $2x^2 = 4x + 3$

46) $2x^2 = -6x - 5$

47) $2a^2 - 5a + 4 = 0$

48) $2x^2 - 7x + 6 = 0$

49) $3x^2 + 6x = 1$

50) $3s^2 + 8s = 3$

51) $2x^2 + 3x + 5 = 0$

52) $3x^2 - 5x + 8 = 0$

CHALLENGE EXERCISES (53–56)

Solve the following by completing the square.

53) $x^2 + \dfrac{1}{3}x - \dfrac{2}{9} = 0$

54) $x^2 + \dfrac{5}{2}x + \dfrac{21}{16} = 0$

55) Check Example 3b.

56) Check Example 4b.

WRITING EXERCISES

57) Why do some quadratic equations have no real number solutions?

58) What is the advantage to solving quadratic equations by completing the square?

CRITICAL THINKING

59) Write a quadratic equation that has no real solutions and explain how you arrived at that equation.

SECTION 9.3

Solving Quadratic Equations by the Quadratic Formula

OBJECTIVE

When you complete this section, you will be able to:

Ⓐ Solve quadratic equations by using the quadratic formula.

INTRODUCTION

• Developing the quadratic formula

Thus far, we have learned two methods of solving quadratic equations: factoring and completing the square. The disadvantage of factoring is that not all quadratic equations can be solved by factoring. The disadvantage of completing the square is that it is complicated and there are many places where errors can be made. In this section we will learn a third method of solving quadratic equations that also allows us to solve *any* quadratic equation. This method involves using the **quadratic formula** which we will now develop.

Let us begin with the general form of a quadratic equation and solve for x by completing the square. This will result in a formula which gives the solutions of the equation in terms of its coefficients.

$ax^2 + bx + c = 0$	Subtract c from both sides.
$ax^2 + bx = -c$	Divide both sides by a to get x^2.
$\dfrac{ax^2 + bx}{a} = -\dfrac{c}{a}$	Simplify the left side.

$$x^2 + \frac{b}{a}x = -\frac{c}{a}$$

Find the expression needed to make the left side a perfect square trinomial and add it to both sides.

$$x^2 + \frac{b}{a}x + \_\_ = -\frac{c}{a} + \_\_$$

$\frac{1}{2} \cdot \frac{b}{a} = \frac{b}{2a}$ and $\left(\frac{b}{2a}\right)^2 = \frac{b^2}{4a^2}$. Therefore, add $\frac{b^2}{4a^2}$ to both sides.

$$x^2 + \frac{b}{a}x + \frac{b^2}{4a^2} = \frac{b^2}{4a^2} - \frac{c}{a}$$

Write the left side as the square of a binomial and write the right side with the LCD of $4a^2$.

$$\left(x + \frac{b}{2a}\right)^2 = \frac{b^2}{4a^2} - \frac{c}{a} \cdot \frac{4a}{4a}$$

Simplify the right side.

$$\left(x + \frac{b}{2a}\right)^2 = \frac{b^2}{4a^2} - \frac{4ac}{4a^2}$$

Add the fractions on the right side.

$$\left(x + \frac{b}{2a}\right)^2 = \frac{b^2 - 4ac}{4a^2}$$

Solve using extraction of roots.

$$x + \frac{b}{2a} = \pm\sqrt{\frac{b^2 - 4ac}{4a^2}}$$

Apply the quotient rule for radicals.

$$x + \frac{b}{2a} = \pm\frac{\sqrt{b^2 - 4ac}}{\sqrt{4a^2}}$$

$\sqrt{4a^2} = 2a$ if $a > 0$.

$$x + \frac{b}{2a} = \pm\frac{\sqrt{b^2 - 4ac}}{2a}$$

Subtract $\frac{b}{2a}$ from both sides.

$$x = -\frac{b}{2a} \pm \frac{\sqrt{b^2 - 4ac}}{2a}$$

Add the fractions.

$$x = \frac{-b \pm \sqrt{b^2 - 4ac}}{2a}$$

Therefore, the solutions of $ax^2 + bx + c = 0$ are $\frac{-b + \sqrt{b^2 - 4ac}}{2a}$ and $\frac{-b - \sqrt{b^2 - 4ac}}{2a}$.

Note: The variables a, b, and c in the quadratic formula come from the equation written in the form $ax^2 + bx + c = 0$. This means that the solutions of a quadratic equation are determined by its coefficients.

THE QUADRATIC FORMULA

The solutions of any equation written in the form $ax^2 + bx + c = 0$ are

$$x = \frac{-b \pm \sqrt{b^2 - 4ac}}{2a}.$$

The first task necessary in using the quadratic formula is to find the values of a, b, and c. This is done by rewriting the given equation, if necessary, in the form of $ax^2 + bx + c = 0$ and comparing the coefficients.

EXAMPLE 1

Find the values of a, b, and c in each of the following.

a) $3x^2 + 2x - 5 = 0$ Compare with $ax^2 + bx + c = 0$.
$a = 3, b = 2, c = -5$

b) $x^2 - 3x = 5$ Subtract 5 from both sides.
$x^2 - 3x - 5 = 0$ Compare with $ax^2 + bx + c = 0$.
$a = 1, b = -3, c = -5$

c) $x^2 = x - 7$ Subtract x from and add 7 to both sides.
$x^2 - x + 7 = 0$ Compare with $ax^2 + bx + c = 0$.
$a = 1, b = -1, c = 7$

d) $3 = x^2 + 7x$ Subtract 3 from both sides.
$0 = x^2 + 7x - 3$ Rewrite in the form $ax^2 + bx + c = 0$.
$x^2 + 7x - 3 = 0$ Compare with $ax^2 + bx + c = 0$.
$a = 1, b = 7, c = -3$

PRACTICE EXERCISES

Find the values of a, b, and c in each of the following.

1) $4x^2 + 5x - 6 = 0$ **2)** $x^2 - x = 9$

3) $6x + 2 = -4x^2$ **4)** $x^2 + 5 = -4x$

To solve quadratic equations using the quadratic formula, we use the following procedure.

SOLVING QUADRATIC EQUATIONS WITH THE QUADRATIC FORMULA

1) If necessary, rewrite the equation in $ax^2 + bx + c = 0$ form.

2) Find a, b, and c.

3) Write the quadratic formula.

4) Substitute the values for a, b, and c into the formula.

5) Simplify the results.

6) Check the solution(s) in the original equation.

In Example 2 below the equations could be solved by factoring, but we solved them using the quadratic formula for illustrative purposes.

EXAMPLE 2

Solve the following equations using the quadratic formula. If there is no real solution, indicate the solution set as $\varnothing$.

a) $x^2 - 2x - 8 = 0$ Find a, b, and c.
$a = 1, b = -2, c = -8$ Write the quadratic formula.

$x = \dfrac{-b \pm \sqrt{b^2 - 4ac}}{2a}$ Substitute for a, b, and c.

$$x = \frac{-(-2) \pm \sqrt{(-2)^2 - 4(1)(-8)}}{2(1)}$$ Simplify. Watch the signs!

$$x = \frac{2 \pm \sqrt{4 + 32}}{2}$$ Add 4 and 32.

$$x = \frac{2 \pm \sqrt{36}}{2}$$ $\sqrt{36} = 6$.

$$x = \frac{2 \pm 6}{2}$$ Write as two separate fractions.

$$x = \frac{2 + 6}{2}, x = \frac{2 - 6}{2}$$ Simplify each fraction.

$$x = \frac{8}{2}, x = \frac{-4}{2}$$ Continue simplifying.

$$x = 4, x = -2$$ Therefore, the solutions are 4 and -2.

Note: These equations are checked as in Section 9.2. Consequently, the checks are omitted.

b) $6x^2 + x = 12$ Rewrite in the form $ax^2 + bx + c = 0$.

     $6x^2 + x - 12 = 0$ Find a, b, and c.

     $a = 6, b = 1, c = -12$ Write the quadratic formula.

$$x = \frac{-b \pm \sqrt{b^2 - 4ac}}{2a}$$ Substitute for a, b, and c.

$$x = \frac{-1 \pm \sqrt{1^2 - 4(6)(-12)}}{2(6)}$$ Simplify.

$$x = \frac{-1 \pm \sqrt{1 + 288}}{12}$$ Continue simplifying.

$$x = \frac{-1 \pm \sqrt{289}}{12}$$ $\sqrt{289} = 17$.

$$x = \frac{-1 \pm 17}{12}$$ Rewrite as two separate fractions.

$$x = \frac{-1 + 17}{12}, x = \frac{-1 - 17}{12}$$ Simplify each fraction.

$$x = \frac{16}{12}, x = \frac{-18}{12}$$ Reduce each to lowest terms.

$$x = \frac{4}{3}, x = -\frac{3}{2}$$ Therefore, the solutions are $\frac{4}{3}$ and $-\frac{3}{2}$.

ADDITIONAL PRACTICE

Solve the following equations using the quadratic formula.

a) $x^2 + x - 12 = 0$

b) $8x^2 + 10x - 3 = 0$

PRACTICE EXERCISES

Solve the following equations using the quadratic formula. If there is no real solution, indicate the solution set as $\varnothing$.

5) $x^2 + 7x + 10 = 0$ **6)** $6x^2 - x - 2 = 0$

     If more practice is needed, do the Additional Practice Exercises in the margin.

The real advantage gained from using the quadratic formula is in solution of quadratic equations that do not factor. Although this type of equation can be solved by completing the square, the quadratic formula is usually easier.

EXAMPLE 3

Solve the following equations using the quadratic formula. If there is no real solution, indicate the solution set as $\varnothing$.

a) $x^2 + 3x - 1 = 0$ Find a, b, and c.

$a = 1$, $b = 3$, $c = -1$ Write the quadratic formula.

$$x = \frac{-b \pm \sqrt{b^2 - 4ac}}{2a}$$

Substitute for a, b, and c.

$$x = \frac{-3 \pm \sqrt{3^2 - 4(1)(-1)}}{2(1)}$$

Simplify.

$$x = \frac{-3 \pm \sqrt{9 + 4}}{2}$$

Continue simplifying.

$$x = \frac{-3 \pm \sqrt{13}}{2}$$

Therefore, the solutions are $\frac{-3 + \sqrt{13}}{2}$ and $\frac{-3 - \sqrt{13}}{2}$.

b) $x(x + 6) = 4$ Distribute x.

$x^2 + 6x = 4$ Subtract 4 from both sides to put in $ax^2 + bx + c = 0$ form.

$x^2 + 6x - 4 = 0$ Find a, b, and c.

$a = 1$, $b = 6$, $c = -4$ Write the quadratic formula.

$$x = \frac{-b \pm \sqrt{b^2 - 4ac}}{2a}$$

Substitute for a, b, and c.

$$x = \frac{-6 \pm \sqrt{6^2 - 4(1)(-4)}}{2(1)}$$

Simplify.

$$x = \frac{-6 \pm \sqrt{36 + 16}}{2}$$

Continue simplifying.

$$x = \frac{-6 \pm \sqrt{52}}{2}$$

Rewrite 52 as $4 \cdot 13$.

$$x = \frac{-6 \pm \sqrt{4 \cdot 13}}{2}$$

Apply the product rule for radicals.

$$x = \frac{-6 \pm 2\sqrt{13}}{2}$$

Factor 2 from the numerator.

$$x = \frac{2(-3 \pm \sqrt{13})}{2}$$

Divide by the common factor of 2.

$$x = -3 \pm \sqrt{13}$$

Therefore, the solutions are $-3 + \sqrt{13}$ and $-3 - \sqrt{13}$.

c) $2x^2 = 7x - 1$ Subtract $7x$ from and add 1 to both sides to put in $ax^2 + bx + c = 0$ form.

$2x^2 - 7x + 1 = 0$ Find a, b, and c.

$a = 2$, $b = -7$, $c = 1$ Write the quadratic formula.

$$x = \frac{-b \pm \sqrt{b^2 - 4ac}}{2a}$$

Substitute for a, b, and c.

$$x = \frac{-(-7) \pm \sqrt{(-7)^2 - 4(2)(1)}}{2(2)}$$

Simplify.

$$x = \frac{7 \pm \sqrt{49 - 8}}{4}$$

Continue simplifying.

$$x = \frac{7 \pm \sqrt{41}}{4}$$

Therefore, the solutions are $\frac{7 + \sqrt{41}}{4}$ and $\frac{7 - \sqrt{41}}{4}$.

 d) $2x^2 + 6 = -2x$

Add $2x$ to both sides and write in $ax^2 + bx + c = 0$ form.

$$2x^2 + 2x + 6 = 0$$

Find a, b, and c.

$$a = 2, b = 2, c = 6$$

Write the quadratic formula.

$$x = \frac{-b \pm \sqrt{b^2 - 4ac}}{2a}$$

Substitute for a, b, and c.

$$x = \frac{-2 \pm \sqrt{2^2 - 4(2)(6)}}{2(2)}$$

Simplify.

$$x = \frac{-2 \pm \sqrt{4 - 48}}{4}$$

Continue simplifying.

$$x = \frac{-2 \pm \sqrt{-44}}{4}$$

Since $\sqrt{-44}$ does not exist as a real number, this equation has no real solutions. Hence the solution set is $\varnothing$.

ADDITIONAL PRACTICE

Solve the following equations using the quadratic formula.

c) $x^2 - 4x = -1$

d) $x(x + 6) = 3$

e) $2x^2 - 6x - 1 = 0$

f) $2x^2 + 5 = -2x$

PRACTICE EXERCISES

Solve the following equations using the quadratic formula. If there is no real solution, indicate the solution set as $\varnothing$.

7) $x^2 + 6x = -6$

8) $x^2 = 4x + 23$

9) $3x^2 - 9x + 3 = 0$

10) $x(3x + 2) = -6$

If more practice is needed, do the Additional Practice Exercises in the margin.

Given a choice of which method to use in solving a quadratic equation, we recommend the following procedure.

SOLVING QUADRATIC EQUATIONS

1) Try to solve by factoring.

2) If $b = 0$, use extraction of roots.

3) If the equation cannot be solved by factoring or if it is difficult to factor, then solve it by using the quadratic formula.

You might ask why we learned to solve quadratic equations by completing the square if we do not use it. Remember, we needed to know how to solve quadratic equations by completing the square in order to derive the quadratic formula. Though completing the square is rarely used in solving quadratic equations, this procedure has many important applications, including graphing of second degree equations.

EXERCISE SET 9.3

Solve the following using the quadratic formula. If there is no real number solution, indicate the solution set as $\varnothing$. Most of the solutions to Exercises 1–20 are rational numbers.

1) $x^2 - 6x + 5 = 0$
2) $r^2 - 8r + 15 = 0$
3) $x^2 = 6x - 8$
4) $x^2 = -8x - 12$
5) $b^2 - 2b = 24$
6) $z^2 - 4z = 12$
7) $a^2 + 6 = -7a$
8) $x^2 - 10 = -3x$
9) $n^2 - 5n = 14$
10) $n^2 + 9n = -20$
11) $2y^2 - 15y + 18 = 0$
12) $2x^2 - 3x - 20 = 0$
13) $3x^2 - 14x = 24$
14) $3x^2 - 11x = 6$
15) $n^2 = -6n - 10$
16) $a^2 = -4a - 6$
17) $8x^2 - 14x - 15 = 0$
18) $9x^2 - 6x - 8 = 0$
19) $15a^2 = 29a + 2$
20) $6x^2 = 25x - 25$

Solve the following quadratic equations using the quadratic formula. If there is no real number solution, indicate the solution set as $\varnothing$. Most of the exercises 21 through 48 have irrational solutions.

21) $x^2 + 3x - 1 = 0$
22) $x^2 + 5x - 2 = 0$
23) $a^2 = -a + 3$
24) $b^2 - 5b = -3$
25) $2c^2 - 3c = 4$
26) $2n^2 - 5n + 1 = 0$
27) $5r^2 - r - 3 = 0$
28) $4y^2 = 5y + 2$
29) $3q^2 + 3 = -2q$
30) $2m^2 + 2 = -2m$
31) $x^2 + 7x + 1 = 0$
32) $2x^2 + 9x = 1$
33) $2a^2 = 9a + 3$
34) $2d^2 = 5d + 4$
35) $x^2 - 2x = 2$
36) $a^2 + 4a = -2$
37) $x^2 - 10x = -20$
38) $x^2 + 12x = -34$
39) $b^2 = 8b - 8$
40) $w^2 + 8w = 11$
41) $2x^2 - 4x - 5 = 0$
42) $2x^2 - 6x = -3$
43) $4r^2 + 10r + 5 = 0$
44) $4x^2 - 12x + 7 = 0$
45) $3x^2 = -2x + 4$
46) $3x^2 = -8x + 2$
47) $5z^2 + 2z + 10 = 0$
48) $6a^2 - 4a + 2 = 0$

Solve the following applications problems.

49) Find two consecutive odd integers with sum of squares equal to 34.

50) Find two consecutive even integers with sum of squares equal to 52.

Exercises 51 through 54 require the use of the formula for the area of a rectangle. The formula for the area of a rectangle is $A = LW$.

51) The length of a rectangle is six inches more than the width. If the area is 55 square inches, find the length and width.

52) The width of a rectangle is three meters less than the length. If the area is 70 square meters, find the length and width.

53) The length of a rectangle is one yard less than twice the width. If the area is 28 square yards, find the length and width.

54) The length of a rectangle is two centimeters more than three times the width. If the area is 33 square centimeters, find the length and width.

CHALLENGE EXERCISES (55–58)

Solve the following quadratic equations using the quadratic formula. If there is no real solution, indicate the solution set as $\varnothing$.

55) $x^2 - \dfrac{7}{2}x + 3 = 0$
56) $x^2 - \dfrac{13}{2}x + 10 = 0$
57) $x^2 - \dfrac{7}{3}x - 2 = 0$
58) $y^2 - \dfrac{24}{5}y - 1 = 0$

WRITING EXERCISES

59) What are the advantages of using the quadratic formula instead of completing the square?

60) In the quadratic formula, $x = \dfrac{-b \pm \sqrt{b^2 - 4ac}}{2a}$, the expression $b^2 - 4ac$ is called the discriminant. For what values (types of numbers) of the discriminant will a quadratic equation have:

a) No real solutions? Why?

b) Two irrational solutions? Why?

c) Two rational solutions? Why?

d) Exactly one solution? Why?

61) Given a quadratic equation in the form $ax^2 + bx + c = 0$. If b is an even integer, then $\sqrt{b^2 - 4ac}$ can always be simplified. Why?

Solving Mixed Equations

When you complete this section, you will be able to solve the following:

Ⓐ Linear equations.

Ⓑ Quadratic equations.

Ⓒ Equations containing rational numbers and/or rational expressions which result in linear or quadratic equations when simplified.

Ⓓ Equations containing radicals which result in linear or quadratic equations when simplified.

INTRODUCTION

Previously we learned to solve linear, quadratic, rational, and radical equations. The techniques that we use to solve each type of equation differ greatly. In this section we will not learn anything new since we are going to review the techniques needed to solve the different types of equations. We have learned many arithmetic and algebraic techniques since we first solved equations, so the initial appearance of the equation may be different than you have seen before. However, all the equations that we have solved, or will solve, will ultimately be quadratic or linear after simplification. Some of these techniques are needed in Section 9.5 when solving applications problems. First, we review the techniques needed to solve linear equations.

SOLVING LINEAR EQUATIONS

1) If necessary, simplify both sides of the equation as much as possible.

2) If necessary, use the Addition Property of Equality to get all the terms with variables on one side of the equation and all the constant terms on the other.

3) If necessary, use the Multiplication Property of Equality to eliminate any coefficient on the variable.

4) Check the answer in the original equation.

EXAMPLE 1

Solve the following equations. All checks are left as exercises for the student.

a) $5(2x - 3) + 2(3x - 7) + 1 = 4 - 6(x - 2)$ Apply the distributive property.

$10x - 15 + 6x - 14 + 1 = 4 - 6x + 12$ Add like terms.

$16x - 28 = 16 - 6x$ Add $6x$ to both sides.

$22x - 28 = 16$ Add 28 to both sides.

$22x = 44$ Divide both sides by 22.

$x = 2$ Therefore, $x = 2$ is the solution.

b) $(x + 4)^2 - x^2 = 48$ Square $x + 4$.

$x^2 + 8x + 16 - x^2 = 48$ Add like terms.

$8x + 16 = 48$ Subtract 16 from both sides.

$8x = 32$ Divide both sides by 8.

$x = 4$ Therefore, $x = 4$ is the solution.

c) $(2x + 3)(x - 2) - 18 = (x - 6)(2x - 1) + 6$ FOIL on each side.

$2x^2 - x - 6 - 18 = 2x^2 - 13x + 6 + 6$ Add like terms on each side.

$2x^2 - x - 24 = 2x^2 - 13x + 12$ Subtract $2x^2$ from both sides.

$-x - 24 = -13x + 12$ Add $13x$ to both sides.

$12x - 24 = 12$ Add 24 to both sides.

$12x = 36$ Divide both sides by 12.

$x = 3$ Therefore, the solution is $x = 3$.

PRACTICE EXERCISES

Solve the following equations.

1) $3(2x + 1) - 4(3x - 8) = 2(4x + 5) - 3$

2) $(a + 3)^2 - a^2 = 33$

3) $(b - 3)(2b + 5) + 15 = (2b - 5)(b + 4) + 4$

In Section 9.3 we solved quadratic equations by using the quadratic formula only. However, near the end of the section we outlined a procedure for solving quadratic equations if the technique for solving is not specified. We repeat the procedure below.

SOLVING QUADRATIC EQUATIONS

To solve a quadratic equation in the form $ax^2 + bx + c = 0$:

1) Try to solve by factoring.

2) If $b = 0$, use extraction of roots.

3) If the equation cannot be solved by factoring or if it is difficult to factor, then solve it by using the quadratic formula.

EXAMPLE 2

Solve the following quadratic equations using any method. The checks are left as exercises for the student.

a) $(2x + 3)(2x + 5) = 35$ Multiply on the left side.

$4x^2 + 16x + 15 = 35$ Subtract 35 from both sides.

$4x^2 + 16x - 20 = 0$ Remove the common factor of 4.

$4(x^2 + 4x - 5) = 0$ Factor $x^2 + 4x - 5$.

$4(x + 5)(x - 1) = 0$ Set each factor equal to 0 $(4 \neq 0)$.

$x + 5 = 0, \ x - 1 = 0$ Solve each linear equation.

$x = -5, \quad x = 1$ Therefore, the solutions are -5 and 1.

b) $x^2 + (2x - 4)^2 = 100$ Square $2x - 4$.

$x^2 + 4x^2 - 16x + 16 = 100$ Add x^2 and $4x^2$.

$5x^2 - 16x + 16 = 100$ — Subtract 100 from both sides.

$5x^2 - 16x - 84 = 0$ — Factor the left side.

$(5x + 14)(x - 6) = 0$ — Set each factor equal to 0.

$5x + 14 = 0, \ x - 6 = 0$ — Solve each equation.

$5x = -14, \quad x = 6$

$x = -\dfrac{14}{5}$ — Therefore, the solutions are 6 and $-\frac{14}{5}$.

c) $2x^2 - 12x + 12 = -3$ — Add 3 to both sides.

$2x^2 - 12x + 15 = 0$ — First try factoring.

$(2x - 1)(x - 15)$

$(2x - 15)(x - 1)$

$(2x - 3)(x - 5)$

$(2x - 5)(x - 3)$ — These are the only possibilities and none work. Therefore, use the quadratic formula. Identify a, b, and c.

$a = 2, \ b = -12, \ c = 15$

$x = \dfrac{-b \pm \sqrt{b^2 - 4ac}}{2a}$ — Substitute for a, b, and c.

$x = \dfrac{-(-12) \pm \sqrt{(-12)^2 - 4(2)(15)}}{2(2)}$ — Simplify.

$x = \dfrac{12 \pm \sqrt{144 - 120}}{4}$ — Continue simplifying.

$x = \dfrac{12 \pm \sqrt{24}}{4}$ — $\sqrt{24} = 2\sqrt{6}$.

$x = \dfrac{12 \pm 2\sqrt{6}}{4}$ — Factor the numerator.

$x = \dfrac{2(6 \pm \sqrt{6})}{4}$ — Reduce.

$x = \dfrac{6 \pm \sqrt{6}}{2}$ — Therefore, the solutions are $\frac{6 + \sqrt{6}}{2}$ and $\frac{6 - \sqrt{6}}{2}$.

d) $6x^2 - 72 = 0$ — Since $b = 0$, use extraction of roots.

Add 72 to both sides.

$6x^2 = 72$ — Divide both sides by 6.

$x^2 = 12$ — Take the square roots of both sides.

$x = \pm\sqrt{12}$ — $\sqrt{12} = 2\sqrt{3}$.

$x = \pm 2\sqrt{3}$ — Therefore, the solutions are $2\sqrt{3}$ and $-2\sqrt{3}$.

PRACTICE EXERCISES

4) $(2x + 2)(2x + 5) = 28$

5) $x^2 + (2x + 2)^2 = 169$

6) $x^2 - 2x - 10 = 0$

7) $5x^2 - 40 = 0$

In Section 5.7 we solved equations that contained rational numbers or expressions. The procedure for solving these types of equations is outlined below.

SOLVING EQUATIONS CONTAINING RATIONAL EXPRESSIONS

1) Multiply both sides of the equation by the LCD in order to eliminate all rational expressions from the equation.

2) Solve the resulting equation using the appropriate techniques depending upon whether it is linear or quadratic.

3) At the minimum, check the solution(s) of the equation in step 2 into the original equation to see if it/they make any denominator equal to 0. If so, that solution of the equation in step 2 is not a solution of the original equation.

EXAMPLE 3

Solve the following equations containing rational expressions.

a) $\dfrac{1}{x} + \dfrac{1}{x + 2} = \dfrac{12}{35}$

Multiply both sides by the LCD $35x(x + 2)$.

$$35x(x + 2)\left(\dfrac{1}{x} + \dfrac{1}{x + 2}\right) = 35x(x + 2)\left(\dfrac{12}{35}\right)$$

Simplify both sides.

$$35x(x + 2)\dfrac{1}{x} + 35x(x + 2)\dfrac{1}{x + 2} = 12x(x + 2)$$

Continue simplifying.

$35(x + 2) + 35x = 12x^2 + 24x$ Multiply 35 and $x + 2$.

$35x + 70 + 35x = 12x^2 + 24x$ Add like terms.

$70x + 70 = 12x^2 + 24x$ Subtract $70x$ from both sides.

$70 = 12x^2 - 46x$ Subtract 70 from both sides.

$0 = 12x^2 - 46x - 70$ Remove the common factor of 2.

$0 = 2(6x^2 - 23x - 35)$ Factor $6x^2 - 23x - 35$.

$0 = 2(6x + 7)(x - 5)$ Set each factor equal to 0 ($2 \neq 0$).

$6x + 7 = 0, x - 5 = 0$ Solve each equation.

$6x = -7, \quad x = 5$

$x = -\dfrac{7}{6}$

Therefore, the solutions are 5 and $-\frac{7}{6}$. Neither solution makes any denominator equal to 0, so both solutions may be solutions of the original equation. The only way to verify that both are solutions is to check both in the original equation.

b) $\dfrac{240}{x+2} = \dfrac{240}{x} - 20$ | Multiply both sides by the LCD of $x(x+2)$.

$$x(x+2)\left(\dfrac{240}{x+2}\right) = x(x+2)\left(\dfrac{240}{x} - 20\right)$$

Simplify both sides.

$$240x = x(x+2)\dfrac{240}{x} - 20x(x+2)$$

Multiply on the right.

$$240x = 240(x+2) - 20x^2 - 40x$$

Multiply 240 and $x+2$.

$$240x = 240x + 480 - 20x^2 - 40x$$

Add like terms.

$$240x = -20x^2 + 200x + 480$$

Write in $ax^2 + bx + c = 0$ form.

$$20x^2 + 40x - 480 = 0$$

Remove the common factor of 20.

$$20(x^2 + 2x - 24) = 0$$

Factor $x^2 + 2x - 24$.

$$20(x+6)(x-4) = 0$$

Set each factor equal to 0 $(20 \neq 0)$.

$$x + 6 = 0, \quad x - 4 = 0$$

Solve each equation.

$$x = -6, \quad x = 4$$

Therefore, the solutions are -6 and 4. Neither solution makes any denominator equal to 0, so both solutions may be solutions of the original equation. The only way to verify that both are solutions is to check both in the original equation.

c) $\dfrac{1}{x-3} + \dfrac{1}{3} = \dfrac{6}{x^2 - 9}$ | Factor $x^2 - 9$.

$$\dfrac{1}{x-3} + \dfrac{1}{3} = \dfrac{6}{(x+3)(x-3)}$$

Multiply both sides by the LCD $3(x+3)(x-3)$.

$$3(x+3)(x-3)\left(\dfrac{1}{x-3} + \dfrac{1}{3}\right) = 3(x+3)(x-3)\left(\dfrac{6}{(x+3)(x-3)}\right)$$

Simplify.

$$3(x+3)(x-3)\dfrac{1}{x-3} + 3(x+3)(x-3)\dfrac{1}{3} = 3 \cdot 6$$

Continue simplifying.

$$3(x+3) + (x+3)(x-3) = 18$$

Continue simplifying.

$$3x + 9 + x^2 - 9 = 18$$

Continue simplifying.

$$x^2 + 3x = 18$$

Subtract 18 from both sides.

$$x^2 + 3x - 18 = 0$$

Factor the left side.

$$(x+6)(x-3) = 0$$

Set each factor equal to 0.

$$x + 6 = 0, \quad x - 3 = 0$$

Solve each equation.

$$x = -6, \quad x = 3$$

Therefore, the solutions are -6 and 3. However, 3 makes the denominators of $\dfrac{1}{x-3}$ and $\dfrac{6}{x^2-9}$ equal to 0. Therefore, the only solution (which can be verified by substituting into the original equation) is -6.

PRACTICE EXERCISES

Solve the following.

8) $\dfrac{1}{x} + \dfrac{1}{x+2} = \dfrac{5}{12}$

9) $\dfrac{120}{x+1} = \dfrac{120}{x} - 10$

10) $\dfrac{x}{x+2} + \dfrac{3}{4x} = \dfrac{3x+1}{2x+4}$

In Section 8.5 we learned to solve equations that contain radicals. We repeat the procedures below. Remember, squaring both sides of an equation can introduce extraneous solutions, so all answers must be checked in the original equation.

SOLVING EQUATIONS CONTAINING ONE SQUARE ROOT

If an equation contains one square root, it can be solved by:

1) Isolating the square root on one side of the equation.
2) Squaring both sides of the equation.
3) Solving the resulting equation.
4) Checking all solutions in the original equation.

SOLVING EQUATIONS CONTAINING TWO SQUARE ROOTS

If an equation contains two square roots, it can be solved by:

1) Isolating one of the radicals on one side of the equation.
2) Squaring both sides of the equation and, if necessary, simplifying each side by combining like terms.
3) Isolating the radical on one side of the equation, if the equation still contains a radical, and squaring both sides of the equation again; then solving the resulting equation.
4) Checking all solutions in the original equation.

EXAMPLE 4

Solve the following.

a) $\sqrt{3x+3} = 2x - 1$ Square both sides of the equation.

$(\sqrt{3x+3})^2 = (2x-1)^2$ Simplify both sides.

$3x + 3 = 4x^2 - 4x + 1$ Subtract $3x$ and 3 from both sides.

$0 = 4x^2 - 7x - 2$ Factor.

$0 = (4x + 1)(x - 2)$ Set each factor equal to 0.

$4x + 1 = 0, \; x - 2 = 0$ Solve each equation.

$4x = -1, \quad x = 2$

$x = -\dfrac{1}{4}$ Therefore, the possible solutions are $-\frac{1}{4}$ and 2.

Check:

$$x = -\frac{1}{4} \qquad\qquad x = 2$$

$$\sqrt{3\left(-\frac{1}{4}\right) + 3} = 2\left(-\frac{1}{4}\right) - 1 \qquad \sqrt{3(2) + 3} = 2(2) - 1$$

$$\sqrt{-\frac{3}{4} + 3} = -\frac{2}{4} - 1 \qquad\qquad \sqrt{6 + 3} = 4 - 1$$

$$\sqrt{\frac{9}{4}} = -\frac{1}{2} - 1 \qquad\qquad\qquad \sqrt{9} = 3$$

$$\frac{3}{2} \neq -\frac{3}{2} \qquad\qquad\qquad\qquad 3 = 3$$

Therefore, 2 is the only solution.

b) $\sqrt{2x} - \sqrt{3x + 10} = -2$ Add $\sqrt{3x+10}$ to both sides.

$\sqrt{2x} = \sqrt{3x+10} - 2$ Square both sides.

$(\sqrt{2x})^2 = (\sqrt{3x + 10} - 2)^2$ Simplify both sides.

$2x = 3x + 10 - 4\sqrt{3x + 10} + 4$ Simplify the right side.

$2x = 3x - 4\sqrt{3x + 10} + 14$ Subtract $3x$ from both sides.

$-x = -4\sqrt{3x + 10} + 14$ Subtract 14 from both sides.

$-x - 14 = -4\sqrt{3x + 10}$ Multiply both sides by -1.

$x + 14 = 4\sqrt{3x + 10}$ Square both sides.

$x^2 + 28x + 196 = 16(3x + 10)$ Multiply on the right.

$x^2 + 28x + 196 = 48x + 160$ Subtract $48x$ and 160 from both sides.

$x^2 - 20x + 36 = 0$ Factor the left side.

$(x - 18)(x - 2) = 0$ Set each factor equal to 0.

$x - 18 = 0, \ x - 2 = 0$ Solve each equation.

$x = 18, \qquad x = 2$ Therefore, the possible solutions are 18 and 2.

Check:

$$x = 18 \qquad\qquad\qquad x = 2$$

$$\sqrt{2 \cdot 18} - \sqrt{3 \cdot 18 + 10} = -2 \qquad \sqrt{2 \cdot 2} - \sqrt{3 \cdot 2 + 10} = -2$$

$$\sqrt{36} - \sqrt{54 + 10} = -2 \qquad\qquad \sqrt{4} - \sqrt{6 + 10} = -2$$

$$6 - \sqrt{64} = -2 \qquad\qquad\qquad 2 - \sqrt{16} = -2$$

$$6 - 8 = -2 \qquad\qquad\qquad\qquad 2 - 4 = -2$$

$$-2 = -2 \qquad\qquad\qquad\qquad\quad -2 = -2$$

Therefore, 18 and 2 are both solutions.

PRACTICE EXERCISES

Solve the following.

11) $\sqrt{3x + 1} = 2x - 6$ 12) $\sqrt{3x + 4} - \sqrt{x + 2} = 2$

E X E R C I S E   S E T   **9.4**

Solve the following equations. Each equation is linear or will be linear when simplified.

1) $3(x + 2) - 12 = 4(2x - 6) - 2(3x + 2)$

2) $4(2x - 1) + 3 = 3(3x - 2) + 2(x - 5)$

3) $(b + 1)^2 = b^2 + 9$

4) $(c + 2)^2 = (c + 1)^2 + 11$

5) $(n + 2)^2 - (n - 2)^2 = 50$

6) $(m + 2)^2 - (m - 2)^2 = 20$

7) $(2z + 3)(z - 4) - 8 = (z - 6)(2z + 5)$

8) $(3x + 2)(x - 1) = (x + 5)(3x - 2) - 6$

Solve the following quadratic equations using any method.

9) $(2x + 3)(2x + 5) = 63$

10) $(2x + 2)(2x + 4) = 80$

11) $(a + 2)(a + 5) - a(a + 3) = 22$

12) $(x + 4)(x + 8) - x(x + 4) = 48$

13) $b^2 + (b + 3)^2 = 225$

14) $x^2 + (2x - 1)^2 = 289$

15) $6c^2 + 8c - 5 = -8$ **16)** $4d^2 + 6d + 2 = 5$

17) $6n + 7 = 5 - 3n^2$ **18)** $-10m + 8 = 4 - m^2$

19) $8a^2 - 192 = 0$ **20)** $3b^2 - 96 = 0$

Solve the following equations containing rational expressions.

21) $\dfrac{1}{x} + \dfrac{1}{x + 2} = \dfrac{3}{4}$ **22)** $\dfrac{1}{x} + \dfrac{1}{x + 2} = \dfrac{3}{4}$

23) $\dfrac{60}{x + 2} = \dfrac{60}{x} - 5$ **24)** $\dfrac{120}{x + 2} = \dfrac{120}{x} - 5$

25) $\dfrac{1}{p - 4} + \dfrac{1}{4} = \dfrac{8}{p^2 - 16}$ **26)** $\dfrac{1}{y - 5} + \dfrac{1}{5} = \dfrac{10}{y^2 - 25}$

27) $\dfrac{6}{2y + 5} = \dfrac{2}{y + 5} + \dfrac{1}{5}$ **28)** $\dfrac{6}{2x + 3} = \dfrac{2}{x - 6} + \dfrac{4}{3}$

Solve the following equations containing radicals.

29) $\sqrt{2a + 5} = 3a - 3$

30) $\sqrt{3b + 1} = 5b - 3$

31) $\sqrt{8m + 1} - 2m + 1 = 0$

32) $\sqrt{6x + 1} - 2x + 3 = 0$

33) $\sqrt{4x + 1} = \sqrt{x + 2} + 1$

34) $\sqrt{3x + 7} = \sqrt{2x + 3} + 1$

35) $\sqrt{2x + 1} - \sqrt{3x + 4} = -1$

36) $\sqrt{2x - 1} - \sqrt{4x + 5} = -2$

CHALLENGE EXERCISES (37–42)

Solve the following.

37) $\dfrac{3}{2n + 5} + \dfrac{2}{n + 1} = \dfrac{8}{3n + 2}$

38) $\dfrac{4}{x + 2} - \dfrac{6}{2x + 4} = \dfrac{2}{x + 7}$

39) $\sqrt{3x + 1} - \sqrt{x - 1} = \sqrt{2x - 6}$

40) $\sqrt{4x + 1} - \sqrt{2x - 3} = \sqrt{x - 2}$

41) $(3x - 2)(x - 4) - (2x + 3)(x + 2) = 4$

42) $(2x + 1)^2 - (x - 5)^2 = x - 14$

S E C T I O N   **9.5**

Applications Involving Quadratic Equations

OBJECTIVES

When you complete this section, you will be able to solve the following types of applications problems involving quadratic equations:

Ⓐ   Numbers.
Ⓑ   Geometric figures.
Ⓒ   Pythagorean theorem.
Ⓓ   Distance, rate, and time.
Ⓔ   Work.
Ⓕ   Applications from science.

INTRODUCTION

Applications problems that resulted in quadratic equations were first introduced in Section 4.8. In that section we were limited to integer problems and areas of squares and rectangles. In later chapters we learned to solve other types of equations. Consequently, in this section we will concentrate on applications problems that involve solving these other types of equations and equations from the sciences.

As in Section 4.8, care must be taken when solving application problems that involve quadratic equations. Quite often a number may solve the equation, but does not satisfy the physical conditions of the problem. For example, the length of a side of a rectangle cannot be negative. Consequently, it is very important that all solutions of the equation be checked in the wording of the original problem. Also, as in Section 4.8, most of the equations in this section can be solved by factoring, though you have the option of using the quadratic formula if you wish.

• **Number problems**

EXAMPLE 1

The sum of the reciprocals of two consecutive even integers is $\frac{7}{24}$. Find the integers.

Solution:

Let x represent the smaller of the two consecutive even integers. Then, $x + 2$ represents the larger of the two consecutive even integers.

$\frac{1}{x}$ represents the reciprocal of the smaller integer.

$\frac{1}{x+2}$ is the reciprocal of the larger integer.

Since the sum of the reciprocals is $\frac{7}{24}$, the equation is:

$$\frac{1}{x} + \frac{1}{x+2} = \frac{7}{24} \qquad \text{Multiply both sides by the LCD } 24x(x+2).$$

$$24x(x+2)\left(\frac{1}{x} + \frac{1}{x+2}\right) = 24x(x+2)\left(\frac{7}{24}\right) \qquad \text{Simplify both sides.}$$

$$24x(x+2)\left(\frac{1}{x}\right) + 24x(x+2)\left(\frac{1}{x+2}\right) = x(x+2)(7) \qquad \text{Multiply.}$$

$$24(x+2) + 24x = 7x(x+2) \qquad \text{Distribute.}$$

$$24x + 48 + 24x = 7x^2 + 14x \qquad \text{Add like terms.}$$

$$48x + 48 = 7x^2 + 14x \qquad \begin{array}{l}\text{Subtract } 48x \text{ and } 48 \\ \text{from both sides.}\end{array}$$

$$0 = 7x^2 - 34x - 48 \qquad \begin{array}{l}\text{Factor} \\ 7x^2 - 34x - 48.\end{array}$$

$$0 = (7x + 8)(x - 6) \qquad \begin{array}{l}\text{Set each factor equal} \\ \text{to 0.}\end{array}$$

$$7x + 8 = 0, \quad x - 6 = 0 \qquad \text{Solve each equation.}$$

$$7x = -8, \qquad x = 6$$

$$x = -\frac{8}{7} \qquad \begin{array}{l}\text{Since } -\frac{8}{7} \text{ is not an} \\ \text{integer, the only} \\ \text{answer is 6.}\end{array}$$

Have we answered the question asked? No, we need the other integer. If $x = 6$, then $x + 2 = 8$. Therefore, the two consecutive even integers are 6 and 8.

Check:

Is the sum of the reciprocals of 6 and 8 equal to $\frac{7}{24}$? Since $\frac{1}{6} + \frac{1}{8} = \frac{4}{24} + \frac{3}{24} = \frac{7}{24}$, our solution is correct.

1) The product of the reciprocals of two consecutive integers is $\frac{1}{12}$. Find the integers.

• Geometric problems

EXAMPLE 2

A rectangular picture, 4 inches by 6 inches, is enclosed by a frame of uniform width. If the area of the picture and the frame is 80 square inches, find the width of the frame.

Solution:

Let x represent the width of the frame.
Draw and label a figure.

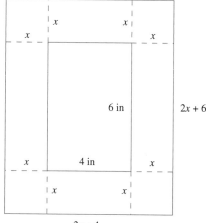

Since there are x inches of frame to the left and the right of the picture and the picture is 4 inches wide, the outside dimension is $2x + 4$ inches. Since there are x inches of the frame above and below the picture and the picture is 6 inches high, the outside dimension is $2x + 6$ inches.

The formula for finding the area of a rectangle is $A = LW$. Since the area of the picture and the frame is 80 square inches, substitute 80 for A, $2x + 6$ for L, and $2x + 4$ for W into the formula.

$A = LW$	Substitute for A, L, and W.
$80 = (2x + 6)(2x + 4)$	Multiply the right side.
$80 = 4x^2 + 20x + 24$	Subtract 80 from both sides.
$0 = 4x^2 + 20x - 56$	Remove the common factor of 4.
$0 = 4(x^2 + 5x - 14)$	Factor $x^2 + 5x - 14$.
$0 = 4(x + 7)(x - 2)$	Set each factor, except 4, equal to 0.
$4 \neq 0,\ x + 7 = 0,\ x - 2 = 0$	Solve each equation.
$x = -7, \quad x = 2$	Since a frame cannot have a width of -7 inches, 2 inches is the only answer.

Since x is the width of the frame, we have answered the question asked.

Check:

If the width of the frame is 2 inches, is the area of the picture and the frame 80 square inches? If the width of the frame is 2 inches, the outside dimensions are $4 + 4 = 8$ inches and $4 + 6 = 10$ inches. The area of a rectangle that is 8 inches by 10 inches is $(8)(10) = 80$ square inches. Therefore, our solution is correct.

PRACTICE EXERCISE

2) The length of a rectangular flower garden is two feet longer than its width. The flower garden is surrounded by a uniform border of mulch 2 feet wide. If the area of the flower garden and the border is 80 square feet, find the length and width of the flower bed.

• **Pythagorean Theorem**

EXAMPLE 3

The longer side of a rectangular mural is three feet less than three times the shorter side. If the mural is 13 feet diagonally across, find the lengths of the sides.

Solution:

Let x represent the length of the shorter side. Then, $3x - 3$ represents the length of the longer side.

Draw and label a figure.

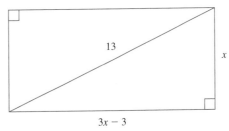

13

x

$3x - 3$

The diagonal divides the rectangle into two right triangles. Remember, the Pythagorean Theorem states that in a right triangle, $a^2 + b^2 = c^2$, where a and b are the lengths of the legs and c is the length of the hypotenuse.

$a^2 + b^2 = c^2$	Substitute for a, b, and c.
$x^2 + (3x - 3)^2 = 13^2$	Raise to powers.
$x^2 + 9x^2 - 18x + 9 = 169$	Add like terms.
$10x^2 - 18x + 9 = 169$	Subtract 169 from both sides.
$10x^2 - 18x - 160 = 0$	Remove the common factor of 2.
$2(5x^2 - 9x - 80) = 0$	Factor the trinomial.
$2(5x + 16)(x - 5) = 0$	Set each factor, except 2, equal to 0.
$5x + 16 = 0,\ x - 5 = 0$	Solve each equation.
$5x = -16, \quad x = 5$	
$x = -\dfrac{16}{5}$	Since a side of a rectangle cannot be $-\frac{16}{5}$, the only answer is 5 feet.

Have we answered the question asked? No, we also need the length.

Since $x = 5$ feet is the length of the shorter side, and $3x - 3$ is the length of the longer side, the length of the longer side $= 3(5) - 3 = 15 - 3 = 12$ feet.

Check:

Do the lengths of the sides satisfy the Pythagorean Theorem? Since $5^2 + 12^2 = 25 + 144 = 169 = 13^2$, our solution is correct.

PRACTICE EXERCISE

3) The distance diagonally across a rectangular vacant lot is 10 feet more than the length of the longer side. If the shorter side is 30 feet long, find the length of the longer side.

EXAMPLE 4

The average speed of a car is 10 miles per hour more than the average speed of a bus. The time required for the bus to travel 200 miles is one hour more than the time required by the car. Find how long it takes the car to travel 200 miles.

Solution:

Remember from Section 6.8 that we use a chart to solve distance, time, rate problems.

	d	r	t
bus			
car			

Let x represent the time for the car to travel 200 miles. Then, $x + 1$ represents the time for the bus to travel 200 miles.
Since both distances equal 200 miles, fill in the distance column also.

	d	r	t
bus	200		$x + 1$
car	200		x

Since $r = \frac{d}{t}$, the distance traveled by the bus is $\frac{200}{x+1}$ and the distance traveled by the car is $\frac{200}{x}$. Put these in the rate column.

	d	r	t
bus	200	$\frac{200}{x+1}$	$x + 1$
car	200	$\frac{200}{x}$	x

The last column filled in was the rate column, so write an equation involving rates. Since the rate of the car is 10 miles per hour faster than the rate of the bus, (the rate of the car) = (the rate of the bus) + 10. Consequently, the equation is:

$$\frac{200}{x} = \frac{200}{x + 1} + 10 \qquad \text{Multiply both sides by the LCD } x(x + 1).$$

$$x(x + 1)\left(\frac{200}{x}\right) = x(x + 1)\left(\frac{200}{x + 1} + 10\right) \qquad \text{Simplify both sides.}$$

$$(x + 1)(200) = x(x + 1)\left(\frac{200}{x + 1}\right) + x(x + 1)(10) \qquad \text{Continue simplifying.}$$

$$200x + 200 = 200x + 10x^2 + 10x \qquad \text{Subtract } 200x \text{ and } 200 \text{ from both sides.}$$

$$0 = 10x^2 + 10x - 200 \qquad \text{Remove the common factor of 10.}$$

$$0 = 10(x^2 + x - 20) \qquad \text{Factor } x^2 + x - 10.$$

$$0 = 10(x + 5)(x - 4)$$ Set each factor, except 10, equal to 0.

$$x + 5 = 0, \; x - 4 = 0$$ Solve each equation.

$$x = -5, \quad x = 4 \text{ hours}$$ Since time cannot be negative, 4 is the only answer. Therefore, it takes the car 4 hours to go 200 miles.

Since x represents the time of the car, we have answered the question asked.

Check:
The rate of the car is $\frac{200}{x} = \frac{200}{4} = 50$ miles per hour. So in 4 hours the car can travel $50(4) = 200$ miles. The time necessary for the bus to travel 200 miles is $x + 1 = 4 + 1 = 5$ hours. The rate of the bus is $\frac{200}{x+1}$. Therefore, the rate of the bus is $\frac{200}{4+1} = \frac{200}{5} = 40$ miles per hour. So in 5 hours the bus can travel $5(40) = 200$ miles. Therefore, our answer is correct.

PRACTICE EXERCISE

4) The average speed of a passenger train is 25 miles per hour more than the average speed of a car. The time required for the car to travel 300 miles is 2 hours more than the time required for the train. Find the average speed of the car.

• **Work problems**

In Section 5.8 we did work problems that resulted in linear equations, but they often result in quadratic equations.

EXAMPLE 5

It would take Ramon 4 hours longer to plant a garden than it would take his wife Aurea. If they can plant the garden in $4\frac{4}{5}$ hours working together, how long would it take each to plant the garden alone?

Solution:
Remember from Section 5.8 that we used a chart to do work problems.
Let $x =$ the number of hours for Aurea to plant the garden working alone. Then, $x + 4 =$ the number of hours for Ramon to plant the garden working alone.

	Time to do the job alone	rate	time worked	fractional part of the total job done
Aurea	x	$\dfrac{1}{x}$	$\dfrac{24}{5}$	$\dfrac{1}{x} \cdot \dfrac{24}{5} = \dfrac{24}{5x}$
Ramon	$x + 4$	$\dfrac{1}{x+4}$	$\dfrac{24}{5}$	$\dfrac{1}{x+4} \cdot \dfrac{24}{5} = \dfrac{24}{5(x+4)}$

The relationship is:
(the fractional part Aurea does) + (the part fractional Ramon does) = 1 (The entire job.).

Hence, the equation is:

$$\frac{24}{5x} + \frac{24}{5(x+4)} = 1$$ Multiply both sides by the LCD $5x(x+4)$.

$$5x(x+4)\left(\frac{24}{5x} + \frac{24}{5(x+4)}\right) = 5x(x+4)(1)$$ Simplify both sides.

$$5x(x+4)\left(\frac{24}{5x}\right) + 5x(x+4)\left(\frac{24}{5(x+4)}\right) = 5x^2 + 20x$$ Continue simplifying.

$$24(x+4) + 24x = 5x^2 + 20x$$ Distribute.

$$24x + 96 + 24x = 5x^2 + 20x$$ Add like terms.

$$48x + 96 = 5x^2 + 20x$$ Subtract $48x$ and 96 from both sides.

$$0 = 5x^2 - 28x - 96$$ Factor $5x^2 - 28x - 96$.

$$0 = (5x+12)(x-8)$$ Set each factor equal to 0.

$$5x + 12 = 0, \quad x - 8 = 0$$ Solve each equation.

$$5x = -12, \quad x = 8$$

$$x = -\frac{12}{5}$$ Since the garden cannot be planted in $-\frac{12}{5}$ hours, 8 is the only answer.

Have we answered the question asked? No, we have found the number of hours for Aurea to plant the garden working alone. We need to find the number of hours for Ramon as well. It would take Ramon $x + 4 = 8 + 4 = 12$ hours to plant the garden alone.

Check:

Since Aurea can plant the garden in 8 hours, she can plant $\frac{1}{8}$ of the garden in one hour. Ramon can plant the garden in 12 hours alone, so he can plant $\frac{1}{12}$ of the garden in one hour. Together they can plant $\frac{1}{8} + \frac{1}{12} = \frac{3}{24} + \frac{2}{24} = \frac{5}{24}$ of the garden in one hour. So in $4\frac{4}{5}$ hours they can plant $4\frac{4}{5} \cdot \frac{5}{24} = \frac{24}{5} \cdot \frac{5}{24} = 1$ garden.

PRACTICE EXERCISE

5) Terri and Tommy run a flea market. It takes Terri 2 hours longer to put the merchandise out than it does Tommy. If they can put the merchandise out together in $1\frac{7}{8}$ hours, how long would it take each working alone?

• Science problems

EXAMPLE 6

A particle moves along a straight line according to the formula $s = 2t^2 + t - 6$, where s represents the distance in inches from the beginning point and t represents the time in seconds. Find the time(s), to the nearest tenth of a second, when the particle is 3 inches from the beginning point.

Solution:

Since *s* represents the distance from the beginning point, we need to find *t* when $s = 3$.

$s = 2t^2 + t - 6$ Substitute 3 for *s*.

$3 = 2t^2 + t - 6$ Subtract 3 from both sides.

$0 = 2t^2 + t - 9$ Solve using the quadratic formula.

$t = \dfrac{-1 \pm \sqrt{1^2 - 4(2)(-9)}}{2(2)}$ Simplify the radical.

$t = \dfrac{-1 \pm \sqrt{1 + 72}}{4}$ Add 1 and 72.

$t = \dfrac{-1 \pm \sqrt{73}}{4}$ Write as two separate answers.

$t = \dfrac{-1 + \sqrt{73}}{4}, t = \dfrac{-1 - \sqrt{73}}{4}$ $\sqrt{73} \approx 8.5440$.

$t = \dfrac{-1 + 8.5440}{4}, t = \dfrac{-1 - 8.5440}{4}$ Simplify.

$t = 1.886, t = -2.386$ Round to the nearest tenth.

$t = 1.9, \quad t = -2.4$ Since we cannot have negative time, the only answer is 1.9 seconds.

Check:

Substitute 3 for *s* and 1.9 for *t* in the formula $s = 2t^2 + t - 6$.

$3 \approx 2(1.9)^2 + 1.9 - 6$

$3 \approx 2(3.61) + 1.9 - 6$

$3 \approx 7.22 + 1.9 - 6$

$3 \approx 3.12$

These are not exactly equal due to the round off error. Since they are approximately equal, our answer is correct.

PRACTICE EXERCISE

6) A ball is thrown upward with an initial velocity of 32 feet per second from the top of a building 64 feet high. The distance, *s*, above the ground *t* seconds after the ball is thrown is given by the formula $s = -16t^2 + 32t + 64$, where *s* is measured in feet. Find the time(s) when the ball is 32 feet above the ground.

EXERCISE SET   **9.5**

Solve the following.

1) The sum of the reciprocals of two consecutive integers is $\frac{9}{20}$. Find the integers.

2) The sum of the reciprocals of two consecutive odd integers is $\frac{12}{35}$. Find the integers.

3) One integer is three more than another. If the sum of the squares of the integers is 117, find the integers.

4) The larger of two integers is one less than twice the smaller. If the difference of the squares of the integers is 56, find the integers.

5) The product of the reciprocals of two consecutive odd integers is $\frac{1}{35}$. Find the integers.

6) The product of the reciprocals of two consecutive even integers is $\frac{1}{48}$. Find the integers.

7) The length of a small rectangular field is 10 yards more than its width. A road five yards wide completely encircles the field. If the area of the road and the field together is 4200 square yards, find the dimensions of the field.

8) A rectangular slab with length two feet more than its width is poured for a utility shed. The shed is placed on the slab so that one foot of the slab is exposed on all sides of the shed. If the area of the floor of the shed is 120 square feet, find the dimensions of the slab.

9) A rectangular sheet of paper is three inches longer than it is wide. Material is printed on the page with a one inch margin at the top and bottom of the page and one-half inch margins on the left and right. If the area of the printed material is 63 square inches, find the length and width of the page.

10) The length of a rectangular picture is two inches more than the width. It is to be mounted with one inch of matting on the sides and one and one-half inches at the top and bottom. If the area of the picture and the matting is 130 square inches, find the length and width of the picture.

11) The sides of a right triangle are three consecutive integers. Find the lengths of the sides.

12) The sides of a right triangle are three consecutive even integers. Find the lengths of the sides.

13) A guy wire is attached to a vertical pole at a point 24 feet above the ground. The length of the wire is six feet more than twice the distance from the pole to the point where the wire is attached to the ground. Find the length of the wire.

14) A telephone pole is supported by two guy wires. The distance from the pole to the points on the ground where the wires are attached is 15 feet. The length of each wire is five feet more than the distance from the ground to the point on the pole where the wires are attached. Find the total length of the wires.

15) Two cars are approaching the same intersection at right angles to each other. The distance from the intersection to one of the cars is 10 feet less than twice the distance from the intersection to the other car. If the distance between the cars is 170 feet, find the distance from each car to the intersection.

16) Two ships are approaching the same port and are traveling on courses that are at right angles to each other. One ship is eight miles from port. The distance of the other ship from port is two miles less than the distance between the ships. If the maximum range of their radios is 20 miles, can the ships communicate with each other?

17) The average rate of a bus is 15 miles per hour more than the average rate of a truck. It takes the truck one hour longer to go 180 miles than it takes the bus. How long does it take the bus to travel 180 miles?

18) The time required for a train to travel 200 miles is two hours more than the time required for a car to travel 180 miles. The average rate of the car is 20 miles per hour more than the average rate of the train. Find the rate of the train.

19) The time required for a bus to travel 360 miles is three hours more than the time required for a motorcycle to travel 300 miles. The average rate of the motorcycle is 15 miles per hour more than the average rate of the bus. Find the average rate of the bus.

20) The time required for a truck to travel 400 miles is three hours more than the time required for a car to travel 300 miles. The average rate of the car is 10 miles per hour more than the average rate of the truck. Find the average rate of the truck.

21) Billy and Jody are commercial fishermen. Working alone it takes Billy two hours longer to run the hoop nets than it takes Jody working alone. Together they can run the hoop nets in $2\frac{2}{5}$ hours. How long does it take each working alone?

22) Using a riding lawn mower, Fran can mow the grass at the campground in four hours less time than it takes Donnie using a push-type mower. Together they can mow the grass in $2\frac{2}{3}$ hours. How long does it take each working alone?

23) It takes Bubba one hour longer to stack a load of hay than it takes Ronnie. Together they can stack a load of hay in $1\frac{1}{5}$ hours. How long does it take each working alone?

24) It takes John, a novice mechanic, five hours longer to repair a transmission than it takes Ramon, an

experienced mechanic. Together they can repair the transmission in $3\frac{1}{3}$ hours. How long does it take each working alone?

25) A particle is moving along a straight line according to the formula $s = t^2 - 2t + 3$, where s is the distance in inches and t is the time in seconds. Find the time(s) when the particle is 18 inches from the beginning point.

26) A particle is moving along a straight line according to the formula $s = t^2 - 6t + 12$, where s is the distance in inches and t is the time in seconds. Find the time(s) when the particle is four inches from the beginning point.

27) The distance a free-falling object falls is given by $s = 16t^2$, where s is in feet and t is in seconds. Find the time(s) to the nearest tenth of a second it takes a free-falling object to fall 48 feet.

28) The distance a free-falling object falls is given by $s = 16t^2$, where s is in feet and t is in seconds. Find the time(s) to the nearest tenth of a second it takes a free-falling object to fall 80 feet.

29) An object is thrown upward with an initial velocity of 48 feet per second from a height of 80 feet. The distance, s, above the ground after t seconds is given by $s = -16t^2 + 48t + 80$, where s is measured in feet. Find the time(s) to the nearest tenth of a second when the object is 40 feet above the ground.

30) An object is thrown upward with an initial velocity of 52 feet per second from a height of 100 feet. The distance, s, above the ground after t seconds is given by $s = -16t^2 + 52t + 100$, where s is measured in feet. Find the time(s) to the nearest tenth of a second when the object is 120 feet above the ground.

WRITING EXERCISES

31) Write an application problem which results in a quadratic equation and involves integers.

32) Write an application problem which results in a quadratic equation and involves geometric figures.

Summary

Solving Incomplete Quadratic Equations: [Section 9.1]

1) Solve incomplete quadratic equations which can be put in the form of $ax^2 + bx = 0$ by factoring.

2) Solve incomplete quadratics which can be put in the form $ax^2 + c = 0$ by extraction of roots. This involves the following:

    a) Solve the equation for x^2. This gives an equation of the form $x^2 = k$.

    b) Find the positive and negative square roots of k.

    c) Write the solutions in simplest form.

3) Solve equations of the form $(ax + b)^2 = c$ by doing the following:

    a) Set $ax + b = \pm\sqrt{c}$.

    b) Rewrite as two equations.

    c) Solve each equation.

Completing the Square: [Section 9.2]

1) To determine the number to be added to $x^2 + bx$ in order to get a perfect square trinomial, find one-half of b and square it. The resulting trinomial is the square of a binomial, with first term equal to the square root of the first term of the trinomial, with last term equal to the square root of the last term of the trinomial, and with sign the same as the middle sign of the trinomial.

Solving Quadratic Equations by Completing the Square: [Section 9.2]

1) Write the equation in the form $x^2 + bx = c$. This may require several steps depending upon the form of the given equation.

2) Determine the number needed to make $x^2 + bx$ a perfect square trinomial and add that number to both sides of the equation.

3) Write one side of the equation as the square of a binomial and add the numbers on the other side of the equation.

4) Solve the equation resulting from step 4 by extraction of roots.

5) Solve each of the resulting equations.

6) Check each solution in the original equation.

Solving Quadratic Equations Using the Quadratic Formula: [Section 9.3]

1) If necessary, rewrite the equation in the form of $ax^2 + bx + c = 0$.

2) Find the values of a, b, and c.

3) Substitute the values of a, b, and c into the quadratic formula which is $x = \frac{-b \pm \sqrt{b^2 - 4ac}}{2a}$.

4) Simplify the results.

5) Check the solutions in the original equation.

Solving Linear Equations: [Section 9.4]

1) If necessary, simplify both sides of the equation as much as possible.

2) If necessary, use the Addition Property of Equality to get all the terms with variables on one side of the equation and all the constants on the other.

3) If necessary, use the Multiplication Property of Equality to eliminate any coefficient on the variable.

4) Check the answer in the original equation.

Solving Equations Containing Rational Expressions: [Section 9.4]

1) Multiply both sides of the equation by the LCD in order to eliminate all rational expressions from the equation.

2) Solve the resulting equation using the appropriate techniques depending upon whether it is linear or quadratic.

3) At the minimum, check the solution(s) of the equation in step 2 in the original equation to see if it/they make any denominator equal to 0. If so, that solution of the equation in step 2 is not a solution of the original equation.

Solving Equations Containing Radicals: [Section 9.4]

If an equation contains one square root, it can be solved by doing the following:

1) Isolate the square root on one side of the equation.

2) Square both sides of the equation.

3) Solve the resulting equation.

4) Check all solutions in the original equation.

If an equation contains two square roots, it can be solved by doing the following:

1) Isolate one of the radicals on one side of the equation.

2) Square both sides of the equation and, if necessary, simplify each side by combining like terms.

3) Isolate the radical on one side of the equation, if the equation still contains a radical, then square both sides of the equation again. Solve the resulting equation.

4) Check all solutions in the original equation.

Solving Applications Problems Involving Quadratic Equations: [Section 9.5]

Solve the applications problems following the same procedure you used earlier. Be sure you answered the question(s) asked. Check the solution(s) in the wording of the problem. Be aware that some solutions of the equation may not satisfy the conditions of the original problem.

Review Exercises

Solve the following incomplete quadratic equations. If there is no real solution, write $\varnothing$. [Section 9.1]

1) $6x^2 + 18x = 0$

2) $5x^2 - 80 = 0$

3) $5a^2 = -13a$

4) $2c^2 = 13c$

5) $n(n + 3) = 5n$

6) $a(a - 5) = 8a$

7) $r^2 = 144$

8) $c^2 = 400$

9) $x^2 - 28 = 0$

10) $c^2 - 45 = 0$

11) $b^2 + 63 = 0$

12) $k^2 + 48 = 0$

13) $3x^2 = 75$

14) $2q^2 = 98$

15) $4x^2 - 80 = 0$

16) $5w^2 - 90 = 0$

17) $(x + 6)^2 = 49$

18) $(a - 3)^2 = 64$

19) $(3x + 4)^2 = 80$

20) $(2x - 5)^2 = 125$

Solve the following quadratic equations by completing the square. If there is no real solution, write $\varnothing$. [Section 9.2]

21) $x^2 + 10x + 24 = 0$

22) $x^2 - 10x + 16 = 0$

23) $c^2 = 5c + 14$

24) $r^2 - 18 = -3r$

25) $2w^2 + 6 = -13w$

26) $3n^2 = 4n + 4$

27) $y^2 + 8y - 2 = 0$

28) $c^2 - 10c + 4 = 0$

29) $a^2 + 7a = 3$

30) $n^2 - 9n + 2 = 0$

31) $x^2 + 3x + 5 = 0$

32) $y^2 + 7y + 9 = 0$

33) $2x^2 + 12x - 5 = 0$

34) $2x^2 - 14x + 3 = 0$

Solve the following quadratic equations using the quadratic formula. If there is no real solution, write $\varnothing$. [Section 9.3]

35) $a^2 + 3a + 2 = 0$

36) $b^2 + 5b + 4 = 0$

37) $a^2 = 8a - 12$

38) $b^2 - 3b = 10$

39) $2x^2 + x - 10 = 0$

40) $3t^2 + t - 2 = 0$

41) $6v^2 = 8v + 3$

42) $6m^2 = -5m + 4$

43) $x^2 + 6x - 4 = 0$

44) $q^2 + 8q - 2 = 0$

45) $2x^2 + 1 = 8x$

46) $4n^2 = -6n + 5$

47) $6r^2 + 2r = 3$

48) $6x^2 = 4x + 1$

49) $3d^2 + 2d + 4 = 0$

50) $4z^2 - 3z + 5 = 0$

51) $2x^2 + 5x - 2 = 0$

52) $3x^2 + 7x = -1$

Solve the following equations. [Section 9.4]

53) $3(4x - 2) - 2(2x + 5) = 4(x - 6) + 4$

54) $2(5x - 6) - 3(4x + 2) = 5(x + 2) - 7$

55) $(x + 2)(x + 3) = x(x + 1) + 26$

56) $(2x + 5)(2x + 6) = 56$

57) $\dfrac{1}{x} + \dfrac{1}{x + 1} = \dfrac{11}{30}$

58) $\dfrac{8}{3x + 1} - \dfrac{5}{4x - 2} = \dfrac{3}{2x + 4}$

59) $\sqrt{5x + 1} = 4x - 8$

60) $\sqrt{x + 2} + \sqrt{4x + 1} = 5$

Solve the following applications problems.

61) One integer is three more than the other. The sum of the reciprocals of the integers is $\frac{7}{10}$. Find the integers.

62) One integer is four more than the other. The sum of the squares of the integers is 40. Find the integers.

63) The sum of the squares of two consecutive integers is 61. Find the integers.

64) One integer is twice the other. The product of the reciprocals of the integers is $\frac{1}{32}$. Find the integers.

65) A rectangular room is four feet longer than it is wide. A rectangular rug is placed on the floor and there is a uniform border of exposed floor one foot wide all the way around the rug. If the area of the rug is 140 square feet, find the dimensions of the room.

66) A rectangular piece of plywood is three feet longer than it is wide. The whole piece of plywood is to be painted with a black rectangle surrounded by a uniform border of red which is one foot wide. If the area of the red border is 30 square feet, find the length and width of the piece of plywood.

67) Two airplanes are approaching the same airport and are flying on courses that are perpendicular to each other. One airplane is one mile less than twice as far from the airport as the other. If the distance between the airplanes is 17 miles, how far is each from the airport?

68) A diver's flag is rectangular in shape and is red with a black stripe going diagonally across. If the width of the flag is five inches less than the length and the length of the black stripe is 25 inches, find the length and width of the flag.

69) The average rate of a recreational vehicle is 20 miles per hour less than the average rate of a car. It takes the recreational vehicle one hour more to travel 200 miles than it takes the car to travel 240 miles. Find the average rate of the recreational vehicle.

70) It takes a bus two hours longer to travel 200 miles than it takes a motorcycle to travel 120 miles. The average rate of the motorcycle is 10 miles per hour more than the average rate of the bus. Find how long it takes the motorcycle to travel 120 miles.

71) Horace is planning to build a house. The lot he has chosen has to be built up with fill dirt. It would take Frank four hours longer to spread the dirt than it would take his worker, Ahmad. If they can spread the dirt in $3\frac{3}{4}$ hours working together, how long would it take each working alone?

72) After Horace built his house in Exercise 71, he and Ahmad painted it. It would take Horace two days longer to paint the house than it would take Ahmad. Together they could paint the house in $3\frac{3}{7}$ days. How long would it take each working alone?

Chapter 9 Test

Solve the following incomplete quadratic equations. If there is no real solution, write $\varnothing$.

1) $6x^2 - 12x = 0$

2) $4x^2 = 20x$

3) $x^2 - 121 = 0$

4) $3x^2 = 81$

5) $(x - 6)^2 = 36$

6) $(3x + 2)^2 = 18$

Find the number needed to make each of the following a perfect square trinomial. Then write each as the square of a binomial.

7) $a^2 - 20a + \underline{}$

8) $b^2 + \dfrac{7}{3}b + \underline{}$

Solve the following quadratic equations by completing the square. If there is no real solution, write $\varnothing$.

9) $x^2 - 6x - 16 = 0$

10) $2x^2 - 6x - 5 = 0$

Solve the following quadratic equations using the quadratic formula. If there is no real solution, write $\varnothing$.

11) $a^2 - 5a + 6 = 0$

12) $2c^2 - 3c = 20$

13) $r^2 - 2r + 6 = 0$

14) $3w^2 = -8w + 3$

Solve the following quadratic equations using the method which seems most appropriate. If there is no real solution, write $\varnothing$.

15) $x^2 = 2x + 35$

16) $a^2 + \dfrac{7}{2}a = 3$

17) $3d^2 + 8 = 5d$

18) $x^2 - 8x - 109 = 0$

Solve the following equations.

19) $5(x - 2) - 3(2x + 1) = -3(x - 5) - 8$

20) $(2x + 4)(2x + 7) = 88$

21) $\dfrac{1}{x} + \dfrac{1}{x + 2} = \dfrac{7}{24}$

22) $\sqrt{5x} - \sqrt{3x+1} = 1$

Solve the following applications problems.

23) Find two consecutive even integers with sum of squares equal to 164.

24) The length of a picture is two inches more than the width. The picture is in a frame which is $1\frac{1}{2}$ inches wide. If the area of the picture and the frame is 99 square inches, find the length and width of the picture.

25) In mountainous country, it takes a recreational vehicle (RV) one hour longer to travel 120 miles than it takes a van. If the rate of the van is 10 miles per hour more than the rate of the RV, find how long it takes the van to travel 120 miles.

TABLE I

POWERS AND ROOTS

n	n^2	n^3	$\sqrt{n}$	$\sqrt[3]{n}$	$\sqrt{10n}$	n	n^2	n^3	$\sqrt{n}$	$\sqrt[3]{n}$	$\sqrt{10n}$
1	1	1	1.000	1.000	3.162	51	2601	132,651	7.141	3.708	22.583
2	4	8	1.414	1.260	4.472	52	2704	140,608	7.211	3.733	22.804
3	9	27	1.732	1.442	5.477	53	2809	148,877	7.280	3.756	23.022
4	16	64	2.000	1.587	6.325	54	2916	157,464	7.348	3.780	23.238
5	25	125	2.236	1.710	7.071	55	3025	166,375	7.416	3.803	23.452
6	36	216	2.449	1.817	7.746	56	3136	175,616	7.483	3.826	23.664
7	49	343	2.646	1.913	8.367	57	3249	185,193	7.550	3.849	23.875
8	64	512	2.828	2.000	8.944	58	3364	195,112	7.616	3.871	24.083
9	81	729	3.000	2.080	9.487	59	3481	205,379	7.681	3.893	24.290
10	100	1000	3.162	2.154	10.000	60	3600	216,000	7.746	3.915	24.495
11	121	1331	3.317	2.224	10.488	61	3721	226,981	7.810	3.936	24.698
12	144	1728	3.464	2.289	10.954	62	3844	238,328	7.874	3.958	24.900
13	169	2197	3.606	2.351	11.402	63	3969	250,047	7.937	3.979	25.100
14	196	2744	3.742	2.410	11.832	64	4096	262,144	8.000	4.000	25.298
15	225	3375	3.873	2.466	12.247	65	4225	274,625	8.062	4.021	25.495
16	256	4096	4.000	2.520	12.649	66	4356	287,496	8.124	4.041	25.690
17	289	4913	4.123	2.571	13.038	67	4489	300,763	8.185	4.062	25.884
18	324	5832	4.243	2.621	13.416	68	4624	314,432	8.246	4.082	26.077
19	361	6859	4.359	2.688	13.784	69	4761	328,509	8.307	4.102	26.268
20	400	8000	4.472	2.714	14.142	70	4900	343,000	8.367	4.121	26.458
21	441	9261	4.583	2.759	14.491	71	5041	357,911	8.426	4.141	26.646
22	484	10,648	4.690	2.802	14.832	72	5184	373,248	8.485	4.160	26.833
23	529	12,167	4.796	2.844	15.166	73	5329	389,017	8.544	4.179	27.019
24	576	13,824	4.899	2.884	15.492	74	5476	405,224	8.602	4.198	27.203
25	625	15,625	5.000	2.924	15.811	75	5625	421,875	8.660	4.217	27.386
26	676	17,576	5.099	2.962	16.125	76	5776	438,976	8.718	4.236	27.568
27	729	19,683	5.196	3.000	16.432	77	5929	456,533	8.775	4.254	27.749
28	784	21,952	5.292	3.037	16.733	78	6084	474,552	8.832	4.273	27.928
29	841	24,389	5.385	3.072	17.029	79	6241	493,039	8.888	4.291	28.107
30	900	27,000	5.477	3.107	17.321	80	6400	512,000	8.944	4.309	28.284
31	961	29,791	5.568	3.141	17.607	81	6561	531,441	9.000	4.327	28.460
32	1024	32,768	5.657	3.175	17.889	82	6724	551,368	9.055	4.344	28.636
33	1089	35,937	5.745	3.208	18.166	83	6889	571,787	9.110	4.362	28.810
34	1156	39,304	5.831	3.240	18.439	84	7056	592,704	9.165	4.380	28.983
35	1225	42,875	5.916	3.271	18.708	85	7225	614,125	9.220	4.397	29.155
36	1296	46,656	6.000	3.302	18.974	86	7396	636,056	9.274	4.414	29.326
37	1369	50,653	6.083	3.332	19.235	87	7569	658,503	9.327	4.431	29.496
38	1444	54,872	6.164	3.362	19.494	88	7744	981,472	9.381	4.448	29.665
39	1521	59,319	6.245	3.391	19.748	89	7921	704,969	9.434	4.465	29.833
40	1600	64,000	6.325	3.420	20.000	90	8100	729,000	9.487	4.481	30.000
41	1681	68,921	6.403	3.448	20.248	91	8281	753,571	9.539	4.498	30.166
42	1764	74,088	6.481	3.476	20.494	92	8464	778,688	9.592	4.514	30.332
43	2849	79,507	6.557	3.503	20.736	93	8649	804,357	9.644	4.531	30.496
44	2936	85,184	6.633	3.530	20.976	94	8836	830,584	9.695	4.547	30.659
45	2025	91,125	6.708	3.557	21.213	95	9025	857,375	9.747	4.563	30.882
46	2116	97,336	6.782	3.583	21.148	96	9216	884,736	9.798	4.579	30.984
47	2209	103,823	6.856	3.609	21.679	97	9409	912,673	9.849	4.595	31.145
48	2304	110,592	6.928	3.634	21.909	98	9604	941,192	9.899	4.610	31.305
49	2401	117,649	7.000	3.659	22.136	99	9801	970,299	9.950	4.626	31.464
50	2500	125,000	7.071	3.684	22.361	100	10,000	1,000,000	10.000	4.642	31.623

TABLE I   631

Answers for Odd Numbered Exercises

1) -8

3) losing 10 lbs

5) 250 mi west

7)

9)

11)

13) 0.875

15) 0.222...

17) $0.\overline{714285}$

19) $-5, -\dfrac{4}{3}, 0, 2.3, 4.\overline{12}, 6$

21) 0, 6

23) $-\dfrac{4}{3}, 2.3, 4.\overline{12}$

25) sometimes

27) always

29) always

31) always

33) always

35) -17

37) 15

39) $-\dfrac{2}{3}$

41) $\dfrac{3}{7}$

43) 3.4

45) 5

47) 9

49) -11

51) $\dfrac{4}{7}$

53) $\dfrac{7}{10}$

55) -9

57) 6

59) $>$

61) $<$

63) $>$

65) $<$

67) $<$

69) $<$

71) $>$

73) $>$

75) $>$

77) $>$

79) $=$

81) $<$

83) $>$

85) $=$

1) prime

3) composite

5) composite

7) prime

9) $2 \cdot 2 \cdot 7$

11) $3 \cdot 3 \cdot 5$

13) $2 \cdot 2 \cdot 5 \cdot 5$

15) $2 \cdot 7 \cdot 11$

17) $2 \cdot 2 \cdot 3 \cdot 3 \cdot 5$

19) $3 \cdot 3 \cdot 5 \cdot 7$

21) $\dfrac{2}{3}$

23) $\dfrac{2}{3}$

25) $\dfrac{3}{2}$

27) $\dfrac{8}{45}$

29) $\dfrac{12}{35}$

31) $\dfrac{22}{25}$

33) $\dfrac{5}{9}$

35) 4

37) $\dfrac{9}{2}$

39) $\dfrac{18}{5}$

41) $\dfrac{48}{5}$

43) $\dfrac{12}{35}$

45) $\dfrac{56}{33}$

47) $\dfrac{5}{7}$

49) 1

51) $\dfrac{2}{15}$

53) $\dfrac{28}{3}$

55) $\dfrac{1}{4}$

57) $\dfrac{1}{6}$

59) $\dfrac{28}{27}$

61) $\dfrac{15}{17}$

63) $\dfrac{5}{13}$

65) $\dfrac{3}{4}$

67) $\dfrac{6}{7}$

69) $\dfrac{9}{13}$

71) $\dfrac{7}{6}$

73) $\dfrac{51}{5}$ or $10\dfrac{1}{5}$

75) $\dfrac{17}{4}$ or $4\dfrac{1}{4}$

77) $\dfrac{2}{3}$

79) $\dfrac{4}{15}$

81) $\dfrac{13}{45}$

83) $\dfrac{51}{50}$

85) $\dfrac{25}{6}$

87) $\dfrac{1}{12}$

89) $\dfrac{7}{4}$

91) $\dfrac{15}{8}$

93) $\dfrac{283}{24}$

95) $\dfrac{27}{350}$

97) 5 cups

99) $\dfrac{5}{8}$ of the pizza

101) $\dfrac{49}{10}$ or $4\dfrac{9}{10}$ ft

103) 18 pieces

105) $\dfrac{13}{3}$ or $4\dfrac{1}{3}$ miles

107) $\dfrac{1}{2}$ of the room

109) 60 square yards

111) $\dfrac{2}{9}$

113) $16

115) 14 lots

117) 25 gallons

EXERCISE SET 1.3

1) 8 squared or 8 to the second power, 64

3) 5 cubed or 5 to the third power, 125

5) 9 to the fourth power, 6561

7) 2 to the fifth power, 32

9) 10 to the sixth power, 1,000,000

11) two-thirds squared of two-thirds to the second power, $\dfrac{4}{9}$

13) 1.3 cubed or 1.3 to the third power, 2.197

15) 9 **17)** 2 **19)** 19

21) 56 **23)** 46 **25)** 6

27) 12 **29)** 288 **31)** 2304

33) $\dfrac{1}{18}$ **35)** 56 **37)** 24

39) 77 **41)** 16 **43)** 21

45) 5 **47)** 3 **49)** 32

51) 4 **53)** 10 **55)** 24

57) 4 **59)** 61 **61)** 2

63) 3 **65)** 4 **C1)** 2187

C3) 53.5824 **C5)** 43.7235 **C7)** 430.204708

67) 18 **69)** 32 **71)** 12

73) 180 **75)** 48 **77)** 8

79) 2 **81)** $\dfrac{4}{75}$ **83)** $\dfrac{4}{25}$

C9) 13.44 **C11)** 181.8432 **C13)** 7.68

85) $654 **87)** $5160 **89)** $396,000

91) $7.38 **93)** $13,514 **95)** 70

97) $\dfrac{29}{48}$ **99)** $\dfrac{7}{48}$ **101)** 5

103) 78 **105)** 1096

EXERCISE SET 1.4

1) 14 m

3) 33 in

5) $P = 42$ m, $A = 54$ m^2

7) $P = 180$ ft, $A = 1400$ ft^2

9) $P = \dfrac{53}{6}$ or $8\dfrac{5}{6}$ yds, $A = \dfrac{55}{12}$ or $4\dfrac{7}{12}$ yds^2

11) $P = 4.8$ yd, $A = 1.44$ yd^2

13) $P = \dfrac{20}{7}$ or $2\dfrac{6}{7}$ cm, $A = \dfrac{25}{49}$ cm^2

15) $P = 26.2$ m, $A = 19.32$ m^2

17) $P = 22$ in, $A = 15$ in^2

19) $C = 14\pi$ m, $A = 49\pi$ m^2

21) $C = 34.5$ ft, $A = 95.0$ ft^2

23) $C = \dfrac{44}{3}$ or $14\dfrac{2}{3}$ in, $A = \dfrac{154}{9}$ or $17\dfrac{1}{9}$ in^2

25) 112 mm^2

27) 1.56 mi^2

29) 42 mi^2

31) 72 in^2

33) 72 ft^3

35) 270.4 m^3

37) 125 mi^3

39) $\frac{27}{125}$ cm^3

41) 300π cm^3

43) 105.84π m^3

45) 42.9 m^3

47) $\frac{1100}{147}$ or $7\frac{71}{147}$ in^3

EXERCISE SET 1.5

1) 208 in   **3)** 69 ft

5) $82.50   **7)** 314 ft

9) 50.24 in   **11)** 30,000, $88,500

13) 18,000 yd^2 **15)** 260, $673.40

17) $1050   **19)** 14.22 yd^3

21) $31,752\pi$ in^3 **23)** $720

25) $500.07   **27)** 219.8 in

29) $2,257.93

EXERCISE SET 1.6

1) 18   **3)** 11   **5)** -7

7) 9   **9)** -11 **11)** -11

13) 48   **15)** 11   **17)** -6

19) 29   **21)** $\frac{4}{11}$ **23)** $\frac{1}{12}$

25) $-\frac{37}{30}$ **27)** 1.8   **29)** -4.7

31) 7   **33)** -5 **35)** -25

37) -70 **39)** 3   **41)** -11

43) -17 **45)** -5 **C1)** -18.59

C3) -942.2249 **C5)** -1086.78

47) commutative   **49)** inverse

51) commutative   **53)** inverse

55) commutative   **57)** associative

59) $6 + (-5)$ **61)** $-4 + (12 + 9)$

63) 0   **65)** $7(8 + 5)$

67) $-11 + 8 = -3$ **69)** $-6 + (-4) + 10 = 0$

71) $8 + 11 = 19$ **73)** $-12 + 7 = -5$

75) $49 + 35 = 84$ **77)** $-18 + 35 = 18$

79) $92° + (-25°)$, $67°$ **81)** $12° + (-19°)$, $-7°$

83) $-12° + 24°$, $12°$ **85)** $-21° + (-7°)$, $-28°$

87) $-68 + 38 = -29$ ft

89) $-65 + -43 = -108$ ft

91) 3rd parking level below ground level

93) 12th floor

95) $143

97) $376

99) $39\frac{1}{4}$

101) $946

103) a) yes, b) Total net gain was 11 yards

105) 18

107) 504

EXERCISE SET 1.7

1) 6   **3)** -6 **5)** 12

7) 4   **9)** -8 **11)** $\frac{3}{5}$

13) $-\frac{1}{2}$ **15)** $-\frac{23}{20}$ **17)** $\frac{29}{24}$

19) -1.7 **21)** -9 **23)** -12

25) 25   **27)** $\frac{5}{4}$ **29)** 2.2

31) 4   **33)** 12   **35)** 1

37) -14 **39)** 39   **41)** $-\frac{1}{2}$

43) 2   **45)** 17   **47)** -26

C1) 16,138.3227   **C3)** -8.93

49) $7 - (-5) = 12$ **51)** $-10 - 6 = -16$

53) $8 - 5 = 3$ **55)** $-5 - 8 = -13$

57) $6 - 9 = -3$ **59)** $9 - (-13) = 22$

61) $26 - 32 = -6$

63) $14 + [6 - (-5)] = 25$

65) $[-4 - (-6)] + 15 = 17$

67) $[9 - (-4)] - (-5 + 6) = 12$

69) $[7 - (-3)] + [-5 - (-2)] = 7$

71) yes   **73)** no   **75)** yes

77) -1 **79)** -2 **81)** 3

83) $-25°$ **85)** $-27°$ **87)** $41°$

89) 19 ft   **91)** 495 ft   **93)** $17,000

95) -375 ft   **97)** 2,687 ft   **99)** $-24\frac{1}{4}$

101) $-1\frac{3}{4}$ **103)** $+$157   **105)** 100

107) 100

EXERCISE SET 1.8

1) monomial   **3)** trinomial

5) binomial   **7)** trinomial

9) none of these   **11)** monomial

13) none of these

15) 3; x; 1

17) 1; x; 2

19) -5; x; 1

21) 4; x; 3

23) -1; x; 3

25) 10; x and y; 6

27) -13; x and z; 6

29) $-5x + 4$; 1

31) $2x^2 + 3x - 4$; 2

33) $-x^4 + x^3 - 4x^2 + 3x$; 4

35) $3x^3 + x$; 3

37) 0

39) 26

41) -1

43) 45

45) 19

47) 4

49) $\dfrac{16}{9}$

51) 100 ft

53) 27 units

55) $9800

57) 5000 sq. ft.

C1) 4.674

EXERCISE SET 1.9

1) $11x$

3) $6y$

5) $\dfrac{5}{7}x$

7) $-\dfrac{1}{15}x$

9) $4z + 3$

11) $-11x$

13) $\dfrac{4}{11}x$

15) $\dfrac{1}{12}x$

17) $-4xy + 8x$

19) $-8x^2 - 2y^2$

21) $12z + x - 12$

23) $7b^2 - 10b + 12$

25) $m^2r + 6r$

27) $\dfrac{2}{5}x^2y + \dfrac{5}{7}xy^2$

29) $-\dfrac{4}{13}m^3n^2 - \dfrac{7}{9}mn^2$

31) $\dfrac{13}{18}x^3y + \dfrac{1}{6}xy^2$

33) $10x - 7$

35) $9x - 3y$

37) $2x + 2$

39) $3x + 13y$

41) $5x^2 - 5x + 14$

43) $10xy + 3y - 2$

45) $5m^2c^2 + 4mc - 9c^2 + 3$

47) $4p^3 - 5p^2k - 13p + 11$

49) $-5r^2 + 6r - 10$

51) $5p^2m^2 + 9mp - 3m^2 + 5$

53) $rp^2 - 5rp - r^3$

55) $h^2 - 9$

57) $\dfrac{3}{5}x^2 - \dfrac{1}{3}x + \dfrac{5}{9}$

59) $-\dfrac{2}{11}r^2 + \dfrac{6}{7}r - \dfrac{2}{5}$

61) $-\dfrac{1}{10}x^2 - \dfrac{1}{8}x + \dfrac{1}{2}$

63) $11x^2 + 6x + 3$

65) $6x^3 - 10x - 12$

67) $-3x^2 - 10x + 4$

69) $2x^3 - 14x^2 + 8x + 1$

71) $8x + 5$

73) $2x + 5$

75) $3x^2 + x - 2$

77) $12z - 11$

79) $x^2 - 11x + 6$

81) $4x - 18$

83) $8x$ cm

85) $8x + 1$ in

87) $16x$ in

89) $7x + 9$ ft

91) $3x - 2$ ft

93) $x - 2$ ft

95) $5x - 2$ m

CHAPTER 1 REVIEW EXERCISES

1) $-\$50$

3) -23

5) $-\dfrac{7}{16}$

7) 8.3

9) 5

11) $<$

13) $>$

15) $<$

17) $>$

19) $>$

21) $>$

23) $\dfrac{1}{3}$

25) $\dfrac{15}{8}$

27) $\dfrac{7}{11}$

29) $\dfrac{58}{13}$

31) $\dfrac{46}{45}$

33) $\dfrac{283}{24}$ or $11\dfrac{19}{24}$

35) 9

37) 56 yds

39) 9^5

41) $3^3 \cdot 5^4$

43) $3^3 \cdot a^3$

45) 324

47) 900

49) $\dfrac{4}{25}$

51) 54

53) 90

55) 2

57) 9

59) 10

61) 31

63) 12

65) 270

67) $\$5.29$

69) $P = 36$ ft, $A = 81$ ft^2

71) $C = 31.4$ m, $A = 78.5$ m^2

73) 81 cm^2

75) 216 in^3

77) 32π m^3

79) 942 ft

81) $\$1820$

83) 64 in^2

85) 120 in^2

87) 35.325 in^3

89) 6

91) -7

93) 23

95) -95

97) $-\dfrac{1}{8}$

99) $-\dfrac{43}{48}$

101) commutative of addition

103) commutative of addition

105) additive inverse

107) $-9 + 6 = -3$

109) $-5 + 16 = 11$

111) $-9 + 8 = -1$

113) $(-5 + 13) + [18 + (-8)] = 18$

115) $[14 + (-8)] - [4 - (-5)] = -3$

117) 9; x; 2

119) -9; x and y; 9

121) trinomial; 2

123) trinomial; 5

125) none of these   **127)** 18

129) 253   **131)** $2x + y$

133) $-5x^2y + 6xy^2 + 4$ **135)** $5a + 11b$

137) $2a^2 - 14a + 5$ **139)** $5x^4 - 4x^2 - 1$

141) $11x^2 - 11x + 4$ **143)** $-2x^2 + 13x - 5$

145) $-2x^3 - 13x^2 + 2x + 5$ **147)** $x + 11$ ft

149) $2x - 9$ dollars

CHAPTER 1 TEST

1) $\dfrac{35}{54}$ **2)** $\dfrac{11}{17}$

3) 12   **4)** $\dfrac{373}{100}$

5) 68   **6)** 64

7) 144   **8)** 33

9) 23   **10)** 256

11) $P = 33$ cm, $A = 42$ cm^2 **12)** $C = 43.96$ m, $A = 153.86$ m^2

13) 90 in^3 **14)** \$36

15) \$72   **16)** $<$

17) $<$ **18)** -11

19) -7 **20)** -1

21) -44 **22)** $\dfrac{11}{60}$

23) $-\dfrac{46}{45}$

24) commutative of addition

25) additive inverse

26) $-7 + 11 = 4$

27) $-12 + [9 - (-6)] = 3$

28) $[4 + (-13)] - (-3 + 9) = -15$

29) $(x^2 - 6x + 2) - (2x^2 + 5x - 3) = -x^2 - 11x + 5$

30) $(4x + 9) - [(3x + 6) + (3x - 3)] = -x + 2$

31) $(3x + 6) - (2x - 2) = x + 8$

32) a) 4;   b) 3;   c) $4x^3 + 3x^2 + 6x$;   d) trinomial

33) -40 ft

34) 497 ft

35) 16

36) $5xy - 4yz$

37) $5x^2 - 3x - 4$

38) $6n^3 - 2n^2 + 7n - 2$

39) \$11,484

40) No. $-x$ is positive if x is negative because the opposite of a negative number is a positive number.

1) -24 **3)** -24 **5)** 24

7) -21 **9)** $-\dfrac{8}{15}$ **11)** $\dfrac{20}{9}$

13) -9.12 **15)** 15   **17)** -48

19) -90 **21)** -120 **23)** $\dfrac{9}{40}$

25) $-\dfrac{3}{10}$ **27)** 64   **29)** -16

31) $\dfrac{4}{9}$ **33)** 108   **35)** 125

C1) -49.651 **C3)** 6,029.6236   **37)** -6

39) 10   **41)** 24   **43)** 40

45) -256 **47)** -128 **49)** -2

51) 2   **53)** -2 **55)** 1 and -1

57) -1 **59)** 2

61) $5(-3) + 6 = -9$

63) $5 - (4)(-7) = 33$

65) $3(-4)^2 = 48$

67) $(-6)(-3) - 8 = 10$

69) $(-5)(4) + (4)(-2) = -28$

71) associative of multiplication

73) commutative of multiplication

75) identity for multiplication

77) commutative of multiplication

79) inverse for multiplication

81) multiplication by 0

83) $[3(-9)](-7)$ **85)** $(-7)(3)$

87) $4(-9) + 4(2)$ **89)** $(-9 + 2)4$

91) -2 **93)** 1

95) 0   **97)** 14

99) -11 **101)** $-12°$

103) -70 ft   **105)** -6 points

1) r^8 **3)** 4^7 **5)** x^9

7) $20x^9$ **9)** $21a^8$ **11)** x^8y^7

13) $8a^{10}b^{11}$ **15)** a^5b^5 **17)** $2^4y^4 = 16y^4$

19) $16x^2$ **21)** $-25x^2$ **23)** $27c^3$

25) $-27b^3$ **27)** y^{18} **29)** 4^{15}

31) x^7y^7 **33)** 4^8x^8 **35)** $c^{24}d^{18}$

37) $4^{20}a^{12}$ **39)** $-4^{20}b^{25}$ **41)** -6^6b^9

43) $108a^{17}b^{12}$ **45)** $4^{20}b^{28}$ **47)** $a^{14}b^{16}c^{22}$

49) $16a^2$ 51) $3x^3$ 53) $89x^6$
55) $-24a^4$ 57) $-2m^4n^6$

1) $-7a^3x^3$ 3) $42f^7k^2b^5$
5) $24p^5q^5r^8$ C1) $-53.7654x^9y^{13}$
C3) $-34.56x^{11}y^9$
7) $2h^5 + 16h^4 - 10h^3$
9) $2b^6 + 14b^5 - 12b^4$
11) $-6r^6 + 8r^5 - 4r^4 + 6r^3$
13) $10x^4y^3 - 20x^3y^4$
15) $-12p^6q^4 + 4p^4q^5 - 4p^2q^4$
C5) $22.79x^5 - 33.54x^4 - 30.96x^3$
17) $a^2 + 9a + 20$ 19) $z^2 - 3z - 40$
21) $x^2 - 16$ 23) $4x^2 - y^2$
25) $x^2 - xy - 2y^2$ 27) $20w^2 - 17w - 24$
29) $6z^2 - 26z + 24$ 31) $6a^2 + 5ab - 4b^2$
33) $x^3 + 2x - 3$ 35) $2x^3 + 2x^2 - 9x + 9$
37) $x^3 + 8$ 39) $6x^3 - x^2 - 18x + 9$
C7) $37.63x^2 - 15x - 17.28$ 41) $2x^2 + 10x + 12$
43) $6x^2 - 25x + 25$ 45) $2a^3 - 6a^2 - 8a + 24$
47) $8y^3 + 8y^2 - 14y + 4$ 49) $2x^2$
51) $4x^2 + 24x$ 53) $2x^2 + x - 6$
55) $3x^3 - x^2 + 4x + 4$ 57) $4y^2$
59) $9z^2$ 61) $2y^2 - 8y$ ft^2
63) $9x^2 + 9x - 10$ ft^2
65) $12.5x^2 + 37.5x$ dollars

1) $x^2 + 5x + 6$ 3) $a^2 - a - 12$
5) $c^2 - 7c + 10$ 7) $a^2 + 7ab + 12b^2$
9) $r^2 - 2rs - 15s^2$ 11) $6a^2 + 13ab + 6b^2$
13) $6x^2 - xy - 15y^2$ 15) $6x^2 - 19xy + 10y^2$
C1) $25.42x^2 + 35.54xy - 32.76y^2$
17) $x^2 - 9$ 19) $p^2 - q^2$
21) $a^2 - 16b^2$ 23) $36a^2 - b^2$
25) $16a^2 - 25$ 27) $4x^2 - 25y^2$
C3) $51.84a^2 - 44.89$ 29) $x^2 + 4x + 4$
31) $a^2 - 8a + 16$ 33) $r^2 + 2rs + s^2$
35) $t^2 - 2ts + s^2$ 37) $4x^2 + 12x + 9$
39) $9a^2 - 6a + 1$ 41) $4x^2 + 20xy + 25y^2$
43) $16x^2 - 24xy + 9y^2$ 45) $25t^4 - 10t^2w + w^2$
C5) $22.09x^2 + 34.78xy + 13.69y^2$

47) $6a^2 + 11ab - 10b^2$ 49) $16a^2 - 9b^2$
51) $9a^2 + 30a + 25$ 53) $8 - 18b + 9b^2$
55) $36 - y^2$ 57) $16 + 24w + 9w^2$
59) $16c^2 - 49d^2$ 61) $(x + 3)^2$
63) $x^2 + 3$ 65) $(x - 5)^2 = 25$
67) $x^2 - 3x - 40$ 69) $12a^2 + 21a - 45$
71) $x^2 + 16x + 64$ 73) $9a^2 - 30ab + 25b^2$
75) $p^3 + q^3$ 77) $x^3 + 27$
79) $b^3 - 64$ 81) $27a^3 + 8b^3$

1) -4 3) -6 5) 6
7) 11 9) -8 11) 5
13) -7 15) 4 17) 2
19) -4 21) -2 23) -2
25) -5 C1) -3.4 C3) -6.73
27) yes 29) yes 31) no
33) yes 35) no 37) -11
39) 8 41) 16 43) 10
45) -5 47) -4 49) -288
51) 10 53) $-7°$ per hour
55) $-\dfrac{3}{50}$ or -0.06 ft per foot 57) -9 ft per sec

1) -2 3) 22 5) -16
7) -33 9) 4 11) 12
13) -22 15) 11 17) -6
19) 1 21) -5 23) -16
25) 27 27) -4 29) 2
31) -2 33) $\dfrac{17}{2}$ 35) $-\dfrac{14}{29}$
C1) 60.9596 C3) 5.8 C5) 9948.2
C7) -2665.18 C9) -2.74 37) 10
39) -89 41) -7 43) 36
45) 90 47) -324 49) 27
51) 60 53) -26 55) -92
57) $\dfrac{5}{3}$ 59) $-\dfrac{5}{18}$ 61) $\dfrac{43}{42}$
63) 6.49 profit 65) 300 loss 67) 1398 profit

1) 2^4 3) c^4 5) z^4
7) $\dfrac{1}{2^4}$ 9) $\dfrac{1}{a^6}$ 11) 5^3

13) y^5 **15)** $\dfrac{1}{x^3}$ **17)** $\dfrac{1}{5^3}$

19) 1 **21)** 3 **23)** 1

25) 2 **27)** -1 **29)** $\dfrac{5}{a^2}$

31) $\dfrac{-9}{c^6}$ **33)** $\dfrac{1}{5^2 a^2}$ **35)** 2^2

37) $\dfrac{1}{5^2}$ **39)** $\dfrac{1}{x^5}$ **41)** $\dfrac{1}{z^2}$

43) $\dfrac{1}{3^4}$ **45)** $\dfrac{1}{y^9}$ **47)** 6^7

49) r^7 **51)** 6 **53)** $\dfrac{1}{z^4}$

55) 8^{18} **57)** r^{20} **59)** $\dfrac{1}{6^{15}}$

61) $\dfrac{1}{y^{21}}$ **63)** $\dfrac{1}{8^{18}}$ **65)** $\dfrac{1}{z^{15}}$

67) 9^{10} **69)** $-\dfrac{108x^3}{y^5}$ **71)** $-\dfrac{2m^{14}}{n^{16}}$

73) $\dfrac{8b^{18}}{a^{21}}$

EXERCISE SET 2.8

1) 35,000 **3)** 0.0095 **5)** 4,790,000

7) 0.00000924 **9)** 100 **11)** 0.00000001

13) 7.6×10^3 **15)** 3.5×10^{-4} **17)** 8.57×10^5

19) 4.98×10^{-7} **21)** 6×10^5

23) 1×10^{-8} **25)** 300,000,000

27) 1376 **29)** 0.0131976

31) 0.000000264388 **33)** 2000

35) 0.00000015 **37)** 1,200,000

39) 50 **41)** 25,000

43) 9,000,000 **45)** 126,500,000

47) 0.252 **49)** 0.00000000000009

51) 23,000 **53)** 0.000025

55) 22,000,000 **57)** 0.18

59) 0.06 **61)** 8

63) 1.5×10^8

65) 1.36×10^4 kg per cubic meter

67) 1.3×10^{-6} meters

69) 299,792,500 meters per second

71) 19,300 kg per cubic meter

73) 0.000000000000001673 grams

75) 4.2 light years

77) 1.5 moles

79) approximately 115,068 years. Yes, since this is considerably longer than our lifespan.

EXERCISE SET 2.9

1) $\dfrac{4}{9}$ **3)** $\dfrac{27}{125}$ **5)** $\dfrac{125}{8}$

7) $\dfrac{4}{9}$ **9)** $\dfrac{2401}{81}$ **11)** $\dfrac{a^3}{b^3}$

13) $\dfrac{q^4}{p^4}$ **15)** $\dfrac{8x^3}{y^3}$ **17)** $\dfrac{27q^3}{p^3}$

19) $\dfrac{a^{12}}{b^{20}}$ **21)** $\dfrac{y^{12}}{x^{24}}$ **23)** $\dfrac{1}{64x^6}$

25) $\dfrac{y^{10}}{4}$ **27)** $\dfrac{25}{a^4 b^{10}}$ **29)** $\dfrac{z^4}{4w^6}$

31) $\dfrac{w^9}{z^{18}}$ **33)** $\dfrac{z^{30}}{x^{20}}$ **35)** $24w^3$

37) $\dfrac{3}{10x}$ **39)** x^2 **41)** $\dfrac{1}{a^{15}}$

43) a^{14} **45)** $\dfrac{16}{y^8}$ **47)** x^4

49) $\dfrac{1}{4x^4}$ **51)** $\dfrac{g^4}{16}$ **53)** $\dfrac{1}{x^{15}}$

55) $\dfrac{1}{q}$ **57)** $\dfrac{1}{a^{12}}$ **59)** $\dfrac{8}{x^{12}}$

61) $81a^3$ **63)** $\dfrac{216a^7}{b^{19}}$

EXERCISE SET 2.10

1) $5x$ **3)** $-3z^2$ **5)** $2x^2$

7) $-\dfrac{5}{a^3}$ **9)** $-6n^2$ **11)** 2

13) $x^2 y^2$ **15)** $-\dfrac{s^4}{r^2}$ **17)** $3x^2 y^2$

19) $-\dfrac{4y^3}{x^3}$ **21)** $\dfrac{xy^2}{z^2}$ **23)** $-\dfrac{4y^3}{x^2 z^2}$

25) $\dfrac{2n^2}{p^3}$ **27)** $9x^2 y^2$ **29)** $-\dfrac{5}{y^2}$

31) $x + 3$ **33)** $a - 1$ **35)** $3 + \dfrac{7}{y}$

37) $2z - 1$ **39)** $3y + \dfrac{2}{y}$ **41)** $4a^3 b - 2ab^4$

43) $\dfrac{2n}{m} + \dfrac{3m^2}{n}$ **45)** $2x^2 + 3x - 1$

47) $y^3 - y^2 - y$ **49)** $3m^3 - m^2 + 2m$

ANSWERS FOR ODD NUMBERED EXERCISES   639

51) $3y^2 - 4y + \dfrac{2}{y}$ **53)** $3m - 2m^2n + 4m^3n^2$

55) $2mn^2$ **57)** $\dfrac{2n}{m^3}$

59) $2a^4 - 3a^2$ **61)** $3y^2 - 4 - \dfrac{2}{y^2}$

63) $2x^2y^2$ **65)** $3y^3 - 5y + 4$

71) $6x^3y^2 - 18x^4y$, because the polynomial equals the divisor times the quotient, i.e., $(3xy)(2x^2y - 6x^3)$.

73) $8x + 12 + \dfrac{4}{x^2}$

CHAPTER 2 REVIEW EXERCISES

1) -8 **3)** -72 **5)** 144

7) -15 **9)** -72 **11)** -25

13) -42

15) associative of multiplication

17) distributive

19) identity for multiplication

21) 4^{14} **23)** $-30a^6b^{12}$

25) $125d^3$ **27)** x^9y^9

29) x^{20} **31)** $a^{16}b^{24}$

33) $a^{37}b^{19}$ **35)** $15x^{10}y^5$

37) $-20x^4 + 25x^2$

39) $-6x^6y^4 + 16x^3y^8 - 8x^5y^3$

41) $12m^2 - 9m - 30$

43) $x^2 - 81$

45) $4s^2 - 25t^2$

47) $9a^2 - 12a + 4$

49) $a^3 - a^2 - 16a + 16$

51) $12x^3 - 26x^2y + 18xy^2 - 4y^3$

53) $27x^3 + 125$ **55)** $10x^2$

57) $36x^2 - 48x + 16$ **59)** -8

61) -9 **63)** 2

65) 4 **67)** -12

69) -61 **71)** -20

73) 16 **75)** -6

77) -1 **79)** 432

81) -41 **83)** $-\dfrac{13}{12}$

85) 2^5 **87)** $\dfrac{1}{3^4}$

89) $\dfrac{1}{5^2}$ **91)** $\dfrac{1}{5}$

93) $\dfrac{1}{x^6}$ **95)** a^3

97) $460{,}000$ **99)** 0.0000703

101) 9.7×10^{10} **103)** 4.78×10^{-6}

105) 0.01395 **107)** 0.03

109) 0.00552 **111)** $50{,}000{,}000$

113) $\dfrac{9}{25}$ **115)** $\dfrac{1}{3^{10}}$

117) $\dfrac{x^{12}}{y^{12}}$ **119)** $\dfrac{1}{5^3a^{12}}$

121) $\dfrac{n^6}{m^8}$ **123)** $8n^3$

125) $\dfrac{6}{m^2}$ **127)** $\dfrac{3}{m^3n^6}$

129) $x - 2$ **131)** $7a - 4$

133) $4x^2 - 3x - 7$ **135)** $-\dfrac{3m^2}{n^2}$

137) $2x^2 - 3x - \dfrac{1}{3}$ **139)** $4x - 2 - \dfrac{2}{x}$

CHAPTER 2 TEST

1) 60 **2)** -72

3) $-6x^5y^5$ **4)** $48x^6y^3$

5) $6x^3 - 8x^2$ **6)** $2p^3q^4 - 4p^4q^2 - 2p^2q$

7) $6p^2 + 5p - 4$ **8)** $6a^2 - 11ab + 4b^2$

9) $3y^3 + 5y^2 + 13y - 5$ **10)** $25x^2 - 9y^2$

11) $4a^2 - 28a + 49$ **12)** -4

13) 3 **14)** $-\dfrac{2z^3}{x^3}$

15) $\dfrac{2a^3}{b^{18}}$ **16)** $\dfrac{2b^6}{a^{10}}$

17) $\dfrac{y^6}{4x^4}$ **18)** $4x - 5 + \dfrac{3}{x}$

19) -14 **20)** -3

21) -5 **22)** a) 4.79×10^9
                  b) 7.49×10^{-8}

23) 1075 **24)** 18

25) $5x^2 - 6x - 10$

EXERCISE SET 3.1

1) $t = 8$ **3)** $r = -6$ **5)** $a = 8$

7) $x = 10.5$ **9)** $m = -1.2$ **11)** $x = 2$

13) $x = -\dfrac{1}{10}$ **15)** $x = \dfrac{13}{12}$ **17)** $w = 9$

19) $x = 7$ **21)** $z = 3$ **23)** $t = -1$

25) $x = 8.1$ **27)** $r = 4$ **29)** $x = 0$

31) $x = -2$ **33)** $x = 6$ **35)** $x = 2$

37) $x = -2.6$ **39)** $y = -7.7$ **41)** $t = 12$

43) $x = 4$ **45)** $x = 10$ **47)** $x = -0.77$

49) Some number minus five is nineteen. A number decreased by five is nineteen.

51) The sum of thirteen and y is seven. Y more than thirteen is seven.

53) Seventeen is some number decreased by six. Seventeen equals the difference of a number and six.

55) $x - 2 = 9$, $x = 11$

57) $x - 14 = 8$, $x = 22$

59) $x - 2 = 15$, $x = 17$

61) $5 + x = 13$, $x = 8$

63) $x - 8 = -10$, $x = -2$

65) $x - 2 = -8$, $x = -6$

67) $275 + x = 550$, $x = 275$

69) $x + 75 = 160$, $x = 85$

71) $x - 8 = 24$, $x = 32$

73) $632 + x = 800$, $x = 168$

75) $x + 46 = 326$, $x = 280$

77) $x - 23 = 118$, 141 lbs

79) $x + 18{,}000 = 40{,}000$, \$22,000

EXERCISE SET 3.2

1) $u = 8$ **3)** $n = -12$ **5)** $x = -3$

7) $p = 4$ **9)** $u = 0.8$ **11)** $x = 50$

13) $t = 110$ **15)** $n = 16$ **17)** $q = -18$

19) $z = -49$ **21)** $w = 9$ **23)** $m = -75$

25) $x = \dfrac{3}{10}$ **27)** $x = -2$ **29)** $y = 7$

31) $x = 3$ **33)** $x = -4$ **35)** $x = -3$

37) $u = 9$ **39)** $x = 3$

41) Sixty-two is equal to four times a number. Four times a number is sixty-two.

43) Negative thirty-nine is three times a number. Three times a number is negative thirty-nine.

45) Two and seven-tenths times a number is equal to five and six-tenths. Five and seven-tenths is two and seven-tenths times a number.

47) $5x = 35$, $x = 7$ **49)** $4x = -28$, $x = -7$

51) $0.4x = 2$, $x = 5$ **53)** $\dfrac{x}{4} = 5$, $x = 20$

55) $\dfrac{x}{-3} = 8$, $x = -24$ **57)** 150

59) \$8.50   **61)** 115

63) 4 miles   **65)** 287

67) \$17,500   **69)** \$500

EXERCISE SET 3.3

1) $x = 2$ **3)** $x = -2$ **5)** $a = 5$

7) $x = 3$ **9)** $u = 2$ **11)** $v = 1$

13) $n = 0$ **15)** $p = -2$ **17)** $v = 2$

19) $k = 1$ **21)** $a = \dfrac{9}{17}$ **23)** $y = 4$

25) $n = -8$ **27)** $r = 5$ **29)** $u = 13$

31) $w = 2$ **33)** $y = \dfrac{5}{2}$ or 2.5   **35)** $x = 10$

37) $n = 1$ **39)** $p = 2$ **41)** $n = -2$

43) $u = 1$ **45)** $b = 1$ **47)** $a = -9$

49) $v = \dfrac{13}{2}$ or 6.5

51) identity, all real numbers

53) identity, all real numbers

55) contradiction, no solutions

57) $x = 2$

59) Four times a number decreased by three is equal to seven. Seven is the difference of four times a number and three.

61) Six times a number decreased by nine is three times that number. Three times a number is equal to six times the number minus nine.

63) Two and eight-tenths more than seven times a number is six and three-tenths. Six and three-tenths is equal to the sum of seven times a number and two and eight-tenths.

65) Twelve times the difference of a number and one is nine. The product of twelve and a number decreased by one is nine.

67) $3x - 8 = 13$, $x = 7$ **69)** $5x - 6 = 9$, $x = 3$

71) $0 = 9(x + 3)$, $x = -3$

73) $4(3x - 2) = 16$, $x = 2$

75) 4   **77)** 5   **79)** \$1500

81) \$160   **83)** \$9.25   **85)** 121

87) \$85.50   **89)** \$4.00   **91)** \$83.33

93) 12 oz salt, 96 oz of water

EXERCISE SET 3.4

1)

3)

5)

7)

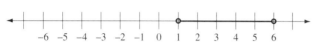

9)

11)

13) $q < -2$

15) $p \leq 2$

17) $p < -2$

19) $s \geq 1$

21) $x > 2$

23) $u > 41$

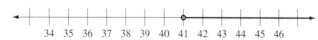

25) $t \leq -3$

27) $y < 5$ **29)** $u < -3$ **31)** $a > 7$
33) $x > -15$ **35)** $r < 2$ **37)** $q > 12$

39) $a \geq 3$ **41)** $p \geq 5$ **43)** $w > -8$
45) $a \leq 1.2$ **47)** $t > 30$ **49)** $s > -1$
51) $p \leq 3$ **53)** $w \geq -\dfrac{5}{7}$ **55)** $m > 4$
57) $x \leq -1$ **59)** $u \leq -4$ **61)** $u > -1$
63) $n > 0$ **65)** $z \leq \dfrac{1}{7}$ **67)** $r > -5$
69) $4 < x < 10$ **71)** $-2 \leq x < 3$ **73)** $-1 < t < 5$
75) $y \geq 5$

77) Five times a number is greater than or equal to ten. Five times a number is at least ten.

79) Three times a number decreased by five is greater than two. Two is less than the difference of three times a number and five.

81) Twenty-eight is less than or equal to the product of four and the difference of a number and two. Twenty-eight is less than or equal to the product of four and two less than a number.

83) $6x < 18, \ x < 3$ **85)** $19 - 4x > 18, \ x < \dfrac{1}{4}$

87) $3(2x + 7) \geq -3, \ x \geq -4$ **89)** $4x - 6 < 18, \ x < 6$

EXERCISE SET 3.5

1) What number is 25% of 36? 25% of 36 is what number?

3) 40% of 60 is what number? What is 40% of 60?

5) 20% of what number is 8? 8 is 20% of what number?

7) 52 is 65% of what number? 65% of what number is 52?

9) What percent of 30 is 21? 21 is what percent of 30?

11) 4.8 is what percent of 80? What percent of 80 is 4.8?

13) 10   **15)** 28   **17)** 80
19) 6000   **21)** 30%   **23)** 12.5%
25) 81%   **27)** 945   **29)** 105 ml
31) 40%   **33)** $85,000   **35)** 190 lb
37) 20,000   **39)** 40   **41)** 100 ml

EXERCISE SET 3.6

1) 3 ft, 5 ft   **3)** 523 mi
5) Richard = $257, Patricia = $300   **7)** $229 and $365
9) $12 per day   **11)** 11, 13
13) 24, 25, 26   **15)** −6, −5, −4
17) 18, 20, 22   **19)** 6, 8, 10
21) 7 hrs   **23)** 8 hrs

25) 12 hrs

27) 6 hrs

29) 5 hrs

31) 9 ft, 12 ft, 15 ft, 18 ft

EXERCISE SET 3.7

1) $400 - x$ miles

3) $25 - x$ dimes

5) $150 - x$ camellias

7) 16 at \$25, 20 at \$32

9) 80 of 36″, 25 of 54″

11) 22 at 18¢, 14 at 25¢,

13) 12 trucks, 15 cars

15) 4 - \$5 bills, 44 - \$20 bills

17) 25 dimes, 30 nickels

19) 20 dimes, 35 quarters

21) \$10,000 at 6%, \$26,000 at 10%

23) \$40,000

25) 4 oz

27) \$3.00 per pound

29) 8 lb peanuts, 4 lb cashews

31) 18 ml

33) 30 tons

35) 3.5 l

37) 40 gallons

39) 25 at \$0.49, 15 at \$1.19

41) 6.25 gallons

43) 12 nickels, 15 dimes, 8 quarters

EXERCISE SET 3.8

1) 3 ft

3) $L = 75$ ft, $W = 35$ ft

5) $L = 22$ ft, $W = 8$ ft

7) 4″, 6″, 8″

9) 8′, 8′, 12′

11) 30°, 60°

13) 75°, 25°, 80°

15) 50°, 40°

17) 40°, 140°

19) $x = 20$, $m\angle A = m\angle B = 78°$

21) $x = 15$, $m\angle A = m\angle B = 57°$

23) $x = 20$, $m\angle A = 70°$, $m\angle B = 110°$

EXERCISE SET 3.9

1) $A = 8$ sq mi

3) $h = 4$ ft

5) 1.5 hours

7) $r = 4$ in

9) $P = 42$ in

11) $A = 24$

13) $V = 15$

15) $P = 60$

17) $W = 2624$

19) $P = 12$

21) $C = K - 273$

23) $T = \dfrac{P}{k}$

25) $m = \dfrac{F}{a}$

27) $t = \dfrac{v}{g}$

29) $V = \dfrac{kT}{P}$

31) $m = \dfrac{E}{c^2}$

33) $r = \dfrac{I}{pt}$

35) $a^2 = c^2 - b^2$

37) $g = \dfrac{2d}{t^2}$

39) $M = DV$

41) $y = -2x + 3$

43) $y = \dfrac{-3x + 6}{2}$

45) $y = \dfrac{-6x + 7}{-3}$

47) $B = \dfrac{2A - hb}{h}$

49) $q_1 = \dfrac{Fr^2}{kq_2}$

CHAPTER 3 REVIEW EXERCISES

1) $x = 4$

3) $v = 2$

5) $z = 40$

7) $u = -3$

9) $a = -12$

11) The difference of twice a number and five is fourteen. Five less than two times a number is 14.

13) $x + 12 = -2$, $x = -14$

15) \$140

17) $-\dfrac{7}{2}$

19) $x = -8$

21) $t = 15$

23) $w = -4.25$

25) Fourteen times a number is negative twenty-eight.

27) $4x = -12$, $x = -3$

29) $w = -14$

31) \$165

33) 75

35) $t = -2$

37) $r = 3$

39) $v = 1$

41) $q = 5$

43) contradiction, no solutions

45) The difference of 3 times a number and 2 is 5.

47) $4x + 3 = 11$, $x = 2$

49) \$45

51)

53)

55) $p < 15$

57) $x > 8$

59) $z > -12$

61) Nineteen is less than or equal to the sum of a number and nine. Nine more than some number is at least nineteen.

63) $t \le -7$ **65)** $x < 2$ **67)** $z > 5$

69) $u \le 6$ **71)** $s \ge -1$ **73)** 88

75) 15% **77)** 15 ml **79)** 138 women

81) $31.82

83) 5, 7, 9

85) 20 nickels, 12 dimes

87) 200 ml at 25%, 300 ml at 50%

89) $L = 100$ ft, $W = 75$ ft

91) 53°, 37°

93) 18.84

CHAPTER 3 TEST

1) $u = -13$ **2)** $v = -3$

3) $k = -6$ **4)** $m = -0.675$

5) $v = -10$ **6)** $t = \dfrac{37}{8}$ or 4.625

7) $y = 5$ **8)** $a = -2.2$

9) $p > 4$ **10)** $q \ge 8$

11) $u \ge 17$ **12)** $x > 7$

13) $y \ge -6$ **14)** $z > -14$

15) Four more than the quotient of a number and three is 6.

16) Two times the difference of a number and five is equal to three times the number.

17) $7x + 5 = 3x - 1$ **18)** $3(x - 6) = x + 1$

19) 62.5% **20)** 80

21) $5250 **22)** $36

23) 90 ml **24)** 2532

25) $-1, 0, 1$

26) 15 incandescent, 10 fluorescent

27) $\dfrac{1}{2}$ hr

28) $W = 31$ cm, $L = 40$ cm

29) $W = \dfrac{P - 2L}{2}$ or $\dfrac{P}{2} - L$

30) $B = \dfrac{2A - bh}{h}$ or $\dfrac{2A}{h} - b$

EXERCISE SET 4.1

1) $2^2 \cdot 5^2$ **3)** $2 \cdot 7 \cdot 11$ **5)** $2^2 \cdot 3^2 \cdot 5$

7) $3^2 \cdot 5 \cdot 7$ **9)** 3 **11)** 5

13) 1 **15)** 3 **17)** 10

19) 90 **21)** 1 **23)** 8

25) 8 **27)** 24 **29)** mn

31) cd^2 **33)** a^3b^2 **35)** 1

37) a^2bc^2 **39)** pr^3 **41)** $6pq$

43) $9s^2t^3$ **45)** $3abc^2$ **47)** $4a^2b^3c^2$

49) 3 **51)** $7b(b + 3)$ **53)** $8cd^2(a - 3)$

55) 2 nickels and 5 dimes

57) 4 Yankees and 8 Braves

59) 3 redfish and 8 trout

61) 48

63) $18x^4y^7(a - 3)^4$

EXERCISE SET 4.2

1) $c(d + f)$ **3)** $r(s - t)$

5) $3(x + 3)$ **7)** $4(2x - 3)$

9) $x(x + 1)$ **11)** $r^2(r^2 - 1)$

13) $7a^2(a + 2)$ **15)** $9p^3(2 - p^2)$

17) $5x(3x^2 + 1)$ **19)** $11r^2(2r^3 - 1)$

21) $4c^2(3c + 2)$ **23)** $6z^2(3 - 2z^4)$

25) $x^2y^2(y + 1)$ **27)** $u^3v^2(u - v)$

29) $cd^2(c^2d^2 + 1)$ **31)** $x^2y^2(1 - xy^2)$

33) $6a^2b^2(3b^2 + 2a)$ **35)** $3xy^2(3y - 4x)$

37) $3xy^3(3xy + 1)$ **39)** $3(3x^2 - 4x + 2)$

41) $5x^2(3x^2 - 2x - 4)$ **43)** $6(a^4 + 2a^2 - 4)$

45) $11x(2x^2 - 3x - 1)$ **47)** $4cd^2(4c^2 - 6cd^2 + 9)$

49) $7xy^2(2x^2y - 3xy^2 - 1)$ **51)** $-3(m - 2n)$

53) $-8(2c - d)$ **55)** $-7(2x - 1)$

57) $-4(x^2 - 2x + 4)$ **59)** $-5x(2x^2 + 3x - 5)$

61) $(m + n)(a + b)$ **63)** $(c + 2)(a - b)$

65) $(t - 3)(t + 6)$ **67)** $(a - 6)(a - 7)$

69) $2x + 3$ and $3x - 5$ **71)** $180x^6y^4(2y - 3x^2)$

73) $3(a + b)^2(a + b + 3)$ **75)** $(x + y)(z + w)$

77) $(n - 4)(m + 3)$ **79)** $(b + 5)(a + 3)$

81) $(x - y)(x + 5)$ **83)** $(b - 3)(a + 1)$

85) $(m + n)(x - y)$ **87)** $(n - 3)(m - 6)$

89) $(d - e)(c - 3)$ **91)** $(x + 3)(2x - y)$

93) $(s + 4)(r - 1)$ **95)** $(d - 3)(c - 1)$

97) $(2b + 5)(2a + 3)$ **99)** $(2z - w)(4x - y)$

101) $(3c + 2d)(2a - 3b)$ **103)** $(2x + 3)(4x - y)$

EXERCISE SET 4.3

1) $(m + n)(m - n)$ **3)** $(x + 2)(x - 2)$

5) $(r + 8)(r - 8)$ **7)** prime

9) $(c + 5d)(c - 5d)$ **11)** $(a + 10b)(a - 10b)$

13) $(2x + 5y)(2x - 5y)$ **15)** $(7p - 9q)(7p + 9q)$

17) prime

19) $(11a + 7b)(11a - 7b)$

21) $(a^2 + b^2)(a + b)(a - b)$

23) $(x^2 + 1)(x + 1)(x - 1)$

25) $(4x^2 + 9)(2x + 3)(2x - 3)$

27) $(x + y + 3)(x + y - 3)$

29) $(6 + x - y)(6 - x + y)$

31) $(2x - 2y + 3)(2x - 2y - 3)$

33) $3(x + 5)(x - 5)$

35) $4(2x + 3y)(2x - 3y)$

37) $3(3x + 4)(3x - 4)$

39) $9(x + 3)(x - 3)$

41) $3(x^2 + 4)(x + 2)(x - 2)$

43) $(3x + 7)(3x - 7)$

45) $(14x - 25)(14x + 25)$

47) $(16c - 19d)(16c + 19d)$

49) $2(13x - 12)(13x + 12)$

51) $(x^4 + y^4)(x^2 + y^2)(x + y)(x - y)$

53) $[9(x + 2) + 4y][9(x + 2) - 4y]$

55) $(x^n + y^n)(x^n - y^n)$

57) $(x + y)(x^2 - xy + y^2)$

59) $(c + 2)(c^2 - 2c + 4)$

61) $(n - 4)(n^2 + 4n + 16)$

63) $(m + 1)(m^2 - m + 1)$

65) $(2x + y)(4x^2 - 2xy + y^2)$

67) $(3a - 2)(9a^2 + 6a + 4)$

69) $(2x + 5y)(4x^2 - 10xy + 25y^2)$

EXERCISE SET 4.4

1) $(x + 2)^2$

3) $(x + 1)^2$

5) $(x + 6)^2$

7) $(y + 10)^2$

9) $(5x - 1)^2$

11) $(a - b)^2$

13) $(2a + b)^2$

15) $(4c - 3)^2$

17) $(3x + 5)^2$

19) prime

21) $(5x + 4y)^2$

23) prime

25) $2(2x - 3)^2$

27) $2(x + 7)^2$

29) $4(x + 10)^2$

31) $x^2(5x - 3)^2$

33) $2xy(5x + 4y)^2$

35) $x - 7$

37) $(9x - 5y)^2$

39) $25(2a^2 + 3b^2)^2$

41) 49

43) 25

EXERCISE SET 4.5

1) $(x + 1)(x + 2)$

3) $(x - 3)(x - 1)$

5) $(a + 2)(a + 3)$

7) $(b - 4)(b - 2)$

9) $(y - 5)(y + 1)$

11) $(a - 5)(a + 7)$

13) $(x + 2)(x + 8)$

15) $(r - 8)(r + 9)$

17) $(a - 9)(a - 3)$

19) prime

21) $(x - 7)(x - 5)$

23) $(y + 7)(y - 6)$

25) $(z + 3)(z + 12)$

27) $(a + b)(a + 3b)$

29) $(r - 7s)(r - s)$

31) $(c - 2d)(c + 9d)$

33) $(r - 11s)(r + 4s)$

35) $(y + 8z)(y - 5z)$

37) $(a - 9b)(a + 6b)$

39) $(r - 12s)(r - 3s)$

41) $3(a + 2)(a + 3)$

43) $5(z - 3)(z - 2)$

45) $x(x - 7)(x + 4)$

47) $y^2(y - 15)(y - 5)$

49) $3a(a^2 - a - 4)$

51) $xy(x - 8)(x + 10)$

53) $y^2(x + 4)(x + 5)$

55) $4s(r - 2)(r + 7)$

57) $3b(a^2 - a - 28)$

59) $(x - 12)(x + 9)$

61) $(y + 8)(y + 16)$

63) $7, -7, 8, -8, 13, -13$

65) $(x + 4)(x + 8)$

67) $x(x + 10)(x - 4)$

EXERCISE SET 4.6

1) $(a + 1)(2a + 5)$

3) $(a + 3)(2a - 1)$

5) $(7m + 1)(m - 5)$

7) prime

9) prime

11) $(11x + 5y)(x + y)$

13) prime

15) $(5x - 4)(x + 2)$

17) $(2x + 1)(2x + 3)$

19) $(2x - 7)(5x + 1)$

21) $(2x + 3)(x + 4)$

23) $(3a - 8)(a - 2)$

25) $(5c - 4)(c + 3)$

27) $(2b - 3)(b + 8)$

29) $(2x + 5)(3x + 1)$

31) $(2x - 5)(4x - 1)$

33) $(4a - 3)(6a + 1)$

35) $(2n + 1)(8n - 5)$

37) $(2x + 3)(3x + 5)$

39) $(3c - 2)(4c - 3)$

41) $(3a + 5)(6a - 7)$

43) $(2w + 3)(4w - 9)$

45) prime

47) prime

49) $(2a + 5b)(3a + 7b)$

51) $(3c - 4d)(6c + 5d)$

53) $(3a + 4b)(4a - 3b)$

55) $(2f - 3g)(6f - 7g)$

57) $3(2a - 3)(a + 2)$

59) $y(4x - 1)(x - 3)$

61) $3x(x - 1)(2x - 5)$

63) $3x(2x - 3)(3x + 4)$

65) $(4x + 5)(9x - 4)$

67) $[(x + 2) - 3][(x + 2) + 2] = (x - 1)(x + 4)$

69) 7, 11

EXERCISE SET 4.7

1) $8(2 - x^2)$

3) $(c + 3)(c + 9)$

5) $3a(a + 3)(a - 6)$

7) $(x + 10)(x - 10)$

9) prime

11) $(a + b)(x + 2)$

13) $6(a - 2)(a - 6)$

15) $(r + 9)(r - 8)$

17) $m^2n^2(m + n^2)$

19) $3a^2(2a - 5)(2a + 5)$

21) $(3a + 5b)^2$

23) $8(r + 4)(2r - 3)$

25) $(ab + 8)(ab - 8)$

27) $(5n + 2)(n + 4)$

29) prime

31) $(y + z)(x + w)$

33) $(a - b + 6)(a - b - 6)$

35) $(2y - 3)(5x - 2)$

37) $(a^2 + 2b)(a^2 - 2b)$

39) prime

41) $(4b^2 + 1)(2b + 1)(2b - 1)$

43) $(6x + 1)(x + 5)$

45) $(2x + 3z)(3x - y)$

47) $(b + 3)(a - 1)$

49) $2u(8u - v)(u + 3v)$

51) $ab(2a - 5ab + 7b)$

53) $(8c + d)(c + 4d)$

55) $3(3a + b)(x - 1)$

57) $(2z + 5)(4z^2 - 10z + 25)$

59) $(3c - 2d)(9c^2 + 6cd + 4d^2)$

61) $3(2c - d)(4c^2 + 2cd + d^2)$

63) $(3x - 8)(4x + 3)$

65) $(9a^2 + 25b^2)(3a + 5b)(3a - 5b)$

67) $(x + 5 - y)(x + 5 + y)$

EXERCISE SET 4.8

1) $6, -4$

3) $5, 9$

5) $0, \dfrac{2}{5}$

7) $0, -\dfrac{11}{7}$

9) $3, -2, 5$

11) $-7, \dfrac{5}{3}, \dfrac{1}{4}$

13) 2

15) $-2, 3$

17) $\dfrac{3}{4}, 5$

19) $0, 2$

21) $0, \dfrac{2}{3}$

23) $2, -2$

25) $5, -5$

27) $3, 4$

29) $-\dfrac{2}{3}, 4$

31) 6

33) $-\dfrac{5}{2}$

35) $12, 14$

37) $-6, -4$

39) -4 or -8

41) 3

43) $w = 4$ ft, $l = 8$ ft

45) $w = 8$ in, $l = 11$ in

47) $b = 6$ mi, $h = 2$ mi

49) a) 7 sec b) 3, 4 sec

51) 4 sec

53) 7

55) 6

57) 9

59) 0, 4, 9

61) $0, -6$

CHAPTER 4 REVIEW EXERCISES

1) composite

3) $2^2 \cdot 3 \cdot 5$

5) 3

7) 14

9) $m^2 n$

11) $6r^2 s^2 t^2$

13) $a(x + y)$

15) $6xy(3x - 4)$

17) $6(3x^2 - 4x + 2)$

19) $(s - 2)(r - 5)$

21) $-5x(3x^2 - x + 5)$

23) $(s - 8)(r + 3)$

25) $(x - 7)(y - 4)$

27) $(2x + 5)(3x - 2y)$

29) $(r + 11)(r - 11)$

31) $(7a + 8b)(7a - 8b)$

33) $16(2c + d)(2c - d)$

35) $(8 + c - d)(8 - c + d)$

37) $(m^2 + n^2)(m + n)(m - n)$

39) $(r - 1)(r^2 + r + 1)$

41) $(4x + 3y)(16x^2 - 12xy + 9y^2)$

43) $(x - 12)^2$

45) $(7a - 1)^2$

47) $(5c + 3d)^2$

49) $4(x - 5)^2$

51) $(x - 5)(x - 7)$

53) $(m - 5)(m + 2)$

55) $(r + 9s)(r - 8s)$

57) $(m - 21n)(m + 2n)$

59) $y(y + 11z)(y - 4z)$

61) $(z + 1)(3z - 5)$

63) $(3a + 2)(a + 8)$

65) $(9a + 5)(2a - 1)$

67) $(5a + 6b)(4a - 3b)$

69) $3c(3c - 2d)(4c - 3d)$

71) $(9x + y)(9x - y)$

73) $(x - 9)(x + 7)$

75) $3(3m - 4)(m - 4)$

77) $(2a + 2b + 7)(2a + 2b - 7)$

79) $(7a - 2b)^2$

81) $2(y - 5)(3x + 1)$

83) $(11a + 13)(11a - 13)$

85) $9(x^2 + 9)$

87) $(r + 10)(r - 4)$

89) $(y^2 + 25)(y + 5)(y - 5)$

91) $(8x - 7)(x - 1)$

93) $0, 6$

95) $\dfrac{8}{3}, -\dfrac{8}{3}$

97) $7, -3$

99) $\dfrac{2}{3}, -4$

101) -9 and -7 or 7 and 9

103) -7 or 4

CHAPTER 4 TEST

1) a) composite b) prime

2) $2^2 \cdot 3^2 \cdot 7$

3) 24

4) $12x^2 y$

5) $x^2(x + 1)$

6) $(m - 9)^2$

7) $(m - 8n)(m + 5n)$

8) $(7a + 3b)(7a - 3b)$

9) $7x^2 y(2x - 4y + 5x^2 y^2)$

10) $(3y - 4)^2$

11) prime

12) $(s + 5)(r - 7)$

13) $(x - y - 10)(x - y + 10)$

14) $3(q - 5)(q + 7)$

15) $(a - 6)(3a + 5)$

16) $(2a - 5b)(4a - 7b)$

17) $3(3x - 2y)(3x + 2y)$

18) $(3b - 2)(2a + 5)$

19) $5(a + 5b)^2$

20) $(x^2 + 25)(x + 5)(x - 5)$

21) $(4a - 3b)(5a + 2b)$

22) $(x - 6)$ and $(x + 4)$

23) $0, -\dfrac{2}{3}$

24) $\dfrac{2}{3}, \dfrac{5}{2}$

25) $w = 4$ in, $l = 8$ in

1) $a \neq 0$

3) $x \neq 0,\ y \neq 0$

5) $x \neq -5$

7) $x \neq -4$

9) $x \neq -2y,\ y \neq -\dfrac{x}{2}$

11) $x \neq 7$ or -3

13) $\dfrac{10}{21}$

15) $\dfrac{2}{3}$

17) $\dfrac{2}{3}$

19) $\dfrac{5}{4}$

21) $\dfrac{21}{11}$

23) $\dfrac{2}{7}$

25) $\dfrac{1}{ab^2}$

27) $4x^3y$

29) $\dfrac{a^2c^2}{b^3}$

31) $\dfrac{16n^2}{9p}$

33) $-\dfrac{3}{5rs}$

35) $-2x^2y$

37) $\dfrac{4}{9}$

39) $\dfrac{2}{3}$

41) $\dfrac{2}{3}$

43) $\dfrac{2}{5}$

45) $-\dfrac{3}{4}$

47) $\dfrac{x+2}{x-3}$

49) $\dfrac{a-2}{a+3}$

51) $\dfrac{1}{x+2}$

53) $\dfrac{x-3}{x-5}$

55) $\dfrac{2x-3y}{3x-5y}$

57) $\dfrac{2a-3b}{3a-2b}$

59) 8

61) -1

63) -1

65) $-a-5$

67) $-x-2$

69) $\dfrac{-1}{x-5}$

EXERCISE SET 5.2

1) 14

3) $\dfrac{4}{3}$

5) $\dfrac{4}{3}$

7) 4

9) $\dfrac{s^2}{r^3}$

11) $\dfrac{1}{m^2}$

13) $\dfrac{a^2x}{b^2y^2}$

15) $\dfrac{1}{rs}$

17) $\dfrac{3xy^2}{2}$

19) $\dfrac{4xy^2}{9}$

21) $\dfrac{8c}{3b^3}$

23) $\dfrac{4a(x+y)}{3}$

25) $\dfrac{xy^2(r+s)}{8}$

27) $\dfrac{6}{35}$

29) $\dfrac{2}{5}$

31) $\dfrac{4(a+1)}{3(a+4)}$

33) $(2x-5)(3x+4)$

35) $-\dfrac{9}{7}$

37) $-\dfrac{3}{2}$

39) $-\dfrac{4y^2}{3x}$

41) $\dfrac{x-2}{x+3}$

43) $\dfrac{x-1}{x+4}$

45) 1

47) $\dfrac{x+2}{x-3}$

49) $-b-5$

51) $\dfrac{(d-5)(a+2)}{(b-4)(d+1)}$

53) $2a-b$

EXERCISE SET 5.3

1) $\dfrac{7}{16}$

3) $-\dfrac{3}{5}$

5) $-\dfrac{8}{5}$

7) $\dfrac{4}{3}$

9) $\dfrac{1}{a^2b^2}$

11) $\dfrac{2}{3ab}$

13) $\dfrac{a^2}{b^3c^2d^2}$

15) $\dfrac{32x^6}{45}$

17) $\dfrac{15}{28}$

19) $\dfrac{4a}{9}$

21) $\dfrac{8}{5}$

23) $\dfrac{1}{4}$

25) $-\dfrac{2}{3}$

27) $\dfrac{1}{(h+7)^2}$

29) $\dfrac{d-4}{2d(3d+4)}$

31) $\dfrac{(m-5)(m+6)}{6}$

33) $\dfrac{x-8}{x+8}$

35) $\dfrac{a-5}{a-6}$

37) $\dfrac{x-2}{x+3}$

39) $x-5$

41) $x+3$

43) $x+2$

45) $2x+5$

47) $2x-5$

49) $x+5+\dfrac{1}{x-3}$

51) $3x+1+\dfrac{9}{4x-3}$

53) $3x-4+\dfrac{12}{2x+7}$

55) $2x^2-6x+3$

57) $2x^2+4x-6$

59) $3x^2-2x+1+\dfrac{4}{2x+5}$

61) $3x^2+15x+2$

63) $2x^2+2x+3+\dfrac{27}{2x-3}$

65) $4x^2-3x+2+\dfrac{4}{4x+3}$

67) $x^2 + 2x + 4$

69) $x^3 + 2x^2 + 4x + 8$

71) $x^2 - 3x + 9 + \dfrac{-18}{x + 3}$

73) $\dfrac{b - 4}{c - 3}$ **75)** 18   **77)** -5

EXERCISE SET 5.4

1) 99

3) 770

5) 252

7) x^3y^3

9) $a^3b^6c^4$

11) $9u^3v^5$

13) $45m^5np^9$

15) $108u^4v^6$

17) $(y + 4)(y + 1)$

19) $(v - 5)(v + 3)$

21) $(x - 4)(x + 3)(x + 5)$

23) $(z + 1)^2(z - 5)$

25) $(t - 3)(t + 3)(t - 5)(t - 4)$

27) 21

29) 243

31) cf

33) 24

35) $99rs^3$

37) $108y^2z^5$

39) $27tx^3y^2$

41) $5y$

43) $16c$

45) $11r + 33$

47) $5v^2 + 10v$

49) $y^2 + 2y - 8$

51) $n^2 - 12n + 27$

53) $14, \dfrac{7}{14}, \dfrac{2}{14}$

55) $15, \dfrac{6}{15}, \dfrac{7}{15}$

57) $30, \dfrac{3}{30}, \dfrac{2}{30}$

59) $120, \dfrac{20}{120}, \dfrac{45}{120}, \dfrac{108}{120}$

61) $bd, \dfrac{ad}{bd}, \dfrac{bc}{bd}$

63) $mstu, \dfrac{rm}{mstu}, \dfrac{ns}{mstu}$

65) $swv^2, \dfrac{sx}{swv^2}, \dfrac{vwy}{swv^2}$

67) $a^3b^5, \dfrac{2at}{a^3b^5}, \dfrac{3b^2z}{a^3b^5}$

69) $24u^2v^4, \dfrac{28}{24u^2v^4}, \dfrac{9uv}{24u^2v^4}$

71) $50w^6z^7, \dfrac{15a}{50w^6z^7}, \dfrac{12bw^4z}{50w^6z^7}$

73) $(u + 8)(u - 10), \dfrac{3u - 30}{(u + 8)(u - 10)}, \dfrac{9u + 72}{(u + 8)(u - 10)}$

75) $12(t - 3), \dfrac{4t}{12(t - 3)}, \dfrac{15t}{12(t - 3)}$

77) $(a + 1)(a - 1)(a + 4), \dfrac{3a^2 + 12a}{(a + 1)(a - 1)(a + 4)},$

$\dfrac{a^2 + a}{(a + 1)(a - 1)(a + 4)}$

79) $(v + 1)(v - 3)(v - 2), \dfrac{v^2 - 7v + 10}{(v + 1)(v - 3)(v - 2)},$

$\dfrac{v^2 + 4v + 3}{(v + 1)(v - 3)(v - 2)}$

81) $(m + 4)^2(m + 1), \dfrac{m^2 + 2m + 1}{(m + 4)^2(m + 1)}, \dfrac{m^2 - 16}{(m + 4)^2(m + 1)}$

83) $x^2(x + 3)(x - 2)(x + 4), \dfrac{x^3 + 9x^2 + 20x}{x^2(x + 3)(x - 2)(x + 4)},$

$\dfrac{x^2 - 5x + 6}{x^2(x + 3)(x - 2)(x + 4)}$

85) $6825x^4y^6z^7$ **87)** $(x + 2)t$

EXERCISE SET 5.5

1) $\dfrac{2}{3}$

3) $\dfrac{15}{17}$

5) $\dfrac{3}{5}$

7) $\dfrac{1}{3}$

9) $\dfrac{u - 5}{v}$

11) $\dfrac{11r + 4m}{21u}$

13 $\dfrac{2a}{3x^2}$

15) $\dfrac{9}{n + 5}$

17) $\dfrac{2}{t - 8}$

19) $\dfrac{3c}{c + 2}$

21) $\dfrac{2a - 1}{a - 5}$

23) $\dfrac{8r - 3}{3r - 2}$

25) $\dfrac{-13}{a - 5}$

27) $\dfrac{m + 5}{2m + 3}$

29) $\dfrac{1}{x + 4}$

31) $\dfrac{1}{a - 2}$

33) $\dfrac{x + 2}{x + 3}$

35 $\dfrac{x - 2}{x + 4}$

37) $\dfrac{t - 1}{t + 7}$

39) $\dfrac{x - 1}{x - 6}$

41) $\dfrac{5}{11}$

43) $\dfrac{8}{4 - t}$ or $\dfrac{-8}{t - 4}$

45) $\dfrac{3m - n - 3}{m - n}$

47) $\dfrac{19}{21}$

49) $\dfrac{7}{10}$

51) $-\dfrac{1}{36}$

53) $-\dfrac{5}{36}$

55) $\dfrac{a^2 + 2b}{ab}$

57) $\dfrac{31z}{20x}$

59) $\dfrac{7x - 7}{10x}$

61) $\dfrac{m + 38}{6m}$

63) $\dfrac{2b^2 + 3b + 21}{14b^2}$

65) $\dfrac{45p^2q - 10pq + 9p + 3}{30p^3q^2}$ **67)** $\dfrac{-2k + 8}{k(k + 2)}$

69) $\dfrac{9w + 13}{(w - 3)(w + 7)}$ **71)** $\dfrac{x - 13}{6(x + 3)}$

73) $\dfrac{8t - 24}{(t - 4)^2}$ **75)** $\dfrac{u + 8}{u - 3}$

77) $\dfrac{h^2 + 15h}{(h + 5)(h - 5)}$ **79)** $\dfrac{6x}{(x + 2)(x + 3)(x - 4)}$

81) $\dfrac{2x - 11}{(x + 4)(x + 2)(x - 1)}$ **83)** $\dfrac{2x^2 - 3x + 19}{(x - 3)(x + 4)(x + 2)}$

85) $\dfrac{v^2 + 16v + 51}{(v + 4)(v - 4)(v + 3)}$ **87)** $\dfrac{2z^2 - z + 9}{(z + 3)^2(z - 2)}$

89) $\dfrac{1}{x}$ **91)** $\dfrac{2x^2 + 13x + 12}{x(x + 6)(x - 6)(x + 3)}$

93) $\dfrac{2x^2 - 13x}{(x + 2)(x - 4)(x + 3)}$ **95)** $\dfrac{a^2 - 5a}{(a + 4)(a - 2)(a - 1)}$

EXERCISE SET 5.6

1) $\dfrac{4}{5}$ **3)** $\dfrac{9}{2}$ **5)** $\dfrac{1}{16}$

7) 4 **9)** $\dfrac{4}{11}$ **11)** $\dfrac{230}{71}$

13) $\dfrac{187}{30}$ **15)** $\dfrac{88}{15}$ **17)** $\dfrac{3m}{2n}$

19) $\dfrac{5f}{3d}$ **21)** $\dfrac{xt}{yr}$ **23)** $\dfrac{g}{4h}$

25) $\dfrac{ac}{b}$ **27)** $\dfrac{a}{xb}$ **29)** $\dfrac{uv}{w}$

31) $\dfrac{1}{b}$ **33)** $\dfrac{u^4}{w^4v^2}$ **35)** $\dfrac{4}{2z - z^2}$

37) $\dfrac{2 + x}{2 - x}$ **39)** $\dfrac{t^2 - t^3}{1 - t^3}$ **41)** $\dfrac{v + 2}{v - 3}$

43) $\dfrac{1}{u + 6}$ **45)** $\dfrac{x - 6}{x(x + 3)}$ **47)** $\dfrac{5r + 44}{(r + 8)^2}$

49) $\dfrac{t^2 - 7t + 9}{11t - 80}$ **51)** $\dfrac{2a^2 + a + 6}{9a^3 + 54a^2 - 3a}$

EXERCISE SET 5.7

1) $x = 2$ **3)** $t = 5$ **5)** $u = 6$

7) $m = -\dfrac{15}{2}$ **9)** $x = 2$ **11)** $t = 5$

13) $x = 2$ **15)** $y = \dfrac{1}{3}$ **17)** $x = -2$

19) $v = 1$ **21)** $q = 1$ **23)** $m = 24$

25) $r = \dfrac{2}{3}$ **27)** $u = 1$ **29)** $w = -2$

31) $x = 4$ **33)** $t = 0$ **35)** $a = 11$

37) $c = 2$ or -5 **39)** $y = \dfrac{1}{3}$ or $y = -2$

41) $m = 0$ **43)** $y = -3$

45) $x = 3$, $x \neq 2$ (makes the denominator 0)

47) $t = 1$ **49)** $w = -\dfrac{35}{4}$

51) $\varnothing$ **53)** $x = \dfrac{9}{4}$

EXERCISE SET 5.8

1) 5 **3)** $\dfrac{10}{7}$

5) $9, 12$ **7)** $6, 12$

9) $4, 6$ **11)** \$18,000

13) $\dfrac{12}{5}$ hr $= 2\dfrac{2}{5} = 2$ hr 24 min

15) $\dfrac{175}{12} = 14\dfrac{7}{12}$ days

17) 12 hrs **19)** $22\dfrac{1}{2}$ min

21) 3 P.M. **23)** 75 mi/hr

25) 60 mi/hr **27)** 441 mi/hr

29) $\dfrac{8}{16}$ **31)** catcher $= 51$, pitcher $= 34$

33) $6\dfrac{2}{3}$ hr $= 6$ hr 40 min **35)** 3

37) 10 mi

EXERCISE SET 5.9

1) $100 : 49$ **3)** $\dfrac{10}{13}$ **5)** yes

7) yes **9)** yes **11)** $x = 15$

13) $t = 12$ **15)** $v = 2$ **17)** $t = 91$

19) $w = 0.6$ **21)** $y = 3$ **23)** $x = 4$

25) $x = 9$ **27)** $x = 5$ **29)** $x = 6$

31) $x = 7$ **33)** $x = -7, 4$ **35)** $x = -\dfrac{5}{2}, 6$

37) $x = 280$ **39)** $x = 7.5$ **41)** $x = 77$

43) $7\dfrac{1}{2}$ in **45)** 400 trout **47)** 20 ft

49) \$248 **51)** $x = -3, 2$ **53)** 65 min

55) 154 lbs, \$48.83

1) $x \neq -9$

3) $x \neq 5y$

5) $\dfrac{6}{7}$

7) $\dfrac{30}{31}$

9) $\dfrac{x}{y^2}$

11) $r^2 s^2$

13) $-\dfrac{14p^3}{9q^2}$

15) $\dfrac{7}{5}$

17) $\dfrac{a-3}{a+2}$

19) $\dfrac{5x-3}{2x-1}$

21) -1

23) $\dfrac{3}{2}$

25) $\dfrac{x^3}{y}$

27) $\dfrac{9n^2}{8m^2(m-n)}$

29) $\dfrac{3(5x+4)}{4}$

31) $\dfrac{x-4}{x+5}$

33) $-y-2$

35) $\dfrac{8}{5}$

37) $\dfrac{40}{27}$

39) $-\dfrac{3}{2}$

41) $\dfrac{(2p+3)(6p-7)}{20}$

43) $4x+5$

45) $2a^2 - 3a + 1 + \dfrac{3}{3a+4}$

47) $4x^2 + 5x - 6$

49) 42

51) 96

53) $a^2 y^4$

55) $42x^5 w^4$

57) $(a+2)^2$

59) 24

61) $20rp^2$

63) $8a$

65) $p^2 - 11p + 24$

67) $\dfrac{8}{14}, \dfrac{35}{14}$

69) $\dfrac{32}{36}, \dfrac{33}{36}$

71) $\dfrac{4}{t^2 v^5}, \dfrac{6tv^2}{t^2 v^5}$

73) $\dfrac{7(t+2)}{(t-4)(t+2)}, \dfrac{9(t-4)}{(t-4)(t+2)}$

75) $\dfrac{28u}{28(2u+3)}, \dfrac{36u}{28(2u+3)}$

77) $\dfrac{3a(a-5)}{(a+3)^2(a-5)}, \dfrac{10a(a+3)}{(a+3)^2(a-5)}$

79) $\dfrac{2}{7}$

81) $\dfrac{12}{13}$

83) $\dfrac{9r+2p}{14x}$

85) $\dfrac{5x}{a^2 - 3a + 4}$

87) $\dfrac{3p^2 + 3p - 6}{p^2 - 16}$

89) $\dfrac{41}{48}$

91) $\dfrac{7n+bm}{mn}$

93) $\dfrac{9y-7}{20y}$

95) $\dfrac{11w-9}{w(w-3)}$

97) $\dfrac{-v^2 + v - 4}{4(v+2)(v-3)}$

99) $\dfrac{8w^2 + 39w + 10}{(w+5)(w-5)(w+2)}$

101) $\dfrac{8}{3}$

103) $\dfrac{8}{9}$

105) $\dfrac{3x}{y}$

107) $\dfrac{t}{5}$

109) $\dfrac{x^4 y}{t^3}$

111) $\dfrac{y-13}{y+2}$

113) $y = 3$

115) $t = \dfrac{12}{55}$

117) $v = \dfrac{17}{6}$

119) $t = \dfrac{7}{4}$

121) $q = -2$ or 1

123) $\dfrac{16}{19}$

125) 18 days

127) 9:06 A.M.

129) yes

131) 2.5

133) 10

135) 5.6

137) 2

139) $x = 5$

141) $1\dfrac{1}{2}$ acres

143) 540 mi

CHAPTER 5 TEST

1) $\dfrac{21a^2}{8b^2}$

2) $\dfrac{3x-2}{2x-3}$

3) $\dfrac{21}{20}$

4) $\dfrac{15b^2 y^2}{4a^3 x^3}$

5) $-\dfrac{13}{21}$

6) $\dfrac{a+2}{a-4}$

7) $\dfrac{y(4y-5)}{(2y-7)(2y-3)}$

8) $4x + 3 + \dfrac{3}{3x-5}$

9) $45x^2 y^4$

10) $15tu^2 w$

11) $6p(q-4)$

12) $\dfrac{7a}{(a+3)(a-6)}, \dfrac{b(a-6)}{(a+3)(a-6)}$

13) $\dfrac{r}{r^2 + 16}$

14) $\dfrac{3v^2 + 26v + 6}{9v^2}$

15) $\dfrac{46w-3}{15(w-4)}$

16) $\dfrac{t^2 + 25t + 10}{2(t+5)(t-5)^2}$

17) $\dfrac{4m}{3}$

18) $\dfrac{12-u}{8u}$

19) $v = \dfrac{19}{17}$ **20)** $p = \dfrac{47}{37}$

21) $t = -7$ **22)** $w = 5$

23) 0 and 4   **24)** 30

25) 2 hr   **26)** 1:100

27) $y = 10$ **28)** $x = 6$

29) 175 minutes   **30)** 54

SECTION 6.1

1) yes   **3)** no   **5)** yes   **7)** no

9) yes   **11)** yes   **13)** no   **15)** yes

17) $(0,3)$, $(-\dfrac{3}{2}, 0)$, $(3,9)$, $(2,7)$

19) $(0,-4)$, $(-2,0)$, $(-3,2)$, $(2,-8)$

21) $(0,-4)$, $(6,0)$, $(-6,-8)$, $(-3,-6)$

23) $(0,4)$, $(5,0)$, $(-10,12)$, $(-5,8)$

25) $(0,-2)$, $(-3,0)$, $(9,-8)$, $(3,-4)$

27) $(0,4)$, $(-3,0)$, $(6,12)$, $(-12,-12)$

29) $(4,0)$, $(4,3)$, $(4,-2)$, $(4,-4)$

31) $(2,4)$,$(3,4)$,$(-4,4)$,$(-1,4)$

33) a) $(1,40)$, $(2,80)$, $(5,200)$,   b) After 8 hours, the motorcycle will travel 320 miles.

35) a) $(2,12)$, $(4,24)$, $(5,30)$,   b) When the length is 9 m, the area is 54 square meters.

37) a) $(1,60)$, $(3,180)$, $(7,420)$,   b) Four candy bars cost 240 cents.

39) a) $(2,10)$, $(5,25)$, $(8,40)$,   b) The value of 10 five-dollar bills is $50.

41) a) $(1,68)$, $(3,4)$, $(6,-92)$,   b) After 2 seconds, the velocity of the ball is 36 feet per second.

43) a) $(1,6)$, $(3,54)$, $(5,150)$,   b) After 2 seconds, the particle is 24 inches from the starting point.

45) a) $(1,334)$, $(2,436)$, $(4,544)$,   b) After 3 seconds, the object is 506 feet above the ground.

47) a) $(1,200)$, $(2,400)$, $(3,800)$,   b) After 4 seconds, there are 1600 bacteria present.

SECTION 6.2

1)

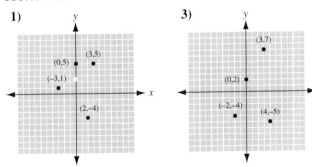

3)

5)

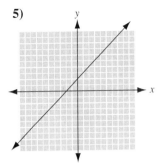

7)

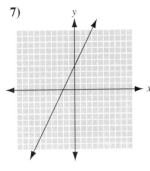

9)

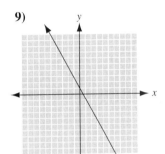

11)

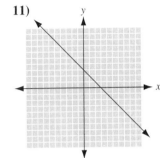

13)

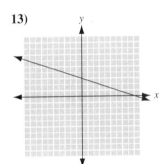

15)

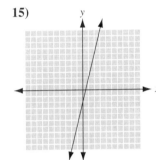

17)

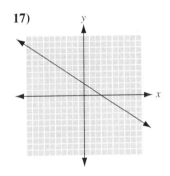

19)

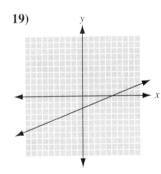

21)

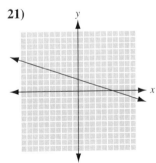

23)

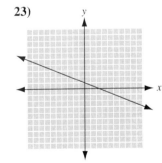

25)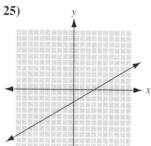

27)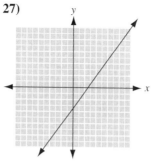

29)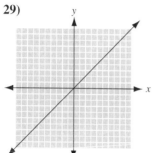

31)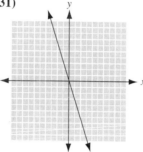

33)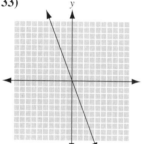

35)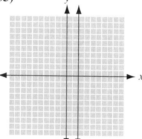

37)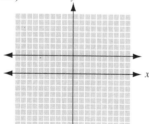

39)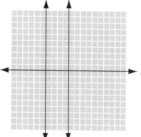

41)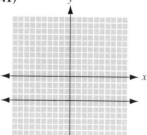

43)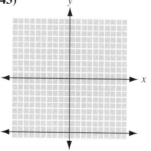

45) $x - y = 4$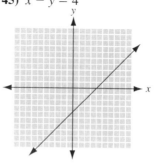

47) $3x - 4y = 12$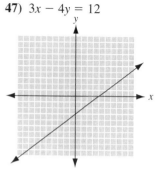

49) $y = 3x + 2$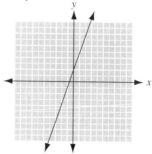

51) $2x + 3y = 24$

53) $5x + 10y = 450$ **55)** $50x + 60y = 500$

57) a) $(0,0), (2,-64), (4,-128)$

b)

c) 3 seconds

59) $a = 3, b = 5$

SECTION 6.3

1) no   **3)** no   **5)** yes

7)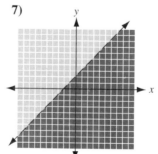

9)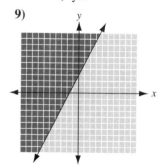

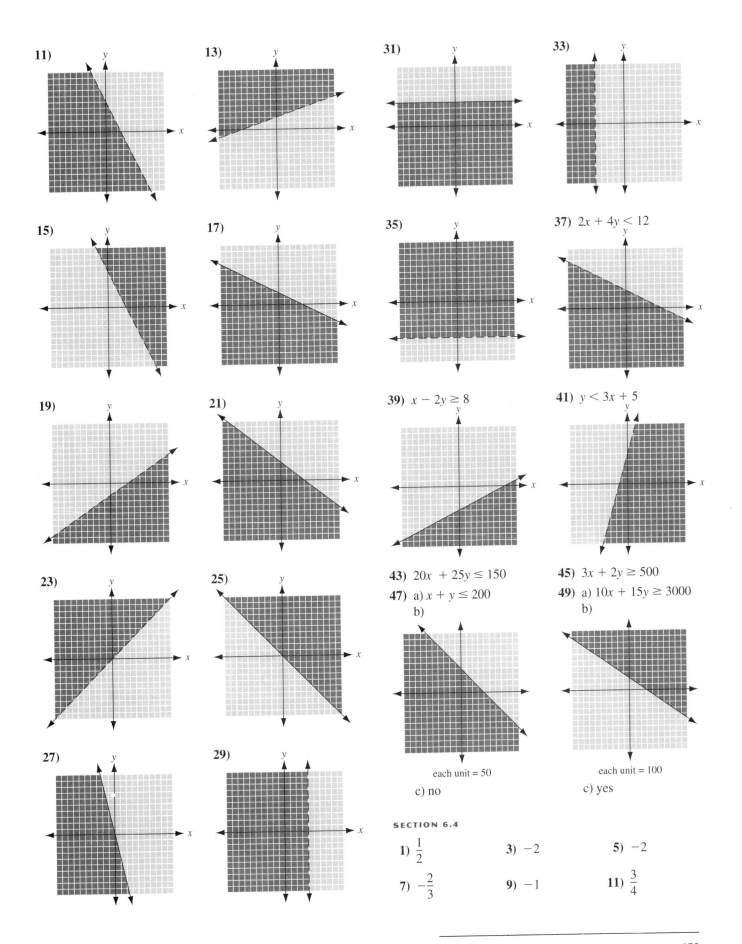

11)

13)

31)

33)

15)

17)

35)

37) $2x + 4y < 12$

19)

21)

39) $x - 2y \geq 8$

41) $y < 3x + 5$

23)

25)

43) $20x + 25y \leq 150$

45) $3x + 2y \geq 500$

47) a) $x + y \leq 200$

b)

each unit = 50

c) no

49) a) $10x + 15y \geq 3000$

b)

each unit = 100

c) yes

27)

29)

SECTION 6.4

1) $\frac{1}{2}$

3) -2

5) -2

7) $-\frac{2}{3}$

9) -1

11) $\frac{3}{4}$

13) $\frac{1}{2}$

15) 0

17) undefined

19) 0

21) 1

23) $-\frac{1}{2}$

25) $\frac{3}{4}$

27) undefined

29) 0

31)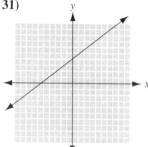

33)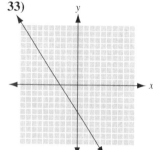

35)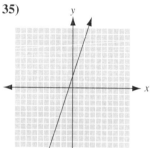

37)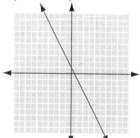

39)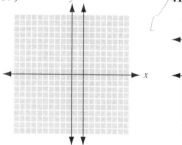

41)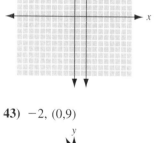

43) -2, $(0,9)$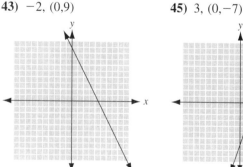

45) 3, $(0,-7)$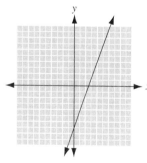

47) $-\frac{2}{3}$, $(0,3)$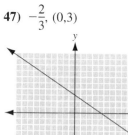

49) $\frac{5}{3}$, $(0,-5)$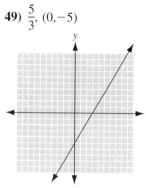

51) -2, $(0,0)$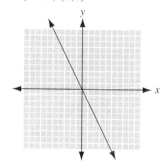

53) $-\frac{3}{2}$, $\left(0,\frac{5}{2}\right)$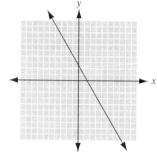

55) undefined

57) undefined

59) 0

61) 0

63) a) $\frac{3}{50}$ b) 6%

65) $\frac{8}{3}$

67) a) $y = 6x$
b)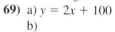
c) same

69) a) $y = 2x + 100$
b)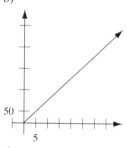
c) same

SECTION 6.5

1) $2x - y = 5$

3) $2x + y = 0$

5) $x - 4y = -23$

7) $x + 3y = -10$

9) $4x - 3y = 17$

11) $2x + 3y = -19$

13) $x = 3$

15) $y = 3$

17) $x = 5$

19) $y = 4$

21) $2x + y = 5$

23) $x - y = -4$

25) $2x + y = -7$

27) $2x + 3y = 12$

29) $x = 5$

31) $y = -4$

33) neither

37) perpendicular

41) parallel

45) parallel

49) parallel

53) perpendicular

57) $x + 2y = -11$

61) $3x - y = -18$

65) $y = -4$

69) $y = \dfrac{3}{5}x - 1$

73) a) $y = 0.15x + 25$
b) \$100
c) 250

77) a) $y = -5000x + 50{,}000$
b) \$10,000
c) 10

35) perpendicular

39) neither

43) perpendicular

47) perpendicular

51) parallel

55) $x + 2y = 1$

59) $4x - y = 17$

63) $x = 6$

67) $y = 2x + 4$

71) $y = -\dfrac{2}{3}x + \dfrac{4}{5}$

75) a) $y = 0.8x + 1.2$
b) \$7.60
c) 5 miles

79) a) $y = 0.05x + 200$
b) \$12,700
c) \$175,000

SECTION 6.6

1) yes, $D = \{-6, -3, 5, 7\}$, $R = \{2, -2, -7, 9\}$

3) yes, $D = \{-4, -1, 3, 5\}$, $R = \{2, 3, 1\}$

5) no, $D = \{-8, -6, 5\}$, $R = \{2, 3, -3, 1\}$

7) yes

9) no

11) yes

13) no

15) yes

17) yes

19) no

21) no

23) no

25) yes

27) yes

29) no

31) yes

33) 5, (0,5)

35) $3a + 5$, $(a, 3a + 5)$

37) -1, $(-1, -1)$

39) 16, (1,16)

41) $16a^2$, $(a, 16a^2)$

43) -6, $(-2, -6)$

45) 7, $(-2, 7)$

47) $|2z - 3|$, $(z, |2z - 3|)$

49) a) \$34, b) \$64, c) The cost of driving 250 miles is \$52

51) a) \$1675 b) \$1175 c) Her salary for \$20,000 of sales is \$2175

53) no

55) yes

CHAPTER 6 REVIEW EXERCISES

1) yes   **3)** yes   **5)** yes

7) (0,3), (1,0), (3,−6), (2,−3)

9) (0,6), (4,0), (−4,12), (2,3)

11) (2,120), (4,240), (7,420), In 3 hours, the motorcycle will travel 180 miles.

13)
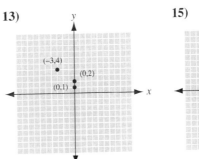

15)
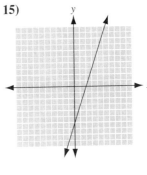

17)
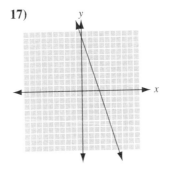

19)
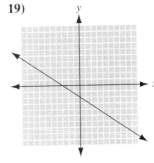

21)
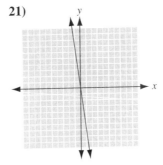

23)
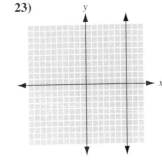

25) $2x + 4y = 16$

27)
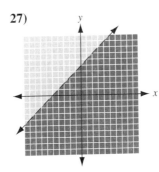

29)
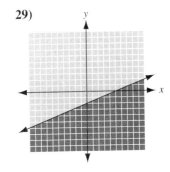

31)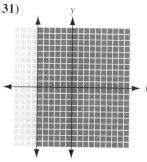

33) $3y - 2x \geq 8$

35) 2

37) $-\dfrac{1}{2}$

39) undefined

41)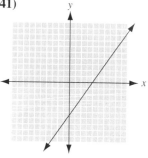

43)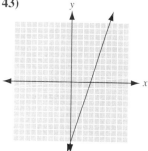

45)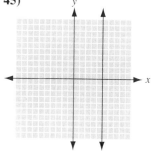

47) $-\dfrac{1}{2}$, (0,4)

49) undefined, none

51) $\dfrac{2}{3}$

53) $3x - y = 0$

55) $5x + 2y = 18$

57) $x = -7$

59) $y = 2$

61) $x + 2y = 5$

63) $3x + 2y = 12$

65) $y = 5$

67) parallel

69) neither

71) parallel

73) parallel

75) $3x + 5y = -2$

77) $2x + 3y = 2$

79) $x = 2$

81) $y = 3$

83) $y = \dfrac{6}{5}x + 4$

85) a) $y = 1.5x + 250$ b) $700 c) 225

87) yes, $D = \{-4, -1, 3, 6\}$, $R = \{2, 4, 5\}$

89) no

91) yes

93) yes

95) yes

97) a) $g(-2) = -18$, $(-2, -18)$
b) $g(0) = 2$, $(0,2)$ $g(b) = b^3 - 3b^2 + 2$, $(b, b^3 - 3b^2 + 2)$

CHAPTER 6 TEST

1) yes

2) $(0, -6)$, $(-8, 0)$, $(-4, -3)$, $(-12, 3)$

3) $(1,2)$, $(2,4)$, $(4,8)$, Ten feet of pipe costs $20.

4)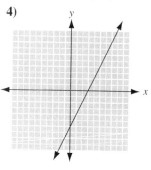

5)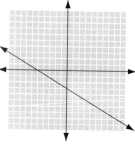

6)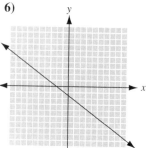

7)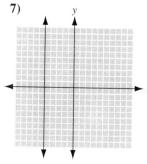

8) $9x + 20y = 480$

9)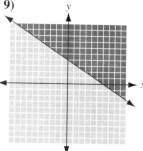

10)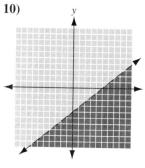

11)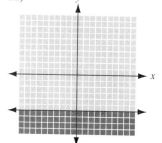

12)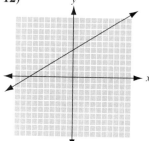

13)

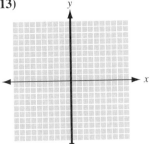

14) −1

15) 0

16) $\frac{1}{4}$

17) $\frac{5}{2}$

18) Undefined

19) perpendicular

20) parallel

21) $5x + 2y = 16$

22) $x - 2y = 8$

23) $x - 3y = -11$

24) $y = -6$

25) The line with slope of −2 slants downward as you go left to right. It drops two units vertically for each horizontal change of one unit to the right. The line with slope of 2 slants upward as you go left to right. It rises two units vertically for each horizontal change of one unit to the right.

26) no, $D = \{-4, 8, 6\}$, $R = \{7, -2, -4, 3\}$

27) no

28) yes

29) a) $f(2) = 0$, $(2,0)$
b) $f(-1) = 6$, $(-1,6)$

30) a) $4500
b) It costs $7000 to operate the business 10 days.

EXERCISE SET 7.1

1) yes

3) no

5) yes

7) no

9) $(0,1)$

11) $(-2,3)$

13) $(-3,4)$

15) $(-3,-2)$

17) $(5,1)$

19) $(3,3)$

21) $(0,0)$

23) inconsistent

25) inconsistent

27) inconsistent

29) consistent $(1,-6)$

31) yes

EXERCISE SET 7.2

1) $(2,0)$

3) $(1,2)$

5) $(-3,-2)$

7) $(-1,4)$

9) $(6,0)$

11) $(4,6)$

13) $(5,-1)$

15) $(4,-4)$

17) $(2,5)$

19) $\left(3, -\frac{1}{2}\right)$

21) $\left(\frac{2}{3}, -4\right)$

23) $\left(-\frac{2}{5}, \frac{2}{3}\right)$

25) $\left(-\frac{1}{4}, \frac{1}{3}\right)$

27) $\left(\frac{3}{13}, \frac{2}{11}\right)$

29) $\left(\frac{47}{70}, -\frac{3}{5}\right)$

31) $\left(\frac{78}{11}, -\frac{28}{11}\right)$

33) $(4,-5)$

35) $\left(\frac{5}{8}, \frac{1}{4}\right)$

37) $(0,4)$

39) $(9,2)$

41) $(3,-1)$

43) dependent

45) $(6,4)$

47) inconsistent

49) $(1,-1)$

51) $\left(2\frac{1}{3}, -\frac{1}{2}\right)$

53) $\left(\frac{2}{3}, -4\right)$

EXERCISE SET 7.3

1) $(2,-1)$

3) $(1,5)$

5) $(2,-3)$

7) $(-4,-1)$

9) $(3,4)$

11) $(3,3)$

13) $(4,-4)$

15) $(-5,2)$

17) $(-3,-2)$

19) $(2,2)$

21) $(-3,-2)$

23) dependent

25) $(7,-9)$

27) inconsistent

29) consistent $(-1,8)$

31) $(4,-1)$

33) $(6,8)$

35) $(4,0)$

EXERCISE SET 7.4

1) 20, 12

3) 11, 4

5) 16 at $25, 20 at $32

7) 80 36-inch fans, 25 54-inch fans

9) 27 dimes, 25 quarters

11) 22 18-cent, 14 25-cent

13) passenger train: 60 mph, freight train: 45 mph

15) 15 mph, 3 mph

17) $2,000, $18,000

19) $150,000, $100,000

21) 8 lb, 4 lb

23) 15 L, 15 L

25) 15, 10

27) 15 cars, 12 trucks

29) 700 quarters, 200 silver dollars

31) 40 mph still water, 5 mph current

33) $1600, $1400

35) 8 cups rice, 12 cups mixture

37) $w = 6$ ft, $l = 10$ ft

39) 5 cm, 7 cm

41) 75°, 105°

1)

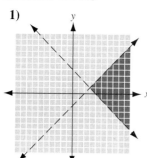

3)

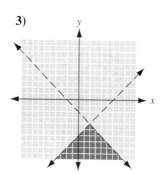

21)

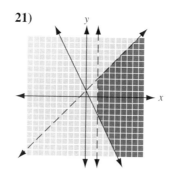

5)

7)

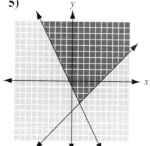

9)

11)

13)

15)

17)

19)

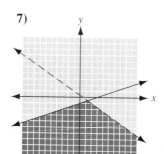

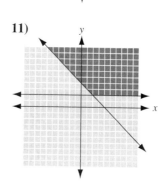

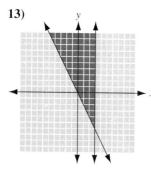

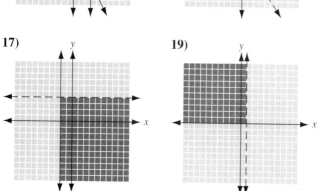

CHAPTER 7 REVIEW EXERCISES

1) no

3) yes

5) $(-1,2)$

7) $(6,7)$

9) dependent

11) $(-3,-14)$

13) $(13,1)$

15) $(-4,6)$

17) $\left(-3,\dfrac{1}{4}\right)$

19) $\left(\dfrac{7}{12},\dfrac{5}{11}\right)$

21) inconsistent

23) dependent

25) $(-2,1)$

27) $(4,6)$

29) $(-5\ -9)$

31) $(-3,-2)$

33) dependent

35) $(-1,4)$

37) $\left(\dfrac{1}{2},2\right)$

39) $(16,9)$

41) 25 nickels, 40 dimes

43) $10,000, $40,000

45)

47)

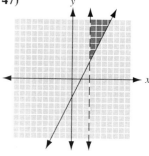

49)

CHAPTER 7 TEST

1) no

2) $(4,-4)$

3) $(-3,2)$

4) dependent

5) $(6,4)$

6) $(5,7)$

7) $(-3,8)$

8) $(2,-4)$

9) $(4,1)$

10) $(-2,5)$ **11)** $(-4,-7)$ **12)** $\left(\dfrac{4}{5}, -2\right)$

13) $(3,4)$ **14)** $(-7,-2)$ **15)** 2 and 14

16) 60 dimes, 30 quarters

17) car = 55 mph, bus = 45 mph

18) \$12,000 at 22% and \$4000 at 7%

19) 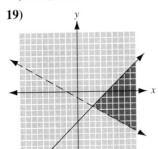   **20)**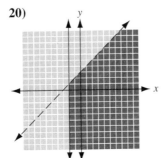

SECTION 8.1

1) 5   **3)** -5

5) -10 **7)** does not exist as a real number

9) 25   **11)** 13

13) 3   **15)** -3

17) -6 **19)** 6

21) does not exist as a real number   **23)** -3

25) irrational, 4.8990   **27)** rational, 12

29) irrational, 3.6840   **31)** rational, 4

33) rational, 1.3   **35)** irrational, 28.4605

37) rational, 0.5   **39)** irrational, 4.6416

41) rational, 46   **43)** rational, 21

45) b **47)** n^2

49) r^4 **51)** n

53) r^2 **55)** c

57) s^2 **59)** a

61) $x - 5$ **63)** $2a + b$

65) $x + 5$ **67)** $2a - 3b$

SECTION 8.2

1) $2\sqrt{3}$ **3)** $7\sqrt{2}$ **5)** $8\sqrt{2}$

7) $15\sqrt{6}$ **9)** $24\sqrt{5}$ **11)** $a\sqrt{a}$

13) xy^2 **15)** $x^3y^4z^5$ **17)** $c^2d^3\sqrt{c}$

19) $r^5s^4\sqrt{rs}$ **21)** $2x^3\sqrt{5}$ **23)** $18x^2\sqrt{2x}$

25) $2\sqrt[3]{4}$ **27)** $8\sqrt[3]{5}$ **29)** $x^2\sqrt[3]{x}$

31) $x^2y\sqrt[3]{y^2}$ **33)** $4z^2\sqrt[3]{2z^2}$ **35)** $4\sqrt[3]{3}$

37) $3\sqrt[4]{2}$ **39)** $ab\sqrt[4]{a^2b^3}$ **41)** $3\sqrt[5]{4}$

43) $ab^3\sqrt[5]{a^4b}$

SECTION 8.3

1) $\sqrt{30}$ **3)** $\sqrt{35}$ **5)** 6

7) 15   **9)** $\sqrt{5x}$ **11)** $\sqrt{15y}$

13) $\sqrt{xy}$ **15)** k **17)** x^2

19) w **21)** x^3 **23)** w

25) $4\sqrt{3}$ **27)** $5\sqrt{3}$ **29)** $24\sqrt{10}$

31) $30\sqrt{14}$ **33)** a^2 **35)** $y^2\sqrt{y}$

37) $m^5\sqrt{m}$ **39)** $x^4y^3\sqrt{y}$ **41)** $a^3b\sqrt{ab}$

43) $24c^4\sqrt{15}$ **45)** $\dfrac{1}{2}$ **47)** $\dfrac{7}{8}$

49) 2   **51)** 7   **53)** $6\sqrt{2}$

55) $6\sqrt{5}$ **57)** x **59)** $c^2\sqrt{d}$

61) $3x$ **63)** $12a^2$ **65)** $6c\sqrt{3c}$

67) $36x^2y^3\sqrt{2y}$ **69)** $\dfrac{2\sqrt{6}}{7}$ **71)** $\dfrac{a^4}{2}$

73) $\dfrac{3x^5\sqrt{5}}{4}$

SECTION 8.4

1) $7\sqrt{7}$ **3)** $-6\sqrt{6}$

5) $9\sqrt{a}$ **7)** $-3\sqrt{n}$

9) $3x\sqrt[3]{9}$ **11)** $18x^2\sqrt[4]{5x}$

13) $5\sqrt{3}$ **15)** $-\sqrt{3}$

17) $8\sqrt{3x}$ **19)** $9x\sqrt{3}$

21) $9y^2\sqrt{2y}$ **23)** $15\sqrt{5}$

25) $-8\sqrt{5}$ **27)** $6\sqrt{6}$

29) $14a\sqrt{3a}$ **31)** $4\sqrt{6}$

33) $18\sqrt{2} - 17\sqrt{3}$ **35)** $28x\sqrt{2y}$

37) $7\sqrt{5}$ **39)** $-15\sqrt{6}$

41) $7\sqrt{2}$ **43)** $-14\sqrt{5}$

45) $3\sqrt{2} + 2$ **47)** $3 - 3\sqrt{5}$

49) $\sqrt{15} + 10\sqrt{3}$ **51)** $24 - 48\sqrt{2}$

53) $12 - 3\sqrt{2} + 4\sqrt{5} - \sqrt{10}$

55) $6 + 5\sqrt{x} + x$

57) $68 - 24\sqrt{6}$

59) $6 + 10\sqrt{2} + 9\sqrt{3} + 15\sqrt{6}$

61) $\sqrt{6} + \sqrt{10} + 3 + \sqrt{15}$

63) $x - \sqrt{xy} - 2y$

65) $-13 - 2\sqrt{6}$

67) $-70 + 24\sqrt{10}$

69) $12\sqrt{14} - 12\sqrt{6} + 6\sqrt{35} - 6\sqrt{15}$

71) $8a + 10\sqrt{ab} - 3b$

73) $22 + 8\sqrt{6}$ **75)** $39 + 12\sqrt{3}$

77) $16 + 8\sqrt{3}$ **79)** 1

81) 4   **83)** $9 - x$

85) -14

87) 25

89) 1

91) $x - y$

93) 50

95) 21

97) $2 + \sqrt{2}$

99) $2 + 4\sqrt{6}$

101) $1 + 2\sqrt{5}$

103) $3 - 2\sqrt{6}$

105) $\dfrac{2 + 3\sqrt{3}}{4}$

107) $\dfrac{4 - 2\sqrt{6}}{3}$

109) $5\sqrt[3]{2}$

111) $9xy\sqrt[3]{2xy^2}$

113) $5\sqrt[4]{3}$

115) $2\sqrt[3]{4} - 2 - 6\sqrt[3]{2}$

SECTION 8.5

1) $\dfrac{3\sqrt{7}}{7}$

3) $\sqrt{5}$

5) $3\sqrt{2}$

7) $2\sqrt{10}$

9) $\dfrac{4\sqrt{x}}{x}$

11) $\dfrac{m\sqrt{n}}{n}$

13) $\dfrac{\sqrt{10}}{2}$

15) $\dfrac{\sqrt{6}}{4}$

17) $\dfrac{\sqrt{10}}{8}$

19) $\dfrac{\sqrt{ab}}{b}$

21) $\dfrac{\sqrt{de}}{e}$

23) $\dfrac{\sqrt{15}}{3}$

25) $\dfrac{\sqrt{21}}{6}$

27) $\dfrac{\sqrt{pq}}{q}$

29) $\dfrac{\sqrt{7x}}{x^3}$

31) $\dfrac{\sqrt{35y}}{5y}$

33) $\dfrac{3\sqrt{7a}}{7a}$

35) $\dfrac{2\sqrt[3]{3}}{3}$

37) $\dfrac{10\sqrt[3]{5}}{5}$

39) $\dfrac{\sqrt[3]{rs^2}}{s}$

41) $\dfrac{\sqrt[3]{xy}}{y}$

43) $\dfrac{\sqrt{10b^2}}{2b}$

45) $\dfrac{\sqrt[3]{28d}}{2d}$

47) $2 - \sqrt{5}$

49) $\dfrac{21 + 7\sqrt{3}}{6}$

51) $\dfrac{7a - a\sqrt{a}}{49 - a}$

53) $\dfrac{15 + 7\sqrt{3}}{6}$

55) $\dfrac{29 - 10\sqrt{5}}{-11}$

57) $\dfrac{20 + 5\sqrt{5} + 4\sqrt{3} + \sqrt{15}}{11}$

59) $\dfrac{14 + 9\sqrt{a} + a}{4 - a}$

61) $4 - \sqrt{15}$

63) $\dfrac{x + 2\sqrt{xy} + y}{x - y}$

65) $\dfrac{\sqrt{10} - 2\sqrt{3}}{2}$

67) $-\dfrac{2\sqrt{10} + 2\sqrt{35} + 6\sqrt{3} + 3\sqrt{42}}{15}$

69) $\sqrt[4]{9}$

71) $y\sqrt[4]{y}$

73) $3\sqrt[5]{4}$

SECTION 8.6

1) 4

3) 49

5) $\varnothing$

7) 16

9) 64

11) 25

13) 6

15) 6

17) $3, 5$

19) 6

21) 7

23) 14

25) 8

27) $4, 5$

29) 6

31) 5

33) 3

35) $0, 4$

37) 5

39) $\varnothing$

41) 4

43) a) 66.6 mph b) 400 ft

45) a) 20 amps b) 27,000 watts

47) a) $\dfrac{\pi}{2}$ or 1.57 seconds b) 1.82 ft

49) 6

51) $x = 8$

53) $x = -11$

55) $x = 16$

57) $x = 41$

SECTION 8.7

1) 15 in

3) 4 ft

5) $4\sqrt{5}$ m

7) $\sqrt{14}$ in

9) $\sqrt{31}$ in

11) $3\sqrt{13}$ ft

13) $2\sqrt{5}$ m

15) $2\sqrt{13}$ in

17) $2\sqrt{2}$ cm

19) 10 in

21) $6\sqrt{2}$ ft

23) $8\sqrt{3}$ mm

25) $4\sqrt{5}$ m

27) $\sqrt{30}$ m

29) $\sqrt{89}$ cm

31) 8 in

33) $7\sqrt{2}$ ft

35) 10 cm

37) $4\sqrt{2}$ in

39) 17 ft

41) $\sqrt{69}$ in

43) 50 in

45) 75 ft

47) $2\sqrt{7}$ ft

CHAPTER 8 REVIEW EXERCISES

1) 4

3) -4

5) 5

7) -5

9) 3

11) -3

13) $n^2\sqrt{n}$

15) $z\sqrt[5]{z^3}$

17) $ab^2\sqrt[3]{a}$

19) irrational, 5.6569

21) irrational, 3.9791

23) rational, 1.2

25) $3\sqrt{7}$

27) $20\sqrt{5}$

29) $3\sqrt[3]{2}$

31) $9x^3\sqrt{2x}$

33) $6z^2y\sqrt{10zy}$

35) $15nm^2\sqrt[3]{4m}$

37) r

39) 5

41) $\sqrt{6r}$

43) 7

45) 15

47) $225\sqrt{2}$

49) c^2

51) $v^2\sqrt{v}$

53) $60v^3\sqrt{3v}$

55) $\dfrac{9x}{2y^2}$

57) 4

59) x^2

61) $6a^2$

63) $\dfrac{\sqrt{15}}{4}$

65) $\dfrac{a\sqrt{21}}{6b}$

67) $18\sqrt{a}$

69) $6\sqrt{3}$

71) $2\sqrt{3}$

73) $-11a\sqrt{5}$

75) $6\sqrt{3}$

77) $3\sqrt{5} - 5\sqrt{6}$

79) $15 + \sqrt{5}$

81) 21

83) $\dfrac{3 - \sqrt{2}}{6}$

85) $\dfrac{6\sqrt{7}}{7}$

87) $\dfrac{\sqrt{15}}{3}$

89) $\dfrac{\sqrt{mn}}{n}$

91) $\dfrac{4\sqrt[3]{3}}{3}$

93) $2\sqrt[3]{4a^2}$

95) $\dfrac{7 - 2\sqrt{7}}{3}$

97) $2\sqrt{2} + 2\sqrt{3}$

99) $\dfrac{5\sqrt{5} + 5\sqrt{3} + \sqrt{15} + 3}{2}$

101) 36   **103)** $\varnothing$ **105)** $\varnothing$

107) 3, 4   **109)** 15 in   **111)** $3\sqrt{6}$ m

113) $\sqrt{89}$ ft   **115)** $\sqrt{181}$ in   **117)** $8\sqrt{2}$ in

119) $10\sqrt{2}$ in   **121)** $5\sqrt{10}$ ft

CHAPTER 8 TEST

1) 9   **2)** -4

3) x^4 **4)** $c^3 d^2$

5) $4\sqrt{6}$ **6)** $21x^2\sqrt{2x}$

7) $8\sqrt[3]{5}$ **8)** $12y^2\sqrt[3]{y}$

9) $30\sqrt{7}$ **10)** $60a^3\sqrt{6a}$

11) 4   **12)** $16c^2\sqrt{2c}$

13) $\dfrac{3c^2\sqrt{7}}{5}$ **14)** $-6\sqrt{6}$

15) $22\sqrt{5}$ **16)** $18a\sqrt{5a}$

17) $3\sqrt{2} + 12\sqrt{6}$ **18)** $3 + 5\sqrt{30}$

19) -12 **20)** $25 - 10\sqrt{x} + x$

21) $\dfrac{4\sqrt{3}}{3}$ **22)** $2\sqrt[3]{4}$

23) $\dfrac{\sqrt{mn}}{n}$ **24)** $\dfrac{21 + 3\sqrt{5}}{11}$

25) $-\dfrac{16 + 7\sqrt{6}}{2}$ **26)** $b = 25$

27) $z = 9$ **28)** $x = -2$

29) $3\sqrt{2}$ ft   **30)** 75 ft

EXERCISE SET 9.1

1) $0, -2$ **3)** $0, -3$ **5)** $0, \dfrac{2}{3}$

7) $0, 2$ **9)** ± 5 **11)** $\varnothing$

13) ± 11 **15)** $\pm\sqrt{7}$ **17)** $\varnothing$

19) $\pm 4\sqrt{3}$ **21)** $\pm 2\sqrt{7}$ **23)** ± 2

25) $\pm\sqrt{5}$ **27)** $\pm 2\sqrt{5}$ **29)** $\pm 2\sqrt{2}$

31) $\varnothing$ **33)** $\pm\dfrac{\sqrt{14}}{2}$ **35)** $\pm\dfrac{\sqrt{55}}{5}$

37) $\pm\dfrac{2\sqrt{3}}{3}$ **39)** $1, -9$

41) $-5 \pm 2\sqrt{11}$ **43)** $1, -\dfrac{11}{3}$

45) $-4, 7$ **47)** $3 \pm \sqrt{7}$

49) $-5 \pm \sqrt{11}$ **51)** $\dfrac{-4 \pm \sqrt{2}}{2}$

53) $2 \pm 7\sqrt{2}$

55) $\dfrac{-2 \pm 2\sqrt{13}}{3}$

57) 2 and -8 **59)** 8 in

61) 3 cm   **63)** $\dfrac{-15 \pm \sqrt{21}}{3}$

65) $\dfrac{-15 \pm \sqrt{55}}{10}$

EXERCISE SET 9.2

1) $9, (a + 3)^2$ **3)** $81, (x - 9)^2$

5) $\dfrac{25}{4}, \left(n + \dfrac{5}{2}\right)^2$ **7)** $\dfrac{49}{4}, \left(b - \dfrac{7}{2}\right)^2$

9) $\dfrac{25}{16}, \left(x + \dfrac{5}{4}\right)^2$ **11)** $\dfrac{9}{196}, \left(a - \dfrac{3}{14}\right)^2$

13) 3, 5   **15)** $-2, -6$

17) $-2, 6$ **19)** $2, -5$

21) $-4, -5$ **23)** $2, \dfrac{3}{2}$

25) $3, -\dfrac{2}{3}$ **27)** $4, -\dfrac{5}{2}$

29) $3, \dfrac{2}{3}$ **31)** $\varnothing$

33) $-2 \pm \sqrt{5}$ **35)** $4 \pm \sqrt{11}$

37) $7 \pm \sqrt{51}$ **39)** $-4 \pm \sqrt{13}$

41) $5 \pm \sqrt{30}$ **43)** $\dfrac{-5 \pm \sqrt{37}}{2}$

45) $\dfrac{2 \pm \sqrt{10}}{2}$ **47)** $\varnothing$

49) $\dfrac{-3 \pm 2\sqrt{3}}{3}$ **51)** $\varnothing$

53) $\dfrac{1}{3}, -\dfrac{2}{3}$

EXERCISE SET 9.3

1) 1, 5   **3)** 2, 4

5) $-4, 6$ **7)** $-1, -6$

9) $-2, 7$ **11)** $6, \dfrac{3}{2}$

13) $6, -\dfrac{4}{3}$ **15)** $\varnothing$

17) $\dfrac{5}{2}, -\dfrac{3}{4}$ **19)** $2, -\dfrac{1}{15}$

21) $\dfrac{-3 \pm \sqrt{13}}{2}$ **23)** $\dfrac{-1 \pm \sqrt{13}}{2}$

25) $\dfrac{3 \pm \sqrt{41}}{4}$ **27)** $\dfrac{1 \pm \sqrt{61}}{10}$

29) $\varnothing$ **31)** $\dfrac{-7 \pm 3\sqrt{5}}{2}$

33) $\dfrac{9 \pm \sqrt{105}}{4}$　　**35)** $1 \pm \sqrt{3}$

37) $5 \pm \sqrt{5}$　　**39)** $4 \pm 2\sqrt{2}$

41) $\dfrac{2 \pm \sqrt{14}}{2}$　　**43)** $\dfrac{-5 \pm \sqrt{5}}{4}$

45) $\dfrac{-1 \pm \sqrt{13}}{3}$　　**47)** $\varnothing$

49) 3 and 5 or -5 and -3

51) $W = 5$ in, $L = 11$ in

53) $W = 4$ yds, $L = 7$ yds

55) $2, \dfrac{3}{2}$

57) $3, -\dfrac{2}{3}$

SECTION 9.4

1) -22　　**3)** 4　　**5)** 6.25

7) -5　　**9)** $2, -6$　　**11)** 3

13) $9, -12$　　**15)** $\varnothing$　　**17)** $\dfrac{-3 \pm \sqrt{3}}{3}$

19) $\pm 2\sqrt{6}$　　**21)** $-\dfrac{4}{3}, 2$　　**23)** $-6, 4$

25) -8　　**27)** $-\dfrac{15}{2}, 5$　　**29)** 2

31) 3　　**33)** 2　　**35)** $0, 4$

37) $-\dfrac{7}{5}, 2$　　**39)** 5　　**41)** $\dfrac{21 \pm \sqrt{449}}{2}$

SECTION 9.5

1) $4, 5$　　**3)** 6 and 9 or -9 and -6

5) 5 and 7　　**7)** $W = 50$ yds, $L = 60$ yds

9) $W = 8$ in, $L = 11$ in　　**11)** 3, 4, and 5

13) 26 ft　　**15)** 80 ft and 150 ft

17) 3 hrs　　**19)** 45 mph

21) Jody = 4 hrs,　　**23)** Ronnie = 2 hrs,
　　Billy = 6 hrs　　　　　Bubba = 3 hrs

25) 5 sec　　**27)** 1.7 sec

29) 3.7 sec

CHAPTER 9 REVIEW EXERCISES

1) $0, -3$　　**3)** $0, -\dfrac{13}{5}$　　**5)** $0, 2$

7) ± 12　　**9)** $\pm 2\sqrt{7}$　　**11)** $\varnothing$

13) ± 5　　**15)** $\pm 2\sqrt{5}$　　**17)** $1, -13$

19) $\dfrac{-4 \pm 4\sqrt{5}}{3}$　　**21)** $-4, -6$　　**23)** $-2, 7$

25) $-6, -\dfrac{1}{2}$　　　　**27)** $-4 \pm 3\sqrt{2}$

29) $\dfrac{-7 \pm \sqrt{61}}{2}$　　**31)** $\varnothing$

33) $\dfrac{-6 \pm \sqrt{46}}{2}$　　**35)** $-1, -2$

37) $2, 6$　　**39)** $2, -\dfrac{5}{2}$

41) $\dfrac{4 \pm \sqrt{34}}{6}$　　**43)** $-3 \pm \sqrt{13}$

45) $\dfrac{4 \pm \sqrt{14}}{2}$　　**47)** $\dfrac{-1 \pm \sqrt{19}}{6}$

49) $\varnothing$　　**51)** $\dfrac{-5 \pm \sqrt{41}}{4}$

53) -1　　**55)** 5

57) $-\dfrac{6}{11}, 5$　　**59)** 3

61) 2 and 5

63) 5 and 6 or -6 and -5

65) $w = 12$ ft and $l = 16$ ft

67) 8 mi and 15 mi

69) 40 mph

71) Ahmad = 6 hrs and Frank = 10 hrs

CHAPTER 9 TEST

1) $0, 2$　　**2)** $0, 5$

3) ± 11　　**4)** $\pm 3\sqrt{3}$

5) $0, 12$　　**6)** $\dfrac{-2 \pm 3\sqrt{2}}{3}$

7) $100, (a - 10)^2$　　**8)** $\dfrac{49}{36}, \left(b + \dfrac{7}{6}\right)^2$

9) $-2, 8$　　**10)** $\dfrac{3 \pm \sqrt{19}}{2}$

11) $2, 3$　　**12)** $4, -\dfrac{5}{2}$

13) $\varnothing$　　**14)** $-3, \dfrac{1}{3}$

15) $-5, 7$　　**16)** $\dfrac{-7 \pm \sqrt{97}}{4}$

17) $\varnothing$　　**18)** $4 \pm 5\sqrt{5}$

19) 10　　**20)** $-\dfrac{15}{2}, 2$

21) $-\dfrac{8}{7}, 6$　　**22)** 5

23) 8 and 10 or -10 and -8

24) $w = 6$ in, $l = 8$ in

25) 3 hrs

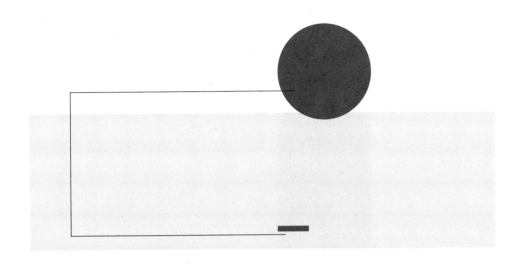

Index

Abscissa, 404
Absolute value, 8
"AC" method for factoring trinomials, 289
Adding
  fractions, 17
    with common denominator, 345
    with unlike denominator, 348
  like terms, 80
  polynomials, 81
  radicals, 544
  rational expressions, 345
    with unlike denominator, 351
  real numbers, 56
  signed numbers, 59
Addition method for solving linear systems, 486
Addition property of equality, 172
Addition property of inequality, 200
Additive identity, 61
  property, 62
Additive inverse, 7, 61
  property, 62, 172
Algebraic expression, 73
  evaluating, 135
  simplifying, 80
Angle
  alternate exterior, 238
  alternate interior, 238
  complementary, 237
  congruent, 237
  supplementary, 237
Angle relationships, 238
Applications
  consecutive integers, 217, 298, 370, 617
  distance, rate, and time, 218, 374, 505, 620

  geometric, 50, 234, 298, 618
  investments, 227, 506
  linear equations in two variables, 452
  linear systems, 501
  mixtures, 229, 507
  money, 224, 247, 504
  motion, 374, 505, 620
  number problems, 369, 502
  proportions, 384
  quadratic equations, 297, 592, 616
  rational expressions, 368
  science problems, 622
  traditional, 214, 223
  work problems, 371, 621
Area
  circle, 39
  definition, 37
  parallelogram, 39
  rectangle, 37, 118, 126
  square, 38, 126
  trapezoid, 39
  triangle, 38, 118
Associative property
  addition, 61
  multiplication, 105

Binomial, 74
  multiplying, 116
    FOIL method, 120
  square of, 122
Binomial products, 116
Braces, 132, 174
Brackets, 59, 68, 132, 174

Calculator exploration
  adding, 56
    radicals, 544
  dividing, 127
  fractions to power, 154
  multiplying, 99
    radicals, 530
  negative exponent, 141
  power rule, 110
  product rule, 108
  quotient rule, 137
  quotient of radicals, 540
  signed numbers to powers, 103
  subtracting, 65
Cartesian coordinate system, 403
Circle
  area, 39
  center, 35
  circumference, 35, 242
  definition, 35
  diameter, 35
  radius, 35
Circumference of a circle, 35
Coefficient, numerical, 74, 259
Combining like terms, 80
Common denominator
  least, 19, 339
Common factor, 254
  binomial, 262
  monomial, 260, 268
  negative, 261
Common multiples, 334
Commutative property
  addition, 61
  multiplication, 101, 105
Complementary angles, 237
Completing the square, 594
Complex fractions, 355
Composite number, 10, 253
Composite polynomial, 258
Compound inequality, 205
Consecutive integers, 216, 298, 370, 617
Constant
  definition, 24
  π as a, 36
Contradiction, 191
Cone
  volume of, 44
Congruent angles, 237
Conjugate of radical, 548, 558
Conjugate pairs, 121, 266
Coordinates of a point, 477
Counting numbers, 2

Cross multiplication, 365
Cross products, 365, 382
Cube
  volume of, 41
Cube root, 524
  rationalizing the denominator, 556
  simplifying, 535
Cubes
  difference of, 269
  sum of, 269

Decimals, 5
Degree
  angle, 237
  monomial, 75
  polynomial, 75
Denominator, 12
  least common, 339
  rationalizing, 552
Dependent system, 480, 493
Descartes, Rene, 403
Descending powers, 75
  in long division, 330
Diameter of a circle, 35
Difference
  of two cubes, 269
  of two squares, 266
Distance formula, 242
Distance-rate-time applications, 218, 374, 505, 620
Distributive property, 79, 115
Dividing
  integers, 126
  polynomial by
    binomial, 327
    monomial, 160
  polynomials with
    long division, 327
  rational expressions, 324
  by zero, 127
Domain, 70
  of relation, 457
Double negative rule, 8

Elements of a set, 2, 457
Empty set, 173, 191
Equation
  addition property of equality, 172
  combining properties in solving, 189
  dependent, 480, 493
  equivalent, 172, 181
  linear, 172
  multiplication property of equality, 181
  percent, 209

quadratic, 295, 586
radical, 560
rational expressions, 364, 612
rational numbers, 362
with one square root, 561, 614
with two square roots, 563, 614
with two variables, 406
Equation of a line, 441
$ax + by = c$ form, 444
containing two given points, 444
horizontal, 446
point-slope form, 442
slope-intercept form, 451
vertical, 446
Equivalent
equations, 172, 181
fractions, 19, 336
rational expressions, 400
Evaluating
algebraic expressions, 29, 135
functions, 463
geometric formulas, 33
polynomials, 75
P(x) notation, 75
Exponent, 25
base of, 25
laws of, 157
expressions combining, 157
power rule, 111, 142
product rule, 109
negative, 139
quotient rule, 137, 143, 155, 259
zero, 139
Expression
exponential, 137
radical, 524
rational, 309
simplifying, 80
translating, 60, 130
variable, 29, 134
Extraneous solutions, 364, 561

Factor, 10, 25
greatest common, 254
Factoring a composite number into prime factors, 11
Factoring a polynomial, 258
"AC" method, 289
aids to complete, 292
binomial, 259
difference of cubes, 299
difference of squares, 266
general trinomials, 276, 285
by grouping, 262

mixed, 292
perfect square trinomial, 272
removing a common binomial factor, 262
removing a common monomial factor, 259
removing a common negative factor, 261
sum of cubes, 269
summary, 282
trinomial, 276, 285
FOIL method, 120
Formulas, using and solving, 242
Function
definition, 458
Fractions, 12
adding, 17
complex, 355
denominator of, 12
dividing, 15
equivalent, 19, 337
fundamental property of, 337
least common denominator, 19, 339
lowest terms, 12, 311
multiplying, 13
numerator of, 12
Function
domain, 458
evaluating, 463
f(x) notation, 463
range, 458
vertical line test, 460
Functional notation, 463
Fundamental property of
fractions, 337
F(x) notation, 463

GCF, 255, 275
General trinomial
factoring, 276
Geometric applications, 234, 618
involving angles, 237
involving quadratic equations, 592
involving shapes, 235, 592
Geometric figures
applications using, 50
Geometric formulas
evaluating, 33
Graphing horizontal line, 413, 436
line given point and slope, 429
linear equation, 406
intercept method, 408
linear inequality with
one variable, 198
two variables, 416
numbers, 3

slope-intercept method, 434
straight lines, 405
summary, 467
system of linear
  equations, 477
  inequalities, 512
vertical line, 412
Greatest common factor
  removing the, 254, 258, 275
Grouping
  factoring by, 262

Horizontal line
  graphing, 413, 436
  slope of, 428, 436, 446
Hypotenuse of a right
  triangle, 568

Identity, 191
  additive, 61
  multiplicative, 99, 105
Imaginary numbers, 525
Inconsistent system, 480, 492
Index of a radical, 524
Inequality
  addition property of, 200
  compound, 199, 205
  graphing, 202
  linear, 197
  multiplication property of, 202
  combining properties in solving, 203
  solutions, 197, 416
  symbols of, 7
  translating from mathematics into English, 206
Integers, 3
  adding, 57, 59
  consecutive, 216, 298
  dividing, 126
  in every day situation, 65
  multiplying, 102
  subtracting, 65
Intercepts of linear equations, 408
Interest
  applications involving, 227, 233, 506, 510
  earned, 227
  rate, 227
Inverse
  additive, 7, 61
  multiplicative, 15, 106, 181
Investment problems, 227, 506
Irrational numbers, 6, 527
  decimal approximation
    cube roots, 527
    square roots, 527

Laws of Exponents, 157
LCD
  finding the, 339
LCM
  finding the, 334
Least common denominator (LCD), 19, 339
  of fractions, 19, 339
  of rational expressions, 339
Least common multiple (LCM), 334
Like terms, 79
  adding, 80
Linear equation
  application, 176
  contradictions, 191
  four-step method to solve, 189
  graphing, 406
  identities, 191
  solving, 172, 609
  systems of, 475
  in two variables, 396
Linear inequality, 197
  graphing, 416
  shading, 417
  systems of, 511
  test point, 417
Lines
  graphing, 406, 408, 413, 429, 434
  parallel, 237
  perpendicular, 447
  writing equations of, 443, 444, 446, 450
Long division, 327
  missing powers, 330
Lowest terms
  definition of, 311
  of rational expressions, 312
    numbers, 12, 309

Mathematical sentences, 176
Measure of angles, 237
Missing powers in long
  division, 330
Mixed factoring, 292
Mixed numbers, 12
Monomial, 74
  degree of, 74
  dividing by a monomial, 160
  product of a polynomial and, 115
Money problems, 224, 247, 504
Motion problems, 374, 505, 620
Multiplication property of
  equality, 181
  inequality,
    negative factor of, 200
    positive factor of, 200

Multiplicative inverse, 15, 106, 181
Multiplying
   cross, 365
   binomials, 116
   fractions, 13, 319
   polynomials, 115
   radicals, 530, 537
   rational expressions, 318, 320
   rational numbers, 318
   real numbers, 99
   by zero, 106

Natural numbers, 2
Negative
   exponents, 141
   numbers, 2, 57
      multiplying, 100
      raised to a power, 104
   slope, 431
Number line, 3
   divided into three parts, 197
Number problems, 369, 502, 617
   quadratic equations, 592
Numbers
   composite, 10, 258
   prime factors of, 11
   directed, 57
   graphing, 3
   imaginary, 525
   integers, 3
   irrational, 6, 527
   mixed, 12
   natural, 2
   negative, 2, 57
   positive, 3, 57
   prime, 10, 253
   rational, 4, 526
   real, 6, 65
   signed, 3
   square root of, 524
   whole, 2
Numerator, 12
Numerical coefficient, 74

Operations
   order of, 26, 129, 132
   with rational expressions, 309
Opposite of a number, 7, 61
Order
   descending, 75
   symbols for, 6
Order of operations, 26, 129, 132
   calculator use, 28
   in everyday situations, 29
   with variable expressions, 29

Ordered pairs, 396, 476
   components of, 397
   finding missing numbers, 397
   solutions to everyday situations written as, 400
Ordinate, 404
Origin, 403

Parallelogram
   area of, 39, 87
   definition of, 39
Parallel lines, 237
   slope of, 447, 481
Parentheses, 26, 59, 68, 174
Percent to decimal converting, 210
Percent Equations, 209
   three cases of, 211
Perfect square number, 272
Perfect square trinomial
   factoring, 272
Perimeter, 33, 87
Perpendicular lines
   slope of, 447
Place holders in long
   division, 330
Plotting Points, 404
Plotting solutions of
   equations in two variables, 405
Points
   plotting, 404
Point-slope formula, 443
Polynomial
   adding, 81
   composite, 258
   defined, 73
   degree, 75
   in descending order, 75
   dividing, 160, 324
   factoring, 258
   multiplying, 115
   evaluating $P(x)$ notation, 75
   in science and the real world, 76
   subtracting, 82
   identifying types, 73
Positive numbers, 3, 57
   multiplying, 100
Positive slope, 431
Power (exponent), 25
   raising a negative number to a, 104
   to a power, 112
   of a product, 111
   for quotients, 155
Power rules for exponents, 111, 142
   quotients, 153
   roots, 539

Prime factorization, 11, 253
   procedure for finding, 254
Prime number, 10, 253
Prime polynomial, 258, 281
Principal root, 525
Product, 10
   of binomials, 116
   negative numbers, 103
   polynomials, 115
     vertically, 117
   radicals, 530, 537
   of form
     $(a + b)(a - b)$, 121
     $(a + b)^2$, 122
     $(a - b)^2$, 122
   signed numbers, 100
Product rule for exponents, 109, 259
Proper fractions, 12
Proportions, 380
   applications, 384
   definition of, 381
   procedure for solving, 382
Pythagorean theorem, 568

Quadrants, 403
Quadratic equations, 295, 610
   applications, 297
   completing the square, 593, 596
   square root method, 587
   solving by factoring, 295, 586
   solving by quadratic formula, 602
   summary, 626
Quadratic formula, 602
Quadrilateral
   perimeter of, 87
Quotient, 4, 126
   power rule, 153, 156
Quotient rule for integer
   exponents, 137, 143, 155

Radical equation, 560
   in real-world situation, 565, 569, 573
   summary, 578
Radical expression, 524
Radical Sign, 534
   index, 524
Radicals, 523
   adding, 544
   equations with, 560
   mixed operations, 546
   multiplying, 530, 537
   products involving sums, 547
   quotients of, 540

   rationalizing denominators, 552
   simplest form, 546
   simplifying, 530
   subtracting, 545
   summary, 578
Radicand, 524
Radius of a circle, 35
Range
   of function, 458
   of relation, 457
Ratio, 380
Rational expressions
   adding, 345
   applications, 368
   definition of, 309
   difference, 346
   dividing, 324
     by factoring, 324
   equations with, 364
   multiplying, 318
   long division, 327
   in lowest terms, 312
   reducing, 309
   subtracting, 345
   sum of, 346
   summary, 388
   undefined, 310
   with zero denominator, 310
Rational numbers, 4, 526
   dividing, 325
   reducing, 309
Real numbers
   adding, 56
   multiplying, 99
   ordering, 6
   subtracting, 65
Reciprocal, 15, 181
Rectangle
   area of, 37, 118, 298
   perimeter of, 34, 242
Rectangular coordinate system, 403
Rectangular solid
   volume of, 42
Relation
   definition, 457
   domain, 457
   range, 457
Replacement set, 70
Right circular cylinder
   volume of, 43
Right triangle, 568
Rise, 426
Root, 523

cube, 524
    definition of, 524
    existence of, 526
    finding, 524
    fourth, 524
    index, 524
    powers of, 539
    principal, 525
    square, 524
    summary, 578
Roots of numbers using
    radicals, definition, 525
Run, 426

Sentences, mathematical, 176, 185, 192
Set
    element of, 2
    empty, 173
    solution, 171
    subset, 2
Set builder notation, 4
Scientific notation, 145
Signed numbers, 3
    adding, 59
    dividing, 126
    multiplying, 100
    to powers, 99
    subtracting, 65
Simplifying
    algebraic expressions, 80
    complex fractions, 356, 357
    radicals, 530
Slope-intercept form, 433
Slope formula, 427
Slope of a line, 424
    equation in standard form, 449
    formula, 426
    horizontal line, 428, 436, 446
    negative, 431
    positive, 431
    steepness, 432
    summary, 438
    undefined, 428
    vertical line, 429, 436, 446
    zero, 428
Slopes of parallel lines, 447
    perpendicular lines, 447
Solution
    equation, 69
        with two variables, 396
    extraneous, 561
    system, 476
Solution problems, 231

Solving
    equations with rational expressions, 612
    equations with one square root, 561, 614
    equations with two square roots, 563, 614
    linear systems, 172, 609
    linear inequalities, 199
    mixed equations, 609
    proportions, 382
        applications, 384
    quadratic equations, 295, 585, 610
        applications, 295
Solving systems of linear equations
    addition, 486
    substitution, 495
Special products of
    binomials, 120
    form $(a + b)(a - b)$, 121
    form $(a + b)^2$, 122
    form $(a + b)(a^2 - ab + b^2)$, 123
    form $(a - b)(a^2 + ab + b^2)$, 124
Square
    area, 38, 592
    definition, 35
    perimeter, 35
Square root, 524
    rationalizing the denominator, 553
    simplifying, 532
Squares
    difference of two, 267
    sum of two, 267
Study tip
    #1, Developing a positive approach, 8
    #2, Motivation—the key to success, 22
    #3, Mathematics is sequential learning, 30
    #4, Reading mathematics critically, 47
    #5, Memorizing and understanding mathematics, 63
    #6, Making sure: confidence building, 77
    #7, Preparing for class, 114
    #8, Making effective use of class time, 125
    #9, Preparing for a test, 136
    #10, Taking a math test, 159
Subscript notation, 424
Subset, 2
Substitution method for
    solving systems of linear equations, 495
Subtracting
    fractions, 17
        with common denominator, 345
        with unlike denominator, 349
    polynomials, 82
    rational expressions, 345
        with unlike denominators, 351
    real numbers, 65
    signed numbers, 65

Sum of
two cubes, 269
two squares, 267
Summary
factoring and quadratic equations, 303
graphing, 467
solving quadratic equations, 626
system of linear equations, 518
system of linear inequalities, 518
Supplementary angles, 237
System of linear equations, 475
applications, 502
consistent, 480
definition, 476
dependent, 480, 493, 499
inconsistent, 480, 499
with infinite number solutions, 480, 493
with no solutions, 480, 492
solution of, 476
solving by addition, 484
solving by graphing, 477
solving by substitution, 494
summary, 518
System of linear inequalities, 511
solving by graphing, 512
summary, 518

Term
adding, 80
combining like, 78
identifying, 73
like, 78
unlike, 78
Translating English to mathematics, 176, 185, 193, 207
mathematics to English, 175, 185, 192, 206
Percent to English, 209
Transversal, 237
Trapezoid
area, 39
definition, 39
Triangle
area, 38, 118, 243
definition, 34
perimeter, 34, 87
sum of angles, 327, 245
Trinomial, 74
factoring, 276
perfect square, 272

Undefined slope, 428
Units
cubic, 41
linear, 33
square, 37

Variable, 4
defined, 4, 24
Variable expression, 29, 134
Variable radicand, 528
Vertical line
graphing, 412, 436
slope of, 429, 436, 446
Volume
cone, 44
cube, 41
geometric solids, 41
rectangular solids, 42
right circular cylinders, 43

Whole numbers, 2
Word problems
consecutive integers, 217, 298, 370, 617
distance, rate, and time, 218, 374, 505, 620
geometric, 50, 234, 298, 618
geometric shapes, 234, 245
investments, 227, 506
mixtures, 229, 507
money, 224, 247, 504
motion, 374, 505, 620
number problems, 369, 502, 617
proportions, 384
Pythagorean theorem, 619
quadratic equations, 297, 592, 616
rational expressions, 368
science, 622
traditional, 214, 223
work, 371, 621
Work problems, 371, 621

X-axis, 403
X-intercept, 408

Y-axis, 403
Y-intercept, 408

Zero
dividing by, 172
exponent, 139
multiplying by, 106
product property, 295
slope, 428